SYMMETRIES AND GROUPS IN CONTEMPORARY PHYSICS

NANKAI SERIES IN PURE, APPLIED MATHEMATICS AND THEORETICAL PHYSICS

Editors: S. S. Chern, C. N. Yang, M. L. Ge, Y. Long

Published:

Vol. 1 Probability and Statistics
eds. Z. P. Jiang, S. J. Yan, P. Cheng and R. Wu

Vol. 2 Nonlinear Anaysis and Microlocal Analysis
eds. K. C. Chung, Y. M. Huang and T. T. Li

Vol. 3 Algebraic Geometry and Algebraic Number Theory
eds. K. Q. Feng and K. Z. Li

Vol. 4 Dynamical Systems
eds. S. T. Liao, Y. Q. Ye and T. R. Ding

Vol. 5 Computer Mathematics
eds. W.-T. Wu and G.-D. Hu

Vol. 6 Progress in Nonlinear Analysis
eds. K.-C. Chang and Y. Long

Vol. 7 Progress in Variational Methods
eds. C. Liu and Y. Long

Vol. 8 Quantized Algebra and Physics
Proceedings of International Workshop
eds. M.-L. Ge, C. Bai and N. Jing

Vol. 9 Operads and Universal Algebra
Proceedings of the International Conference
eds. C. Bai, L. Guo and J.-L. Loday

Vol. 10 Emerging Topics on Differential Equations and Their Applications
Proceedings on Sino-Japan Conference of Young Mathematicians
eds. H. Chen, Y. Long and Y. Nishiura

Vol. 11 Symmetries and Groups in Contemporary Physics
Proceedings of the XXIX International Colloquium on Group-Theoretical Methods in Physics
eds. C. Bai, J.-P. Gazeau and M.-L. Ge

Nankai Series in Pure, Applied Mathematics and Theoretical Physics

Vol. 11

Proceedings of the XXIX International Colloquium on Group-Theoretical Methods in Physics

Tianjin, China 20–26 August 2012

SYMMETRIES AND GROUPS IN CONTEMPORARY PHYSICS

Edited by

Chengming Bai
Nankai University, China

Jean-Pierre Gazeau
Paris Diderot University, France

Mo-Lin Ge
Nankai University, China

World Scientific

NEW JERSEY • LONDON • SINGAPORE • BEIJING • SHANGHAI • HONG KONG • TAIPEI • CHENNAI

Published by

World Scientific Publishing Co. Pte. Ltd.
5 Toh Tuck Link, Singapore 596224
USA office: 27 Warren Street, Suite 401-402, Hackensack, NJ 07601
UK office: 57 Shelton Street, Covent Garden, London WC2H 9HE

British Library Cataloguing-in-Publication Data
A catalogue record for this book is available from the British Library.

Nankai Series in Pure, Applied Mathematics and Theoretical Physics — Vol. 11
SYMMETRIES AND GROUPS IN CONTEMPORARY PHYSICS
Proceedings of the XXIX International Colloquium on Group-Theoretical Methods in Physics

ISBN 978-981-4518-54-3

Printed in Singapore by World Scientific Printers.

PREFACE

The 29th session of the International Colloquia on Group-Theoretical Methods in Physics ("Group 29") was held at the Chern Institute of Mathematics, Nankai University, Tianjin, on August 20-26, 2012.

It was the first time that the venue was in China since the beginning of the series in 1972, after so many places all around the world.

The series of International Colloquia on Group-Theoretical Methods in Physics was initiated in 1972 following the efforts of Henri Bacry in Marseille and Aloysio Janner in Nijmegen. The aim was to provide a forum for physicists interested in group-theoretical methods which were mostly represented at that time by two large communities, namely elementary-particle theorists and phenomenologists on the one hand and by solid-state specialists on the other hand. There were also mathematicians who were eager to apply newly-developed group-theoretical techniques and algebraic structures. Since then the colloquia have been held in various places and have earned a strong scientific reputation around the world. Currently the ICGTMP attract theoretical and experimental physicists, mathematicians and scientists in related disciplines who are interested in the latest methods and applications of group theory in its widest sense.

More than 280 scientists, from over 150 universities and institutes and 32 countries and districts (of China) have participated in Group 29. The conditions and organization were unanimously acknowledged as excellent: all the objectives and goals that had been set out when the Chern Institute made a bid for this meeting four years back have been fully attained.

This Colloquium was one of the most important events for the community in physics and mathematics in China: 197 lectures, including 12 plenary talks and 9 poster contributions, 13 parallel sessions with 176 talks, and a round table on *the Nature of Space and Time.*

88 written contributions are gathered in the present volume, which have given by themselves a vivid idea of the scientific richness of the conference.

Whether written or oral, all contributions cover a wide spectrum of topics pertaining to *group theory in its broadest sense and applications in different parts of physics, mathematics and in other natural sciences*, as is shown by the list of plenary talks,

- *The surprising connection between discrete holomorphicity and integrability*, Murray Batchelor (Australian National University, Australia)
- *Building a quantum computer: a Microsoft approach*, Zhenghan Wang (Microsoft Station Q, USA)
- *Loop quantizable gravity*, Jerzy Lewandowski (Uniwersytet Warszawski, Poland)
- *Yang-Baxter system and quantum information*, Mo-Lin Ge (Nankai University, China)
- *Tensor field theories*, Vincent Rivasseau (University Paris-Sud XI, France)
- *Towards the origin?*, Public Lecture by Daniel Treille (CERN and ETHZ, Switzerland)
- *Permutation symmetry for molecular systems with identical nuclei*, Wigner Medal Conference by Alden Mead (University of Minnesota, USA)
- *The large N limit of Tensor Models*, Weyl Prize Conference by Razvan Gurau (Perimeter Institute for Theoretical Physics, Canada)
- *Aspects of defects in integrable quantum field theory*, Edward Corrigan (University of York, UK)
- *Dynamical symmetries and algebraic methods in Nuclear Structure Physics*, Akito Arima (Shizuoka University of Art and Culture, Japan and Musashi Gakuen, Japan)
- *Quantum invariants of knots and the hyperbolic volume*, Jun Murakami (Waseda University, Japan)
- *An overview of coherent states and coherent state quantization*, Syed Twareque Ali (Concordia University, Canada)

and the topics of the parallel sessions,

(1) Nonlinear Sciences (chairs: Q. Ouyang and W. Wang)
(2) Quantum Algebras (chair: N. Jing)
(3) Superintegrable Hamiltonians (chairs: W. Miller Jr., P. Winternitz and K.B. Wolf)
(4) Applications in Biology (chairs: R. Kerner and R. Twarock)
(5) Quantum Information and Representation Theory (chairs: G. Chiribella, M. Christandl, A. Harrow and L. Vinet)

(6) Finite Quantum Systems and Combinatorial Physics (chairs: A. Solomon and A. Vourdas)
(7) Symmetries in String Theory and Supergravity (chair: F. Toppan)
(8) Loop Quantum Gravity (chair: J. Lewandowski)
(9) Tensor and Group Field Theories for Quantum Gravity (chair: V. Rivasseau)
(10) Conformal Field Theories (chairs: Y. Kawahigashi, L. Kong and C. Schweigert)
(11) Supersymmetry and Quantum Groups (chairs: V. Dobrev and A. Isaev)
(12) Applications in Particle and Nuclear Physics (chairs: P. Isacker and Y.-M. Zhao)
(13) General Quantum Mechanics and its Mathematical Methods Involving Group Theory (chairs: S.T. Ali and G. Goldin)

The meeting was an intense period of scientific exchanges between researchers in those various fields. It perfectly fulfilled the exigency required of any colloquium, namely to occasion a mutual irrigation of ideas, methods, and objectives, between specialists.

We express gratitude for the help given by Professor Yiming Long and Professor Lei Fu, respectively former and current Directors of the Chern Institute, in hosting the colloquium.

We thank all the Chairpersons and anonymous referees for their great efforts, which have made this proceedings possible.

The conference was graciously assisted by the staff members at Chern Institute of Mathematics. We warmly thank all of them for their help.

We are grateful to

- the Chern Institute of Mathematics
- the Mathematical Center of Ministry of Education, China
- the National Natural Science Foundation of China

for financial support.

The editors

Chengming Bai

Jean-Pierre Gazeau

Mo-Lin Ge

XXIXth INTERNATIONAL COLLOQUIUM ON GROUP-THEORETICAL METHODS IN PHYSICS

INTERNATIONAL ADVISORY COMMITTEE

M. Atiyah (Edinburgh, UK)
H.-D. Doebner (Clausthal, Germany)
L. Faddeev (RAS, Russia)
C. Gu (1926 – 2012) (Fudan, China)
V. Kac (MIT, USA)
Yu.I. Manin (MPI, Germany)
C. Vafa (Harvard, USA)
C.N. Yang (Tsinghua, China)

ORGANIZING COMMITTEE

Chengming Bai (Scientific Secretary)
Mo-Lin Ge (Chair)
Yiming Long
Weiping Zhang

ICGTMP STANDING COMMITTEE

The ***International Colloquia on Group-Theoretical Methods in Physics*** *focus on developments in the field of group theory in its broadest sense and applications in different parts of physics, mathematics and in other natural sciences. The* ***Standing Committee*** *is the organizing body of the Colloquium series. It decides where future Colloquia will be held and who will be the local organizers. Its current members are*

S.T. Ali (Concordia U., Montreal)
L. J. Boya (Zaragoza U.)
C. Burdik (Czech Technical U., Prague)
E. Celeghini (Firenze U., INFN)
V. Dobrev (Bulgarian Acad. Sciences)
H.D. Doebner (Honorary Member, Clausthal U.)
J.-P. Gazeau (Chairman, Paris Diderot U.)
M.-L. Ge (Nankai U.)
G. Goldin (Rutgers U.)
F. Iachello (Yale U.)
P. Van Isacker (GANIL, Caen)
J. Van der Jeugt (Ghent U.)
R. Kerner (P&M Curie U., Paris)
P. Kielanowski (Cinvestav, Mexico)
A. Kostelecky (Indiana U.)
M. del Olmo (Valladolid U.)
G. Pogosyan (UNAM, Mexico, JINR, Dubna)
C. Schweigert (Hamburg U.)
A. Solomon (Honorary Member, Open U., UK)
F. Toppan (CBPF, Rio de Janeiro)
R. Twarock (York U.)
L. Vinet (Montral U.)
A. Vourdas (Bradford U.)
K. Wendland (Freiburg U.)
K.B. Wolf (UNAM, Cuernavaca)

CONTENTS

OPENING ADDRESS

H.-D. DOEBNER
Former Chairman and Honorary Member of the ICGTMP Standing Committee
Technical University of Clausthal,
Germany

Dear Friends and Colleagues
Ladies and Gentlemen

I'm delighted to be here at the Chern Institute of Mathematics of the Nankai University and to address the Opening of the 29th Colloquium on Group-Theoretical Methods in Physics. This is due to two reasons.

The first reason is connected with the scientific exchange in our fields: Four years ago, in 2008, the Standing Committee of our Colloquia series decided in a meeting during the 27th Colloquium in Yerevan, Armenia, under my chairmanship to give the 29th Colloquium in 2012 to the Chern Institute following the excellent proposal of Professor Mo-Lin Ge and his colleagues and supported in particular by our honorary member Alan Solomon. I remember that this meeting was arranged at a local restaurant with Armenian music during and after a very lively dinner. I collected the secret votes individually as one of my last activities during the long term of my chairmanship.

The decision in favor for Nankai was motivated by the character of our series which reflects the international scientific community working in our field. Hence the Colloquium jumps from region to region and from country to country. This international character gained further momentum in the last decade. Connected with the world wide economic and political development in Asiatic, African and South American countries, their scientific institutions are more and more dynamic and productive. They contributed to the development of science together with the traditional scientific centers in Europe and North-America. This includes also to our field. After the first Colloquium 40 years ago 25 out of 29 Colloquia were organized in these European and North- American centers, only two in Asia: 1985 South Korea/Seoul and 1994 Japan/Toyonaka, one 1998 in the Australian region: Hobart/Tasmania, and one 1979 in the Middle East region: Israel/Kiriat Anavim. I expect that this distribution will change in the future. The 29th Colloquium is now organized in China, the next will be again in Europe, Belgium/Ghent, and the following one in South America: Brazil/Rio de Janeiro.

My second reason is related to a research project of the former Arnold Sommerfeld Institute at the Technical University of Clausthal, Germany. We were interested in quantization methods for systems localized on smooth manifolds and with motions modeled through representations of their kinematical algebra. A realization of this approach is connected with the work of S.-S. Chern on principal fiber bundles. We classified these quantizations through topological quantum numbers and showed, how physical observables depend on properties of the manifold and the kinematic of the system.

I mention Chern's work and its relevance for physical models to indicate the scientific background of our conference series. The traditional name 'Group-Theoretical Methods in Physics' refers to a special mathematical structure: 'group theory' which is applied to a special field: 'physics'. The name was coined 40 years ago; it reflects the situation in the first period of the series. In those days the Colloquia were centered on crystallographic groups, space-time groups, their representations and applications to atomic and molecular systems; sub-centers dealt with physical systems or mathematical questions with a geometrical background, like general relativity and symmetries of partial differential equations. The following second period was characterized through the fast development of particle physics and successful group-theoretical models. Group theory was considered as an, or even the, essential tool not only for understanding and description of atoms and molecules, but also for elementary particles and nuclei. This progress was in retrospect a convincing justification of the name of our series. With the development of mathematics and physics the situation changed slowly. The last Colloquia are already in the third period of our series. Mathematical fields beyond or related group theory came into play. Topological and differential geometric methods were applied to model properties of quantum systems, e.g. like anyons with possible application to computer science. Algebraic structures like supersymmetry, quantum-groups, quantum-algebras and non-commutative geometry were developed - also with the aim to enlarge the somehow narrow traditional group theory. These new structures open the door for innovative ideas and applications and for special subfields which are presented in the third period of the series. I mention string theory, including the AdS string theory viewed as a quantum field theory, and their application to models in solid state physics. I add quantum gravity. I continue this list with cosmology, quantum information, mathematical biology including mathematical virology and engineering science. Parts of these growing subfields are covered in the corresponding proceedings. In this connection I should add that future scientific tendencies of the Colloquia will depend not only on the development of our field but also on our activity and interest of their organizers.

The Standing Committee is aware of these developments. We kept the traditional name, but interpreted the notion 'group theory' now as 'group theory in its widest sense', mentioned also in the talk of Jean-Pierre Gazeau. This notion includes concepts and methods which are related to or emerged from traditional 'group theory'. It is not meant as definition but as a paraphrase, indicating that group theory in its widest sense is an integral part of the present 'mathematical technology' in theoretical and mathematical physics.

Our Colloquia are now a biannual series; they started 1972 as annual series and shifted 1988 to a biannual modus. To fill the odd years between to Colloquia we launched 1999 in Goslar, Germany, a biannual sister series 'International Symposia on Quantum Theory and Symmetries'. This QTS series is centered on mathematical concepts and technologies based on the plethora of notions of symmetry, including group theory in its widest sense, and with application to quantum systems and to systems and structures which are governed by quantum mechanical objects and rules. The 7th Symposium will be organized 2013 in Mexico City.

Hence the Colloquia and the Symposia span a network of annual discussions of group theory in its widest sense and the plethoric set of symmetries, of their applications to real systems and, in our anthropomorphic limitation, of an understanding of parts of our nature.

After my experience in international scientific exchanges I have great respect for the very effective and successful organization of the present edition of our series; the excellent scientific program presents important trends of the developments in our field; the performance of the organizational team behind the Colloquia, especially of Professor Mo-Lin Ge and of Professor Chengming Bai, was outstanding. With a look into the history of our series, I think that the 29th Colloquium on Group-Theoretical Methods in Physics at the Chern Institute in Nankai was a remarkable edition of our series.

THE WIGNER MEDAL

21 AUGUST 2012

WIGNER MEDAL

The Wigner Medal was established in 1977/8 and was awarded for the first time at the Integrative Conference on Group Theory and Mathematical Physics (7th International Group Theory Colloquium 1978) to Eugene P. Wigner and Valentine Bargmann. The purpose of the Wigner Medal is to recognize outstanding contributions to the understanding of physics through group theory.

The Wigner Medal is administered by the Group Theory and Fundamental Physics Foundation, a publicly supported organization.

WIGNER MEDALISTS

1978 E. P. Wigner
1978 V. Bargmann
1980 I. M. Gel'fand
1982 L. Michel
1984 Y. Ne'eman
1986 F. Gürsey
1988 I.M. Singer
1990 F. Iachello
1992 J. Wess and B. Zumino
1996 V. Kac and R.V. Moody
1998 M. Moshinsky
2000 L. O'Raifeartaigh
2002 H.J. Lipkin
2004 E. Inonü
2006 S. Okubo
2010 M. Jimbo
2012 A. Mead

THE WIGNER MEDAL AT THE CHERN INSTITUTE OF MATHEMATICS, NANKAI UNIVERSITY, TIANJIN, CHINA

ARNO BOHM*

Department of Physics
The University of Texas at Austin
Austin, TX 78712-1081
**E-mail: bohm@physics.utexas.edu*

The Wigner Medal was established in 1977/78 and is administered by the Group Theory and Fundamental Physics Foundation. The first Wigner Medal was awarded to Eugene P. Wigner and Valentine Bargmann at the Integrative Conference on Group Theory and Mathematical Physics /VII International Group Theory Colloquium, 1978 in Austin, Texas, USA.

The purpose of the Wigner Medal is to recognize outstanding contributions to the understanding of physics through group theory.

The Wigner Medal for 2012 was awarded at the XXIX International Colloquium on Group Theoretical Methods to the quantum chemist, C. Alden Mead for the *Gauge Theory of Molecules and its Application to Molecular Spectra and Scattering*. Of the 19 scientists who have received the Wigner Medal in the past, Alden Mead is the second awardee who has his Dr. degree in Chemistry. The first awardee with a Chemistry degree was Eugene Wigner himself, who obtained in 1925 a Dr. Ing. degree in chemical engineering from the Technical University Berlin. That was long before he had become a group theoretician, theory of groups and their representation was not much used at that time.

In the meanwhile to a large extent on the basis of Wigner's work, groups and their algebras have become major tools not only for mathematicians and physicists, but also for theoretical chemists.

Gauge groups are not only important for relativistic field theoreticians but also for quantum chemists who investigate spectra and structure of molecules. Starting with the Born-Oppenheimer approximation, but not considering the slow variables of the nuclei as fixed, one is led to the gauge potential and a geometric (Berry) phase effect. It was in molecular physics where the effect of the geometric phase was first noticed as a sign ambiguity of the electronic wave function and explained by the gauge theory for molecules by Alden Mead and collaborators.

LAUDATIO FOR THE 2012 WIGNER MEDALIST

C. ALDEN MEAD

"For the Gauge Theory of Molecules and its Application to Molecular Spectra and Scattering"

BRIAN K. KENDRICK

Theoretical Division
Group T-1, Mail Stop B221
Los Alamos National Laboratory
Los Alamos, NM USA 87545
E-mail: bkendric@lanl.gov

1. Mead's Background and Research Interests

C. Alden Mead was born in St. Louis, Missouri on December 9, 1932. He graduated from Webster Groves High School in 1950 and received his BS in Chemistry at Carleton College, Northfield, MN in 1954. He earned his PhD in Chemistry from Washington University, St. Louis in 1957. His advisor was Professor Sam Weissman. He pursued postdoctoral study at Brookhaven National Laboratory from 1957 through 1958 after which he became Professor of Physical Chemistry at the University of Minnesota in 1958. Alden took two sabbaticals: one in Professor David Bohm's group at Birbeck College, London from 1964 to 1965, and another in Professor Ernst Ruch's group at the Institut für Quantenchemie der Freien Universität, Berlin, from 1971-1972.

Alden's research interests span a variety of important problems in theoretical physics and chemistry. His 1957 thesis title was "Quantum Theory of the Refractive Index" in which he investigated the formulation of classical electromagnetic and optical properties of matter in terms of quantized elecromagnetic fields and the theory of spectral line shapes. During his postdoctoral studies at Brookhaven and also later during his sabbatical in London, Alden studied gravity theory and its connection to a fundamental length scale (now commonly referred to as the Planck scale).[1,2] During his sabbatical in Berlin, he pursued the study of permutation groups and their applications to molecular properties such as symmetry and chirality. In 1974 Alden wrote a review on this subject for the Springer series "Topics in Current Chemistry" entitled "Permutation Groups, Symmetry and Chirality in Molecules".[3] In the late 1970's, Alden became interested in degenerate Born-Oppenheimer electronic potential energy surfaces in molecules. In particular, he rederived a more general formulation of the "noncrossing" rule using time-reversal invariance.[4] This work laid the foundation for the "gauge theory of molecules" (to be discussed in Sect. 3) which originated from the investigation of the permutation symmetry of identical nuclei in molecules in the presence of an electronic degeneracy. Eugene Wigner's work[5–7] played an influential role in several of Alden's research interests mentioned above.

2. The Anomalous Sign Change in Molecules

An anomalous sign change associated with degenerate Born-Oppenheimer electronic wave functions in molecules was first noticed by Longuet-Higgins[8,9] in 1958 and later in a more general treatment by Herzberg and Longuet-Higgins[10] in 1963. They showed that in the neigborhood of a conical intersection (i.e., electronic degeneracy, see Fig. 1), the electronic Hamiltonian can be expanded in terms of a 2x2 matrix of the form

$$h(\mathrm{x}) = \begin{pmatrix} -u & v \\ v & u \end{pmatrix} = r \begin{pmatrix} -\cos\theta & \sin\theta \\ \sin\theta & \cos\theta \end{pmatrix} \tag{1}$$

where the Cartesian coordinates (u, v) and corresponding polar coordinates (r, θ) denote the displacement from the degeneracy point (origin) and are defined in Fig. 1. The electronic eigenvectors for the Hamiltonian in Eq. 1 are given by

$$|+\rangle = \begin{pmatrix} \sin\frac{\theta}{2} \\ \cos\frac{\theta}{2} \end{pmatrix} \quad , \quad |-\rangle = \begin{pmatrix} \cos\frac{\theta}{2} \\ -\sin\frac{\theta}{2} \end{pmatrix} . \tag{2}$$

The eigenvectors in Eq. 2 are obviously double-valued functions of θ. Thus, these eigenvectors change sign for any curve C which encircles the conical intersection (i.e., the origin in Fig. 1). The sign change or double-valuedness of the electronic wave functions was correctly included in early treatments of Jahn-Teller vibronic systems using a 2x2 matrix approach[8,9,11] and a single electronic potential energy surface.[12] However, this effect was not

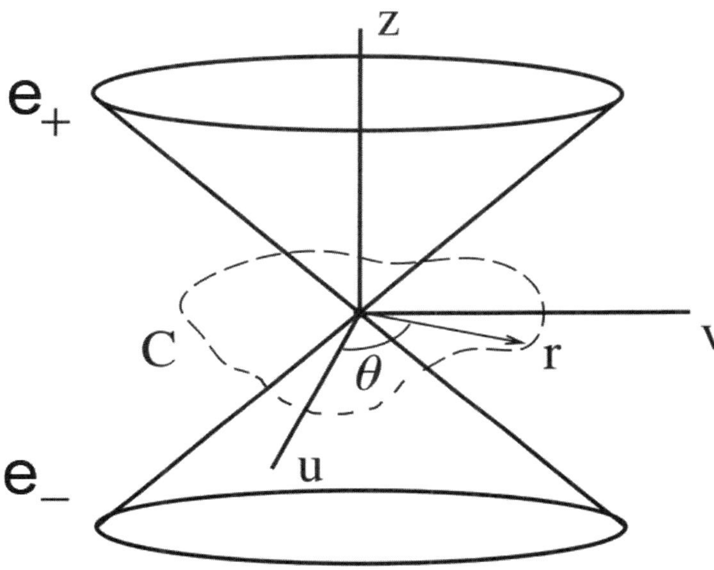

Fig. 1. Conical intersection between two Born-Oppenheimer electronic states with energies e_+ and e_- which become degenerate at the origin ($u = 0$, $v = 0$). The dashed curve denotes a closed path C in the uv-plane which encircles the origin.

widely known in the mid-1970's within the physics and chemistry communities. For example, it was neglected in early quantum reactive scattering calculations for $H + H_2$.[13] At that time, it was not clear how to include the effects of this sign change and make it consistent with the identical particle permutation symmetry in a scattering calculation on a **single** (adiabatic) Born-Oppenheimer electronic potential energy surface.

3. The Gauge Theory of Molecules

The anamolous sign change problem was solved in 1979 by Mead and Truhlar[14] in their paper entitled: *On the determination of Born-Oppenheimer nuclear motion wave functions including complications due to conical intersections and identical nuclei.* In this paper, Mead and Truhlar uncovered the underlying $U(1)$ gauge group symmetry associated with the anomalous sign change in molecular electronic wave functions. They showed that a consistent treatment of both the sign change and identical particle permutation symmetry is possible by transforming the real double-valued Born-Oppenheimer electronic eigenvector (Eq. 2) into a *complex* single-valued eigenvector $|-\rangle_C$ via the transformation

$$|-\rangle_C = \exp\big[i\,l\,\theta/2\big]\,|-\rangle \tag{3}$$

where l is an odd integer and θ is a phase angle which changes by $2\,\pi$ for any path which encircles the conical intersection (see Fig. 1). However, the resulting Hamiltonian for the nuclear motion now contains a "magnetic interaction" term given by $\mathbf{p} \to \mathbf{p} - \mathbf{a}$ where $\mathbf{a}$ is a non-trivial $U(1)$ gauge (vector) potential. Quoting Mead and Truhlar:[14] *"We see that the extra terms behave mathematically exactly as if there were a magnetic interaction between the nuclei, with vector potential proportional to* **a**. *Moreover, since* $\mathbf{a} = -\frac{3}{2}\nabla\theta$*, the pseudomagnetic field* $\mathbf{h} = \nabla \times \mathbf{a} = 0$ *everywhere except along the curve of intersection of the two potential surfaces.... The pseudomagnetic field* **h***, therefore, points ... along the intersection curve, and is zero excecpt along that curve, where it has a delta-function singularity"*. For a triatomic molecule there are three internal (nuclear) degrees of freedom, three rotational degrees of freedom, and three center of mass degrees of freedom. The "non-crossing" rule for this case (no spin) specifies **two** constraints ($u = 0, v = 0$) on the three internal degrees of freedom which must be satisfied in order for two electronic potential energy surfaces to become degenerate.[4] Thus, the degeneracy subspace $\mathcal{D}$ is a one-dimensional curve imbedded in the the three-dimensional internal space $\mathcal{M}$, and the effective "magnetic field" **h** points in the direction along the intersection curve $\mathcal{D}$ (see Fig. 2).

In his first 1980 paper entitled *The Molecular Aharonov-Bohm Effect in Bound States*[15] Mead named the effect the *"Molecular Aharonov-Bohm (MAB) Effect"* and showed that it can lead to appreciable effects on the vibrational energies in molecules: (1) it reverses the symmetry of the lowest vibrational state $E \leftrightarrow A$, (2) it leads to significant shifts and reordering of the higher-lying vibrational energy levels, and (3) it must be included in order to correctly predict the symmetry allowed states consistent with Bose/Fermi statistics. Mead also verified that gauge invariance was satisfied by the numerical calculations (see Fig. 3). The MAB effect in molecular spectra has since been confirmed experimentally in several molecules: Cu_3,[16–21] Li_3,[22,23] and Na_3.[24] Recent theoretical calculations for N_3 have reported the largest MAB effect in molecular vibrational spectra to date.[25]

In his second 1980 paper entitled *Electronic spin-orbit interaction and*

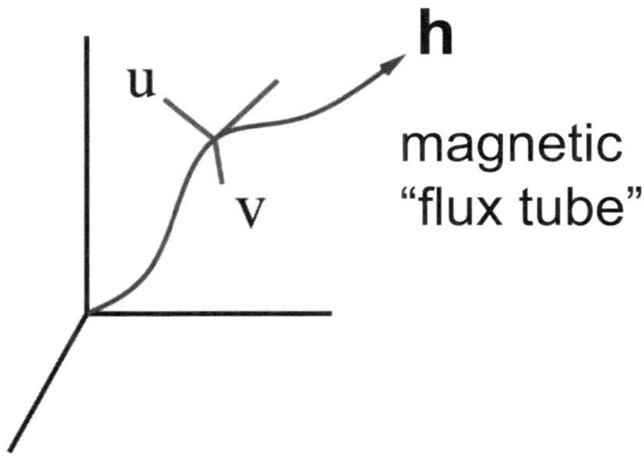

Fig. 2. The one-dimensional degeneracy subpsace $\mathcal{D}$ (determined by the contraints $u = 0$, $v = 0$) is imbedded in the internal three-dimensional space $\mathcal{M}$. The non-trivial $U(1)$ gauge field $\mathbf{h}$ is mathematically equivalent to a "magnetic flux tube" (i.e., it is non-zero only on $\mathcal{D}$ and points along the direction of $\mathcal{D}$).

the Molecular Aharonov-Bohm Effect[26] Mead showed that by including spin-orbit interaction with half-odd integer spin (Kramer's doublets), the electronic degeneracy is lifted but that the MAB effect is qualitatively the same in regions far from the region where the curves now nearly intersect. Quoting Mead: *"Instead of a vector potential corresponding to a delta-function pseudo-magnetic field, the field is now spread out over a finite region where the curves nearly intersect."* and *"In all cases, it is shown that the results of earlier studies in which the spin-orbit interaction was neglected are not qualitatively altered."*

In his third 1980 paper entitled *"Superposition of reactive and nonreactive scattering amplitudes in the presence of a conical intersection"*[27] Mead showed that the MAB effect can also significantly alter scattering cross sections. Quoting Mead: *"The conclusion is that the interference between reactive and nonreactive scattering amplitudes for identical nuclei is exactly opposite to what it would be without consideration of the conical intersection [MAB effect]."* For example, the MAB effect changes the sign of the interference term in the cross section calculations for the ortho-ortho and para-para transitions in $H + H_2$ scattering. Recent calculations have shown that this leads to significant "phase shifts" in the differential cross sections when plotted as a function of scattering angle.[28] Due to experimental difficulties, these cross sections have not yet been measured experimentally, but

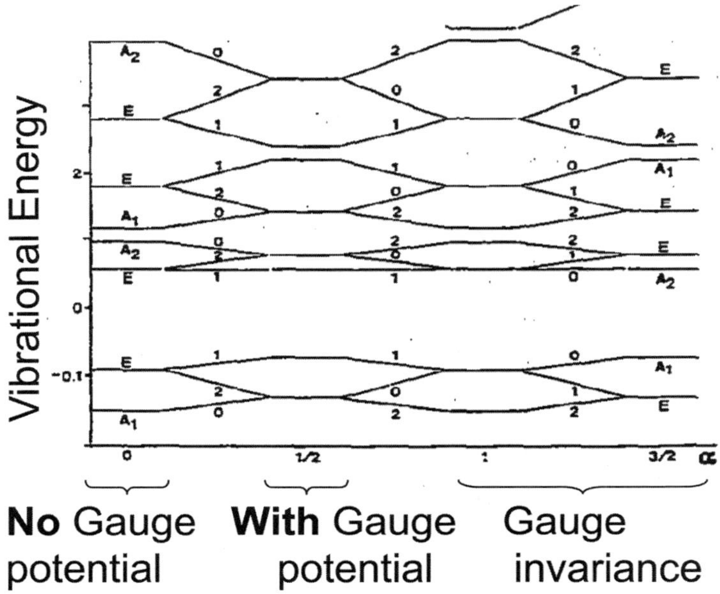

Fig. 3. The vibrational spectra for Li_3 computed by Mead[15] including the MAB effect (i.e., the results "**With** Gauge potential"). *Reproduced with permission (note: the x and y labels were added to aid discussion).*

when they are measured, it is fully expected that the experimental results will confirm Mead's prediction.

The non-abelian case was also studied by Mead and Truhlar[29] in their 1982 paper entitled *Conditions for the definition of strictly diabatic electronic basis for molecular systems* and further by Mead[30] in his 1983 paper entitled *Electronic Hamiltonian, wave functions, and energies, and derivative coupling between Born-Oppenheimer states in the vicinity of a conical intersection.* In the non-abelian case more than one Born-Oppenheimer electronic wave function is considered (i.e., an $N > 1$ dimensional electronic subspace). Mead and Truhlar showed that the effective $U(N)$ non-abelian gauge potential is given by the "derivative coupling matrix"

$$A_i^{nm} = i \langle n(x)| \partial_i |m(x)\rangle \tag{4}$$

where $n = 1, \ldots, N$ and $m = 1, \ldots, N$ label the electronic states and the subscript i denotes the internal nuclear coordinate (x_i). The corresponding "magnetic interaction" (i.e., non-abelian gauge field) is given by

$$\mathbf{F}_{ij} = \partial_i \mathbf{A}_j - \partial_j \mathbf{A}_i - i\big[\mathbf{A}_i, \mathbf{A}_j\big] \tag{5}$$

where the last term on the right-hand-side of Eq. 5 is a non-zero matrix commutator. The non-abelian gauge field in Eq. 5 can be written in the following form[31,32]

$$\mathbf{F} = \mathrm{d}\mathbf{W}^{\dagger} \wedge (1 - \Lambda_{\mathcal{H}_N})\,\mathrm{d}\mathbf{W} \tag{6}$$

where $\Lambda_{\mathcal{H}_N}$ denotes the projection operator onto the finite N-dimensional electronic subspace, d is the exterior derivative, and $\mathbf{W}$ is a complex matrix satisfying $\mathbf{W}^{\dagger}\,\mathbf{W} = \mathbf{I}$ and $\mathbf{W}\,\mathbf{W}^{\dagger} = \Lambda_{\mathcal{H}_N}$.[32] Mead and Truhlar[29] showed that if the electronic subspace is complete (i.e., $\Lambda_{\mathcal{H}_N} = 1$), then $\mathbf{F} = 0$ and therefore the non-abelian gauge potential can be removed by a $U(N)$ gauge transformation. The resulting electronic basis is referred to as a "strictly diabatic electronic basis" (i.e., a basis for which $\mathbf{A}' = 0$). However, in most molecular applications, an incomplete finite dimensional electronic basis is used (i.e., $\Lambda_{\mathcal{H}_N} \neq 1$). In this case $\mathbf{F} \neq 0$ and it is impossible to completely remove the non-abelian gauge potential $\mathbf{A}$ by a $U(N)$ gauge transformation. Thus, the truncation to an incomplete finite dimensional electronic subspace induces "curvature". Using perturbation theory, Mead[30] further investigated the behavior of the non-abelian gauge potential and gauge field in the neighborhood of an electronic degeneracy (conical intersection) and explicitly identified the "removable" and "non-removable" components of $\mathbf{A}$. In later work, Mead[33] also investigated the non-abelian "phase factor" associated with a degenerate Kramer's doublet.

The underlying mathematical structure is represented by the diagram in Fig. 4 where the top row denotes the universal $U(N)$ principal fiber fundle

$$\begin{array}{ccccc} U(N) & \xrightarrow{\;i\;} & V(k+1,N) & \xrightarrow{\;\pi\;} & G(k+1,N) \\ & & & & \uparrow f \\ U(N) & \xrightarrow{\;i_f\;} & Q_N & \xrightarrow{\;\pi_f\;} & M \end{array}$$

Fig. 4. The universal $U(N)$ principal fiber bundle over G (top row) and the induced $U(N)$ principal fiber bundle over the nuclear parameter space M (bottom row). The mapping f is determined by the electronic structure of the molecule.

with total space $V(k+1,N)$ (a complex Stiefel manifold) and base space $G(k+1,N)$ (a complex Grassmann manifold).[31,32,34,35] The dimensionality of the full electronic Hilbert space is $k+1$. The bottom row in Fig. 4

denotes the "pullback" or induced bundle over the nuclear parameter space M which is defined by the mapping $f = \Lambda_{\mathcal{H}_N}$ from M to G. In molecular applications, the projection operator $\Lambda_{\mathcal{H}_N}$ is computed from the electronic wave functions of the N dimensional subspace (i.e., it is determined by the electronic structure of a given molecule). Therefore, the non-abelian gauge potential (connection one-form) **A** and gauge field (curvature two-form) **F** given above in Eqs. 4 and 5, respectively, are the pullbacks with respect to f of the universal connection one-form and curvature two-form of the universal $U(N)$ principal fiber bundle over G.[32]

In summary, between 1979 and 1983 Mead published seven papers on the gauge theory of molecules and made experimentally verifiable predictions of the MAB effect in both molecular spectra and scattering. This was *before* the 1984 paper by Michael Berry.[36] In fact, Berry became aware of Mead's work while drafting his 1984 paper and giving a seminar on the topic.[37] Berry recognized the generality of Mead's work. Quoting Berry:[36] *"Mead (1980a,b) employs the term 'molecular Aharonov-Bohm effect' in a different context, to describe how degeneracies in electronic energy levels affect the spectrum of nuclear vibrations..... and this introduces a vector potential into the Schrödinger equation for the nuclear motion. In general one may expect such effects whenever an isolated system is considered as being divided into two interacting parts, each slaved to a different aspect of the other .... The systems considered in this [Berry's] paper might be regarded as a special case, ...".* Other researchers also recognized the originality and general implications of Mead's work. Quoting Moody, Shapere, and Wilczek:[38] *"To compensate for these and other more general, phase ambiguities and thus to make the Born-Oppenheimer wave functions single valued, Mead[6] suggested adding a 'vector-potential-type term' to the nuclear Hamiltonian. He refers to the effects of this term on the nuclear motion as the molecular Bohm-Aharonov effect."* In 1992, Mead published a comprehensive review article in Reviews of Modern Physics entitled: *The geometric phase in molecular systems*[39] which has been cited as "one of the pillars of the contemporary theory of nonadiabatic chemistry".[40] Indeed, the gauge theory of molecules represents a signficant generalization of the Born-Oppenheimer method of molecules and is sometimes also referred to as "non-Born-Oppenheimer theory" or "generalized Born-Oppenheimer theory". The well known Born-Oppenheimer method has been the bedrock theory of molecules which chemists and physicists have relied on for 85 years.[41] Yet, from Mead's pioneering work we know that it represents only a special case within the more general gauge theory of molecules. The appearance of gauge potentials in molecules is more the rule than the exception, and their treatment remains an exciting challenge rich with opportunities for new discoveries and applications.

Mead's work on the gauge theory of molecules has received praise from numerous experts in the field, such as: *"In the past three decades, nobody has delved deeper into the fundamentals of electron-nuclear couplings [gauge potentials] in polyatomic molecules than Alden Mead."* and *"Due to their fundamental nature, Alden's contributions to the theory of non-Born-Oppenheimer dynamics in molecules will be of lasting importance."*.[40] The gauge theory of moleules has also expanded into many other fields, and is actively growing today. For example, the phonon-electronic coupling in condensed matter physics gives rise to gauge potential interactions which are mathematically identical to Mead's molecular gauge potentials. The analogue of a molecular conical intersection, the potential energy surfaces associated with a topological insulator, was featured recently on the cover of Physics Today.[42] Recently, magnetic solenoid type gauge potentials have been "induced" experimentally to "punch holes" in a Bose-Einstein Condensate of ^{87}Rb.[43] Related applications include topological phases of matter which may form the basis of a fault tolerant quantum computer.[44] The gauge theory of molecules has also been featured in text books[45] and continues to be an active area of research within the physical chemistry community. A recent international conference on this topic was held in Tsukuba, Japan entitled: *XXIst International Symposium on the Jahn-Teller Effect*, August 26-31, 2012, which featured a special session on the role of gauge potentials in molecules and solids including important applications such as high-T_c superconductors.

The gauge theory of molecules represents the first realization of a much more general gauge field framework which appears anytime a system can be decomposed into "fast" and "slow" degrees of freedom. The couplings between these degrees of freedom give rise to non-trival gauge potentials which in turn alter the dynamics of the overall system. Mead's gauge theory of molecules will continue to influence both chemists and physicists for generations to come. Congratulations Alden Mead!

References

1. C. A. Mead, *Phys. Rev.* **135**, B849 (1964).
2. C. A. Mead, *Phys. Rev.* **143**, 990 (1966).
3. C. A. Mead, *Permutation Groups, Symmetry and Chirality in Molecules* in the Springer series "Topics in Theoretical Chemistry" **49** (1974).
4. C. A. Mead, *J. Chem. Phys.* **70**, 2276 (1979).

5. E. P. Wigner, *Rev. Mod. Phys.* **29**, 255 (1957); *Phys. Rev.* **120**, 643 (1960).
6. E. Wigner, *Gruppentheorie und ihre Anwendung auf die Quantenmechanik der Atomspektren*, Braunschweig: Friedr. Vieweg and Sohn (1931). English translation: *Group Theory and its Application to the Quantum Mechanics of Atomic Spectra.* New York: Academic Press (1959).
7. J. von Neumann and E. Wigner, *Phys. Z.* **30**, 467 (1929).
8. H. C. Longuet-Higgins, U. Öpik, M. H. L. Pryce, and R. A. Sack, *Proc. R. Soc. London, Ser. A* **244**, 1 (1958).
9. H. C. Longuet-Higgins, *Adv. Spectrosc.* **2**, 429 (1961).
10. G. Herzberg and H. C. Longuet-Higgins, *Discuss. Faraday Soc.* **35**, 77 (1963).
11. R. Englman, *The Jahn-Teller Effect in Molecules and Crystals* (Wiley-Interscience, New York, 1972).
12. M. C. M. O'Brien, Proc. R. Soc. London Ser. A **281**, 323 (1964).
13. G. C. Schatz and A. Kuppermann, *J. Chem. Phys.* **65**, 464 (1976); **65** 466 (1976).
14. C. A. Mead and D. G. Truhlar, *J. Chem. Phys.* **70**, 2284 (1979); **78**, 6344 (1983).
15. C. A. Mead, *Chem. Phys.* **49**, 23 (1980).
16. M. D. Morse, J. B. Hopkins, P. R. R. Langridge-Smith, and R. E. Smalley, *J. Chem. Phys.* **79**, 5316 (1983).
17. D. P. DiLella, K. V. Taylor, and M. Moskovits, *J. Phys. Chem.* **87**, 524 (1983).
18. E. A. Rohlfing and J. J. Valentini, *Chem. Phys. Lett.* **126**, 113 (1986).
19. T. C. Thompson and C. A. Mead, *J. Chem. Phys.* **82**, 2408 (1985).
20. T. C. Thompson, D. G. Truhlar, and C. A. Mead, *J. Chem. Phys.* **82**, 2392 (1985).
21. D. G. Truhlar, T. C. Thompson, and C. A. Mead, *Chem. Phys. Lett.* **127**, 287 (1986).
22. J. P. Wolf, G. Delacretaz, and L. Woste, *Phys. Rev. Lett.* **63**, 1946 (1989).
23. M. Keil, H.-G. Krämer, A. Kudell, M. A. Baig, J. Zhu, W. Demtröder, and W. Meyer, *J. Chem. Phys.* **113**, 7414 (2000).
24. H. von Busch, V. Dev, H. A. Eckel, S. Kasahara, J. Wang, W. Demtröder, P. Sebald and W. Meyer, *Phys. Rev. Lett.* **81** 4584 (1998).
25. D. Babikov, B. K. Kendrick, P. Zhang, and K. Morokuma, *J. Chem. Phys.* **122**, 044315 (2005).
26. C. A. Mead, *Chem. Phys.* **49**, 33 (1980).
27. C. A. Mead, *J. Chem. Phys.* **72**, 3839 (1980).
28. F. Bouakline, B. Lepetit, S. C. Althorpe, and A. Kuppermann, in Springer Series in Chemical Physics **97**, 201 (2009).
29. C. A. Mead and D. G. Truhlar, *J. Chem. Phys.* **77**, 6090 (1982).
30. C. A. Mead, *J. Chem. Phys.* **78**, 807 (1983).
31. T. Eguchi, P. R. Gilkey, and A. J. Hanson, *Phys. Rep.* **66**, 213 (1980).
32. M. S. Narashimhan and S. Ramanan, *Am. J. Math.* **83**, 563 (1961); **85**, 223 (1963).
33. C. A. Mead, *Phys. Rev. Lett.* **59**, 161 (1987).

34. C. Nash and S. Sen, *Topology and Geometry for Physicists* (Academic Press, San Diego, CA, 1983).
35. N. Steenrod, *The Topology of Fiber Bundles* (Princeton U.P., 1970).
36. M. V. Berry, *Proc. R. Soc. London Ser. A* **392**, 45 (1984).
37. M. V. Berry, in *Geometric Phases in Physics*, edited by A. Shapere and F. Wilczek (World Scientific, Singapore), p. 7 (1989).
38. J. Moody, A. Shapere, and F. Wilczek, *Phys. Rev. Lett.* **56**, 893 (1986).
39. C. A. Mead, *Rev. Mod. Phys* **64**, 51 (1992).
40. Quotes from a nomination letter.
41. M. Born and R. Oppenheimer, Ann. Phys. (Leipzig) **84**, 457 (1927).
42. X. L. Qi and S. C. Zhang, in Physics Today Vol. **63**, No. **1**, p. 33 (2010).
43. Y. J. Lin, R. L. Compton, K. Jiménez-Garci, J. V. Porto, and I. B. Spielman, *Nature* **462** 628 (2009).
44. C. Nayak, *Nature* **464** 693 (2010).
45. A. Bohm, *Quantum Mechanics: Foundations and Applications*, 3rd Edition, (Springer-Verlag, New York), Chapters XII and XIII, (1993), (2001).

Ceremonies for Wigner Medal and Weyl Prize

Ceremonies for Wigner Medal and Weyl Prize

Wigner Medal and Weyl Prize

WIGNER MEDAL ACCEPTANCE

C. ALDEN MEAD[†*]

92 Bartram Road
Savannah, GA 31411, U.S.A
[†] *Professor Emeritus of Chemistry, University of Minnesota*
[*] *E-mail: aldykar@att.net*

In accepting this award, I feel very honored, and also very humble, because I had no idea that anyone would consider me to be worthy of it. I offer my heartfelt thanks to the foundation and to the organizers of this conference for the chance to be here tonight. I would also like to thank Brian Kendrick for that very kind and flattering speech. The St. Louis Cardinals logo was also a nice touch, and much appreciated.

The award is all the more pleasing because of being named after Eugene Wigner, who has always been an idol of mine, and whose book 'Gruppentheorie' was a bible to me and to many theoreticians of my generation. I would like to quote from the introduction to this book, written by Wigner in 1931:

This book emerges from the wish to make the applications of group theoretical methods in quantum mechanics available to a larger audience. The actual solution of the quantum mechanical differential equations encounters such great difficulties that at most only a rough approximation can be obtained. Thus it is even more satisfying that such a large part of the quantum mechanical results can be obtained by pure symmetry considerations. One has often raised the objection to the group theoretical treatment of the Schrödinger equation that it is 'not physical'. But it seems to me that the conscious application of basic symmetry properties corresponds more to one's physical intuition than does the more purely computational method. This objection may be traceable to the fact that group theory has a significantly different character from the mathematical methods most familiar to physicists, so that it requires some time before one feels comfortable with it.

When Wigner wrote this, the computer age was not even in its infancy, but even now it rings largely true: Even though much better and more exact solutions of the equations are available now than then, great insight can still be achieved through the use of group theory.

My involvement with group theory entered another chapter in about 1970, when, through my contact with Prof. Ernst Ruch of Berlin, I became interested in the properties of the permutation groups and their representations in terms of Young diagrams and Young operators.

Another important property of the Young operators is that without them I would never have met my wife Karin, whom I met in Berlin in 1971, on sabbatical leave with Prof. Ruch's group. She is here tonight, along with our daughter, son-in-law, and granddaughter.

My work on the gauge theory of molecules has been by no means a single-handed effort. The first paper on the subject that I was involved in was a collaboration with my very distinguished colleague Prof. Don Truhlar of the University of Minnesota, without whom my involvement in this field almost certainly would not have taken place. In subsequent papers, there has been important collaboration with Prof. Truhlar, also with Sean Keating, Todd Thompson, Maynard Brandt, and, last but far from least, Brian Kendrick, who is here tonight.

In summary, whatever I have achieved is owing in great part to the help of others, with whom I have been most fortunate to be associated.f Without all these, I would not be standing here tonight.

Again, I wish to express how highly honored I feel, and to express most sincere thanks to the foundation and to the organizers of this conference.

THE HERMANN WEYL PRIZE

21 AUGUST 2012

HERMANN WEYL PRIZE

The Hermann Weyl Prize was created in 2000 by the Standing Committee of the International Group Theory Colloquium. The purpose of the Weyl Prize is to provide recognition for young scientists (younger than 35 years of age) who have performed original work of significant scientific quality in the area of understanding physics through symmetries. The Hermann Weyl prize consists of a certificate citing the accomplishments of the recipient, prize money and an allowance towards attendance at the bi-annual International Group Theory Colloquium at which the award is presented. Candidates are nominated by or through members of the Standing Committee.

H. WEYL PRIZE AWARDEES

2002 Edward Frenkel
2004 Nikita A. Nekrasov
2006 Boyko Bakalov
2008 Mohammad M. Sheikh-Jabbari
2010 Giulio Chiribella
2012 Razvan Gurau

PRESENTATION OF THE 2012 WEYL PRIZE AWARDEE: RAZVAN GURAU

E. CORRIGAN
Chairman of the Hermann Weyl Prize Committee

The Hermann Weyl Prize Committee received four very strong nominations this year and after carefully considering the nominations and supporting papers it decided to award the Prize to Razvan Gurau.

Razvan Gurau was born in Romania in 1980, began his university education in Bucharest and continued it in Paris, first at the École Normale Supérieure and afterwards at Paris-Sud XI, where he worked for his PhD under the supervision of Vincent Rivasseau. Since 2008 he has been a postdoctoral researcher at the Perimeter Institute. His talent and exceptional potential were evident from his student days and he has subsequently written or co-authored many papers on a variety of topics in the general area of quantum field theory. In 2009 he discovered and has subsequently developed the theory of coloured random tensors, which generalises beyond two dimensions the theory of random matrices. This unexpected discovery has opened up new avenues for investigation, with potential applications in probability theory, statistical mechanics, integrable systems, quantum field theory and random discrete geometries. It has led already to significant new steps including a novel generalisation of the Virasoro algebra beyond two dimensions and the first exact calculation of exponents in three dimensional critical statistical mechanics.

For these reasons it is an honour and great pleasure to introduce Razvan Gurau to you.

Ceremonies for Wigner Medal and

PLENARY SESSION

20 & 21 & 23 & 24 AUGUST 2012

THE SURPRISING CONNECTION BETWEEN EXACTLY SOLVED LATTICE MODELS AND DISCRETE HOLOMORPHICITY

MURRAY T. BATCHELOR

Mathematical Sciences Institute, and
Department of Theoretical Physics, Research School of Physics and Engineering,
The Australian National University,
Canberra, ACT 0200, Australia
E-mail: Murray.Batchelor@anu.edu.au

Over the past few years it has been discovered that an "observable" can be set up on the lattice which obeys the discrete Cauchy-Riemann equations. The ensuing condition of discrete holomorphicity leads to a system of linear equations which can be solved to yield the Boltzmann weights of the underlying lattice model. Surprisingly, these are the well known Boltzmann weights which satisfy the star-triangle or Yang-Baxter equations at criticality. This connection has been observed for a number of exactly solved models. I briefly review these developments and discuss how this connection can be made explicit in the context of the Z_N model. I also discuss how discrete holomorphicity has been used in recent breakthroughs in the rigorous proof of some key results in the theory of planar self-avoiding walks.

Keywords: statistical mechanics; exactly solved models, integrability; discrete holomorphicity; self-avoiding walks; polymer adsorption transition.

1. Introduction

The study of exactly solved lattice models in statistical mechanics has been a richly rewarding area of activity over many decades. Among the pantheon of exactly solved models is the two-dimensional Ising model, which is fundamental to the theory of phase transitions and critical phenomena.[1–3] Two-dimensional lattice models with local interactions like the Ising model possess remarkable properties at criticality. The fluctuations in such systems are conformally invariant at the critical point. One of the triumphs of statistical physics is that this property entails the classification of universality classes of second order phase transitions in terms of conformal field theories.[3] Roughly speaking, there is an exactly solved model for each universality class. The beauty of exactly solved models is that their critical Boltzmann weights satisfy the star-triangle or Yang-Baxter equation.[1–3]

A remarkable link has been uncovered between Yang-Baxter integrability and the condition of discrete holomorphicity.[4–6] This involves the establishment of a local identity which is satisfied by an operator defined on the lattice. This operator is an example of a discrete parafermionic observable appearing in conformal field theory. The observable obeys a lattice version of the Cauchy-Riemann equations and is thus known to be discretely holomorphic. The condition of discrete holomorphicity leads to a system of linear equations which can be solved to yield the Boltzmann weights of the lattice model. These are the Boltzmann weights which are known to satisfy the Yang-Baxter equations at criticality. This connection has been observed for a number of exactly solved models,[4–6] including the Z_N model,[4] the Potts model[5] and the O(n) loop model.[5]

Here we will discuss these developments in terms of the O(n) loop model.

2. The O(n) loop model

The O(n) loop model on the square lattice was originally constructed by Nienhuis.[7,8] It is exactly solved by means of the Bethe Ansatz, with periodic[9] and open[10,11] boundaries. As an integrable model, its underlying R-matrix is that of the Izergin-Korepin,[12] or $A_2^{(2)}$ vertex model.[10] The partition function is given by

$$Z_{\text{loop}} = \sum_{\mathcal{G}} \rho_1^{m_1} \ldots \rho_9^{m_9}\, n^P \tag{1}$$

which is a sum over all possible loop configurations $\mathcal{G}$. Here m_i is the number of occurrences of Boltzmann weight ρ_i and P is the total number of closed loops of fugacity n in any given configuration. The nine non-zero Boltzmann weights ρ_i are shown in Fig. 1. These weights are given by[7,10]

$$\begin{aligned}
\rho_1 &= \sin(3\lambda - u)\sin u + \sin 2\lambda \sin 3\lambda \\
\rho_2 &= \rho_3 = \epsilon_1 \sin(3\lambda - u)\sin 2\lambda \\
\rho_4 &= \rho_5 = \epsilon_2 \sin u \sin 2\lambda \\
\rho_6 &= \rho_7 = \sin(3\lambda - u)\sin u \\
\rho_8 &= \sin(3\lambda - u)\sin(2\lambda - u) \\
\rho_9 &= -\sin(\lambda - u)\sin u
\end{aligned} \tag{2}$$

with $\epsilon_1^2 = \epsilon_2^2 = 1$, $0 < u < 3\lambda$ and $n = -2\cos 4\lambda$.

Fig. 1. The Boltzmann weights of the O(n) loop model.

The parafermion operator is defined on the lattice as[5]

$$F(z) = \sum_{\mathcal{G}\in\Gamma(0,z)} P(\mathcal{G}) e^{-i\sigma W(z)}. \tag{3}$$

Here $P(\mathcal{G})$ is the probability of configuration $\mathcal{G}$, $\Gamma(0,z)$ is a set of loop configurations for which points 0 and z belong to the same loop, $W(z)$ is the winding angle of this loop from points 0 and z, and σ is the spin of the parafermion $F(z)$. For the O(n) loop model

$$F(z) = \sum_{\gamma(a\to z)\in\Omega} e^{-i\sigma W(\gamma(a\to z))} \rho_1^{m_1} \dots \rho_9^{m_9} n^P \tag{4}$$

for some path $\gamma(a \to z)$ across a domain Ω.

What Ikhlef and Cardy[5] have shown is that the parafermion $F(z)$ is discretely holomorphic, i.e., it satisfies

$$F(p) - e^{i\theta} F(q) - F(r) + e^{i\theta} F(s) = 0 \tag{5}$$

on each vertex as defined in Fig. 2. This condition is equivalent to requiring that F is divergence and curl free.

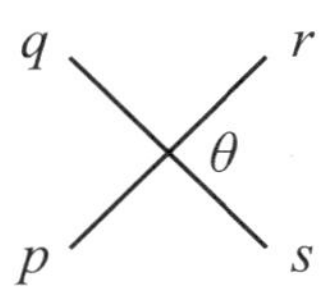

Fig. 2. The labelling of a vertex.

To establish this, one chooses a particular mid edge p. There are four possible different external connectivities, as shown in Fig. 3. These possibilities result in four equations

$$
\begin{aligned}
\rho_1 + \nu\rho_2 - \nu\zeta^{-1}\rho_4 - \rho_7 &= 0 \\
-\zeta^{-1}\rho_2 + n\rho_5 + \nu\zeta\rho_7 - \nu\zeta^{-1}(\rho_8 + n\,\rho_9) &= 0 \\
n\,\rho_3 - \zeta\rho_4 - \nu\zeta^{-2}\rho_7 + \nu(n\,\rho_8 + \rho_9) &= 0 \\
-\nu\zeta^{-2}\rho_2 + \nu\zeta\rho_4 + n\,\rho_6 - \zeta^{-2}\rho_8 - \zeta^2\rho_9 &= 0
\end{aligned}
\tag{6}
$$

where $\nu = \mathrm{e}^{\mathrm{i}\theta(\sigma+1)}$ and $\zeta = \mathrm{e}^{\mathrm{i}\sigma\pi}$. The remarkable point is that Ikhlef and Cardy[5] were able to solve this set of linear equations to yield the Boltzmann weights $\rho_1, \ldots, \rho_9$ and the conformal spin

$$
\sigma = 1 - \frac{3\lambda}{\pi}
\tag{7}
$$

provided $u = \sigma(\theta - \pi) + \theta$. In particular, at $\theta = \frac{\pi}{2}$, $u = 3\lambda/2$, the isotropic point. Substitution of the Boltzmann weights (2) into equations (6) shows that these equations are indeed satisfied.

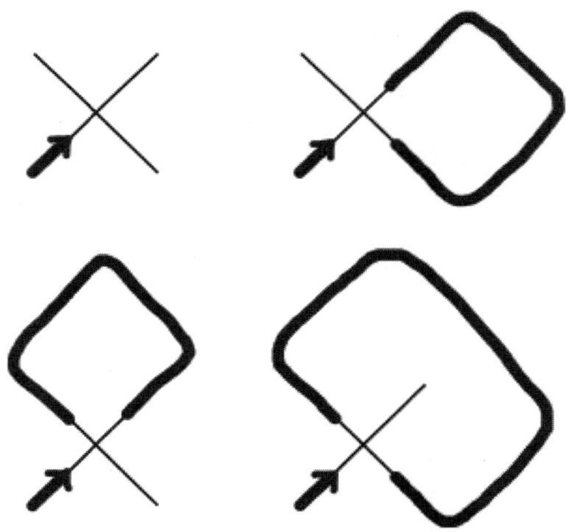

Fig. 3. The four possible external connectivities from mid edge p.

The surprising point is that the Yang-Baxter equations are *cubic* in the Boltzmann weights, whereas imposing discrete holomorphicity led to a set of equations which are *linear* in the Boltzmann weights. There thus appears to be a simpler route to uncovering integrable lattice models at criticality.

Why is this? And what is the precise link between discrete holomorphicity and integrability? These questions will be taken up further in Section 4 in the context of the Z_N model, which can be discussed from an algebraic point of view.

We now turn to the related developments in the theory of self-avoiding random walks.

3. Self-avoiding random walks

The random walk, also known as the drunkard's walk, was introduced over 100 years ago.[13] The study of random walks has since developed into the central pillar of the theory of probability and random processes.[14] A key feature of the random walk is that, like the drunkard, it has no memory of where it has been. This absence of memory is the defining property of a Markovian process.

A more challenging class of mathematical problems arise when a random walk is not allowed to revisit a site it has already visited. Such self-avoiding random walks were originally proposed as models of long polymer chains in a good solvent. Despite the simplicity of their definition, the non-Markovian nature of self-avoiding random walks makes them notoriously difficult to study. Nevertheless much is known about the theory of self-avoiding walks, particularly in higher dimensions.[15] Much is also known in two and three dimensions from extensive computer enumeration. The self-avoiding constraint is more restrictive in two dimensions and thus self-avoiding walks on planar lattices are arguably more interesting to study.

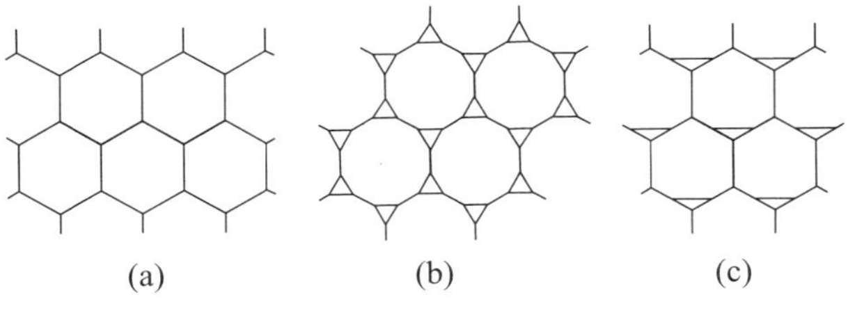

Fig. 4. The (a) honeycomb, (b) 3-12 and (c) martini lattices.

In 1982 Nienhuis[16] was able to obtain some key information for self-avoiding walks by considering the O(n) loop model on the honeycomb (hexagonal) lattice depicted in Fig. 4(a). In general, the O(n) model has a connection to self-avoiding walks in the limit $n = 0$.[17] The critical Boltzmann weights of the model on the honeycomb lattice were obtained from an exact mapping to the Potts model.[16] The Boltzmann weights defining the critical line of the O(n) loop model on the honeycomb lattice also follow as a condition for the model to be solved exactly in terms of the Bethe Ansatz.[18–20] More generally the O(n) loop model on the honeycomb lattice follows as a limiting case of the exactly solved O(n) loop model on the square lattice.[a]

The critical Boltzmann weights of the model give the value for the connective constant μ, which is a measure of the growth in the number of allowed self-avoiding walks with increasing length. The Nienhuis result,

$$\mu = \sqrt{2+\sqrt{2}} = 1.847\,759\ldots \tag{8}$$

was generally believed to be exact, having been confirmed to high precision from extensive enumeration of self-avoiding walks on the honeycomb lattice. No such exact results are known for the other regular planar lattices.

Similar exact results are known for self-avoiding walks on related semi-regular lattices. These are the 3-12 lattice (Fig. 4(b)) and the martini lattice (Fig. 4(c)). For the 3-12 lattice, the exact value[21,22] $\mu = 1.711\,041\ldots$ follows from the solution of

$$\frac{1}{\mu^2} + \frac{1}{\mu^3} = \frac{1}{\sqrt{2+\sqrt{2}}}. \tag{9}$$

And for the martini lattice, the exact value[23] $\mu = 1.750\,564\ldots$ follows from the solution of

$$\frac{1}{\mu^3} + \frac{1}{\mu^4} = \frac{1}{\sqrt{2+\sqrt{2}}}. \tag{10}$$

More generally, the O(n) model on the honeycomb lattice can be mapped to the O(n) model on the 3-12[22] and martini lattices.[23]

Up until recently, a rigorous mathematical proof of the result (8) had seemed out of reach. In a remarkable paper, Duminil-Copin and Smirnov[24] have given an ingenious proof of the Nienhuis result. The first part of the proof involves defining a parafermionic observable on the honeycomb lattice which is shown to be discretely holomorphic. The Boltzmann weights of the model follow from the condition of discretely holomorphicity. In this sense the value (8) also appears as a consequence of discrete holomorphicity. The second part of the proof builds on previously hard won results in the general

[a]Specifically, when either $u = \lambda$ or $u = 2\lambda$ in (2), corresponding to the two orientations in Fig. 5.

theory of self-avoiding walks to rigorously establish that this value is indeed the connective constant.

3.1. *The polymer adsorption transition*

These arguments have been extended[25,26] to provide a rigorous proof of the critical surface fugacity y_c of self-avoiding walks on the honeycomb lattice in the presence of a boundary. This is the critical adsorption temperature at which the polymer becomes attached to the surface. For $y > y_c$ the polymer is adsorbed and for $y < y_c$ the polymer is desorbed. The critical value depends on the orientation of the lattice (see Fig. 5).

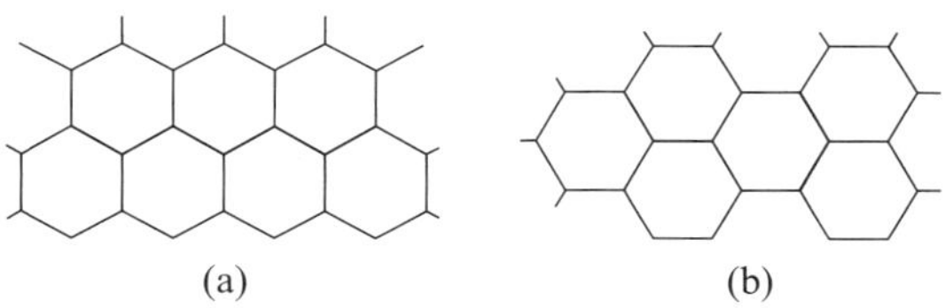

Fig. 5. The two orientations of the honeycomb lattice.

For the orientation shown in Fig. 2(a), the exact value[27]

$$y_c = 1 + \sqrt{2} = 2.414\ldots \tag{11}$$

had been obtained earlier using the underlying integrability of the $O(n)$ loop model with a boundary.[10] For the orientation shown in Fig. 2(b), the exact value[28]

$$y_c = \sqrt{\frac{2+\sqrt{2}}{1+\sqrt{2}-\sqrt{2+\sqrt{2}}}} = 2.455\ldots \tag{12}$$

also followed from the integrable boundary weights of the $O(n)$ loop model.[10]

Building on the approach of Duminil-Copin and Smirnov, these results have been shown by Beaton *et al.*[25,26] to also follow from the discrete holomorphicity of a parafermionic observable. Their mathematical proof that these values are indeed the critical adsorption temperatures is an impressive *tour de force*, involving the adaptation of previous rigorous work on self-avoiding walks and the rigorous treatment of self-avoiding bridges in a finite strip.

We turn now to further examine the connection between discrete holomorphicity and integrability.

4. The Z_N model

The Z_N model is an N-state model in which the spins s_r at vertex r take values from the Nth roots of unity, $s_r = \omega^{q_r}$, with $q_r \in \{0, 1, \ldots, N-1\}$ and $\omega = \exp(2\pi \mathrm{i}/N)$. The model contains a number of well known models as special cases, namely the Ising model ($N = 2$), the three-state Potts model ($N = 3$) and the Ashkin-Teller model ($N = 4$). The Z_N model has a well known set of critical points obtained by Fateev and Zamolodchikov (FZ) as solutions to the star-triangle relations.[29] By defining lattice parafermions as products of neighbouring order and disorder variables with a suitable phase factor, it was shown by Rajapour and Cardy that the FZ Boltzmann weights ensure that the lattice parafermions obey the discrete version of the Cauchy-Riemann equations, i.e., they enjoy the property of discrete holomorphicity.[4,6] These results were extended to lattices with anisotropic interactions.

The underlying motivation for this work on the Z_N model is the construction of discretely holomorphic parafermions, which are expected to become the holomorphic parafermions of the conformal theory. Here our interest is in the connection with integrability. From that perspective, the key finding is that the contour sum around each elementary face or rhombi of the lattice vanishes, the key ingredient for which is the set of FZ critical Boltzmann weights. This approach was extended by explicitly considering the condition of discrete holomorphicity on two and three adjacent rhombi.[30] For two rhombi this leads to a quadratic equation for the Boltzmann weights which implies the known inversion relations for the Z_N model. For three rhombi it leads to a cubic equation from which the star-triangle relation follows. The simplicity of the discrete holomorphic approach is that the two- and the three-rhombus equations are equivalent to the one-rhombus equation.

5. Discussion

As alluded to in the previous section, the original motivation behind the recent progress on various aspects of discrete holomorphicity covered in this article was to set up the necessary discrete building blocks for the passage to the continuous scaling limit. There is ample motivation for doing this. For example, the geometry of the curves traced out by self-avoiding walks is conjectured to be described by the mathematical theory of Schram-Loewner evolution ($\mathrm{SLE}_{8/3}$).[31] Basically, if the scaling limit of two-dimensional self-avoiding walks exists and has a certain conformal invariance property, then the scaling limit must be described by the particular stochastic process $SLE_{8/3}$. However, establishing the existence of the scaling limit (and therefore also its conformal invariance) is a difficult open problem. Nevertheless, significant progress in this direction has been made for other two-dimensional models, most notably percolation and the Ising model.[32–34]

Although the programme to establish the scaling limit of self-avoiding walks in a mathematically rigorous way is not yet complete, the first step, involving establishing analytic properties on the lattice via the discrete Cauchy-Riemann equations, has led to some remarkable developments. These include the surprising mathematical proofs of the results (8), (11) and (12) for the connective constant and critical surface fugacity for self-avoiding walks on the honeycomb lattice.[24–26]

There are also ongoing developments with regard to the connection with integrability, particularly at a boundary. A boundary condition can also be imposed on a discretely parafermionic observable.[35,36] This approach has led to a set of new integrable boundary weights for the Z_N model.[35] Several new sets of boundary weights have also been obtained in this way for the $O(n)$ and $C_2^{(1)}$ loop models.[36]

More developments are eagerly awaited.

Acknowledgments

This work has been partly supported by the Australian Research Council.

References

1. R. J. Baxter, *Exactly Solved Models in Statistical Mechanics* (Academic Press, London, 1982).
2. B. M. McCoy, *Advanced Statistical Mechanics* (Oxford University Press, Oxford, 2010).
3. G. Mussardo, *Statistical Field Theory* (Oxford University Press, Oxford, 2010).
4. M. A. Rajapour and J. Cardy, *J. Phys. A* **40**, 14703 (2007).
5. Y. Ikhlef and J. Cardy, *J. Phys. A* **42**, 102001 (2009).
6. J. Cardy, *J. Stat. Phys.* **137**, 814 (2009).
7. B. Nienhuis, *Int. J. Mod. Phys.* **4**, 929 (1990).
8. H. W. J. Blöte and B. Nienhuis, *J. Phys. A* **22**, 1415 (1989).
9. M. T. Batchelor, B. Nienhuis and S. O. Warnaar, *Phys. Rev. Lett.* **62**, 2425 (1989).
10. C. M. Yung and M. T. Batchelor, *Nucl. Phys. B* **435**, 430 (1995).
11. M. T. Batchelor and C. M. Yung, *J. Phys. A* **28**, L421 (1995).
12. A. G. Izergin and V. E. Korepin, *Commun. Math. Phys.* **79**, 303 (1981).

13. K. Pearson, *Nature* **72**, 294 (1905).
14. B. D. Hughes, *Random Walks and Random Environments, Volume 1* (Oxford University Press, Oxford, 1995).
15. N. Madras and G. Slade, *The Self-Avoiding Walk* (Birkhauser, Boston, 1993).
16. B. Nienhuis, *Phys. Rev. Lett.* **49**, 1062 (1982).
17. P. G. de Gennes, *Phys. Lett. A* **38**, 339 (1972).
18. R. J. Baxter, *J. Phys. A* **19**, 2821 (1986).
19. M. T. Batchelor and H. W. J. Blöte, *Phys. Rev. Lett.* **61**, 138 (1988); *Phys. Rev. B* **39**, 2391 (1989).
20. J. Suzuki, *J. Phys. Soc. Japan* **57**, 2966 (1988).
21. I. Jensen and A. J. Guttmann, *J. Phys. A* **31**, 8137 (1998).
22. M. T. Batchelor, *J. Stat. Phys.* **92**, 1203 (1998).
23. C.-X. Deng, Z. Fu and W.-A. Guo, *Phys. Rev. E* **85**, 062101 (2012).
24. H. Duminil-Copin and S. Smirnov, *Ann. Math.* **175**, 1653 (2012).
25. N. R. Beaton, M. Bousquet-Melou, J. de Gier, H. Duminil-Copin and A. J. Guttmann, arXiv:1109.0358v3.
26. N. R. Beaton, arXiv:1210.0274
27. M. T. Batchelor and C. M. Yung, *Phys. Rev. Lett.* **74**, 2026 (1995).
28. M. T. Batchelor, D. Bennett-Wood and A. L. Owczarek, *Eur. Phys. J. B* **5**, 139 (1998).
29. V. A. Fateev and A. B. Zamolodchikov, *Phys. Lett. A* **92**, 37 (1982).
30. I. T. Alam and M. T. Batchelor, *J. Phys. A* **45**, 494014 (2012).
31. G. F. Lawler, O. Schramm and W. Werner, *Proc. Symposia Pure Math.* **72**, 339 (2004).
32. S. Smirnov, *C. R. Math. Acad. Sci. Paris* **333**, 239 (2001).
33. S. Smirnov, *Ann. Math.* **172**, 101 (2010).
34. H. Duminil-Copin and S. Smirnov, *Clay Mathematics Proceedings* **15**, 213 (2012).
35. Y. Ikhlef, *J. Phys. A* **45**, 265001 (2012).
36. J. de Gier, A. Lee and J. Rasmussen, arXiv:1210.5036.

QUANTUM COMPUTING: A QUANTUM GROUP APPROACH

ZHENGHAN WANG

Microsoft Station Q
CNSI Bldg Rm 2237
University of California
Santa Barbara, CA 93106-6105
U.S.A.
Email: zhenghwa@microsoft.com

There is compelling theoretical evidence that quantum physics will change the face of information science. Exciting progress has been made during the last two decades towards the building of a large scale quantum computer. A quantum group approach stands out as a promising route to this holy grail, and provides hope that we may have quantum computers in our future.

Keywords: Quantum group, unitary modular category, topological quantum field theory, topological phases of matter, quantum computing.

1. Quantum group and quantum computing

Working at Microsoft Station Q, I am interested in building a quantum computer by braiding non-abelian anyons. The mathematical underpin of anyon theory is unitary modular tensor category theory— an abstraction of the representation theory of quantum groups[13].[21] A quantum group commonly refers to a certain Hopf algebra such as a quasi-triangular one, though the terminology is used only loosely. I would like to advocate the use of a quantum group to mean a modular category. Modular categories are elaborate algebraic structures that arise in a variety of subjects: Hopf algebras, von Neumann algebras, topological quantum field theories (TQFTs), conformal field theories (CFTs), and topological phases of matter. They are quantum analogues of finite abelian groups in a sense that I will explain below. We could use unitary fusion categories to describe quantum symmetries of quantum phases of matter just as finite groups are used to describe symmetries of crystals.

Qubits are just the modern abstraction of a two-level quantum system and its composites. Theoretically, there is compelling evidence that quantum physics can change the face of information science: the BB84 private key protocol,[1] Shor's factoring algorithm,[19] and Hastings' additivity counterexamples.[7] But for the last two decades, the most interesting challenge has been to realize quantum computation in the real world. A key post-Shor idea is to use topology to protect quantum information—a revolutionary idea articulated by A. Kitaev as follows: "if a physical system were to have quantum topological (necessarily nonlocal) degrees of freedom, which were insensitive to local probes, then information contained in them would be automatically protected against errors caused by local interactions with the environment. This would be fault tolerance guaranteed by physics at the hardware level, with no further need for quantum error correction, i.e., topological protection."[5 12]

Materials with such topological degrees of freedom are called topological phases of matter.[16] The prospect of utilizing topological phases of matter to build a quantum computer has been greatly enhanced by recent experimental progress. This approach to quantum computing at the triple juncture of mathematics, computer science, and physics is referred to as topological quantum computation (TQC). The low energy effective theories of topological phases of matter are TQFTs, and their algebraic data are modular categories. Therefore, unitary modular categories (UMCs) are algebraic models of anyons in topological phases of matter. TQC leads to new perspectives on TQFTs and modular categories.

2. Modular categories as finite abelian groups

A modular category $\mathcal{C}$ is a non-degenerate ribbon fusion category[4].[2] A ribbon fusion category is a fusion category with compatible braiding and pivotal structures. A practical way to describe such a complicated structure $\mathcal{C}$ is to use a representative set $\mathrm{Irr}(\mathcal{C}) = \{X_i\}_{i\in I}$ of the isomorphism classes of simple objects of $\mathcal{C}$, and the various structures on $\mathrm{Irr}(\mathcal{C})$. The finite index set I of $\mathrm{Irr}(\mathcal{C})$ is the isomorphism classes of simple objects of $\mathcal{C}$, and an element of I will be called a label. The cardinality $|I|$ of the label set I is called the *rank* of the modular category. It is important to keep in mind the difference between labels and the simple object representatives. To make an analogy with a finite group, we will think of each simple object X_i as an analogue of a group element in the modular category $\mathcal{C}$. Then there will be a binary operation on $\mathrm{Irr}(\mathcal{C})$ which satisfies the analogues of associativity, commutativity, and inverse of a group.

The binary operation on $\mathrm{Irr}(\mathcal{C})$ is called the tensor product. Just as tensor products of irreps of algebraic structures such as Hopf algebras, the tensor product $X_i \otimes X_j$ is not necessarily an irrep, so it could decompose into a direct sum $\oplus_{k=1}^{|I|} N_{i,j}^k X_k$, i.e., $X_i \otimes X_j \cong \oplus_{k=1}^{|I|} N_{i,j}^k X_k$ for some non-negative integers $N_{i,j}^k$ (called multiplicities). This is some kind of quantum multiplication as the outcome is a superposition. Actually the outcomes $\{X_k\}$ are endowed with an intrinsic probability distribution related to the quantum dimensions of the simple objects.

The associativities between $(X_i \otimes X_j) \otimes X_k$ and $X_i \otimes (X_j \otimes X_k)$ are weakened to be functorial isomorphisms that satisfy the pentagon axioms.

The dual X_i^* of X_i is quantum analogue of an inverse. As such we would only require that the tensor unit $\mathbf{1}$ appears in $X_i \otimes X_i^*$ as a sub-object.

Finally, the commutativity of $X_i \otimes X_j$ and $X_j \otimes X_i$ is relaxed to be a functorial isomorphism that satisfies the hexagon axioms.

There is a very convenient picture calculus for a modular category. Basically, the axioms of a ribbon fusion category is to define topological invariants of colored braided trivalent framed graphs in the plane. In particular, we have invariants of colored framed links.

Important numerical data for a modular category $\mathcal{C}$ are the fusion multiplies $\{N_{i,j}^k\}, i, j, k \in I$, the S-matrix and T-matrix. The fusion multiplies are organized into matrices $N_i, i \in I$ such that $(N_i)_{j,k} = N_{i,j}^k$. The unnormalized $\tilde{S} = (S_{ij})$-matrix consists of link invariants of the colored Hopf link: S_{ij} is the link invariant of the Hopf link with the two components colored by X_i and X_j. We always identify I with $\{0, 1, ..., |I|-1\}$, and assume X_0 is the tensor unit $\mathbf{1}$. The T-matrix T is diagonal with the i-th diagonal entry equal to the twist θ_i of X_i. The first row of the $\tilde{S}$ matrix consists of the invariants of the unknot colored by X_i, denoted as d_i, called the quantum dimension of the simple object X_i. Let $D = \sqrt{\sum_{i \in I} d_i^2}$, $p_\pm = \sum_{i \in I} \theta_i^\pm d_i^2$. Then an important identity for a modular category is:

$$p_+ p_- = D^2.$$

The S-matrix S is $\frac{1}{D}\tilde{S}$, and S, T together leads to a projective representation of the modular group $SL(2, \mathbb{Z})$.

Any finite abelian group G can be endowed with modular category structures in many different ways so that G becomes the Irr($\mathcal{C}$)'s. One way to do it is as follows: choose an $n \times n$ even integral matrix B so that the cokernel of the map $B : \mathbb{Z}^n \to \mathbb{Z}^n$ is isomorphic to G, i.e., $\mathbb{Z}^n/\text{Im}(B) \cong G$. Each element $g \in G$ can be identified as an equivalence class of vectors $v_g \in \mathbb{Q}^n$ such that $Bv_g \in \mathbb{Z}^n$. Two such v_g's are equivalent if their difference is in

$\mathbb{Z}^n$. Then there is a modular category G_B with all $d_i = 1$, fusion rules as G, and $\tilde{S}$-matrix $\tilde{S}_{ij} = e^{2\pi\sqrt{-1}<v_{g_i},v_{g_j}>_B}$, $T_i = e^{\pi\sqrt{-1}<v_{g_i},v_{g_i}>_B}$, where $< v, w >_B = v^t B w$.

It is obvious that for a fixed order, there exist only finitely many groups. The analogous statement for modular categories would be for a fixed rank, there are only finitely many equivalence classes of modular categories. This fundamental conjecture implies that modular categories are amenable to a classification, especially the low rank cases. Modular categories up to rank 4 have been classified.[18]

3. Modular categories as topological phases of matter

Solid, liquid, and gas are all familiar states of matter. But by a more refined classification, each state consists of many different phases of matter. For example, within the crystalline solid state, there are many different crystals distinguished by their different lattice structures. All those phases are classical in the sense they depend crucially on the temperature. More mysterious and challenging to understand are quantum phases of matter: phases of matter at zero temperature (in reality very close to zero). Quantum phases of matter whose low energy physics can be modeled well by TQFTs are called topological phases of matter. There are two classes of topological phases of matter in real materials: two dimensional electron systems that exhibit the fractional quantum Hall effect (FQHE), and topological insulators and superconductors. The universal properties of those materials can be modeled well by TQFTs, and their variations—symmetry enriched TQFTs.

Traditionally, phases of matter are classified by classical symmetries modeled by groups. Quantum phases of matter cannot be fit into this Landau paradigm. Intensive effort in physics right now is to develop a post-Landau paradigm to classify quantum phases of matter. The well-studied classes of quantum phases of matter are topological phases of matter. To characterize topological phases of matter, we focus on both the ground states and first excited states. The algebraic models of elementary excitations of the topological phases are unitary modular categories (UMCs). If a classical symmetry modeled by a group G is also present, then the UMC is G-graded. An interesting example is the trivial case with symmetry including the topological insulators discovered in the last few years—symmetry protects topological phases of matter.

Anyons are localized quantum fields. The mathematical model of an anyon under UMCs is a simple object X in a UMC, and in picture calculus, described as a small arrow. Therefore its world line in $\mathbb{R}^3$ is a ribbon. The isomorphism class x of an anyon X is called the anyon type of X, or the anyonic charge.

A dictionary of terminologies between anyon theory and UMC theory is given in the following table.

UMC	*Anyonic system*
simple object	anyon
label	anyon type or anyonic charge
tensor product	fusion
fusion rules	fusion rules
triangular space V_{ab}^{c} or V_{c}^{ab}	fusion/splitting space
dual	antiparticle
birth/death	creation/annihilation
mapping class group representations	generalized anyon statistics
nonzero vector in $V(Y)$	ground state vector
unitary F-matrices	recoupling rules
twist $\theta_x = e^{2\pi i s_x}$	topological spin
morphism	physical process or operator
colored braided framed trivalent graphs	anyon trajectories
quantum invariants	topological amplitudes

A UMC leads to an Atiyah-Segal type unitary TQFT.[20] An $(n+1)$-Atiyah-Segal-type TQFT is a symmetric monoidal functor from the $(n+1)$-bordism category Bord$(n+1)$ of n and $n+1$-manifolds to the category of finitely dimensional vector spaces. An important question is when such a TQFT has a Hamiltonian realization on triangulations such as the quantum double of finite groups and Turaev-Viro type $(2+1)$-TQFTs[12].[14]

4. Modular categories as anyonic quantum computers

Given n anyons of type x in a disk D^2, their ground state degeneracy $\dim(V(D, x, \cdots, x))$ has the asymptotic growth rate d_x—the quantum dimension of x. An anyon X with $d_X = 1$ is called an abelian anyon, and an anyon with $d_X > 1$ a non-abelian anyon.

Given a non-abelian anyon of type x, we can construct an anyonic quantum computation model as follows: choose $V_{n,x}^{a}$ as the computational subspace—the Hilbert space of n anyons of type x and boundary label a—the total anyonic charge.

To simulate a traditional quantum circuit $U_L : (\mathbb{C}^2)^{\otimes n} \to (\mathbb{C}^2)^{\otimes n}$ we need to find a braid b making the square

$$\begin{array}{ccc} (\mathbb{C}^2)^{\otimes n} & \xrightarrow{\iota} & V^a_{n,x} \\ {\scriptstyle U_L}\downarrow & & \downarrow{\scriptstyle \rho(b)} \\ (\mathbb{C}^2)^{\otimes n} & \xrightarrow[\iota]{} & V^a_{n,x} \end{array}$$

commute, where $\rho(b)$ is the braid matrix and ι is an efficient embedding of the n-qubit space $(\mathbb{C}^2)^{\otimes n}$ into the ground states $V^a_{n,x}$. This is rarely achievable. Therefore we seek b making the square commute up to arbitrary precision. To achieve universality for quantum computation, we need to implement a universal gate set of unitary matrices. Then universality for anyonic quantum computing becomes the question: can we find braids b whose $\rho(b)$'s approximate a universal gate set efficiently to arbitrary precision? For non-abelian anyons, since the Hilbert spaces always grow exponentially, universality is usually guaranteed if the braid group representations afforded by the UMC have a dense image in the special unitary groups $\mathrm{SU}(V^a_{n,x})$.

4.0.1. *Braiding universal anyons*

To simulate the standard quantum circuit model with braidings alone, we choose a non-abelian anyon type x and use m and n anyons respectively to encode one and two qubits. If the braid group representations $B_m \to \mathrm{SU}(V^a_{m,x})$ and $B_n \to \mathrm{SU}(V^b_{n,x})$ are independently dense for all anyon types a, b, then the anyonic quantum computing model is universal. Anyon with such properties will be called braiding universal. An example of a braiding universal anyon is the Fibonacci anyon τ.

4.0.2. *Resource assisted universal anyons*

It is conjectured that anyons with quantum dimensions $d_x^2 \notin \mathbb{Z}$ are essentially braiding universal. But current experimentally accessible anyons are those $d_x^2 \in \mathbb{Z}$. An example of such anyons is the Ising anyon σ with $d_\sigma = \sqrt{2}$. More examples are the metaplectic anyons.[8] It is conjectured that such anyons always lead to representations of the braid groups with finite images, hence not braiding universal.[17] Therefore, to obtain a universal braiding gate set, we need to supplement the braidings with other gates such as measurement, magic state distillation, ... We will call an anyon R-assisted universal if the braidings supplemented with the resource R give rise to a universal gate set. For example, the Ising anyon is (magic-state-distillation)-assisted universal.[3]

5. Majorana physics and its realization

At the writing, we are the closest to detecting and harnessing Majoranas. Majoranas mean various things in the literature, and sometimes are very misleading. We will distinguish between Majorana fermions from Majorana zero modes. By Majorana fermions, we will mean abelian anyons with twist $\theta = -1$. Therefore, Majorana fermions are never non-abelian. An example of a Majorana fermion is the ψ particle in the Ising theory. By Majorana zero modes, we will mean the zero modes $\{\sum_\alpha c_\alpha|\psi_\alpha >\}$ obtained from the Majorana algebras $\mathcal{M}_n$ through the state-operator correspondence $\gamma = \sum_\alpha |\psi_\alpha >< \psi_\alpha|$. In the Ising theory, the ground state manifolds of many Ising anyons σ's consist of Majorana zero modes. However, Majorana zero modes can arise without the presence of Ising anyons.

The Majorana algebra $\mathcal{M}_n$ is generated by the operators $\{\gamma_i\}, i = 1, \cdots, n-1$ such that $\gamma_i^\dagger = \gamma_i, \{\gamma_i, \gamma_j\} = 2\delta_{ij}$. $\mathcal{M}_n$ supports a representation $\rho_{\mathcal{M}_n}$ of the braid group B_n by sending the braid generator σ_i to an automorphism $\rho_{\mathcal{M}_n}(\sigma_i) = \frac{1}{\sqrt{2}}(1 - \gamma_i\gamma_{i+1})$ of the Majorana zero modes:

$$\rho_{\mathcal{M}_n}(\sigma_i)\gamma = \frac{1}{\sqrt{2}}(1 - \gamma_i\gamma_{i+1}) \cdot \gamma \cdot \frac{1}{\sqrt{2}}(1 - \gamma_i\gamma_{i+1})^\dagger,$$

for any $\gamma \in \mathcal{M}_n$.

$$\text{It is easy to check: } \begin{cases} \rho_{\mathcal{M}_n}(\sigma_i)\rho_{\mathcal{M}_n}(\sigma_i)^\dagger = \text{Id}, \\ \rho_{\mathcal{M}_n}(\sigma_i)\gamma_i = \gamma_{i+1}, \\ \rho_{\mathcal{M}_n}(\sigma_i)\gamma_{i+1} = -\gamma_i, \\ \rho_{\mathcal{M}_n}(\sigma_i)\gamma_j = \gamma_j, \text{ if } j \neq i, i+1. \end{cases}$$

This projective braid group representation $\rho_{\mathcal{M}_n}(\sigma_i) = \frac{1}{\sqrt{2}}(1 - \gamma_i\gamma_{i+1})$ is essentially the Jones representation of the braid groups at the 4-th root of unity $q = e^{\frac{2\pi i}{4}}$ because the Temperley-Lieb Jones algebra at $q = e^{\frac{2\pi i}{4}}$ is the Majorana algebra $\mathcal{M}_n$[10].[21]

There are important experimental progress towards the confirmation of Majorana physics in real materials. Interferometry experiments for the FQH liquids at filling fraction $\mu = \frac{5}{2}$ agree with the Ising physics,[22] and nanowires experiments are consistent with Majorana zero modes.[15]

Recently, a generalization of Majorana physics to metaplectic physics is proposed.[8] Potentially, metaplectic anyons can be used to perform universal quantum computation more easily than Majorana anyons.

6. Open problems and future directions

There are new problems in mathematics, physics, and computer science inspired by the quantum group approach to quantum computing. The classification of modular categories is both an interesting question in mathematics and physics. As beautiful mathematical structures, modular category theory could be regarded as quantum version of group theory. In physics, a classification is very beneficial to the identification of topological phases of matter.[9]

As the theory of modular category reaches certain maturity, many new directions emerge. UMCs capture topological properties of bosonic physical systems in 2-spatial dimension. Three directions that we would like to explore are for fermion systems, for boson systems with group symmetries, and for 3-spatial dimension.

6.0.3. *SPT and SET*

Topological phases of matter can have conventional group symmetries G. Such topological phases of matter are called symmetry enriched topological (SET) phases. When the intrinsic topological order is trivial, SET is called symmetry protect topological (SPT) phase. Important examples of SPTs are topological insulators and topological superconductors. In 2-spatial dimension, SETs are described by G-equivariant TQFTs/modular categories. Classification of SETs then involves the interplay between group theory and modular category theory.

6.0.4. *Fermion systems*

Fermion systems share similar Hilbert space as bosonic spin systems, but the meaning of locality is different.[6] A local Hamiltonian for a fermion system is actually non-local. Therefore, fermionic topological phases of matter are strictly richer than their bosonic counterparts.

6.0.5. *(3+1)-TQFTs*

Topological phases of matter behave differently in different spatial dimensions. In $1D$, there are no intrinsic non-trivial topological order. In $2D$, there is a richness of topological orders. In $3D$, the situation is more intrigue. $(3+1)$-TQFTs whose partitions that can detect smooth structures are very rare.[11] I would conjecture that all unitary Atiyah-Segal-type $(3+1)$-TQFTs give rise to partition functions that are simple homotopy invariants. for $3D$, an algebraic framework analogous to modular category is lacking, at least we do not understand higher category for $3D$ to have a good understanding.

Quantum computers are still hypothetical models that promise revolutionary advance in information science. The quantum algebra/topology approach provides substantial evidence that we may see quantum computers in our future, so the time of machines is coming.

Acknowledgments

The author is partially supported by NSF DMS 1108736.

References

1. Bennett, C.H. and Brassard,G. (1984) Quantum Cryptography: Public key distribution and coin tossing, in Proceedings of the IEEE International Conference on Computers, Systems, and Signal Processing, Bangalore, p. 175 (1984).
2. B. Bakalov and A. Kirillov, Jr., *Lectures on Tensor Categories and Modular Functors*, University Lecture Series **21**, Amer. Math. Soc., 2001.
3. S. Bravyi and A. Kitaev, *Universal Quantum Computation with ideal Clifford gates and noisy ancillas*, Phys. Rev. A 71, 022316 (2005), arXiv:quant-ph/0403025.
4. P. Etingof, D. Nikshych, and V. Ostrik, *On fusion categories*, Ann. of Math. (2) **162**, 2 (2005), 581–642, arXiv:math/0203060.
5. M. Freedman, *P/NP, and the quantum field computer*, Proc. Natl. Acad. Sci. USA **95**, 1 (1998), 98–101.
6. Z.-G. Gu, Z. Wang, X.-G. Wen, A classification of 2D fermionic and bosonic topological orders, arXiv:1010.1517.
7. Hastings, M.B. (2009), Superadditivity of communication capacity using entangled inputs, Nature Physics 5, 255 - 257 (2009).
8. M.B. Hastings, C. Nayak, and Z. Wang, *Metaplectic Anyons, Majorana Zero Modes, and their Computational Power*, arXiv:1210.5477.
9. H.C. Jiang, L. Balents, and Z. Wang, *Identifying topological order by entanglement entropy*, Nature Physics 8, 902–905(2012), doi:10.1038/nphys2465, arXiv:1205.4289.
10. V. Jones, *Braid groups, Hecke algebras and type* II_1 *factors*, in Geometric Methods in Operator Algebras (Kyoto, 1983), Pitman Res. Notes Math. Ser. **123**, Longman Sci. Tech., Harlow, 1986, 242–273. Harlow, 1986.
11. K. Walker and Z. Wang, *(3+1)-TQFTs and Topological Insulators*, Front. Phys., 2012, 7(2): 150-159, arXiv:1104.2632.

12. A. Kitaev, *Fault-tolerant quantum computation by anyons*, Ann. Physics **303**, 1 (2003), 2–30, `arXiv:quant-ph/9707021`.
13. A. Kitaev, *Anyons in an exactly solved model and beyond*, Ann. Physics **321**, 1 (2006), 2–111, `arXiv:cond-mat/0506438`.
14. M. Levin and X.-G. Wen, *String-net condensation: A physical mechanism for topological phases*, Phys. Rev. B **71**, 4 (2005), 045110, `arXiv:cond-mat/0404617`.
15. V. Mourik, K. Zuo, S. Frolov, S. Plissard, E. Bakkers, and L. P. Kouwenhoven, *Signatures of Majorana Fermions in Hybrid Superconductor-Semiconductor Nanowire Devices*, Science 1222360 DOI:10.1126/science.1222360.
16. C. Nayak, S. Simon, A. Stern, M. Freedman, and S. Das Sarma, *Non-abelian anyons and topological quantum computation*, Rev. Mod. Phys. **80**, 3 (2008), 1083–1159, `arXiv:0707.1889`.
17. D. Naidu and E. C. Rowell; *A finiteness property for braided fusion categories*, Algebr. Representation Theory. **15** (2011) no. 5, 837–855.
18. E. Rowell, R. Stong, and Z. Wang, *On classification of modular tensor categories*, Communications in Mathematical Physics **292**, 2 (2009), 343–389, `arXiv:0712.1377`.
19. Shor, P. W. (1994), Algorithms for quantum computation: discrete logarithms and factoring. 35th Annual Symposium on Foundations of Computer Science (Santa Fe, NM, 1994), 124–134, IEEE Comput. Soc. Press, Los Alamitos, CA, 1994.
20. V. Turaev, *Quantum Invariants of Knots and 3-Manifolds*, De Gruyter Studies in Mathematics **18**, Walter de Gruyter & Co., Berlin, 1994.
21. Z. Wang, *Topological quantum computation*, CBMS monograph vol. 112, American Mathematical Society, 2010.
22. R. L. Willett, L. N. Pfeiffer, and K. W. West, *Measurement of filling factor 5/2 quasiparticle interference with observation of charge e/4 and e/2 period oscillations,* PNAS (2009), vol. 106, no. 22, 8853—8858.

YANG-BAXTER EQUATION AND QUANTUM INFORMATION

MO-LIN GE*

Chern Institute of Mathematics, Nankai Univ, Tianjin, 300071, China
** E-mail: geml@nankai.edu.cn*

KANG XUE

Dept. of Physics, Northeast Normal Univ,, ChangChun, Jilin, 120024, China

Two types of (4×4)-dimensional solutions of YBE can be mapped to 2-dimensional YBE by acting the Temperely-Lieb algebra on the topological bases $|e_1\rangle$ and $|e_2\rangle$. The fermionic expressions for $|e_1\rangle$ and $|e_2\rangle$ have been shown. The 3-d topological base have been found for Birman-Wenzl-Murakami algebra. The extreme of L_1-norm of Wigner's D-function is introduced to explain why there are only the two types in Physics that are related to entanglement in quantum information.

Keywords: Braiding Matrices; Yang-Baxter Equations; Topological base.

1. Introduction

Braid relation

$$B_i B_{i+1} B_i = B_{i+1} B_i B_{i+1} \quad (B_i \equiv B_{i,i+1}) \tag{1.1}$$

$$B_i = I \otimes \cdots \underset{ii+1}{B} \cdots \otimes I \tag{1.2}$$

has $(N^2 \times N^2)$-d solution:

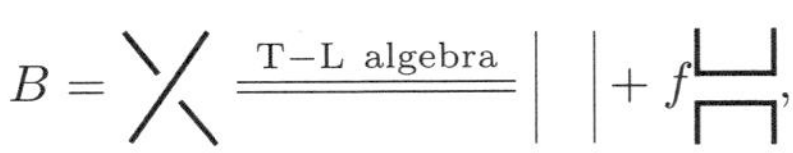

(1.3)

The simplest realizations: Type-I (Spin chain models, NSE, δ-interaction model$\cdots$),

$$B_{\rm I} = P, \quad P^2 = I \quad \text{(permutation)} \tag{1.4}$$

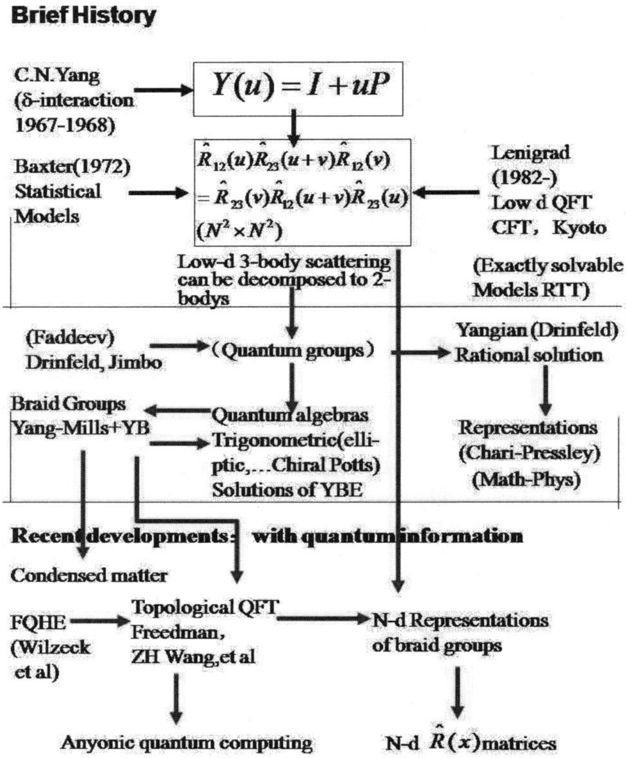

There were many remarkable works in this field. We could only list a few part of them in the refs. [1-5].

Type-II (Quantum Information for Bell states[6])(Kauffman-Lomonaco):

$$
\begin{aligned}
&|\Psi\rangle = \left(|\Psi_+\rangle\,, |\Phi_+\rangle\,, |\Psi_-\rangle\,, |\Phi_-\rangle\right)^T \\
&|\Psi_+\rangle = \frac{1}{\sqrt{2}}\left(|\uparrow\uparrow\rangle + |\downarrow\downarrow\rangle\right), \quad |\Phi_+\rangle = \frac{1}{\sqrt{2}}\left(|\downarrow\uparrow\rangle + |\uparrow\downarrow\rangle\right) \\
&|\Psi_-\rangle = \frac{1}{\sqrt{2}}\left(|\downarrow\uparrow\rangle - |\uparrow\downarrow\rangle\right), \quad |\Phi_+\rangle = \frac{1}{\sqrt{2}}\left(|\uparrow\uparrow\rangle - |\downarrow\downarrow\rangle\right)
\end{aligned}
\tag{1.5}
$$

$$
\begin{aligned}
&|\Phi_0\rangle = \left(|\downarrow\downarrow\rangle\,, |\uparrow\downarrow\rangle\,, |\downarrow\uparrow\rangle\,, |\uparrow\uparrow\rangle\right)^T \text{ (natural basis)} \\
&|\Psi\rangle = B_{\text{II}}\,|\Psi_0\rangle
\end{aligned}
\tag{1.6}
$$

$$B_{\mathrm{II}} = \frac{1}{\sqrt{2}} \begin{bmatrix} 1 & 0 & 0 & 1 \\ 0 & 1 & 1 & 0 \\ 0 & -1 & 1 & 0 \\ -1 & 0 & 0 & 1 \end{bmatrix} = \frac{1}{\sqrt{2}} (I + M) , \quad M^2 = -1 \tag{1.7}$$

$M = i\gamma_1$ (Dirac's γ−matrix)

The B_{II} satisfies (1.2) and M satisfies

$$\begin{aligned} &M_i M_{i\pm 1} = -M_{i\pm 1} M_i \\ &M_i M_j = M_j M_i \qquad |i-j| \geq 2 \end{aligned} \tag{1.8}$$

that form "special extra 2-group".
The unified form for both type-I and type-II can be viewed as the realizations of the operator

$$\hat{B} =$$  $$= \rho \left(I + f\hat{U} \right) \quad (\rho = \text{const}) \tag{1.9}$$

$$f = \frac{1}{2}(-d \pm \sqrt{d^2 - 4}) \quad d =$$  (1.10)

where $\hat{U}$ satisfies the T-L algebra:

$$\hat{U}_i \hat{U}_{i\pm 1} \hat{U}_i = \hat{U}_i, \quad \hat{U}_i^2 = d\hat{U}_i \tag{1.11}$$

$$\hat{U}_i =$$  $$, \quad d =$$  (value of loop) (1.12)

Explicitly the (4×4) representation of $\hat{U}$ for the type-I takes the form

$$U_{\mathrm{I}} = \begin{bmatrix} 0 & 0 & 0 & 0 \\ 0 & q & \eta & 0 \\ 0 & \eta^{-1} & q^{-1} & 0 \\ 0 & 0 & 0 & 0 \end{bmatrix} \quad (d = q + q^{-1}) \tag{1.13}$$

or

$$U_{\mathrm{I}}' = \begin{bmatrix} 1 & 0 & 0 & e^{i\varphi'} \\ 0 & 0 & 0 & 0 \\ 0 & 0 & 0 & 0 \\ e^{-i\varphi'} & 0 & 0 & 1 \end{bmatrix} \quad d = 2 \tag{1.14}$$

whereas for the type-II

$$U_{\mathrm{II}} = \frac{1}{\sqrt{2}} \begin{bmatrix} 1 & 0 & 0 & e^{i\varphi} \\ 0 & 1 & i\epsilon & 0 \\ 0 & -i\epsilon & 1 & 0 \\ e^{-i\varphi} & 0 & 0 & 1 \end{bmatrix} \quad d = \sqrt{2} \quad (\epsilon^2 = 1) \tag{1.15}$$

When $\eta = 1$, $q = 1$ in (1.14) it leads to the permutation, whereas for $\epsilon = 1$, $\varphi = \pi/2$, (1.15) reduces to (1.7). Eq. (1.9) is our starting point.
With the operator form of (1.9) we shall find the 2×2 representations of braid group by acting $\hat{U}$ on the topological basis with two components.

2. 2-dimensional representations of $\hat{U}$ and topological basis

In accordance with the (4×4) represention of both (1.13) (type-I) and (1.15) (type-II) we want to find the corresponding 2-dimensional representations of $\hat{U}$.
Freedman, Z.H. Wang et al. introduced the topological basis (anyonic). For both type-I and type-II,[7,8] with T-L algebra we have

$|e_1\rangle = \frac{1}{d}$  , $\quad |e_2\rangle = \frac{1}{\sqrt{d^2-1}}\Big($  $- \frac{1}{d}$  $\Big)$ (2.1)

Acting $\hat{B} = \rho(1 + f\hat{U})$ on $|e_1\rangle$ and $|e_2\rangle$, we obtain

$$B_{12}\,|e_1\rangle = B_{34}\,|e_1\rangle = \rho(1 + df)\,|e_1\rangle$$
$$B_{12}\,|e_2\rangle = B_{34}\,|e_2\rangle = \rho\,|e_2\rangle$$
$$B_{23}\,|e_1\rangle = B_{41}\,|e_1\rangle = \rho\left[(1 + \frac{f}{d})\,|e_1\rangle + \frac{f\sqrt{d^2-1}}{d}\,|e_2\rangle\right]$$
$$B_{23}\,|e_2\rangle = B_{41}\,|e_2\rangle = \rho\sqrt{d^2-1}\left[d^{-1}f\,|e_1\rangle + \left(\frac{1}{\sqrt{d^2-1}} + \frac{f}{d}\sqrt{d^2-1}\right)|e_2\rangle\right] \tag{2.2}$$

With (2.2) we shall find the 2-dimensional representations corresponding to (1.13),(1.14) and (1.15), respectively. The operator $\hat{U}$ for both of types can be written in terms of spin operators at the neighbour sites. Based on (2.2) the topological bases $|e_1\rangle$ and $|e_2\rangle$ can then be expressed through four fermions.
For the type-I (1.14) the U_{I}' can be written in terms of the spin operators:

$$U_{i,\mathrm{I}}' = \frac{1}{2}(1_{i,i+1} + 4S_i^3 S_{i+1}^3) + e^{i\varphi} S_i^+ S_{i+1}^+ + e^{-i\varphi} S_i^- S_{i+1}^- \tag{2.3}$$

Suppose, in general, $|e_i\rangle$ $(i = 1, 2)$ is expressed in terms of 4-spin states[9] $|e_i\rangle = a_i^{(+)}\,|\uparrow\uparrow\uparrow\uparrow\rangle + a_i^{(-)}\,|\downarrow\downarrow\downarrow\downarrow\rangle + b_i^{(+)}\,|\uparrow\uparrow\downarrow\downarrow\rangle + b_i^{(-)}\,|\downarrow\downarrow\uparrow\uparrow\rangle + c_i^{(+)}\,|\uparrow\downarrow\downarrow\uparrow\rangle + c_i^{(-)}\,|\downarrow\uparrow\uparrow\downarrow\rangle + d_i^{(+)}\,|\uparrow\downarrow\uparrow\downarrow\rangle + d_i^{(-)}\,|\downarrow\uparrow\downarrow\uparrow\rangle$. The coefficients are to be determined by acting $\hat{U}_{\mathrm{I}}'$ on $|e_i\rangle$ $(i = 1, 2)$. After a lengthy calculation we find for the type-I:

$$|e_1\rangle = |\psi_{12}\rangle\,\langle\psi_{34}| \tag{2.4}$$

$$|e_2\rangle = \frac{1}{\sqrt{2}}\left(2\,|\psi_{41}\rangle\,\langle\psi_{23}| + |\psi_{12}\rangle\,\langle\psi_{34}|\right) \tag{2.5}$$

where

$$|\psi_{ij}\rangle = \frac{1}{\sqrt{2}}\left(|\uparrow\uparrow\rangle_{ij} + e^{-i\varphi}\,|\downarrow\downarrow\rangle_{ij}\right) \quad (j = i+1) \tag{2.6}$$

with φ arbitrary phase factor. Hence

$$= |\uparrow\uparrow\rangle_{ij} + e^{-i\varphi}\,|\downarrow\downarrow\rangle_{ij} = \sqrt{2}\,|\psi_{ij}\rangle \tag{2.7}$$

and

$$\hat{U}^{I}_{ij} =$$

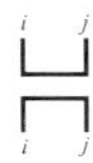

$$= 2\,|\psi_{ij}\rangle\,\langle\psi_{ij}| \tag{2.8}$$

The (2×2) matrix elements can be given by (2.2). Denoting

$$A_{ij} = (B_{12})_{ij}, \quad A$$

$$B_{ij} = (B_{23})_{ij}, \quad B$$

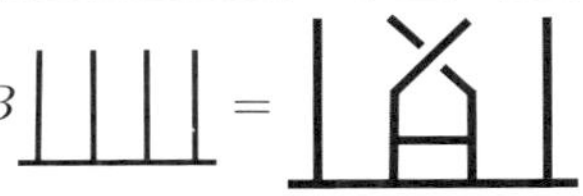

(2.9)

we find

$$A^{\mathrm{I}} = \rho\begin{bmatrix} 1+df & 0 \\ 0 & 1 \end{bmatrix} = \rho\begin{bmatrix} -1 & 0 \\ 0 & 1 \end{bmatrix} \tag{2.10}$$

and

$$B^{\mathrm{I}} = \rho\begin{bmatrix} 1+\frac{f}{d} & \frac{f\sqrt{d^2-1}}{d} \\ \frac{f\sqrt{d^2-1}}{d} & 1+\frac{f}{d} \end{bmatrix} = \frac{\rho}{2}\begin{bmatrix} -1 & -\sqrt{3} \\ -\sqrt{3} & 1 \end{bmatrix} \tag{2.11}$$

satisfying $ABA = BAB$. This is the 2-d form of braiding relation corresponding to the permutation in spin chain models, i.e. for the type-I.

Fermionic expression of braiding matrix for the type-II

The U^{II} is related to the Bell basis. Starting from

$$U^{\mathrm{II}} = \frac{1}{\sqrt{2}}\left(I + \begin{bmatrix} 0 & 0 & 0 & e^{i\varphi} \\ 0 & 0 & i\epsilon & 0 \\ 0 & i\epsilon & 0 & 0 \\ e^{-i\varphi} & 0 & 0 & 0 \end{bmatrix}\right) \tag{2.12}$$

which can be expressed in terms of the spin operators:[9]

$$\hat{U}_{ij} = \sqrt{2}\left(|\psi_{ij}\rangle\,\langle\psi_{ij}| + |\varphi_{ij}\rangle\,\langle\varphi_{ij}|\right) \tag{2.13}$$

where

$$|\psi_{ij}\rangle = \frac{1}{\sqrt{2}}\left(|\uparrow\uparrow\rangle_{ij} + e^{-i\varphi}|\downarrow\downarrow\rangle_{ij}\right) \tag{2.14}$$

the (2.2) together with (2.13) lead to

$$\begin{aligned}|e_1\rangle &= (1+|\alpha|^2)^{-1/2}\left(|\psi_{12}\rangle\langle\psi_{34}| + \alpha|\varphi_{12}\rangle\langle\psi_{34}|\right)\\ |e_2\rangle &= (1+|\alpha|^2)^{-1/2}\left\{(1-i\epsilon\alpha e^{i\varphi})|\psi_{23}\rangle\langle\psi_{41}|\right.\\ &\qquad \left.-\alpha(1+i\alpha^{-1}e^{-i\varphi})|\psi_{23}\rangle\langle\psi_{41}| - |e_1\rangle\right\}\end{aligned} \tag{2.15}$$

i.e. they can be expressed through entangled states. Finally, because $A_{ij} = \langle e_i|\hat{B}_{12}|e_j\rangle$ and $B_{ij} = \langle e_i|\hat{B}_{23}|e_j\rangle$ we obtain

$$A = \rho\begin{bmatrix} i\epsilon & 0\\ 0 & 1\end{bmatrix}, \quad B = \frac{\rho}{2}\begin{bmatrix} 1+i\epsilon & -1+i\epsilon\\ -1+i\epsilon & 1+i\epsilon\end{bmatrix}, \quad (\epsilon^2 = 1) \tag{2.16}$$

As $\epsilon = -1$ we have

$$A = -i\rho\begin{bmatrix} 1 & 0\\ 0 & i\end{bmatrix}, \quad B = -\frac{i\rho}{\sqrt{2}}e^{i\pi/4}\begin{bmatrix} 1 & -i\\ -i & 1\end{bmatrix} \tag{2.17}$$

They satisfy

$$ABA = BAB \tag{2.18}$$

which is 2-d braiding relation and was known in the fractional QHE presented by Nayak-Wilczek long time ago.

3. Yang-Baxterizaton (parametrization of braiding matrices)

The Yang-Baxter equation (YBE) reads (Faddeev, Sklynian, Kulish ...)

$$\breve{R}_{12}(x)\breve{R}_{23}(xy)\breve{R}_{12}(y) = \breve{R}_{23}(y)\breve{R}_{12}(xy)\breve{R}_{12}(x) \tag{3.1}$$

where x is spectral parameter. We are looking for the rational form of $\breve{R}(x)$ satisfying YBE:

$$\begin{aligned}\breve{R}_{12}(u) &= a_1(u)I_{12} + b_1(u)U_{12}\\ \breve{R}_{23}(u) &= a_2(u)I_{12} + b_2(u)U_{23}\end{aligned} \tag{3.2}$$

The (4×4)-solution for the type-I is well known ($x = e^{iu}$):

$$\breve{R}_{\rm I}(u) = 1 + uP \tag{3.3}$$

that can be extended to $q-$deformation forms. The new one is for the Bell basis (type-II)

$$\breve{R}_{\rm II}(x) = \begin{bmatrix} \cos\frac{\theta}{2} & 0 & 0 & \sin\frac{\theta}{2}e^{-i(\varphi-\pi/2)} \\ 0 & \cos\frac{\theta}{2} & \sin\frac{\theta}{2} & 0 \\ 0 & -\sin\frac{\theta}{2} & \cos\frac{\theta}{2} & 0 \\ -\sin\frac{\theta}{2}e^{i(\varphi-\pi/2)} & 0 & 0 & \cos\frac{\theta}{2} \end{bmatrix} \tag{3.4}$$

where $\cos\theta = (1-x)/\sqrt{2(1+x^2)}$. When $\theta = \pi/2$, $\varphi = \pi/2$, it reduces to the braiding matrix related to the Bell basis.

$$\begin{aligned} \text{Type}-\text{I}: \quad & \breve{R}(u) = 1 + uP, \quad P^2 = I\,, \\ \text{Type}-\text{II}: \quad & \breve{R}_{\rm II}(x) = \cos\frac{\theta}{2} + M\sin\frac{\theta}{2} = \cos\frac{\theta}{2}(1+uM)\,, \end{aligned} \tag{3.5}$$

where $M^2 = -I$ and $u = \tan\frac{\theta}{2}$.
With the type-II solution the Berry's phase can be shown[12] and a chain model had been also solved.[13] The type-II solution can be simulated through quantum optics.[14] Acting the operator $\hat{U}$ on the topological basis $|e_1\rangle$ and $|e_2\rangle$ and taking (3.2) into account, after calculations we find for the type-II:

$$A(\theta) = D^{\frac{1}{2}\frac{1}{2}}(\theta, \varphi = 0) = \begin{bmatrix} \cos\frac{\theta}{2} & -\sin\frac{\theta}{2} \\ \sin\frac{\theta}{2} & \cos\frac{\theta}{2} \end{bmatrix} \quad (u = \tan\frac{\theta}{2}) \tag{3.6}$$

$$B(\theta, \varphi) = D^{\frac{1}{2}\frac{1}{2}}(\theta, \varphi) = \begin{bmatrix} \cos\frac{\theta}{2} & -\sin\frac{\theta}{2}e^{-i\varphi} \\ \sin\frac{\theta}{2}e^{i\varphi} & \cos\frac{\theta}{2} \end{bmatrix} \tag{3.7}$$

where $D^{\frac{1}{2}\frac{1}{2}}(\theta, \varphi)$ is the Wigner rotation function and satisfies YBE. For satisfying the braiding relation $ABA = BAB$, θ and φ should be constrained by[15]

$$\cos\varphi = \frac{\cos\theta}{1-\cos\theta} \tag{3.8}$$

The YBE reads in 2-d for the type-I

$$A(u)B(u+v)A(u) = B(u)A(u+v)B(u) \tag{3.9}$$

whereas for the type-II

$$A(u)B\left(\frac{u+v}{1+uv}\right)A(v) = B(u)A\left(\frac{u+v}{1+uv}\right)B(u) \tag{3.10}$$

where

$$\tan\frac{\theta}{2} = u \tag{3.11}$$

In general, the (2J+1)-d YBE takes the form

$$A(\theta_1)B(\theta_2,\varphi)A(\theta_3) = B(\theta_3,\varphi)A(\theta_2)B(\theta_1,\varphi) \tag{3.12}$$

which leads to[9]

$$\cos\varphi = \frac{1}{2}\left[\frac{(\tan\frac{\theta_1}{2}+\tan\frac{\theta_3}{2})-\tan\frac{\theta_2}{2}}{\tan\frac{\theta_1}{2}\cdot\tan\frac{\theta_2}{2}\cdot\tan\frac{\theta_3}{2}}-1\right] \tag{3.13}$$

for $A(\theta) = D^J(\theta,\varphi=0)$ and $B(\theta) = D^J(\theta,\varphi)$. The (2J+1)-d YEB takes the form shown by (3.12). When $\theta_1 = \theta_2 = \theta_3 = \theta$ (3.12) reduces to the braiding relation. The strategy for (2×2)-d representation of A and B can be extended to any $D^J(\theta,\varphi)$ functions for J=half (even integers) integers. Eq(3.13) still holds for any J in (3.12).

Why are there such two possibilities? It is related to the extremization of L_1-norm of the D-functions: Acting A or B on the topological basis the wave function becomes, say[9]

$$\Psi = \cos\frac{\theta}{2}\,|e_1\rangle - \sin\frac{\theta}{2}e^{i\varphi}\,|e_2\rangle \tag{3.14}$$

Extremum of L_1-norm $|\cos\frac{\theta}{2}| + |\sin\frac{\theta}{2}|$ reads

$$||\Psi||_{L_1} = \text{Extremum}\left(\left|\cos\frac{\theta}{2}\right| + \left|\sin\frac{\theta}{2}\right|\right) \tag{3.15}$$

Minimum $\theta = \pm\pi$ (for spin chain), and Maximum $\theta = \pm\pi/2$ (for Bell basis).

It is known well that in information theory the equivalence between L_0-norm and the L_1-minimization was introduced by Donoho-Candes-Tao in the compressed sensing theory for a sparse signal. In quantum mechanics, as usual only L_2 norm is interesting to normalize the probability. Here we see that the L_1-norm in quantum mechanics shows the entanglement in information.

For any J =half-integer, the common extremes occur at $\theta = \pm\pi/2$ and $\theta = \pm\pi$. For the general discussion see.[9]

Topological basis related to Birman-Wenzl-Marakami algebra

Suppose a braiding matrix S has three distinct eigenvalues

$$\prod_{i=1}^{3}(S-\lambda_i) = 0 \tag{3.16}$$

It can always be set $\lambda_1\lambda_2 = -1$. Denoting $\lambda_3 = \sigma$ and $w = \lambda_1 + \lambda_2$, we then have

$$S - S^{-1} = w(I - E) \tag{3.17}$$

Denoting by

$S_{12} =$  and $E_i =$ 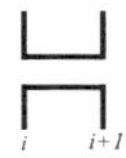 (3.18)

where for BWM algebra:

 $= w(I -$ $) = w(I - E)$ (3.19)

E_i satisfies T-L algebra, and there are many other relations for BWM algebra. Denoting the eigenvalues of a braiding matrix S with three eigenvalues by λ_1, λ_2, and λ_3, where S satisfies braid relation

$$S_{12}S_{23}S_{12} = S_{23}S_{12}S_{23} \qquad (S_1 \equiv S_{12} = S \otimes I, \quad S_2 = S_{23} = I \otimes S) \tag{3.20}$$

and without loss of generality by setting the eigenvalues of S to be[17] λ_1, λ_2 and λ_3 with

$$\lambda_1\lambda_2 = -1, \quad W = \lambda_1 + \lambda_2, \quad \lambda_3 = \sigma \tag{3.21}$$

we have

$$S - S^{-1} = WI + \frac{1}{\sigma}(I + WS - S^2). \tag{3.22}$$

Defining

$$E = \frac{1}{\sigma W}(S^2 - WS - I) \tag{3.23}$$

(3.22) becomes

$$S - S^{-1} = W(I - E). \tag{3.24}$$

Where S and E occupy k-th and $(k+1)$-th sites and satisfy all the relations of BWM algebra. Noting that a loop takes the value

$$d = 1 + \frac{1}{W}(\sigma^{-1} - \sigma). \tag{3.25}$$

3-*d Topological Basis of BWMA for S unitary*

When S is unitary with q a root of unity, namely

$$\lambda_\mu^* = \lambda_\mu^{-1} \text{ (e.g. } \lambda = q^m, \; q \text{ root of unity, with } m = \pm \text{ integer)}, \tag{3.26}$$

the basis reads $(i = 1, 2)$

$$|e_i\rangle = f_i \left(\quad + \alpha_i \quad + \beta_i \quad \right), \tag{3.27}$$

$$\langle e_i| = f_i^* \left(\quad + \alpha_i^* \quad + \beta_i^* \quad \right). \tag{3.28}$$

The 3-D matrix form is found:

$$A = \begin{bmatrix} \lambda_1 & & \\ & -\lambda_1^{-1} & \\ & & \lambda_3 \end{bmatrix}, \tag{3.29}$$

$B =$

$$\begin{bmatrix} \lambda_1^{-1}[\lambda_3(d-1) - \lambda_1^{-1}] & (1 - \lambda_1^{-1}\lambda_3 - \lambda_1\lambda_3 d) f_1^{-1} f_2 & \lambda_1^{-1} f_1^{-1} \\ -(1 + \lambda_1\lambda_3 + \lambda_1^{-1}\lambda_3 d) f_1 f_2^{-1} & \lambda_1[\lambda_3(d-1) + \lambda_1] & \lambda_1 f_2^{-1} \\ \lambda_1^{-1} f_1^{*-1} & -\lambda_1 f_2^{*-1} & (\lambda_1 + \lambda_1^{-1})\lambda_3^{-1} \end{bmatrix} \tag{3.30}$$

Following Ref.[17] for q a root of unity it allows 'non-standard' braiding matrices, say, for B_n some of $\lambda_k = q$, and for the others $\lambda_i = -q^{-1}$, then the general form of σ is

$$\sigma = \lambda_3 = q^m \quad (|m| \leq 2n). \tag{3.31}$$

It leads to $d = 1 - [m]q$, where m is the difference between the positive power number and negative ones in the power of q in the third eigenvalues of S for the fundamental representations of B_n, D_n and C_n.[16,17]

4. Conclusion Remarks

(1) There are two types of solutions of YBE and the corresponding braiding matrices. One is the most familiar ones associated with chain models and extensions that take the permutation as the rational limit. The other is related to the quantum information, say Bell basis and extensions. For both T-L and BWM algebraic structures by acting the braiding operators on the topological base we find that the N-dimensional solutions of YBE for both of the types are nothing but the Wigner's

D-functions. The "velocity additivity" in the YBE is Galileo's for the type-I as we have been got used to, whereas for the type-II it is Lorentz-like.

(2) We have shown that the two types of braiding ways in physics may be related to the extreme of L_1- norm of the D-functions. This provides a strong hint that the L_1- norm in quantum mechanics may explore the entanglement of two individual entries. As is known well that the entanglement exists neither entry 1 nor entry 2, but in the interaction between 1 and 2. If we use two lines to represent 1 and 2 (say,locations of two particles), then a braiding operation naturally describes the entanglement. Now a new knowledge is that the extreme of L_1- norm of D-functions describes the entangling degrees. Therefore YBE does not only favor what we have been familiar, say integrable models and remarkable progress in mathematical physics, but also quantum information theory.
The behavior of L_1-norm has been intensively studied in the information theory.[19] It deserves to apply the idea to quantum world.[20,21]

(3) The topological base can be viewed as a "bulk" of fermions or other particles. By acting the braiding operation on the topological base, a $(N^2 \times N^2)$-d YBE can be mapped onto $(N \times N)$-d YBE. It is natural to ask whether a new solvable system of Hamiltonian can be set up in parallel to the $(N^2 \times N^2)$-d cases. We hope the powerfulness of YBE in quantum information will be much developed in the future.

Acknowledgments

The authors appreciate Joan Birman, Z.H. Wang, M. Spera, K Niu, X.B. Peng, Y.S. Wu for the enlighten discussions.This work was supported in part by NSF of China No. 11075077.

References

1. R.J. Baxter, Exactly Solved Models in Statistical Mechanics, Academic Press, London, 1982.
2. D.C. Mattis, The Many-body Problem, World Sci. Publ., Singapore, 1993.
3. L. Faddeev, M. Henneaux, R. Kashaev, F. Lambert, A. Volkov (Eds) Bethe Ansatz: 75 Years Later, Proceedings, Solvay Workshops and Symposia, Université Libre de Bruxelles-Vrije Univ.Brussel, International Solvay Institute for Physics and Chemistry, 2006.
4. C.N. Yang and M.L. Ge (Eds), Braid Group, Knot Theory and Statistical Mechanics, World Sc. Publ, Singapore, 1990; M. Jimbo (ED), Yang-Baxter Equation in Integrable Systems, World Scientific Pub. Singapre, 1990.

5. L. Kauffman, Knots in Physics, World Scientific Pub. Singapore, 1991.
6. L. Kauffman and S. Lomonaco Jr, New Journal of Physics 6, 134 (2004).
7. Z. Wang, Topologization of Electron Liquids with Chern-Simons Theory and Quantum Computation, Differential Geometry an Physics, 106-120, Nankai Tracts. Math.10, World Sci. Publ, Hackensack, NJ, 2006; M.H.Freedman, M. Larsen, Z. Wang, Comm. Math. Phys., 227, 605 (2002); S. Das Sarma, M. Freedman, C. Nayak, Phys. Rev. Lett. 94, 166802 (2005).
8. J. Preskill, Lecture Notes on Topological Quantum Computation, (http://www.theory.caltech.edu/preskill/ph219).
9. K. Niu, K. Xue, Q. Zhao and M.-L. Ge, The $l_1$norm in Quantum Information via the Approach of Yang-Baxter Equation, J. Phys. A: Math. Theor. 44, 265304 (2011).
10. C. Nayak, F. Wilczek, Nucl. Phys. B479, 529 (1996).
11. E.H. Rezayi, N. Read, Nucl. Phys. B56, 16864 (1996).
12. J. L. Chen, K. Xue, and M. L. Ge, Berry Phase and Quantum Criticality in Yang-Baxter Systems, Annals of Physics 323, (2008)2614-2623; Phys Rev. A76,042324 (2007).
13. M.G. Hu, K. Xue and M.-L. Ge, Exact Solution of Yang-Baxter Spin-1/2 Chain Model and Quantum Entanglement, Phys. Rev. A 78, 052324 (2008).
14. S.W. Hu, K. Xue, and M. L. Ge, Phys. Rev. A 78, 022319 (2008).
15. A. Benveanu, M. Spera, Rev. Math. Phys. 18, 1075 (2006)(Singapore).
16. M. Jimbo, Comm. Math. Phys. 102, 537 (1986), Lett. Math. Phys.10,63 (1985)
17. Y. Cheng, M. L. Ge, K. Xue, Comm. Math. Phys. 136, 195 (1991).
18. T.D.M. Haldane, Phys. Rev. Lett. 50,1153 (1983).
19. D. Donoho, IEEE Transactions on Information Theory 52, 1289 (2006); E.Candes and T.Tao, IEEE Transactions on Information Theory 52, 5406 (2006)
20. R. Baraniuk, J. Romberg, and M. Wakin, "Tutorial on compressive sensing", 2008 Information Theory and Applications Workshop, (2008), http://www.dsp.ece.rice.edu/richb/talks/cstutorial-ITA-feb08-complete.pdf.
21. J. Altepeter, E. Jeffrey, and P. Kwiat, "Quantum states tomography", http://research.physics.illinois.edu/QI/Photonics/Tomography/.R.L.Kosut, Arxiv preprint arXiv:0812.4323 (2009).

THE TENSOR TRACK: AN UPDATE

VINCENT RIVASSEAU

Laboratoire de Physique Théorique, CNRS UMR 8627,
Université Paris-Sud, 91405 Orsay, France
and Perimeter Institute for Theoretical Physics
31 Caroline St. N., ON, N2L 2Y5, Waterloo, Canada
**E-mail: rivass@th.u-psud.fr*

The *tensor track* approach to quantum gravity,[1] is based on a new class of quantum field theories, hereafter called tensor group field theories (TGFTs).[2–6] We provide a brief review of recent progress and list some desirable properties of TGFTs. In order to narrow the search for interesting models, we also propose a set of guidelines for TGFT's loosely inspired by the Osterwalder-Schrader axioms of ordinary Euclidean QFT.

Keywords: Tensor models, quantum gravity, group field theory, axiomatic field theory.

1. Introduction

String theory and loop quantum gravity (LQG), the two leading approaches to quantum gravity, are currently stuck by a common problem: the lack of a convincing second-quantized non-perturbative formulation.

About twenty years ago, d-branes with $d \geq 3$ were recognized as key features of string theory. The non-perturbative framework that should explain the presence of branes and their beautiful associated dualities was called M-theory, where M means *matrix*, *m*ystery or *m*agic. But a simple action for this M-theory is still missing. This problem may be related to the huge and puzzling *landscape* of perturbative ordinary string vacua. The way forward may require some radical simplification.

A candidate for a non-perturbative second quantized formulation of LQG was quite early identified[7] as group field theory (GFT).[8,9] But GFT developed slowly and is not yet the mainstream formulation of LQG; in particular its correct combinatorics and renormalization have been found only recently.

The *tensor track* is a generalization of the random matrix approach to the quantization of two-dimensional gravity. It rebuilds early tensor models[10] and GFT around new principles derived from the universal properties of general random tensors.[11] It leads to a new class of quantum field theories which successfully renormalize GFT divergences, now correctly interpreted as ultraviolet rather than infrared in the Wilsonian sense*.

Properly supplemented with standard-model matter fields (and possibly supersymmetry?), this approach may hopefully some day relate different approaches such as LQG and superstrings through a framework that we could nickname T-theory (T like *tensor* or *t*otal). Indeed tensor models contain many embedded matrix models (their *jackets*[13]). They have therefore at least in principle the potential to quantize strings and higher dimensional branes on the same footing, leading to simpler models.

2. Basic Hypotheses

In the absence of direct experimental evidence, we expect the search for a good theory of quantum gravity to remain speculative and based on analogies for quite a while. The tensor track emphasizes quantum field theory, Feynman functional integrals, phase transitions and the Wilsonian renormalization group. Hence it reflects certain prejudices. Other approaches emphasize other concepts, such as the unification of all interactions, extended symmetries, canonical quantization, lattice regularizations etc...

Nevertheless the tensor track is rooted in deep convictions. Quantum field theory, functional integrals, phase transitions and the Wilsonian renormalization group together form our most advanced and most successful tools to understand physical systems with *many degrees of freedom*†. Only quantum field theory together with renormalization can compute accurately (more than ten digits!) physical effects which involve radiative corrections. Only quantum field theory has successfully renormalized all other interactions. Although gravity around a flat Minkovski background is not renormalizable in the perturbative sense, still the most conservative option seems to enlarge quantum field theory in a suitable minimal way to quantize it.

The tensor track bets upon the idea that quantum gravity should be background independent, even topology-independent, and that classical

---

*See[12] for renormalization of spin-foams based on lattice-like coarse-graining.

†We should in particular certainly not consider quantum field theory as just a way to combine special relativity and quantum mechanics, nor renormalization as a way to hide infinities. They are far more universal, as exemplified by their great success, first of course in particle physics but also in condensed matter, which is not relativistic, in statistical mechanics, which is not quantum, etc.

space-time and general relativity are effective concepts emerging from a more fundamental theory through one or several *phase transitions*, nicknamed geometrogenesis.[14]

Indeed phase transitions, whose modern understanding is provided by the Wilsonian renormalization group, are generic features of physical systems with many degrees of freedom. In fact the myriad of phase transitions leading to composite structures is perhaps the most obvious characteristic of our universe. Why would the geometry of the universe itself, with its huge number of degrees of freedom, not follow this trend? Another observation is that phase transitions occur at particular scale, hence they could provide an explanation for the existence of the Planck and cosmological constant scales in our universe. Finally of course geometrogenesis nicely fits with the big-bang, as they could be just the same thing.

The main *new* (hence perhaps most controversial) *bet of the tensor track* is to replace the ordinary principle of local interactions by an extended notion based on tensor invariance[‡]. But again this choice is not arbitrary. We believe that quantizing gravity is essentially the same as correctly randomizing geometry. Since our universe is very large, we need a robust tool to perform a statistical analysis of *large geometries in three and four dimensions*. The most fundamental tool in probability theory is the law of large numbers and the central limit theorem. The theory of random matrices and of their invariant interactions provides the equivalent of this tool to analyze two dimensional geometries. The recently discovered theory of higher rank random tensors and of their invariant interactions[11] is the natural candidate for their generalization to dimensions three and four.

3. GFT's and TGFT's

Consider a (simple, compact) Lie group G, endowed with its natural Haar measure and metric. Complex valued square integrable functions on G form an associated Hilbert space $H(G)$. The group structure on G allows Fourier analysis. $H(G)$ is infinite-dimensional, and admits various approximation

[‡]One of the few consensual ideas on the subject is that ordinary locality should be extended to quantize gravity. Spatial distances or areas smaller than the Planck scale cannot be measured in the usual way, as the measuring probes would disappear into the black hole created by their own gravitational field. However this does not mean that meaningful physics necessarily stops at that scale. Transplanckian *scales* could exist in the renormalization group sense even when there is no longer any well-defined notion of *distance*;[1] and transplanckian *physics* might be detected through indirect but convincing effects, which may lead to future predictions and physical discoveries.

schemes through finite N-dimensional vector spaces $H_N(G)$. Rank d tensor fields are defined as elements of the external tensor product of d copies of $H(G)$, or of $H_N(G)$, in which case we are interested in letting the cutoff N tend to infinity[§]. The scalar product in $H(G)$ or $H_N(G)$ allows to raise and lower tensor indices, hence to contract indices at identical positions between a tensor and its complex conjugate.

Equivalently a tensor field can be considered as a function on the product G^d of d copies of G, but this erases its tensor aspect. Group field theory (GFT) nevertheless emphasizes this second point of view; it is defined by an action for fields living on G^d.[9] In the initial example of group field theory, the Boulatov model,[8] the group G is $SO(d)$ or its universal covering group and the field incorporates a projection which averages over a common group translation of the d variables. This projection trivializes the holonomies along the faces of any Feynman amplitude, hence it implements the BF action on the 2-complex corresponding to the Feynman graph.[8] However the usual vertex envisioned by ordinary GFT is not a tensor invariant (in the precise sense defined below) and does not correspond to a stable action. The theory triangulates very singular pseudo-manifolds in addition to regular manifolds. No $1/N$ expansion has been found to organize the amplitudes of this theory, and although many amplitudes become infinite in the no-cutoff limit, they could not be properly *renormalized.*

This situation changed with the discovery of *colored* group field theories[15] and of their associated simpler random tensor models.[13] They triangulate better behaved spaces[16] and admit a 1/N expansion[17] (where N is the ultraviolet cutoff). Their uncolored formulation[11,18] rests on the classification of all the $U(N)^{\otimes d}$-*invariant interactions* of a pair of complex conjugate random tensors. It generalizes the standard invariance of (Wishart) matrix models. A welcome property is that such interactions are also often *stable* for a suitable sign of the coupling constant, curing one of the main problems of GFT.

The tensor track proposal conjectures that *this* tensor invariance should be the proper extension of the ordinary notion of *locality* needed to quantize gravity. Rebuilding GFT to incorporate this tensorial aspect of the field. we obtain a new class of quantum field theories, namely the TGFTs.[2,6]

However just as locality in quantum field theory is fundamental but is only an exact property of interactions, not of propagators, we expect in-

---

[§]As usual in QFT, tensor fields may be in fact *distributions* rather than functions but we skip this technicality here.

teresting TGFT's to have non-$U(N)^{\otimes d}$-invariant propagators. It is in fact the interplay between approximately local propagators and local interactions which launches the renormalization group flow of coupling constants in quantum field theory, and the same happens in TGFTs.

4. Desirable Properties

4.1. *Renormalization*

Just renormalizability is a property shared by all physical interactions except (until now!) gravity. In the renormalization group sense it is *natural.* Indeed just renormalizable interactions survive long-lived RG flows. They can be considered as the result of a kind of *Darwinian selection* associated to such flows. Therefore if quantum gravity can be renormalized as proposed in[1,19] it will rely on the same powerful technique that applies successfully to all other interactions of the standard model. There will be no longer any need for a teleological or anthropic interpretation.

The simplest renormalizable TGFT has been found in dimensions three and four for the $U(1)$ group.[2,4] In dimension 4 it has two unexpected ϕ^6-like marginal interactions, hence a richer RG flow than the usual ϕ^4 models.[3]

4.2. *Asymptotic Freedom*

Again asymptotic freedom is a property shared by all physical interactions except (until now!) gravity. Indeed QCD is asymptotically free and the electromagnetic sector inherits at high energy the asymptotic freedom of the unified electroweak theory¶.

Asymptotic freedom is desirable to build a geometrogenesis scenario for TGFTs,[14,20–22] and in fact may be generic in the world of tensors of rank greater or equal to three. It has been *already established* for the simplest renormalizable TGFTs in dimension 3 and 4.[3,4] The new locality axiom allows wave function renormalization to compete with coupling constant radiative corrections, and typically to win in the case of rank ≥ 3 tensors. Recall that absence of asymptotic freedom is the rule for the simplest models of scalar, vector and matrix type (except of course non-Abelian gauge theories) and that asymptotic safety is barely reached for natural matrix field theories such as the Grosse-Wulkenhaar model.[23] In addition, the infrared growth of the coupling constant occurs for the *stable* sign of the

¶The ultraviolet behavior of the Higgs sector is a subtle issue not considered here.

interaction, hence may lead to the discovery of singularities which could represent *unitary matter*. This would improve on single-matrix model singularities which lead to (non-unitary) Lee-Yang type singularities.

4.3. *Constructibility*

Constructibility of a quantum field theory means that its perturbative series can be uniquely resummed (typically through a kind of Borel resummation).[26,27] Physically it is related to *stability and uniqueness of the vacuum*. It guarantees that at least the perturbative phase of the theory is unique and mathematically well-defined at small coupling.

TGFTs with stable positive interactions should be constructible, and the corresponding proofs seem doable, thanks to a new constructive tool called the loop vertex expansion (LVE),[28] adapted to the extended notions of locality that govern matrix or tensor models. Significant results have been already obtained in this direction.[11,29] We expect the full Borel summability of renormalizable asymptotically free TGFTs to be more difficult than those of infrared ϕ^4_4 and of the Gross-Neveu model,[27] but much simpler than the corresponding study for non-Abelian gauge theories.

The existence of such a constructive perspective is a very important *long term asset* of the tensor track program, which (to our knowledge) is missing in all other current approaches to quantum gravity.

4.4. *Geometricity*

We are ultimately interested in models whose effective infrared physics leads naturally (under suitable boundary conditions) to our universe, namely a large quasi-flat four dimensional space-time with a metric obeying the (classical) Einstein equations.

Recently models were developed which incorporate the constraint projector of the BF theory. This could lead to a geometrogenesis with a smoother metric. The first four-dimensional models of this type have been proved superrenormalizable on the $U(1)$ group.[6] We expect ϕ^6 models in dimension 3 and on the $SU(2)$ group to be just renormalizable.[25]

To guide geometrogenesis towards the desired outcome in four dimensions we may have to decorate the most natural renormalizable TGFT's with additional *geometric conditions*. Spin foam models, in particular the 4d models which incorporate Plebanski simplicity constraints[24] could inspire such decorations. We are open to other possibilities, as the only rule is to find the *simplest* such models with gravity as their effective limit.

4.5. *Dualities, holography*

Dualities such as Born duality, Langmann-Szabo dualities in the Grosse-Wulkenhaar matrix models or the many dualities of string theory may have interesting analogs in the TGFT world. Such dualities could allow integrability and exact solvability of particular TGFT models. This possibility should be systematically investigated.

Similarly it might be interesting to incorporate some kind of holographic principle in TGFT's. The structure of the boundaries of TGFT amplitudes, which are themselves lower rank TGFT vacuum amplitudes, suggests some principle of this kind.

5. Rules for TGFT's

We sketch now tentative rules for TGFTs that could later evolve into a true axiomatic scheme. Axioms embody the long term reflection of the scientific community on the most fundamental aspects of quantum field theory and are therefore a valuable source of inspiration into unexplored territory such as quantum gravity. But at this early stage we intentionally formulated our proposal in a non-technical language. It should not be considered rigid nor exclude interesting future developments (for instance Fermionic axioms etc...). The hard work, which remains entirely to be done, requires a more precise mathematical formulation of these rules and the proof that interesting interacting TGFTs indeed obey them.

According to our conservative analogy-based approach we search for natural analogs of the main axioms of Euclidean quantum field theory. These new rules should imply a new kind of constructive program, for TGFTs. The initial constructive program[26,27] is far from complete, as it does not include yet the full construction of the four dimensional Yang-Mills theories. However a constructive program for TGFTs could actually progress faster in the coming years, since interesting asymptotically free models may be free of subtle constructive issues such as Gribov ambiguities which plague the ordinary non-Abelian gauge theories.

Our rules are formulated in terms of approximation schemes based on limits of functional integrals with cutoffs. In ordinary quantum field theory we know that each cutoff violates some axiom; but uniqueness of the limit typically ensures that the theory without cutoffs satisfies all of them.

Rule 1: Tensor Invariance and Positivity of Interactions

This rule replaces locality. The fields are considered both as functions on G^d, the *pre-space*, and as rank-d tensors on $H(G)$. The bare functional measure $d\nu(\phi, \bar{\phi})$ of a TGFT should be formally of the form

$$d\nu(\phi, \bar{\phi}) = \frac{1}{Z} d\mu_C(\phi, \bar{\phi}) e^{-S_{int}(\phi, \bar{\phi})}, \quad S_{int}(\phi, \bar{\phi}) = \sum_{b \in \mathcal{B}} \lambda_b I_b(\phi, \bar{\phi}), \tag{1}$$

where C is the covariance or bare propagator, $\mathcal{B}$ is a finite set of *connected positive tensor invariants* labeled by b, and the coupling constants $\lambda_b \in \mathbb{C}$, should have positive real parts. The non formal definition requires as usual to introduce an ultraviolet cutoff N, then to control the limit $N \to \infty$.

Tensor invariants are obtained by convolution of a set of fields φ and $\overline{\varphi}$, in such a way that the k-th index of a field φ is always contracted with the k-th index of a conjugate field $\overline{\varphi}$, resulting in a polynomial invariant under $U(N)^{\otimes d}$. They are canonically represented by closed bipartite *d-colored graphs*: each field φ (resp. $\overline{\varphi}$) is represented by a white (resp. black dot), and each contraction of a k-th index between two fields is pictured as a line with *color* label k linking the two relevant dots (see Figure 1). *Connected* invariants correspond to connected graphs. *Positive* invariants admit a mirror symmetry allowing to write them as sums of moduli squares. For instance in Figure 1 the first two tensor invariants admit such a symmetry, but not the last one. The conditions on positive real parts for the couplings λ_b ensure stability of the corresponding action.

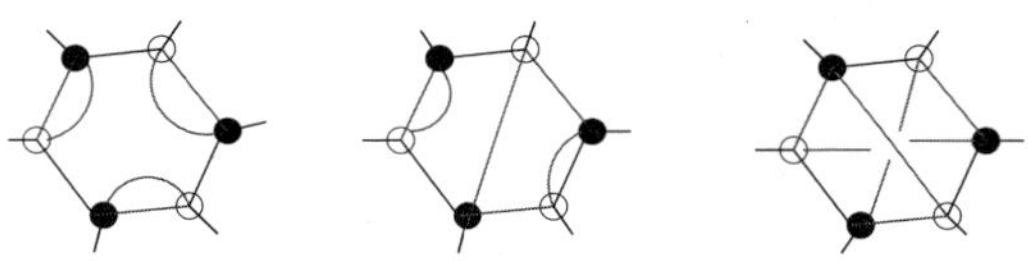

Fig. 1. Some connected tensor invariants in $d = 3$

Rule 2: Discrete Permutational Symmetry

We suggest to replace continuous rotational and translational invariance by a discrete permutation invariance, as appropriate in case of geometric discretizations. Hence we require that the Schwinger functions should be invariant under the discrete symmetry group Σ_d with $d!$ elements. In particular tensorial interactions should be symmetrized over the discrete permutations of the d space-time colors. This implies constraints on the coupling constants: they should be equal for invariants which differ only by a permutation of colors.

Rule 3: Clustering

The clustering axiom in ordinary QFT requires the Schwinger functions to decay as their external arguments are taken apart. In the pregeometric tensor world external arguments of rank d models represent boundaries which are themselves colored tensor models of rank $d-1$. There is not any good notion of distance yet, hence we should rather ask for a decay in the defining parameters of the sum $S_r(c, d_1, \cdots d_c)$ of Feynman amplitudes which have r external legs defining a boundary with c connected components, each having *degree* d_i.

In case of exponential decay, this would mean that there exist constants $K > 0$ and $\epsilon > 0$ such that

$$|S_r(c, d_1, \cdots d_c)| \leq K e^{-\epsilon(r+c+\sum_{i=1}^{c} d_i)} \tag{2}$$

The number of connected components of the boundary and the sum of their degrees gives some measure of the complexity both of the topology and of the cellular structure of that boundary. Decay of this type holds at the perturbative level for the models considered so far.[2] Beware however that the notion of *connectedness* may depend on the models considered.[6]

Rule 4: Positivity and Mirror-Positivity of Propagator

The only quadratic invariant for tensors is the mass term which is also local. Hence *locality coincides with tensor invariance for the 2-point function.* We want to consider a bare propagator which softly breaks both of these invariances, in order to launch a renormalization group flow. We require that it should have a non-trivial positive spectrum, allowing its parametric representation as $C = \int_0^\infty e^{-\alpha C^{-1}} d\alpha$. For renormalization group analysis to work we also need that it should become approximately local in the ultraviolet regime $\alpha \to 0$.

Most controversially perhaps, we suggest that an analog of Osterwalder-Schrader positivity or of the Markov property of ordinary Euclidean fields should also hold for TGFTs. Indeed this key property of Euclidean fields allows the continuation to Lorentzian time and the unitary time evolution of states. Having an analog of that in the tensor world could hopefully also leads to a time interpretation of the resulting effective theory. Without it we could build arbitrarily many convergent quantum field theories in four dimensions, namely non-unitary theories with *ultraviolet cutoffs*. We certainly want to exclude these.

Assuming convergence problems solved through constructive theory, any Euclidean quantum field theory with OS-positive bare propagator and *local interactions* is fully OS positive.[26] Since the interactions we consider are non-local, we have not found yet any analog of this result for tensors. But we think interesting, at least as a heuristic and tentative rule, to suggest the bare propagator of our theory to be mirror-positive in the following sense. The pre-space G^d, thanks to the group structure of G, comes equipped with d fundamental involutions, or mirror symmetries

$$(g_1, ..., g_i, ...g_d) \to S_i(g) = (g_1, ..., g_i^{-1}, ...g_d).$$

We could then define a generalized OS-positivity of the propagator C, which means that the p by p matrix with matrix elements $C(g_k, S_i(g_l))$ should be positive for any $i \in \{1, \cdots d\}$ and any finite collection of p arguments $\{g_k\}$ in G^d. This property holds for propagators admitting a Euclidean Källen-Lehmann representation

$$C = \int_0^\infty \frac{d\rho(m)}{-\Delta + m}, \tag{3}$$

where Δ is the Laplacian on G^d and $d\rho(m)$ a positive measure. This representation excludes better ultraviolet behavior than the one of the Laplacian.

Another strong argument for the Laplacian as propagator in the pre-space comes from the Taylor expansion around divergent two-point functions required by renormalization.[30] Ultimately the Laplacian is a natural choice on G, which as a (simple, compact) Lie group comes equipped with a differential structure and a metric, so TGFTs should use it.

6. Conclusion

Renormalizable 3D and 4D TGFTs *exist*. The most natural models are asymptotically free, an encouraging fact for geometrogenesis scenarios. Adding some pre-geometric content is possible at least in some cases (eg $U(1)$ in $D = 4$, probably SU(2) in $D = 3$). Axiomatic schemes can be considered for TGFTs, leading to a new constructive program. Simplified models have been proved Borel-summable using the LVE.[28]

The main open problem is to analyze geometrogenesis of natural renormalizable TGFTs and to find the right pre-geometric content that would lead to general relativity as effective lower energy physics in dimension 4.

Success, for probably a long time to come, will be measured in terms of how far these attempts can be pushed on the mathematical level, how convincingly they imply general relativity as effective limit and how many applications to different domains they bring. In this last respect the tensor track is promising, as it is linked to a growing list of applications of random tensors to statistical mechanics.[31]

Acknowledgments

This paper has greatly benefitted from many discussions with J. Ben Geloun, S. Carrozza, R. Gurau and D. Oriti. I would like to thank the organizers of the XXIX International Colloquium on Group-Theoretical Methods in Physics and the Perimeter Institute for support.

References

1. V. Rivasseau, "Quantum Gravity and Renormalization: The Tensor Track," AIP Conf. Proc. **1444**, 18 (2011) [arXiv:1112.5104 [hep-th]].
2. J. Ben Geloun and V. Rivasseau, "A Renormalizable 4-Dimensional Tensor Field Theory," Commun. Math. Phys. 2012, (DOI) 10.1007/s00220-012-1549-1, arXiv:1111.4997 [hep-th].
3. J. Ben Geloun, "Two and four-loop β-functions of rank 4 renormalizable tensor field theories," arXiv:1205.5513 [hep-th].
4. J. Ben Geloun and D. O. Samary, "3D Tensor Field Theory: Renormalization and One-loop β-functions," arXiv:1201.0176 [hep-th].
5. J. B. Geloun and E. R. Livine, "Some classes of renormalizable tensor models," arXiv:1207.0416 [hep-th].
6. S. Carrozza, D. Oriti and V. Rivasseau, "Renormalization of Tensorial Group Field Theories: Abelian U(1) Models in Four Dimensions," arXiv:1207.6734 [hep-th].
7. M. Reisenberger, C. Rovelli, Class. Quant. Grav. 18 (2001) 121-140, gr-qc/0002095
8. D. V Boulatov, Mod.Phys.Lett. A**7**: 1629-1646 (1992).
9. L. Freidel, "Group field theory: An overview," Int. J. Theor. Phys. **44**, 1769 (2005) [arXiv:hep-th/0505016]; D. Oriti, "The microscopic dynamics of quantum space as a group field theory," arXiv:1110.5606 [hep-th].
10. M. Gross, Nucl. Phys. Proc. Suppl. **25A**, 144-149, (1992); J. Ambjorn, B. Durhuus, T. Jonsson, Mod. Phys. Lett. **A6**, 1133-1146, (1991); N. Sasakura, Mod.Phys.Lett. A6 (1991) 2613-2624
11. R. Gurau, "Universality for Random Tensors," arXiv:1111.0519 [math.PR].
12. B. Bahr, B. Dittrich, F. Hellmann and W. Kaminski, "Holonomy Spin Foam Models: Definition and Coarse Graining," arXiv:1208.3388 [gr-qc].
13. R. Gurau and J. P. Ryan, "Colored Tensor Models - a review," SIGMA **8**, 020 (2012) [arXiv:1109.4812 [hep-th]].
14. T. Konopka, F. Markopoulou and S. Severini, "Quantum Graphity: a model of emergent locality," Phys. Rev. D **77**, 104029 (2008) [arXiv:0801.0861 [hep-th]].
15. R. Gurau, "Colored Group Field Theory," Commun. Math. Phys. **304**, 69 (2011) [arXiv:0907.2582 [hep-th]].
16. R. Gurau, "Lost in Translation: Topological Singularities in Group Field Theory," Class. Quant. Grav. **27**, 235023 (2010) [arXiv:1006.0714 [hep-th]].
17. R. Gurau, "The 1/N expansion of colored tensor models," Annales Henri Poincaré **12**, 829 (2011) [arXiv:1011.2726 [gr-qc]]; R. Gurau and V. Ri-

vasseau, "The 1/N expansion of colored tensor models in arbitrary dimension," Europhys. Lett. **95**, 50004 (2011) [arXiv:1101.4182 [gr-qc]]; R. Gurau, "The complete 1/N expansion of colored tensor models in arbitrary dimension," arXiv:1102.5759 [gr-qc].

18. V. Bonzom, R. Gurau and V. Rivasseau, "Random tensor models in the large N limit: Uncoloring the colored tensor models," Phys. Rev. D **85**, 084037 (2012) [arXiv:1202.3637 [hep-th]].
19. V. Rivasseau, "Towards Renormalizing Group Field Theory," PoS CNCFG **2010**, 004 (2010) [arXiv:1103.1900 [gr-qc]].
20. D. Oriti, "A quantum field theory of simplicial geometry and the emergence of space-time," J. Phys. Conf. Ser. **67**, 012052 (2007) [arXiv:hep-th/0612301].
21. D. Oriti, "Group field theory as the microscopic description of the quantum space-time fluid: a new perspective on the continuum in quantum gravity," arXiv:0710.3276 [gr-qc].
22. L. Sindoni, "Emergent models for gravity: an overview," arXiv:1110.0686 [gr-qc].
23. M. Disertori, R. Gurau, J. Magnen and V. Rivasseau, "Vanishing of Beta Function of Non Commutative Phi**4(4) Theory to all orders," Phys. Lett. B **649**, 95 (2007) [hep-th/0612251].
24. L. Freidel, K. Krasnov, Class. Quant. Grav. **25**, 125018 (2008) [arXiv: 0708.1595]; J. Engle, R. Pereira, C. Rovelli, Nucl. Phys. B **798**, 251 (2008), [arXiv: 0708.1236]; J. Engle, E. Livine, R. Pereira, C. Rovelli, Nucl. Phys. B **799**, 136 (2008), [arXiv:0711.0146]; J. Ben Geloun, R. Gurau, V. Rivasseau, Europhys.Lett. 92 (2010) 60008, arXiv:1008.0354 [hep-th] A. Baratin and D. Oriti, Phys. Rev. D **85**, 044003 (2012), arXiv:1111.5842 [hep-th].
25. S. Carrozza, D. Oriti and V. Rivasseau, work in preparation
26. J. Glimm, A. Jaffe, "Quantum Physics: A Functional Integral Point of View", Springer-Verlag, 1987.
27. V. Rivasseau, "From perturbative to constructive renormalization," Princeton University Press, 1991.
28. V. Rivasseau, "Constructive Matrix Theory," JHEP **0709**, 008 (2007) [arXiv:0706.1224 [hep-th]].
29. J. Magnen, K. Noui, V. Rivasseau and M. Smerlak, "Scaling behaviour of three-dimensional group field theory," Class. Quant. Grav. **26**, 185012 (2009) [arXiv:0906.5477 [hep-th]].
30. J. Ben Geloun and V. Bonzom, "Radiative corrections in the Boulatov-Ooguri tensor model: The 2-point function," Int. J. Theor. Phys. **50**, 2819 (2011) [arXiv:1101.4294 [hep-th]].
31. V. Bonzom, R. Gurau, A. Riello and V. Rivasseau, "Critical behavior of colored tensor models in the large N limit," Nucl. Phys. B **853**, 174 (2011) [arXiv:1105.3122 [hep-th]]; V. Bonzom, R. Gurau and V. Rivasseau, "The Ising Model on Random Lattices in Arbitrary Dimensions," Phys. Lett. B **711**, 88 (2012) [arXiv:1108.6269 [hep-th]]; V. Bonzom and H. Erbin, "Coupling of hard dimers to dynamical lattices via random tensors," arXiv:1204.3798 [cond-mat.stat-mech]; V. Bonzom, R. Gurau and M. Smerlak, "Universality in p-spin glasses", arXiv:1206.5539.

TOWARDS THE ORIGIN?

DANIEL TREILLE
Physics Department, ETH Zurich and CERN
Geneva, 1223, Switzerland

Traveling backward in time along the timeline of the Universe, we underline what is known experimentally, how sound is our understanding at various moments of its evolution and, towards the origin, at which point hypotheses start to replace certainties.

Keywords: Universe, cosmology, particle, Standard Model, Brout-Englert-Higgs boson, Supersymmetry, Grand Unification, inflation.

Introduction

The title may seem a bit presumptuous, but the goal is simply to underline what is actually known experimentally about the Universe, how sound is our understanding at various moments of its evolution and at which point hypotheses start to replace certainties. To do so, we will travel backward in time along the timeline of the Universe in the expanding model which is now the prevailing one. Due to size constraints, this exposé is a short version of a more complete one, containing figures and references.[1]

Starting from the present time, one can learn about the history of the Universe by observing it, using various techniques, all of them relying presently, as messengers of the past, on electromagnetic radiation, i.e. photons of a large variety of energies. This information is accessible as long as the Universe was transparent to this radiation, namely after the formation of atoms.

For ages younger than this epoch of recombination (at 380000 years of age), when atoms formed from light nuclei and electrons, the only way to get information is to re-create in a microscopic way samples of the physics which was governing the Universe at a given time.

This is the role of particle accelerators and colliders, providing the required energies. We will make a few stops at some crucial early times of the evolution of the Universe and a longer and more detailed one at the present high-energy frontier of particle physics, dominated by the Large Hadron Collider (LHC) at CERN. We will describe its recent discovery of a new particle, explain what this could mean and comment on the main other searches under way at the energy and intensity frontiers of present machines.

For energies which will likely stay for ever beyond the reach of colliders, i.e. for times earlier than a picosecond (10^{-12} s) or so, one must then extrapolate the results obtained so far, and this can only be done in the frame of various hypothetical models of particle physics. We will introduce briefly some of these models and present the hints they suggest at high energy. We will then discuss the main ideas concerning the very first moments of the Universe, and, with many question marks, reflecting both the general limits of our understanding and the author's own ones, we will present the speculations about its origin.

1. The expanding Universe

The model of an expanding Universe, evolving from small to big, from hot to cold, from dense to diluted, from simple to complex, is presently unanimously recognized as the most credible one. About this model, what one should keep in mind is the representation of an expansion, as the uniform expansion of the arms of Escher's lattice (as for many references, please see[1]) would suggest, rather than the confusing idea of an explosion (of what into what?).

Let us first introduce the key authors of the expanding Universe model. Einstein, with its General Relativity, wrote the relevant equations. Friedmann was the first to get the solution corresponding to an expanding universe, an idea shared by Lemaître, but which Einstein first refused. Hubble brought the first experimental proof of the expansion by showing that galaxies recede from one another at a speed proportional to their distance, the Hubble law ($v = H_0 D$), which became the first pillar of the model. Several other great names, de Sitter, Weyl, Schwarzschild, etc, played important roles in this grandiose construction. The present value of the Hubble constant H_0 is 72.1 ± 1.7 (km/s) / Mpc (a parsec, pc, is 3.09 10^{13} km).

Presently, the model is exploited in the simplifying frame of the Cosmological Principle, namely the assumption of a homogeneous and isotropic space. While this view is probably valid at very large scales, the Universe is obviously not homogeneous at small and intermediate scales, since for instance a void of a billion light-years across is known. It is always good therefore, before drawing conclusions, to remember the simplification assumed.

2. The timeline of the Universe

As already said, one learns about the Universe first by "looking" at it, as long as it is transparent, then, when it becomes opaque, by "recreating" microscopic samples of its physics.

Looking far in space means also looking far in the past. The unit of distance will be the light-year (l.y., $9.46\ 10^{15}\,\mathrm{m}$). An observer looking for instance at the Sloan Great Wall, 1 billion l.y. away, sees it as it was about 1 billion years ago.

The messengers are photons of all wavelengths and energies, from TeV ($10^{12}\,\mathrm{eV}$) gamma rays to radio waves. A crucial fact is that the expansion of the Universe expands the wavelength of a given radiation. A number, the z factor or red shift, measures this dilatation. For large values, z gives simply the ratio of the observed and emitted wavelengths. Visible light emitted at $z = 1000$ is thus detected at a millimeter wavelength. The relation between z and age is complex.

It may be that in the future other messengers from the past, as gravitational waves (GW), ripples of space-time itself, which do not care about the transparency of the Universe, will bring further information. Large terrestrial interferometers as LIGO, VIRGO are taking data, with their advanced versions under way, which will increase much the volume of the Universe they can scan for sources of GW. The LISA interferometer, three satellites forming a triangle of $5\ 10^6$ km arms, is foreseen, after its Pathfinder exploratory mission in 2015 or so. As for the re-creation of samples, it is obtained by colliding accelerated particles of the relevant energy. In the collision, their kinetic energy is transformed into newborn particles. An electron-positron collision at LEP can for instance lead to a pair of Z^0 bosons, each decaying in a quark and an antiquark, which manifest themselves as jets of ordinary particles.

Obviously, with subatomic physics, we enter a world obeying to the laws of quantum theory and special relativity, synthesized for instance in the formulation of the Dirac equation.

A quick remark about the term "high energy": the energy of a proton-proton collision at LHC is no more than the one of a little mosquito in flight. The most energetic cosmic ray recorded is less energetic than a service ball of a tennis champion. However these tiny energies are delivered in an infinitesimal volume and this leads to huge densities of energy.

Fig 1 gives in logarithmic scales the timeline of the Universe in the expanding model. In abscissa is the age of the Universe. The log scale shrinks dramatically the domain of observation, although it covers nearly the full time span, over thirteen billions of years during which the force of gravity was at work, shaping the present universe. In ordinate is temperature, also in log scale. In the upper scale we indicate the typical energy of particles at a given time.

With this choice of coordinates, the timeline is nearly exactly the straight line drawn, with an imperceptible change of slope when the universe changes from radiation to matter domination, near the time of recombination, and the "recent" acceleration of the expansion to which we come back later. We also give, for information, an idea of the size of the presently observable Universe at the corresponding age. These huge sizes illustrate well the fact that, contrary to hasty statements, high-energy physics is not "re-doing the Big Bang".

Before the recombination, the important transitions are:

- the synthesis of light nuclei, in the first few minutes, when the Universe was acting as a thermonuclear reactor, and which is therefore the domain of nuclear physics;
- the quark-hadron transition, at 1 to 10 microsecond, when quarks, free until then, got trapped inside hadrons. This is the domain of hadronic physics, at a typical energy of a GeV (the mass of the nucleon) and, experimentally, of the heavy ion collisions performed at colliders, in particular at LHC;
- the transition corresponding to electroweak symmetry breaking, which occurred at a picosecond or so. This is the time where elementary particles are supposed to have got their masses. This transition is the focus

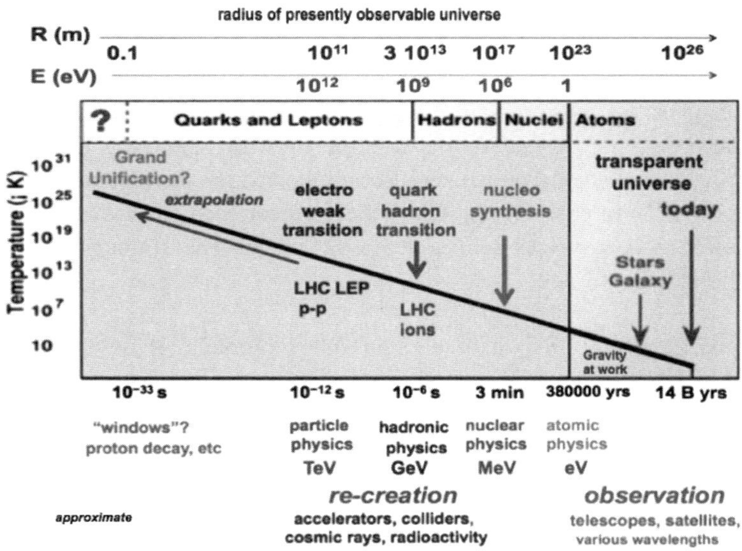

Fig. 1. The timeline of the Universe.

of the experimentation at the high energy frontier, dominated presently by the LHC proton-proton collisions. Given the important results obtained recently at LHC, our stop there will be longer and abundantly documented.

The first observational glimpse at the young universe, at 380000 years of age, when the atoms formed and after which photons could travel freely, is a pivotal information which deserves a special attention. Indeed the cosmic microwave background (CMB) emitted at this time of recombination exhibits, as we will see, some fluctuations of temperature, indicating fluctuations of density, which are the seeds of the large scale structures of the present Universe. It is also suspected that these fluctuations may themselves result from the quantum fluctuations of a field during an hypothetical very early phase of the Universe, called inflation: a brief exponential growth of space, with a super-cooling and a reheating phases, sudden and dramatic changes of parameters, quite different from the ulterior smooth timeline.

3. Observing the Universe

The set of observational pieces of information about the Universe have been obtained by a collection of large instruments operating in a variety of wavelengths: let us quote the Hubble telescope, which should hopefully be replaced by the James Webb telescope in 2018, the Very Large Telescope (VLT) of Atacama which should be followed by the Extremely Large Telescope (ELT) in early 2020s. The oldest object seen by Hubble is a tiny galaxy seen as it was 13.2 billion years ago, when the Universe had only 480 millions years of age.

The ground based LSST, a camera of 3.2 gigapixels, is being prepared for 2015 in Chile. The satellite EUCLID mission is under study for about 2019.

These observatories explore the far universe, as far as their resolution allows, and offer a variety of information about large scale structure of galaxies, supernovae and the distribution of mass, including dark matter, through lensing effects.

Before reaching (still going backward) the recombination time, one encounters the Dark Ages of the Universe, a period where the Universe was transparent but had no light emitted yet, since no source of light (galaxies, stars) already existed. After some delay, the atomic hydrogen from recombination started emitting its typical 21cm line, corresponding to an atomic transition between spin states. Emitted at a z of about 10, this radiation has now a 2m wavelength, in the domain of radio waves. When the first light sources occurred, they re-ionized progressively the hydrogen around them and stopped the emission. These important studies of the re-ionization period will be led in particular by giant radiotelescopes like the distributed Square Kilometer Array (SKA). China has already such telescopes at work or under realization, like FAST, a huge "bowl" of diameter 500 m, to be installed in Guizhou, and PAST, a spectacular deployment of 10000 TV antennas, looking at the polar region of the sky and at work in the Tian Shan mountains.

4. The Cosmic Microwave background (CMB)

If your eyes were sensitive to millimeter scale radiation and very high quality instruments, the sky would look like on Fig 2, as observed by WMAP.

We are indeed bombarded by microwaves coming from all directions of space. This is the fossil Cosmic Microwave background (CMB), second pillar of our vision of the expanding universe. It corresponds to a perfect Planck spectrum of 2.725 K. This was emitted at 3000 K at a z of 1100, hence the present millimeter wave length. It was predicted, approximately, in 1948 by Gamow and discovered in 1965 (Nobel 1978). Its temperature is very homogeneous over the full sky and it took some time and effort to spot its inhomogeneities at the level of 10^{-5}, what the COBE satellite did first (Nobel 2006), followed, with much better resolution, by WMAP.

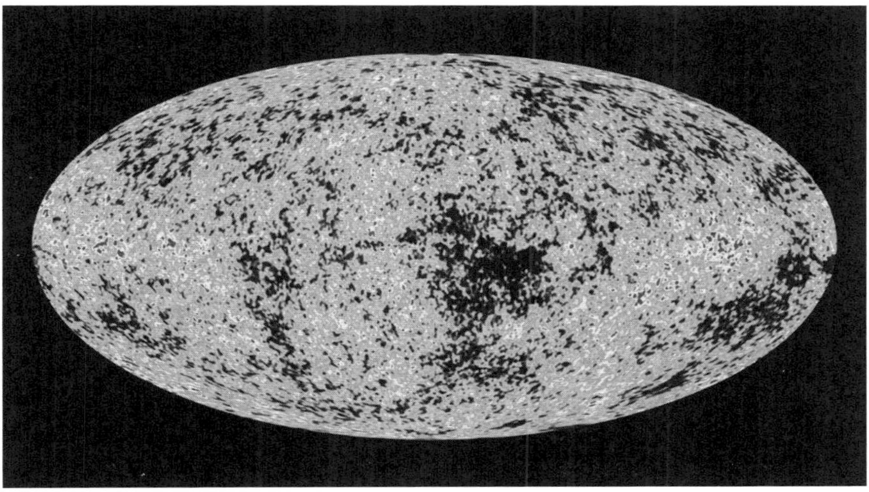

Fig. 2. The CMB as seen by WMAP. Color spots indicate small temperature fluctuations.

The power spectrum of the CMB, namely the distribution in angular opening of the patterns it exhibits, is dominated by a series of well-understood acoustic peaks. They reflect the oscillations of the plasma of electrons and baryons at the time of last scattering, under the counteracting forces of gravity and the outward pressure due to photon-matter interactions before decoupling. The main peak corresponds to an angular scale of about 1°. The power spectrum has delivered an amazing quantity of information about the properties and content of the universe.

The CMB is also polarized and the study of its modes of polarization is the present focus of the observations, as well as its "non-gaussian" behaviour, not expected from the simplest versions of inflation, i.e. 3-point correlations, instead of the 2-point ones studied so far.

Above our heads, the PLANCK satellite has already accomplished its mission and, after many publications on the foregrounds, should deliver soon its first results on the CMB itself. Ground based observations (in particular from Antartica) are also in the game.

5. Bilan of observation

Let us briefly summarize what these observations of our Universe, up to the Last Scattering time, told about its properties.

From the CMB, one learned first that the Universe is spatially flat, i.e. euclidean. Indeed, through the study of acoustic peaks, i.e. the observation of understood patterns of a known size at a known distance from the observer, proved that the photons of the Last Scatter came to us along straight paths.

From the study, as a function of z, of the brightness of Supernovae, exploding stars used as standard candles, one concludes that our universe is not only expanding, but also in a phase of accelerated expansion (Nobel 2011). This was confirmed by WMAP. From the basic formulae of cosmology, one deduces easily that some kind of repulsive gravity, with negative pressure, is required. Lacking a better name, it was called dark energy.

The need for dark matter being recognized since long, this leads to the present "cocktail" of the Universe content (fig 3), generally accepted and called the Concordance Model. The figure tells where the information come from. Baryon acoustic oscillations (BAO) correspond to fluctuations in the density of visible baryonic matter caused by the acoustic waves at the last scattering time. They provide a "standard ruler" (500 M l.y. in the present Universe) measured by looking at the large scale structure of the Universe by astronomical surveys.

The cocktail shows that known forms of matter represent only a tiny part of it. One should however recall that "heretic" alternatives are still being considered, as a possible modification of General Relativity at very large distances or the effect of the non-homogeneity of the present Universe at intermediate scales.

6. Re-creating samples

Let us now cross the wall of the "visible" and enter the early times, for which the only method of investigation is to re-create samples of the physics prevailing then.

At the high-energy frontier (the TeV scale, and high energy physics at LHC), the objects of our studies and the observables in our detectors are leptons, propagating freely, and quarks, confined, manifesting themselves as collimated jets of ordinary particles. These objects do not show any structure, down to 10^{-17} cm, and are for the time being considered as point like and elementary.

On various sites, one can find photos of past (starting with little ADA which was the first e^+e^- collider) and present colliders, and of LHC detectors and sub-detectors. No picture can replace a visit to the underground experimental zones of LHC, which will be quite accessible next year, during a long shutdown of the machine

7. The first few minutes

Our first stop will be at the first few minutes, when the Universe was functioning as a thermonuclear reactor, synthesizing light nuclei from protons and neutrons. This is a fairly well documented period.

To estimate the outcome of this transition, two "terrestrial" pieces of information are needed. Since the neutron is part of the game, an accurate knowledge of its lifetime is necessary. The accuracy of this measurement is progressively improving, in particular due to the availability of ultra cold neutrons sources. One needs also to know how many species of light active neutrinos were participating at that time. Since LEP, it is clear that the answer is three.

Then one can plot the expected rates of various light nuclei, starting with helium, as a function of the key parameter, the density of ordinary matter relative to photons, another result from the study of the CMB. For the value obtained by WMAP, the expectation is 25% of helium 4 (for mass abundance), which is exactly what astrophysicists measure.

Let us recall that nuclei heavier than Be will be synthesized later in stars,

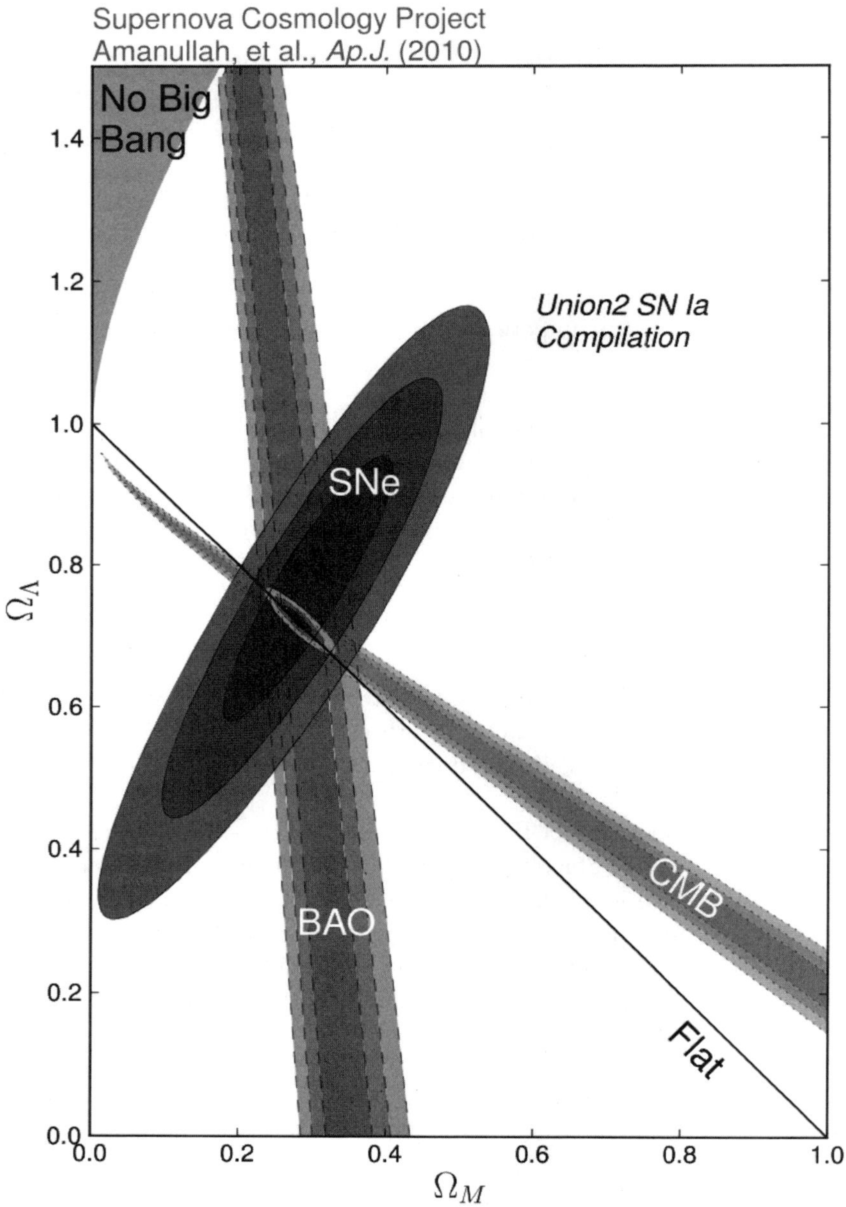

Fig. 3. The cocktail making the Universe and information about where it comes from.

accrediting the fact that we are indeed "star dust". This primordial nucleosynthesis is the 3rd pillar of the expanding Universe model. It has even turned into a probe of physics/cosmolology beyond the Standard Model: new hypothetical ingredients should not "spoil" its success.

8. The quark-hadron transition

Our second stop concerns the transition from quarks to hadrons, between one and ten microsecond. Quarks and gluons will never be freed again. The question is whether one can, at least partly, observe the inverse process, i.e. get some evidence for a transient quark gluon plasma giving the proof of some de-confinement.

The method is to collide big nuclei, like gold, accelerated at high energy. This gives a medium of high temperature, which lasts long enough to reach thermalization. After the CERN SPS and the US RHIC collider, this is now performed at higher energy at LHC, in particular in the dedicated ALICE experiment. The final state of such collisions, especially head-on ones, is densely populated. To get informations about the medium (temperature, chemical potential, hydrodynamical flow, equation of state), one studies all possible messengers emerging from it, more or less easily, leptons, jets, hadrons of identified flavour, Z bosons, quarkonia (i.e. quark-antiquark bound states), and their emission patterns. Photons and weak vector bosons leave unaffected and serve as standard candles. Jets and hadrons, including heavy ones (charm, beauty) do interact with the medium. One observes also the partial disappearance of loosely bound states, like the heavy states of the Upsilon family (bound states of beauty quark and antiquark), as if they were melting. For quarkonia, there may be however competition with mechanisms regenerating them.

The present status of these studies is that this medium corresponds to a nearly perfect frictionless fluid, with viscosity (i.e. inability to transport momentum and sustain a wave) near the lowest theoretically possible. This implies strong interaction or coupling between the constituents of the medium, explaining its opacity to jets and heavy charm quarks.

It is amusing to realize that physics has now identified two nearly perfect fluids, one, the Bose Einstein condensate of very cold atoms, close to the absolute zero at 10^{-7} K, and this one at 10^{12} K.

At a few microsecond of age the presently observable universe had the size of the solar system while the Hubble radius (i.e. ct) was a few kilometers. One does not expect that this transition may have created fluctuations of cosmic relevance, nor wiped out pre-existing ones.

9. The Standard Model (SM) of particles

The SM summarizes our knowledge of the elementary constituents and our good and unified understanding of the forces acting between them. It is extremely successful, but incomplete (no account of gravity) and full of mysteries (e.g. why three families?). Limited to its minimal ingredients, constituents and force carriers, the SM becomes incorrect when extrapolated to higher energies. Even supplemented by the Brout-Englert-Higgs (BEH) boson, it calls for an explanation of the hierarchy of masses observed, e.g. the huge disparity between the electroweak and the Planck scales.

The SM also splits the world of particles in two quite different species, the constituents, fermions of spin 1/2, and the force carriers, bosons of spin 1.

Let us introduce readily an extension of the SM which cures some of the problems underlined: Supersymmetry (SUSY), which, to each particle of the SM, associates a new partner of the opposite species. The existence of such doublets at the same mass being ruled out, SUSY must be a broken symmetry, with the partners somewhat heavier than the SM particles. Mathematically, SUSY is a deeply founded theory. Superstrings require its existence. However, only experiment will tell whether SUSY exists in a sufficiently light version to offer partners accessible to present and foreseeable colliders.

10. A quantum world

A first obvious remark is that, in particle physics, we are dealing with a quantum world. This means that particles can act as virtual states, appearing in the loops of the diagrams representing interactions. They are created and destroyed, according to the Heisenberg relations, but nevertheless change the rate of the processes in which such "loops" intervene. This has several crucial consequences.

First, "constants" of the theory are actually not constant, but evolve with the resolving power of the observation, hence with the energy scale. This is quite manifest for the coupling of the strong force, decreasing with energy and announcing the "asymptotic freedom" of QCD (Nobel 2004). The couplings of the three forces, accurately measured at LEP, can be extrapolated to higher energies, using renormalization group equations (RGE), the mathematical expression of the dependence on the resolving power. In the frame of SUSY, this extrapolation gives a perfect convergence of the couplings at a very high scale of about 10^{16} GeV, suggesting the possibility of a Supersymmetric Grand Unification (GUT).

Another consequence of the quantum nature of subatomic physics is that particles can be "felt" before being "really" produced, through their role as virtual states. At LEP, accurate measurements of the vector bosons properties revealed quite rapidly that the top quark was heavy, before its discovery at Fermilab in USA. With the top effect duly taken into account, LEP could also detect a small residual effect attributed to the BEH boson and predicting its mass in the 100 GeV ballpark.

Another way to search for new particles is to consider rare or forbidden processes in the SM. Rare decay modes of beauty particles (e.g. of the strange-and-beautiful B^0_s in two muons), the rate of particle-antiparticle oscillation, the decay of the muon into electron and gamma are examples. These studies, requiring a huge number of the particles under scrutiny, are performed at the high intensity frontier of the present programs. Evidence for an electric dipole moment of any particle (e.g. muon or neutron) would also be a major discovery.

11. The SM in brief

An idea to keep is that, besides existing in space-time, constituents live in other spaces, called "internal". Quarks do so in the color space. All fermions do so in the weak isospin space. Left handed fermions (their spin opposite to their momentum) live as couples, or doublets, right handed ones as bachelors, or singlets, another odd feature of the SM.

In these spaces one can perform "rotations" (true rotations or unitary transformations), or substitutions of particle species, which leave the physics invariant, for instance changing a red quark to a green quark, or a lepton to its neutrino inside a weak isospin left-handed doublet. These substitutions actually occur by emission or absorption of the bosons.

In brief, the Standard Model is the mathematical description of what was just said in sentences. It rests on two great principles

1. its formulation is invariant when one interchanges particles, which are a priori different, as explained above
2. this can only be done in a coherent way if one introduces the required bosons, which will operate the rotations mentioned

This gives to the theory a local invariance, namely the freedom to perform, in each point of space-time, arbitrary "rotations". It is the locality of the invariance which gives all its power to the theory. Mathematicians will recognize Yang-Mills theories. Unfortunately this beautiful construction applies only to massless particles, which is not our world, and this delayed the advent of the SM for a while.

But there is a happy end to the story

12. The Brout-Englert-Higgs (BEH) mechanism in brief

Indeed there may be something in nature that refuses this freedom of performing arbitrarily such kinds of "rotations".

Inside a superconductor, the field of Cooper pairs, bosons acting coherently over macroscopic distances, forbids the freedom of rotating the phase of the wave function of an electron arbitrarily in each point. By interacting with this field the photon gets a mass m. It is the Meissner effect. The magnetic field penetrates the superconductor only over a distance $1/m$.

The BEH mechanism mimics such a situation, with however big differences. A field acting in the weak isospin space gives an orientation and suppresses the freedom to rotate arbitrarily in this space. The weak bosons then get a mass, e.g. $m_w = 0.5\, g\, v$, where v is the "vacuum expectation" of the Higgs field (246 GeV) and g is one of the two electroweak couplings.

The BEH field is a complex doublet (4 components) in weak isospace. Spontaneous symmetry breaking is crucial. The resulting Nambu-Goldstone bosons are "eaten" by the gauge bosons W and Z and become their longitudinal modes (3 components). One component is left, a Lorentz scalar, the BEH boson.

The "how" of the BEH mechanism in the SM is the following. The energy added by the Higgs field is $E = \mu^2\phi^2 + \lambda\phi^4$. With λ (quartic coupling) > 0 and μ^2 chosen negative, quite arbitrarily, one gets the "mexican hat" potential, ensuring that the minimal energy is obtained for ϕ non zero. The boson mass is $\approx \sqrt{\lambda}$ times v, therefore unknown, as is λ. The only requirements are that it should be neither too heavy, to avoid for instance a Landau singularity (i.e. the coupling becoming infinite), nor too light, as we will see later.

However SUSY tells more about the "why" of the BEH mechanism. One can choose a normal potential, with $\mu^2 > 0$, at very high energy, and the evolution makes it "mexican" at the electroweak scale, provided that the top quark, dominating the evolution, is heavy enough, which is indeed the case.

Actually the Higgs sector of SUSY is more complicated. But the "smoking gun" is that one neutral boson at least must be lighter than about 130 GeV, a fact known since about 20 years.

The exact nature of the electroweak phase transition (which order?) is an important question, since some authors associate a first order transition to the possibility of electroweak baryogenesis.

13. Electroweak symmetry breaking: the BEH boson

This brings us to our 3rd and longer stop: the time of electroweak symmetry breaking and the Brout-Englert-Higgs (BEH) boson, the "goddamn" particle that became "God's particle"

It is the simplest version of the origin of mass, although one can imagine other mechanisms, as Technicolor, now disfavoured by experiment.

The boson gives to elementary particles a mass more or less heavy by interacting more or less with them. We will come back later to the case of neutrinos.

Let us note also that this has almost nothing to do with our mass, nor the mass of the Universe. We weight the mass of our nucleons, which do not weight the mass of their constituents, gluons, massless, and u and d quarks, nearly so (few MeV), but their "dance".

While the mass of the SM BEH is undetermined, within the limits described, one knows everything else about it. In brief the boson couples to the heaviest states available. This is the case for its decay modes, as well as for its production. However the dominant production mode, through gluon-gluon fusion, as well as its decay mode into a pair of photons, seems to contradict that statement. It is nevertheless true, since the boson communicates with these massless particles through a loop a virtual heavy particles involving the top quark and the vector bosons (and eventually other non SM particles if we go beyond the SM). It is thus remarkable that, as we shall see next, the boson will have been discovered purely through quantum effects.

Knowing the production cross section and branching ratios of the SM boson, as well as the SM backgrounds, one can then, as a function of its mass and the integrated luminosity at LHC, anticipate what would be the significance of its observation. The p-factor, more precisely its local value, is defined as the probability that a similar or more extreme effect would be seen when the signal does not exist. One sees that, for a mass of 125 GeV and for the presently set of data, CMS expects a local p-value of about 10^{-8}, corresponding to a 6 sigma effect, and dominated by the gamma-gamma, the Z-Z* and the W-W* decay modes.

Fig 4 [2] show the results obtained by the experiments ATLAS and CMS at LHC in the modes gamma-gamma and Z-Z*. In both channels and both experiments, a signal is visible above background at a mass of about 125 GeV.

Overall, ATLAS observes a 6 sigma effect (with 5 sigma expected) and CMS a 5 sigma effect (with 6 sigma expected). Within errors, the masses of the effects agree. The observed branching ratios are still compatible with a SM BEH boson. More data are eagerly waited for.

The mass of 125 GeV found for what could be the SM BEH boson is particularly interesting. For too light a mass of the boson, the Higgs potential would not be bounded from below or would develop at a higher scale a minimum lower than the electroweak one, leading to the instability, or metastability, of the vacuum. The instability mass limit is estimated to be 127 GeV, with a few GeV uncertainty due to the uncertainties on the top mass and the strong coupling. Some authors, on the basis of arguments concerning the behaviour of the Higgs quartic coupling at high scale, have

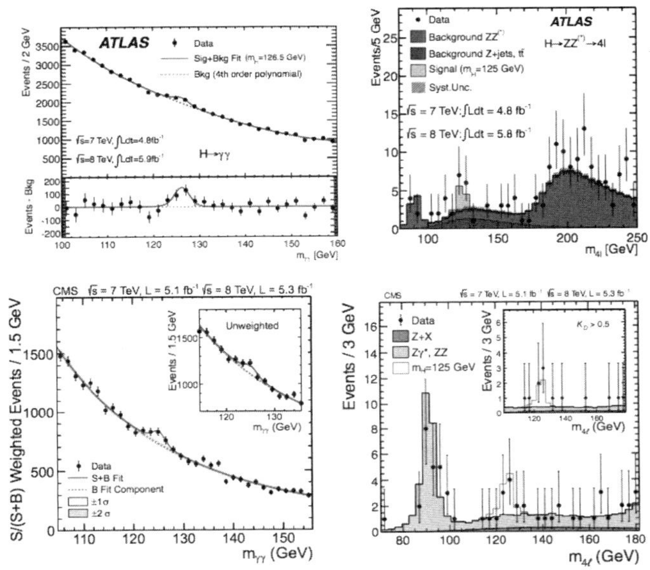

Fig. 4. Evidence for a new particle in gamma-gamma and 4 leptons as observed in ATLAS and CMS experiments.

actually predicted that the SM boson mass would be found close to this limit, which was indeed the case. These authors also consider that the SM is valid up to arbitrary high scale, with no need for new physics (see later for precisions). This domain of mass is however also a region favoured by SUSY

14. Supersymmetry and Dark Matter

Then, what about SUSY?

The observed boson at 125 GeV, still possibly SM like, is compatible with SUSY, may be slightly heavier than expected. As we said already, a light elementary scalar as the BEH boson needs new physics to stabilize its mass against radiative corrections, and already against the effect of the top. In SUSY indeed, the effect of partners cancel each other, provided they, or some of them, are not too heavy. We saw also that SUSY offers an exact convergence of couplings at the high scale of a presumed Grand Unification scale.

But SUSY offers another bonus, namely an ideal candidate for dark matter. Indeed SUSY particles are produced in pairs, any of them cascades down to another one of lower mass, except the lightest SUSY particle (LSP), which cannot and therefore is stable. If partners were produced at earlier times, the LSP can be a fossil massive particle, weakly interacting (a WIMP), and we could be immersed in a cloud of WIMPs.

Furthermore the "WIMP miracle" is that the amount of dark matter needed by the "cocktail" shown previously corresponds to WIMPs of masses near the electroweak scale, interacting through a weak cross section with ordinary matter.

One can then conceive various search methods to look for them.

In very sensitive detectors (bolometers, liquid argon or xenon), well shielded from cosmic rays, hence deep underground, one can look for tiny signals deposited in the collision of a WIMP with a nucleus of the detector. Presently, a detector with 100 kg of liquid xenon seems to set the strongest limit on the WIMPs relic density times cross section, although other experiments still claim for a possible signal.

Another way is to look for an anomalous rate of ordinary particles (photons, positrons, antiprotons) which could result from the annihilation of two WIMPS in regions of the Cosmos where WIMPs have good reason to accumulate, e.g. near the Galactic Center. For instance, the MAGIC detector, in Canaries islands, looks for cosmic gamma rays. One should follow the analysis of the FERMI satellite data to see if a reported "gamma line" at 130 GeV is real or not.

If SUSY does not exist, the next candidate for dark matter, as an effect of a new particle, is the axion, which would deserve a coordinated and strong search program.

SUSY at LHC?

Clearly, SUSY can also appear at LHC and searches for super-symmetric particles is presently a priority.

If supersymmetric particles are produced in a collision, they will appear in pair and each of them will decay, cascading down to the stable LSP which will stay unobserved. So the event will be characterized by missing energy and momentum carried by the invisible LSPs. However the SM also contains such a type of events, for instance when a Z boson is produced and decays to a pair of undetectable neutrinos, or if some mis-measurement of jets produces fake missing quantities. SUSY therefore can only appear as a signal over background, and the heavier the partners, the smaller the signal.

After two years of searches, there is nothing yet in sight. These searches were concerning mostly the partners of the quarks of the first two generations and of the gluon, and excluded them up to a mass of about 1 TeV, provided however the LSP mass is less than 400 GeV or so.

But they don't tell much yet about other partners, in particular those of the quarks of the third generation, bottom and top. These are "closer relatives" of the BEH boson since they govern the value of its mass, and the observed mass suggests that they may be within reach. Looking for them is the next objective of the searches at LHC.

No signal of new physics is reported either from the studies of rare events, as described earlier, where the contribution of virtual SUSY particles in the loops governing their production could have led to an enhancement of their production rate.

SUSY was not the only possible new physics sector searched for. In particular, the possibility of effects due to the existence of "large" extra dimensions (ED) in various models has been explored: gravitons disappearing in the ED and leaving unbalanced events, resonances corresponding to recurrences ("echoes") of particles in the extra space. Nothing new has been observed yet. Special attention is given to the top quark, because of its intriguing very high mass (172 GeV). While some anomalies were reported at the Tevatron, the LHC, up to now, does not see any in the properties of the top.

So the SM seems OK. However we are only at the start of LHC studies. Running near the design energy of 14 TeV and accumulating few hundred inverse femtobarn, i.e. 100 times more than up to now, is the next goal.

15. Grand Unification?

We already alluded to the possibility of a Grand Unification near 10^{16} GeV. This would imply that the symmetry group was then larger than the SM one and including it. The first one which was considered was SU(5). SO(10) would allow to accommodate all fermions in the same multiplet, with a slot left for a right-handed neutrino. This would offer an explanation for the small mass of the three active neutrinos, through a simple mechanism called the see-saw. Heavy RH neutrinos offer also the possibility of leptogenesis, i.e. the generation of an asymmetry between leptons and antileptons, transmitted later to the world of baryons, a plausible scenario, however difficult to prove.

Seen from the point of view of superstrings (SS), the GU group should also be contained in the large exceptional groups relevant to SS, which have generally to do with octonions.

A smoking gun for the validity of GU would be the observation of proton decay, as a far "remembrance" of this time where being quark or lepton was indifferent, since they were components of the same multiplet. The present limit is of about 10^{34} years, obtained from the 20 ktons water Cerenkov detector SuperKamiokande (SK). This already excludes the simplest version of SU(5) GUT. Exploring the next order of magnitude would require a megaton detector, which, if one judges from the size of SK, would be a very big enterprise

16. Inflation?

Inflation is a hypothetical brief period of exponential expansion. It solves several mysteries: spatial flatness, as one can intuitively guess by considering the surface of an inflating balloon, homogeneity of the CMB over the full sky, since inflation made the observable Universe from a single causally connected initial region, etc, and it is compatible with the data of WMAP, showing a nearly scale invariant spectrum of the fluctuations of CMB, as expected from inflation.

However the questions are many. What is the actual process? Many variants exist, including eternal inflation, which expresses the fact that, once started, inflation cannot stop. One can quote the zero energy version of A. Linde.

Another question is the nature of the agent of inflation? Which kind of scalar field could it be? In this respect, some authors propose that the Higgs field is indeed a possible agent of inflation. These authors are the same who predicted that the Higgs mass would be close to the stability limit, as we discussed. With an adequate coupling of the Higgs field with gravity and the addition of three sterile neutrinos (see later) of mass below the electroweak scale, they claim also that this minimal set is sufficient to explain all known phenomena, a radical and challenging conclusion, which is however under debate.

Other concerns are the way to ensure a "graceful exit" from inflation, the time where the matter content of the Universe was created, as well as the proper initial conditions which made inflation possible and with an outcome in agreement with WMAP results.

Such speculations about the pre-inflation era raise the far-reaching question of the origin that we will formulate in our conclusion. Further studies of the CMB will hopefully bring more information

17. Why something rather than nothing?

In all processes one observes an equal production of particles and antiparticles. If this was the case at all stages in the history of the Universe, a mystery appears. Why is our Universe made only of matter? Where did the antimatter go? There is no sign of antimatter accumulation anywhere in the Universe, which would annihilate violently with surrounding matter. The cosmic radiation does not seem to contain primordial antiparticles, other than what interaction of cosmic particles or known phenomena (pulsars, etc) can explain.

Or if matter and antimatter annihilated, what explains that something is left, making the observable Universe? Why something rather than nothing?

The situation can be described by considering the following numerology. At the beginning of the Universe let us consider, say, a billion of particles, plus one, and exactly the same number of antiparticles. It happened as if, shortly after, one antiparticle transformed into one particle, giving thus an excess of two particles. After mutual annihilation of the billion of both species, this excess of two particles is left, and this is what makes the

Universe. That particles and antiparticles can transform into each other is possible, as a quantum effect already quoted. But why is it asymmetrical?

18. The broken mirrors

In 1956 it was found, at the general surprise, that the weak interaction "violates parity". The P mirror image of a "good" physical objet may not be a "good" physical objet. The extreme case is the neutrino: the mirror image of the left-handed neutrino is a right-handed neutrino which does not exist In the mirror CP, where C is charge conjugation, one seems to get a good image, except that the image is antimatter. A right-handed antineutrino does exist. However in 1964 it was shown, in the world of neutral strange mesons K^0, that the weak interaction also violates CP. This tiny violation has been observed since then in other sectors, most noticeably in the world of beauty particles.

The possibility of CP violation is explained in the SM by quark mixing between the three families (Nobel 2008), since the unitary 3×3 mixing matrix, besides three mixing angles, contains one phase, key of CP violation. Let us note that this requires the existence of three families, which a priori seemed a useless replication.

Furthermore CP violation appears as a most welcome phenomenon. Indeed the famous Sakharov's conditions, to explain the mysterious and short victory of matter over antimatter are:

1. baryonic number is not conserved, i.e. proton can decay
2. the C and CP symmetries are violated, i.e. changing a particle to an antiparticle has a rate different from the reverse
3. these violations happened when the Universe was not in equilibrium.

So, the questions are: when did it happen? And how? One knows that the level of the observed violation of CP, well accounted for by the SM, is not sufficient to explain the asymmetry. The hope that physics beyond the SM, as SUSY, could boost it to the required level has been deceived. It is generally admitted that the study of CP in the world of neutrinos may give further hints, although there is no guarantee of an explanation.

This brings us to the next question. Can one have access experimentally to the measurement of CP violation for neutrinos?

19. The neutrinos

Needless to say, neutrino physics is not an easy one, given the tiny interaction cross-section of neutrinos with matter (of the order of 10^{-38} E (GeV) cm2). A 10 Gev neutrino has 10^{-8} probability to interact in 100 m of iron Huge fluxes and detectors are needed.

A basic fact in the neutrino world is that the flavor eigenstates (electron-neutrino, etc) are different from the mass eigenstates, i.e. those propagating, both being related by a 3×3 unitary mixing matrix. The result of this mixing is the phenomenon of neutrino oscillations, i.e. neutrinos changing their flavor while propagating. For a neutrino of energy E and a flight L, the probability of appearance of a new flavor is

$$\sin^2(2\theta)\,\sin^2(1.267\,\Delta m^2\,L/E\,\mathrm{GeV/eV^2\,km})\,,$$

where θ is a mixing angle and Δm^2 is the difference of mass squared of the two species. The study of oscillations tells therefore about mass differences, not the absolute values of the masses.

By studying solar, atmospheric, reactor and accelerator neutrinos, the pattern observed offers two different Δm^2, the "small" $\Delta m^2_{\mathrm{sol}}$, (7.6 10^{-5} eV^2), between 1 and 2, and the "large" $\Delta m^2_{\mathrm{atm}}$ (2.4 10^{-3} eV^2), between (1 or 2) and 3. But, in absolute masses, one does not know whether 1-2 is below 3 (normal hierarchy) or above (inverted hierarchy). Determining which is the correct one is the next question to be answered.

Most interesting is the question of an eventual CP violation in the world of neutrinos. It is possible due the presence in the 3×3 unitary matrix of a phase, the effect of which is however always factorized with the mixing angle θ_{13}.

A key was therefore the determination of this mixing angle. Very recently the result came, first from the Daya Bay reactor experiment in China, involving nuclear reactors and near and far detectors. The disappearance of electron-antineutrinos with distance from the reactors was clearly observed. Daya Bay finds $\sin^2(2\theta_{13}) = 0.1$, a comfortably large value, confirmed by other measurements.

This opens the road to further programs aiming at the search for CP violation in the neutrino world. However this may not be an easy road, since, for instance, matter effects along the flight are different between electron-neutrinos and antineutrinos. Various long base line experiments (up to 2300 km) are under study, requiring very high power proton sources to produce enough neutrinos.

Another important question is the exact nature of the neutrino: is it a Dirac particle, or a Majorana particle, i.e. identical to its antiparticle? The way the neutrino may acquire mass is different in the two cases. The observation of a process called neutrino-less double beta decay, actively searched for by a series of experiments, would prove its Majorana nature.

Another question is whether there is room for more neutrino species, called sterile, i.e. not interacting as the three we know, in particular with the Z boson, since LEP has shown that the light active ones are of three and only three species. The possible masses of these hypothetical sterile neutrinos span a vast domain, from zero to the GUT scale. Possible anomalies spotted in oscillation, reactor or source experiments suggest light sterile, with $\Delta m^2 = 1\,\mathrm{eV}^2$. As we said, the authors previously quoted need three heavier sterile neutrinos (one in the keV region and two at a few GeV) to complete the set of elementary particles, with which, they claim, all known phenomena are accounted for.

Concerning the absolute value of neutrino masses, the measurement of the end point of the spectrum of beta decay of tritium (3H$\rightarrow$3He + e^- + electron anti-neutrino) is still considered. But the KATRIN spectrometer, with its 70m length, is no more a table top experiment. It will bring the mass limit to 0.2 eV/c^2 (90% CL) or discover the actual mass if it is larger than 0.35 eV/c^2. It is again cosmology (CMB, large surveys) which sets the best limit on the sum of neutrino masses, $\sum m_\nu < 440\,\mathrm{meV}$.

20. Some final questions

Let me end by asking, as a layman, a set of questions. It is often recalled, for instance by M.Rees in his book "Just six numbers", that the value of several quantities, from the number of large dimensions to the 10^{36} ratio between nuclear force and gravity or the value of the electromagnetic coupling constant, could not be different, even slightly, otherwise the Universe would have failed. The question is then: will science offer ONE solution leading to such numbers? Or will it happen another way?

It could be that inflation, studied through the CMB, may be in the future through gravitational waves, marks the limit of observational science. However one cannot refrain from asking what conditions made it possible, i.e. speculate about the period around and before Planck time. Then general relativity and quantum theory have to be reconciled.

To achieve that, do we need a radical change of our vision of the world, as the Superstring (SS) theory proposes, by building it not from points but from extended objects, strings, branes? Such a theory is only viable with more dimensions of space. At each point of our space-time, one must attach a supplementary 6-dimensional space (7-dim for M-theory). The physics laws are supposed to depend of its topology. But the number of possible "compactifications" of such spaces has been shown to be gigantic, $\sim 10^{500}$ or so, leading to the hypothesis of the so called SS "landscape".

The idea of a permanent inflation then suggests that as many universes, with all kinds of different sets of parameters, have been "tried", populating the Multiverse. We are in a successful universe. With so many trials, one should not worry about "fine tunings". This is the so-called anthropic point of view.

How can one reject this possibility? Can we falsify any one of the two elements, landscape and permanent inflation? Can one find apparently fine tuned quantities which have no anthropic implication?

What about if the anthropic argumentation was winning? This may sound as a big turn or defeat for science. However we must anyway understand the coherence of OUR universe and the scientific approach is maintained.

At this meeting in particular, another approach is considered: loop quantum cosmology, which seems also to avoid initial singularities and allow naturally for inflation. As far as I understand, in both approaches some authors consider the possible emergence of space and time, while others pursue the idea of no-beginning, as in bouncing universe cosmology.

Will the scientific proof ever come or not? Concerning inflation, one can hope it is at least falsifiable. But one will probably never decide on the existence of the Multiverse. Nevertheless it is legitimate to ask such basic questions about the origin and fascinating to have in mind ideas like the world as an hologram, a zero energy universe emerging from a tiny fluctuation or some version of a "cosmological darwinism".

Acknowledgments

I thank Robert Barate and Felicitas Pauss for their careful reading and many corrections and suggestions.

References

1. For an extended version with figures and references, see http://ihp-pw1.ethz.ch/treille/origin_tianjin_corr2.pdf and D.Treille, CERN-OPEN-2012-026 http://cds.cern.ch/record/1498725/files/origin_tianjin_corr2.pdf
2. For ATLAS: *Phys.Lett. B* **716** (2012) 1-29; for CMS: *Phys.Lett. B* **716** (2012) 30-61

GAUGE POTENTIAL AND PERMUTATION SYMMETRY FOR MOLECULAR SYSTEMS WITH IDENTICAL NUCLEI

C. ALDEN MEAD[†*]

92 Bartram Road
Savannah, GA 31411, U.S.A
[†] *Professor Emeritus of Chemistry, University of Minnesota*
[*] *E-mail: aldykar@att.net*

In this article we discuss some aspects of the geometric phase in molecular systems, starting with the Herzberg - Longuet-Higgins sign change appearing when a Born-Oppenheimer (BO) electronic wave function is carried through a closed path surrounding a conical intersection, and the gauge potential that arises when one rephases the BO function so as to make it single-valued. The discussion is generalized to the nonabelian gauge potential arising when the electronic wave function is a Kramers doublet. We also discuss the behavior of electronic wave functions under permutations of identical nuclei.

Keywords: molecular systems, electronic wave function, geometric phase, Born-Oppenheimer, Kramers doublet.

1. Introduction

In this article we discuss various aspects of the so-called geometric phase in molecular systems. In Sec. 2 we discuss some simple aspects of the sign change experienced by a Born-Oppenheimer (BO) electronic wave function when the nuclei are carried around a closed path while keeping the wave function real. This effect was defined definitively by Herzberg and Longuet-Higgins[1] following up on an earlier treatment by Longuet-Higgins et al,[2] and is often referred to as the Herzberg - Longuet-Higgins or HLH sign change. Section 3 takes up the methods that can be used to deal with the multiple-valuedness of BO electonic wave funcions, both in the usual non-degenerate case and the case in which the electronic wave function is a Kramers doublet.

The geometric phase, a generalization of the HLH sign change, has been further developed for molecular systems by the present author and

Truhlar,[3] and for quantum mechanical systems in general in the famous paper by Berry[4] The phase in molecular systems has been reviewed by the present author,[5] and more recently by Wittig.[6]

2. The Herzberg - Longuet-Higgins Sign Change

Consider two BO electronic states, labeled 1 and 2, which become degenerate at some nuclear configuration or manifold, and their behavior in the near neighborhood of such a manifold. Define coordinates $x = rcos\theta$ along the gradient of $(H_{11} - H_{22})/2$ and $y = rsin\theta$ along the gradient of H_{12}. Then the Hamiltonian is proportional to

$$\begin{pmatrix} \cos\theta & \sin\theta \\ \sin\theta & -\cos\theta \end{pmatrix} \tag{1}$$

with eigenvalues ± 1 and eigenfunctions

$$\begin{pmatrix} \cos\frac{\theta}{2} \\ \sin\frac{\theta}{2} \end{pmatrix} ; \begin{pmatrix} -\sin\frac{\theta}{2} \\ \cos\frac{\theta}{2} \end{pmatrix} . \tag{2}$$

One sees immediately that the eigenfunctions undergo a sign change when the angle θ increases by 2π, i.e. when a path is traversed around the intersection. Such an intersection, in which the energy separation behaves linearly with the separation from the point (manifold) of degeneracy is called a conical intersection (CI).

Of course, the situation can also be looked at from another direction, starting with the wave functions and eigenvalues and deducing the Hamiltonian. Suppose we have two functions, with eigenvalues ± 1 and defined with respect to some angle τ as

$$|a_+\rangle = \begin{pmatrix} \cos\tau \\ \sin\tau \end{pmatrix} ; |a_-\rangle = \begin{pmatrix} -\sin\tau \\ \cos\tau \end{pmatrix} . \tag{3}$$

Then for the Hamiltonian matrix we have

$$H = |a_+\rangle\langle a_+| - |a_-\rangle\langle a_-| = \begin{pmatrix} \cos 2\tau & \sin 2\tau \\ \sin 2\tau & -\cos 2\tau \end{pmatrix} . \tag{4}$$

From this point of view the sign change appears quite natural.

We are usually interested in degeneracies that arise from symmetry, where coordinate changes that break the symmetry may also break the degeneracy, but in this case one needs to be careful. It is well known that, for the case where the electronic Hamiltonian can be considered real (time reversal invariance, total electronic spin integer), the number of conditions on the nuclear coordinates required for degeneracy is just 2. In general, however, the number of requirements for symmetry is greater than this, so there will often be coordinates that break the symmetry without breaking the degeneracy. For example, consider the case of an atomic p-state placed in the base of an equilateral-triangular pyramid, so that its symmetry is reduced to C_{3v}, which can be removed by distorting the base or by moving the atom located at the apex, as shown in Fig. 1.

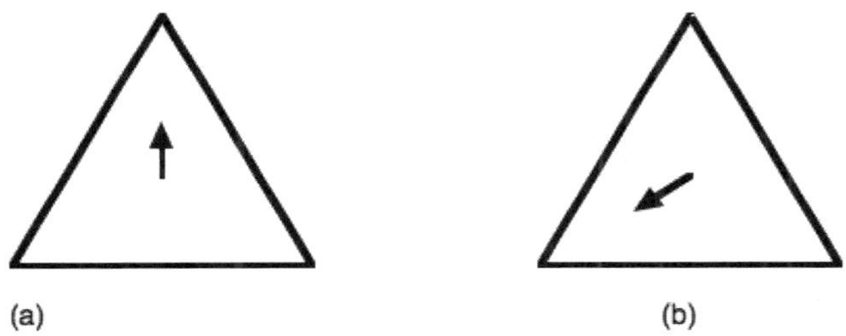

Figure1 (a) Atom at apex of pyramid moved off center toward top corner; (b) Apex atom moved toward bottom left

Clearly, in both parts of Fig. 1, the symmetry has been reduced from that of an equilateral triangle to just a symmetry plane. In Fig. 1(a), obviously one of the eigenfunctions will be symmetric under reflection through this plane. In Fig. 1(b) the symmetry-breaking coordinate has been rotated through $2\pi/3$, and from symmetry so has the corresponding eigenfunction, *not* by just $\pi/3$ as one might expect from eq. (2). With respect to this symmetry-breaking coordinate, the intersection is glancing, not conical, and there is no HLH sign change.

On the other hand, consider a triangle abc, with electronic wave functions linear combinations of functions φ_a, φ_b, φ_c located on the nuclei a, b, c:

$$\psi = q_a\varphi_a + q_b\varphi_b + q_c\varphi_c\,. \tag{5}$$

Define coordinates representing the distortion from C_{3v} symmetry:

$$x = \frac{1}{\sqrt{6}}(r_{ab} + r_{ac} - 2r_{bc}) \equiv \rho\cos\theta\,;$$
$$y = \frac{1}{\sqrt{2}}(r_{ab} - r_{ac}) \equiv \rho\sin\theta\,. \tag{6}$$

For small distortions from equilateral triangle we get

$$\delta r_{bc} = -\sqrt{\frac{2}{3}}\rho\cos\theta \equiv -\varepsilon\cos\theta\,;$$
$$\delta r_{ab} = -\varepsilon\cos(\theta + 2\pi/3)\,;\ \delta r_{ac} = -\varepsilon\cos(\theta - 2\pi/3)\,. \tag{7}$$

Thus, for $\theta = 0$ we have

$$\delta r_{bc} = -\varepsilon\,;\ \delta r_{bc} = \delta r_{bc} = \varepsilon/2\,. \tag{8}$$

Thus atom a is at the apex of an isosceles triangle. For $\theta = 2\pi/3$, atom b is at the apex.

Setting H_{jk} proportional to δr_{jk} for each pair of atoms, a short calculation leads to an eigenfunction of the form (5) with

$$q_a = \sqrt{\frac{2}{3}}\cos\left(\frac{\theta}{2}\right)\,;\ q_b = \sqrt{\frac{2}{3}}\cos\left(\frac{\theta}{2} + \frac{2\pi}{3}\right)\,;$$
$$q_c = \sqrt{\frac{2}{3}}\cos\left(\frac{\theta}{2} - \frac{2\pi}{3}\right)\,. \tag{9}$$

For $\theta = 0$, with atom a at the apex, q_a has the largest value and is positive. When θ advances to $2\pi/3$, atom b now has pseudorotated to the apex, and q_b now has the largest absolute value, but is now negative, i.e. there has been a sign change for the pseudorotation compared with a simple rotation. Note that this represents a smaller change in the value of q_b than would be the case if it had to assume the large positive value, so the sign change can be seen as the result of changing the coefficient by as little as possible.

3. Rephasing of Electronic Wave Functions; Gauge Potentials

3.1. *Integer Spin*

In this section, we will always assume that the electronic Hamiltonian commutes with time reversal (actually motion reversal) for all nuclear configurations. The properties of the time reversal operator have been extensively treated by Wigner,[7] and also more recently by Truhlar et al.[8] The time reversal operator T is antiunitary, with the properties:

$$\langle T\psi|T\varphi\rangle = \langle\psi|\varphi\rangle^* \,;\, T\left(a|\psi\rangle\right) = a^*|T\psi\rangle\,, \tag{10}$$

where $|\psi\rangle$ and $|\varphi\rangle$ are arbitrary kets and a is a complex number. T commutes with quantities such as energy and position that are unchanged by motion reversal, and anticommutes with those that change sign, such as velocity and angular momentum components. Also

$$T^2 = \pm 1\,, \tag{11}$$

with the plus sign applying if the total spin is integer, minus if it is half-odd integer.[7,8]

In this subsection we consider the case of integer total electronic spin, which has the consequence that BO electronic wave functions can be chosen to be unchanged by time reversal, $|T\alpha\rangle = |\alpha\rangle$, which is equivalent to the possibility of choosing them to be real, also to requiring that

$$\langle\alpha|\delta\alpha\rangle = 0 \tag{12}$$

for infinitesimal transport along some path, since normalization requires the inner product to be pure imaginary.

However, as we have seen in Sec. 2, the real wave function is subject to the HLH sign change, meaning that it is not single-valued as function of the nuclear coordinates, and also it changes sign under cyclic permutations of an odd number of identical nuclei.

There are basically two ways of dealing with this situation: Method A is to choose the nuclear wave function ψ to be also double-valued and thus to cancel the double-valuedness of the electronic function, so that the product $|\alpha\rangle\psi$ is single-valued. Alternatively, one can make use of the fact that the BO eigenvalue equation determines the electronic wave function only up to a phase factor that can depend on the nuclear coordinates, and choose this phase factor so that the revised electronic function $|\bar{\alpha}\rangle = |\alpha\rangle e^{i\gamma}$ is single-valued and, if desired, also invariant under permutations of identical nuclei. This we call method B. Note that, under method B, the electronic wave function is no longer real, and thus no longer unchanged by time reversal. We will return to this point later.

In principle, method B solves the problem of multiple-valuedness, but its practical implementation can be difficult. One must first locate all conical intersections and for each intersection j identify the coordinates

$$u_j = r_j \cos\theta_j \,;\; v_j = r_j \sin\theta_j$$

that break the degeneracy, and then replace the real electronic ket $|\alpha\rangle$ by

$$|\bar{\alpha}\rangle = |\alpha\rangle \exp\left(i\sum_j \nu_j \frac{\theta_j}{2}\right) , \tag{13}$$

where the sum goes over all intersections and each ν_j is the number of cyclic permutations of identical nuclei that make up a closed path around intersection j. This is obviously a very difficult procedure.

Another way of accomplishing the same thing, however, is to define a reference ket $|\alpha_r\rangle$ which is chosen to be single-valued and invariant under permutations of identical nuclei (perhaps even independent of the nuclear coordinates), then simply phase the electronic ket so that $\langle\alpha_r|\bar{\alpha}\rangle$ is real and positive everywhere. Thus, if

$$\langle\alpha_r|\alpha\rangle = qe^{-i\gamma} \tag{14}$$

with q a positive number, then define

$$|\bar{\alpha}\rangle = |\alpha\rangle e^{i\gamma} . \tag{15}$$

The rephased ket will now be single-valued and invariant under permutations, as desired.

There is one apparent defect with this approach: If the reference ket is itself real, then the product $\langle\alpha_r|\alpha\rangle$ will also be real, and the phase factor can only be ± 1, changing abruptly at certain places. The solution to this, though, is easy: Just use a reference ket that is not real, e.g.

$$|\alpha_r\rangle = \frac{1}{\sqrt{2}}\left(|u\rangle + i|v\rangle\right) , \tag{16}$$

where $|u\rangle$ and $|v\rangle$ are real invariant kets.

Now following the usual BO procedure, one defines the BO nuclear-electronic wave function Ψ as

$$\Psi = |\bar{\alpha}\rangle\psi \tag{17}$$

and obtains the effective nuclear Schrödinger equation

$$-\frac{1}{2M}(\nabla + i\nabla\gamma)^2\psi + U\psi = E\psi \tag{18}$$

where $\nabla\gamma$ plays the role of a gauge potential.

Taking the complex conjugate of (18) we find

$$-\frac{1}{2M}(\nabla - i\nabla\gamma)^2\psi^* + U\psi^* = E\psi^* \,. \tag{19}$$

Defining $\psi^* = e^{2i\gamma}\varphi$ we find that φ also satisfies (18). On the other hand, the solution to (18) is unique in the absence of degeneracy, so we have

$$\psi = \psi^* e^{-2i\gamma} \; ; \; \psi = e^{-i\gamma}\chi_{\text{real}} \tag{20}$$

so our total wave function can also be written

$$\Psi = |\bar{\alpha}\rangle\psi = |\alpha\rangle\chi_{\text{real}} \tag{21}$$

i.e. the end result is the same as if we had used method A.

3.2. *Half-odd Integer Spin*

If the total electronic spin is half-odd integer, i.e. if the number of electrons is odd, then the electronic state will be a Kramers doublet[7,8]

$$|A\rangle \equiv (|\alpha\rangle \; |T\alpha\rangle) \; ; \; \langle A| \equiv \begin{pmatrix} \langle\alpha| \\ \langle T\alpha| \end{pmatrix} . \tag{22}$$

The equivalent of (12), requiring the doublet to be "real" is

$$\langle A|\delta A\rangle = 0 \,. \tag{23}$$

Similarly, we define a reference doublet $|A_r\rangle$, $\langle A_r|$.

Now if

$$\begin{aligned} \langle A_r|A\rangle &= \begin{pmatrix} \langle\alpha_r|\alpha\rangle & \langle\alpha_r|\tilde{T}\alpha\rangle \\ \langle\tilde{T}\alpha_r|\alpha\rangle & \langle\tilde{T}\alpha_r|\tilde{T}\alpha\rangle \end{pmatrix} \\ &= q\begin{pmatrix} \cos\beta e^{i\lambda} & \sin\beta e^{i\tau} \\ -\sin\beta e^{-i\tau} & \cos\beta e^{-i\lambda} \end{pmatrix} \equiv q\hat{U} \equiv qe^{-i\hat{\gamma}} \end{aligned} \tag{24}$$

we transform the doublet as

$$\begin{aligned} |\bar{A}\rangle &= |A\rangle\hat{U}^\dagger = |A\rangle e^{i\hat{\gamma}} \; ; \\ \langle A_r|\bar{A}\rangle &= q = q\begin{pmatrix} 1 & 0 \\ 0 & 1 \end{pmatrix} . \end{aligned} \tag{25}$$

This will be single-valued and invariant under permutations. The gauge potential $\nabla\hat{\gamma}$ will now be nonabelian, an operator in the two-dimensional space of the doublet instead of just a number.

This concludes our brief discussion of these aspects of the geometric phase in molecules.

Acknowledgments

The author is grateful to the organizers of this conference for the kind invitation. He is also grateful to Prof. Donald G. Truhlar for the fruitful collaboration that has been indispensible for whatever the author may have achieved in this field.

References

1. G. Herzberg and H. C. Longuet-Higgins, *Discuss. Faraday Soc.* **35**, 77 (1963).
2. H. C. Longuet-Higgins, U. Öpik, M. H. L. Pryce, and R. A. Sack, *Proc. R. Soc. London Ser. A* **244**, 1 (1958).
3. C. A. Mead and D. G. Truhlar, *J. Chem. Phys.* **70**, 2284 (1979).
4. M. V. Berry, *Proc. R. Soc. London Ser A* **392**, 45 (1984).
5. C. A. Mead, *Rev. Mod. Phys.* **64**, 51 (1992).
6. C. Wittig, *Phys. Chem. Chem. Phys.* **14**, 6409 (2012).
7. E. P. Wigner, *Group Theory*, translated by J. J. Griffin (Academic Press, New York, 1959), Chapter 26.
8. D. G. Truhlar, C. A. Mead, and M. A. Brandt, *Adv. Chem. Phys.* **33**, 295 (1975).

A REVIEW OF THE LARGE N LIMIT OF TENSOR MODELS

R. GURAU*

Perimeter Institute, 31 Caroline St. N., Waterloo, ON, N2L 2Y5, Canada
** E-mail: rgurau@perimeterinstitute.ca*

Random matrix models encode a theory of random two dimensional surfaces with applications to string theory, conformal field theory, statistical physics in random geometry and quantum gravity in two dimensions. The key to their success lies in the $1/N$ expansion introduced by 't Hooft. Random tensor models generalize random matrices to theories of random higher dimensional spaces. For a long time, no viable $1/N$ expansion for tensors was known and their success was limited.

A series of recent results has changed this situation and the extension of the $1/N$ expansion to tensors has been achieved. We review these results in this paper.

Keywords: Random Tensors; $1/N$ expansion; Critical behavior.

1. Introduction

In many theories space appears as a fixed background, a manifold with a fixed metric. Length scales are defined with respect to this background. The scales encode the causality: fundamental physics at short distances determines the effective phenomena at large distances. It is however clear that the length scale are too crude to be applied in all instances: not only general relativity promotes the metric to a dynamical variable but also quantum field theory suggests that at the fundamental level space should become random, quantum. A fundamental theory of space is for now out of reach. Nevertheless we know that, no matter what this theory is, it must answer three major questions: how to define a statistical theory of random geometries, how to define an appropriate notion of scale and how to recover the usual space time as an effective phenomenon.

In this paper we present a framework to deal with random geometries in arbitrary dimensions. In this framework we answer precisely each of the three question formulated above.

Invariant probability distributions for random $N \times N$ matrices[a] encode the most successful (but restricted to two dimensions) theory of random geometry we have so far. The moments and partition functions of such a probability distribution evaluate in terms of ribbon Feynman graphs. Ribbon graphs are in one to one correspon-

[a]Random matrices were introduced by Wishart[1] in statistics and used for the first time in physics by Wigner[2] for the study of the spectra of heavy atoms.

dence with triangulated surfaces[3] providing the bridge between random matrices and random geometries.

In his seminal work[4] 't Hooft realized that matrix models have a built in notion of scale: the size of the matrix, N. The perturbative expansion can be reorganized in powers of $1/N$ (indexed by the genus) and in the large N limit only planar graphs[5] contribute. Matrix models undergo a phase transition to continuum infinitely refined surfaces[6,7] because the planar graphs form a summable family (i.e. a power series with a finite radius of convergence). Random matrices provide the quantization of two dimensional gravity coupled to conformal matter[8–15] and through the KPZ correspondence[16–19] they relate to conformal field theory in fixed geometry. In the double scaling limit matrix models provide a quantization of the string world sheet[20–22] in string theory.

The resounding success of matrix models has inspired their generalization in higher dimensions to random tensor models.[23–29] Until recently however tensor models have failed to provide an analytically controlled theory as, for a long time, no generalization of the $1/N$ expansion to tensors has been found.

The discovery of colored[30–32] tensor models[33–42] has drastically changed this situation. The colored models support a $1/N$ expansion[43–45] indexed by the *degree*. The leading order (melonic[46]) graphs, triangulate the sphere in any dimension.[43–45] They map to trees[46] hence are a summable family (different form the planar family). Colored tensor models undergo a phase transition[46] to a theory of continuous infinitely refined random spaces.

These results generalize to all invariant models for a random complex tensor.[47] The colors arise as a canonical bookkeeping device labeling the indices of the tensor. The resulting universal[48–51] theory of random tensors is the generalization of invariant matrix models to higher dimensions. The $1/N$ expansion can be realized dynamically as a renormalization group flow in tensor group field theories[52–57] which are, in at least two simple cases, asymptotically free.[53,54]

2. Tensor Models

Let us consider a covariant complex tensor of rank D transforming under the external tensor product of D fundamental representations of the unitary group $\otimes_{i=1}^{D} U(N_i)$

$$\begin{aligned} T'_{a_1\dots a_D} &= \sum_{n_1,\dots,n_D} U_{a_1 n_1} \cdots V_{a_D n_D} \, T_{n_1\dots n_D} \\ \bar{T}'_{a_1\dots a_D} &= \sum_{n_1,\dots,n_D} \bar{U}_{a_D n_D} \cdots \bar{V}_{a_1 n_1} \, \bar{T}_{n_1\dots n_D} \, . \end{aligned} \tag{1}$$

We stress that each unitary group $U(N_i)$ acts independently on its corresponding index. The tensor T has *no symmetry properties* under permutation of its indices. The dimensions $N_1, \dots N_D$ can be different but for simplicity we set $N_i = N$, $\forall i$. We denote by $\bar{n}_i$ the indices of the complex conjugated tensor $\bar{T}$ (which is contravariant

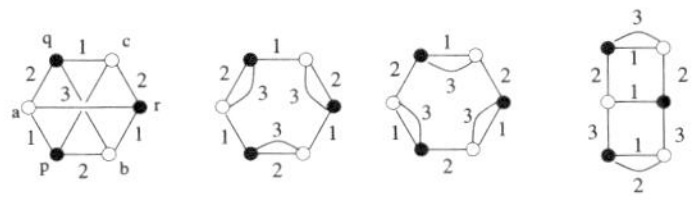

Fig. 1. Some trace invariants for $D = 3$.

of rank D), and by $\vec{n}$ the D-uple of integers $(n_1, \ldots, n_D)$. We define the *color* of an index as its position: n_1 (and $\bar{n}_1$) has color 1, n_2 (and $\bar{n}_2$) has color 2 and so on.

Invariants and action. An invariant tensor model is a probability measure defined by an invariant polynomial $S(T, \bar{T})$, which we call the "action",

$$d\nu = \frac{1}{Z} e^{-N^{D-1} S(T,\bar{T})} \left(\prod dT_{n_1 \ldots n_D} d\bar{T}_{n_1 \ldots n_D} \right) . \tag{2}$$

By the fundamental theorem of classical invariants of the unitary group (see[58] for a modern proof) any invariant polynomial is a linear combination of *trace invariants* obtained by contracting pairs of covariant and contravariant indices in a product of tensor entries

$$\mathrm{Tr}(T, \bar{T}) = \sum \prod \delta_{n_1, \bar{n}_1} \, T_{n_1 \ldots} \ldots \bar{T}_{\bar{n}_1 \ldots} \, , \tag{3}$$

where all indices are saturated. Note that, in order to obtain an invariant, we always contract the first index n_1 of a T with the first index $\bar{n}_1$ on some $\bar{T}$, the second index n_2 of T with the second index $\bar{n}_2$ on some $\bar{T}$ and so on. That is we only contract indices of the *same color*.

The trace invariants can be represented graphically as D-colored graphs. We represent every $T_{\ldots n_i \ldots}$ by a white vertex v and every $\bar{T}_{\ldots n_i \ldots}$ by a black vertex $\bar{v}$. The contraction of the two indices n_i and $\bar{n}_i$, $\delta_{n_i \bar{n}_i}$, is represented by a line $l^i = (v, \bar{v})$ of color i connecting the two vertices (see figure 1). For example the invariant

$$\begin{aligned} \sum \delta_{a_1 p_1} \delta_{a_2 q_2} \delta_{a_3 r_3} \quad \delta_{b_1 r_1} \delta_{b_2 p_2} \delta_{b_3 q_3} \quad \delta_{c_1 q_1} \delta_{c_2 r_2} \delta_{c_3 p_3} \\ T_{a_1 a_2 a_3} T_{b_1 b_2 b_3} T_{c_1 c_2 c_3} \bar{T}_{p_1 p_2 p_3} \bar{T}_{q_1 q_2 q_3} \bar{T}_{r_1 r_2 r_3} \, , \end{aligned} \tag{4}$$

is represented by the leftmost graph in figure 1 (the vertex a in the drawing represents $T_{a_1 a_2 a_3}$ and so on). The trace invariant associated to the graph $\mathcal{B}$ writes

$$\mathrm{Tr}_{\mathcal{B}}(T, \bar{T}) = \sum_{\{\vec{n}^v, \vec{\bar{n}}^v\}_{v, \bar{v} \in \mathcal{B}}} \left(\prod_{i=1}^{D} \prod_{l^i = (v, \bar{v}) \in \mathcal{B}} \delta_{n_i^v \bar{n}_i^{\bar{v}}} \right) \prod_{v, \bar{v} \in \mathcal{B}} T_{\vec{n}^v} \bar{T}_{\vec{\bar{n}}^{\bar{v}}} \, , \tag{5}$$

where v and $\bar{v}$ run over the white and black vertices of $\mathcal{B}$ and l^i runs over the lines of color i of $\mathcal{B}$. The (unique) graph with two vertices connected by D lines is called the D-dipole and is denoted $\mathcal{B}_1$. It represents the unique quadratic invariant

$$\mathrm{Tr}_{\mathcal{B}_1}(T, \bar{T}) = \sum_{\vec{n}, \vec{\bar{n}}} T_{\vec{n}} \bar{T}_{\vec{\bar{n}}} \left[\prod_{i=1}^{D} \delta_{n_i \bar{n}_i} \right] . \tag{6}$$

The most general "single trace" invariant tensor model is defined by the action

$$S(T,\bar{T}) = \mathrm{Tr}_{\mathcal{B}_1}(T,\bar{T}) + \sum_{\mathcal{B}} t_{\mathcal{B}} N^{-\frac{2}{(D-2)!}\omega(\mathcal{B})} \mathrm{Tr}_{\mathcal{B}}(T,\bar{T}) \ , \tag{7}$$

where the sum over invariants includes only *connected* graphs $\mathcal{B}$. The parameters $t_{\mathcal{B}}$ are the coupling constants of the model. We have added in eq. (7) a scaling $N^{-\frac{2}{(D-2)!}\omega(\mathcal{B})}$ for each invariant as it simplifies some formulae. Up to trivial modifications all the results presented below hold in its absence.

Feynman graphs. We will discuss in the sequel the partition function

$$Z(t_{\mathcal{B}}) = \int \Big(\prod_{\vec{n}=\vec{\bar{n}}} dT_{\vec{n}} d\bar{T}_{\vec{\bar{n}}} \Big) \, e^{-N^{D-1} S(T,\bar{T})} \ , \tag{8}$$

as the expectations of the invariant observables (which are also trace invariants represented by D-colored graphs) are treated similarly.

By Taylor expanding with respect to $t_{\mathcal{B}}$ and evaluating the Gaussian integral in terms of Wick contractions, $Z(t_{\mathcal{B}})$ becomes a sum over Feynman graphs. The graphs are made of *effective vertices* connected by *effective propagators*. The effective vertices are the invariants $\mathrm{Tr}_{\mathcal{B}}(T,\bar{T})$, that is they are themselves graphs $\mathcal{B}$ with colors $1,\dots,D$. The effective propagators correspond to pairings (Wick contractions) of $T_{a_1\dots a_D}$'s and $\bar{T}_{\bar{p}_1\dots\bar{p}_D}$'s. A Wick contraction with the quadratic part eq. (6) amounts to replacing the two tensors by $\frac{1}{N^{D-1}}\prod_{i=1}^{D}\delta_{a_i\bar{p}_i}$. We represent the effective propagators by dashed lines to which we assign a new color, 0. The Feynman graphs, denoted from now on $\mathcal{G}$, are then $D+1$ colored graphs (see figure 2).

We label $\mathcal{B}_{(\rho)}$, $\rho = 1,\dots|\rho|$ the effective vertices (subgraphs with colors $1,\dots D$) of $\mathcal{G}$ and denote l^0 the effective propagators (lines of color 0) of $\mathcal{G}$. The free energy is a sum over connected Feynman graphs $\mathcal{G}$

$$F(t_{\mathcal{B}}) = -\ln Z(t_{\mathcal{B}}) = \sum_{\mathcal{G}} \frac{(-1)^{|\rho|}}{s(\mathcal{G})} \Big(\prod_{\rho=1}^{|\rho|} t_{\mathcal{B}_{(\rho)}} \Big) A(\mathcal{G}) \ , \tag{9}$$

where $s(\mathcal{G})$ is a symmetry factor and $A(\mathcal{G})$ is the amplitude of $\mathcal{G}$

$$A(\mathcal{G}) = \sum_{\{\vec{n}^v,\vec{\bar{n}}^{\bar{v}}\}} \Big[\prod_{\rho} N^{D-1-\frac{2}{(D-2)!}\omega(\mathcal{B}_{(\rho)})} \Big(\prod_{i=1}^{D} \prod_{l^i=(v,\bar{v})\in\mathcal{B}_{(\rho)}} \delta_{n_i^v \bar{n}_i^{\bar{v}}} \Big) \Big] \\ \Big[\prod_{l^0=(v,\bar{v})} \frac{1}{N^{D-1}} \prod_i \delta_{n_i^v,\bar{n}_i^{\bar{v}}} \Big] \ . \tag{10}$$

Colored Graphs and triangulated spaces. Matrix models generate ribbon graphs which represent triangulated surfaces. The colored graphs of tensor models represent triangulated spaces:[32]

Theorem 2.1. *A closed connected $D+1$ colored graph is a D dimensional normal simplicial pseudo manifold.*

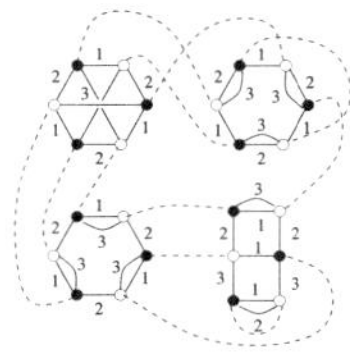

Fig. 2. A Feynman graph.

A pseudo manifold is a generalization of a manifolds having a finite number of conical singularities. This pseudo manifold can be built by gluing simplices. Consider $D = 3$, thus the $D + 1$ colored graphs have lines of colors 0, 1, 2 and 3.

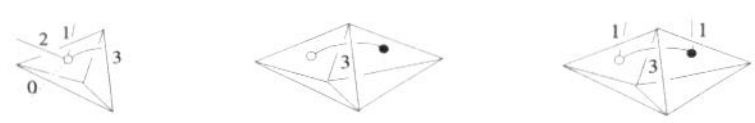

(a) (b) (c)

Fig. 3. Gluing of tetrahedra associated to a graph. (a) Vertex. (b) Line. (c) Face.

Every four valent white (or black) vertex is dual to a positive (or negative) oriented tetrahedron. The lines emanating from a vertex are dual to the triangles bounding the tetrahedron (see figure 3(a)). The four triangles inherit the color of the lines 0, 1, 2 and 3. All the lower dimensional simplices are then canonically colored: the edge common to the triangles 1 and 2 is colored by the couple of colors 12, and the apex common to the triangles 1, 2 and 3 is colored by the triple of colors 123. A line in the graph represents the *unique* gluing of two tetrahedra which respects *all* the colorings.

Take the example presented in fig. 3(b). The line of color 3 represents the gluing of the two tetrahedra along triangles of color 3 such that the edges 13, 23 and 03 as seen from the positive tetrahedron are glued on the edges 13, 23 and 03 as seen from the negative tetrahedron (and similarly for apices).

The cellular structure of the resulting gluing of tetrahedra is encoded in the colors: an edge ($D - 2$ simplex for $D + 1$ colored graphs) 13 corresponds to a subgraph with colors 13 (see fig. 3(c)). Such subgraphs are called the *faces* of $\mathcal{G}$.

A classical result[59] guarantees that tensor models generate all manifolds:

Theorem 2.2 (Pezzana's Existence Theorem). *Any piecewise linear D dimensional manifold admits a representation as a $D + 1$ colored graph.*

An initial trace invariant (subgraph with colors $1, \ldots D$) represents a "chunk" of a $D + 1$ dimensional space. Set $D = 3$. As a graph with 3 colors (see fig. 4) an invariant represents a surface. When seen as a subgraph in a $3 + 1$ colored Feynman graph, the invariant is decorated by lines of color 0. This amounts to taking the topological cone over the surface and obtain the $3 + 1$ dimensional "chunk" of

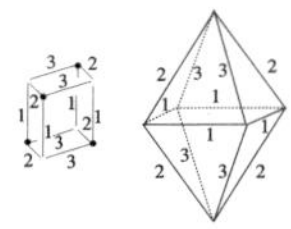

Fig. 4. A trace invariant and associated chunk.

space. Note that, if the surface represented by the invariant is non planar, taking the topological cone leads to a conical singularity in the $3+1$ dimensional space.

The $1/N$ expansion. The amplitude of a closed connected ribbon graph of a matrix model is[4] $A(\mathcal{G}) = N^{2-2g(\mathcal{G})}$, where $g(\mathcal{G})$ is the *genus* of the graph. What replaces the genus in higher dimensions? We answer this question below.

The genus arises in matrix models because the numbers of vertices, lines and faces in a ribbon graph are not independent. A closed connected ribbon graph with $V = 2p$ trivalent vertices has $L = 3p$ lines and $F = p + 2 - 2g(\mathcal{G})$ faces, where $g(\mathcal{G})$ is the genus of $\mathcal{G}$. This generalizes in higher dimensions for D colored graphs. Consider a closed connected D colored graph $\mathcal{B}$ with $2p$ black and white vertices (and Dp lines). The number of faces (subgraphs with two colors) of $\mathcal{B}$, is[43–45]

$$F = \frac{(D-1)(D-2)}{2}p + (D-1) - \frac{2}{(D-2)!}\omega(\mathcal{B}) \,, \tag{11}$$

where the *degree* of $\mathcal{B}$, $\omega(\mathcal{B})$ is a *non negative* integer. Of course a similar relation holds by shifting D to $D+1$ for $D+1$ colored graphs $\mathcal{G}$.

Although, like the genus, the degree is a non negative integer it is *not* a topological invariant. It is an intrinsic number one can compute starting from the graph which mixes information about the topology and the cellular structure[60] of $\mathcal{B}$. Some examples of $D+1$ colored graphs and their degrees are presented in figure 5.

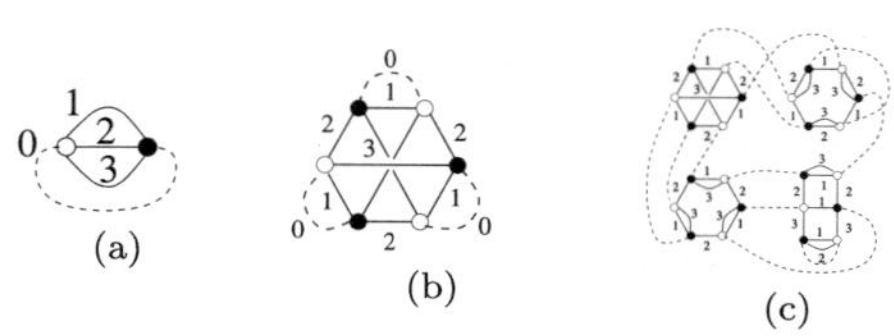

Fig. 5. $3+1$ colored graphs of degree (a) $\omega(\mathcal{G}) = 0$. (b) $\omega(\mathcal{G}) = 4$. (c) $\omega(\mathcal{G}) = 10$.

The idea is that, embedded in the graph $\mathcal{B}$, one can identify some special ribbon graphs $\mathcal{J}$, called jackets.[43–45,61] One then counts the number of faces of each jacket in terms of its genus $g(\mathcal{J})$. Summing over the jackets one obtains the total number of faces of the colored graph in terms of the sum of these genera, the degree of the colored graph $\omega(\mathcal{B}) = \sum_{\mathcal{J}} g(\mathcal{J})$.

We chose the scaling of the invariants in the action eq. (7), $\omega(\mathcal{B})$, to be precisely

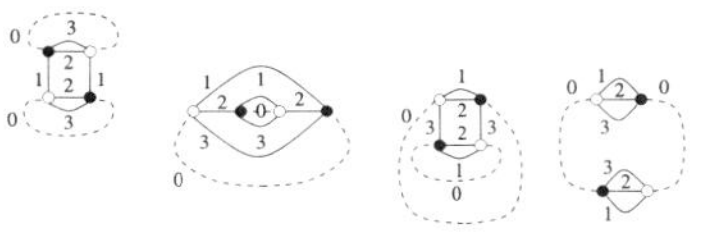

Fig. 6. Melons with $p = 2$ and $3 + 1$ colors.

the degree of $\mathcal{B}$. To evaluate the amplitude of a Feynman graph $\mathcal{G}$ (eq. (10)), one needs to count the number of independent sums over indices. Each solid line (of colors $1, \dots D$) represents the identification of one index $\delta_{n_i \bar{n}_i}$. The dashed lines (of color 0) represent the identifications of D indices $\prod_{i=1}^{D} \delta_{a_i \bar{p}_i}$. An index n_i is identified first along a line of color i, then along a line of color 0, then along a line of color i and so on until the cycle of colors 0 and i closes. We obtain a free sum over an index (hence a factor N) for every face (subgraph with two colors) of colors $0i$. Combining this with the explicit scalings in eq. (10) we obtain the $1/N$ expansion in tensor models in arbitrary dimension:

Theorem 2.3. *The amplitude of a closed connected Feynman ($D+1$ colored) graph generated by the action* (7) *is*

$$A(\mathcal{G}) = N^{D - \frac{2}{(D-1)!}\omega(\mathcal{G})} . \tag{12}$$

For $D = 2$ the $2 + 1$ colored graphs are ribbon graphs, the degree is the genus and the $1/N$ expansion of tensor models reduces to the one of matrices.

The leading order graphs. In matrix models the planar graphs (of genus zero) dominate the $1/N$ expansion. They represent spherical surfaces and form a summable family. What generalizes planar graphs to arbitrary dimensions? We answer this question below.

The leading order graphs are the $D + 1$ colored graphs of degree zero. The structure of such graphs is quite different for $D = 2$ (matrices) and $D \geq 3$ (tensors). However, for all D, the graphs of degree zero represent spherical topologies and form a summable family.

A first example of a graph of degree zero is the $D + 1$ dipole (see fig. 5(a)). It has 2 vertices and $\frac{D(D-1)}{2}$ faces (subgraphs with two colors ij) hence degree 0. For $D \geq 3$, the $D + 1$ colored graphs of degree 0 with $2p + 2$ vertices are obtained by inserting two vertices connected by D lines arbitrarily on any line of a $D+1$ colored graph of degree zero with $2p$ vertices.[46] This clearly fails for $D = 2$. We call these graphs *melons* (see figure 6).

The $D + 1$ dipole represents a sphere in D dimensions obtained by identifying coherently two D simplices along their boundary. Two $D + 1$ valent vertices connected by D-lines represent a D dimensional ball. As the insertion of balls into spheres preserves the topology[43–45] we have:

Theorem 2.4. *For any D, the graphs of degree* 0 *have spherical topology.*

The melonic graphs are generated by iterative insertions which can be mapped onto abstract (colored, $D+1$-ary) trees. Melonic graphs thus form a summable family with a finite radius of convergence. They are weighted by the coupling constants $t_{\mathcal{B}}$ of the model. When tuning the coupling constants to criticality, tensor models undergo a phase transition to infinitely refined continuous random spaces.

The algebra of constraints. The observables are graphs with D colors. At leading order only observables corresponding to melonic graphs $\mathcal{B}$ contribute. To each observable we associate a Schwinger Dyson equation[49,50] and at leading order only the equations corresponding to D melons survive. Each equation translates into a constraint satisfied by the partition function $\mathcal{L}_{\mathcal{T}} Z = 0$ for an explicit partial differential operator $\mathcal{L}_{\mathcal{T}}$. The operators form a Lie algebra. As the melons are indexed by D colored trees, the Lie algebra is indexed by colored rooted D-ary trees.[49]

A colored rooted D-ary tree $\mathcal{T}$ is a tree such that all its vertices (including the root) have at most D descendants and all its lines have a color index, $1, \dots, D$. Furthermore the *direct descendants* of any vertex are connected by lines with different colors.

Two trees $\mathcal{T}_1$ and $\mathcal{T}_2$ can be *joined* at a vertex $V \in \mathcal{T}_1$. The joined tree $\mathcal{T}_1 \star_V \mathcal{T}_2$ is obtained as follows (see fig. 7, where V is the bold vertex). We first cut the successors of V in $\mathcal{T}_1$ (and get a collection of branches, represented as dashed in figure 7(b)). We then glue the tree $\mathcal{T}_2$, with its root on top of V. Finally we reattach the branches at the end of the branches of the appropriate color starting at V (see fig. 7(c)).

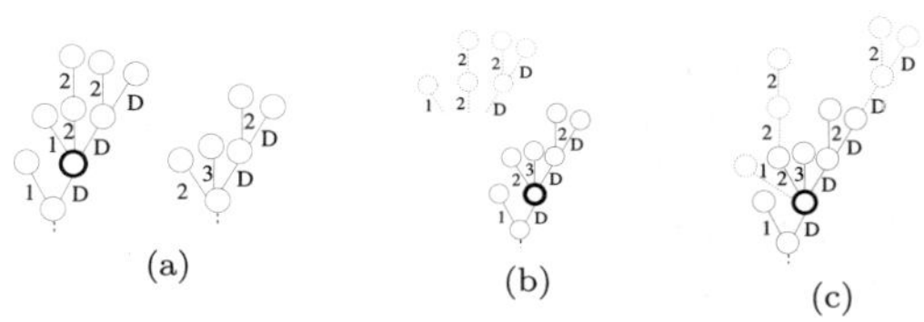

Fig. 7. Joining of two trees.

All the vertices of $\mathcal{T}_1$ except V and all the vertices in $\mathcal{T}_2$ except its root, which we denote ( ), are mapped into an unique vertex in $\mathcal{T}_1 \star_V \mathcal{T}_2$, while V and ( ) are both mapped into V in $\mathcal{T}_1 \star_V \mathcal{T}_2$. The constraint operators form the Lie algebra with Lie bracket

$$\Big[\mathcal{L}_{\mathcal{T}_2}, \mathcal{L}_{\mathcal{T}_1}\Big] = \sum_{V \in \mathcal{T}_2} \mathcal{L}_{\mathcal{T}_2 \star_V \mathcal{T}_1} - \sum_{V \in \mathcal{T}_1} \mathcal{L}_{\mathcal{T}_1 \star_V \mathcal{T}_2} \,. \tag{13}$$

The fact that the bracket is antisymmetric is clear. What is less obvious is that it respects the Jacobi identity. All the more so because the composition of trees is *not* associative. Indeed, $(\mathcal{T}_1 \star_V \mathcal{T}_2) \star_W \mathcal{T}_3 = \mathcal{T}_1 \star_V (\mathcal{T} \star_W \mathcal{T}_3)$ only if $W \in \mathcal{T}_2$ (or $W = V$). If $W \in \mathcal{T}_1 \setminus V$ the right hand side is not defined. However the Jacobi identity holds due to the presence of the sums. To see this we evaluate

$$\begin{aligned}
&\Big[\mathcal{L}_{\mathcal{T}_1}, \big[\mathcal{L}_{\mathcal{T}_2}, \mathcal{L}_{\mathcal{T}_3}\big]\Big] = \Big[\mathcal{L}_{\mathcal{T}_1}, \sum_{V\in\mathcal{T}_2} \mathcal{L}_{\mathcal{T}_2\star_V\mathcal{T}_3} - \sum_{V\in\mathcal{T}_3} \mathcal{L}_{\mathcal{T}_3\star_V\mathcal{T}_2}\Big] \\
&= \sum_{V\in\mathcal{T}_2}\Big(\sum_{V'\in\mathcal{T}_1} \mathcal{L}_{\mathcal{T}_1\star_{V'}(\mathcal{T}_2\star_V\mathcal{T}_3)} - \sum_{V'\in\mathcal{T}_2\backslash V} \mathcal{L}_{(\mathcal{T}_2\star_V\mathcal{T}_3)\star_{V'}\mathcal{T}_1} \\
&\quad - \sum_{V'\in\mathcal{T}_3\backslash(\,)} \mathcal{L}_{(\mathcal{T}_2\star_V\mathcal{T}_3)\star_{V'}\mathcal{T}_1} - \mathcal{L}_{(\mathcal{T}_2\star_V\mathcal{T}_3)\star_V\mathcal{T}_1}\Big) \\
&\quad - \sum_{V\in\mathcal{T}_3}\Big(\sum_{V'\in\mathcal{T}_1} \mathcal{L}_{\mathcal{T}_1\star_{V'}(\mathcal{T}_3\star_V\mathcal{T}_2)} - \sum_{V'\in\mathcal{T}_3\backslash V} \mathcal{L}_{(\mathcal{T}_3\star_V\mathcal{T}_2)\star_{V'}\mathcal{T}_1} \\
&\quad - \sum_{V'\in\mathcal{T}_2\backslash(\,)} \mathcal{L}_{(\mathcal{T}_3\star_V\mathcal{T}_2)\star_{V'}\mathcal{T}_1} - \mathcal{L}_{(\mathcal{T}_3\star_V\mathcal{T}_2)\star_V\mathcal{T}_1}\Big)\,, \qquad (14)
\end{aligned}$$

hence $\Big[\mathcal{L}_{\mathcal{T}_1}, \big[\mathcal{L}_{\mathcal{T}_2}, \mathcal{L}_{\mathcal{T}_3}\big]\Big] + \Big[\mathcal{L}_{\mathcal{T}_2}, \big[\mathcal{L}_{\mathcal{T}_3}, \mathcal{L}_{\mathcal{T}_1}\big]\Big] + \Big[\mathcal{L}_{\mathcal{T}_3}, \big[\mathcal{L}_{\mathcal{T}_1}, \mathcal{L}_{\mathcal{T}_2}\big]\Big]$ writes

$$\begin{aligned}
&\sum_{V\in\mathcal{T}_2}\sum_{V'\in\mathcal{T}_1} \mathcal{L}_{\mathcal{T}_1\star_{V'}(\mathcal{T}_2\star_V\mathcal{T}_3)} - \sum_{V\in\mathcal{T}_2}\sum_{V'\in\mathcal{T}_2\backslash V} \mathcal{L}_{(\mathcal{T}_2\star_V\mathcal{T}_3)\star_{V'}\mathcal{T}_1} \\
&\quad - \sum_{V\in\mathcal{T}_2}\sum_{V'\in\mathcal{T}_3\backslash(\,)} \mathcal{L}_{(\mathcal{T}_2\star_V\mathcal{T}_3)\star_{V'}\mathcal{T}_1} - \sum_{V\in\mathcal{T}_2} \mathcal{L}_{(\mathcal{T}_2\star_V\mathcal{T}_3)\star_V\mathcal{T}_1} \\
&\quad - \sum_{V\in\mathcal{T}_3}\sum_{V'\in\mathcal{T}_1} \mathcal{L}_{\mathcal{T}_1\star_{V'}(\mathcal{T}_3\star_V\mathcal{T}_2)} + \sum_{V\in\mathcal{T}_3}\sum_{V'\in\mathcal{T}_3\backslash V} \mathcal{L}_{(\mathcal{T}_3\star_V\mathcal{T}_2)\star_{V'}\mathcal{T}_1} \\
&\quad + \sum_{V\in\mathcal{T}_3}\sum_{V'\in\mathcal{T}_2\backslash(\,)} \mathcal{L}_{(\mathcal{T}_3\star_V\mathcal{T}_2)\star_{V'}\mathcal{T}_1} + \sum_{V\in\mathcal{T}_3} \mathcal{L}_{(\mathcal{T}_3\star_V\mathcal{T}_2)\star_V\mathcal{T}_1} \\
&+ \sum_{V\in\mathcal{T}_3}\sum_{V'\in\mathcal{T}_2} \mathcal{L}_{\mathcal{T}_2\star_{V'}(\mathcal{T}_3\star_V\mathcal{T}_1)} - \sum_{V\in\mathcal{T}_3}\sum_{V'\in\mathcal{T}_3\backslash V} \mathcal{L}_{(\mathcal{T}_3\star_V\mathcal{T}_1)\star_{V'}\mathcal{T}_2} \\
&\quad - \sum_{V\in\mathcal{T}_3}\sum_{V'\in\mathcal{T}_1\backslash(\,)} \mathcal{L}_{(\mathcal{T}_3\star_V\mathcal{T}_1)\star_{V'}\mathcal{T}_2} - \sum_{V\in\mathcal{T}_3} \mathcal{L}_{(\mathcal{T}_3\star_V\mathcal{T}_1)\star_V\mathcal{T}_2} \\
&\quad - \sum_{V\in\mathcal{T}_1}\sum_{V'\in\mathcal{T}_2} \mathcal{L}_{\mathcal{T}_2\star_{V'}(\mathcal{T}_1\star_V\mathcal{T}_3)} + \sum_{V\in\mathcal{T}_1}\sum_{V'\in\mathcal{T}_1\backslash V} \mathcal{L}_{(\mathcal{T}_1\star_V\mathcal{T}_3)\star_{V'}\mathcal{T}_2} \\
&\quad + \sum_{V\in\mathcal{T}_1}\sum_{V'\in\mathcal{T}_3\backslash(\,)} \mathcal{L}_{(\mathcal{T}_1\star_V\mathcal{T}_3)\star_{V'}\mathcal{T}_2} + \sum_{V\in\mathcal{T}_1} \mathcal{L}_{(\mathcal{T}_1\star_V\mathcal{T}_3)\star_V\mathcal{T}_2} \\
&+ \sum_{V\in\mathcal{T}_1}\sum_{V'\in\mathcal{T}_3} \mathcal{L}_{\mathcal{T}_3\star_{V'}(\mathcal{T}_1\star_V\mathcal{T}_2)} - \sum_{V\in\mathcal{T}_1}\sum_{V'\in\mathcal{T}_1\backslash V} \mathcal{L}_{(\mathcal{T}_1\star_V\mathcal{T}_2)\star_{V'}\mathcal{T}_3} \\
&\quad - \sum_{V\in\mathcal{T}_1}\sum_{V'\in\mathcal{T}_2\backslash(\,)} \mathcal{L}_{(\mathcal{T}_1\star_V\mathcal{T}_2)\star_{V'}\mathcal{T}_3} - \sum_{V\in\mathcal{T}_1} \mathcal{L}_{(\mathcal{T}_1\star_V\mathcal{T}_2)\star_V\mathcal{T}_3} \\
&\quad - \sum_{V\in\mathcal{T}_2}\sum_{V'\in\mathcal{T}_3} \mathcal{L}_{\mathcal{T}_3\star_{V'}(\mathcal{T}_2\star_V\mathcal{T}_1)} + \sum_{V\in\mathcal{T}_2}\sum_{V'\in\mathcal{T}_2\backslash V} \mathcal{L}_{(\mathcal{T}_2\star_V\mathcal{T}_1)\star_{V'}\mathcal{T}_3} \\
&\quad + \sum_{V\in\mathcal{T}_2}\sum_{V'\in\mathcal{T}_1\backslash(\,)} \mathcal{L}_{(\mathcal{T}_2\star_V\mathcal{T}_1)\star_{V'}\mathcal{T}_3} + \sum_{V\in\mathcal{T}_2} \mathcal{L}_{(\mathcal{T}_2\star_V\mathcal{T}_1)\star_V\mathcal{T}_3}\,, \qquad (15)
\end{aligned}$$

which cancels due to identities like

$$- \sum_{V\in\mathcal{T}_2}\sum_{V'\in\mathcal{T}_2\backslash V} \mathcal{L}_{(\mathcal{T}_2\star_V\mathcal{T}_3)\star_{V'}\mathcal{L}_{\mathcal{T}_1}} + \sum_{V\in\mathcal{T}_2}\sum_{V'\in\mathcal{T}_2\backslash V} \mathcal{L}_{(\mathcal{T}_2\star_V\mathcal{T}_1)\star_{V'}\mathcal{T}_3} = 0\,,$$

$$
\begin{aligned}
&\sum_{V\in\mathcal{T}_2}\sum_{V'\in\mathcal{T}_1}\mathcal{L}_{\mathcal{T}_1\star_{V'}(\mathcal{T}_2\star_V\mathcal{T}_3)}-\sum_{V\in\mathcal{T}_1}\sum_{V'\in\mathcal{T}_2\setminus(\,)}\mathcal{L}_{(\mathcal{T}_1\star_V\mathcal{T}_2)\star_{V'}\mathcal{T}_3}\\
&-\sum_{V\in\mathcal{T}_1}\mathcal{L}_{(\mathcal{T}_1\star_V\mathcal{T}_2)\star_V\mathcal{T}_3}=\sum_{V'\in\mathcal{T}_1}\mathcal{L}_{\mathcal{T}_1\star_{V'}(\mathcal{T}_2\star_{(\,)}\mathcal{T}_3)}-\sum_{V\in\mathcal{T}_1}\mathcal{L}_{(\mathcal{T}_1\star_V\mathcal{T}_2)\star_V\mathcal{T}_3}=0\,,\\
&-\sum_{V\in\mathcal{T}_3}\sum_{V'\in\mathcal{T}_1}\mathcal{L}_{\mathcal{T}_1\star_{V'}(\mathcal{T}_3\star_V\mathcal{T}_2)}+\sum_{V\in\mathcal{T}_1}\sum_{V'\in\mathcal{T}_3\setminus(\,)}\mathcal{L}_{(\mathcal{T}_1\star_V\mathcal{T}_3)\star_{V'}\mathcal{T}_2}\qquad(16)\\
&+\sum_{V\in\mathcal{T}_1}\mathcal{L}_{(\mathcal{T}_1\star_V\mathcal{T}_3)\star_V\mathcal{T}_2}=-\sum_{V'\in\mathcal{T}_1}\mathcal{L}_{\mathcal{T}_1\star_{V'}(\mathcal{T}_3\star_{(\,)}\mathcal{T}_2)}+\sum_{V\in\mathcal{T}_1}\mathcal{L}_{(\mathcal{T}_1\star_V\mathcal{T}_3)\star_V\mathcal{T}_2}=0\,.
\end{aligned}
$$

This algebra generalizes to all orders to an algebra indexed by D-colored graphs.[50] As it contains the dilation operator,[49,50] in order to connect with conformal field theories in arbitrary dimensions it remains to properly identify the rotation operators. This requires the completion of this Lie algebra by the appropriate generalization of the negative part of the Virasoro (Witt) algebra.

References

1. J. Wishart, "The generalised product moment distribution in samples from a normal multivariate population," Biometrika **20** A, 32, (1928)
2. E. Wigner, "Characteristic vectors of bordered matrices with infinite dimensions," Ann. Of Math. **62** (3), 548 (1955).
3. P. Di Francesco, P. H. Ginsparg and J. Zinn-Justin, "2-D Gravity and random matrices," Phys. Rept. **254**, 1 (1995) [arXiv:hep-th/9306153].
4. G. 't Hooft, "A Planar Diagram Theory for Strong Interactions," Nucl. Phys. B **72**, 461 (1974).
5. E. Brezin, C. Itzykson, G. Parisi and J. B. Zuber, "Planar Diagrams," Commun. Math. Phys. **59**, 35 (1978).
6. V. A. Kazakov, "Bilocal Regularization of Models of Random Surfaces," Phys. Lett. B **150**, 282 (1985).
7. F. David, "A Model Of Random Surfaces With Nontrivial Critical Behavior," Nucl. Phys. B **257**, 543 (1985).
8. V. A. Kazakov, "Ising model on a dynamical planar random lattice: Exact solution," Phys. Lett. A **119**, 140 (1986).
9. D. V. Boulatov and V. A. Kazakov, "The Ising Model on Random Planar Lattice: The Structure of Phase Transition and the Exact Critical Exponents," Phys. Lett. **186B**, 379 (1987).
10. E. Brezin, M. R. Douglas, V. Kazakov and S. H. Shenker, "The Ising Model Coupled To 2-d Gravity: A Nonperturbative Analysis," Phys. Lett. B **237**, 43 (1990).
11. V. A. Kazakov, "The Appearance of Matter Fields from Quantum Fluctuations of 2D Gravity," Mod. Phys. Lett. A **4**, 2125 (1989).
12. J. Ambjorn, J. Jurkiewicz and Yu. M. Makeenko, "Multiloop correlators for two-dimensional quantum gravity," Phys. Lett. B **251**, 517 (1990).
13. M. Fukuma, H. Kawai and R. Nakayama, "Continuum Schwinger-Dyson Equations and universal structures in two-dimensional quantum gravity," Int. J. Mod. Phys. A **6**, 1385 (1991).
14. Yu. Makeenko, "Loop equations and Virasoro constraints in matrix models," arXiv:hep-th/9112058.
15. R. Dijkgraaf, H. L. Verlinde and E. P. Verlinde, "Loop equations and Virasoro constraints in nonperturbative 2-D quantum gravity," Nucl. Phys. B **348**, 435 (1991).

16. V. G. Knizhnik, A. M. Polyakov and A. B. Zamolodchikov, "Fractal Structure of 2D Quantum Gravity," Mod. Phys. Lett. A **3**, 819 (1988).
17. F. David, "Conformal Field Theories Coupled to 2D Gravity in the Conformal Gauge," Mod. Phys. Lett. A **3**, 1651 (1988).
18. J. Distler and H. Kawai, "Conformal Field Theory and 2D Quantum Gravity Or Who's Afraid of Joseph Liouville?," Nucl. Phys. B **321**, 509 (1989).
19. Bertrand Duplantier "Conformal Random Geometry" Les Houches, Session LXXXIII, 2005, Mathematical Statistical Physics, A. Bovier, F. Dunlop, F. den Hollander, A. van Enter and J. Dalibard, eds., pp. 101-217, Elsevier B. V. (2006); [arXiv:math-ph/0608053].
20. E. Brezin and V. A. Kazakov, "Exactly Solvable Field Theories Of Closed Strings," Phys. Lett. B **236**, 144 (1990).
21. M. R. Douglas and S. H. Shenker, "Strings in Less Than One-Dimension," Nucl. Phys. B **335**, 635 (1990).
22. D. J. Gross and A. A. Migdal, "Nonperturbative Two-Dimensional Quantum Gravity," Phys. Rev. Lett. **64**, 127 (1990).
23. J. Ambjorn, B. Durhuus and T. Jonsson, "Three-Dimensional Simplicial Quantum Gravity And Generalized Matrix Models," Mod. Phys. Lett. A **6**, 1133 (1991).
24. N. Sasakura, "Tensor model for gravity and orientability of manifold," Mod. Phys. Lett. A **6**, 2613 (1991).
25. M. Gross, "Tensor models and simplicial quantum gravity in > 2-D," Nucl. Phys. Proc. Suppl. **25A**, 144 (1992).
26. N. Sasakura, "Tensor models and hierarchy of n-ary algebras," Int. J. Mod. Phys. A **26**, 3249 (2011), arXiv:1104.5312 [hep-th].
27. N. Sasakura, "Super tensor models, super fuzzy spaces and super n-ary transformations," Int. J. Mod. Phys. A **26**, 4203 (2011), arXiv:1106.0379 [hep-th].
28. D. Oriti, "The microscopic dynamics of quantum space as a group field theory," arXiv:1110.5606 [hep-th].
29. A. Baratin and D. Oriti, "Ten questions on Group Field Theory (and their tentative answers)," arXiv:1112.3270 [gr-qc].
30. R. Gurau, "Colored Group Field Theory," Commun. Math. Phys. **304**, 69 (2011), arXiv:0907.2582 [hep-th].
31. R. Gurau, "Topological Graph Polynomials in Colored Group Field Theory," Annales Henri Poincaré **11**, 565 (2010), arXiv:0911.1945 [hep-th].
32. R. Gurau, "Lost in Translation: Topological Singularities in Group Field Theory," Class. Quant. Grav. **27**, 235023 (2010), arXiv:1006.0714 [hep-th].
33. R. Gurau and J. P. Ryan, "Colored Tensor Models - a review," SIGMA **8**, 020 (2012) [arXiv:1109.4812 [hep-th]].
34. J. B. Geloun, J. Magnen and V. Rivasseau, "Bosonic Colored Group Field Theory," Eur. Phys. J. C **70**, 1119 (2010), arXiv:0911.1719 [hep-th].
35. J. P. Ryan, "Tensor models and embedded Riemann surfaces," Phys. Rev. D **85**, 024010 (2012) [arXiv:1104.5471 [gr-qc]].
36. S. Carrozza and D. Oriti, "Bounding bubbles: the vertex representation of 3d Group Field Theory and the suppression of pseudo-manifolds," Phys. Rev. D **85**, 044004 (2012) [arXiv:1104.5158 [hep-th]].
37. S. Carrozza and D. Oriti, "Bubbles and jackets: new scaling bounds in topological group field theories," JHEP **1206**, 092 (2012) [arXiv:1203.5082 [hep-th]].
38. V. Bonzom, R. Gurau and V. Rivasseau, "The Ising Model on Random Lattices in Arbitrary Dimensions," arXiv:1108.6269 [hep-th].
39. D. Benedetti and R. Gurau, "Phase Transition in Dually Weighted Colored Tensor

Models," Nucl. Phys. B **855**, 420 (2012) arXiv:1108.5389 [hep-th].
40. R. Gurau, "The Double Scaling Limit in Arbitrary Dimensions: A Toy Model," arXiv:1110.2460 [hep-th], Phys. Rev. D **84**, 124051 (2011)
41. V. Bonzom, "Multicritical tensor models and hard dimers on spherical random lattices," arXiv:1201.1931 [hep-th].
42. V. Bonzom and H. Erbin, "Coupling of hard dimers to dynamical lattices via random tensors," arXiv:1204.3798 [cond-mat.stat-mech].
43. R. Gurau, "The 1/N expansion of colored tensor models," Annales Henri Poincaré **12**, 829 (2011), arXiv:1011.2726 [gr-qc].
44. R. Gurau and V. Rivasseau, "The 1/N expansion of colored tensor models in arbitrary dimension," Europhys. Lett. **95**, 50004 (2011), arXiv:1101.4182 [gr-qc].
45. R. Gurau, "The complete 1/N expansion of colored tensor models in arbitrary dimension," Annales Henri Poincaré **13**, 399 (2012) [arXiv:1102.5759 [gr-qc]].
46. V. Bonzom, R. Gurau, A. Riello and V. Rivasseau, "Critical behavior of colored tensor models in the large N limit," Nucl. Phys. **B853**, 174-195 (2011), arXiv:1105.3122 [hep-th]
47. V. Bonzom, R. Gurau and V. Rivasseau, "Random tensor models in the large N limit: Uncoloring the colored tensor models," arXiv:1202.3637 [hep-th].
48. R. Gurau, "Universality for Random Tensors," arXiv:1111.0519 [math.PR].
49. R. Gurau, "A generalization of the Virasoro algebra to arbitrary dimensions," Nucl. Phys. B **852**, 592 (2011) [arXiv:1105.6072 [hep-th]].
50. R. Gurau, "The Schwinger Dyson equations and the algebra of constraints of random tensor models at all orders," Nucl. Phys. B **865**, 133 (2012) [arXiv:1203.4965 [hep-th]].
51. V. Bonzom, "Revisiting random tensor models at large N via the Schwinger-Dyson equations," arXiv:1208.6216 [hep-th].
52. J. Ben Geloun and V. Rivasseau, "A Renormalizable 4-Dimensional Tensor Field Theory," arXiv:1111.4997 [hep-th].
53. J. Ben Geloun and D. O. Samary, "3D Tensor Field Theory: Renormalization and One-loop β-functions," arXiv:1201.0176 [hep-th].
54. J. Ben Geloun, "Two and four-loop β-functions of rank 4 renormalizable tensor field theories," arXiv:1205.5513 [hep-th].
55. J. B. Geloun and E. R. Livine, "Some classes of renormalizable tensor models," arXiv:1207.0416 [hep-th].
56. S. Carrozza, D. Oriti and V. Rivasseau, "Renormalization of Tensorial Group Field Theories: Abelian U(1) Models in Four Dimensions," arXiv:1207.6734 [hep-th].
57. V. Rivasseau, "Quantum Gravity and Renormalization: The Tensor Track," arXiv:1112.5104 [hep-th].
58. B. Collins, "Moments and cumulants of polynomial random variables on unitary groups, the Itzykson-Zuber integral, and free probability," Int. Math. Res. Not. **17**, (2003) 953 [arXiv:math-ph/0205010].
59. M. Pezzana, "Sulla struttura topologica delle varieta compatte", Atti Sem. Mat. Fis. Univ. Modena **23** (1974), 269.
60. V. Bonzom and M. Smerlak, "Bubble divergences: sorting out topology from cell structure," Annales Henri Poincaré **13**, 185 (2012) [arXiv:1103.3961 [gr-qc]].
61. J. Ben Geloun, T. Krajewski, J. Magnen and V. Rivasseau, "Linearized Group Field Theory and Power Counting Theorems," Class. Quant. Grav. **27**, 155012 (2010) [arXiv:1002.3592 [hep-th]].

ASPECTS OF INTEGRABLE DEFECTS

E. CORRIGAN*
Department of Mathematics, University of York,
York, YO10 5DD, United Kingdom
**E-mail: edward.corrigan@york.ac.uk*

1. Outline

Defects and impurities are ubiquitous in nature and essential for the understanding of phenomena in many domains of physics ranging from classical continuum mechanics to condensed matter physics, or cosmology. In every case, a defect is likely to be modeled by a discontinuity in a field variable, such as the abrupt change in fluid velocity across a shock, or the discontinuity in the derivative of a wave function in the presence of an impurity. From a mathematical point of view it is natural to ask what kinds of defect can be supported by an integrable field theory without destroying integrability. Some integrable models do appear to be useful in certain special physical situations but it remains to be seen if the defects they can support are themselves relevant. The purpose of this talk is exploratory, to discover what might be possible and review a collection of ideas and questions. The main points to be addressed are: examples of integrable defects and the special role played by energy-momentum conservation and Bäcklund transformations; the description of solitons scattering with defects and some curious unanticipated effects; the description of defects in integrable quantum field theory and associated transmission matrices; integrable field theories with boundaries revisited.

2. Modeling an integrable discontinuity

A convenient description of the setup is described in Ref. 1. There is a single component scalar field and one selected point x_0 on the x-axis, say $x_0 = 0$, and denote the field to the left ($x < 0$) by u, and to the right ($x > 0$) by v:

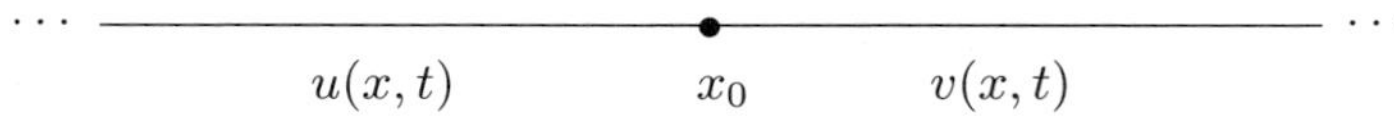

There will be field equations in separated domains:

$$\partial^2 u = -\frac{\partial U}{\partial u}, \quad x < 0, \quad \partial^2 v = -\frac{\partial V}{\partial v}, \quad x > 0,$$

where U, V represent integrable potentials. Then the mathematical question is simply, how can the fields u, v and their derivatives be 'sewn' together at $x = x_0 = 0$ to preserve integrability in the whole system?

One natural choice (a δ-impurity) would be to put

$$u(0,t) = v(0,t), \quad u_x(0,t) - v_x(0,t) = F(u(0,t)).$$

However, unless the system is linear, besides having interesting properties, integrability is generally lost.[2]

Clearly, because there is a distinguished point, translation symmetry is broken and conservation laws - at least some of them - will be violated unless the impurity contributes compensating terms. Integrable field theories enjoy a (countably) infinite number of conserved conservation laws that Poisson commute with each other and at least a significant subset of these should be preserved by the defect. These have been investigated for several models but, curiously, integrable models are selected in the presence of a defect by very simple considerations involving energy-momentum alone.

Consider the field contributions to energy-momentum:

$$P^\mu = \int_{-\infty}^{0} dx\, T^{0\mu}(u) + \int_0^\infty dx\, T^{0\mu}(v).$$

Using the field equations, can we arrange

$$\frac{dP^\mu}{dt} = \left[T^{1\mu}(u)\right]_{x=0} - \left[T^{1\mu}(v)\right]_{x=0} = -\frac{dD^\mu(u,v)}{dt},$$

with the right hand side depending only on the fields at $x = 0$ (defined by limits from the left and right domains)? If so, $P^\mu + D^\mu$ is conserved. Surprisingly, only a few possible sewing conditions (and bulk potentials $U,\ V$) are permitted for this to work.

Consider the field contribution to energy and calculate

$$\frac{dP^0}{dt} = [u_x u_t]_0 - [v_x v_t]_0.$$

Setting $u_x = v_t + X(u,v)$, $v_x = u_t + Y(u,v)$, we find $\dot{P}^0 = u_t X - v_t Y$, which is a total time derivative if

$$X = -\frac{\partial D^0}{\partial u}, \quad Y = \frac{\partial D^0}{\partial v},$$

for some D^0, meaning $\dot{P}^0 = -\dot{D}^0$. This is not really surprising since time translation is not expected to be violated by a defect, which may itself contribute to the energy of the system.

On the other hand, for momentum,

$$\frac{dP^1}{dt} = -\left[\frac{u_t^2 + u_x^2}{2} - U(u)\right]_{x=0} + \left[\frac{v_t^2 + v_x^2}{2} - V(v)\right]_{x=0}$$

$$= \left[-v_t X + u_t Y - \frac{X^2 - Y^2}{2} + U - V\right]_{x=0} = -\frac{dD^1}{dt}.$$

This is a total time derivative provided the first piece is a perfect differential and the second piece vanishes. Thus

$$X = -\frac{\partial D^0}{\partial u} = \frac{\partial D^1}{\partial v}, \quad Y = \frac{\partial D^0}{\partial v} = -\frac{\partial D^1}{\partial u}.$$

In other words, assembling the requirements at $x = 0$,

$$\frac{\partial^2 D^0}{\partial v^2} = \frac{\partial^2 D^0}{\partial u^2}, \quad \frac{1}{2}\left(\frac{\partial D^0}{\partial u}\right)^2 - \frac{1}{2}\left(\frac{\partial D^0}{\partial v}\right)^2 = U(u) - V(v),$$

a surpisingly strong set of conditions that not only restrict the defect contributions D^μ but also the bulk potentials U, V.

It is not difficult to check that the only possible choices for the latter are sine-Gordon, Liouville, massless free, or massive free. With one exception, these are the only single field integrable relativistic field theories, the exception being the Tzitzéica model (also known as the Bullough-Dodd, Mikhailov-Zhiber-Shabat, or $a_2^{(2)}$ affine Toda model), with potential

$$U(u) = e^u + 2e^{-u/2},$$

which will be discussed further later.

For example, if $U(u) = m^2u^2/2$, $V(v) = m^2v^2/2$, D^0 turns out to be

$$D^0(u,v) = \frac{m\sigma}{4}(u+v)^2 + \frac{m}{4\sigma}(u-v)^2,$$

and σ is a free parameter. This is an interesting sewing condition because it implies plane waves are purely transmitted with a phase change; quite unlike a linear δ-impurity, which is always accompanied by both transmission and reflection.

It is also worth noting in passing that there is a Lagrangian description of this type of defect (call it type I):

$$\mathcal{L} = \theta(-x)\mathcal{L}(u) + \delta(x)\left(\frac{uv_t - u_t v}{2} - D^0(u,v)\right) + \theta(x)\mathcal{L}(v).$$

As a second example, consider sine-Gordon by choosing u, v to be sine-Gordon fields (scaling the coupling and mass parameters to unity for simplicity), the allowed possibilities are:

$$D^0(u,v) = -2\left(\sigma \cos\frac{u+v}{2} + \sigma^{-1}\cos\frac{u-v}{2}\right),$$

where σ is a constant, to find

$$\begin{aligned}
x<0: &\quad \partial^2 u = -\sin u,\\
x>0: &\quad \partial^2 v = -\sin v,\\
x=0: &\quad u_x = v_t - \sigma\sin\frac{u+v}{2} - \sigma^{-1}\sin\frac{u-v}{2},\\
x=0: &\quad v_x = u_t + \sigma\sin\frac{u+v}{2} - \sigma^{-1}\sin\frac{u-v}{2}.
\end{aligned}$$

It is worth emphasising that the final two equations are a 'Bäcklund transformation' frozen at $x = 0$, and also that the defect could be placed anywhere along the x-axis. The process described above narrows the possibilities though it is necessary to confirm the integrability by other means.[1,3,4]

3. Solitons and defects

The sine-Gordon model allows solitons and antisolitons so consider a soliton incident from $x < 0$. It will not be possible to satisfy the sewing conditions (in general) unless a similar soliton emerges into the region $x > 0$. Thus, using detailed expressions for the well-known soliton solutions, the following is expected:[1]

$$\begin{aligned}
x<0: &\quad e^{iu/2} = \frac{1+iE}{1-iE},\\
x>0: &\quad e^{iv/2} = \frac{1+izE}{1-izE},\\
E = e^{ax+bt+c}, &\quad a=\cosh\theta, \quad b = -\sinh\theta,
\end{aligned}$$

where z is to be determined. It is also useful to set $\sigma = e^{-\eta}$ to find

$$z = \coth\left(\frac{\eta - \theta}{2}\right).$$

This result has a number of interesting consequences:

- $\theta > \eta$ implies $z < 0$; that is, the soliton emerges as an anti-soliton, meaning the final field configuration will contain a discontinuity of magnitude 4π at $x = 0$;
- $\theta \to \eta$ implies $z \to \infty$; that is, there is **no** emerging soliton, meaning the energy-momentum and topological charge of the soliton are captured by the 'defect', meaning the final field configuration has a discontinuity 2π;
- $\theta < \eta$ implies $z > 0$; that is, the soliton retains its character.

Defects at $x = x_1 < x_2 < x_3 < \cdots < x_n$ behave independently - each contributing a factor z_i for a total $z = z_1 z_2 \ldots z_n$. Each component of a multisoliton is affected separately - thus, since the components of a multisoliton have distinct rapidities, at most one can be 'filtered out' by the defect. The idea that defects could be used to control solitons is fascinating.[6] Since a soliton can be absorbed, it is natural to ask if a starting configuration with $u = 0$, $v = 2\pi$ would decay into a soliton? This needs quantum mechanics to provide the probability for decay after a given time. Defects can also move (with constant speed), and scatter.[5]

4. Generalisations

To complete the story for single component fields it is necessary to return to consider the Tzitzéica model mentioned before for which there can be no type I defect. There is no time to consider multi-component fields here in any detail but it is worth mentioning that type I defects are restricted to the $a_r^{(1)}$ collection of affine Toda field theories.[7] These are the only field theories of this type possessing Bäcklund transformations generalising the Bäcklund transformation of sine-Gordon.[8] Thus new ideas are necessary.

This time, start with a Lagrangian description with relativistic fields u and v on either side of the defect, as before, but add a new degree of freedom $\lambda(t)$ on the defect in a minimal way:[9]

$$\mathcal{L} = \theta(-x)\mathcal{L}_u + \theta(x)\mathcal{L}_v + \delta(x)\left((u - v)\lambda_t - D^0(\lambda, p, q)\right),$$

where

$$q = \left.\frac{u - v}{2}\right|_0, \quad p = \left.\frac{u + v}{2}\right|_0.$$

In fact, λ is the conjugate variable to the discontinuity across the defect. Then, the usual steps lead to the equations of motion in each domain together with sewing conditions at $x = 0$:

$$2q_x = -D^0_p, \quad 2p_x - 2\lambda_t = -D^0_q, \quad 2q_t = -D^0_\lambda.$$

As before, demanding momentum conservation requires a functional $D^1(u, v, \lambda)$ such that $\dot{P}^1 = -\dot{D}^1$, with the latter defined at $x = 0$. This leads to conditions implying,

$$D^0 = f(p+\lambda, q) + g(p-\lambda, q), \quad D^1 = f(p+\lambda, q) - g(p-\lambda, q)$$

with

$$\frac{\partial f}{\partial \lambda}\frac{\partial g}{\partial q} - \frac{\partial g}{\partial \lambda}\frac{\partial f}{\partial q} = U(u) - V(v).$$

The nonlinear constraint is very powerful since λ does not enter the right hand side, and takes the form of a Poisson bracket with respect to the defect discontinuity and its 'conjugate momentum' λ, which has to match the potential difference across the defect - an intriguing condition. This defect will be called type II

It is now possible to choose f, g with potentials U, V given by any of sine-Gordon, Liouville, Tzitzéica, or free massive or massless. For example, in the case of Tzitzéica with

$$U(u) = (e^u + 2\,e^{-u/2} - 3), \quad V(v) = (e^v + 2\,e^{-v/2} - 3),$$

the defect potential $D^0(\lambda, p, q)$ is given by[10]

$$\begin{aligned} D^0 = 2\sigma &\left(e^{(p+\lambda)/2} + e^{-(p+\lambda)/4}\left(e^{q/2} + e^{-q/2}\right)\right) \\ &+ \frac{1}{\sigma}\left(8\,e^{-(p-\lambda)/4} + e^{(p-\lambda)/2}\left(e^{q/2} + e^{-q/2}\right)^2\right). \end{aligned}$$

This looks a bit messy but it is interesting to see how solitons traverse the defect and, as a by-product, this expression and accompanying sewing conditions provide a new Bäcklund transformation for the Tzitzéica model.

In the sine-Gordon model the type II defect is new with two free parameters; in a sense it is two type-I defects fused together.[9,10] It remains to be seen whether this idea generalises to all the affine Toda models. It seems to work for a subset but possibly other new ideas are required to obtain the full picture; this work is ongoing.

5. Defects in quantum field theory

This will have to be brief because the story is too lengthy to be comprehensive. The general approach is algebraic, meaning one seeks a transmission matrix to describe the scattering of solitons by defects (remember, within free massive theories the defects described as type I/II are purely transmitting), and demands compatibility with the bulk scattering of solitons with each other. This provides expressions that may then be checked in various ways, for example using perturbation theory or TBA. It is expected scattering will respect topological charge conservation because a defect can itself carry topological charge. The latter means transmission matrices are intrinsically infinite-dimensional and may come in several varieties. For example, in the sine-Gordon model, since topological charge can change by $\pm$ two, as a soliton converts to an anti-soliton, or vice-versa, there will be two, labeled either by even or by odd integers. In fact, this could be paraphrased by saying the transmission matrix is labeled by weights of a_1, either spin weights (odd) or vector weights (even). In generalisations to other models this is precisely the situation. The transmission matrices must respect a form of crossing and may be unitary.[11,12] In the sine-Gordon model, if the even-labeled transmission matrix is unitary, the odd-labeled matrix is not - the latter can be regarded as a resonance of the former.

Finally, an open question is to try to understand the scattering of defects themselves, which really requires a description of scattering matrices labelled by whole weight lattices (rather than individual finite representations), constructed to be compatible with the transmission matrices. There are already some ideas but not, so far, a clear picture.[5,13]

Thus for the sine-Gordon model it is convenient to write,

$$T_{a\alpha}^{b\beta}(\theta,\eta), \quad a+\alpha=b+\beta,$$

where α, β (even or odd integers) and a, b (±1) are defect and soliton labels, respectively, and η is a (set of) defect parameter(s). Then, the components of the transmission matrix are required to be compatible with the (already known) bulk soliton-soliton scattering:[11]

$$S_{ab}^{cd}(\Theta)\,T_{d\alpha}^{f\beta}(\theta_a)T_{c\beta}^{e\gamma}(\theta_b)=T_{b\alpha}^{d\beta}(\theta_b)T_{a\beta}^{c\gamma}(\theta_a)S_{cd}^{ef}(\Theta),$$

where $\Theta=\theta_a-\theta_b$, and sums over the 'internal' indices β, c, d are to be understood. Effectively, given S, these are a set of quadratic recurrence relations for the components of the transmission matrix. A solution has been known for some time.[12] It is not the most general but will be described first.

Zamolodchikov's sine-Gordon soliton-soliton S-matrix[14] is given conveniently by

$$S_{ab}^{cd}(\Theta) = \rho(\Theta) \begin{pmatrix} A & 0 & 0 & 0 \\ 0 & C & B & 0 \\ 0 & B & C & 0 \\ 0 & 0 & 0 & A \end{pmatrix}$$

where

$$A(\Theta) = \frac{qx_2}{x_1} - \frac{x_1}{qx_2}, \; B(\Theta) = \frac{x_1}{x_2} - \frac{x_2}{x_1}, \; C(\Theta) = q - \frac{1}{q}$$
$$x_a = e^{\gamma\theta_a}, \; a = 1, 2, \; \Theta = \theta_1 - \theta_2, \; q = e^{i\pi\gamma}, \; \gamma = \frac{8\pi}{\beta^2} - 1,$$

and

$$\rho(\Theta) = \frac{\Gamma(1+z)\Gamma(1-\gamma-z)}{2\pi i} \prod_1^\infty R_k(\Theta) R_k(i\pi - \Theta)$$
$$R_k(\Theta) = \frac{\Gamma(2k\gamma + z)\Gamma(1 + 2k\gamma + z)}{\Gamma((2k+1)\gamma + z)\Gamma(1 + (2k+1)\gamma + z)}, \; z = i\gamma/\pi.$$

It is useful to define the variable $Q = e^{4\pi^2 i/\beta^2} = \sqrt{-q}$. and then the Konik-LeClair solutions take the form (apologies for overusing β),

$$T_{a\alpha}^{b\beta}(\theta) = f(q,x) \begin{pmatrix} Q^\alpha\, \delta_\alpha^\beta & q^{-1/2} e^{\gamma(\theta-\eta)}\, \delta_\alpha^{\beta-2} \\ q^{-1/2}\, e^{\gamma(\theta-\eta)}\, \delta_\alpha^{\beta+2} & Q^{-\alpha}\, \delta_\alpha^\beta \end{pmatrix}$$

where $f(q,x)$ is not uniquely determined but, for a unitary transmission matrix, should satisfy

$$\bar{f}(q,x) = f(q,qx), \quad f(q,x)f(q,qx) = \left(1 + e^{2\gamma(\theta-\eta)}\right)^{-1}.$$

A 'minimal' solution has the following form

$$f(q,x) = \frac{e^{i\pi(1+\gamma)/4}}{1 + ie^{\gamma(\theta-\eta)}} \frac{r(x)}{\bar{r}(x)},$$

where it is convenient to put $z = i\gamma(\theta - \eta)/2\pi$ and

$$r(x) = \prod_{k=0}^\infty \frac{\Gamma(k\gamma + 1/4 - z)\Gamma((k+1)\gamma + 3/4 - z)}{\Gamma((k+1/2)\gamma + 1/4 - z)\Gamma((k+1/2)\gamma + 3/4 - z)}.$$

At this point a few remarks can be made (supposing $\theta > 0$). First, it is tempting to suppose η (possibly renormalized) is the same parameter as in the type I classical model. Clearly

- $\eta < 0$ - the off-diagonal entries dominate;
- $\theta > \eta > 0$ - the off-diagonal entries dominate;
- $\eta > \theta > 0$ - the diagonal entries dominate,

all similar features to those found in the classical soliton-defect scattering. Notice, $\theta = \eta$ is not special but there is a simple pole (in $f(q,x)$) nearby, at

$$\theta = \eta - \frac{i\pi}{2\gamma} \to \eta, \ \beta \to 0.$$

This pole is like a resonance, with complex energy,

$$E = m_s \cosh\theta = m_s(\cosh\eta\cos(\pi/2\gamma) - i\sinh\eta\sin(\pi/2\gamma))$$

and possesses a 'width' proportional to $\sin(\pi/2\gamma)$, which vanishes as $\beta \to 0$.

On the other hand, the type I defect is not the only possibility for the sine-Gordon model and therefore there should exist transmission matrices for type II defects as well. These were overlooked originally but correspond to more general solutions[10]

$$\rho(\theta) \begin{pmatrix} (a_+Q^\alpha + a_-Q^{-\alpha}\,x^2)\,\delta_\alpha^\beta & x\,(b_+Q^\alpha + b_-Q^{-\alpha})\,\delta_\alpha^{\beta-2} \\ x\,(c_+Q^\alpha + c_-Q^{-\alpha})\,\delta_\alpha^{\beta+2} & (d_+Q^\alpha\,x^2 + d_-Q^{-\alpha})\,\delta_\alpha^\beta \end{pmatrix}$$

where the free constants satisfy the two constraints

$$a_\pm\, d_\pm - b_\pm\, c_\pm = 0.$$

These and $\rho(\theta)$ are constrained further by crossing and unitarity. With the choice $a_- = d_+ = 0$ and $b_+ = c_- = 0$ or $b_- = c_+ = 0$ (after a similarity transformation), the expression reduces to the type I solution. Intriguingly, for certain other choices of parameters it reduces to a direct sum of the Zamolodchikov S-matrix and two infinite dimensional pieces (suggesting, in a manner to be explored further, the type II defect should be closely related to a soliton).

6. Boundaries revisited: the sine-Gordon model

Start with a single selected point on the x-axis, say $x_0 = 0$, and denote the field to the left ($x < 0$) by u:

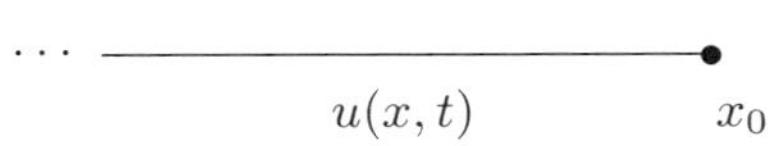

The sine-Gordon model on a half-line with a general (two-parameter) integrable boundary condition was analyzed some time ago.[15] However, a defect (or several defects) could be placed in front of the boundary to generate a new boundary effect (as seen from $x \ll 0$) (for example see Ref. 16). It is then natural to think of the boundary as carrying topological charge (of any even or odd number of units), since typically the charge changes as solitons reflect. Taking this into account means considering reflection matrices of the form

$$R^{b\,\beta}_{a\,\alpha}(\theta) = \begin{pmatrix} r_+(\alpha,x)\,\delta^{\beta}_{\alpha} & s_+(\alpha,x)\,\delta^{\beta-2}_{\alpha} \\ s_-(\alpha,x)\,\delta^{\beta+2}_{\alpha} & r_-(\alpha,x)\,\delta^{\beta}_{\alpha} \end{pmatrix}$$

and solving the Boundary Yang-Baxter equation[17]

$$R^{q\,\beta}_{a\,\alpha}(\theta_a)\,S^{p\,s}_{b\,q}(\Theta_+)R^{r\,\gamma}_{p\,\beta}(\theta_b)\,S^{d\,c}_{s\,r}(\Theta_-) = S^{p\,q}_{b\,a}(\Theta_-)R^{r\,\beta}_{p,\alpha}(\theta_b)S^{s\,c}_{q\,r}(\Theta_+)R^{d\,\gamma}_{s\,\beta}(\theta_a),$$

with $\Theta_+ = (\theta_b + \theta_a)$ and $\Theta_- = (\theta_b - \theta_a)$. One solution, the Ghoshal-Zamolodchikov solution reformulated, in which the matrix elements do not depend on the state of the boundary, is

$$R^{b\,\beta}_{a\,\alpha}(\theta) = \sigma(\theta)\begin{pmatrix} (r_1 x + r_2/x)\,\delta^{\beta}_{\alpha} & k_0\left(x^2 - 1/x^2\right)\delta^{\beta-2}_{\alpha} \\ l_0\left(x^2 - 1/x^2\right)\delta^{\beta+2}_{\alpha} & (r_2 x + r_1/x)\,\delta^{\beta}_{\alpha} \end{pmatrix}$$

and $l_0 = k_0$, $r_1 r_2 = 1$, but the general solution is[18]

$$\begin{aligned}
r_+(\alpha,x) &= \left(x^2 - 1/x^2\right)\left(r_3 q^{\alpha+1}x - r_4 q^{-\alpha-1}/x\right) + r_1 x + r_2/x,\\
r_-(\alpha,x) &= \left(x^2 - 1/x^2\right)\left(r_4 q^{-\alpha+1}x - r_3 q^{\alpha-1}/x\right) + r_2 x + r_1/x,\\
s_+(\alpha,x) &= \left(x^2 - 1/x^2\right)\left(k_0 + k_1 q^{\alpha} + k_2 q^{-\alpha}\right),\\
s_-(\alpha,x) &= \left(x^2 - 1/x^2\right)\left(l_0 + l_1 q^{\alpha} + l_2 q^{-\alpha}\right),
\end{aligned}$$

$$k_1 l_1 = -r_3^2, \quad k_2 l_2 = -r_4^2, \quad k_1 l_0 + q^2 k_0 l_1 = q r_2 r_3, \quad k_0 l_2 + q^2 k_2 l_0 = q r_1 r_4.$$

A defect placed in front of a boundary generalises R according to

$$R^{b\,\beta\,\tilde\beta}_{a\,\alpha\,\tilde\alpha}(\theta) = T^{c\,\tilde\gamma}_{a\,\tilde\alpha}(\theta)R^{d\,\beta}_{c\,\alpha}(\theta)\hat{T}^{b\,\tilde\beta}_{d\,\tilde\gamma}(\theta)$$

where $\hat{T}(\theta) = T^{-1}(-\theta)$.

Using a diagonal R matrix corresponding to a Dirichlet boundary condition,

$$R^{(0)}{}^{d\,\beta}_{c\,\alpha}(\theta) = \sigma(\theta)\begin{pmatrix} (rx + x^{-1}r^{-1})\,\delta^{\beta}_{\alpha} & 0 \\ 0 & (rx^{-1} + xr^{-1})\,\delta^{\beta}_{\alpha} \end{pmatrix}$$

$TR^{(0)}\hat{T}$ is equivalent to the general solution given above when T is the general type II transmission matrix, but equivalent to the GZ solution when T is restricted to the type I transmission matrix. It is natural to seek a Lagrangian description for the generalized setup.[18]

7. Remarks

These examples point the way to further questions. The transmission matrix for the Tzitzéica model defect has been obtained recently, together with transmission matrices for the $a_r^{(1)}$ class of affine Toda models.[20] It has proved profitable to consider the charges carried by a defect in terms of a tower of states related by generalised annihilation and creation operators,[19] leading to succinct expressions for transmission matrices.[13,20] Indeed, the transmission matrix can be considered as intertwining the tensor products of finite and infinite dimension representations of a Borel subalgebra of a relevant quantum group. In turn, this has prompted an investigation into the nature of infinite-dimensional representations of quantum groups. Interestingly, this makes contact with work motivated by the analysis of spin chains.[21] Beyond this, there has been work on other types of integrable model,[22–25] and models with fermions.[26]

Acknowledgements

I am grateful to the organisers of Group29 for the opportunity to present these ideas, and to Peter Bowcock, Craig Robinson and Cristina Zambon for enjoyable collaborations and discussions. I also apologise to several authors for the lack of space in this proceedings to describe or refer to their work.

References

1. P. Bowcock, E. Corrigan and C. Zambon, *Classically integrable field theories with defects*, Int. J. Mod. Physics A19 (Sup) (2004) 82-91; hep-th/0305022.
2. R. H. Goodman, P. J. Holmes and M. I. Weinstein, *Interaction of sine-Gordon kinks with defects: phase transport in a two mode model*, Physica D161 (2002), 21.
3. P. Bowcock, E. Corrigan and C. Zambon, *Affine Toda field theories with defects*, JHEP01(2004)056; hep-th/0401020.
4. I. Habibullin and A. Kundu, *Quantum and classical integrable sine-Gordon model with defect*, Nucl. Phys. B795 (2008) 549; arXiv:0709.4611 [hep-th].
5. P. Bowcock, E. Corrigan and C. Zambon, *Some aspects of jump-defects in the quantum sine-Gordon model*, JHEP08(2005)023; hep-th/0506169.
6. E. Corrigan and C. Zambon, *Aspects of sine-Gordon solitons, defects and gates*, J. Phys. A: Math. Gen. 37 (2004) L471-L477; hep-th/0407199.
7. E. Corrigan and C. Zambon, *On purely transmitting defects in affine Toda field theories*, JHEP07(2007)001; arXiv:0705.1066
8. A. P. Fordy and J. Gibbons, *Integrable nonlinear Klein-Gordon equations and Toda lattices,* Commun. Math. Phys. 77 (1980) 21.
9. E. Corrigan and C. Zambon, *A new class of integrable defects*, J. Phys. A: Math. Theor. 42 (2009) 475203; arXiv:0908.3126.

10. E. Corrigan and C. Zambon, *A transmission matrix for a fused pair of integrable defects in the sine-Gordon model*, J. Phys. A: Math. Theor. 43 (2010) 345201; arXiv:1006.0939.
11. G. Delfino, G. Mussardo and P. Simonetti, *Scattering theory and correlation functions in statistical models with a line of defect*, Nucl. Phys. B432 (1994) 518; hep-th9409076.
12. R. Konik and A. LeClair, *Purely transmitting defect field theories*, Nucl. Phys. B538 (1999) 587; hep-th9703085.
13. R. Weston, *An algebraic setting for defects in the XXZ and sine-Gordon models*, arXiv:1006.1555.
14. A. B. Zamolodchikov and Al. B. Zamolodchikov, *Factorized S-Matrices in Two Dimensions as the Exact Solutions of Certain Relativistic Quantum Field Theory Models*, Ann. Phys. 120 (1979) 253.
15. S. Ghoshal, A. B. Zamolodchikov, *Boundary S matrix and boundary state in two-dimensional integrable quantum field theory,* Int. J. Mod. Phys. A9 (1994) 3841-3886; hep-th/9306002.
16. Z. Bajnok and Z. Simon, *Solving topological defects via fusion*, Nucl. Phys. B802 (2008) 307; arXiv:0712.4292.
17. I. V. Cherednik, *Factorizing Particles on a Half Line and Root Systems*, Theor. Math. Phys. 61 (1984) 977-983.
18. E. Corrigan and C. Zambon, *Infinite dimension reflection matrices in the sine-Gordon model with a boundary*, JHEP06(2012)050; arXiv:1202.6016
19. A.J. Macfarlane, *On q-analogues of the quantum harmonic scillator and the quantum group $SU(2)_q$*, J. Phys. A: Math. Gen. 22 (1989) 4581;
L.C. Biedenharn, *The quantum group $SU_q(2)$ and a q-analoque of the boson operators*, J. Phys. A: Math. Gen. 22 (1989) L873.
20. E. Corrigan and C. Zambon, *Integrable defects in affine Toda field theory and infinite dimensional representations of quantum groups*, Nuclear Physics B848 (2011), 545-577; arXiv:1012.4186.
21. H. Boos, F. Göhmann, A. Klümper, K. S. Nirov and A. V. Razumov, *Exercise with the universal R-matrix*, J. Phys. A: Math. Theor. 43 (2010) 415208; arXiv:1004.5342v3.
22. E. Corrigan and C. Zambon, *Jump-defects in the nonlinear Schrödinger model and other non relativistic field theories*, Nonlinearity 19 (2006) 1447; nlin.si/0512038
23. V. Caudrelier, *On a systematic approach to defects in classical integrable field theories*, Int. J. Geom. Meth. Mod. Phys. 5 (2008) 1085.
24. F. Nemes, *Semiclassical analysis of defect sine-Gordon theory*, Int. J. Mod. Phys. A25 (2010) 4493; arXiv:0909.3268.
25. J. Avan and A. Doikou, *Liouville integrable defects: the non-linear Schrodinger paradigm*, JHEP1201 (2012) 040; arXiv:1110.4728.
26. A. R. Aguirre, J. F. Gomes, L. H. Ymai and A. H. Zimerman, *Grassmannian and bosonic Thirring models with jump defects*, JHEP1102 (2011) 017; arXiv:1012.1537.

DYNAMICAL SYMMETRIES AND ALGEBRAIC METHODS IN NUCLEAR STRUCTURE PHYSICS

A. ARIMA

Shizuoka University of Art and Culture, Hamamatsu City, Shizuoka, Japan
and
Musashi Academy of Nezu Foundation
Nerima Ku, Tokyo, Japan

In this paper I start from the early history of application of group theory to nuclear physics, then I discuss the seniority scheme in nuclear shell model under strong pairing interaction. Then I go to the Interacting Boson Model (IBM) and its applications. As the microscopic foundation, I describe nucleon pair approximation and its application to complex nuclei. Finally I discuss the microscopic method to derive the IBM Hamiltonian in strongly deformed nuclei from the mean-field theory.

Keywords: Dynamical Symmetries; group theory; nuclear Structure.

1. Early history

1.1. *The pioneering works by Wigner and Racah*

The use of group theoretical methods in nuclear physics has a long history starting from the pioneering work by E. P. Wigner.[1] He used $U(4)$ group to describe spin-isospin invariance in light nuclei. Between ^{4}He and ^{16}O, both protons and neutrons occupy the $0p$-shell. The Coulomb interaction does not affect so much those p-shell nuclei. If we neglect the Coulomb interaction, the isospin invariance works well. In those light nuclei, the influence of non-central interactions, such as tensor and LS interactions, is still very weak. Furthermore, among central interactions, the Wigner and Majorana interactions dominate. Those Wigner and Majorana interactions are invariant under the U(4) transformation. This is called Wigner's supermultiplet theory. Racah[2] developed convenient scheme to classify atomic and nuclear states and technique to calculate atomic and nuclear quantities such as energies and electro-magnetic transitions by using the O(3) group.

1.2. *Nuclear Shell Model*

1.2.1. *Magic numbers*

Some nuclei are very stable. They have special numbers of protons and/or neutrons. Those numbers are called the nuclear magic numbers. They are $2, 8, 20, 28, 50, 82$ for protons, and $2, 8, 20, 28, 50, 82, 126$ for neutrons. In atomic physics, the magic numbers are 2(He), 10(Ne), 18(Ar), 36(Kr), 54(Xe) and 86(Rn), called inert-gas atoms. In such atoms, the electronic shell orbits are closed. Therefore the existence of the magic numbers in nuclei indicates that nucleons rotate in their orbits similarly as electrons do in atoms. Differently from atoms, however, the nuclear mean field must have a strong spin-orbit force to explain the magic numbers larger than 20, which was discovered by Mayer and Jensen in 1949.[3]

1.2.2. *Seniority scheme*

All even-even nuclei have spin 0^+ without any exceptions, and almost all the first excited states of the even-even nuclei have spin 2^+. In order to explain that all even-even nuclei have spin 0^+ in their ground states, a strong attractive pairing interaction must work in nuclei. A realistic nucleon-nucleon interaction has such a property.

Now let us simplify the argument by assuming that all nucleons are in a single j-shell. Let a creation operator $a_m^{\dagger j}$ satisfy the commutation relations which define spherical tensor operators of half-integer rank j, $[J_Z, a_m^{\dagger j}] = m a_m^{\dagger j}$, and $[J_x \pm iJ_y, a_m^{\dagger j}] = \sqrt{(j \pm m + 1)(j \mp m)}\, a_{m\pm 1}^{\dagger j}$. If we define $\tilde{a}_m^j = (-1)^{j-m} a_{-m}^j$, the anti-commutation relations are given by $\{a_m^{\dagger j}, a_{m'}^{\dagger j'}\} = \{\tilde{a}_m^j, \tilde{a}_{m'}^{j'}\} = 0, \{\tilde{a}_m^j, a_{m'}^{\dagger j'}\} = (-1)^{j-m}\delta(j, j')\delta(m, -m')$. The three operators, $S_+ = \sqrt{\frac{\Omega}{2}}[a^{\dagger j} \times a^{\dagger j}]_0^{(0)}, S_0 = \frac{1}{2}(N - \Omega)$, and $S_- = \sqrt{\frac{\Omega}{2}}[a^{\dagger j} \times a^{\dagger j}]_0^{(0)}$, satisfy the commutation relations appropriate to angular momentum operators. In the above relation, $\Omega = (2j + 1)/2$ and N is the number operator.

The pairing Hamiltonian H_p can be written as $H_p = 2GS_+S_-$. The eigenvalues of this Hamiltonian GS_+S_- are given by $2G(S - S_z)(S - S_z + 1)$. The quasi-spin S is given by the seniority quantum number ν as $S = (\Omega - \nu)/2$, while the relation between $S_z = S_0$ and the number of nucleons is $S_z = (N - \Omega)/2$.

We can now easily calculate the matrix elements of any two body interactions,[4] $\langle j^n \nu\alpha J|V|j^n \nu'\alpha' J\rangle, \quad \nu' = \nu, \nu \pm 2, \nu \pm 4$. The algebraic reduction formula for example when $\nu = \nu'$ results in

$$
\begin{aligned}
&\langle j^n \nu\alpha J|V|j^n \nu\alpha' J\rangle \\
&= \left\{ \frac{(\Omega - 2\nu)(2\Omega - n - \nu}{4(\Omega - \nu)(\Omega - \nu - 1)}(G_0 + 2F_0) - F_0 \right\} (n-\nu)\delta(\alpha, \alpha') \\
&+ \frac{(\Omega - \nu)(\Omega - \nu - 2}{2(\Omega - \nu)(\Omega - \nu - 1)} \langle j^\nu \nu\alpha J|V|j^\nu \nu\alpha' J\rangle \\
&+ \frac{(n - \nu)(2\Omega - n - \nu}{2(\Omega - \nu)(\Omega - \nu - 1)} \langle j^\nu \nu\alpha J|\bar{V}|j^\nu \nu\alpha' J\rangle.
\end{aligned} \tag{1}
$$

Here

$$
\begin{aligned}
G_J &= \langle j^2 J|V|j^2 J\rangle, F_k = \sum_J (2J+1)W(jjjj;JK)G_J, \\
V &= \sum_J G_J \left([a^{\dagger j} \times a^{\dagger j}]^{(J)} \cdot [\tilde{a}^j \times \tilde{a}^j]^{(J)} \right), \\
\bar{V} &= \sum_K F_K \left([a^{\dagger j} \times a^{\dagger j}]^{(K)} \cdot [\tilde{a}^j \times \tilde{a}^j]^{(K)} \right).
\end{aligned}
$$

We often call V particle-particle interactions and $\bar{V}$ particle-hole interactions. This simple formula was derived for the first time by Kawarada and Arima 1964.[5] Please notice that the matrix element in a n-body system is reduced to that in the ν-body system. The ground state has $\nu = 0$ in even-even nuclei and $\nu = 1$ in even-odd and odd-even nuclei when the pairing interaction dominates. Similarly, the first excited states of even-even nuclei have $\nu = 2$ when the pairing interaction is very strong. The diagonal matrix element of any two body interaction can be reduced to that of the two body system with $\nu = 2$.

1.2.3. *Quadrupole-Quadrupole and Quadrupole-pairing interactions*

Bohr and Mottelson[6] discovered the collective motions in nuclear structure. Most importantly, they discovered that some nuclei are deformed. The best evidence is the rotational motion in those deformed nuclei. The other evidence of the deformation is given by large electric quadrupole moments of those nuclei.[7]

Now we need an interaction which causes to deform nuclei. Because quadrupole moments are large in deformed nuclei, it is very natural to introduce the quadrupole-quadrupole interaction, $V_q = Q \cdot Q$, where $Q = r^2 Y^{(2)}_m(\theta, \phi)$. The constancy of the first 2^+ excitation energies of even-even Sn isotopes is the evidence that the seniority scheme works well.[8] If we use an interaction $V = g_0 S_+ S_- + \kappa Q \cdot Q$, one obtains that the 2^+_1 energies for nuclei with more valence nucleons are systematically larger than those with less valence nucleons parabolically up to the middle of the shell.

In 1965 I was very surprised by this fact.[9] This observation is contrary to our expectation that the attractive $Q \cdot Q$ interaction decreases the 2^+ excitation energy to induce the deformation where the 2^+ state comes down to very low energy. Many other calculations those days also recognized this problem. Many of them have used very strong attractive $Q \cdot Q$ interaction to explain nuclear deformation. These days one group of nuclear theorists, who favor the seniority scheme particularly in nuclei such as the Sn isotopes in which protons form stable closed shell core with Z=50, said that the $Q \cdot Q$ interaction must be very weak. Therefore a compromise is that the $Q \cdot Q$ interaction among the same kind of nucleons (κ_{nn} and κ_{pp}) should be weak, but the $Q \cdot Q$ interaction can be strong between neutrons and protons ($\kappa_{pn} \gg \kappa_{nn}$ and κ_{pp}). Indeed I believed this interpretation even in 1980 when I gave a talk on nuclear effective interactions in San Francisco. This was one of my great mistake. Later I realized what I have done in much earlier time 1964 and 1966. This was the formula which I have just discussed.[5]

Let me introduce the attractive quadrupole pairing interaction together with the monopole pairing interaction. The spectrum of ${}^{210}_{82}\mathrm{Pb}_{128}$ shows that the 2^+ first excited state is much lower than other excited states. This means that the two-body attractive interaction in the 2^+ state is also strong, as well as in the 0+ ground state. Thus we need to have also an attractive quadrupole-pairing interaction, $V_2 = G_2([a^{+j} \cdot a^{+j}]^{(2)} \cdot [\tilde{a}^j \cdot \tilde{a}^j]^{(2)})$. Numerical calculation shows that for an effective interaction $V = g_0 S_+ S_- + V_2$, the quadrupole-pairing V_2 pushes down the 2^+ state as n increases, contrary to the $Q \cdot Q$ interaction. It is now very apparent how to explain the stable 2^+ excitation energies in the Sn isotopes—we need the following combination, $V = g_0 S_+ S_- + V_2 + \kappa Q \cdot Q$. Adjusting the strengths G_2 and κ, we can obtain the constant value of the 2^+ excitation energies.

I can now say clearly that the effective interactions should contain the $Q \cdot Q$ interaction and the quadrupole pairing interaction together with the strong monopole pairing interactions between the same kind of nucleons. Therefore κ_{pn} can be of the same order of magnitude as $\kappa_{nn} = \kappa_{pp}$. The role of each interaction in $V = g_0 S_+ S_- + V_2 + \kappa Q \cdot Q$ is as follows. The pairing interaction $S_+ S_-$ stabilize 0^+ pairs of nucleons, and the sum of the quadrupole pairing interaction and the quadrupole-quadrupole interaction $V_2 + \kappa Q \cdot Q$ stabilize 2^+ pairs of nucleons. These days more realistic nucleon-nucleon interactions are available. They must keep the 2^+ excitation energies of the Sn isotopes constant. We, however, still need schematic interactions such as $V = g_0 S_+ S_- + V_2 + \kappa Q \cdot Q$ to simplify calculations and to understand more intuitively the structure of complicated nuclei.

1.2.4. *The strong coupling limit in the BCS theory*

Bardeen, Cooper and Schrieffer[10] discovered an epoch-making explanation of the superconductivity, that is the BCS theory, in 1957. In the BCS theory the Hamiltonian is given by $H = \sum \omega_k n_{ks} - V \sum b_k^* b_k$, where ω_k is the Bloch energy and n_{ks} is the single-particle number operator defined by $n_{ks} = c_{ks}^+ c_{ks}$, in terms of electron creation and annihilation operators c_{ks}^+ and c_{ks}. Here k and s are momentum and spin of an electron. The average coupling constant V is inversely proportional to the size of the system, and b_k means the annihilation operator and the creation operator of an electron pair with opposite momenta and spins, and is defined by $b_k = c_{k\downarrow} c_{k\uparrow}$ and $b_k^* = c_{k\uparrow} c_{k\downarrow}$. The Cooper pair is the sum of them: $\sum b_n^*$.

We define the following three operators, $M_+ = \sum b_k^*/\sqrt{2}, M_- = \sum b_k/\sqrt{2}, M_z = (N - \frac{N(0)\omega}{2})/2$. N is the total number of electrons involved in the Cooper pair, $N(0)$ is the density of the Bloch states of one spin per unit energy at the Fermi surface, and ω is the energy of phonon responsible to produce the pairing interaction. The three operators satisfy the same commutation relations as the angular momentum operators.

We can rewrite the pairing-interaction H_p, $H_p = -2VM_+M_- = 2V(M^2 - M_z^2 + M_z)$. When V is very large, we can treat the kinetic energy by perturbation. The strong coupling of the BCS theory is very similar as the seniority scheme in the nuclear structure theory. The electron pair, produced by the pairing interaction, is nothing but the Cooper pair. The reason why I discussed these very old stories both in the nuclear structure and the condensed matter physics is that I wanted to tell you that nuclear theorists, including myself, missed a great chance to discover at least the strong coupling limit of the BCS theory, if we could understand the role of the pairing interaction in the superconductor. In condensed matter physics it is the momentum which play an essential role, but in nuclear physics it is the angular momentum and spin. However, we can recognize that the $SU(2)$ and $O(3)$ algebra can commonly be used. The seniority scheme was developed before the BCS theory. Therefore if nuclear physicists were more interested in the condensed matter physics, they had a chance to contribute at least to develop the strong coupling limit of the BCS theory. I am very sorry that I was too busy in nuclear physics to study other fields such as condensed matter physics.

1.2.5. *The Elliott $SU(3)$ model in the sd shell of nuclei*

Nuclear collective motions were discovered and predicted by Bohr and Mottelson 1951-1953.[11] Bohr predicted a rotational motion of deformed nuclei with large mass numbers. On the other hand the many-body shell model calculations have been accomplished, for example in ^{20}Ne at that time. The observed spectrum of ^{20}Ne shows the rotational feature which can be well explained by the shell model calculation.[12] There are two familiar so-called accidental degeneracy in energy levels. One of them is the Hamiltonian with the Coulomb potential. The other is that of the three dimensional harmonic oscillator. Those degeneracies are not accidental but due to symmetries of those systems. The Coulomb potential has $SU(2) \times SU(2)$ symmetry found by Pauli and Elliott found $SU(3)$ symmetry in the harmonic oscillator. Elliott found furthermore that all states made of nucleons in the $1s$ and $0d$ shells can be classified in terms of $O(6)$ group, which has $SU(3)$ as a subgroup. Then Elliott introduced the $SU(3)$ model of the $1s - 0d$ shell nuclei. His Hamiltonian is $H(SU3) = H_{core} + (a - 3/4b)L(L+1) + C_2(SU3)$. It is very easy to obtain eigen-value as the following expression, $E = E_{core} + \varepsilon n + (a - 3/4b)L(L+1) + b(\lambda^2 + \lambda\mu + \mu^2 + 3\lambda + 3\mu)$. This expression contain a term $L(L+1)$. This is indeed the rotational energy. Elliott succeeded in understanding why nuclear shell model can explain the nuclear rotational motion in light nuclei such as $1s0d$-nuclei.[13]

2. The Interacting Boson Model (IBM)

2.1. *The IBM-I*

I was very inspired by the phonon model, in which the spin of the phonon is 2^+. By this model several low-lying states of even-even nuclei are explained as multi-phonon states. I started to develop further the phonon model. One of my ideas was that low-lying states of even-even nuclei are made of 2^+ phonons which I named d bosons and 0^+ phonons which I named s bosons, and these bosons interact with each other. I wrote only a Japanese paper on this idea. Franco Iachello also developed the phonon model in his Ph D thesis, the title of which was indeed the Interacting Boson Model.[14]

In 1974, I was invited by the KVI in the Groningen University. Franco Iachello and I started our cooperation to develop the Interacting Boson Model. Our basic idea is that basic building blocks in even-even nuclei are the 0^+ (mono-pole) pair and the 2^+ (quadrupole) pair. Those are analogues of the Cooper pair in the condensed matter physics. We further approximate those fermion pairs by bosons. They are s and d bosons, which interact each other. This model is now called the Interacting Boson Model.[15]

We assumed that all states are made of s bosons and d bosons. The building blocks of all operators of this model are creation operators of bosons $s^+, d_\mu^+(\mu = 0, \pm1, \pm2)$, and their annihilation operators. $s, d_\mu(\mu = 0, \pm1, \pm2)$. Using these six operators, we can construct $36 = 6 \times 6$ infinitesimal operators of the $U(6)$ group. The group $U(6)$ has many subgroups, such as $U(5)$, $U(3)$ and $U(2)$ and $O(6)$, $O(5)$ and $O(3)$. Using linear (one-body) and quadratic (two-body) Casimir operators, we can write the most general Hamiltonian of the interacting boson model as

$$\begin{aligned} H = \ & e_0 + e_1 C_1(U6) + e_2 C_2(U6) + \varepsilon C_1(U5) + \alpha C_2(U5) \\ & + \beta C_2(O5) + \gamma C_2(O3) + \delta C_2(SU3) + \eta C_2(O6). \end{aligned}$$

2.1.1. *Chain I: the $U(5)$ symmetry*

Considering the Hamiltonian

$$H^{(I)} = e_0 + e_1 C_1(U6) + e_2 C_2(U6) + \varepsilon C_1(U5) + \beta C_2(O5) + \gamma C_2(O3),$$

we have the eigenvalues of $H^{(I)}$ as follows.

$$\begin{aligned} & E^{(I)}(N, nd, \nu, n_\Delta, L, M_L) \\ & \quad = \langle [N], n_d, \nu, n_\Delta, L, M_L | H^{(I)} | [N], n_d, \nu, n_\Delta, L, M_L \rangle \\ & \quad = E_0 + \varepsilon n_d + \alpha n_d(n_d + 4) + \beta 2\nu(\nu + 1) + \gamma 2L(L + 1), \end{aligned}$$

where $E_0 = e_0 + e_1 N + e_2 N(N + 5)$. The chain I corresponds to the vibrational nucleus in the Bohr-Mottelson model. The agreement between the theoretically predicted spectrum and the observed spectrum is very good in the vibrational motion, e.g., the nucleus ${}^{110}_{48}\mathrm{Cd}_{62}$.

2.1.2. *Chain II: $SU(3)$ Symmetry*

For the Hamiltonian

$$H^{(II)} = e_0 + e_1 C_1(U6) + e_2 C_2(U6) + \gamma C_2(O3) + \delta C_2(SU3),$$

the quantum numbers are the total number of bosons, λ, μ, L and M, the eigenvalues of $H^{(II)}$ are given by $E^{(II)}(N, \lambda, \mu, L, M) = E_0 + \gamma 2L(L+1) + \delta 2/3(\lambda^2 + \mu^2 + \lambda\mu + 3\lambda + 3\mu)$, under the groups $U(6)$, $SU(3)$, $O(3)$ and $O(2)$. A typical example of nuclei showing the $SU(3)$ symmetry is ${}^{156}_{64}\mathrm{Gd}_{92}$. The $SU(3)$ nuclei correspond to β stable deformed rotational nuclei of the Bohr - Mottelson collective model.

2.1.3. *Chain III: O(6) Symmetry*

Let us consider the Hamiltonian

$$H^{(III)} = e_0 + e_1 C_1(U6) + e_2 C_2(U6) + \beta C_2(O5) + \gamma C_2(O3) + \eta C_2(O6), \tag{2}$$

which has eigenvalues $E^{(III)}(\omega, \delta, \tau, \nu, L, M_L) = E_0 + \beta 2\tau(\tau+3) + \gamma 2L(L+1)$. An example of a spectrum with $O(6)$ symmetry is ${}^{196}_{87}\mathrm{Pt}_{118}$. The predicted levels of the IBM are not in as good an agreement with the observed levels as the previous cases with $O(5)$ and $SU(3)$ symmetries. Nevertheless, the typical feature of unstable nuclei of the Bohr-Mottelson Model is well explained. In this symmetry, $E2$ transitions with $\Delta\tau = 2$ are forbidden. Indeed, observed data are all of order 10^{-2}.

2.1.4. *Transitional cases*

As I have discussed, the eigenvalue problem for H can be easily solved in closed form in three chains of groups. Most nuclei, however, are not in this ideal situation, and one has to treat the full Hamiltonian H. Before going to the most general cases, it is instructive to study three transitional cases. There are three classes, (A) nuclei with properties intermediate between I and II limits, (B) nuclei between II and III limits, and (C) nuclei between III and I limits. This can be visualized by using Castens triangle.[16] After studying those three classes, we can attack the most general and complicated cases, nuclei with properties intermediate between all three limits. Let us take the transitional class (A). One can study the Hamiltonian $H^{(I)+(II)} = \varepsilon C_1(U5) + \alpha C_2(U5) + \beta C_2(O5) + \gamma C_2(O3) + \delta C_2(SU3)$. By neglecting the inessential terms α and β, the Hamiltonian can be rewritten as $H^{(I)+(II)} = n_d - \kappa Q \cdot Q - \kappa L \cdot L$. Numerical calculations of low-lying states for the ${}_{62}\mathrm{Sm}$ isotopes by using this Hamiltonian show good agreement with experimental data. We see that 0_2^+ and 2_2^+ jump up after neutron number 92. This indicates that the Sm isotopes change their shapes from spherical to deformed. Namely the Sm isotopes move from the region of $O(5)$ symmetry to that of $SU(3)$. This is an example of quantum phase transitions. We can study systematically other transitional cases. Not only energies, but also other properties such as $B(E2)$ can be studied in the same way. These studies of transitional nuclei indicate that the IBM is a very good tool to study quantum phase transitions.

2.2. *The IBM-II and application to molecular physics*

In the IBM-1, we do not differentiate between neutrons and protons. In the interacting boson model-2, we introduce proton bosons s_π and d_π, and neutron bosons s_ν and d_ν.[15,17] The Hamiltonian has three parts, $H = H_\pi + H_\nu + V_{\pi\nu}$, where H_π is the Hamiltonian of proton bosons, H_ν is the Hamiltonian of neutron bosons, and $V_{\pi\nu}$ is an interaction between a proton boson and a neutron boson.

The algebra of proton bosons is $U_\pi(6)$ and that of neutrons is $U_\nu(6)$. Then the algebra for the whole system made of proton and neutron bosons is the product of them, $U_\pi(6) \times U_\nu(6)$. It is physically clear that both protons and neutrons have rotation algebras, $O_\pi(3)$ and $O_\nu(3)$, and their generators are nothing but the angular momenta L_π and L_ν. They can be physically added. This corresponds to the familiar addition of angular momenta for protons and neutrons, This is the generator of $O_{\pi+\nu}(3)$. Since each of the $U(6)$ algebra has a rich subalgebra structure, there are a variety of ways in which the algebra $U_\pi(6) \times U_\nu(6)$ can be reduced to $O_{\pi+\nu}(3)$.

From the practical point of view, this case is not particularly interesting, since it does not produce anything new. In order to diagonalize the Hamiltonian, we must use computers. There are two types of lattices of particular importance. The first is that in which the algebras of $U_\pi(6)$ and $U_\nu(6)$ are jointed at the first step. It is easy to obtain compact expressions of eigenvalues of the Hamiltonian in each chains. Many numerical calculations of nuclear properties using the IBM-2 have been accomplished.

One of the most exciting facts concerning the IBM-2 is the discovery of the scissors mode of M1 excitations. It can occur that the group of protons behaves like one blade and that of neutron behaves like another blade of a scissors. The two blades oscillate oppositely to each other. This kind of motion cannot be described by the IBM-1 and neither by ordinary Bohr-Mottelson Model. It is easy to treat such states by the IBM-2. Such states can be excited by magnetic dipole excitations. Indeed, Richter[18] discovered in 1985 the scissors-mode of M1 excitations.

There are many other version of interacting boson models. Elliott and White introduced in the IBM-3 the isospin degree of freedom, and Elliott and Evans expanded this idea further in the IBM-4.[19] There are also other kinds of extensions of the interacting boson model. In those extensions, higher spin bosons such as $f(l = 3)$ and $g(l = 4)$ bosons are added in order to explain negative parity states and states with alignment.

Algebraic methods for molecular physics was introduced by F. Iachello[20] in 1981. His model is called the Vibron model of molecules. In the Vibron model, the building blocks are s-boson with $l^\pi = 0^+$ and p-boson with $l^\pi = 1^-$. The Vibron model[21] is going now to polymers such as 9-paraffine molecule C_9H_{20}.

3. The pair approximation and the foundation of the IBM

In the pair approximation of the shell model, one assumes that low-lying states consist of pairs of nucleons, such as $0^+, 2^+$, and generally spin-J pairs. I have been interested in the pair approximation, because this is powerful to study complicated nuclei which have many valence nucleons outside a doubly magic nucleus. The shell model Hamiltonians of those nuclei have very large dimensions, which are very hard even for supercomputers to handle. By this reason, approximations such as the pair approximation are necessary. The pair approximation plays also an important role to give a microscopic foundation to the IBM.

The pair approximation method[22] has been developed by J. Q. Chen and his collaborators, particularly Y. M. Zhao. The most important technique developed by Chen is the factorization of commutators. By recursive application of Wick's theorem for coupled operators, one can use computers to evaluate the overlaps. Zhao calculated the energy spectrum of the system made of six nucleons in a single-j ($j = 19/2$) shell. Up to 6^+ states, we see a good agreement between the shell model calculation, and the SD-pair approximation. This gives a microscopic foundation of the sd bosons in the IBM. In order to explain 8^+ and 10^+ states, we need definitely G pair and I pair. Zhao and his collaborators are working very hard these days by using the pair approximation to study the structure of many complicated nuclei.[23] I myself have been cooperating with Zhao to pursue this project.

As I have discussed, the s and d bosons of the IBM represent, respectively, collective pairs of valence nucleons of $J_\pi = 0^+$ (S-pair) and $J_\pi = 2^+$ (D-pair). In the OAI mapping,[24] the boson Hamiltonian is derived from the S-pair condensate and D pair condensate. Shell model states are calculated by using S pairs and D pairs, which are mapped onto s-boson condensate and d-boson excitations. The parameters of the IBM Hamiltonian are determined so that matrix elements of former state for the nucleon Hamiltonian are reproduced by the corresponding matrix elements of latter states for the IBM Hamiltonian. The OAI method has been successful for spherical and modestly deformed nuclei, including nuclei with γ-unstable shape. It has, however, been unclear how to derive the Hamiltonian of the IBM in strongly deformed nuclei.

T. Otsuka and his collaborators have developed a way to solve the problem of microscopic foundation of the IBM in deformed nuclei.[25] They calculate first the intrinsic states by using mean field model with some nucleon-nucleon interactions such as the Skyrme and Gogny interactions. They map this intrinsic states onto those of the interacting boson model. For ^{134}Ba and ^{208}W, theoretical energies are calculated by using the IBM Hamiltonian which parameters are obtained by their mapping approach, and a remarkable agreement to experiment is obtained. Therefore the IBM gains predictive power even in well deformed nuclei. Because the calculation of energy levels is rather difficult and complicated with mean-field models, the IBM-II with the Hamiltonian derived by using this new mapping method gives a handy tool in this respect.

4. Conclusion

Now I wish to tell you the conclusion of my talk. I have discussed the following points: (1) strong attractive pairing interaction and seniority scheme $SU(2)$-Symmetry, (2) the Interacting Boson Model (IBM) in which the $U(6)$ Symmetry plays an essential role, (3) the IBM-II, in which both proton and neutron degrees of freedom are taken into account, and in which the scissors mode of deformed nuclei is explained, (4) Nucleon pair approximation which have been developped by Y. M. Zhao and collaborators. (5) Microscopic basis of IBM in deformed nuclei which have been developped by T. Otsuka and collaborators.

Dynamical Symmetry is defined as following: This is not the rigorous symmetry found in the space and time; such as time reversality and Lorentz invariance. Dynamical symmetry is symmetry which the Hamiltonian and/or Lagraigian of a physical system has. Finding dynamical symmetry in a system, of which we are interested in, we can very easily obtain algebraic expressions of eigenvalues of the Hamiltonian of the system. We can understand intuitively and physically, the structure of the system.

Why is the discovery of symmetry important? First, even if we will have extremely powerful computers, we will be unable to solve secular equations with huge dimension. Monte Carlo shell model calculations goes to 10^{10} at present. Dy and other nuclei require us to solve secular equations which dimensions are more than $10^{16} \sim 10^{20}$. Second, only several numerical results out of huge numbers of numerical data are important. Third, in order to understand the essence of physics in a system under consideration, we must apply our knowledge of symmetry.

I wish to thank F. Iachello, I. Talmi, P. Van Isacker, Y. M. Zhao, T. Otsuka and N. Yoshinaga for their helps and cooperation.

References

1. E. Wigner, *Phys. Rev.* **51**, 106 (1937).
2. G. Racah, *Phys. Rev.* **61**, 186 (1942). ,
3. M. G. Mayer, *Phys. Rev.* **74**, 235 (1948); *ibid.* **75**, 1969 (1949); *ibid.* **78**, 16 (1950); M. G. Mayer and J. Jensen, *Elementary theory of nuclear shell structure* (Wiley, 1960).
4. M. Ichimura and A. Arima, *Nucl. Phys. A* **204**, 225 (1973).
5. A. Arima and H. Kawarada *J. Phys. Soc. Japan* **19**, 1768 (1964).
6. A. Bohr and BR. Mottelson, *Nuclear structure I and II* (Benjamin, New York, 1975).
7. J. Rainwater, *Phys. Rev.* **79**, 432 (1950).
8. A. de Shalit and I. Talmi, *Nuclear Shell Theory* (Academic Press, New York, 1963); I. Talmi, Simple Models of Complex Nuclei (Harwood Academic Publishers, 1993).
9. A. Arima, M. Nomura and H. Kawarada *Phys. Lett.* **19**, 400 (1965).
10. J. Bardeen, L. N. Cooper and J. R. Schrieffer, *Phys. Rev.* **106**, 162 (1957); *ibid.* **108**, 1175 (1957).
11. A. Bohr and BR. Mottelson, *Mat. Fys. Medd. Dan. Vid. Selsk.* **27**, 16 (1953).
12. T. Inoue, T. Sebe, H. Hagiwara and A. Arima, *Nucl. Phys.* **59**, 1 (1964); *ibid.* **85**, 184 (1966); A. Arima, S. Cohen, R. D. Lawson, M. H. MacFarlane, *ibid.* **108**, 94 (1968).
13. H. A. Elliot, *Proc. Roy. Soc. A* **245**, 128 (1958); *ibid.* **245**, 562 (1958).
14. F. Iachello, MIT PhD Thesis (1969).
15. A. Arima and F. Iachello, *Phys. Rev. Lett.* **35**, 1069 (1975); *Ann. Phys. (NY)* **99**, 253(1976); *ibid.* **111**, 201 (1978); *ibid.* **123**, 468 (1979); F. Iachello and A. Arima, *The Interacting Boson Model* (Cambridge Univ. Press, 1987).
16. R. F. Casten, *Interacting Boson-Fermion Systems in Nuclei* (Plenum, 1981).
17. A. Arima, T. Otsuka, F. Iachello and I. Talmi, *Phys. Lett. B* **66**, 205 (1977); T. Otsuka, A. Arima, F. Iachello, and I. Talmi, *Phys. Lett. B* **76**, 139 (1978).
18. D. Bohle, A. Richter, W. Steffen, A. E. L. Dieperink, N. Lo Iudice, F. Palumbo, O. Scholten, *Phys. Lett. B* **137**, 27 (1984).
19. J. P. Elliot and A. P. White, *Phys. Lett. B* **97**, 169(1980); J. P. Elliot and J. A. Evans, *Phys. Lett. B* **101**, 216 (1981);
20. F. Iachello, *Chem. Phys. Lett.* **78**, 581 (1981).
21. F. Iachello and P. Truini, *Ann. Phys (N.Y)* **299**, 35 (1999); T. Marinkovic and S. Oss, *Phys. Chem. Comm.* **5**, 66 (2002).
22. J. Q. Chen and Y. M. Zhao, *Nucl. Phys. A* **554**, 61 (1992); J. Q. Chen, *Nucl. Phys. A* **563**, 218 (1993); *ibid.* **626**, 686 (1997); Y. M. Zhao, N. Yoshinaga, S. Yamaji, J. Q. Chen, and A. Arima, *Phys. Rev. C* **62**, 014304 (2000).
23. H. Jiang, J. J. Shen, Y. M. Zhao and A. Arima, *J. Phys. G* **38**, 045103 (2011).
24. T. Otsuka, A. Arima and F. Iachello, *Nucl. Phys. A* **309**, 1 (1978).
25. K. Nomura, N. Shimizu, and T. Otsuka, *Phys. Rev. Lett.* **101**, 142501 (2008).

AN OVERVIEW OF COHERENT STATES AND COHERENT STATE QUANTIZATION

S. TWAREQUE ALI*

Department of Mathematics and Statistics,
Concordia University, Montréal, Québec,
CANADA H3G 1M8
**E-mail: twareque.ali@concordia.ca*

We attempt in this paper to present a quick overview of coherent states and coherent state quantization, emphasizing the role of group theory. Coherent states feature in many areas of quantum physics and mathematics, but here the focus is on the mathematical aspects.

Keywords: coherent states, quantization

1. Introduction

Very generally, a family of coherent states is a set of continuously labeled quantum states, with specific mathematical and physical properties, in terms of which arbitrary quantum states can be expressed as linear superpositions. Since coherent states are continuously labeled, they form overcomplete sets of vectors in the Hilbert space of states.

The literature on coherent states is vast; we only mention here the two books,[1,3] the review article[2] and the recent special issue[4] of the Journal of Physics, which contains a detailed bibliography and a wide selection of papers, some of a review nature, touching on many aspects of the subject.

What we call coherent states today, were originally these introduced into physics by Schrödinger (1926), as a family of quantum states in terms of which the transition from quantum to classical mechanics could be conveniently studied. These states, which were not called coherent states by Schrödinger, have the minimal uncertainty property, in the sense that they saturate the Heisenberg uncertainty relations. The name coherent state was applied when these states were rediscovered in the context of quantum optical radiation by Glauber, Klauder and Sudarshan. It was demonstrated that in these states the correlation functions of the quantum optical field factorize as they do in classical optics, so that the optical field has a near classical behaviour, with the optical beam being coherent. We shall refer to these states as the canonical coherent states.

The canonical coherent states, apart from their use in quantum optics, have also been found to be extremely useful in computations in atomic and molecular physics, in quantum statistical mechanics, and in certain areas of mathematics and mathematical physics, including harmonic analysis, symplectic geometry and quantization theory.

In view of the of the wide applicability of the canonical coherent states, it is natural to look for generalizations sharing similar mathematical and physical properties. These other families of states are usually called generalized coherent states, even when there is no link to optical coherence in such studies.

2. Some properties of the canonical coherent states

One starts with a set of quantum state vectors

$$|0\rangle, |1\rangle, \ldots, |n\rangle, \ldots ,$$

corresponding to states of $0, 1, \ldots, n, \ldots,$ excitons, respectively. The Hilbert space of these states, in which they form an orthonormal basis, is often known as Fock space. The canonical coherent states are then defined in terms of this basis, for each complex number z, by the analytic expansion:

$$|z\rangle = e^{-\frac{|z|^2}{2}} \sum_{n=0}^{\infty} \frac{z^n}{\sqrt{n!}} |n\rangle . \tag{1}$$

The states $|z\rangle$ are normalized to unity: $\langle z|z\rangle = 1$ and satisfy the formal eigenvalue equation

$$a|z\rangle = z|z\rangle , \tag{2}$$

where a is the annihilation operator for excitons which, together with its adjoint $a^\dagger$, acts on the basis vectors (Fock states) as

$$a|n\rangle = \sqrt{n}|n-1\rangle, \quad a^\dagger|n\rangle = \sqrt{n+1}|n+1\rangle, \quad [a, a^\dagger] = aa^\dagger - a^\dagger a = I . \tag{3}$$

Writing

$$Q = \frac{a + a^\dagger}{\sqrt{2}} , \qquad P = \frac{a - a^\dagger}{i\sqrt{2}} , \tag{4}$$

one can derive the relation ($\hbar = 1$):

$$\langle \Delta Q \rangle \langle \Delta P \rangle = \frac{1}{2} , \tag{5}$$

where for any observable A,

$$\langle \Delta A \rangle = \left[\langle z|A^2|z\rangle - \langle z|A|z\rangle^2 \right]^{\frac{1}{2}} .$$

There also follows the resolution of the identity,

$$\int_{\mathbb{C}} |z\rangle\langle z| \; \frac{dq \; dp}{2\pi} = I \; , \tag{6}$$

where $z = \frac{1}{\sqrt{2}}(q - ip)$. Equation (6) incorporates the mathematical fact that the set of vectors $|z\rangle$ is overcomplete in the Hilbert space.

One obtains analytical properties of these states by looking at $\langle \overline{z}|\phi\rangle$ for an arbitrary vector $|\phi\rangle$ in Fock space. From (1) it is clear that

$$F(z) = \langle \phi|z\rangle = e^{-\frac{|z|^2}{2}} f(z) \; ,$$

where f is an entire analytic function in the complex variable z. Moreover, the mapping $\phi \mapsto f$ is an isometric embedding of the Fock space onto the Hilbert space of analytic functions, with respect to the norm

$$\|f\| = \left[\int_{\mathbb{C}} |f(z)|^2 \; d\mu(z, \overline{z}) \right]^{\frac{1}{2}} \; , \tag{7}$$

defined by the measure $d\mu(z, \overline{z}) = \dfrac{1}{2\pi} e^{-|z|^2} dq \; dp$.

Group theoretical properties of the canonical coherent states, emerge upon noting that

$$|z\rangle = e^{-\frac{|z|^2}{2}} \; e^{za^\dagger} |0\rangle = U(z)|0\rangle \; , \qquad U(z) = e^{za^\dagger - \overline{z}a} \; . \tag{8}$$

The vectors $|z\rangle$ and the unitary operator $U(z)$ can be re-expressed in terms of the real variables q, p and the operators Q, P as

$$|z\rangle = |q, p\rangle = U(q, p)|0\rangle \; , \qquad U(q, p) = e^{i(pQ - qP)} \; . \tag{9}$$

The operators $U(q, p)$ realize a (projective) unitary, irreducible representation of the Weyl-Heisenberg group, which is the group whose Lie algebra has the generators Q, P and I, obeying the commutation relations

$$[Q, P] = iI \; .$$

The existence of the resolution of the identity (6) is the statement of the fact that this representation is square-integrable. This leads to another method for building general coherent states, this time by the action, on a fixed vector, of the unitary operators of a square-integrable representation of a locally compact group.

3. A possible general definition of coherent states

Let $\mathfrak{H}$ be an abstract, separable Hilbert space over the complexes, X a locally compact space and $d\nu$ a measure on X. Let $|x, i\rangle$ be a family of vectors in $\mathfrak{H}$, defined for each x in X and $i = 1, 2, 3, \ldots, N$, where N is usually a finite integer, although it could also be infinite. We assume that this set of vectors possesses the following properties.

(1) For each i, the mapping $x \mapsto |x, i\rangle$ is weakly continuous, i.e., for each vector $|\phi\rangle$ in $\mathfrak{H}$, the function $\Psi_i(x) = \langle x, i|\phi\rangle$ is continuous (in the topology of X).

(2) For each x in X, the vectors $|x, i\rangle$, $i = 1, 2, \ldots, N$, are linearly independent.

(3) The resolution of the identity

$$\sum_{i=1}^{N} \int_X |x, i\rangle\langle x, i| \, d\nu(x) = I_{\mathfrak{H}} \tag{10}$$

holds in the weak sense on the Hilbert space $\mathfrak{H}$, i.e., for any two vectors $|\phi\rangle, |\psi\rangle$ in $\mathfrak{H}$, the following equality holds:

$$\sum_{i=1}^{N} \int_X \langle\phi|x, i\rangle\langle x, i|\psi\rangle \, d\nu(x) = \langle\phi|\psi\rangle \, .$$

A set of vectors $|x, i\rangle$, satisfying the above three properties is called a family of generalized vector coherent states. In case $N = 1$, the set is called a family of generalized coherent states.

Sometimes the resolution of the identity condition is replaced by a weaker condition, with the vectors $|x, i\rangle$ simply forming a total set in $\mathfrak{H}$ and the functions $F_i(x) = \langle x, i|\phi\rangle$, as $|\phi\rangle$ runs through $\mathfrak{H}$.

Equation (10) is immediately seen to imply that

$$|\phi\rangle = \sum_{i=1}^{N} \int_X \Psi_i(x)|x, i\rangle \, d\nu(x) \, , \tag{11}$$

where $\Psi_i(x) = \langle x, i|\phi\rangle$.

Associated to a family of generalized coherent states on a Hilbert space $\mathfrak{H}$, there is an intrinsic isomorphism between this space and a Hilbert space of (in general vector valued) continuous functions over X. Using this isomorphism, it is always possible to look upon coherent states as a family of continuous functions which are square integrable with respect to the measure $d\nu$. To demonstrate this we note that, in view of (10), for each vector $|\phi\rangle$ in $\mathfrak{H}$, the vector-valued function $\mathbf{\Psi}(x)$ on x, with components $\Psi_i(x) = \langle x, i|\phi\rangle$, $i = 1, 2, \ldots, N$, satisfies the norm condition

$$\sum_{i=1}^{N} \int_X |\Psi_i(x)|^2 \, d\nu(x) = \|\phi\|^2_{\mathfrak{H}} \, .$$

This means that the set of vectors $\mathbf{\Psi}$, as $|\phi\rangle$ runs through $\mathfrak{H}$, is a closed subspace of the Hilbert space $L^2_{\mathbb{C}^N}(X, d\nu)$ of N-vector valued functions on x. Let us denote this subspace by $\mathfrak{H}_K$ and note that this space is a reproducing kernel Hilbert space with a matrix-valued kernel $\mathbf{K}(x, y)$ having matrix elements,

$$K(x, y)_{ij} = \langle x, i|y, j\rangle, \qquad i, j = 1, 2, \ldots, N, \tag{12}$$

and enjoying the properties,

$$K(x, y)_{ij} = \overline{K(y, x)_{ji}} \, , \qquad K(x, x)_{ii} > 0 \, , \tag{13}$$

and

$$\sum_{\ell=1}^{N} \int_X K(x, z)_{i\ell} \, K(z, y)_{\ell j} \, d\nu(z) = K(x, y)_{ij} \, . \tag{14}$$

If $\mathbf{e}^i$, $i = 1, 2, \ldots, N$, are the vectors constituting the canonical basis of $\mathbb{C}^N$, then for each x in X and $i = 1, 2, \ldots, N$, the vector-valued function $\boldsymbol{\xi}^i_x$ on X, defined by $\boldsymbol{\xi}^i_x(y) = \mathbf{K}(y, x)\mathbf{e}^i$, is the image in $\mathfrak{H}_K$ of the generalized vector coherent state $|x, i\rangle$, under the above mentioned isometry. The vectors $\boldsymbol{\xi}^i_x$ span the space $\mathfrak{H}_K$ and for an arbitrary element $\mathbf{\Psi}$ of this Hilbert space, the reproducing property (14) of the kernel implies the relation

$$\int_X \mathbf{K}(x, y)\mathbf{\Psi}(y) \, d\nu(y) = \mathbf{\Psi}(x) \, . \tag{15}$$

Conversely, given any reproducing kernel Hilbert space, with a kernel satisfying the relations (13) and (14), generalized coherent states can be constructed as above in terms of this kernel.

4. A few examples of generalized coherent states

A simple modification of the analytic structure yields a large class of generalizations of the canonical coherent states (1). Indeed, let $x_1 \leq x_2 \leq \ldots \leq x_n \leq \ldots$ be an infinite sequence of positive numbers ($x_1 \neq 0$). Set $x_n! = x_1 x_2 \ldots x_n$ and by convention set $x_0! = 1$. In the same Fock space in which the canonical coherent states were described, we now define the related deformed or nonlinear coherent states via the analytic expansion

$$|z\rangle = \mathcal{N}(|z|^2)^{-\frac{1}{2}} \sum_{n=0}^{\infty} \frac{z^n}{\sqrt{x_n!}} |n\rangle \ . \tag{16}$$

The normalization factor $\mathcal{N}(|z|^2)$ is chosen so that $\langle z|z\rangle = 1$.

These generalized coherent states are overcomplete in the Fock space and satisfy a resolution of the identity, of the type

$$\int_{\mathcal{D}} |z\rangle\langle z| \, \mathcal{N}(|z|^2) \, d\nu(z, \overline{z}) = I \ , \tag{17}$$

$\mathcal{D}$ being an open disc in the complex plane of radius L, the radius of convergence of the series $\sum_{n=0}^{\infty} \frac{z^n}{\sqrt{x_n!}}$. In the case of the canonical coherent states, $L = \infty$.

Generally the measure $d\nu$ has the form $\frac{1}{2\pi} d\theta \, d\lambda(r)$ (for $z = re^{i\theta}$), where $d\lambda$ is related to the $x_n!$ through the moment condition:

$$x_n! = \int_0^L r^{2n} \, d\lambda(r), \qquad n = 0, 1, 2, \dots \ . \tag{18}$$

Once again, we see that for an arbitrary vector $|\phi\rangle$ in the Fock space, the function $F(z) = \langle \phi|z\rangle$, of the complex variable z, is of the form

$$F(z) = \mathcal{N}(|z|^2)^{-\frac{1}{2}} f(z),$$

where f is an analytic function on the domain $\mathcal{D}$. The reproducing kernel associated to these coherent states is

$$K(\overline{z}, z') = \langle z|z'\rangle = \left[\mathcal{N}(|z|^2)\mathcal{N}(|z'|^2)\right]^{-\frac{1}{2}} \sum_{n=0}^{\infty} \frac{(\overline{z}z')^n}{x_n!} \ . \tag{19}$$

By analogy with (2), one can define a generalized annihilation operator A by its action on the vectors $|z\rangle$,

$$A|z\rangle = z|z\rangle \quad \text{and} \quad A|n\rangle = \sqrt{x_n}|n-1\rangle \ , \qquad A^\dagger|n\rangle = \sqrt{x_{n+1}}|n+1\rangle \ . \tag{20}$$

Depending on the exact values of the quantities x_n, these two operators, together with the identity I and all their commutators, could generate a wide range of algebras including various deformed quantum algebras.

The term non-linear, as often applied to these generalized coherent states, comes again from quantum optics where many such families of states are used in studying the interaction between the radiation field and atoms, where the strength of the interaction itself depends on the frequency of radiation. Of course, these coherent states will not in general have either the group theoretical or the minimal uncertainty properties of the CCS.

The following is an example of generalized coherent states of the above type, built over the unit disc, $\mathcal{D} = \{z \in \mathbb{C} | |z| < 1\}$: on the Fock space, we define the states

$$|z\rangle = (1-r^2)^{\kappa} \sum_{n=0}^{\infty} \left[\frac{(2\kappa)_n}{n!}\right]^{\frac{1}{2}} z^n |n\rangle \,, \qquad r = |z| \,, \tag{21}$$

where $\kappa = 1, \frac{3}{2}, 2, \frac{5}{2}, \ldots$, and

$$(a)_m = \frac{\Gamma(a+m)}{\Gamma(a)} = a(a+1)(a+2)\ldots(a+m-1) \,.$$

Comparing (21) with (16) we see that $x_n = \dfrac{n}{2\kappa+n-1}$ so that $\lim_{n\to\infty} x_n = 1$. Thus, the infinite sum is convergent for any z lying in the unit disc. These generalized coherent states arise from representations of the group SU(1, 1) belonging to the discrete series, each irreducible representation being labelled by a specific value of the index κ. The associated Hilbert space of functions, analytic on the unit disc, is a subspace of $L^2(\mathcal{D}, d\mu_\kappa)$, with

$$d\mu_\kappa(z,\overline{z}) = (2\kappa-1)\frac{(1-r^2)^{2\kappa-2}}{\pi}\, r\, dr\, d\theta \,, \qquad z = re^{i\theta} \,,$$

which can be obtained by solving the moment problem (18). The resolution of the identity satisfied by these states is

$$\frac{2\kappa-1}{\pi} \int_{\mathcal{D}} |z\rangle\langle z| \, \frac{r\, dr\, d\theta}{(1-r^2)^2} = I \,. \tag{22}$$

The associated generalized creation and annihilation operators are,

$$A|n\rangle = \sqrt{\frac{n}{2\kappa+n-1}}|n-1\rangle \,, \qquad A^\dagger|n\rangle = \sqrt{\frac{n+1}{2\kappa+n}}|n+1\rangle \,, \tag{23}$$

so that clearly, $[A, A^\dagger] \neq I$.

Operators A and $A^\dagger$ of the general type defined in (20) are also known as ladder operators. When such operators appear as generators of representations of Lie algebras, their eigenvectors [see (20)] are usually called Barut-Girardello coherent states. As an example, the representation of the Lie algebra of $SU(1,1)$ on the Fock space is generated by the three operators K_+, K_- and K_3, which satisfy the commutation relations

$$[K_3, K_\pm] = \pm K_\pm, \qquad [K_-, K_+] = 2K_3 \,. \tag{24}$$

They act on the vectors $|n\rangle$ in the manner,

$$K_-|n\rangle = \sqrt{n(2\kappa+n-1)}|n-1\rangle\ , \quad K_+ = K_-^\dagger\ , \quad K_3|n\rangle = (\kappa+n)|n\rangle\ . \tag{25}$$

Thus $K_-|0\rangle = 0$ and

$$|n\rangle = \frac{1}{\sqrt{n!(2\kappa)_n}}\, K_+^n|0\rangle\ .$$

The Barut-Girardello coherent states $|z\rangle$ are now defined as the formal eigenvectors of the ladder operator K_- :

$$K_-|z\rangle = z|z\rangle\ , \qquad z \in \mathbb{C}\ . \tag{26}$$

They have the analytic form:

$$|z\rangle = \frac{|z|^{2\kappa-1}}{\sqrt{I_{2\kappa-1}(2|z|)}} \sum_{n=0}^{\infty} \frac{z^n}{\sqrt{n!(2\kappa+n-1)!}}\, |n\rangle\ , \tag{27}$$

where $I_\nu(x)$ is the order-ν modified Bessel function of the first kind.

A non-analytic extension of the expression (16) is often used to define generalized coherent states associated to physical Hamiltonians having pure point spectra. These coherent states, known as Gazeau-Klauder coherent states, are labelled by action-angle variables. Suppose that we are given the physical Hamiltonian $H = \sum_{n=0}^{\infty} E_n|n\rangle\langle n|$, with $E_0 = 0$, i.e., it has the energy eigenvalues E_n and eigenvectors $|n\rangle$, which we assume to form an orthonormal basis for the Hilbert space of states $\mathfrak{H}$. Let us write the eigenvalues as $E_n = \omega\epsilon_n$ by introducing a sequence of dimensionless quantities $\{\epsilon_n\}$ ordered as: $0 = \epsilon_0 < \epsilon_1 < \epsilon_2 < \ldots$. Then, for all $J \geq 0$ and $\gamma \in \mathbb{R}$, the Gazeau-Klauder coherent states are defined as

$$|J,\gamma\rangle = \mathcal{N}(J)^{-\frac{1}{2}} \sum_{k=0}^{\infty} \frac{J^{n/2} e^{-i\epsilon_n\gamma}}{\sqrt{\epsilon_n!}} |n\rangle\ , \tag{28}$$

where again $\mathcal{N}$ is a normalization factor, which turns out to be dependent on J only.

These coherent states satisfy the *temporal stability* condition,

$$e^{-iHt}|J,\gamma\rangle = |J,\gamma+\omega t\rangle\ , \tag{29}$$

and the *action identity*,

$$\langle J,\gamma|H|J,\gamma\rangle_{\mathfrak{H}} = \omega J\ . \tag{30}$$

To construct the second set of examples of generalized coherent states, let G be a locally compact group and suppose that it has a continuous, irreducible representation on a Hilbert space $\mathfrak{H}$ by unitary operators

$U(g),\ g \in G$. This representation is called square integrable if there exists a non-zero vector $|\psi\rangle$ in $\mathfrak{H}$ for which the following integral converges:

$$c(\psi) = \int_G |\langle\psi|U(g)\psi\rangle|^2 \, d\mu(g) \tag{31}$$

Here $d\mu$ is a Haar measure of G, which for definiteness, we take to be the left invariant measure. A vector $|\psi\rangle$, satisfying (31), is said to be admissible, and it can be shown that the existence of one such vector guarantees the existence of an entire dense set of such vectors in $\mathfrak{H}$. Moreover, if the group G is unimodular, i.e., if the left and the right invariant measures coincide, then the existence of one admissible vector implies that every vector in $\mathfrak{H}$ is admissible.

Given a square integrable representation and an admissible vector $|\psi\rangle$, let us define the vectors

$$|g\rangle = \frac{1}{\sqrt{c(\psi)}}\, U(g)|\psi\rangle\ , \tag{32}$$

for all g in the group G. These vectors are to be seen as the analogues of the canonical coherent states (9), written there in terms of the representation of the Weyl-Heisenberg group. Next, it can be shown that a resolution of the identity holds on $\mathfrak{H}$:

$$\int_G |g\rangle\langle g| \, d\mu(g) = I_{\mathfrak{H}} \tag{33}$$

Thus, the vectors $|g\rangle$ constitute a family of generalized coherent states.

A typical example of the above construction is provided by the affine group, G_{Aff}. This is the group of all 2×2 matrices of the type,

$$g = \begin{pmatrix} a & b \\ 0 & 1 \end{pmatrix}\ , \tag{34}$$

a and b being real numbers with $a \neq 0$. We shall also write $g = (b, a)$. This group is non-unimodular, with the left invariant measure being given by $d\mu(b, a) = \dfrac{1}{a^2}\, db\ da$. (The right invariant measure is $\dfrac{1}{a}\ db\ da$.) The affine group has a unitary irreducible representation on the Hilbert space $L^2(\mathbb{R}, dx)$. The (unitary) operators $U(b, a)$ of this representation act as

$$(U(b,a)\phi)(x) = \frac{1}{\sqrt{|a|}}\, \phi\left(\frac{x-b}{a}\right) \qquad \phi \in L^2(\mathbb{R}, dx). \tag{35}$$

If ψ is a function in $L^2(\mathbb{R}, dx)$ such that its Fourier transform $\widehat{\psi}$ satisfies the condition

$$\int_{\mathbb{R}} \frac{|\widehat{\psi}(k)|^2}{|k|}\, dk < \infty\ , \tag{36}$$

then it can be shown to be an admissible vector, i.e.,

$$c(\psi) = \int_{G_{\mathrm{Aff}}} |\langle\psi|U(b,a)\psi\rangle|^2 \, \frac{db\, da}{a^2} < \infty \, .$$

Thus, following the general construction outlined above, the vectors

$$|b,a\rangle = \frac{1}{\sqrt{c(\psi)}} \, U(b,a)\psi \, , \qquad (b,a) \in G_{\mathrm{Aff}} \tag{37}$$

define a family of generalized coherent states and one has the resolution of the identity

$$\int_{G_{\mathrm{Aff}}} |b,a\rangle\langle b,a| \, \frac{db\, da}{a^2} = I \tag{38}$$

on $L^2(\mathbb{R}, dx)$.

In the signal analysis literature a vector satisfying the admissibility condition (36) is called a mother wavelet and the generalized coherent states (37) are called wavelets. The function

$$F(b,a) = \langle b,a|\phi\rangle \, , \tag{39}$$

is called the continuous wavelet transform of the signal ϕ.

There exist alternative ways of constructing generalized coherent states using group representations. For example, the Perelomov method is based on the observation that the vector $|0\rangle$, appearing in the construction of the canonical coherent states in (8) and (9) using the representation of the Weyl-Heisenberg group, is invariant up to a phase, under the action of its centre. Consequently, the coherent states $|z\rangle$, as written in (8), are labelled, not by elements of the group itself, but only by the points in the quotient space of the group by its (central) phase subgroup.

Generally, let G be a locally compact group and U a unitary irreducible representation of it on the Hilbert space $\mathfrak{H}$. We do not assume U to be square integrable. We fix a vector $|\psi\rangle$ in $\mathfrak{H}$, of unit norm and denote by H the subgroup of G consisting of all elements h for which

$$U(h)|\psi\rangle = e^{i\omega(h)}|\psi\rangle \, , \tag{40}$$

where ω is a real-valued function of h. Let $X = G/H$ be the left coset space and x an arbitrary element in X. Choosing a coset representative $g(x) \in G$, for each coset x, we define the vectors in $\mathfrak{H}$:

$$|x\rangle = U(g(x))|\psi\rangle \tag{41}$$

The dependence of these vectors on the specific choice of the coset representative $g(x)$, is only through a phase. Vectors $|x\rangle$, defined in this manner, are called Gilmore-Perelomov coherent states. In this definition of generalized coherent states, no resolution of the identity is postulated. The Perelomov construction can be used to define coherent states for any locally compact group.

On the other hand, there exist other constructions of generalized coherent states, using group representations, which generalize the notion of square integrability to homogeneous spaces of the group. Briefly, in this approach one starts with a unitary irreducible representation U and attempts to find a vector $|\psi\rangle$, a subgroup H and a section $\sigma : G/H \longrightarrow G$ such that

$$\int_{G/H} |x\rangle\langle x| \, d\mu(x) = T \ , \tag{42}$$

where $|x\rangle = U(\sigma(x))|\psi\rangle$, T is a bounded, positive operator with bounded inverse and $d\mu$ is a quasi-invariant measure on $X = G/H$. It is not assumed that $|\psi\rangle$ be invariant up to a phase under the action of H and clearly, the best situation is when T is a multiple of the identity.

Although somewhat technical, this general construction is of enormous versatility for semi-direct product groups of the type $\mathbb{R}^n \rtimes K$, where K is a closed subgroup of $GL(n, \mathbb{R})$. Thus, it is useful for many physically important groups, such as the Poincaré or the Euclidean group, which do not have square integrable representations in the sense of the earlier definition (see Eq. (31)).

The canonical coherent states satisfy the minimal uncertainty relation (5). It is possible to build families of coherent states by generalizing from this condition also,

5. Quantization using coherent states

Using a family of coherent states $|x\rangle := \eta_x$ on the Hilbert space $\mathfrak{H}$, we now introduce a mapping from functions f on the measure space X (which now will typically be a physical phase space $\mathbf{\Gamma}$) to operators on $\mathfrak{H}$.

Suppose $f : X \to \mathbb{R}$ is a reasonably decent function, then the corresponding Berezin-Toeplitz operator $\widehat{f}$, on the state space $\mathfrak{H}$ is defined in the manner:

$$\widehat{f} = \int_X f(x) \, |\eta_x\rangle\langle\eta_x| \, d\mu(x)$$

provided this integral makes sense. It is this correspondence $f \longmapsto \widehat{f}$, defined over some appropriate algebra $\mathcal{A}(X)$ of functions on X to operators on $\mathfrak{H}$, that we would like to call a quantization map:

$$Q(f) = \widehat{f} \, .$$

Clearly, this map is linear and $Q(\mathbb{I}) = I_{\mathfrak{H}}$. But we would also like to verify if the asymptotic correspondence principle

$$[Q(f), Q(g)] = i\hbar Q(\{f, g\}) + O(h^2).$$

is satisfied. This can be done in specific cases, and generally for a large class of domains, in the context of Berezin-Toeplitz quantization.

Thus, the quantization procedure that we have adopted here is to:

- isolate an appropriate algebra $\mathcal{A}(X)$ of (generally complex-valued) functions on the measure space X;
- determine a Hilbert space $\mathfrak{H}$ and a family of coherent states $\eta_x\ ,\quad x \in X$, in it;
- implement the quantization $f \longmapsto Q(f) := \widehat{f}$;
- verify the asymptotic correspondence principle and identify the set of functions f for which this correspondence holds exactly.

But as soon as the family of coherent states η_x has been obtained, there comes with it a reproducing kernel Hilbert space $\mathfrak{H}_{\text{rk}} \subset L^2(X, \mu)$, which most often is a space of continuous (even holomorphic functions), and a unitary map $W : \mathfrak{H} \longrightarrow \mathfrak{H}_{\text{rk}}$. We write $\mathbb{P}$ for the projection operator from $L^2(X, \mu)$ to $\mathfrak{H}_{\text{rk}}$ and $W\eta_x = \xi_x$. Let $\widehat{f}_{\text{rk}} = W\widehat{f}W^*$ be the image of this operator on $\mathfrak{H}_{\text{rk}}$. Then it is easy to see that, as an operator on $L^2(X, d\mu)$

$$\widehat{f}_{\text{rk}} = \mathbb{P} f_{\text{op}} \mathbb{P} = \int_X f(x)\ |\xi_x\rangle\langle\xi_x|\ d\mu(x) = W\widehat{f}W^*,$$

where f_{op} is the operator of multiplication by the function f on $L^2(X, d\mu)$.

Thus, it would appear that one could start with $L^2(X, \mu)$ itself, identify a reproducing kernel subspace of it and introduce a quantization by the correspondence $f \longmapsto \widehat{f}_{\text{rk}}$.

References

1. S.T. Ali, J-P. Antoine and J-P. Gazeau, *Coherent States, Wavelets and Their Generalizations*, (Springer-Verlag, New York, 2000).
2. S.T, Ali, *Coherent states*, in *Encyclopedia of Mathematical Physics*, Eds., Jean Pierre Françoise, Gregory L. Naber and Tsou Sheung Tsun, Elsevier Academic Press (2006), pp. 19.
3. A. Perelomov, *Generalized Coherent States and their Applications*, Springer-Verlag, Berlin (1986).
4. S.T. Ali, J.-P. Antoine, F. Bagarello and J.-P. Gazeau, (guest editors), Special Issue on *Coherent States: Mathematical and Physical Aspects* of the *Journal of Physics A*. Volume **45**, No. 24 (June 2012).

PARALLEL SESSIONS

20 – 24 AUGUST 2012

PARALLEL SESSION
Nonlinear Sciences

20 & 22 & 24 AUGUST 2012

EXACTLY SOLVABLE BCS-BEC CROSSOVER HAMILTONIANS

A. BIRRELL*, P. S. ISAAC, and J. LINKS

School of Mathematics and Physics, The University of Queensland, Brisbane, Queensland 4072, Australia.
**E-mail: a.birrell@uq.edu.au*

We demonstrate a novel approach that allows the determination of very general classes of exactly solvable Hamiltonians via Bethe ansatz methods. This approach combines aspects of both the co-ordinate Bethe ansatz and algebraic Bethe ansatz. The eigenfunctions are formulated as factorisable operators acting on a suitable reference state. Yet, we require no prior knowledge of transfer matrices or conserved operators. By taking a variational form for the Hamiltonian and eigenstates we obtain general exact solvability conditions. The procedure is conducted in the framework of Hamiltonians describing the crossover between the low-temperature phenomena of superconductivity, in the Bardeen-Cooper-Schrieffer (BCS) theory, and Bose-Einstein condensation (BEC).

Keywords: Exactly Solvable Models; Bethe Ansatz; Condensed Matter Theory.

1. Introduction

Recent work suggests an adequate understanding of the nature of the BCS-BEC crossover requires the study of Hamiltonians exhibiting coupling between Cooper pairs of atoms and bosonic molecular modes.[1,2] Further, the necessity of exact solutions to properly explain the physics of superconducting devices such as ultrasmall superconducting grains[3] motivates the search for exactly solvable generalisations of the BCS theory.

A standard technique in this field is the implementation of the Quantum Inverse Scattering Method (QISM)[4–6] to construct quantum Hamiltonians with multiple conserved operators, and then find their exact solution by Bethe ansatz methods. Typically, a solution of the Yang-Baxter equation (YBE)[7–9] is used to construct a *transfer matrix* which generates the conserved operators of the Hamiltonian. The exact solution may then be obtained via Bethe ansatz techniques. We note that in principle the implementation of the co-ordinate Bethe ansatz[10] does not depend on any prior knowledge of an associated solution of the YBE, nor the conserved operators of the Hamiltonian that it generates. A more modern approach in the framework of the QISM is the algebraic Bethe ansatz.[4–6]

The strategy demonstrated here enables the direct determination of general classes of exactly solvable Hamiltonians, with both a bosonic and Cooper pairing degrees of freedom, such that they model BCS-BEC crossover behaviour. By contrast with standard techniques, this gives a unified construction for classes of exactly solvable Hamiltonians with multiple free coupling parameters. By taking appropriate limits we recover seven exactly solvable subcases from the literature.[11–17]

In Section 2 a general Hamiltonian that models the BCS-BEC crossover behaviour is introduced. In Section 3 the method of obtaining exact solvability constraints for the variational Hamiltonian is outlined. Calculations that have been omitted for brevity appear in the literature.[18]

2. Variational BCS-BEC crossover Hamiltonian

We consider the general Hermitian family of pairing Hamiltonians coupled to a bosonic degree of freedom

$$H = H_0 - H_1 \tag{1}$$

where

$$H_0 = \alpha N_0 + \kappa N_0^2 + \sum_{k=1}^{L} f(z_k) N_k, \tag{2}$$

$$H_1 = \beta \sum_{k=1}^{L} g(z_k) b_0 b_k^\dagger + \beta \sum_{k=1}^{L} \overline{g(z_k)} b_0^\dagger b_k + \sigma \sum_{k,s}^{L} g(z_k)\overline{g(z_s)} b_k^\dagger b_s, \tag{3}$$

for some complex-valued function $f(z)$ and real-valued function $g(z)$, and real-valued parameters α, κ, β and σ, which will be subject to certain solvability constraints yet to be determined. An overline is used to denote complex conjugation. The operators $b_k^\dagger$ and $N_k = b_k^\dagger b_k$ for $k > 0$ are hard-core Cooper pair creation and number operators[a]. There is a single bosonic mode with operators $b_0^\dagger$ and N_0. The particle operators satisfy the following commutation relations:

$$[b_j, b_k] = [b_j^\dagger, b_k^\dagger] = 0 \quad \forall\, j,k \geq 0, \quad [b_j, b_k^\dagger] = \begin{cases} I & j,k = 0 \\ (I - 2N_k)\delta_{jk} & \text{else} \end{cases}$$

[a]The Cooper pair operators are related to the fermionic operators $c_k^\dagger$ via the identity $b_k^\dagger = c_{-k}^\dagger c_k^\dagger$ where fermions are paired such that there is zero total momentum.

where δ_{jk} denotes the standard Kronecker delta. The Hamiltonian commutes with the total number operator $N = N_0 + \sum_k N_k$. Hamiltonians in this family describe a system of bosonic molecules, condensed into a single bosonic degree of freedom, coupled to L Cooper pairs which are bosonic-like pairs of fermions that must observe an exclusion principle. The Hamiltonian consists of a diagonal part H_0, given in Eq. 2, describing the allowed Cooper pair energy levels and the self-interacting bosonic mode, and a cross-interaction part H_1, given in equation Eq. 3, describing level dependent molecule-pair coupling and pair-pair couplings. For the special case $\alpha = \beta = \kappa = 0$, which suppresses any action of the Hamiltonian on the bosonic part of the underlying Hilbert space, the Hamiltonian Eq. 1 reduces to a general form of the BCS Hamiltonian. It is for this reason we can refer to the general pairing Hamiltonian as a BCS-BEC crossover Hamiltonian.

We will attempt to directly solve the Hamiltonian by formulating the wave-functions as factorisable operators acting on a suitable reference state, analogous to the algebraic Bethe ansatz.[4] On the other hand, we note that a co-ordinate Bethe ansatz approach[10,19,20] need not resort to any prior knowledge of a transfer matrix or a set of conserved operators. We do not expect that the Hamiltonian has an exact solution in general, however, by combining these aspects of the co-ordinate and algebraic Bethe ansatz methods, this technique allows us to obtain exactly solvable models in a very general fashion. In particular, we have found solvability conditions for two sub-classes of the Hamiltonian, namely the cases *i)* $\kappa = 0$ of no self-interaction term and *ii)* $\sigma = 0$ of no BCS pair-pair scattering term.

3. Variational Approach for Exact Solvability

Motivated by the approach of Richardson,[11] we assume the ansatz,

$$|\Psi\rangle = \prod_{j=1}^{M} C(y_j)|0\rangle, \quad C(y) = \gamma(y)b_0^\dagger + \sum_{k=1}^{L} h(y, z_k)b_k^\dagger, \tag{4}$$

for the eigenstates of Eq. 1, where $|0\rangle$ denotes the vacuum state, $h(y,z)$ is yet to be determined, and $y \in \mathbb{C}$. In general, acting the Hamiltonian H on the state $|\Psi\rangle$ will lead to terms that are linearly independent of $|\Psi\rangle$:

$$H|\Psi\rangle = E|\Psi\rangle + |\Phi\rangle, \quad \langle\Psi|\Phi\rangle = 0.$$

The objective is to choose appropriate constraints that allow these terms to be isolated. The ansatz will be an eigenstate if the coefficients of the terms linearly independent of $|\Psi\rangle$ cancel. The Hamiltonian is furnished with an exact solution on the manifold in the coupling parameter space obtained through compatibility of the corresponding constraints. Technical details of the calculation can be found in the literature.[18] We introduce the notation

$$|\Psi_j\rangle = \prod_{l\neq j}^{M} C(y_l)|0\rangle, \qquad |\Psi_{ij}\rangle = \prod_{l\neq i,j}^{M} C(y_l)|0\rangle.$$

By direct calculation the action of H on the state $|\Psi\rangle$ is of the form

$$H|\Psi\rangle = \sum_{j=1}^{M}\sum_{k=0}^{L} O_1(y_j, z_k) b_k^\dagger |\Psi_j\rangle + \sum_{j,l\neq j}^{M}\sum_{k,s=0}^{L} O_2(y_j, y_l, z_k, z_s) b_k^\dagger b_s^\dagger |\Psi_{jl}\rangle.$$

for some functions O_i depending on the various functions introduced above.

In order to determine the exact solution we look to solve the eigenvalue problem $H|\Psi\rangle = E|\Psi\rangle$ for some scalar E. The set of *solvability constraints* required to find such a solution define a manifold in the coupling parameter space along which the Hamiltonian has an exact solution. We briefly remark on the specific choices for the constraints. Imposing a constraint of the form

$$O_2(x, y, w, z) = g(w)k(x, y)h(y, z) + g(w)k(y, x)h(x, z),$$

for some function $k(x, y)$ to be determined later, allows the reduction of linear combinations of $b_k^\dagger b_s^\dagger |\Psi_{ij}\rangle$ and $b_k^\dagger b_0^\dagger |\Psi_{ij}\rangle$ terms in to linear combinations of $b_k^\dagger |\Psi_i\rangle$ and $b_0^\dagger |\Psi_i\rangle$ terms by use of the definition $C(y_l)|\Psi_{jl}\rangle = |\Psi_j\rangle$. To keep equations concise we have introduced the parameter z_0, however, we take $h(y, z_0) = \gamma(y)$, $g(z_0) = 1$. For $\kappa \neq 0 \neq \sigma$, this constraint yields a trivial exact solution due to the incompatibility of the coefficients of $b_k^\dagger b_s^\dagger |\Psi_{ij}\rangle$ and $b_k^\dagger b_0^\dagger |\Psi_{ij}\rangle$ terms. Thus the $\kappa = 0$ and $\sigma = 0$ cases are treated separately. A constraint is chosen such that the eigenvalues are of a standardised form,

$$h(y, z) = \beta^{-1}(\sigma(\alpha - y) + \beta^2)\frac{g(z)\gamma(y)}{f(z) - y}, \qquad E = \sum_j y_j.$$

Compatibility of these constraints defines $k(x, y)$. A final constraint is obtained by imposing that the remaining coefficients cancel. These are analogous to the Bethe ansatz equations and, subject to the compatibility of all constraints, the Hamiltonian is exactly solvable if the y_j are their roots.

For the case $\sigma = 0$ compatibility of the constraints leads to the constraining relations for exact solvability:

$$h(y_j, z_k) = \frac{\beta g(z_k)\gamma(y_j)}{f(z_k) - y_j}, \quad f(z_k) = \kappa^{-1}\beta^2 g(z_k)\overline{g(z_k)} + \kappa^{-1}c_1$$

$$y_j - (\alpha + \kappa) + \sum_{k=1}^{L} \frac{\kappa f(z_k) - c_1}{f(z_k) - y_j} = 2\sum_{l\neq j}^{M} \frac{c_1 - c_2 y_l}{y_j - y_l}, \qquad c_2 = \kappa.$$

for arbitrary constants β(or α), c_1, and c_2.

In the case $\kappa = 0$ the solvability conditions are

$$c_1\sigma\beta = c_2(\sigma\alpha+\beta^2)\beta, \quad c_1 = c_2 f(z_k) - \overline{g(z_k)}g(z_k),$$

$$h(y_j, z_k) = \beta^{-1}(\sigma(\alpha - y_j) + \beta^2)\frac{g(z_k)\gamma(y_j)}{f(z_k) - y_j},$$

$$\frac{y_j - \alpha}{\sigma(\alpha - y_j) + \beta^2} + \sum_{k=1}^{L} \frac{c_2 f(z_k) - c_1}{f(z_k) - y_j} = 2\sum_{l\neq j}^{M} \frac{c_1 - c_2 y_l}{y_j - y_l}.$$

for arbitrary constants β(or α), c_1, and c_2. In either case the $\gamma(y_j)$ will be fixed by any normalisation of $|\Psi\rangle$.

4. Final Remarks

We have determined manifolds in the coupling parameters of Eq. 1 for which an exact solution exists. As shown in Fig. 1, in appropriate limits of the general exact models found above eight subcases are recovered, seven of which are known.[11–17] Remarkably, this method has enabled a unified construction for exact solutions of models that were previously independent. There is opportunity to extend the scope of this approach to other models[21–23] by adopting an ansatz of a different form and relaxing the assumed separability of the pair-pair interaction.

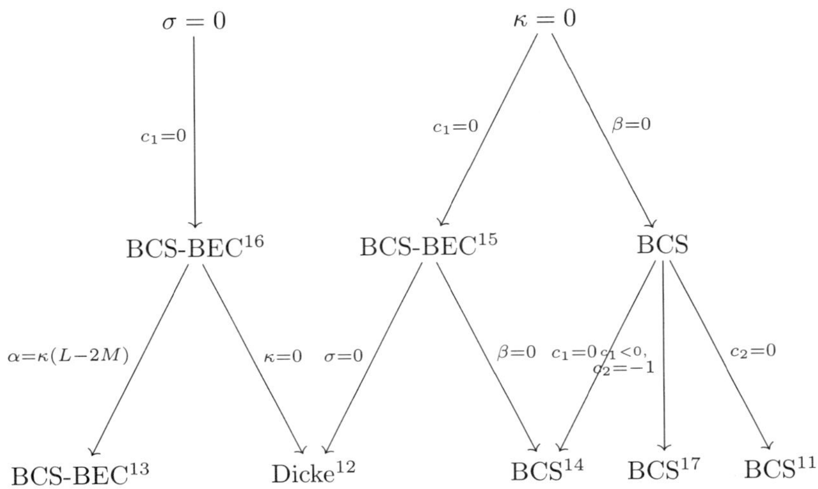

Fig. 1. Hierarchy of exactly solvable models that have been recovered. Specific models can be obtained from the references, excluding three previously unknown cases.

Acknowledgements

This work was supported in part by the Australian Research Council under Discovery Project DP110101414. A. Birrell acknowledges the support of an Australian Postgraduate Award.

References

1. M. Holland, S. J. J. F. Kokkelmans, M. L. Chiofalo and R. Walser, *Phys. Rev. Lett.* **87**, 120406 (2001).
2. Y. Ohashi and A. Griffin, *Phys. Rev. Lett.* **89**, 130402 (2002).
3. G. Sierra, J. Dukelsky, G. G. Dussel, J. von Delft and F. Braun, *Phys. Rev. B* **61**, R11890 (2000).
4. E. K. Sklyanin, L. A. Takhtadzhyan and L. D. Faddeev, *Theor. and Math. Phys.* **40**, 688 (1979).
5. L. A. Takhtadzhyan and L. D. Faddeev, *Russ. Math. Surv.* **34**, 11 (1979).
6. P. P. Kulish and E. K. Sklyanin, *Phys. Lett. A* **70**, 461 (1979).
7. J. B. McGuire, *J. Math. Phys.* **5**, 622 (1964).
8. C. N. Yang, *Phys. Rev. Lett.* **19**, 1312 (1967).
9. R. J. Baxter, *Ann. Phys.* **70**, 193 (1972).
10. H. Bethe, *Z. Phys. A* **71**, 205 (1931).
11. R. Richardson, *Phys. Lett.* **3**, 277 (1963).
12. M. Gaudin, *J. Physique* **37**, 1087 (1976).
13. C. Dunning, K. E. Hibberd and J. Links, *Nucl. Phys. B* **748**, 458 (2006).
14. M. Ibañez, J. Links, G. Sierra and S.-Y. Zhao, *Phys. Rev. B* **79**, 180501 (2009).
15. C. Dunning, P. S. Isaac, J. Links and S.-Y. Zhao, *Nucl. Phys. B* **848**, 372 (2011).
16. S. H. Lerma, S. M. A. Rombouts, J. Dukelsky and G. Ortiz, *Phys. Rev. B* **84**, 100503 (2011).
17. J. Dukelsky, S. Lerma, L. M. Robledo, R. Rodriguez-Guzman and S. M. A. Rombouts, *Phys. Rev. C* **84**, 061301 (2011).
18. A. Birrell, P. S. Isaac and J. Links, *Inverse Problems* **28**, 035008 (2012).
19. E. H. Lieb and F. Y. Wu, *Phys. Rev. Lett.* **20**, 1445 (1968).
20. P. Schlottmann, *Phys. Rev. B* **36**, 5177 (1987).
21. L. Amico, A. D. Lorenzo and A. Osterloh, *Nucl. Phys. B* **614**, 449 (2001).
22. J. Dukelsky, C. Esebbag and P. Schuck, *Phys. Rev. Lett.* **87**, 066403 (2001).
23. I. Marquette and J. Links, *Nucl. Phys. B* **866**, 378 (2013).

NONLINEAR WAVE DYNAMICS OF THE TRANSPORT PROCESS ON NONLINEAR LATTICES

ZONGQIANG YUAN and ZHIGANG ZHENG*

Department of Physics, Beijing Normal University,
Beijing, 100875, China
** Corresponding author: zgzheng@bnu.edn.cn*

The spatiotemporal propagation of a momentum excitation traveling along the Fermi-Pasta-Ulam β (FPU-β) lattice is investigated. The lifetime of the solitary wave against the anharmonic parameter β exhibits a hierarchical multiple-peak fractal structure for the case of an open boundary condition. The energy of the solitary wave decays exponentially during the final stage of the collapse process. The decay rate against β is found to exhibit a crossover between two different scaling laws. The hierarchical multiple-peak structure corresponds to a weak stochasticity where the KAM tori are dominant. Our studies of the solitary-wave dynamics support the discussions of energy carriers in the heat conduction processes of low-dimensional lattices, where phonons play a dominant role.

Keywords: Nonlinear wave dynamics; Heat conduction; Solitary wave.

1. Introduction

Recently, heat conduction in low-dimensional materials became a hot topic for the reason that classical one-dimensional lattices frequently exhibit anomalous heat conduction behaviors.[1,2] This further arouse a tide of interest in the fundamental issue of the microscopic foundation of a normal heat conduction.

In studies of heat conductions of nonlinear lattices, the thermal baths usually contact with the system by coupling the particles at two ends. The influence of the thermal bath on the lattice can be considered as a series of stochastic perturbations. Therefore, exploring the evolution behavior of a single pulse excited at one end of the lattice is of fundamental importance. In this aspect, the energy propagation for an excitation on infinitely long nonlinear lattices has been well discussed.[3–5] Practically, most of the systems have finite sizes, and the propagations of energy on materials usually possesses finite time scales.

2. The Fermi-Pasta-Ulam Lattice Model

The FPU model was introduced by Fermi, Pasta, and Ulam to investigate the energy equipartition problem and the ergodic hypothesis in nonlinear systems.[6] Recently the FPU lattice was studied in relating to heat conductions in low-dimensional systems.[7–11] The Hamiltonian of the FPU-β lattice consisting of N particles with open boundary condition is

$$H = \sum_{i=1}^{N} \frac{p_i^2}{2} + \sum_{i=1}^{N-1} \left[\frac{k}{2} \left(q_{i+1} - q_i \right)^2 + \frac{\beta}{4} \left(q_{i+1} - q_i \right)^4 \right], \tag{1}$$

where p_i and q_i denotes the momentum and the displacement from the equilibrium position of the i-th particle, respectively. The presence of the anharmonic terms breaks the integrability of the harmonic chain and brings forth the intermingling of regular and chaotic motions in phase space.[12]

We may impart an initial momentum excitation to the first particle of an initially quiescent lattice and fix $k = 0.5$ throughout the simulation.

3. Long Transient Excitation Propagation

Here we are concerned with the role of the anharmonic coupling, and we focus on the destiny of an initial pulse on a lattice. We first keep the excitation energy $E = 50$ and study the dynamical evolution of the initially local excitation on a lattice with $N = 50$ particles by adjusting the parameter β.

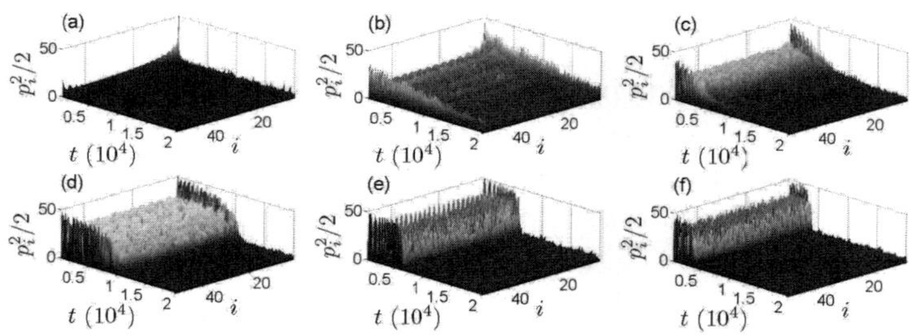

Fig. 1. Spatiotemporal propagation of the kinetic energy pulse with $N = 50$ and $E = 50$. (a) $\beta = 0.001$; (b) $\beta = 0.004$; (c) $\beta = 0.01$; (d) $\beta = 0.04$; (e) $\beta = 0.4$; (f) $\beta = 0.7$.

In Fig. 1, We depict the spatiotemporal propagation of the excitation for some typical values of β. With increasing β, different propagation behaviors can be observed. In Figs. 1(d)-(f), we give the evolution of the kinetic energy of each particle for relatively larger values of $\beta = 0.04, 0.4,$ and 0.7, respectively. One finds the evolution consists of two distinctively different stages. In the first stage, the kinetic energy pulse initiated at the first particle can be coherently transferred to other particles, and this pulse forms a solitary wave. The second stage comes when the solitary wave loses its stability and collapses in a rather short duration.

To give the long transient solitary wave state a more precise description, it is necessary to study the lifetime τ of the solitary wave. The lifetime τ is sensitive to β, as shown in Fig. 2(a), especially for $\beta \in (0.05, 0.2)$ where a multiple-peak structure is found. By adopting the method proposed in Ref. 13, the fractal dimension of the structure is estimated at $D \approx 1.66$.

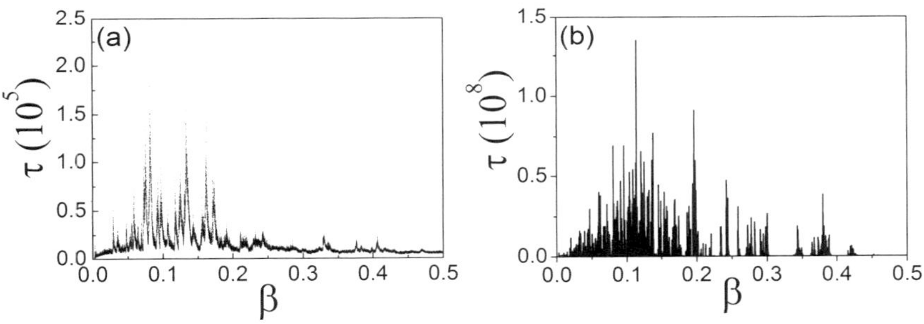

Fig. 2. The lifetime τ of the momentum excitation for $N = 50$ and $E = 50$. (a) and (b) correspond to the cases of open and periodic boundary conditions, respectively.

The lifetime τ against β shown in Fig. 2(a) is for the case of open boundary condition. In this case, both the reflection effect at two ends of the lattice and the interaction between the solitary wave and the phonon waves play important roles. It should be valuable to consider the periodic boundary condition, where there is no reflection of the wave at the boundary. The lifetime τ against β for the periodic boundary condition is presented in Fig. 2(b), which is found to increase drastically as compared with that for the open boundary condition cases.

4. Instability of the Solitary Wave

A sufficiently large pulse excites not only a solitary wave, but also a small-amplitude tail. In Figs. 3(a), (c) and (e), the snapshots of propagations of the energy waves for the open boundary condition case are plotted. The solitary wave moves with a hierarchy of lower pulses that move slower with different speeds, and the heights of these small pulses decrease when they move forward. As the solitary wave moves to the boundary of the lattice, it will be bounced back with a radiation of additional phonon waves. The multiple collisions between the solitary and phonon waves give rise to the instability of the solitary wave.

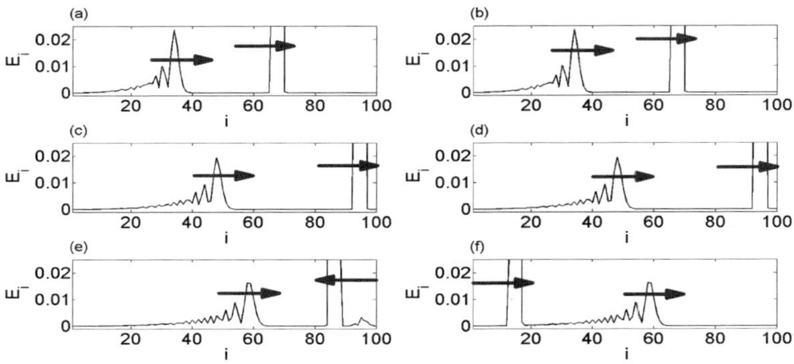

Fig. 3. The energy distribution among the particles with $\beta = 0.08$, $N = 100$ and $E = 50$ at $t = 500$ [(a), (b)], $t = 700$[(c), (d)], and $t = 850$[(e), (f)], respectively. The left and right columns correspond to cases of open and periodic boundary conditions, respectively.

This phonon-soliton-interaction mechanism is also valid for the case of periodic boundary condition, where no reflections occur at the boundary. Therefore the solitary wave moves unidirectionally, as shown in Figs. 3(b), (d) and (f). Due to the lack of additional excitations of phonons at the boundary, the lifetime of the solitary wave moving on a circular topology of the lattice should be much longer than that on an open lattice.

To understand the dynamics of the long-transient state, we focus on the collapse process of the solitary wave by resorting to the evolution of the energy of the small-amplitude tails. The energy of the tails is defined as the residual energy of the solitary wave $E_{\text{tails}} = E - \sum_{i=i_c-i_n}^{i_c+i_n} E_i(t)$, where $i_c(t)$ is the center position of the solitary wave at t. Numerically $i_n = 2$ is enough due to the energy localization of the solitary wave.

Although the lifetime varies for different boundary conditions, the collapse processes exhibit essentially the same scenario, i.e., the energy of the solitary wave decays exponentially during the final stage of the collapse process. The system exhibits similar behaviors for different boundary conditions, and the collapse process is only related to the intrinsic dynamical property of the lattice.

In Fig. 4(a), the decay rate γ of the solitary wave is computed for the case of open boundary condition. It can be seen that γ against β displays two different scaling laws $\gamma \propto \beta^{\kappa}$ for weak and strong nonlinearities, respectively. For lower values of β, $\kappa \approx 2/3$. For larger β, $\kappa \approx 1/4$. It is instructive to note that a similar result was obtained in studies of nonlinear Hamiltonian dynamics of the FPU-β lattice.[14,15] It was found that the largest Lyapunov exponent of the system varying with the β exhibits a two-segment behavior, and the same power-law exponents were given. The existence of a threshold separating two scalings corresponds to the transition from weak to strong stochasticity. In the weak chaos regime, the phase space of the Hamiltonian system is composed of chaotic trajectories intermingled with KAM tori. Then the multiple-peak structure observed in Fig. 2 can be explained as transiently quasi-regular motions in the phase space induced by weak stochasticity, where KAM invariant tori are dominant.

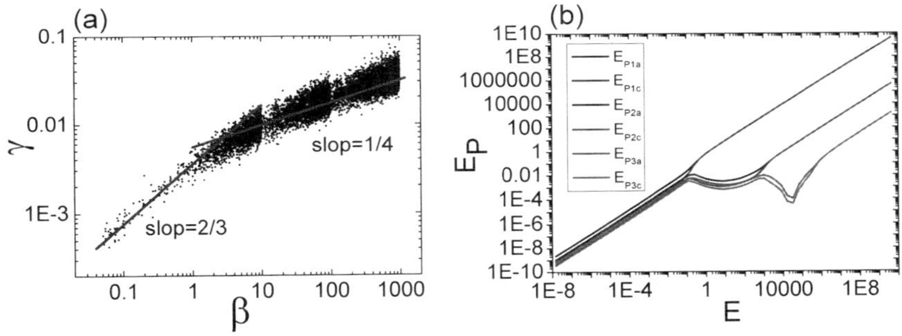

Fig. 4. (a) The decay rate γ against β for $N = 50$ and $E = 50$ for the case of open boundary condition. (b) The hight of the energy packet E_{pij} for $a = 50$, $c = 250$, and $\beta = 0.5$.

5. Emergence of a Series of Solitary Waves

The harmonic and anharmonic couplings induce the dispersion and localization of a energy packet on the lattice, respectively. Since a higher energy packet samples more anharmonic portions of the potential, one expects that the bifurcation of a solitary wave takes place as the energy of the packet exceeds a threshold. We examine the packet hight E_{pij} for the i-th wave packet passing the j-th particle on the lattice, as shown in Fig. 4(b). When the initial excitation energy E is low, all the packets are dispersing during their traveling along the lattice due to the dominating role of the harmonic coupling. A solitary wave emerges from the first energy packet when E is

Sufficiently large. The second and other solitary waves may also emerge from the energy packets as the total E increases to high enough values over thresholds.

6. Concluding Remarks

The results we discussed are based on a microscopic viewpoint of energy transfer process. The interaction between solitary wave and phonon waves leads to the instability of the energy wave packet. Li *et al.*[11] studied the energy carriers in the heat conduction processes of low-dimensional lattices. They argued through the analysis of the maximum sound velocity in terms of renormalized phonon theory and numerical computations that the dominant energy carrier of heat transport is the renormalized/effective phonons rather than solitons. Our microscopic studies of the solitary-wave dynamics support this judgement. The dynamical instability of the energy solitary wave is the microscopic mechanism of the dominance of phonons in thermal conductions.

Acknowledgments

Project supported by the National Natural Science Foundation of China (Grant No. 11075016), the Fundamental Research Funds for the Central Universities of China, and the Research Fund for the Doctoral Program of Higher Education of China (Grant No. 20100003110007).

References

1. S. Lepri, R. Livi and A. Politi, *Phys. Rev. Lett.* **78**, 1896 (1997).
2. S. Lepri, R. Livi and A. Politi, *Physics Reports* **377**, 1 (2003).
3. G. S. Zavt, M. Wagner and A. Lütze, *Phys. Rev. E* **47**, 4108 (1993).
4. A. Sarmiento, R. Reigada, A. H. Romero and K. Lindenberg, *Phys. Rev. E* **60**, 5317 (1999).
5. A. Rosas and K. Lindenberg, *Phys. Rev. E* **69**, 016615 (2004).
6. E. Fermi, J. Pasta and S. Ulam, *Los Alamos Document No. LA-1940, 1955* .
7. B. Li, J. H. Lan and L. Wang, *Phys. Rev. Lett.* **95**, 104302 (2005).
8. T. Mai, A. Dhar and O. Narayan, *Phys. Rev. Lett.* **98**, 184301 (2007).
9. A. Dhar and K. Saito, *Phys. Rev. E* **78**, 061136 (2008).
10. O. V. Gendelman and A. V. Savin, *Phys. Rev. E* **81**, 020103 (2010).
11. N. Li, B. Li and S. Flach, *Phys. Rev. Lett.* **105**, 054102 (2010).
12. C. Antonopoulos and T. Bountis, *Phys. Rev. E* **73**, 056206 (2006).
13. S. R. Forrest and T. A. W. Jr, *J. Phys. A* **12**, p. L109 (1979).
14. L. Casetti, R. Livi and M. Pettini, *Phys. Rev. Lett.* **74**, 375 (1995).
15. L. Casetti, C. Clementi and M. Pettini, *Phys. Rev. E* **54**, 5969 (1996).

PARALLEL SESSION
Quantum Algebras

21 & 22 & 23 & 24 AUGUST 2012

POSITIVE REPRESENTATIONS OF SPLIT REAL QUANTUM GROUPS

IVAN CHI-HO IP

Kavli IPMU, The University of Tokyo,
Kashiwa, Chiba 277-8583, Japan
E-mail: ivan.ip@ipmu.jp

We give a summary of the recent progress in the theory of positive representations of split real quantum groups. We outline the motivation of the research program and briefly describe the main ingredients of the construction, the appearance of Langlands duality, and future aspects of the theory.

Keywords: Positive operators, Modular double, Langlands duality.

1. Introduction

The present article summarizes the research program and some recent results described in Ref. 4,6 and 7. Let $\mathfrak{g}$ be a simple Lie algebra, and $\mathfrak{g}_c$ its compact form. For example, in type A_n, $\mathfrak{g} = \mathfrak{sl}(n+1, \mathbb{C})$ and $\mathfrak{g}_c = \mathfrak{su}(n+1)$. It is known that the category of irreducible finite dimensional representations $Rep(\mathfrak{g}_c)$ possesses several nice properties. Let $V_\lambda \in Rep(\mathfrak{g}_c)$, then:

- V_λ is parametrized by the cone of the positive weights $\lambda \in P^+ \subset \mathfrak{h}_\mathbb{R}^*$, where $\mathfrak{h}_\mathbb{R}$ is the real form of the Cartan subalgebra $\mathfrak{h} \subset \mathfrak{g}$,
- We have Peter-Weyl theorem:

$$\mathbb{C}[G] \simeq \bigoplus_\lambda V_\lambda \otimes V_\lambda^* \tag{1}$$

  where $\mathcal{U}(\mathfrak{g}_c)$ acts on $\mathbb{C}[G]$ by the regular representations,
- It is closed under tensor product:

$$V_\lambda \otimes V_\mu \simeq \bigoplus_\nu c_{\lambda\mu}^\nu V_\nu \tag{2}$$

  for some multiplicities $c_{\lambda\mu}^\nu \in \mathbb{N}$.

It is known that the above properties also hold in general for the Drinfeld-Jimbo compact quantum group $\mathcal{U}_q(\mathfrak{g}_c)$,[2,8] and together with the existence of the universal R matrix, $Rep(\mathcal{U}_q(\mathfrak{g}_c))$ can be given a braided tensor category structure.

It is natural to consider the other real form of $\mathfrak{g}$, most notably the split real form $\mathfrak{g}_{\mathbb{R}}$ (e.g. in type A_n, $\mathfrak{g}_{\mathbb{R}} = \mathfrak{sl}(n+1,\mathbb{R})$), and address the q-deformation of its irreducible representations. We know that the classical situation is not as nice compared to the compact case. The class of irreducible representations consists of, at least, the continuous and discrete series, and they do not behave as uniformly under the tensor products as well as the decomposition of $L^2(G)$.

However, in the case of $\mathcal{U}_q(\mathfrak{sl}(2,\mathbb{R}))$, where $|q|=1$, $q := e^{\pi i b^2}$ with $b^2 \in (0,1) \setminus \mathbb{Q}$, Teschner *et al.*[1,10,11] have constructed from Liouville conformal field theory a special class of representations $\mathcal{P}_\lambda$, parametrized by $\lambda \in \mathbb{R}_{>0}$, acting as unbounded operators in the space $L^2(\mathbb{R})$, and at the same time demonstrating the above nice properties in parallel to the compact case:

Theorem 1.1. *The class of representations is closed under tensor product (in the continuous sense):*[11]

$$\mathcal{P}_\alpha \otimes \mathcal{P}_\beta \simeq \int_{\mathbb{R}_+}^{\oplus} P_\lambda d\mu(\lambda). \tag{3}$$

Define the modular double $\mathcal{U}_{q\widetilde{q}} := \mathcal{U}_q(\mathfrak{sl}(2,\mathbb{R})) \otimes \mathcal{U}_{\widetilde{q}}(\mathfrak{sl}(2,\mathbb{R}))$ *with* $\widetilde{q} := e^{\pi i b^{-2}}$. *Then we have a similar Peter-Weyl theorem*[5,10]

$$L^2(SL^+_{q\widetilde{q}}(2,\mathbb{R})) \simeq \int_{\mathbb{R}_+}^{\oplus} P_\alpha \otimes P_\alpha d\mu(\alpha) \tag{4}$$

as left-right regular representations of $\mathcal{U}_{q\widetilde{q}} \otimes \mathcal{U}_{q\widetilde{q}}$. *In both equations,* $\mu(\lambda)$ *is expressed in terms of the quantum dilogarithm function* $g_b(x)$. *Furthermore, there exists a universal* R *operator, unitary on* $L^2(\mathbb{R} \otimes \mathbb{R})$,

$$R = q^{\frac{H\otimes H}{4}} g_b(e \otimes f) q^{\frac{H\otimes H}{4}}, \tag{5}$$

such that it is invariant under the change $b \longleftrightarrow b^{-1}$.

The Hilbert space $L^2(SL^+_{q\widetilde{q}}(2,\mathbb{R}))$ is constructed from the GNS representations of a C^*-algebraic version of $SL_q^+(2,\mathbb{R})$ by applying the quantum double construction to the quantum plane, while the action of $\mathcal{U}_{q\widetilde{q}}$ is obtained by dualizing the regular corepresentation of $SL_q^+(2,\mathbb{R})$. Details of the construction can be found in Ref. 5.

To better formulate the generalization of this class of representations to arbitrary $\mathcal{U}_q(\mathfrak{g}_{\mathbb{R}})$, let us observe the following defining properties of $\mathcal{P}_\lambda$:

(1) $\mathcal{P}_\lambda$ is parametrized by $\lambda \in \mathbb{R}_{>0}$.

(2) The generators E, F, K are represented by positive essentially self-adjoint operators on $L^2(\mathbb{R})$. Specifically, they are expressed in terms of Laurent polynomials in $u := e^{2\pi bx}$ and $v := e^{2\pi bp}$ with positive coefficients, where $p = \frac{1}{2\pi i}\frac{d}{dx}$.
(3) Define $e := 2\sin(\pi b^2)E, f := 2\sin(\pi b^2)F$ and similarly for $\widetilde{E}, \widetilde{F}$ with b replaced by b^{-1}. We have the *transcendental relations*

$$\widetilde{e} := e^{\frac{1}{b^2}}, \qquad \widetilde{f} := f^{\frac{1}{b^2}}, \qquad \widetilde{K} := K^{\frac{1}{b^2}}, \tag{6}$$

such that $\widetilde{E}, \widetilde{F}$ and $\widetilde{K}$ give the same representation of $\mathcal{U}_{\widetilde{q}}(\mathfrak{sl}(2,\mathbb{R}))$ with b replaced by b^{-1} acting on the same space $\mathcal{P}_\lambda$, and commute weakly with E, F, K. Hence this gives a representation of the modular double $\mathcal{U}_{q\widetilde{q}}$.

In parallel to the above properties, we define the class of *positive principal series representations* of $\mathcal{U}_q(\mathfrak{g}_\mathbb{R})$ for $|q| = 1$, generated by E_i, F_i, K_i subjected to the usual quantum relations, such that

(1) It is parametrized by the $\mathbb{R}_{>0}$-span $P_\mathbb{R}^+$ of the discrete cone P^+, or equivalently, by $(\mathbb{R}_{>0})^{rank(\mathfrak{g})}$,
(2) The action of the generators are represented by positive essentially self-adjoint operators,
(3) Define $e_i := 2\sin(\pi b^2)E_i$ and $f_i := 2\sin(\pi b^2)F_i$. We have the transcendental relations as before with

$$\widetilde{e_i} := e_i^{\frac{1}{b^2}}, \qquad \widetilde{f_i} := f_i^{\frac{1}{b^2}}, \qquad \widetilde{K_i} := K_i^{\frac{1}{b^2}}, \tag{7}$$

such that E_i, F_i, K_i together with $\widetilde{E_i}, \widetilde{F_i}, \widetilde{K_i}$ gives a representation of the modular double $\mathcal{U}_{q\widetilde{q}}(\mathfrak{g}_\mathbb{R})$.

This class of representation has been constructed for $\mathcal{U}_q(\mathfrak{g}_\mathbb{R})$ corresponding to simple Lie algebra $\mathfrak{g}$ of any type. In the next section we will outline the main ingredients of the construction.

2. Construction

To obtain representations by positive operators, the basic idea is to express the action of the generators in terms of combinations of Weyl pairs $u := e^{2\pi bx}$ and $v := e^{2\pi bp}$, which is essentially multiplication and shifting operators. We know that classically, $P_\mathbb{R}^+$ parametrizes the irreducible unitary representations associated to the Borel subalgebra $\mathfrak{b}_\mathbb{R}$, called the minimal principal series, which act on the functions of the (upper) unipotent subgroup U^+ by differential operators. Hence the crucial step of our construction is to apply Mellin transform which converts differential operators into difference operators. In order to do this in the L^2 sense, we are required to work on the functions of *totally positive* unipotent matrices $U_{>0}^+$. Hence the natural choice for a good coordinate system is given by the Lusztig's data for total positivity:[9]

Theorem 2.1. *Fix a longest element $w_0 = s_{i_1}...s_{i_m} \in W$ in the Weyl group with reduced expression. Then the totally positive upper unipotent semi-subgroup $U_{>0}^+$ of the totally positive semigroup $G_{>0}$ is parametrized by*

$$\begin{aligned} \mathbb{R}_{>0}^m &\longrightarrow U_{>0}^+ \\ (a_1, ..., a_m) &\mapsto x_{i_1}(a_1)...x_{i_m}(a_m), \end{aligned}$$

where $x_i(t)$ is the root subgroup corresponding to the i-th root.

Write the Gauss decomposition as $G_{>0} = U_{>0}^- T_{>0} U_{>0}^+$. From this we can apply the regular representation of G on functions of $U_{>0}^+$:

$$g \cdot f(g_+) := [f(g_+ g)]_+ \chi_\lambda(g_+ g), \tag{8}$$

where $[-]_+$ denotes the projection onto $U_{>0}^+$, and χ_λ is the character on $T_{>0}$ parametrized by $\lambda \in P_{\mathbb{R}}^+ \simeq \mathbb{R}_{>0}^{rank(\mathfrak{g})}$. We then obtain the actions of the generators of $\mathfrak{g}_{\mathbb{R}}$ in terms of differential operators by infinitesimal actions.

Now we apply the Mellin transform

$$L^2(U_{>0}^+) \simeq L^2(\mathbb{R}_{>0}^{dim(U^+)}) \longrightarrow L^2(\mathbb{R}^{dim(U^+)}),$$

which converts the actions into difference operators with weights of the form

$$f(\mathbf{u}) \mapsto P(\mathbf{u}) f(\mathbf{u} + i\mathbf{e}),$$

where $P(\mathbf{u})$ is linear in the new coordinates $\mathbf{u}$, and $\mathbf{e}$ is a constant integral vector. To obtain a representation of $\mathcal{U}_q(\mathfrak{g}_{\mathbb{R}})$ we simply quantize the weight

$$P(\mathbf{u}) \mapsto [P(\mathbf{u})]_q,$$

where $[n]_q := \frac{q^n - q^{-n}}{q - q^{-1}}$ is the quantum number [a]. Then one can check, despite non-trivially, that all quantum relations of $\mathcal{U}_q(\mathfrak{g})$ are satisfied.

After rescaling by b, the final step is to introduce a twist $\frac{Q}{2b}$ in the weight with $Q = b + b^{-1}$, so that the action is represented by positive self-adjoint operators. More precisely, we obtain sums of operators of the form

$$\left[\frac{Q}{2b} - i\frac{P(\mathbf{u})}{b}\right]_q e^{-2\pi b \mathbf{p_e}} = \frac{1}{2\sin(\pi b^2)} (e^{\pi b (P(\mathbf{u}) - \mathbf{p_e})} + e^{\pi b(-P(\mathbf{u}) - \mathbf{p_e})}). \tag{9}$$

[a]This is done in type A_n, and carry over to other types by the unitary transformations given in Theorem 2.2.

Under this twist however, we can no longer take the classical limit $b \longrightarrow 0$.

So far we have fixed a reduced expression for the longest element $w_0 \in W$ to obtain a representation. It turns out that the representation is independent of the choice of reduced expression in the following sense, hence it can be considered canonical:

Theorem 2.2. *The transformation of the operators $\mathcal{U}_q(\mathfrak{g}_\mathbb{R})$ corresponding to the change of words $...s_is_js_i... = ...s_js_is_j...$ of w_0 is given by*

$$X \mapsto \Phi X \Phi^{-1}, \tag{10}$$

where Φ is a unitary transformation on $L^2(\mathbb{R}^{dim(U^+)})$ defined in terms of ratios of the quantum dilogarithm function $g_b(x)$.

3. Non-simply-laced case

In the non-simply-laced case, the definition of $\mathcal{U}_q(\mathfrak{g}_\mathbb{R})$ depends on different q-parameters $q_i := e^{\pi i b_i^2} := q^{\frac{1}{2}(\alpha_i,\alpha_i)}$ for different simple root α_i, where $(-,-)$ is the inner product on the root lattice. Representing the generators by similar constructions above, if we let as before $e_i := 2\sin\pi b_i^2 E_i$, $f_i := 2\sin\pi b_i^2 F_i$, and define

$$\widetilde{e_i} := e_i^{\frac{1}{b^2}}, \qquad \widetilde{f_i} := f_i^{\frac{1}{b^2}} \qquad \widetilde{K_i} := K_i^{\frac{1}{b^2}}, \tag{11}$$

then we have

Theorem 3.1. *The operators $\widetilde{E_i}, \widetilde{F_i}, \widetilde{K_i}$ gives a positive representation of $\mathcal{U}_{\widetilde{q}}({}^L\mathfrak{g}_\mathbb{R})$. More precisely, if E_i correspond to short root in $\mathcal{U}_q(\mathfrak{g}_R)$, then $\widetilde{E_i}$ corresponds to long root in $\mathcal{U}_{\widetilde{q}}({}^L\mathfrak{g}_\mathbb{R})$ and vice versa.*

Hence this provides a very direct analytic relation between $\mathfrak{g}$ and its Langlands dual in the operator algebraic setting. This correspondence between modular duality and Langlands duality is expected to have deep consequences.[3,12]

4. Future Perspectives

The existence of positive representations for arbitrary split real quantum groups strongly suggests that one can try to find the analogues of various remarkable results and constructions that were discovered and studied in relation to the braided tensor category of the finite-dimensional representations of the quantum group $\mathcal{U}_q(\mathfrak{g}_c)$. We conjecture the closure of tensor products of this class of representations, and consequently we envision applications in several area of mathematics and physics where split real Lie group $G_{\mathbb{R}}$ is used in recent development. To name a few, these includes for example Chern-Simons-Witten (CSW) model and topological quantum field theories (TQFT's) related to the quantization of Teichmüller spaces, geometrization and categorification of $\mathcal{U}_{q\widetilde{q}}(\mathfrak{g}_{\mathbb{R}})$ based on super-symmetric gauge theory, and equivalence of categories of affine Lie algebras and quantum groups, which in the case of $\mathcal{U}_{q\widetilde{q}}(\mathfrak{sl}(2,\mathbb{R}))$, is related to representations of Virasoro algebra associated to Liouville conformal field theory. A more detailed description of these future projects can be found in Ref. 4 and references therein.

Acknowledgments

I would like to thank the committee of Group 29 for their invitation and the session chairs for organizing the event and generating interesting discussions.

References

1. A.G. Bytsko, K. Teschner, *R-operator, co-product and Haar-measure for the modular double of* $\mathcal{U}_q(\mathfrak{sl}(2,\mathbb{R}))$, Comm. Math. Phys., **240**, (2003), 171-196
2. V. G. Drinfeld, *Hopf algebras and the quantum Yang-Baxter equation*,
3. V.V. Fock, A.B. Goncharov, *The quantum dilogarithm and representations of quantum cluster varieties*, Invent. Math. **175**, (2009), 223286
4. I. B. Frenkel, I. Ip, *Positive representations of split real quantum groups and future perspectives*, arXiv:1111.1033, (2011)
5. I. Ip, *Representation of the quantum plane, its quantum double and harmonic analysis on* $GL_q^+(2,R)$, arXiv:1108.5365, (2011)
6. I. Ip, *Positive representations of split real simply-laced quantum groups*, arXiv:1203:2018, (2012)
7. I. Ip, *Positive representations of split real quantum groups of type* B_n, C_n, F_4, *and* G_2, arXiv:1205.2940, (2012)
8. M. Jimbo *A q-difference analogue of* $\mathcal{U}(\mathfrak{g})$ *and the Yang-Baxter equation*, Lett. Math. Phys., **10**, (1985), 63-69
9. G. Lusztig *Total positivity in reductive groups*, in: "Lie theory and geometry: in honor of B. Kostant", Progr. in Math., **123**, Birkhauser, (1994), 531-568
10. B. Ponsot, J. Teschner, *Liouville bootstrap via harmonic analysis on a noncompact quantum group*, arXiv: hep-th/9911110, (1999)
11. B. Ponsot, J. Teschner, *Clebsch-Gordan and Racah-Wigner coefficients for a continuous series of representations of* $\mathcal{U}_q(\mathfrak{sl}(2,\mathbb{R}))$, Comm. Math. Phys., **224**, (2001), 613-655
12. J. Teschner, *Quantization of the Hitchin moduli spaces, Liouville theory, and the geometric Langlands correspondence I*, arXiv:1005.2846, (2010)

A SOLUTION OF THE 3D REFLECTION EQUATION FROM QUANTIZED ALGEBRA OF FUNCTIONS OF TYPE B

ATSUO KUNIBA

Institute of Physics,
Graduate School of Arts and Sciences, University of Tokyo,
Komaba, Tokyo 153-8902, Japan
E-mail: atsuo@gokutan.c.u-tokyo.ac.jp

MASATO OKADO

Department of Mathematical Science,
Graduate School of Engineering Science, Osaka University,
Toyonaka, Osaka 560-8531, Japan
E-mail: okado@sigmath.es.osaka-u.ac.jp

Let $A_q(\mathfrak{g})$ be the quantized algebra of functions associated with simple Lie algebra $\mathfrak{g}$ defined by generators obeying the so called RTT relations. We describe the embedding $A_q(B_2) \hookrightarrow A_q(C_2)$ explicitly. As an application, a new solution of the Isaev-Kulish 3D reflection equation is constructed by combining the embedding with the previous solution for $A_q(C_2)$ by the authors.

Keywords: 3D reflection equation, quantized algebra of functions.

1. Introduction

The reflection equation[1,2] is a companion system of the Yang-Baxter equation[3] in 2 dimensional (2D) integrable systems with boundaries. A 3D analogue of such a system is known by Isaev and Kulish[4] as the 3D reflection equation which accompanies the Zamolodchikov tetrahedron equation.[5] In Ref. 6, the first solution of the 3D reflection equation was constructed by invoking the representation theory[7] of the quantized algebra of functions $A_q(\mathfrak{g})$[8,9] for $\mathfrak{g}$ of type C. It was done by succeeding the approach in Ref. 10, where the solution of the tetrahedron equation[11] was obtained (up to misprint) as the intertwiner for $A_q(\mathfrak{g})$-modules with $\mathfrak{g}$ of type A.

In this paper we extend such results to $\mathfrak{g}$ of type B. The algebra $A_q(B_n)$ is formulated following Ref. 9, and its fundamental representations $\pi_i (i = 1, 2, 3)$ are presented for $n = 3$. According to the general theory,[7] one has the equivalence of the tensor products $\pi_{121} \simeq \pi_{212}$ and $\pi_{2323} \simeq \pi_{3232}$ ($\pi_{i_1,\dots,i_r} := \pi_{i_1} \otimes \cdots \otimes \pi_{i_r}$) according to the Coxeter relations $s_1s_2s_1 = s_2s_1s_2$ and $s_2s_3s_2s_3 = s_3s_2s_3s_2$ among the simple reflections in the Weyl group $W(B_3)$. We determine their intertwiners $\mathcal{S}$ and $\mathcal{J}$ explicitly and show that they yield a new solution to the 3D reflection equation, confirming a conjecture in Ref. 6. In fact, our strategy is to attribute the analysis to the $A_q(C_3)$ case.[6] The key to this is the embedding $A_q(B_2) \hookrightarrow A_q(C_2)$ in Theorem 2.1. It enable us to bypass the complicated intertwining relation (11) for type B and to utilize the result on the type C case. The embedding originates in a formulation of $A_q(\mathfrak{g})$ by the RTT relations[9] and the fact that the vector representation of C_2 is the spinor one for B_2.

The paper is organized as follows. In Section 2 we recall the Hopf algebra $A_q(B_n)$ following Ref. 9. The embedding $A_q(B_2) \hookrightarrow A_q(C_2)$ in Theorem 2.1 will be the key to the proof of our main Theorem 4.2. In Section 3 we present the fundamental representations $\pi(i = 1, 2, 3)$ of $A_q(B_3)$. The π_2 and π_3 corresponding to the subalgebra $A_q(B_2)$ actually come from the fundamental representations of $A_q(C_2)$ through the embedding (Remark 3.1). In Section 4 we formulate the intertwiners along the scheme parallel to $A_q(C_3)$[6] and determine them. In Section 5 we explain that these intertwiners yield a new solution to the 3D reflection equation, which is distinct from the type C case in Ref. 6.

2. Algebra $A_q(B_n)$

Let $N = 2n + 1$ with $n \in \mathbb{Z}_{\geq 2}$. We define $A_q(B_n)$ following Ref. 9, where it was denoted by $\mathrm{Fun}(\mathrm{SO}_q(N))$. $A_q(B_n)$ is a Hopf algebra generated by t_{ij} $(1 \leq i, j \leq N)$ with the relations

$$R_{ij,mp}t_{mk}t_{pl} = t_{jp}t_{im}R_{mp,kl}, \quad C_{jk}C_{lm}t_{ij}t_{lk} = C_{ij}C_{kl}t_{kj}t_{lm} = \delta_{im}, \tag{1}$$

where the repeated indices are summed over $\{1, 2, \dots, N\}$. The structure constants are specified by $C_{ij} = \delta_{i,N+1-j}q^{\varrho_j}$ with $(\varrho_1, \dots, \varrho_N) = (2n-1, \dots, 3, 1, 0, -1, -3, \dots, -2n+1)$, and $\sum_{i,j,m,l} R_{ij,ml}E_{im} \otimes E_{jl} = q^2 \lim_{x\to\infty} x^{-2}R(x)|_{k=q^{-2}}$, where $R(x)$ is the quantum R matrix[12,13] for the vector representation of $U_q(B_n^{(1)})$ given in Ref. 13, eq.(3.6). The former one in (1) is the so called RTT relation on the generators $T = (t_{ij})$. The coproduct is given by $\Delta(t_{ij}) = \sum_k t_{ik} \otimes t_{kj}$. We omit the antipode and counit as they are not necessary in this paper.

Reflecting the equivalence $U_q(B_2) \simeq U_q(C_2)$, the simplest case $A_q(B_2)$ is related to $A_q(C_2)$. The latter was introduced in Ref. 9 and detailed in Ref. 6, sec. 3.1, where it was denoted by $A_q(\mathrm{Sp}_4)$. The $A_q(C_2)$ is the Hopf algebra defined similarly to $A_q(B_2)$ by replacing the structure constants $R_{ij,mp}$ and C_{jk} by those associated with the quantum R matrix of the vector representation of $U_q(C_2^{(1)})$. The generators t_{ij} $(1 \le i,j \le 4)$ of $A_q(C_2)$ (Ref. 6, sec. 3.1) will be denoted by s_{ij} here for distinction. The vector representation of $U_q(C_2)$ is the spinor one in terms of $U_q(B_2)$. This fact is reflected in the following.

Theorem 2.1. *There is an embedding of the Hopf algebra* $\iota : A_q(B_2) \hookrightarrow A_q(C_2)$ *which maps the generators* $\{t_{ij} \mid 1 \le i,j \le 5\}$ *as follows:*

$$
\begin{aligned}
t_{ij} &\mapsto (-1)^{\delta_{i5}+\delta_{j5}}(\sqrt{r})^{\delta_{i3}}(s_{\xi_i\xi_j}s_{\eta_i\eta_j} - q s_{\xi_i\eta_j}s_{\eta_i\xi_j}) \quad (j \neq 3),\\
t_{i3} &\mapsto (-1)^{\delta_{i5}}\sqrt{r}\,(-q^{-1}s_{\xi_i 1}s_{\eta_i 4} + q s_{\xi_i 4}s_{\eta_i 1}) \quad (i \neq 3),\\
t_{33} &\mapsto s_{22}s_{33} - q s_{21}s_{34} + q s_{24}s_{31} - q^2 s_{23}s_{32},
\end{aligned}
$$

where $r = 1+q^2$ *and* ξ_i, η_i *are given by* $\left(\begin{smallmatrix}\xi_1\xi_2\xi_3\xi_4\xi_5\\ \eta_1\eta_2\eta_3\eta_4\eta_5\end{smallmatrix}\right) = \left(\begin{smallmatrix}11223\\ 23344\end{smallmatrix}\right)$.

We have checked that under ι, all the relations on $\{t_{ij}\}$ in $A_q(B_2)$ are guaranteed by those on $\{s_{ij}\}$ in $A_q(C_2)$ by tedious but direct calculations. We note that in the different formulation of $A_q(\mathfrak{g})$ of Ref. 14, one has the isomorphism $A_q(B_2) \simeq A_q(C_2)$ by definition rather than the embedding.

3. Fundamental representations of $A_q(B_3)$

Let $\mathrm{Osc}_q = \langle \mathbf{1}, \mathbf{a}^+, \mathbf{a}^-, \mathbf{k}\rangle$ be the q-oscillator algebra, i.e. an associative algebra with the center $\mathbf{1}$ and the relations

$$\mathbf{k}\,\mathbf{a}^+ = q\,\mathbf{a}^+\,\mathbf{k}, \;\; \mathbf{k}\,\mathbf{a}^- = q^{-1}\mathbf{a}^-\,\mathbf{k}, \;\; \mathbf{a}^-\mathbf{a}^+ = \mathbf{1} - q^2\,\mathbf{k}^2, \;\; \mathbf{a}^+\mathbf{a}^- = \mathbf{1} - \mathbf{k}^2.$$

It has a representation on the Fock space $\mathcal{F}_q = \oplus_{m\ge 0}\mathbb{C}(q)|m\rangle$:

$$\mathbf{1}|m\rangle = |m\rangle, \; \mathbf{k}|m\rangle = q^m|m\rangle, \; \mathbf{a}^+|m\rangle = |m+1\rangle, \; \mathbf{a}^-|m\rangle = (1-q^{2m})|m-1\rangle.$$

We shall also use Osc_q with q replaced by q^2. For distinction it is denoted by $\mathrm{Osc}_{q^2} = \langle \mathbf{1}, \mathbf{A}^+, \mathbf{A}^-, \mathbf{K}\rangle$, which acts on $\mathcal{F}_{q^2}$.

From now on we consider $A_q(B_3)$ which includes $A_q(B_2)$ as a subalgebra. Set $(q_1, q_2, q_3) = (q^2, q^2, q)$, which reflects the squared root length of the simple roots of B_3. Consider the maps $\pi_i\,(i=1,2,3) : A_q(B_3) \to \mathrm{Osc}_{q_i}$ that send the generators $(t_{ij})_{1\le i,j\le 7}$ to the following:

$$\pi_1 : \begin{pmatrix} \mu_1\mathbf{A}^- & \alpha_1\mathbf{K} & & & & & \\ \beta_1\mathbf{K} & \nu_1\mathbf{A}^+ & & & & & \\ & & \kappa_1\mathbf{1} & & & & \\ & & & \sigma_1\mathbf{1} & & & \\ & & & & \kappa_1^{-1}\mathbf{1} & & \\ & & & & & \nu_1^{-1}\mathbf{A}^- & q^2\beta_1^{-1}\mathbf{K} \\ & & & & & q^2\alpha_1^{-1}\mathbf{K} & \mu_1^{-1}\mathbf{A}^+ \end{pmatrix}, \tag{2}$$

$$\pi_2 : \begin{pmatrix} \kappa_2\mathbf{1} & & & & & & \\ & \mu_2\mathbf{A}^- & \alpha_2\mathbf{K} & & & & \\ & \beta_2\mathbf{K} & \nu_2\mathbf{A}^+ & & & & \\ & & & \sigma_2\mathbf{1} & & & \\ & & & & \nu_2^{-1}\mathbf{A}^- & q^2\beta_2^{-1}\mathbf{K} & \\ & & & & q^2\alpha_2^{-1}\mathbf{K} & \mu_2^{-1}\mathbf{A}^+ & \\ & & & & & & \kappa_2^{-1}\mathbf{1} \end{pmatrix}, \tag{3}$$

$$\pi_3 : \begin{pmatrix} \kappa_{31}\mathbf{1} & & & & & & \\ & \kappa_{32}\mathbf{1} & & & & & \\ & & \mu_3(\mathbf{a}^-)^2 & \alpha_3\mathbf{k}\,\mathbf{a}^- & -(r\mu_3)^{-1}\alpha_3^2\mathbf{k}^2 & & \\ & & -r\alpha_3^{-1}\mu_3\mathbf{a}^-\mathbf{k} & \mathbf{a}^-\mathbf{a}^+ - \mathbf{k}^2 & -(q\mu_3)^{-1}\alpha_3\mathbf{k}\,\mathbf{a}^+ & & \\ & & -r(q\alpha_3^{-1})^2\mu_3\mathbf{k}^2 & r\alpha_3^{-1}\mathbf{k}\,\mathbf{a}^+ & \mu_3^{-1}(\mathbf{a}^+)^2 & & \\ & & & & & \kappa_{32}^{-1}\mathbf{1} & \\ & & & & & & \kappa_{31}^{-1}\mathbf{1} \end{pmatrix}, \tag{4}$$

where r is defined in Theorem 2.1 and blanks mean 0. The symbols $\alpha_i, \beta_i, \mu_i, \nu_i, \sigma_i, \kappa_i, \kappa_{31}, \kappa_{32}$ are parameters. They obey the constraint

$$\alpha_i\beta_i = -q^2\mu_i\nu_i, \quad \sigma_i = \pm 1 \quad (i = 1, 2). \tag{5}$$

Proposition 3.1. *The maps $\pi_i\,(i = 1, 2, 3)$ are naturally extended to the algebra homomorphisms. The resulting representations of $A_q(B_3)$ on $\mathcal{F}_{q_i}$ are irreducible.*

We call $\pi_i\,(i = 1, 2, 3)$ the *fundamental representations* of $A_q(B_3)$. It is easy to infer the fundamental representation π_i of $A_q(B_n)$ for general n.

Remark 3.1. The symbols $(t_{ij})_{2\le i,j\le 6}$ generate $A_q(B_2)$. The 5×5 submatrices of (3) and (4) without the first and the last rows and columns provide the two fundamental representations of the $A_q(B_2)$. Denote them by $\pi_1^{B_2}$ and $\pi_2^{B_2}$ respectively. Similarly one can extract the fundamental representations of $A_q(C_2)$ from the 4×4 submatrices of Ref. 6,eq.(3.4) and eq.(3.5). Denote them by $\pi_1^{C_2}$ and $\pi_2^{C_2}$, respectively. Then $\pi_i^{B_2}$ coincides with the representation induced from $\pi_{3-i}^{C_2}$ via Theorem 2.1 with a suitable adjustment of the parameters, i.e. $\pi_i^{B_2}(f) = \pi_{3-i}^{C_2}(\iota(f))$ for any $f \in A_q(B_2)$.

4. Intertwiners

The Weyl group $W(B_3) = \langle s_1, s_2, s_3 \rangle$ is generated by the simple reflections with the relations $s_1^2 = s_2^2 = s_3^2 = 1$, $s_1 s_3 = s_3 s_1$, $s_1 s_2 s_1 = s_2 s_1 s_2$ and $s_2 s_3 s_2 s_3 = s_3 s_2 s_3 s_2$. According to Ref. 7 one should have the equivalence

$$\pi_{13} \simeq \pi_{31}, \quad \pi_{121} \simeq \pi_{212}, \quad \pi_{2323} \simeq \pi_{3232} \tag{6}$$

under an appropriate tuning of the parameters. Here and in what follows we use the shorthand $\pi_{i_1,\ldots,i_r} = \pi_{i_1} \otimes \cdots \otimes \pi_{i_r}$. It is easy to see that $\pi_{13} \simeq \pi_{31}$ holds if and only if

$$\kappa_1 = \sigma_1, \quad \kappa_{31} = \kappa_{32} \tag{7}$$

and the intertwiner is just the transposition $P(x \otimes y) = y \otimes x$. The equivalence $\pi_{121} \simeq \pi_{212}$ holds if and only if

$$\kappa_1 = \kappa_2 = \sigma_2, \quad \alpha_1\beta_1 = \alpha_2\beta_2 \tag{8}$$

are further satisfied. Assuming them we introduce the intertwiner $\Phi \in \mathrm{End}(\mathcal{F}_{q^2} \otimes \mathcal{F}_{q^2} \otimes \mathcal{F}_{q^2})$ characterized by

$$\pi_{212}(\Delta(f)) \circ \Phi = \Phi \circ \pi_{121}(\Delta(f)) \quad (\forall f \in A_q(B_3))$$

and the normalization $\Phi(|0\rangle \otimes |0\rangle \otimes |0\rangle) = |0\rangle \otimes |0\rangle \otimes |0\rangle$. Set $S = \Phi P_{13}$, where $P_{13}(x \otimes y \otimes z) = z \otimes y \otimes x$. The S is regarded as a matrix $S = (S_{ijk}^{abc})$ whose elements are specified by $S(|i\rangle \otimes |j\rangle \otimes |k\rangle) = \sum_{a,b,c \in \mathbb{Z}_{\geq 0}} S_{ijk}^{abc} |a\rangle \otimes |b\rangle \otimes |c\rangle$. We use the notation $\delta_i^a = 1 (a = i)$, $\delta_i^a = 0 (a \neq i)$ and

$$(q)_i = \prod_{j=1}^{i} (1 - q^j), \quad \begin{bmatrix} i_1, \ldots, i_r \\ j_1, \ldots, j_s \end{bmatrix}_q = \begin{cases} \frac{\prod_{k=1}^{r} (q)_{i_k}}{\prod_{k=1}^{s} (q)_{j_k}} & \forall i_k, j_k \in \mathbb{Z}_{\geq 0}, \\ 0 & \text{otherwise.} \end{cases}$$

The following result confirms the conjecture stated in Ref. 6, eq.(4.2).

Theorem 4.1. *Under (5), (7) and (8), the following formulas are valid:*

$$S_{ijk}^{abc} = (-\alpha_1\beta_1 q^{-2})^j \mu_1^{a-j+k} \mu_2^{b-a-k} \sigma_1^{b+j} \mathcal{S}_{ijk}^{abc},$$

$$\mathcal{S}_{ijk}^{abc} = \delta_{i+j}^{a+b} \delta_{j+k}^{b+c} \sum_{\lambda+\mu=b} (-1)^\lambda q^{2i(c-j)+2(k+1)\lambda+2\mu(\mu-k)} \begin{bmatrix} i, j, c+\mu \\ \mu, \lambda, i-\mu, j-\lambda, c \end{bmatrix}_{q^4}.$$

This $\mathcal{S}^{abc}_{ijk}$ is equal to $\mathcal{R}^{abc}_{ijk}$ in Ref. 6, eq.(2.20) with q replaced by q^2. Our notation here is taken consistently with Ref. 6, eq.(4.2). This $\mathcal{S}$ coincides with the 3D R matrix[11] which follows as the intertwiner of $A_{q^2}(\mathrm{SL}_3)$-modules.[10] See Ref. 6, sec. 2 for an exposition. It satisfies the tetrahedron equation

$$\mathcal{S}_{356}\mathcal{S}_{246}\mathcal{S}_{145}\mathcal{S}_{123} = \mathcal{S}_{123}\mathcal{S}_{145}\mathcal{S}_{246}\mathcal{S}_{356} \tag{9}$$

and $\mathcal{S} = \mathcal{S}^{-1}$, $\mathcal{S}_{ijk} = \mathcal{S}_{kji}$. Let us turn to $\pi_{2323} \simeq \pi_{3232}$ in (6). It holds if and only if

$$\kappa_{31} = \kappa_{32} = \pm 1, \quad \alpha_2\beta_2 = \pm q^2. \tag{10}$$

The constraints (5), (7), (8) and (10) are summarized as

$$\sigma = \kappa_i = \sigma_i, \;\; \rho = \kappa_{3i}, \;\; \alpha_i\beta_i = -\varepsilon q^2, \;\; \mu_i\nu_i = \varepsilon \quad (i = 1, 2)$$

in terms of the three independent sign factors $\sigma, \rho, \varepsilon \in \{1, -1\}$.

Let $\Xi : \mathcal{F}_{q^2} \otimes \mathcal{F}_q \otimes \otimes \mathcal{F}_{q^2} \otimes \mathcal{F}_q \to \mathcal{F}_q \otimes \otimes \mathcal{F}_{q^2} \otimes \mathcal{F}_q \otimes \mathcal{F}_{q^2}$ be the intertwiner characterized by

$$\pi_{3232}(\Delta(f)) \circ \Xi = \Xi \circ \pi_{2323}(\Delta(f)) \quad (\forall f \in A_q(B_3)) \tag{11}$$

and the normalization $\Xi(|0\rangle \otimes |0\rangle \otimes |0\rangle \otimes |0\rangle) = |0\rangle \otimes |0\rangle \otimes |0\rangle \otimes |0\rangle$. Introduce further $J = \Xi P_{14}P_{23} \in \mathrm{End}(\mathcal{F}_q \otimes \otimes \mathcal{F}_{q^2} \otimes \mathcal{F}_q \otimes \mathcal{F}_{q^2})$, where $P_{14}P_{23}(x \otimes y \otimes z \otimes w) = w \otimes z \otimes y \otimes x$. The J is regarded as a matrix $J = (J^{abcd}_{ijkl})$ whose elements are specified by

$$J(|i\rangle \otimes |j\rangle \otimes |k\rangle \otimes |l\rangle) = \sum_{a,b,c,d \in \mathbb{Z}_{\geq 0}} J^{abcd}_{ijkl} |a\rangle \otimes |b\rangle \otimes |c\rangle \otimes |d\rangle.$$

Setting $J^{abcd}_{ijkl} = \varepsilon^{b+i+l}\rho^{b+c}(\sigma\mu_2)^{c-k}\mu_3^{b-j}\mathcal{J}^{abcd}_{ijkl}$, it can easily be checked that $\mathcal{J}^{abcd}_{ijkl}$ depends only on q. In this sense we call $\mathcal{J} = (\mathcal{J}^{abcd}_{ijkl})$ the parameter-free part of the intertwiner for $A_q(B_3)$.

Let $\mathcal{K} = (\mathcal{K}^{abcd}_{ijkl})$ be the parameter-free part in the similar sense of the intertwiner for the quantized function algebra $A_q(C_3)$. The elements $\mathcal{K}^{abcd}_{ijkl}$ are polynomials in q and given explicitly in Ref. 6, Th. 3.4. The $\mathcal{J}$ and $\mathcal{K}$ will simply be referred to as type B and type C, respectively. Now we present the main result of the paper, which confirms the conjecture in Ref. 6, eq.(4.1).

Theorem 4.2. *The type B and C intertwiners are related by the transposition of the components as $\mathcal{J} = P_{14}P_{23}\mathcal{K}P_{14}P_{23}$, i.e. $\mathcal{J}^{abcd}_{ijkl} = \mathcal{K}^{dcba}_{lkji}$ holds.*

Proof. The $\mathcal{J}$ and $\mathcal{K}$ are characterized as the intertwiners of the representations of the subalgebras $A_q(B_2)$ and $A_q(C_2)$, respectively. Thus the assertion follows from Theorem 2.1 and Remark 3.1. □

As a corollary of Ref. 6, Th. 3.5, $\mathcal{J}^{abcd}_{ijkl}$ is a polynomial in q^2 vanishing unless $(a+2b+c, b+c+d) = (i+2j+k, j+k+l)$. Moreover $\mathcal{J}$ has the birational and combinatorial counterparts as shown in Ref. 6, Table 1.

Example 4.1. The following is the list of all the nonzero $\mathcal{J}^{1102}_{ijkl}$.

$$\mathcal{J}^{1102}_{3003} = q^8(1-q^6)(1-q^{12}), \qquad \mathcal{J}^{1102}_{2012} = q^4(1-q^4)(1-q^2+q^4-q^6-q^{10}),$$
$$\mathcal{J}^{1102}_{1021} = -q^6(1-q^6), \qquad \mathcal{J}^{1102}_{0030} = -q^2(1-q^6),$$
$$\mathcal{J}^{1102}_{1102} = q^2(1-q^8+q^{14}), \qquad \mathcal{J}^{1102}_{0111} = 1-q^4+q^{10}.$$

5. 3D reflection equation

The intertwiners $\mathcal{S}$ and $\mathcal{J}$ yield a solution of the 3D reflection equation proposed by Isaev and Kulish.[4] Since its derivation is the same as the type C case in Ref. 6, sec. 3.6, we shall only describe the outline.

Let w_0 be the longest element of the Weyl group $W(B_3)$ and consider the two reduced expressions $w_0 = s_1s_2s_3s_2s_1s_2s_3s_2s_3 = s_3s_2s_3s_2s_1s_2s_3s_2s_1$. According to Ref. 7, Th. 5.7, Cor. 1, we have the equivalence $\pi_{123212323} \simeq \pi_{323212321}$ of the representations of $A_q(B_3)$. The intertwiner for this can be constructed by composing the 'basic ones' P, Φ and Ξ for (6). Moreover there are two ways to do so, the results of which must coincide since $\pi_{123212323} \simeq \pi_{323212321}$ is irreducible. This postulate leads to the equality

$$\mathcal{S}_{456}\mathcal{S}_{489}\mathcal{J}_{3579}\mathcal{S}_{269}\mathcal{S}_{258}\mathcal{J}_{1678}\mathcal{J}_{1234} = \mathcal{J}_{1234}\mathcal{J}_{1678}\mathcal{S}_{258}\mathcal{S}_{269}\mathcal{J}_{3579}\mathcal{S}_{489}\mathcal{S}_{456} \qquad (12)$$

in $\mathrm{End}(\mathcal{F}_q \otimes \mathcal{F}_{q^2} \otimes \mathcal{F}_q \otimes \mathcal{F}_{q^2} \otimes \mathcal{F}_{q^2} \otimes \mathcal{F}_{q^2} \otimes \mathcal{F}_q \otimes \mathcal{F}_{q^2} \otimes \mathcal{F}_{q^2})$, where the indices signify the positions of the tensor components on which the intertwiners act nontrivially. The equation (12) is the 3D reflection (or "tetrahedron reflection") equation introduced in Ref. 4 without a spectral parameter. In view of Theorem 4.2, one may substitute $\mathcal{J}_{ijkl} = \mathcal{K}_{lkji}$. The resulting relation confirms the conjecture stated in Ref. 6, eq.(4.3).

A pair $(\mathcal{S}, \mathcal{J})$ is called a solution to the 3D reflection equation if it satisfies (12) and $\mathcal{S}$ satisfies the tetrahedron equation (9) by itself. Thus we have established a new solution of the 3D reflection equation distinct from the previous one $(\mathcal{R}, \mathcal{K})$ in Ref. 6. The two solutions $(\mathcal{S}, \mathcal{J})$ and $(\mathcal{R}, \mathcal{K})$ are associated with $A_q(B_3)$ and $A_q(C_3)$, respectively. Although they are simply related as $(\mathcal{S}, \mathcal{J}) = (\mathcal{R}|_{q\to q^2}, P_{14}P_{23}\mathcal{K}P_{14}P_{23})$, the fact that these replacements do keep (12) is highly nontrivial. The main achievement of this paper is to have established it by exploiting the embedding in Theorem 2.1.

Let us finish by giving a diagram for the 3D reflection equation

$$\mathcal{S}_{489}\mathcal{J}_{3579}\mathcal{S}_{269}\mathcal{S}_{258}\mathcal{J}_{1678}\mathcal{J}_{1234}\mathcal{S}_{654} = \mathcal{S}_{654}\mathcal{J}_{1234}\mathcal{J}_{1678}\mathcal{S}_{258}\mathcal{S}_{269}\mathcal{J}_{3579}\mathcal{S}_{489}. \qquad (13)$$

This is $\mathcal{S}_{456}^{-1} \times \text{eq.}(12) \times \mathcal{S}_{456}$ with the properties mentioned after (9) applied. It is depicted in Fig.1.

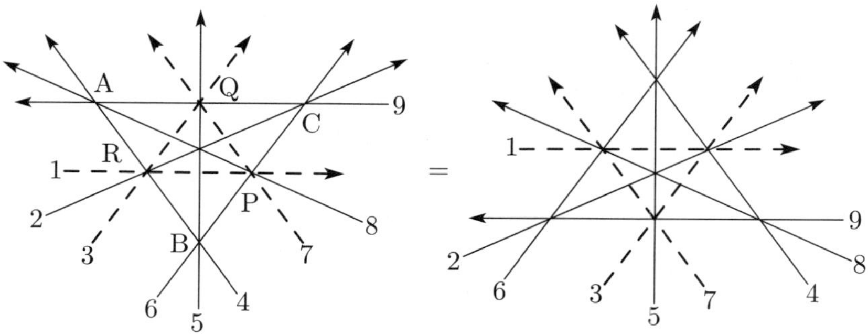

Fig. 1. The 2D projection diagram of the 3D reflection equation (13).

To draw the LHS, first put three lines 8, 5 and 2 that intersect at one point, the center. Take generic points A, B and C on every second half-infinite line emanating from the center along 8, 5 and 2. The points P, Q and R are the crossings of the triangle ABC with the lines 8, 5 and 2. The broken lines are formed by connecting P, Q and R. The RHS is obtained by changing A, B and C in such a way that the lines 6, 9 and 4 are shifted and the triangle ABC is reversed.

There is a natural geometric interpretation of Fig.1. Let us call the plane containing all the lines the *boundary plane*. Fig. 1 shows the projection of the three planes reflected by the boundary plane. A physical interpretation is the world sheets of three straight strings in 3D exhibiting boundary reflections.

The intersections of the three and four arrows correspond to $\mathcal{S}$ and $\mathcal{J}$ in (13), respectively. The indices 1, 3 and 7 are attached only to $\mathcal{J}$, hence represent the boundary degrees of freedom. In Fig.1 they are depicted by the broken lines. The other solid lines are the projection of the configurations (or events) in 3D explained in what follows onto the boundary plane.

To each broken line, associate a pair of half-planes in 3D sharing the broken line as the common boundary. They represent the world sheet of a straight string moving in 3D and reflected by the boundary plane exactly at the broken line. The six in total half-planes should be located on the same side of the boundary plane. A broken line is like a spine of an open (by a generic angle) book and the half-planes are like the front and back of the book. The books should be "upright" on the boundary plane since the incident and reflection angles coincide. There are three such books with spines $1,3$ and 7. At the intersection R of the spines 1 and 3, the fronts and backs of the two books generate the four intersecting half-lines on them. Their projections are the solid lines 2 and 4. The crossing of these four lines $1,2,3,4$ yields $\mathcal{J}_{1234}$. The other $\mathcal{J}$ can be understood similarly. The intersections of three planes without a spine take place off the boundary plane and correspond to three string scatterings encoded by $\mathcal{S}$. By an elementary geometry one can prove that the 2D projection of the 3D configuration explained so far exactly forms the pattern depicted in Fig.1. Finally the orientations of the lines indicated by the arrows signify the time ordering of the scattering and reflection events. They specify the order of the products of various $\mathcal{S}$ and $\mathcal{J}$ in (13).

Acknowledgments

The authors thank the organizers of The XXIX International Colloquium on Group-Theoretical Methods in Physics, August 20-26, 2012 at Chern Institute of Mathematics, Tianjin, China for invitation and warm hospitality. This work is supported by Grants-in-Aid for Scientific Research No. 23340007, No. 24540203 and No. 23654007 from JSPS.

References

1. I. V. Cherednik, *Theor. Math. Phys.* **61**, 977 (1984).
2. E. K. Sklyanin, *J. Phys. A: Math. Gen.* **21**, 2375 (1988).
3. R. J. Baxter, *Exactly solved models in statistical mechanics* (Dover, 2007).
4. A. P. Isaev and P. P. Kulish, *Mod. Phys. Lett. A* **12**, 427 (1997).
5. A. B. Zamolodchikov, *Soviet Phys. JETP* **52**, 325 (1980).
6. A. Kuniba and M. Okado, *J. Phys. A: Math. Theor.* **45** 465206, 27pp (2012).
7. Y. S. Soibelman, *Leningrad Math. J.* **2**, 161 (1991).
8. V. G. Drinfeld, Quantum groups, in *Proceedings of the International Congress of Mathematicians*, pp798–820 (Berkeley, 1986).
9. N. Yu. Reshetikhin, L. A. Takhtadzhyan and L. D. Faddeev, *Leningrad Math. J.* **1**, 193 (1990).
10. M. M. Kapranov and V. A. Voevodsky, *Proc. Symposia in Pure Math.* **56**, 177 (1994).
11. V. V. Bazhanov and S. M. Sergeev, *J. Phys. A: Math. Theor.* **39**, 3295 (2006).
12. V. V. Bazhanov, *Phys. Lett. B* **159**, 321 (1985).
13. M. Jimbo, *Commun. Math. Phys.* **102**, 537 (1986).
14. M. Kashiwara, *Duke Math. J.* **69**, 455 (1993).

DISUCCESSORS AND DUPLICATORS OF OPERADS, MANIN PRODUCTS AND OPERATORS

JUN PEI

Department of Mathematics, Lanzhou University,
Lanzhou, Gansu 730000, China
**E-mail: peitsun@163.com*

CHENGMING BAI

Chern Institute of Mathematics & LPMC, Nankai University,
Tianjin 300071, China
**E-mail: baicm@nankai.edu.cn*

LI GUO

Department of Mathematics and Computer Science, Rutgers University,
Newark, New Jersey 07102, USA
**E-mail: liguo@rutgers.edu*

XIANG NI

Department of Mathematics, Caltech,
Pasadena, California 91125, USA
**E-mail: xni@caltech.edu*

We briefly summarize recent constructions of the disuccessor and duplicator of a quadratic operad and relate them to the Manin products, and Rota-Baxter and average operators.

Keywords: Operad, successor, duplicator, Manin products, Rota-Baxter operator, average operator.

1. Introduction

Several algebraic structures are important in physics related studies. The group, associative algebra and Lie algebra are some of them. A less well-known one is the pre-Lie algebra[3] which gives the Lie algebra by anti-symmetrization as in the case of the associative algebra. There are other structures that also have this property and it is natural to ask what made the pre-Lie algebra stand out. Another algebraic structure that recently found physical application is the dendriform algebra[8] with two binary operations whose sum is associative. Thus the two operations can be regarded as a "splitting" of the associative operation. Many structures appeared in recent years (see[12]) with similar "splitting" property, such as the tridendriform, quadri, ennea and octo algebras for the associative algebra, the PostLie and L-dendriform algebras, in addition to the pre-Lie algebra, for the Lie algebra, and the Zinbiel algebra for the commutative algebra. In the opposite direction, there were also new structures that doubled or tripled some known structures, such as the diassociative and triassociative algebras for the associative algebra, the Leibniz algebra for the Lie algebra and the permutative (or commutative diassociative) algebra for the commutative algebra. Each of these structures was discovered individually, often in quite complicated manner, with their connections found only afterwards. For instance, they have some relationship with the Manin products of operads and the action of a Rota-Baxter operator or an average operator. Thus it is desirable that general procedures be developed to connect the dots and put these constructions in a uniform framework.

In[1,10] we conceptualized the phenomena of "splitting" and "doubling" by the notions of successors (namely the disuccessor and trisuccessor) and replicators (namely the duplicator and triplicator) of a binary operad. This paper is a survey of these results. For the limit of space, only the disuccessor and duplicator are discussed. Further details can be found in the original papers and the references therein (in particular[7]).

In Section 2, the concepts of the disuccessor and duplicator are introduced, together with some examples of the two constructions. The duality between the disuccessor and the duplicator, and the relationship of the disuccessor and duplicator with the Manin products and linear operators are given in Section 3.

2. Successors and duplicators of operads

To fix notations we start with giving some background on operads. For further details see.[9] Then we will give the definition of a successor (resp. duplicator) and conclude this section with some examples about the two constructions of operads.

An **$\mathbb{S}$-module** over $\mathbf{k}$ is a family $\mathcal{P} := \{\mathcal{P}(n)\}_{n\geq 0}$ of $\mathbf{k}[\mathbb{S}_n]$-modules $\mathcal{P}(n)$. An **(symmetric) operad** is an $\mathbb{S}$-module $\mathcal{P}$ equipped with **partial compositions**:

$$\circ_i := \circ_{m,n,i} : \mathcal{P}(m) \otimes \mathcal{P}(n) \longrightarrow \mathcal{P}(m+n-1), 1 \leq i \leq m,$$

such that, for $\lambda \in \mathcal{P}(\ell), \mu \in \mathcal{P}(m)$, $\nu \in \mathcal{P}(n)$, $\sigma \in \mathbb{S}_m$ and $\tau \in \mathbb{S}_n$,

$$\lambda \circ_i (\mu \circ_j \nu) = (\lambda \circ_i \mu) \circ_{i+j-1} \nu, \quad 1 \leq i \leq \ell, 1 \leq j \leq m, \tag{1}$$

$$(\lambda \circ_k \nu) \circ_i \mu = (\lambda \circ_i \mu) \circ_{k+m-1} \nu, 1 \leq i \leq k \leq \ell, \tag{2}$$

$$\mu^\sigma \circ_i \nu = (\mu \circ_{\sigma(i)} \nu)^{\sigma'} \text{ for a well-defined } \sigma' \in \mathbb{S}_{m+n}, \tag{3}$$

$$\mu \circ_i \nu^\sigma = (\mu \circ \nu)^{\sigma''} \text{ for a well-defined } \sigma'' \in \mathbb{S}_{m+n}. \tag{4}$$

These conditions are modeled after the properties of compositions of multilinear maps with permutations of the arguments. When the $\mathbb{S}$-module in the definition of operad is replaced by a graded **k**-vector space $\mathcal{P} = \{\mathcal{P}_n\}_{n\geq 0}$ and the partial composition only satisfies (1) and (2), we call $\mathcal{P}$ a **non-symmetric operad**.

An operad is called **binary** if $\mathcal{P}(1) = \mathbf{k}\,\mathrm{id}$ and $\mathcal{P}(n)$, $n \geq 3$, are induced from $\mathcal{P}(2)$ by compositions. Let V be an $\mathbb{S}_2$-module, regarded as an $\mathbb{S}$-module concentrated at arity two. Let $\mathbb{T}$ denote the isomorphic classes of reduced binary rooted trees and with symmetric group action on the leafs. For such a tree t, let $\mathrm{V}in(\mathsf{t})$ denote the set of labels of the vertices of t and $\mathrm{I}n(v)$ the input set of a vertex v. Define

$$\mathsf{t}[V] := \bigotimes_{v \in \mathrm{V}in(\mathsf{t})} V(|\mathrm{I}n(v)|)\,.$$

The free operad $\mathcal{T}(V)$ on an $\mathbb{S}_2$-module $V = V(2)$ is given by the $\mathbb{S}$-module

$$\mathcal{T}(V) := \bigoplus_{\mathsf{t}\in\mathbb{T}} \mathsf{t}[V]\,.$$

An associative algebra can be presented as the quotient of a free algebra modulo some relations. Likewise, a binary operad can be presented as the quotient $\mathcal{P} = \mathcal{T}(V)/(R)$ of the free operad $\mathcal{T}(V)$ modulo some relations R. Then $\mathcal{P}$ is called **quadratic** if all relations R among the binary operations in V are derived from $\mathcal{T}(V)(3)$. Further $\mathcal{T}(V)(3)$ is isomorphic to $3V \otimes V$, represented by the following three types of trees ($\mu, \nu \in V$):

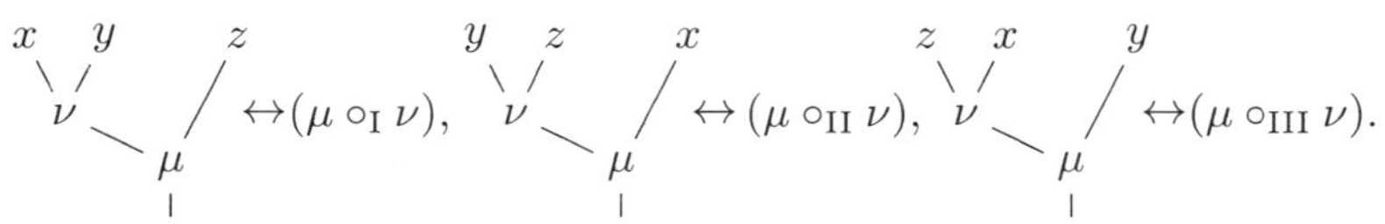

We first give the definition of the disuccessor (resp. duplicator) of a binary planar tree and then of a binary operad.

Definition 2.1. For $\tau \in \mathcal{T}(V)$ and a leaf $x \in \mathrm{L}in(\tau)$, the set of labels of the leaves of τ. The **disuccessor** $\mathrm{Su}_x(\tau)$ (resp. **duplicator** $\mathrm{Du}_x(\tau)$) of τ is defined by relabeling a vertex ω of $Vin(\tau)$ by

$$\begin{cases} \binom{\omega}{\prec}\ (resp.\ \binom{\omega}{\dashv}), \text{ the path from the root of } \tau \text{ to } x \text{ turns left at } \omega; \\ \binom{\omega}{\succ}\ (resp.\ \binom{\omega}{\vdash}), \text{ the path from the root of } \tau \text{ to } x \text{ turns right at } \omega; \\ \binom{\omega}{\star}\ (resp.\ \binom{\omega}{\dagger}), \ \omega \text{ is not on the path from the root of } \tau \text{ to } x, \end{cases}$$

where $\binom{\omega}{\star} := \left\{\binom{\omega}{\prec} + \binom{\omega}{\succ}\right\}$ (resp. $\binom{\omega}{\dagger} := \left\{\binom{\omega}{\dashv}, \binom{\omega}{\vdash}\right\}$).

Note that $\mathrm{Du}_x(\tau)$ might be a set of several labeled trees.

Definition 2.2. Let $\mathcal{P} = \mathcal{T}(V)/(R)$ be a binary operad with space of operations V and space of relations generated by

$$R := \left\{ r_s = \sum_i c_{s,i}\tau_{s,i} \;\middle|\; c_{s,i} \in \mathbf{k},\ \tau_{s,i} \in \mathcal{T}(V),\ 1 \leq s \leq k \right\}.$$

For $1 \leq s \leq k$ and $x \in Lin(\tau_{s,i})$, denote

$$\mathrm{Su}_x(r_s) := \sum_i c_{s,i}\mathrm{Su}_x(\tau_{s,i}),\ \ \mathrm{Du}_x(r_s) := \sum_i c_{s,i}\mathrm{Du}_x(\tau_{s,i}),$$

where, for nonempty subsets $W_i \subseteq \mathcal{T}(V)$ and $c_i \in \mathbf{k}, 1 \leq i \leq r$, denote

$$\sum_{i=1}^{r} c_i W_i := \left\{ \sum_{i=1}^{r} c_i w_i \;\middle|\; w_i \in W_i, 1 \leq i \leq r \right\}.$$

The **disuccessor** $\mathrm{Su}(\mathcal{P})$ and the **duplicator** $\mathrm{Du}(\mathcal{P})$ of $\mathcal{P}$ are respectively the binary operad defined by

$$\mathrm{Su}(\mathcal{P}) := \mathcal{T}(\mathrm{Su}(V))/(\mathrm{Su}(R)), \quad \mathrm{Du}(\mathcal{P}) := \mathcal{T}(\mathrm{Du}(V))/(\mathrm{Du}(R)),$$

where $\mathrm{Su}(V) = V \otimes (\mathbf{k} \prec \oplus\, \mathbf{k} \succ)$, $\mathrm{Du}(V) = V \otimes (\mathbf{k} \dashv \oplus\, \mathbf{k} \vdash)$ with

$\binom{\omega}{\dashv}^{(12)} := \binom{\omega^{(12)}}{\vdash}$, $\binom{\omega}{\vdash}^{(12)} := \binom{\omega^{(12)}}{\dashv}$, $\binom{\omega}{\prec}^{(12)} := \binom{\omega^{(12)}}{\succ}$, $\binom{\omega}{\succ}^{(12)} := \binom{\omega^{(12)}}{\prec}$,

for all $\omega \in V$, as the $\mathbb{S}_2$-actions and

$$\mathrm{Su}(R) := \{\mathrm{Su}_x(r_s) \mid x \in Lin(\tau_{s,i}), 1 \leq s \leq k\},$$
$$\mathrm{Du}(R) := \bigcup_{s=1}^{k} \left(\bigcup_{x \in Lin(\tau_{s,i})} \mathrm{Du}_x(r_s) \right).$$

The notion of the disuccessor $\mathrm{Su}(\mathcal{P})$ of $\mathcal{P}$ makes precise the process of splitting an operad into two pieces, while the notion of the duplicator $\mathrm{Du}(\mathcal{P})$ makes precise the process of duplicating an operad into two pieces. The above definitions are also valid for non-symmetric operads.

Example 2.1. If we take $\mathcal{P}$ to be the non-symmetric operad As for the associative algebra, then we have $\mathrm{Su}(As) = Dend$ and $\mathrm{Du}(As) = Dias$, where $Dend$ and $Dias$ are the operads for the dendriform algebra and diassociative algebra respectively.

Example 2.2. If we take $\mathcal{P}$ to be the operad Lie (resp. $Comm$, resp. $Pois$) for the Lie algebra (resp. commutative associative algebra, resp. Poisson algebra), then we have

$$\mathrm{Su}(Lie) = PreLie, \quad \mathrm{Su}(Comm) = Zinb, \quad \mathrm{Su}(Pois) = PrePois,$$
$$\mathrm{Du}(Lie) = Leib \quad \mathrm{Du}(Comm) = Perm, \quad \mathrm{Du}(Pois) = dualprePois.$$

where $PreLie, Zinb, prePois, Leib, Perm, dualprePois$ are the operads for the pre-Lie algebra, Zinbiel algebra, pre-Poisson algebra, Leibniz algebra, permutative algebra, dualprePoisson algebra respectively.

More information about these operads can be found in.[12] As these examples show, the two constructions provide a general framework to study the well-known di-type (resp. dendriform-type) algebras which are analogues of a diassociative algebra (resp. dendriform algebra) structure associated to an associative algebra. For example, it gives another reasonable explanation of the definition of pre-Lie dialgebra.[5] More importantly, it provides a general recipe to construct new di-type (resp. dendriform-type) algebraic structures associated to any binary operads.

3. Duality, products and operators of operads

The similarity between the definitions of the disuccessor and the duplicator suggests that there is a close relationship between the two constructions. This is indeed the case. For binary quadratic operads $\mathcal{P}$ and $\mathcal{Q}$, let $\mathcal{P}^!$ denote the Koszul dual of $\mathcal{P}$ and let $\mathcal{P} \bullet \mathcal{Q}$ (resp. $\mathcal{P} \bigcirc \mathcal{Q}$) denote the Manin black (resp. white) product.[6,9]

Theorem 3.1.

(1) Let $\mathcal{P} = \mathcal{T}(V)/(R)$ be a binary quadratic operad with $R \neq 0$. Then $\mathrm{Du}(\mathcal{P})^! = \mathrm{Su}(\mathcal{P}^!)$.

(2) Let $\mathcal{P}, \mathcal{Q}$ be binary quadratic operads. Then $\mathrm{Su}(\mathcal{P}) \cong PreLie \bullet \mathcal{P}, \quad \mathrm{Du}(\mathcal{P}) \cong Perm \bigcirc \mathcal{P}$.

The analogue of Theorem 3.1.(2) for nonsymmetric operads can be found in.[4,11] Theorem 3.1 provides a general method to prove the duality for certain operads. For example, $\mathrm{Du}(Leib)$ is Koszul dual to $\mathrm{Su}(Zinb)$

which is the commutative quadri-algebra, and Du($Perm$) is Koszul dual to Su($preLie$) which is the L-dendriform algebra.[2] Theorem 3.1 also provides convenient ways to compute Manin black and while products of $PreLie$ and $Perm$ with other operads.

We also define the notion $\mathrm{RB}_0(\mathcal{P})$ (resp. $\mathrm{DA}(\mathcal{P})$) of the action of the Rota-Baxter (resp. average) operator of weight 0 on a binary operad $\mathcal{P}$ and show that there are morphisms of binary quadratic operads

$$PreLie \bullet \mathcal{P} \cong \mathrm{Su}(\mathcal{P}) \to \mathrm{RB}_0(\mathcal{P}), \;\; Perm \bigcirc \mathcal{P} \cong \mathrm{Du}(\mathcal{P}) \to \mathrm{DA}(\mathcal{P}).$$

This conceptualizes the phenomenon that a Rota-Baxter (resp. average) operator on an algebraic structure gives "splitting" (resp. "doubling") of the operations.

Acknowledgements

C. Bai thanks the support by NSFC (10920161, 11271202). L. Guo acknowledges support from NSF grant DMS 1001855.

References

1. C. Bai, O. Bellier, L. Guo and X. Ni, Spliting of operations, Manin products and Rota-Baxter operators, *IMRN* (2012). DOI: 10.1093/imrn/rnr266.
2. C. Bai, L. Liu and X. Ni, Some results on L-dendriform algebras, *J. Geom. Phys.* **60** (2010) 940-950.
3. D. Burde, Left-symmetric algebras, or pre-Lie algebras in geometry and physics, *Cent. Eur. J. Math.* **4** (2006) 323-357.
4. K. Ebrahimi-Fard and L. Guo, On products and duality of binary, quadratic, regular operads, *J. Pure. Appl. Algebra,* **200**, (2005), 293-317.
5. R. Felipe, A brief foundation of the left-symmetric dialgebras, *Comunicación del CIMAT,* NO I, (2011).
6. V. Ginzburg and M. M. Kapranov, Koszul duality for operads, *Duke Math. J.* **76** (1995) 203-272.
7. V. Yu. Gubarev and P. S. Kolesnikov, On embedding of dendriform algebras into Rota-Baxter algebras, *Cent. Eur. Jour. Math* **11** (2013), 226-245.
8. J.-L. Loday, Dialgebras, in Dialgebras and related operads, Lecture Notes in Math. 1763 (2002) 7-66.
9. J.-L. Loday, B. Vallette, Algebraic operads, *Grundlehren Math. Wiss.* **346**, Springer, Heidelberg, 2012.
10. J. Pei, C. Bai, L. Guo and X. Ni, Replicating of binary operads, Koszul duality, Manin products and average operators, arXiv:1212.0177v2.
11. B. Vallette, Manin products, Koszul duality, Loday algebras and Deligne conjecture, *J. Reine Angew. Math.* **620** (2008) 105-164.
12. G.W. Zinbiel, Encyclopedia of types of algebras 2010, in: "Operads and Universal Algebra", *Nankai Series in Pure, Applied Mathematics and Theoretical Physics* **9**, 217-296, World Scientific, Singapore, 2011.

NOVEL TENSOR NETWORK RENORMALIZATION GROUP APPROACHES FOR QUANTUM LATTICE MODELS

GANG SU*

Theoretical Condensed Matter Physics and Computational Materials Physics Laboratory, School of Physics, University of Chinese Academy of Sciences, Beijing 100049, China
**E-mail: gsu@ucas.ac.cn*
tcmp2.ucas.ac.cn

Two novel tensor network-based algorithms, say, the linearized tensor renormalization group and the optimized decimation of tensor network with super-orthogonalization methods, as well as their applications to quantum spin lattice models, are discussed. Both methods can be applied to explore not only the thermodynamic but also the ground state properties of low-dimensional quantum lattice systems.

Keywords: tensor network; renormalization group; modified Tucker decomposition; quantum lattice models.

1. Introduction

It is known that the strongly correlated quantum spin and electron systems are very fundamental for understanding the magnetism and superconductivity in condensed matter and materials physics. The efficient and accurate numerical methods are of great importance for these systems from both theoretical and experimental aspects. The powerful quantum Monte Carlo (QMC) method suffers from a notorious negative sign problem that hinders applications to the two-dimensional (2D) Hubbard model away from the half-filling and frustrated quantum spin models; the exact diagonalization with Lanczos, Jacobi-Davidson algorithms, etc., are always subjected to the constraint of computer's capacity; the density matrix renormalization group-based methods are powerful for one-dimensional (1D) quantum correlated electron and spin lattice models but they are still under progress for the 2D quantum systems; the recently developed tensor renormalization group approaches are promising and achieve successes for the ground state properties of some 2D quantum lattice models, but they are difficult for studying the thermodynamic properties due to low efficiency and accuracy. Developing new methods beyond them is a challenge.

Recently, we developed two methods, the linearized tensor renormalization group (LTRG)[1] and the optimized decimation of tensor networks with super-orthogonalization (ODTNS)[2] algorithms, that can be applicable to explore with better accuracy and efficiency not only the thermodynamic but also ground state properties of low-dimensional quantum lattice systems. The LTRG and ODTNS methods are successfully applied to several 1D and 2D quantum spin models, and very good results are obtained.

2. Linearized Tensor Renormalization Group

Our purpose is to develop a novel algorithm for calculating the thermodynamic properties of low-dimensional quantum lattice models. The strategy we adopted is the following:[1] First, one can transform the D-dimensional quantum lattice model to a (D + 1)-dimensional classical tensor network by means of the Trotter-Suzuki decomposition; and then, decimate linearly the tensor network first along the Trotter direction to reach a given temperature and then contract the tensor network along the lattice site direction to obtain the partition function and thus the thermodynamics of the original quantum many-body system can be explored. This algorithm is coined as the LTRG.

We take the 1D quantum many-body lattice system as an example. The Hamiltonian can be written as

$$H = \sum_{i=1}^{N} h_{i,i+1} = H_1 + H_2,$$
$$H_1 = \sum_{i=1}^{N/2} h_{2i-1,2i}, \; H_2 = \sum_{i=1}^{N/2} h_{2i,2i+1}, \tag{1}$$

where N (even) is the number of sites. By inserting $2K$ (large K) complete sets of states $\{|\sigma_i^j\rangle\}(\sigma_i^j = 1, \cdots, D)$ with i the site index and j the Trotter index, the partition function of this model can be denoted as

$$Z_N \simeq Tr[e^{-\beta H_1/K} e^{-\beta H_2/K}]^K$$
$$= \sum_{\{\sigma_i^j\}} \prod_{i=1}^{N/2} \prod_{j=1}^{K} v_{\sigma_{2i-1}^{2j-1}\sigma_{2i}^{2j-1},\sigma_{2i-1}^{2j}\sigma_{2i}^{2j}} v_{\sigma_{2i}^{2j}\sigma_{2i+1}^{2j},\sigma_{2i}^{2j+1}\sigma_{2i+1}^{2j+1}}, \tag{2}$$

where $v_{\sigma_1\sigma_4,\sigma_2\sigma_3} \equiv \langle\sigma_1\sigma_4|\exp(-\beta h_{i,i+1}/K)|\sigma_2\sigma_3\rangle$ is the transfer matrix (a

4th-order tensor), and β is the inverse temperature. The partition function Z_N can be represented graphically as a classical tensor network.

To obtain Z_N, we need to sum over all intermediate states $|\sigma_i^j\rangle$ or contract all bonds σ in the tensor network. To do so, we first make a local transformation by making use of the singular value decomposition (SVD) on tensor v, which transforms the original square tensor network into a honeycomb network, and then contract the bonds between the bottom lines, which gives rise to a brick-wall structure of the tensor network. Subsequently, we project layer by layer the brick-wall structure of transfer-matrix tensor network until the fixed temperature is reached. Note that during this process every two projections accomplish a complete Trotter step, and after the projection, the density matrix of the present system is represented in the form of matrix product operator (MPO). By tracing the physical indices the remaining 1D tensor network can be evaluated by utilizing the coarse-graining procedure. To avoid divergence during the renormalization group process, we need to normalize the tensors. Two normalization factors n_i and m_j are involved, where n_i comes from the linear decimation of 2D classical tensor network, and m_j is from the coarse-graining procedure of 1D MPO. Consequently, the free energy per site f at inverse temperature $\beta = K\tau$ can be determined by

$$f = -\frac{1}{K\tau}\left(\sum_{i=1}^{2K-2}\frac{\ln n_i}{2} + \sum_{j=1}^{p}\frac{\ln m_j}{2^j}\right), \tag{3}$$

where p is the coarse-graining steps. The other thermodynamic quantities can be obtained from the free energy.

To examine the accuracy of the LTRG method, we take the exactly solvable model — spin-1/2 XY chain as a benchmark. The LTRG result on this model gives a very remarkable agreement with the exact solution. For instance, at $\beta = 120$ and $D_c = 150$, the energy per site approaches the exact solution $-1/\pi$. The linear behavior of the specific heat at extremely low temperature is also clearly shown by LTRG calculations. Besides, the error curves of LTRG and transfer matrix renormalization group (TMRG) calculations coincide at high temperatures, with relative errors about 10^{-5}, where the errors are contributed mainly from the Trotter-Suzuki decomposition. For low temperature, the LTRG can achieve comparable or even better accuracy than TMRG.

The LTRG method is successfully applied to the spin-1 bond-alternating Heisenberg aniferromagnetic (AF) chain, which gives a very nice fitting to the experimental data of susceptibility and magnetic curves of model material $Ni(C_2H_{24}N_4)NO_2(ClO_4)$ (NTENP). In particular, we identified that, at extreme low temperatures, this material shows clearly a Tomonaga-Luttinger liquid behavior in a longitudinal magnetic field, in consistent with the experimental observation, while it reveals a gapful excitation in a transverse field.[3] Note that we adopted the TMRG method to study this issue a few years ago and failed to access to such low temperatures.[4] In addition, we also utilized the LTRG method to explore the 1D extended quantum compass model, and got very accurate results of the specific heat even at extremely low temperatures.[5] So far, the LTRG approach is quite efficient and accurate for 1D and quasi-1D quantum spin systems, and its computational cost is lower than other numerical methods, and it can treat the case with spin frustrations. The extensive application of LTRG method to quantum spin systems, boson models and so on is emerging.

3. Optimized Decimation of Tensor Network with Super-orthogonalization

For 2D quantum lattice systems, we developed the ODTNS algorithm. The primary strategy is:[2] we first represent the finite temperature density operator of the 2D quantum lattice system by a three-dimensional (3D) closed tensor network comprised of the tensor product density operator (TPDO) and the 3D brick-wall tensor network. By linearly contracting the brick-wall tensor network with the corresponding imaginary time length to get the TPDO, the finite temperature properties of the 2D quantum systems can hence be obtained. In this course, we propose the so-called network Tucker decomposition (NTD) and transform the tensor network into a super-orthogonal form to get an optimal lower-dimensional approximation for the bond space in terms of the robust network singular value spectrum. The efficiency and accuracy of the ODTNS scheme are checked satisfactorily with 2D unfrustrated as well as frustrated quantum Heisenberg spin systems.

Let us take the quantum spin system on a honeycomb lattice as an example. The Hamiltonian of the system reads $H = \sum_{i,j} \hat{H}_{ij}$ with $\hat{H}_{ij}$ the local Hamiltonian of pairs of spins on site i and j. By means of the Trotter-Suzuki decomposition, the density matrix can be denoted as $\rho \simeq [\exp(-\tau \sum_{i,j} \hat{H}_{ij})]^{K+1} \simeq [\prod_{i,j} \hat{U}_{ij}]^{K+1} = \prod_{t=1}^{K+1} \prod_{i,j} \hat{U}^t_{ij}$, where $\hat{U}_{ij} = \exp(-\tau \hat{H}_{ij})$ is a local evolution operator, $\beta = (K+1)\tau$ with τ the infinitesimal imaginary time slice, and t is the Trotter index. The density matrix can thus be represented by a tensor network. By performing a SVD on the local evolution operator $\hat{U}$, it becomes $U^{ij}_{i'j'} = \sum_s G^L_{ii',s} \lambda^0_s G^R_{jj',s}$, where G^L and G^R are two local evolution tensors, and λ^0 is the singular value vector. By contracting the shared bonds among G^L and G^R, we can represent the density operator ρ by

$$\rho_{\cdots ii'jj'\cdots} = Tr_G(\cdots \lambda_y^{II}\lambda_z^{III}T^L_{ii',xyz}\lambda_x^I T^R_{jj',xy'z'}\lambda_{y'}^{II}\lambda_{z'}^{III}\cdots), \tag{4}$$

where Tr_G stands for the trace over all contracted geometrical bonds, and λ^I, λ^{II}, λ^{III} are three inequivalent singular value vectors with the initial value λ^0, and

$$T^L_{il,xyz} = \sum_{jk} G^L_{ij,x}G^L_{jk,y}G^L_{kl,z}, \; T^R_{il,xyz} = \sum_{jk} G^R_{ij,x}G^R_{jk,y}G^R_{kl,z}, \tag{5}$$

where x, y and z denote three inequivalent bonds on a honeycomb lattice. It is seen that the density operator is represented by the TPDO that consists of an infinite product of two inequivalent tensors T^L and T^R for two sublattices of the honeycomb lattice as well as λ^I, λ^{II} and λ^{III} for three inequivalent bonds. The main computational cost is around $O(tzD_g^3)$, where t is the transformation steps, z is the coordination number and D_g is the dimension of the geometrical bond. The cost related to the coordination number is much lower.

The free energy per site of the quantum system can be obtained by

$$f(\beta) = \frac{1}{2\beta}(\sum_{q=1}^{K}\sum_{S=I,II,III} \ln r_q^S + 2\ln \bar{r}), \tag{6}$$

where $r_q^S = \sqrt{\sum_g \lambda_g^S}$ is the normalization factor with q the Trotter step, and $\bar{r}$ is the factor obtained when contracting the TPDO. The thermodynamic properties of the 2D quantum system can thus be studied in light of the free energy. On the other hand, the ground state properties of the 2D quantum spin lattice systems can also be explored by the ODTNS method. By taking $K \to \infty$ and $\tau \to 0$, the ground state energy per site e_0 can be obtained by

$$e_0 = \lim_{K\to\infty}\lim_{\tau\to 0}\frac{1}{2\tau}\ln\prod_{S=I,II,III} r^S. \tag{7}$$

Consequently, the ODTNS algorithm can be utilized to investigate both the thermodynamic and ground state properties of the 2D quantum lattice systems, which would be a powerful numerical method in the field of correlated quantum many-body systems.

To check the efficiency and accuracy of the ODTNS method, we have studied the thermodynamics of the spin-1/2 Heisenberg antiferromagnet on a honeycomb lattice. The results from the ODTNS calculations are found to be in well agreement with the QMC results. To show the scalability of the ODTNS scheme, we have also studied the thermodynamics of a spin-1/2 frustrated Heisenberg model on a bilayer honeycomb lattice with alternating interlayer coupling. Note that this model cannot be accessed by the QMC calculations. By means of the ODTNS algorithm, the sublattice magnetization as well as the specific heat as functions of temperature for this frustrated quantum system can be readily obtained, revealing that the ODTNS method is also applicable to the frustrated quantum spin systems.

4. Summary

In summary, we have discussed the two tensor network-based methods for low-dimensional quantum spin lattice systems, say, LTRG and ODTNS methods. The LTRG algorithm is developed for better calculating the thermodynamics of 1D quantum lattice systems, and applications of the LTRG approach to quantum spin chains are successful. The ODTNS method is proposed for calculating the thermodynamic and ground state properties of 2D quantum lattice systems with good accuracy and efficiency, and its application to a 2D frustrated bilayer honeycomb Heisenberg model was shown to be practical. These methods are as well expected to have applications in the correlated boson and fermion quantum systems.

Acknowledgments

This work is collaborated with, in particular, Wei Li, Shi-Ju Ran, and also Xin Yan, Bin Xi, Shou-Shu Gong, Zhe Zhang, Fei Ye, and Yang Zhao. It is supported in part by the NSFC (Grant Nos. 90922033 and 10934008), the MOST of China (Grant No. 2012CB932901 and 2013CB933401) and the CAS.

References

1. W. Li, S. J. Ran, S. S. Gong, Y. Zhao, B. Xi, F. Ye, and G. Su, Phys. Rev. Lett. **106**, 127202 (2011).
2. S. J. Ran, W. Li, B. Xi, Z. Zhang and G. Su, Phys. Rev. B **86**, 134429 (2012).
3. X. Yan, W. Li, Y. Zhao, S. J. Ran, and G. Su, Phys. Rev. B **85**, 134425 (2012).
4. B. Gu, Ph.D. Thesis, (Graduate University of Chinese Academy of Sciences, Beijing, 2006).
5. G. H. Liu, W. Li, W. L. You, G. S. Tian and G. Su, Phys. Rev. B **85**, 184422 (2012).

ON EXPLICIT FREE FIELD REPRESENTATIONS OF CURRENT (SUPER) ALGEBRAS

WEN-LI YANG*

Institute of Modern Physics, Northwest University, Xian, 710069, China
* *E-mail: wlyang@nwu.edu.cn*

We construct the explicit free field representations of the current (super) algebras at a generic level k. The corresponding energy−momentum tensors from the Sugawara construction is also given in terms of free fields.

Keywords: The WZNW model; current algebras; free field realization.

1. Introduction

The interest in two-dimensional non-linear sigma models with supermanifold as target space, such as the Wess-Zumino-Novikov-Witten (WZNW) models associated with super-groups, has drastically increased over the last ten years due to their applications in superstring theory,[1,2] logarithmic conformal field theory (CFT) and condensed matter physics.[3–8] Free field realization[9–13] has been proved to be a powerful method in analyzing CFTs such as WZNW models. In the construction of a free field realization of the currents, one needs[9–12] firstly to construct the differential operator representation of the corresponding finite-dimensional superalgebra. However, it might be very involved, if not impossible, to apply the general procedure proposed in[9,10,12] to derive explicit expressions of differential operator representation for higher-rank algebras. It was shown[14] that a particular ordering of all positive roots of a higher-rank (super)algebra could drastically simplify the computation of its explicit differential operator expression. In this paper, we shall use such an ordering to construct the free field realization of the $D(2,1;\alpha)$ current algebra. It was shown that the superalgebra $D(2,1;\alpha)$ has played an important role in describing the origin of the Yangian symmetry of Ads/CFT[15] and the symmetry of string theory on $AdS_3 \times S^3 \times S^3 \times S^1$.[16] Moreover, To our knowledge, it might be the first result on the current superalgebra associated with exceptional superalgebra.

2. $D(2,1;\alpha)$ current algebra

Let us start with some basic notation of the $D(2,1;\alpha)$ current algebra, i.e. the $\widehat{D(2,1;\alpha)}$ affine superalgebra.[17] The exceptional Lie superalgebra $D(2,1;\alpha)$ with $\alpha \neq 0,\ -1,\ \infty$ forms a one-parameter (i.e. α) family of superalgebras of rank 3 and dimension 17. The bosonic (or even) part is a direct sum of three $su(2)$ and the fermionic (or odd) part is a spinorial representation $(2,2,2)$ of the even part. Let $\Pi := \{\alpha_1,\ \alpha_2,\ \alpha_3\}$ be the simple roots with α_1 being fermionic and α_2, α_3 being bosonic. The Cartan matrix a_{ij} is given by $a = \begin{pmatrix} 0 & 1 & \alpha \\ -1 & 2 & 0 \\ -1 & 0 & 2 \end{pmatrix}$. The positive even roots Δ_0^+ and the positive odd roots Δ_1^+ are then given respectively by $\Delta_0^+ = \{\alpha_2,\ \alpha_3,\ 2\alpha_1+\alpha_2+\alpha_3\}$ and $\Delta_1^+ = \{\alpha_1,\ \alpha_1+\alpha_2,\ \alpha_1+\alpha_3,\ \alpha_1+\alpha_2+\alpha_3\}$, the set of all positive roots Δ^+ is a union of the positive even and odd roots, i.e. $\Delta^+ = \Delta_0^+ \bigcup \Delta_1^+$. Associated with each positive root δ, there is a raising operator E_δ, a lowering operator $F_\delta \equiv E_{-\delta}$ and a Cartan generator H_δ. These operators have definite $\mathbb{Z}_2$-gradings: $[H_\delta] = 0,\ [E_\delta] = [F_\delta] = \begin{cases} 0,\ \delta \in \Delta_0^+, \\ 1,\ \delta \in \Delta_1^+. \end{cases}$ For any two homogenous elements (i.e. elements with definite $\mathbb{Z}_2$-gradings) $a, b \in D(2,1;\alpha)$, the Lie bracket is defined by $[a,\ b] = ab - (-1)^{[a][b]}ba$, this (anti)commutator extends to inhomogenous elements through linearity. Then the commutation relations of $D(2,1;\alpha)$ are[17] [a]

$$[E_{\alpha_i},\ F_{\alpha_j}] = \delta_{ij}\, H_{\alpha_i}, \qquad [H_{\alpha_i},\ H_{\alpha_j}] = 0, \qquad i,j = 1,2,3, \tag{1}$$

$$\begin{aligned}
&[H_{\alpha_i},\ E_{\alpha_j}] = a_{ij} E_{\alpha_j}, \qquad [H_{\alpha_i},\ F_{\alpha_j}] = -a_{ij} F_{\alpha_j},, \qquad i,j = 1,2,3,\\
&[E_{\alpha_1},\ E_{\alpha_2}] = -E_{\alpha_1+\alpha_2}, \qquad [E_{\alpha_1},\ E_{\alpha_3}] = -E_{\alpha_1+\alpha_3},\\
&[E_{\alpha_1},\ E_{\alpha_1+\alpha_2+\alpha_3}] = -(1+\alpha) E_{2\alpha_1+\alpha_2+\alpha_3},\\
&[E_{\alpha_2},\ E_{\alpha_1+\alpha_3}] = [E_{\alpha_3},\ E_{\alpha_1+\alpha_2}] = E_{\alpha_1+\alpha_2+\alpha_3},\\
&[E_{\alpha_1+\alpha_2},\ E_{\alpha_1+\alpha_3}] = (1+\alpha) E_{2\alpha_1+\alpha_2+\alpha_3},\\
&[F_{\alpha_1},\ E_{2\alpha_1+\alpha_2+\alpha_3}] = -E_{\alpha_1+\alpha_2+\alpha_3}, \qquad [F_{\alpha_1},\ E_{\alpha_1+\alpha_2}] = -E_{\alpha_2},\\
&[F_{\alpha_1},\ E_{\alpha_1+\alpha_3}] = -\alpha E_{\alpha_3}, \qquad [F_{\alpha_2},\ E_{\alpha_1+\alpha_2+\alpha_3}] = E_{\alpha_1+\alpha_3},\\
&[F_{\alpha_2},\ E_{\alpha_1+\alpha_2}] = E_{\alpha_1}, \qquad [F_{\alpha_3},\ E_{\alpha_1+\alpha_3}] = E_{\alpha_1},\\
&[F_{\alpha_3},\ E_{\alpha_1+\alpha_2+\alpha_3}] = E_{\alpha_1+\alpha_2}.
\end{aligned} \tag{2}$$

---

[a] Here we only give the relevant relations for our purpose, the others can be found in[17]

One may introduce a nondegenerate and invariant supersymmetric metric or bilinear form which was given in,[18] denoted by (X,Y) for any $X,Y \in D(2,1;\alpha)$. Let $X(z)$ be current associated with generator $X \in D(2,1;\alpha)$. For an example, denote the currents corresponding to E_δ, H_δ, F_δ ($\forall\delta \in \Delta^+$) by $E_\delta(z)$, $H_\delta(z)$, $F_\delta(z)$ respectively. The $D(2,1;\alpha)$ current algebra is generated by a set of currents $\{X(z)|\, X \in D(2,1;\alpha)\}$, which obey the following OPEs,[17]

$$X(z)\,Y(w) = k\frac{(X,Y)}{(z-w)^2} + \frac{[X,\,Y]\,(w)}{(z-w)}, \tag{3}$$

where k is the level (or central charge) of the current algebra.

3. Free field realization

To obtain a free field realization of the $D(2,1;\alpha)$ currents, one needs[9–13] firstly to construct the differential operator representation of the corresponding finite-dimensional superalgebra $D(2,1;\alpha)$. It was shown[14] that a particular ordering of all positive roots of a higher-rank (super)algebra could drastically simplify the computation of its explicit differential operator expression. Here we find that for the $D(2,1;\alpha)$ case the very ordering of the positive roots is given by[18]

$$\alpha_3,\ \alpha_2,\ \alpha_1+\alpha_2+\alpha_3,\ \alpha_1+\alpha_3,\ 2\alpha_1+\alpha_2+\alpha_3,\ \alpha_1+\alpha_2,\ \alpha_1. \tag{4}$$

Such an ordering allows us to construct the explicit expression of the differential operator representation of $D(2,1;\alpha)$ by the method developed in Ref.[14] With the help of the differential representation we can obtain the free field realization of the $D(2,1;\alpha)$ current algebra in terms of 3 bosonic β-γ pairs $((\beta_{\alpha_2}(z),\ \gamma_{\alpha_2}(z)),\ (\beta_{\alpha_3}(z),\ \gamma_{\alpha_3}(z))$ and $(\beta_{2\alpha_1+\alpha_2+\alpha_3}(z),\ \gamma_{2\alpha_1+\alpha_2+\alpha_3}(z)))$, 4 fermionic b-c pairs $((\Psi^\dagger_{\alpha_1}(z),\ \Psi_{\alpha_1}(z))$, $(\Psi^\dagger_{\alpha_1+\alpha_2}(z),\ \Psi_{\alpha_1+\alpha_2}(z))$, $(\Psi^\dagger_{\alpha_1+\alpha_3}(z),\ \Psi_{\alpha_1+\alpha_3}(z))$ and $(\Psi^\dagger_{\alpha_1+\alpha_2+\alpha_3}(z),\ \Psi_{\alpha_1+\alpha_2+\alpha_3}(z)))$ and 3 free scalar fields ($\phi_i(z)$, for $i = 1,2,3$). These free fields obey the following OPEs:

$$\beta_i(z)\,\gamma_j(w) = -\gamma_j(z)\,\beta_i(w) = \frac{\delta_{ij}}{z-w}, \qquad i,j \in \Delta_0^+, \tag{5}$$

$$\Psi_i(z)\Psi_j^\dagger(w) = \Psi_j^\dagger(z)\Psi_i(w) = \frac{\delta_{ij}}{z-w}, \qquad i,j \in \Delta_1^+, \tag{6}$$

$$\phi_1(z)\phi_1(w) = -\ln(z-w), \tag{7}$$

$$\phi_i(z)\phi_j(w) = \delta_{ij}\ln(z-w), \qquad i,j = 2,3. \tag{8}$$

and the other OPEs are trivial. Here we present the results for the currents associated with the simple roots,

$$E_{\alpha_1}(z) = \Psi_{\alpha_1}(z), \tag{9}$$

$$\begin{aligned}
E_{\alpha_2}(z) &= -\Psi^+_{\alpha_1}(z)\Psi_{\alpha_1+\alpha_2}(z) - \Psi^+_{\alpha_1+\alpha_3}(z)\Psi_{\alpha_1+\alpha_2+\alpha_3}(z) + \beta_{\alpha_2}(z), \\
E_{\alpha_3}(z) &= -\Psi^+_{\alpha_1}(z)\Psi_{\alpha_1+\alpha_3}(z) - (1+\alpha)\Psi^+_{\alpha_1}(z)\Psi^+_{\alpha_1+\alpha_2}(z)\beta_{2\alpha_1+\alpha_2+\alpha_3}(z) \\
&\quad -\Psi^+_{\alpha_1+\alpha_2}(z)\Psi_{\alpha_1+\alpha_2+\alpha_3}(z) + \beta_{\alpha_3}(z),
\end{aligned} \tag{10}$$

$$\begin{aligned}
H_{\alpha_1}(z) &= -\Psi^+_{\alpha_1+\alpha_2}(z)\Psi_{\alpha_1+\alpha_2}(z) - (1+\alpha)\gamma_{2\alpha_1+\alpha_2+\alpha_3}(z)\beta_{2\alpha_1+\alpha_2+\alpha_3}(z) \\
&\quad -\alpha\Psi^+_{\alpha_1+\alpha_3}(z)\Psi_{\alpha_1+\alpha_3}(z) - (1+\alpha)\Psi^+_{\alpha_1+\alpha_2+\alpha_3}(z)\Psi_{\alpha_1+\alpha_2+\alpha_3}(z) \\
&\quad -\gamma_{\alpha_2}(z)\beta_{\alpha_2}(z) - \alpha\gamma_{\alpha_3}(z)\beta_{\alpha_3}(z) \\
&\quad +\sqrt{k}\left(\sqrt{\frac{1+\alpha}{2}}\,\partial\phi_1(z) + \frac{\sqrt{2}}{2}\,\partial\phi_2(z) + \sqrt{\frac{\alpha}{2}}\,\partial\phi_3(z)\right),
\end{aligned} \tag{11}$$

$$\begin{aligned}
H_{\alpha_2}(z) &= \Psi^+_{\alpha_1}(z)\Psi_{\alpha_1}(z) - \Psi^+_{\alpha_1+\alpha_2}(z)\Psi_{\alpha_1+\alpha_2}(z) + \Psi^+_{\alpha_1+\alpha_3}(z)\Psi_{\alpha_1+\alpha_3}(z) \\
&\quad -\Psi^+_{\alpha_1+\alpha_2+\alpha_3}(z)\Psi_{\alpha_1+\alpha_2+\alpha_3}(z) - 2\gamma_{\alpha_2}(z)\beta_{\alpha_2}(z) + \sqrt{2k}\,\partial\phi_2(z), \\
H_{\alpha_3}(z) &= \Psi^+_{\alpha_1}(z)\Psi_{\alpha_1}(z) + \Psi^+_{\alpha_1+\alpha_2}(z)\Psi_{\alpha_1+\alpha_2}(z) - \Psi^+_{\alpha_1+\alpha_3}(z)\Psi_{\alpha_1+\alpha_3}(z) \\
&\quad -\Psi^+_{\alpha_1+\alpha_2+\alpha_3}(z)\Psi_{\alpha_1+\alpha_2+\alpha_3}(z) - 2\gamma_{\alpha_3}(z)\beta_{\alpha_3}(z) + \sqrt{\frac{2k}{\alpha}}\,\partial\phi_3(z),
\end{aligned}$$

$$\begin{aligned}
F_{\alpha_1}(z) &= \Psi^+_{\alpha_1+\alpha_2}(z)\Psi^+_{\alpha_1+\alpha_3}(z)\Psi_{\alpha_1+\alpha_2+\alpha_3}(z) - \Psi^+_{\alpha_1+\alpha_2}(z)\beta_{\alpha_2}(z) \\
&\quad +\gamma_{2\alpha_1+\alpha_2+\alpha_3}(z)\Psi_{\alpha_1+\alpha_2+\alpha_3}(z) - \alpha\Psi^+_{\alpha_1+\alpha_3}(z)\beta_{\alpha_3}(z) \\
&\quad -\Psi^+_{\alpha_1}(z)\left\{\Psi^+_{\alpha_1+\alpha_2}(z)\Psi_{\alpha_1+\alpha_2}(z) + \alpha\Psi^+_{\alpha_1+\alpha_3}(z)\Psi_{\alpha_1+\alpha_3}(z)\right\} \\
&\quad -\Psi^+_{\alpha_1}(z)\left\{\gamma_{\alpha_2}(z)\beta_{\alpha_2}(z) + (1+\alpha)\gamma_{2\alpha_1+\alpha_2+\alpha_3}(z)\beta_{2\alpha_1+\alpha_2+\alpha_3}(z)\right\} \\
&\quad -\Psi^+_{\alpha_1}(z)\left\{(1+\alpha)\Psi^+_{\alpha_1+\alpha_2+\alpha_3}(z)\Psi_{\alpha_1+\alpha_2+\alpha_3}(z) + \alpha\gamma_{\alpha_3}(z)\beta_{\alpha_3}(z)\right\} \\
&\quad +\sqrt{k}\,\Psi^+_{\alpha_1}(z)\left\{\sqrt{\frac{1+\alpha}{2}}\,\partial\phi_1(z) + \frac{\sqrt{2}}{2}\,\partial\phi_2(z) + \sqrt{\frac{\alpha}{2}}\,\partial\phi_3(z)\right\} \\
&\quad -k\,\partial\Psi^+_{\alpha_1}(z),
\end{aligned} \tag{12}$$

$$\begin{aligned}
F_{\alpha_2}(z) &= -\Psi^+_{\alpha_1+\alpha_2}(z)\Psi_{\alpha_1}(z) - \Psi^+_{\alpha_1+\alpha_2+\alpha_3}(z)\Psi_{\alpha_1+\alpha_3}(z) - \gamma^2_{\alpha_2}(z)\beta_{\alpha_2}(z) \\
&\quad +\sqrt{2k}\,\gamma_{\alpha_2}(z)\,\partial\phi_2(z) + (k-2)\,\partial\gamma_{\alpha_2}(z),
\end{aligned} \tag{13}$$

$$\begin{aligned}
F_{\alpha_3}(z) &= (1+\alpha)\Psi^+_{\alpha_1+\alpha_2+\alpha_3}(z)\Psi^+_{\alpha_1+\alpha_3}(z)\beta_{2\alpha_1+\alpha_2+\alpha_3}(z) \\
&\quad -\Psi^+_{\alpha_1+\alpha_3}(z)\Psi_{\alpha_1}(z) - \Psi^+_{\alpha_1+\alpha_2+\alpha_3}(z)\Psi_{\alpha_1+\alpha_2}(z) - \gamma^2_{\alpha_3}(z)\beta_{\alpha_3}(z) \\
&\quad +\sqrt{\frac{2k}{\alpha}}\,\gamma_{\alpha_3}(z)\,\partial\phi_3(z) + (\frac{k}{\alpha}-2)\,\partial\gamma_{\alpha_3}(z).
\end{aligned} \tag{14}$$

Here normal ordering of the free field expressions is implied. The free field realization for currents associated with the non-simple roots can be obtained from the OPEs of the simple ones. It is straightforward to check that (9)-(14) satisfy the OPEs of the $D(2,1;\alpha)$ current algebra given in the last section.

It is well-known that the energy-momentum tensor associated with a current algebra can be obtained by means of the Sugawara's construction. After tedious calculation, we obtain that the corresponding energy-momentum tensor $T(z)$ can be expressed in terms of the free fields

$$\begin{aligned} T(z) = &-\frac{1}{2}\partial\phi_1(z)\partial\phi_1(z) - \sqrt{\frac{1+\alpha}{2k}}\,\partial^2\phi_1(z) \\ &+\frac{1}{2}\partial\phi_2(z)\partial\phi_2(z) - \sqrt{\frac{1}{2k}}\,\partial^2\phi_2(z) \\ &+\frac{1}{2}\partial\phi_3(z)\partial\phi_3(z) - \sqrt{\frac{\alpha}{2k}}\,\partial^2\phi_3(z) \\ &+\beta_{\alpha_2}(z)\partial\gamma_{\alpha_2}(z) + \beta_{\alpha_3}(z)\partial\gamma_{\alpha_3}(z) + \beta_{2\alpha_1+\alpha_2+\alpha_3}(z)\partial\gamma_{2\alpha_1+\alpha_2+\alpha_3}(z) \\ &+\partial\Psi^+_{\alpha_1}(z)\,\Psi_{\alpha_1}(z) + \partial\Psi^+_{\alpha_1+\alpha_2}(z)\,\Psi_{\alpha_1+\alpha_2}(z) \\ &+\partial\Psi^+_{\alpha_1+\alpha_3}(z)\,\Psi_{\alpha_1+\alpha_3}(z) + \partial\Psi^+_{\alpha_1+\alpha_2+\alpha_3}(z)\,\Psi_{\alpha_1+\alpha_2+\alpha_3}(z), \end{aligned} \tag{15}$$

where the normal ordering of the free field expressions is implicit. We find that the Sugawara-type stress tensor $T(z)$ satisfies the Virasoro algebra

$$T(z)T(w) = \frac{c/2}{(z-w)^4} + \frac{2T(w)}{(z-w)^2} + \frac{\partial T(w)}{(z-w)}, \tag{16}$$

with central charge $c = 1$. Moreover, it is easy to check that

$$T(z)X(w) = \frac{X(w)}{(z-w)^2} + \frac{\partial X(w)}{z-w} = \partial_w\left\{\frac{X(w)}{z-w}\right\}, \ \forall X \in D(2,1;\alpha). \tag{17}$$

Namely, all the currents are primary fields of conformal dimension one.

Acknowledgments

The financial supports from the National Natural Science Foundation of China (Grant Nos. 11075126,11191240050 and 11291240037), the State Education Ministry of China (Grant No. 20116101110017 and SRF for ROCS) are gratefully acknowledged. I would like to thank thank the organizers, especially Profs. M.-L. Ge, N. Jing and C.-M Bai, of the conference for giving me the opportunity to present the work.

References

1. N. Berkovits, C. Vafa and E. Witten, *JHEP* **03**, 018 (1999).

2. M. Bershadsky, S. Zhukov and A. Vaintrob, *Nucl. Phys.* **B 559**, 205-234 (1999).
3. V. Gurarie, *Nucl. Phys.* **B 410**, 535-549 (1993).
4. L. Rozansky and H. Saleur, *Nucl. Phys.* **B 376**, 461-509 (1992).
5. D. Bernard, (Perturbed) Conformal field theory applied to 2D disordered systems: an introduction, `hep-th/9509137`.
6. C. Mudry, C. Chamon and X.-G. Wen, *Nucl. Phys.* **B 466**, 383-443 (1996).
7. Z. Maassarani and D. Serban, *Nucl. Phys.* **B 489**, 603-625 (1997).
8. M. J. Bhaseen, J.-S. Caux, I. I. Kogan and A. M. Tsveilk, *Nucl. Phys.* **B 618**, 465-499 (2001).
9. B. Feigin and E. Frenkel, *Commun. Math. Phys.* **128**, 161 (1990).
10. P. Bouwknegt, J. McCarthy and K. Pilch, *Prog. Phys. Suppl.* **102**, 67 (1990).
11. K. Ito, *Phys. Lett.* **B 252**, 69 (1990).
12. J. Rasmussen, *Nucl. Phys.* **B 510**, 688 (1998).
13. X.-M. Ding, M. Gould and Y.-Z. Zhang, *Phys. Lett.* **A 318**, 354 (2003).
14. W.-L. Yang, Y.-Z. Zhang and S. Kault, *Nucl. Phys.* **B 823**, 372 (2009); W.-L. Yang, *Nankai Series in Pur. Appl. Math. and Phys.* **Vol. 9** 173 (2010).
15. T. Matsumoto and S. Moriyama, *JHEP* **04** (2008), 022; *JHEP* **09**, 097 (2009).
16. O. Ohlsson Sax and B. Stefanski, jr., *arXiv:1106.2558.*
17. L. Frappat, P. Sorba and A. Scarrino, Dictionary on Lie algebras and superalgebras, Academic Press, New York, 2000.
18. X. Chen, X.-M. Ding, X.-H. Wang and W.-L. Yang, *Commun. Theor. Phys.* **57**, 825-832 (2012); X. Chen, W.-L. Yang, X.-M. Ding, J. Feng, K.-S. Ming, K. Wu and Y.-Z. Zhang, *J. Phys.* **A** 45, 405204 (2012).

PARALLEL SESSION
Superintegrable Hamiltonians

23 & 24 AUGUST 2012

SUPERINTEGRABLE QUANTUM OSCILLATOR AND KEPLER-COULOMB SYSTEMS ON CURVED SPACES

A. BALLESTEROS[1], A. ENCISO[2], F. J. HERRANZ*[1], O. RAGNISCO[3] and D. RIGLIONI[4]

[1] *Departamento de Física, Universidad de Burgos, E-09001 Burgos, Spain*
[2] *Instituto de Ciencias Matemáticas, CSIC, E-28049 Madrid, Spain*
[3] *Dipartimento di Fisica, Università di Roma Tre and Instituto Nazionale di Fisica Nucleare sezione di Roma Tre, I-00146 Roma, Italy*
[4] *Centre de Recherches Mathématiques, Université de Montréal, H3C 3J7, Canada*
* *E-mail: fjherranz@ubu.es*

An overview of maximally superintegrable classical Hamitonians on spherically symmetric spaces is presented. It turns out that each of these systems can be considered either as an oscillator or as a Kepler–Coulomb Hamiltonian. We show that two possible quantization prescriptions for all these curved systems arise if we impose that superintegrability is preserved after quantization, and we prove that both possibilities are gauge equivalent.

Keywords: Integrable Systems; Exact Solvability; Quantization; Curvature.

1. Bertrand spacetimes

Bertrand's theorem asserts that any three-dimensional (3D) spherically symmetric natural Hamiltonian system $H = \frac{1}{2}\mathbf{p}^2 + V(|\mathbf{q}|)$ that has a stable circular trajectory passing through each point of $\mathbb{R}^3$ and all whose bounded trajectories are closed is either a harmonic oscillator, $V(r) = \omega^2\mathbf{q}^2 + a$, or a Kepler–Coulomb (KC) system, $V(r) = \frac{k}{|\mathbf{q}|} + b$. An extension of such a classical theorem was found by Perlick[1] in a General Relativity framework. Consider a $(3+1)$D spherically symmetric spacetime $(\mathcal{M}\times\mathbb{R}, \eta)$, where $\mathcal{M}$ is a 3-manifold. Then the Lorentzian metric η can be written as

$$\eta = h(r')^2\,\mathrm{d}r'^2 + r'^2\big(\mathrm{d}\theta^2 + \sin^2\theta\,\mathrm{d}\varphi^2\big) - \frac{\mathrm{d}t^2}{V(r')} = g - \frac{\mathrm{d}t^2}{V(r')},$$

where g is a Riemannian metric on $\mathcal{M}$ and $h(r'), V(r')$ are smooth functions. A *trajectory* in the spacetime is the projection of an inextendible timelike geodesic to a constant time leaf $\mathcal{M}\{t_0\}$. The Lorentzian (3+1)-manifold

$(\mathcal{M}\times\mathbb{R},\eta)$ is a *Bertrand spacetime* if it is verified that: (i) there is a circular (r' = constant) trajectory passing through each point of $\mathcal{M}$, and (ii) such circular trajectories are stable. Under these assumptions, Perlick undertook the classification of all Bertrand spacetimes finding two multiparametric families of metrics by obtaining explicitly the functions $h(r'), V(r')$.

These results can be extended to arbitrary dimension and expressed in a conformally flat form. Let us consider an $(N+1)$D spherically symmetric spacetime $(\mathcal{M}\times\mathbb{R},\eta)$, where hereafter $\mathcal{M}$ is an N-manifold. The Lorentzian metric η can be written as

$$\eta = g - \frac{\mathrm{d}t^2}{V(r)}, \qquad g = f(|\mathbf{q}|)^2\,\mathrm{d}\mathbf{q}^2 = f(r)^2(\mathrm{d}r^2 + r^2\mathrm{d}\Omega^2), \tag{1}$$

where $\mathbf{q} = (q_1,\dots,q_N)$, $\mathrm{d}\mathbf{q}^2 = \sum_{i=1}^N \mathrm{d}q_i^2$, $r = |\mathbf{q}| = \sqrt{\mathbf{q}^2}$ and $\mathrm{d}\Omega^2$ is the standard metric on the unit $(N-1)$D sphere. Hence, g defines a Riemannian metric on $\mathcal{M}$ with $f(r)$ playing the role of *conformal factor* of the Euclidean metric $g_0 = \mathrm{d}\mathbf{q}^2$. The scalar curvature of g is generally nonconstant and turns out to be

$$R = -(N-1)\left(\frac{(N-4)f'(r)^2 + f(r)\left(2f''(r) + 2(N-1)r^{-1}f'(r)\right)}{f(r)^4}\right). \tag{2}$$

By starting from Perlick's classification of (3+1)D Bertrand spacetimes, if we change the initial radial coordinate in the form $r' = \mathcal{F}(r)$ and we accordingly transform the initial Perlick's parameters, then it can be proven through very cumbersome computations that the metric η of an $(N+1)$D Bertrand spacetime belongs to one of the following classes:[2]

• **Type I**. This is a *three-parametric* family of metrics depending on $(\beta;\kappa,\xi)$ where β is a positive rational number and κ,ξ are real constants:

$$\eta_{\rm I} = \frac{1}{r^2\left(r^{-\beta}+\kappa r^{\beta}\right)^2}\,\mathrm{d}\mathbf{q}^2 - \frac{\mathrm{d}t^2}{\left(r^{-\beta}-\kappa r^{\beta}\right)+\xi}. \tag{3}$$

• **Type II**. This is a *four-parametric* family depending on $(\gamma;\lambda,\delta,\chi)$ where γ is a positive rational number and λ^2,δ,χ are real constants:

$$\eta_{\rm II} = \frac{\left(r^{-2\gamma}+\lambda^2 r^{2\gamma}-2\delta\right)}{r^2\left(r^{-2\gamma}-\lambda^2 r^{2\gamma}\right)^2}\,\mathrm{d}\mathbf{q}^2 - \frac{\mathrm{d}t^2}{\left(r^{-2\gamma}+\lambda^2 r^{2\gamma}-2\delta\right)^{-1}+\chi}. \tag{4}$$

The sets of parameters $(\beta;\kappa,\xi)$ and $(\gamma;\lambda,\delta,\chi)$ can be, in fact, regarded as *deformation parameters* of the *flat* Minkowskian metric, since this is recovered by a contraction limit of (3) and (4) as follows:

$$\eta_{\rm I} = \mathrm{d}\mathbf{q}^2 - \mathrm{d}t^2 \quad \text{for} \quad \beta=1, \quad \kappa=0, \quad \xi=1/\varepsilon^2; \quad t\to t/\varepsilon, \quad \varepsilon\to 0.$$
$$\eta_{\rm II} = \mathrm{d}\mathbf{q}^2 - \mathrm{d}t^2 \quad \text{for} \quad \gamma=1, \quad \lambda=\delta=0, \quad \chi=1/\varepsilon^2; \quad t\to t/\varepsilon, \quad \varepsilon\to 0.$$

2. Classical Bertrand Hamiltonians

Timelike geodesics in an $(N+1)$D spacetime with metric η (1) are naturally related to trajectories of the ND classical Hamiltonian in $\mathcal{M}$ given by

$$H = \frac{\mathbf{p}^2}{2f(|\mathbf{q}|)^2} + V(|\mathbf{q}|) = \frac{p_r^2 + r^{-2}\mathbf{L}^2}{2f(r)^2} + V(r), \tag{5}$$

where $\mathbf{p} = (p_1, \ldots, p_N)$ and p_r are, in this order, the conjugate momenta of the coordinates $\mathbf{q}$ and r, while $\mathbf{L}$ is the total angular momentum.

Thus, Perlick's study of Bertrand spacetimes provides a complete classification of the central potentials $V(r)$ and spherically symmetric metrics for which the statement of the classical Bertrand theorem remains true. Moreover, we showed that these Hamiltonian systems are still superintegrable.[3] Therefore, by considering the classification of Bertrand spacetimes (3) and (4), we write the corresponding *classical Bertrand Hamiltonians* which are cast into the following two families:[2]

• **Type I**. *Two-parametric* family of Hamiltonians depending on $(\beta;\kappa)$:

$$H_{\rm I} = \frac{1}{2} r^2 \left(r^{-\beta} + \kappa r^{\beta}\right)^2 \mathbf{p}^2 + A\big(r^{-\beta} - \kappa r^{\beta}\big). \tag{6}$$

• **Type II**. *Three-parametric* family depending on $(\gamma;\lambda,\delta)$:

$$H_{\rm II} = \frac{r^2\left(r^{-2\gamma} - \lambda^2 r^{2\gamma}\right)^2}{2\left(r^{-2\gamma} + \lambda^2 r^{2\gamma} - 2\delta\right)}\,\mathbf{p}^2 + \frac{B}{\left(r^{-2\gamma} + \lambda^2 r^{2\gamma} - 2\delta\right)}. \tag{7}$$

We remark that the parameters ξ, χ, which are rather relevant in the metrics, become additive constants in the Hamiltonians. The coupling constants A, B of the potentials do not appear in the metrics since they can be set equal to 1 through a time scaling. Note also that Bertrand Hamiltonians are, as expected, generalizations (deformations) of the flat harmonic oscillator and KC systems, which are recovered as particular cases:

$$H_{\rm I} = \tfrac{1}{2}\mathbf{p}^2 + A/r \quad \text{for} \quad \beta = 1, \quad \kappa = 0.$$
$$H_{\rm II} = \tfrac{1}{2}\mathbf{p}^2 + Br^2 \quad \text{for} \quad \gamma = 1, \quad \lambda = \delta = 0.$$

The main properties of Bertrand Hamiltonians are the following:

(i) *Superintegrability.* The spherical symmetry of any classical Hamiltonian of the type (5) provides $2N-3$ integrals of motion which are quadratic in the momenta. *Maximally superintegrable* systems are distinguished cases for which an additional constant of the motion exists (which is a component of a Runge–Lenz N-vector), so that H is endowed with $2N-2$ functionally independent integrals. This has been explicitly proved for 3D Bertrand Hamiltonians,[3] finding that the integrals of motion are not, in general, quadratic in the momenta (such higher order is determined by β, γ).

(ii) *Geometric interpretation.* The *intrinsic KC and oscillator potentials* on the ND space $\mathcal{M}$ are defined by means of the radial symmetric Green function $U(r)$ of the Riemannian metric g (1) as

$$U(r) = \int^r \frac{\mathrm{d}r}{r^2 f(r)}, \qquad \mathcal{U}_{\mathrm{KC}}(r) := A\,U(r), \qquad \mathcal{U}_{\mathrm{O}}(r) := \frac{B}{U(r)^2}.$$

If we apply these definitions to the two families of Bertrand Hamiltonians it can be shown[4] that type I (6) and type II Hamiltonians (7) always define, in this order, intrinsic KC systems and oscillator potentials on curved spaces. Consequently, the extension of the classical Bertrand's theorem to spherically symmetric spaces only provides either oscillators or KC systems.

(iii) *Stäckel transform.* Both types of Bertrand Hamiltonians are equivalent via the Stäckel transform.[5] If we denote (6) as $H_{\mathrm{I}} = T_{\mathrm{I}} + V_{\mathrm{I}}$ and consider $H_{0,\mathrm{I}} = T_{\mathrm{I}} + B$ and $U_{\mathrm{I}} = V_{\mathrm{I}}/A + C$, then $H_{0,\mathrm{I}}/U_{\mathrm{I}} \equiv H_{\mathrm{II}}$ provided that

$$\beta = 2\gamma, \quad \kappa = -\lambda^2, \quad C = -2\delta.$$

Conversely, Hamiltonians of type I can be obtained from those of type II, so establishing a *"duality"* between curved oscillators and KC systems.[2]

We illustrate the above results by displaying in Table 1 some relevant specific Bertrand Hamiltonians that arise for $\beta = 1, 2 \leftrightarrow \gamma = \frac{1}{2}, 1$; all of them have quadratic integrals.[2] The remaining parameters, κ and (λ, δ) are related with the curvature of the space. In each row we present a KC system with its Stäckel equivalent ("dual") oscillator. The names "Euclidean", "Taub–NUT"..., refer to the underlying Riemannian space. Notice that the KC system on the *three* spaces of *constant* sectional curvature κ ($>, <, = 0$ for the sphere, hyperbolic space, Euclidean space, respectively) is recovered in the cases 1A and 1B of type I ($\beta = 1$), meanwhile their corresponding oscillator potential appears in the cases 2A.1 and 2B.1 of type II ($\gamma = 1$), with $\lambda = \delta$ now playing the role of the constant curvature.

3. Quantum Bertrand Hamiltonians

In order to obtain the quantization of the classical Bertrand Hamiltonians (6) and (7), the ordering ambiguity coming from the kinetic term can be fixed by imposing the resulting quantum Hamiltonians $\hat{\mathcal{H}}$ to be maximally superintegrable. This means that the classical integrals are promoted to $(2N - 2)$ algebraically independent operators that commute with $\hat{\mathcal{H}}$. We consider the quantum position and momenta operators, $\hat{\mathbf{q}}$, $\hat{\mathbf{p}}$, such that $[\hat{q}_i, \hat{p}_j] = \mathrm{i}\hbar\delta_{ij}$, $\hat{q}_i = q_i$, $\hat{p}_i = -\mathrm{i}\hbar\frac{\partial}{\partial q_i}$, and we present two quantization prescriptions that fulfil the superintegrability condition:[6] the so-called Schrödinger quantization and the conformal Laplace–Beltrami (LB) one.

Table 1. Examples of N-dimensional classical Bertrand Hamiltonians.

Type I: KC Hamiltonians $(\beta;\kappa)$	Type II: Oscillator Hamiltonians $(\gamma;\lambda,\delta)$
• 1A $(\beta=1;\kappa=0)$ Euclidean $H=\frac{1}{2}\mathbf{p}^2+\frac{A}{r}$	• 1A $(\gamma=\frac{1}{2};\lambda=0,\delta)$ Taub–NUT $H=\frac{r}{2(1-2\delta r)}\,\mathbf{p}^2+\frac{Br}{(1-2\delta r)}$
• 1B $(\beta=1;\kappa)$ Sphere/Hyperbolic $H=\frac{1}{2}(1+\kappa r^2)^2\mathbf{p}^2+A\,\frac{(1-\kappa r^2)}{r}$	• 1B $(\gamma=\frac{1}{2};\lambda,\delta)$ Darboux IV $H=\frac{(1-\lambda^2r^2)^2r}{2(1+\lambda^2r^2-2\delta r)}\,\mathbf{p}^2+\frac{Br}{(1+\lambda^2r^2-2\delta r)}$
• 2A $(\beta=2;\kappa=0)$ $H=\frac{\mathbf{p}^2}{2r^2}+\frac{A}{r^2}$	• 2A.1 $(\gamma=1;\lambda=0,\delta=0)$ Euclidean $H=\frac{1}{2}\mathbf{p}^2+Br^2$
	• 2A.2 $(\gamma=1;\lambda=0,\delta)$ Darboux III $H=\frac{\mathbf{p}^2}{2(1-2\delta r^2)}+\frac{Br^2}{(1-2\delta r^2)}$
• 2B $(\beta=2;\kappa)$ $H=\frac{(1+\kappa r^4)^2}{2r^2}\,\mathbf{p}^2+A\,\frac{(1-\kappa r^4)}{r^2}$	• 2B.1 $(\gamma=1;\lambda,\delta=\lambda)$ Sphere/Hyperbolic $H=\frac{1}{2}(1+\lambda r^2)^2\mathbf{p}^2+\frac{Br^2}{(1-\lambda r^2)^2}$
	• 2B.2 $(\gamma=1;\lambda,\delta)$ $H=\frac{(1-\lambda^2r^4)^2}{2(1+\lambda^2r^4-2\delta r^2)}\,\mathbf{p}^2+\frac{Br^2}{(1+\lambda^2r^4-2\delta r^2)}$

• *"Direct" or Schrödinger quantization.*[7] This prescription provides maximally superintegrable Hamiltonians whose spectra presents, as expected, accidental degeneracy.[8] They are given by

$$\hat{\mathcal{H}}_{\rm I}=\frac{1}{2}\hat{\mathbf{q}}^2\left(|\hat{\mathbf{q}}|^{-\beta}+\kappa|\hat{\mathbf{q}}|^{\beta}\right)^2\hat{\mathbf{p}}^2+A\big(|\hat{\mathbf{q}}|^{-\beta}-\kappa|\hat{\mathbf{q}}|^{\beta}\big),$$

$$\hat{\mathcal{H}}_{\rm II}=\frac{\hat{\mathbf{q}}^2\left(|\hat{\mathbf{q}}|^{-2\gamma}-\lambda^2|\hat{\mathbf{q}}|^{2\gamma}\right)^2}{2\left(|\hat{\mathbf{q}}|^{-2\gamma}+\lambda^2|\hat{\mathbf{q}}|^{2\gamma}-2\delta\right)}\,\hat{\mathbf{p}}^2+\frac{B}{\left(|\hat{\mathbf{q}}|^{-2\gamma}+\lambda^2|\hat{\mathbf{q}}|^{2\gamma}-2\delta\right)}.$$

• *Conformal LB quantization.* The quantum free Hamiltonian could be constructed by means of the usual LB operator $\Delta_{\rm LB}$ for the metric g (1):

$$\hat{\mathcal{H}}_{\rm LB,I}=-\tfrac{1}{2}\hbar^2\Delta_{\rm LB}+A\big(|\hat{\mathbf{q}}|^{-\beta}-\kappa|\hat{\mathbf{q}}|^{\beta}\big),$$

$$\hat{\mathcal{H}}_{\rm LB,II}=-\tfrac{1}{2}\hbar^2\Delta_{\rm LB}+B\big(|\hat{\mathbf{q}}|^{-2\gamma}+\lambda^2|\hat{\mathbf{q}}|^{2\gamma}-2\delta\big)^{-1}.$$

However, such quantum Hamiltonians are not maximally superintegrable and their spectrum does not convey accidental degeneracy. Therefore these operators define *different* physical systems with respect to the above ones. Nevertheless, maximal superintegrability can be restored in the LB quantization by considering the so-called conformal LB operator $\Delta_{\rm CLB}$,

$$\Delta_{\rm CLB}=\Delta_{\rm LB}-\frac{(N-2)}{4(N-1)}\,R(|\mathbf{q}|),$$

where R is the scalar curvature (2). Then it can be shown that[8]

$$\hat{\mathcal{H}}_{\mathrm{CLB,I}} = -\tfrac{1}{2}\hbar^2\Delta_{\mathrm{CLB}} + A\big(|\hat{\mathbf{q}}|^{-\beta} - \kappa|\hat{\mathbf{q}}|^{\beta}\big),$$
$$\hat{\mathcal{H}}_{\mathrm{CLB,II}} = -\tfrac{1}{2}\hbar^2\Delta_{\mathrm{CLB}} + B\big(|\hat{\mathbf{q}}|^{-2\gamma} + \lambda^2|\hat{\mathbf{q}}|^{2\gamma} - 2\delta\big)^{-1},$$

are superintegrable Hamiltonians and present accidental degeneracy.

To end with, we stress that the Schrödinger and the conformal LB quantizations are related by means of a gauge transformation. Thus they have the same spectrum and differ in the explicit expressions of the eigenfunctions. In particular, if we consider the Schrödinger equations $\hat{\mathcal{H}}\Phi = E\Phi$ and $\hat{\mathcal{H}}_{\mathrm{CLB}}\Phi_{\mathrm{CLB}} = E\Phi_{\mathrm{CLB}}$ for types I and II, then[8]

$$\hat{\mathcal{H}}_{\mathrm{CLB}} = f(|\hat{\mathbf{q}}|)^{(2-N)/2}\hat{\mathcal{H}}f(|\hat{\mathbf{q}}|)^{(N-2)/2}, \quad \Phi_{\mathrm{CLB}} = f(|\hat{\mathbf{q}}|)^{(2-N)/2}\Phi,$$

where we have used the conformal factor of the metrics (6) and (7), namely,

$$f_{\mathrm{I}}(|\hat{\mathbf{q}}|)^2 = \frac{1}{\hat{\mathbf{q}}^2\left(|\hat{\mathbf{q}}|^{-\beta} + \kappa|\hat{\mathbf{q}}|^{\beta}\right)^2}, \qquad f_{\mathrm{II}}(|\hat{\mathbf{q}}|)^2 = \frac{\left(|\hat{\mathbf{q}}|^{-2\gamma} + \lambda^2|\hat{\mathbf{q}}|^{2\gamma} - 2\delta\right)}{\hat{\mathbf{q}}^2\left(|\hat{\mathbf{q}}|^{-2\gamma} - \lambda^2|\hat{\mathbf{q}}|^{2\gamma}\right)^2}.$$

Acknowledgments

This work was partially supported by the Spanish MINECO through the Ramón y Cajal program (A.E.) and under grants MTM2010-18556 (A.B and F.J.H.), AIC-D-2011-0711 (MINECO-INFN) (A.B, F.J.H. and O.R.) and FIS201-22566 (A.E.), by the ICMAT Severo Ochoa under grant SEV-2011-0087 (A.E.), by Banco Santander-UCM under grant GR35/10-A-910556 (A.E.), and by a postdoctoral fellowship by the Laboratory of Mathematical Physics of the CRM, Université de Montréal (D.R.).

References

1. V. Perlick, *Class. Quant. Grav.* **9**, 1009 (1992).
2. A. Ballesteros, A. Enciso, F. J. Herranz, O. Ragnisco and D. Riglioni, *preprint* (2012).
3. A. Ballesteros, A. Enciso, F. J. Herranz and O. Ragnisco, *Comm. Math. Phys.* **290**, 1033 (2009).
4. A. Ballesteros, A. Enciso, F. J. Herranz and O. Ragnisco, *Class. Quant. Grav.* **25**, 165005 (2008).
5. E. G. Kalnins, J. M. Kress and W. Jr. Miller, *J. Math. Phys.* **46**, 053510 (2005); *ibid J. Math. Phys.* **47**, 043514 (2006).
6. A. Ballesteros, A. Enciso, F. J. Herranz, O. Ragnisco and D. Riglioni, *Ann. Phys.* **326**, 2053 (2011).
7. T. Iwai, Y. Uwano and N. Katayama, *J. Math. Phys.* **37**, 608 (1996).
8. A. Ballesteros, A. Enciso, F. J. Herranz, O. Ragnisco and D. Riglioni, in preparation.

A FINITE MODEL OF THE OSCILLATOR IN TWO-DIMENSIONS WITH $SU(2)$ SYMMETRY

HIROSHI MIKI

Department of Applied Mathematics and Physics,
Graduate School of Informatics, Kyoto University,
Sakyo-Ku, Kyoto 606 8501, Japan
E-mail: miki@amp.i.kyoto-u.ac.jp

SARAH POST

Department of Mathematics
University of Hawaii,
Honolulu, HI 96822, USA
E-mail: spost@hawaii.edu

LUC VINET

Centre de Recherches Mathématiques and Département de Physique,
Université de Montréal,
Montréal, QC H3C 3J7, Canada
E-mail: luc.vinet@umontreal.ca

ALEXEI ZHEDANOV

Donetsk Institute for Physics and Technology,
Donetsk 83 114, Ukraine
E-mail: zhedanov@fti.dn.ua

A superintegrable finite model of the oscillator in two-dimensions is presented. It is defined on a uniform lattice of triangular shape. The wavefunctions are expressed in terms of bivariate Krawtchouk polynomials. The constants of motion form an $SU(2)$ symmetry algebra.

Keywords: Discrete model, harmonic oscillator, orthogonal polynomials, Lie algebras.

1. Introduction

Superintegrable Hamiltonian systems have a number of independent constants of motion that is higher than the number of dimensions in which they are defined. The recent years have seen considerable advances in the classification of these superintegrable models.[1] In spite of this, little is known in the discrete realm. Finite systems are interesting in their own right, they are used for instance in optics and signal processing.[2] The purpose of this note is to report the identification[3] of a finite model of the two-dimensional oscillator which is superintegrable and like its continuum, isotropic version has $SU(2)$ as its symmetry algebra.

2. The model and its $SU(2)$ symmetry algebra

We wish to construct a time-independent finite-difference Schrödinger equation in two-dimensions whose continuum limit will be that of the standard isotropic oscillator. The model will be defined on a uniform lattice of triangular shape with coordinates $\{(x,y) \in \mathbb{Z}^2_{>0} | 0 \leq x+y \leq N\}$. The integer N determines the size of the grid and the continuum limit corresponds to $N \to \infty$. We shall use the following notation for difference operators

$$\Delta_{\pm x, \pm y} f(x,y) = f(x \pm 1, y \pm 1) - f(x,y). \tag{1}$$

The following two seven-point difference operators[4] Λ_1^N and Λ_2^N, play a central role:

$$\begin{aligned}
\Lambda_1^N = & \frac{(N-x-y)p_1p_3\sum p_i}{(p_1+p_3)(p_1p_4-p_2p_3)}\left(\frac{p_2}{p_1+p_2}\Delta_x - \frac{p_4}{p_3+p_4}\Delta_y\right) \\
& + \frac{p_1p_4-p_2p_3}{(p_1+p_2)(p_3+p_4)}\left(x\frac{p_3}{p_1+p_3}\Delta_{-x} - y\frac{p_1}{p_1+p_3}\Delta_{-y}\right) \\
& - x\frac{p_3p_4}{(p_1+p_3)(p_3+p_4)}\Delta_{-x,y} - y\frac{p_1p_2}{(p_1+p_2)(p_1+p_3)}\Delta_{x,-y} \\
\Lambda_2^N = & \frac{(N-x-y)p_2p_4\sum p_i}{(p_2+p_4)(p_1p_4-p_2p_3)}\left(-\frac{p_1}{p_1+p_2}\Delta_x + \frac{p_3}{p_3+p_4}\Delta_y\right) \\
& + \frac{p_1p_4-p_2p_3}{(p_1+p_2)(p_3+p_4)}\left(-x\frac{p_4}{p_2+p_4}\Delta_{-x} + y\frac{p_2}{p_2+p_4}\Delta_{-y}\right) \\
& - x\frac{p_3p_4}{(p_2+p_4)(p_3+p_4)}\Delta_{-x,y} - y\frac{p_1p_2}{(p_1+p_2)(p_2+p_4)}\Delta_{x,-y}.
\end{aligned} \tag{2}$$

A striking observation[3] is that these operators factorize, namely

$$\Lambda_1^N = A_+^{(R,N-1)}A_-^{(R)}, \qquad \Lambda_2^N = A_+^{(L,N-1)}A_-^{(L)}, \tag{3}$$

with

$$\begin{aligned}
A_-^{(R)} &= \frac{p_1p_2p_3p_4\sum p_i}{(p_1+p_3)(p_1p_4-p_2p_3)}\left(\frac{1}{p_4(p_1+p_2)}\Delta_x - \frac{1}{p_2(p_3+p_4)}\Delta_y\right),\\
A_-^{(L)} &= \frac{p_1p_2p_3p_4\sum p_i}{(p_2+p_4)(p_1p_4-p_2p_3)}\left(\frac{-1}{p_3(p_1+p_2)}\Delta_x + \frac{1}{p_1(p_3+p_4)}\Delta_y\right),\\
A_+^{(R,N)} &= \frac{p_1p_4-p_2p_3}{p_1+p_2+p_3+p_4}\left(\frac{x}{p_1}T_x^{-1} - \frac{y}{p_3}T_y^{-1}\right) + (N+1-x-y)\\
A_+^{(L,N)} &= \frac{p_1p_4-p_2p_3}{p_1+p_2+p_3+p_4}\left(\frac{-x}{p_2}T_x^{-1} + \frac{y}{p_4}T_y^{-1}\right) + (N+1-x-y), \quad (4)
\end{aligned}$$

where T is the shift operator defined by $T_lf(l) = f(l+1)$. One can verify that the operators given in Eq. (4) satisfy commutation relations of the Heisenberg algebra type:

$$\begin{aligned}
A_-^{(i)}A_+^{(i,N)} - A_+^{(i,N-1)}A_-^{(i)} &= 1, \qquad i = R, L\\
A_-^{(R)}A_-^{(L)} - A_-^{(L)}A_-^{(R)} &= 0,\\
A_+^{(R,N+1)}A_+^{(L,N)} - A_+^{(L,N+1)}A_+^{(R,N)} &= 0,\\
A_-^{(R)}A_+^{(L,N)} - A_+^{(L,N-1)}A_-^{(R)} &= 0,\\
A_+^{(R,N+1)}A_-^{(L)} - A_-^{(L)}A_+^{(R,N)} &= 0. \qquad (5)
\end{aligned}$$

This hence makes clear that they act as creation and annihilation operators and that Λ_1^N and Λ_2^N are commuting and have linear spectra. In other words, Λ_1^N and Λ_2^N have joint eigenfunctions $K_{m,n}^N$ satisfying

$$\Lambda_1^N K_{m,n}^N = mK_{m,n}^N, \qquad \Lambda_2^N K_{m,n}^N = nK_{m,n}^N. \qquad (6)$$

This suggests taking

$$\mathfrak{h} = \Lambda_1^N + \Lambda_2^N. \qquad (7)$$

as the Hamiltonian of a finite model of the isotropic oscillator in two-dimensions. The spectrum $E_{m,n} = m+n$ of $\mathfrak{h}$ is that of the continuum version. Furthermore, $\mathfrak{h}$ is readily seen to be superintegrable. Indeed, from the commutation relations Eq. (5), it is observed that the quantities

$$\begin{aligned}
J_X &= \frac{1}{2}\left(A_+^{(R,N-1)}A_-^{(L)} + A_+^{(L,N-1)}A_-^{(R)}\right),\\
J_Y &= \frac{i}{2}\left(A_+^{(R,N-1)}A_-^{(L)} - A_+^{(L,N-1)}A_-^{(R)}\right),\\
J_Z &= \frac{1}{2}\left(A_+^{(R,N-1)}A_-^{(R)} - A_+^{(L,N-1)}A_-^{(L)}\right), \qquad (8)
\end{aligned}$$

commute with $\mathfrak{h}$ exactly as in the Schrödinger model (of which this is a finite version). The constants of the motion (8) are seen to realize the $SU(2)$ commutation relations:

$$[J_X, J_Y] = iJ_Z, \quad [J_Y, J_Z] = iJ_X, \quad [J_Z, J_X] = iJ_Y. \tag{9}$$

A basis for the energy eigenstates of $\mathfrak{h}$ is given by the functions $K_{m,n}^N$ satisfying Eq. (6) or equivalently

$$\mathfrak{h}K_{m,n}^N = (m+n)K_{m,n}^N, \qquad J_Z K_{m,n}^N = \frac{1}{2}(m-n)K_{m,n}^N. \tag{10}$$

Here is thus a superintegrable finite model in two-dimensions with $SU(2)$ symmetry that has attributes of the isotropic oscillator. We shall indicate momentarily that it also has the correct continuum limit.

3. The wavefunctions: bivariate Krawtchouk polynomials

The wavefunctions $K_{m,n}^N(x,y)$ of $\mathfrak{h}$ can be built through repeated action of the creation operators. Let us write

$$K_{m,n}^N(x,y) = \frac{(N-m-n)!}{N!}\Pi_{j=0}^{m-1} A_+^{R,N-m-j}\Pi_{i=0}^{n-1} A_+^{L,N-m-n-i}\cdot 1, \tag{11}$$

$m+n \leq N$. As shown in,[3] this corresponds to the Rodrigues formula for the two-variable version of the Krawtchouk polynomials introduced by Rahman.[5] These are given as follows in terms of Aomoto-Gelfand hypergeometric functions:

$$K_{m,n}^N = \sum_{0\leq i+j+k+\ell\leq N} \frac{(-m)_{i+j}(-n)_{k+\ell}(-x)_{i+k}(-y)_{j+\ell}}{i!j!k!\ell!(-N)_{i+j+k+\ell}} u_1^i v_1^j u_2^k v_2^\ell, \tag{12}$$

with parameters

$$\begin{aligned} u_1 &= \frac{(p_1+p_2)(p_1+p_3)}{p_1\sum p_i}, \; u_2 = \frac{(p_1+p_2)(p_2+p_4)}{p_2\sum p_i} \\ v_1 &= \frac{(p_1+p_3)(p_3+p_4)}{p_3\sum p_i}, \; v_2 = \frac{(p_2+p_4)(p_3+p_4)}{p_4\sum p_i}, \end{aligned} \tag{13}$$

and $(a)_n = a(a+1)\cdots(a+n-1)$. They are orthogonal with respect to the trinomial distribution:

$$\omega(x,y) = \binom{N}{x,y}\eta_1^x\eta_2^y(1-\eta_1-\eta_2)^{N-x-y}, \tag{14}$$

with

$$\binom{N}{x,y} \equiv \frac{N!}{x!y!(N-x-y)!}, \text{ and } \quad \begin{aligned} \eta_1 &= \frac{p_1p_2(p_1+p_2+p_3+p_4)}{(p_1+p_2)(p_1+p_3)(p_2+p_4)}, \\ \eta_2 &= \frac{p_3p_4(p_1+p_2+p_3+p_4)}{(p_2+p_4)(p_3+p_4)(p_1+p_3)}. \end{aligned} \tag{15}$$

Given that A_-^L and A_-^R annihilate constants, the action of the ladder operators on the Rahman polynomials is directly obtained from Eq. (5) and Eq. (11); one finds:

$$A_-^{(R)} K_{m,n}^N = \frac{m}{N} K_{m-1,n}^{N-1}, \qquad A_-^{(L)} K_{m,n}^N = \frac{n}{N} K_{m,n-1}^{N-1},$$
$$A_+^{(R,N)} K_{m,n}^N = (N+1) K_{m+1,n}^{N+1}, \qquad A_+^{(L,N)} K_{m,n}^N = (N+1) K_{m,n+1}^{N+1}. \tag{16}$$

Moreover, in view of the $SU(2)$ symmetry of $\mathfrak{h}$, the Rahman polynomials with fixed $j = (m+n)/2$ form a basis for the $(2j+1)$-dimensional irreducible representations of $SU(2)$.

4. The continuum limit

Let us now briefly indicate how the standard isotropic oscillator is recovered in its continuum limit. Introduce the variables s and t related to x and y as follows

$$x = N\eta_1 + \sqrt{N}\,(c_1 s + c_2 t)\,, \qquad y = N\eta_2 + \sqrt{N}\,(c_3 s + c_4 t)\,, \tag{17}$$

with

$$c_1 = \frac{-p_2}{(p_1+p_2)(p_1+p_3)} \sqrt{\frac{2(p_1+p_2+p_3+p_4)p_1 p_3}{p_2+p_4}}$$
$$c_2 = \frac{p_1}{(p_1+p_2)(p_2+p_4)} \sqrt{\frac{2(p_1+p_2+p_3+p_4)p_2 p_4}{p_1+p_3}}$$
$$c_3 = \frac{p_4}{(p_3+p_4)(p_1+p_3)} \sqrt{\frac{2(p_1+p_2+p_3+p_4)p_2 p_4}{p_1+p_3}}$$
$$c_4 = \frac{-p_3}{(p_3+p_4)(p_2+p_4)} \sqrt{\frac{2(p_1+p_2+p_3+p_4)p_2 p_4}{p_1+p_3}}. \tag{18}$$

Note that as $N \to \infty$, the range of s and t becomes the whole real line. With this change of variables, as $N \to \infty$ the operators Λ_1^N and Λ_2^N have the following limits:

$$\lim_{N\to\infty} \Lambda_1^N = -\frac{1}{2}\partial_s^2 + s\partial_s, \qquad \lim_{N\to\infty} \Lambda_2^N = -\frac{1}{2}\partial_t^2 + t\partial_t. \tag{19}$$

Thus, in the limit $N \to \infty$, the Hamiltonian $\mathfrak{h}$ tends to the Hamiltonian

$$\lim_{N\to\infty} \mathfrak{h} = -\frac{1}{2}\left(\partial_s^2 + \partial_t^2\right) + s\partial_s + t\partial_t \tag{20}$$

of the two-dimensional oscillator which has been conjugated by the ground state wavefunction.

As for the wavefunctions, if we let

$$\widehat{K}^N_{m,n}(x,y) = \alpha^N_{m,n} K^N_{m,n}(x,y)$$
$$\alpha^N_{m,n} = \frac{(-\sqrt{2\sum p_i})^{m+n}}{(p_1p_4 - p_2p_3)^{m+n}} \sqrt{\frac{N!\,(p_1p_3(p_2+p_4))^m\,(p_2p_4(p_1+p_3))^n}{(N-m-n)!}} \quad (21)$$

it can be shown[3] that

$$\lim_{N\to\infty} \widehat{K}^N_{m,n} = \epsilon H_m(s)H_n(t), \qquad \epsilon^2 = 1. \quad (22)$$

That is, the $N \to \infty$ limit of the Rahman polynomials is, as expected, the product of Hermite polynomials.

5. Concluding remarks

This rare example we presented of a superintegrable discrete model, has in particular provided context for an algebraic description of the Rahman polynomials. There is another family of bivariate Krawtchouk polynomials introduced by Tratnik.[6] They also form a different basis for the eigenfunctions of $\mathfrak{h}$. The model we developed also allowed to establish the relationship between the Tratnik polynomials and those of Rahman.

Acknowledgments

The authors are grateful to W. Miller Jr, P. Winternitz, and K. B. Wolf for their interest in this model. The research of HM is supported by the Japan Society for the Promotion of Science (JSPS) and that of LV by the Natural Science and Engineering Research Council of Canada.

References

1. W. Miller, S. Post, P. Winternitz and K. B. Wolf (Eds.) *Special Issue on Superintegrability, Exact Solvability, and Special Functions* (SIGMA, 2012) http://www.emis.de/journals/SIGMA/SESSF2012.html
2. N. Atakishiyev, G. Pogosyan and K. B. Wolf *Physics of Particles and Nuclei* **36** 247–265 (2005)
3. H. Miki, S. Post, L. Vinet, and A. Zhedanov, (2012) arXiv: 1208.4142.
4. P. Iliev and P. Terwilliger *Trans. Amer. Math. Soc.* **364** 4225–4238 (2012)
5. M. R. Hoare and M. Rahman M *SIGMA* **4** 089 (2008)
6. M. Tratnik M *J. Math. Phys.* **30** 2740–2749 (1989)

STRUCTURE THEORY FOR EXTENDED KEPLER–COULOMB 3D QUANTUM SUPERINTEGRABLE SYSTEMS

WILLARD MILLER, JR.

School of Mathematics, University of Minnesota,
Minneapolis, MN 55455, USA
** E-mail: mille003@umn.edu*

A quantum superintegrable system is an integrable n-dimensional Hamiltonian system with potential: $H = \Delta_n + V$ that admits $2n-1$ algebraically independent partial differential operators commuting with the Hamiltonian, the maximum number possible. The system is of order ℓ if the maximum order of the symmetry operators, other than H, is ℓ. Typically, the algebra generated by the symmetry operators and their commutators has been proven to close polynomially. However the degenerate 3-parameter potential for the 3D extended Kepler-Coulomb system (2nd order superintegrable) appeared to be an exception as Kalnins et al. (2007) showed that it didn't close polynomially. The 3D 4-parameter potential for the extended Kepler-Coulomb system is not even 2nd order superintegrable. However, Verrier and Evans (2008) showed it was 4th order superintegrable, Tanoudis and Daskaloyannis (2011) showed it closed polynomially. We consider an infinite class of quantum extended Kepler-Coulomb systems that we show to be superintegrable of arbitrarily high order, compute the structure algebras and demonstrate that algebraic closure is the norm, whereas polynomial closure requires extra symmetry. This is a report on joint work with E. G. Kalnins (Waikato) and J. M. Kress (New South Wales).

Keywords: Superintegrable systems; Schrödinger equation.

1. Definitions

A quantum system is defined by a Schrödinger operator $H = \Delta + V(\mathbf{x})$ where $\Delta = \frac{1}{\sqrt{g}} \sum_{ij} \partial_{x_i}(\sqrt{g} g^{ij})\partial_{x_j}$ is the Laplace-Beltrami operator on an n-dimensional Riemannian manifold, in local coordinates x_j. The system is *superintegrable* of order ℓ if it admits $2n-1$ algebraically independent globally defined differential symmetry operators (the maximal number possible) $S_j, \quad 1 \le j \le 2n-1, \quad n \ge 2$, with $S_1 = H$ and $[H, S_j] \equiv HS_j - S_jH = 0$, such that ℓ is the maximum order of the generating symmetries (other than H) as a differential operator. Systems associated with Lie algebras ($\ell = 1$)

and separation of variables ($\ell = 2$) are the simplest and best studied. An integrable system has n algebraically independent commuting symmetry operators whereas a superintegrable system has $2n-1$ independent symmetry operators which cannot all commute and this nonabelian structure is critical for finding the spectral resolution of H by algebraic methods alone. The importance of these systems is that they can be solved exactly.

2. The 4-parameter Kepler system

We shall start with the 4-parameter extended Kepler Coulomb system

$$H = \partial_x^2 + \partial_y^2 + \partial_z^2 + \frac{\alpha}{r} + \frac{\beta}{x^2} + \frac{\gamma}{y^2} + \frac{\delta}{z^2}, \tag{1}$$

where x, y, z are Cartesian coordinates and $r = \sqrt{x^2+y^2+z^2}$. H is 4th order superintegrable. We use the idea introduced in[1,2] and extend the system by passing to spherical coordinates and replacing each angular coordinate θ_i by $k_i\theta_i$ where k_i is a fixed rational number. The extended Kepler-Coulomb operator is $H\Psi = E\Psi$, where $[L_2, L_3] = [L_j, H] = 0$ and

$$H = \partial_r^2 + \frac{2}{r}\partial_r + \frac{\alpha}{r} + \frac{1-k_1^2}{4r^2} + \frac{L_2}{r^2}, \quad L_3 = \partial_{\theta_2}^2 + \frac{\beta}{\cos^2(k_2\theta_2)} + \frac{\gamma}{\sin^2(k_2\theta_2)},$$

$$L_2 = \partial_{\theta_1}^2 + k_1\cot(k_1\theta_1)\partial_{\theta_1} + \frac{L_3}{\sin^2(k_1\theta_1)} + \frac{\delta}{\cos^2(k_1\theta_1)}.$$

Here the 2nd order operators L_2, L_3 are just those that determine multiplicative separation of the Schrödinger equation. The scalar potential is

$$\tilde{V} = \frac{\alpha}{r} + \frac{1-k_1^2}{4r^2} + \frac{1}{r^2}\left(\frac{\beta}{\sin^2(k_1\theta_1)\cos^2(k_2\theta_2)} + \frac{\gamma}{\sin^2(k_1\theta_1)\sin^2(k_2\theta_2)} + \frac{\delta}{\cos^2(k_1\theta_2)}\right).$$

It differs from the classical potential V by the term $\frac{1-k_1^2}{4r^2}$ which corresponds to $R/8$ where R is the scalar curvature of the manifold, just the correction term needed for the conformally invariant Laplacian. It is also needed for superintegrability. Note that for $k_1 \neq 1$ the space isn't flat.

Our basic tasks are 1) to demonstrate that this generalized 4-parameter Kepler-Coulomb problem is superintegrable for all rational k_1, k_2 by constructing the two remaining basic generating symmetries explicitly, and 2) to investigate the closure of the symmetry algebra generated by the basic symmetries and compute the structure equations. The fundamental difficulty is the construction and manipulation of symmetry operators of arbitrarily high order.

The separated solutions for the eigenvalue equations $H\Psi = E\Psi$, $L_3\Psi = -\mu^2\Psi$, $L_2\Psi = \lambda\Psi$ are $\Xi_{p,m,n} = R_p^{k_1\rho}(r)\Phi_m^{(c,b)}(\cos(2k_2\theta_2))\frac{\Psi_n^{(\mu/k_1,d)}(\cos k_1\theta_1)}{\sqrt{\sin(k_1\theta_1)}}$,
$\Phi_m^{(c,b)} = \sin^{c+1/2}(k_2\theta_2)\cos^{b+1/2}(k_2\theta_2)P_m^{(c,b)}(\cos(2k_2\theta_2))$,
$\Psi_n^{(\mu/k_1,d)} = \sin^{\mu/k_1+1/2}(k_1\theta_1)\cos^{d+1/2}(k_1\theta_1)P_n^{(\mu/k_1,d)}(\cos(2k_1\theta_1))$,
$R_p^{k_1\rho} = \frac{e^{-\sqrt{E}r}r^{(k_1\rho+1)/2}}{2\sqrt{E}r}L_p^{k_1\rho}(2\sqrt{E}r)$.
Here $\beta = k_2^2(\frac{1}{4}-b^2)$, $\gamma = k_2^2(\frac{1}{4}-c^2)$, $\delta = k_2^2(\frac{1}{4}-d^2)$, $\mu = k_2(2m+b+c+1)$, $\beta = k_2^2(\frac{1}{4}-b^2)$, $\lambda = \frac{k_1^2}{4}(1-\rho^2)$; $P_n^{(\mu/k_1,d)}$, $P_m^{(c,b)}$ are Jacobi functions, $L_p^{k_1\rho}$ are associated Laguerre functions; $\rho = 2(2n+\frac{\mu}{k_1}+d+1)$ and the energy is $E = \alpha^2/(2[p+2k_1n+2k_2m]+2k_1[d+1]+2k_2[b+c+1]+1)^2$.

A basic insight here is that formal eigenspaces of the Hamiltonian are invariant under action of any symmetry operator, so the operator must induce recurrence relations for the basis of separated eigenfunctions. Our strategy is to use the known recurrence relations for hypergeometric functions to reverse this process and determine a symmetry operator from the recurrence relations. We look for recurrence operators that change the eigenvalues of L_2, L_3 but preserve E. For $k_1 = p_1/q_1, k_2 = p_2/q_2$ we use the mappings 1) : $p \to p+2p_1$, $m \to m$, $n \to n-q_1$, 2) : $p \to p-2p_1$, $m \to m$, $n \to n+q_1$, 3) : $p \to p$, $m \to m-p_1q_2$, $n \to n+q_1p_2$, 4) : $p \to p$, $m \to m+p_1q_2$, $n \to n-q_1p_2$. We first look at recurrences for associated Laguerre functions. To effect the r-dependent transformations 1) and 2) we use $Y(1)^{\pm p}$: ($\tau = -\frac{4\alpha}{2p+k_1\rho+1}$)

$$Y(1)^p_- R_p^{k_1\rho} = [2(k_1\rho+1)\partial_r + (2\alpha + \frac{1-k_1^2\rho^2}{r})]R_p^{k_1\rho} = \tau R_{p-1}^{k_1\rho+2},$$

$$Y(1)^p_+ R_p^{k_1\rho} = [2(-k_1\rho+1)\partial_r + (2\alpha + \frac{1-k_1^2\rho^2}{r})]R_p^{k_1\rho} = \tau(p+1)(p+k_1\rho)R_{p+1}^{k_1\rho-2}.$$

Similarly for Jacobi functions with $z = \cos(2k_1\theta_1)$, we can find first order parameter–dependent recursion operators $Z(1)^n_-$, $Z(1)^n_+$ such that

$$Z(1)^n_- \frac{\Psi_n^{\mu/k_1,d}}{\sqrt{\sin(k_1\theta_1)}} = -2(\frac{\mu}{k_1}+n)(d+n)\frac{\Psi_{n-1}^{\mu/k_1,d}}{\sqrt{\sin(k_1\theta_1)}},$$

$$Z(1)^n_+ \frac{\Psi_n^{\mu/k_1,d}}{\sqrt{\sin(k_1\theta_1)}} = -2(\frac{\mu}{k_1}+d+n+1)(n+1)\frac{\Psi_{n+1}^{\mu/k_1,d}}{\sqrt{\sin(k_1\theta_1)}}.$$

We now form the operators

$$J^{\pm} = \left(Y(1)^{p\pm 2p_1-1}_{\pm} Y(1)^{p\pm 2p_1-2}_{\pm}\cdots Y(1)^{p\pm 1}_{+} Y(1)^{p}_{\pm}\right)\left(Z(1)^{n\mp(q_1-1)}_{\mp}\cdots Z(1)^{n}_{\mp}\right),$$

We can observe that J^+ and J^- switch places under $\rho \to -\rho$ so $J_2 = J^+ + J^-$, $J_1 = (J^- - J^+)/\rho$ are even functions of both ρ and μ and can be interpreted as differential operators. Similarly we can find raising and lowering recurrences satisfying properties 3) and 4), acting on basis functions $\Psi_n^{\mu/k_1,d}$, $\Phi_m^{(c,b)}$ to construct ladder operators K^+, K^-, even functions of ρ^2 that switch places under the reflection $\mu \to -\mu$. Thus $K_2 = K^+ + K^-$, $K_1 = (K^- - K^+)/\mu$ are even functions in both ρ and μ so are differential operators. This proves superintegrability.

The action of J^- on a generalized eigenbasis is given in terms of the rising factorial $(\alpha)_n$ by:

$$J^-\Xi_{p,m,n} = \frac{(2)^{4p_1+q_1}(-1)^{q_1}\alpha^{2p_1}(n+1)_{q_1}(\mu/k_1+d+n+1)_{q_1}}{(2p+k_1\rho+1)^{2p_1}}\Xi_{p-2p_1,m,n+q_1}$$

with similar formulas for $J^+, K^\pm$. Recalling the actions of H, L_2, L_3 on this eigenbasis we have $J^+J^-\Xi_{p,m,n} = 4^{4p_1+q_1}E^{2p_1}(\frac{\rho/2-\frac{\mu}{k_1}-d+1}{2})_{q_1} \times (\frac{\rho/2+\frac{\mu}{k_1}+d+1}{2})_{q_1}(\frac{\rho/2+\frac{\mu}{k_1}-d+1}{2})_{q_1}(\frac{\rho/2-\frac{\mu}{k_1}+d+1}{2})_{q_1}(\frac{k_1\rho-\frac{\alpha}{\sqrt{E}}+1}{2})_{2p_1}(\frac{k_1\rho+\frac{\alpha}{\sqrt{E}}+1}{2})_{2p_1}$ $\times\Xi_{p,m,n}$ and similarly for $J^-J^+\Xi_{p,m,n}$, $K^+K^-\Xi_{p,m,n}$, $K^-K^+\Xi_{p,m,n}$. Thus, each of J^+J^-, J^-J^+ is a polynomial in μ^2 and E and since these operators switch places under the reflection $\rho \to -\rho$, $P_1 = J^+J^- + J^-J^+$, $P_2 = \frac{J^+J^- - J^-J^+}{\rho}$ are polynomials in H, L_2, L_3. Also K^+K^-, K^-K^+ are polynomials in ρ^2 and in μ and switch places under the reflection $\mu \to -\mu$ so $P_3 = K^+K^- + K^-K^+$, $P_4 = \frac{K^+K^- - K^-K^+}{\mu}$ are polynomials in L_2, L_3.

To determine the structure relations it is sufficient to establish them on the generalized eigenbases. The results can then be shown valid independent of basis,[3] For example $[J_1, J_2] = -2q_1J_1^2 - 2P_2(H, L_2, L_3)$, $[K_1, K_2] = -2p_1p_2K_1^2 - 2P_4(L_2, L_3)$, and the unsymmetrized structure relations $J_1^2(\frac{1}{4} - k_1^{-2}L_2) = J_2^2 - 2P_1(H, L_2, L_3) - 2q_1J_1J_2$, $K_1^2 = -K_2^2 + 2P_3(L_2, L_3) + 2p_1p_2K_1K_2$. Eventually, we find for $h, \ell = 1, 2$,

$$[K_\ell, J_h]Q = \{J_1, K_1\}P_{11}^{h\ell} + \{J_1, K_2\}P_{12}^{h\ell} + \{J_2, K_1\}P_{21}^{h\ell} + \{J_2, K_2\}P_{22}^{h\ell},$$

where Q, $P_{jk}^{h\ell}$ are explicit polynomials in L_2, L_3 and $\{A, B\} \equiv AB - BA$. Here, Q is defined by $B(d,\mu)B(d,-\mu)B(-d,\mu)B(-d,-\mu)$ where $B(d,\mu) = (-\frac{d}{2} - \frac{\mu}{k_1} - 2q_1p_2 + \frac{1}{2})_{q_1} + (-\frac{d}{2} - \frac{\mu}{k_1} + \frac{1}{2})_{q_1}$ on a generalized eigenbasis. These identities can be symmetrized, but they show clearly that double commutators of algebra generators cannot be expressed as polynomials in the generators. Rather the double commutators satisfy polynomial equations in the generators. In this sense the algebra closes, but it closes algebraically.

However, there is another problem: the generators $\{H, L_2, L_3, J_1, K_1\}$ are not of minimal order. In particular, for the case $k_1 = k_2 = 1$ we have found a system of generators of orders $(2,2,2,3,5)$ whereas it is known that a generating set of orders $(2,2,2,2,4)$ exists. It is easy to see that J_1 is of order 5 and J_2 is of order 6, whereas K_1 is of order 3 and K_2 is of order 4. We know that there must be other lower order symmetries J_0, K_0 for this case, such that $[L_2, J_0] = J_1,\ [L_3, J_0] = 0,\ L_3, K_0] = K_1,\ [L_2, K_0] = 0$. In[3] we show how to obtain these symmetries from the raising and lowering operators $J^{\pm}, K^{\pm}$, for all rational k_1, k_2. Now consider the symmetry algebra generated by H, L_2, L_3, K_0, J_0. We can find relations between $[J_1, K_0]$ and the generators, so that the algebra closes algebraically.

3. The caged Kepler problem: $k_1 = k_2 = 1$

In the case $k_1 = k_2 = 1$, (1), we are in Euclidean space and our system has additional symmetry. Note that in this case any permutation of the ordered pairs $(x,\beta), (y,\gamma), (z,\delta)$ leaves the Hamiltonian unchanged. This leads to additional structure in the symmetry algebra. This system was shown to be superintegrable by Verrier and Evans,[4] The basic symmetries are: $L_3 = I_{xy} = (x\partial_y - y\partial_x)^2 + \frac{\beta(x^2+y^2)}{x^2} + \frac{\gamma(x^2+y^2)}{y^2}$. and by permutation symmetry, I_{xz}, I_{yz}, and $L_2 = I_{xy} + I_{xz} + I_{yz} - (\beta+\gamma+\delta)$. The constant of the motion K_0 is 2nd order: $K_0 = -\frac{1}{32}(I_{yz} - I_{xz})$, and J_0 is 4th order: $J_0 = -4M_3^2 - \frac{1}{2}\times$

$$\{(\{x,\partial_x\}+\{y,\partial_y\}+\{z,\partial_z\})^2, \frac{\delta}{z^2}\}+2H\left(I_{xz}+I_{yz}-(\beta+\gamma+\frac{3}{4})\right)+5\frac{\delta}{z^2}+\frac{\alpha^2}{2},$$

$$M_3 = \frac{1}{2}\{(y\partial_z - z\partial_y), \partial_y\} - \frac{1}{2}\{(z\partial_x - x\partial_z), \partial_x\} - z\left(\frac{\alpha}{2r} + \frac{\beta}{x^2} + \frac{\gamma}{y^2} + \frac{\delta}{z^2}\right).$$

The symmetries H, L_2, L_3, J_0, K_0 form a generating (algebraic) basis for the symmetry operators. Under the transposition $(x,\beta) \leftrightarrow (z,\delta)$ the operators H, L_2, L_3, K_0 transform to linear combinations of themselves. However, J_0 is mapped to a new symmetry J_0' linearly independent of the original basis symmetries. Tanoudis and Daskaloyannis[5] show that the algebra generated by the 6 algebraically dependent symmetries H, L_2, L_3, J_0, K_0 and J_0' closes polynomially, in the sense that all double commutators of the generators are again expressible as polynomials in the generators. However more is true. *All* formally self-adjoint symmetry operators, such as $\{K_1, J_1\}$, K_1^2, J_1^2, $[K_1, J_1]$ are expressible as symmetrized polynomials in the generators. Further we can deduce an explicit polynomial relation among the 6 generators, an operator identity of order 12, quadratic in J_0': $0 \equiv Q_5 + \{Q, J^2\} + 32\{K_0, J_0\}^2 -$

$16\{L_2+L_3-\delta-1, \{J, \{K_0, J_0\}\}\}+\{Q_1, K_0^2\}+\{Q_2, J_0^2\}+\{Q_3, K_0\}+\{Q_4, J_0\},$

where $J = -8(J_0 + 2J_0' + \frac{1}{2}H - \frac{\alpha^2}{2})$, and Q_1, Q_2, Q_3, Q_4, Q_5 are explicit polynomials in H, L_2, L_3 of orders 2,4,5,4,6, respectively.

4. Conclusions and Outlook

The recurrence relation method developed by Kalnins, Kress and Miller[6] for proving superintegrability and determining the structure equations for families of 2D quantum superintegrable systems can be extended to the 3D case. The construction appears to be quite general and not restricted to Kepler-Coulomb analogs. For these higher order superintegrable systems it appears that algebraic closure is the norm. For polynomial closure, extra symmetry is needed. We have not proved in all cases that there do not exist other generators of lower order but, if they exist, they must also be obtainable in terms of recurrence relations of hypergeometric functions. A crucial role is played by the raising and lowering operators. They are not defined independent of eigenbasis and are not even symmetries, but all symmetries are built from them.

Acknowledgment

This work was partially supported by a grant from the Simons Foundation (# 208754 to Willard Miller, Jr).

References

1. F. Tremblay, A.Turbiner and P. Winternitz. An infinite family of solvable and integrable quantum systems in the plane. J.Phys.A: Math. Theor., **42**, (2009) 242001.
2. F. Tremblay, A. Turbiner and P. Winternitz. Periodic orbits for an infinite family of classical superintegrable systems. J. Phys. A: Math. Theor. **43** (2010) 015202.
3. E. G. Kalnins, J. M. Kress and W. Miller, Jr. The 3 and 4-parameter extended Kepler-Coulomb quantum systems in 3 dimensions (submitted)
4. N. W. Evans and P. E. Verrier. Superintegrability of the caged anisotropic oscillator. *J. Math. Phys.* **49**, 092902, 10 pages, arXiv:0808.2146, (2008).
5. Y Tanoudis and C. Daskaloyannis. Algebraic calculation of the energy eigenvalues for the non degenerate three dimensional Kepler Coulomb potential. arXiv:11020397v1, (2011)
6. E. G. Kalnins, J. M. Kress and W. Miller Jr. A recurrence relation approach to higher order quantum superintegrability, SIGMA 7 (2011), 031, 24 pages arXiv:1011.6548v3 [math-ph]

PARALLEL SESSION
Applications in Biology

21 & 23 & 24 AUGUST 2012

ON THE ION-MEDIATED INTERACTION BETWEEN PROTEIN AND DNA

M. BARBI

CNRS LPTMC UMR 7600, Université Pierre et Marie Curie-Paris 6
4 place Jussieu, 75252 Paris Cedex 05, France
E-mail: barbi@lptmc.jussieu.fr

F. PAILLUSSON

Department of Chemistry, University of Cambridge, Lensfield Road
CB2 1EW, Cambridge, UK
E-mail: fp286@cam.ac.uk

The mechanism allowing a protein to search for a target sequence on DNA is currently described as an intermittent process made of 3D diffusion in bulk and 1D diffusion along the DNA molecule. Due to the relevant charge of protein and DNA, electrostatic interaction should play a crucial role during this search. In this paper, we explicitly derive the mean field theory allowing for a description of the protein-DNA electrostatics in solution. This approach leads to a unified model of the search process, where 1D and 3D diffusion appear as a natural consequence of the diffusion on an extended interaction energy profile.

Keywords: DNA; Proteins; ionic liquids; modelling.

1. Introduction

Many proteins in living cells have to search specific, short sequences on long DNA molecules in order to perform their biological task. Such *DNA-binding proteins* have proven to be very efficient in searching their target: their association constants can be two orders of magnitude higher than what is expected from a simple 3D diffusion.[1,2] It has been suggested[3,4] that such a rapid reaction rate can results from an *intermittent diffusion*, swapping between a 1D diffusion along DNA – or *sliding* – and a 3D diffusion in solution – or *jumping*. An increasing number of single molecule experiments has been able to evidence sliding, confirming this scenario.[5–8] Experiments also show that both sliding and jumping result to be sensitive to the salt concentration.[1,6,8] This supports the idea that electrostatics is involved in the intermittent behaviour, with a probable role for the ions in solution.

In this paper, we shall first recall the statistical mechanics of ions in solution for a given fixed charge distribution. We will then introduce a toy model for a DNA-protein system[9,10] for which we will discuss some unexpected features. This will finally allow us to gain more insights about the physics at play during the search of target by a protein and to propose a method to get further insights on protein physical properties.

2. Statistical mechanics of electrolytes

In this section we consider two fixed macro-ions confined in a domain $\Sigma \subset \mathbb{R}^3$ that contains an electrolyte solution. We denote ρ_f the charge density carried by these macro-ions. Within the framework of statistical mechanics,[11] the position $\mathbf{R}_j$ of an ion j is a random vector that can take any value $\mathbf{r}_i$ belonging to a subset Ω of Σ. The set of N particle positions $\{\mathbf{r}_1, .., \mathbf{r}_N\}$ gotten at each *trial* for N ions in the system will be called an *ionic configuration* and denoted $\mathcal{C}$. In thermodynamic equilibrium, the conditional probability density to get a specific configuration $\mathcal{C}$ knowing that the system is at temperature β^{-1}/k_B and confined in the domain Ω of volume V reads:

$$p(\mathcal{C}|\beta, \mathbf{N}, V) \equiv \mathbf{1}_{\mathcal{C}\in\Omega^N}\, e^{-\beta H(\mathcal{C})}/Q_{[\beta,\mathbf{N},V]} \tag{1}$$

where the characteristic function $\mathbf{1}_{\mathcal{C}\in\Omega^N}$ is zero if any of the ions is outside Ω and one otherwise. In Eq. (1), we introduced the normalization functional:

$$Q_{[\beta,\mathbf{N},V]} \equiv \frac{1}{\prod_\alpha^m N_\alpha! \Lambda_\alpha^{3N_\alpha}} \int_{\Omega^N} d^{3N}(\mathcal{C})\, e^{-\beta H(\mathcal{C})} \tag{2}$$

with $d^{3N}(\mathcal{C})$ being the 3N-dimensional Lebesgue measure on $\mathbb{R}^{3N}$, m is the number of different ionic species in solution, Λ_α is the de Brooglie wavelength of the species α and the sets $\mathbf{N} = \{N_\alpha\}_{\alpha=1..m} \in \mathbb{N}^m$ and $\mathbf{q} = \{q_\alpha\}_{\alpha=1..m} \in \mathbb{R}^m$ characterize the *ionic composition* of the mixture. The real valued function $H(\mathcal{C})$ in Eqs. (1) and (2) is the energy of the system for a given configuration $\mathcal{C}$ that defines the model used. Here, we rely on the so called *Restricted Primitive Model* (RPM),[12] defined as:

$$H(\mathcal{C}) = \frac{1}{2} \int_\Sigma d^6 rr' \frac{\rho_\mathcal{C}(\mathbf{r})\rho_\mathcal{C}(\mathbf{r}')}{4\pi\varepsilon|\mathbf{r}-\mathbf{r}'|} + \sum_{j<k} v_{HS}(j,k) \tag{3}$$

where ε is the dielectric permittivity of water and where we use the Dirac delta "function" to define the charge density $\rho_\mathcal{C}(\mathbf{r}) \equiv \sum_j eq_j\delta(\mathbf{r}-\mathbf{r}_j)+\rho_f(\mathbf{r})$.

The second term in Eq. (2) stands for a hard-sphere repulsion such that $e^{-\beta v_{HS}(j,k)}$ behaves as $\Theta(|\mathbf{r_j} - \mathbf{r_k}| - D)$ where $\Theta(x)$ is the Heavyside step function and D the diameter of the ions[a]. From Eq. (2), one can then perform a *Hubbard-Stratonovich* transform that effectively breaks Coulomb pair interactions into a one-body potential[12,13] ϕ [b]:

$$e^{-\frac{\beta}{2}\int_\Sigma d^6rr' \frac{\rho_C(\mathbf{r})\rho_C(\mathbf{r}')}{4\pi\varepsilon|\mathbf{r}-\mathbf{r}'|}} \equiv \int \frac{\mathcal{D}[\phi]}{\mathcal{Z}[0]} e^{-\frac{\beta}{2}\int_\Sigma d^6rr'\, \phi(\mathbf{r})G^{-1}_{\mathbf{r},\mathbf{r}'}\phi(\mathbf{r}') - i\beta\int d^3r\phi(\mathbf{r})\rho_C(\mathbf{r})} \tag{4}$$

where $i^2 = -1$ and $G^{-1}_{\mathbf{r},\mathbf{r}'} = -\varepsilon\Delta\delta(\mathbf{r} - \mathbf{r}')$.[13] The factor $\mathcal{Z}[0]$ is the normalization factor for the free field ϕ in absence of ρ_C. If we introduce the Gaussian average over configurations of the field ϕ, $\langle . \rangle_\phi$, then the r.h.s. of Eq. (4) reads $\langle e^{-i\beta\int d^3r\,\rho_C\phi}\rangle_\phi$. It is now convenient to swap to a grand canonical ensemble where the composition $\mathbf{N}$ is a random vector taking any value in $\mathbb{N}^m$ with a probability weight $e^{\beta\mathcal{M}\cdot\mathbf{N}} \equiv e^{\beta\sum_\alpha \mu_\alpha N_\alpha}$ set by m *chemical potentials* $\mathcal{M} = \{\mu_\alpha\}_{\alpha=1..m} \in \mathbb{R}^m$, each of which corresponds to a particular ionic species. In this case, the normalization functional writes:

$$Q_{[\beta,\mathcal{M},V]} \equiv \sum_{\mathbf{N}} e^{\beta\mathcal{M}\cdot\mathbf{N}} Q_{[\beta,\mathbf{N},V]} \tag{5}$$

Inserting the partition function (2) (after performing (4)) into Eq. (5) yields:

$$Q_{[\beta,\mathcal{M},V]} = \left\langle Q^{HS}_{[\beta,\mathcal{M}(i\phi),V]} e^{-i\beta\int d^3r\,\rho_f\phi} \right\rangle_\phi \tag{6}$$

where $Q^{HS}_{[\beta,\mathcal{M}(i\phi),V]}$ is the grand partition function[11] of a mixture of *bare* hard spheres with the set of chemical potentials $\mathcal{M}(i\phi) = \mathcal{M} - ie\mathbf{q}\phi$. We now use the fact that the mixture is dilute by approximating $Q^{HS}_{[\beta,\mathcal{M}(i\phi),V]}$ by the first term of its Mayer expansion[11] i.e. $Q^{HS}_{[\beta,\mathcal{M}(i\phi),V]} \approx \exp(\int d^3r \sum_\alpha e^{\beta(\mu_\alpha - ieq_\alpha\phi)}/\Lambda^3_\alpha)$. Eq. (6) becomes then explicitely:

$$Q_{[\beta,\mathcal{M},V]} = \int \frac{\mathcal{D}[\phi]}{\mathcal{Z}[0]} e^{-\beta\int_\Sigma d^3r\,[\varepsilon\frac{(\nabla\phi)^2}{2} - \beta^{-1}\mathbf{1}_{\mathbf{r}\in\Omega}\sum_{\alpha=1}^m e^{\beta(\mu_\alpha - ieq_\alpha\phi)}/\Lambda^3_\alpha + i\rho_f\phi]} \tag{7}$$

Note that the quadratic term in ϕ from Eq. (4) has now become quadratic in $\nabla\phi$, by applying twice the divergence theorem[c].

---

[a]The diameter D is such that we can assume each ionic species to be in a stable gas phase in solution.

[b]The measure $\mathcal{D}[\phi]$ can be thought of as the limit of the measure $\prod_k^{(L/\epsilon+1)^3} d\gamma(\phi_k)$ — γ being a complex measure — characterizing field configurations on a 3D-lattice of size L and lattice spacing ϵ when the latter tends to zero.

[c]The boundary terms arising from this theorem do not contribute since global electroneutrality is assumed in Σ.

The so called *Poisson-Boltzmann* (PB) theory can now be readily gotten from Eq. (7) by using a functional saddle point or *mean field* approximation.[12,14] The grand potential defined as $\mathcal{G}_{[\beta,\mathcal{M},V]} \equiv -\beta^{-1} \ln Q_{[\beta,\mathcal{M},V]}$ reads then:

$$\mathcal{G}_{[\beta,\mathcal{M},V]} \stackrel{MF}{=} -\int_{\Sigma} d^3r \, [\varepsilon \frac{(\nabla\varphi)^2}{2} + \beta^{-1} \mathbf{1}_{\mathbf{r}\in\Omega} \sum_{\alpha=1}^{m} \frac{e^{\beta(\mu_\alpha - h - eq_\alpha\varphi)}}{\Lambda_\alpha^3} - \rho_f \varphi] \quad (8)$$

where the sign $\stackrel{MF}{=}$ stands for an equality within the mean field approximation and where $\varphi \equiv i\phi_s$, ϕ_s being the field evaluated at the saddle point. With the new field h introduced in (8) and ρ_f, $\mathcal{G}_{[\beta,\mathcal{M},V]}$ can be seen as a generating functional from which one can get grand canonical averages of meaningful quantities. In particular $\langle i\phi \rangle_{\beta,\mathcal{M},V} \equiv -(\delta\mathcal{G}_{\beta,\mathcal{M},V}/\delta\rho_f)_{h=0} \stackrel{MF}{=} \varphi$ and $\langle \rho_C^\alpha \rangle_{\beta,\mathcal{M},V} \equiv -eq_\alpha(\delta\mathcal{G}_{\beta,\mathcal{M},V}/\delta[\beta(\mu - h)])_{h=0} \stackrel{MF}{=} eq_\alpha \mathbf{1}_{\mathbf{r}\in\Omega} e^{\beta\mu_\alpha} e^{-\beta eq_\alpha\varphi}/\Lambda_\alpha^3$. The average charge density $\langle \rho_C^\alpha \rangle_{\beta,\mathcal{M},V}$ has to be a real number which implies then that φ has to be a real field. It is therefore common to use Eq. (8) as a functional of the real valued function φ called the *Poisson-Boltzmann functional* that has to be extremalized numerically to find the most probable electrostatic field φ and its corresponding charge density.[15] An equivalent way to look at it is to realize that if φ maximizes the PB functional: the functional derivative of the latter with respect to the former has therefore to be zero. This gives rise to an Euler-Lagrange type of equation called the *Poisson-Boltzmann equation* to be solved for φ:

$$\Delta\varphi = -\frac{1}{\varepsilon} \left(\mathbf{1}_{\mathbf{r}\in\Omega} \sum_{\alpha=1}^{m} eq_\alpha \frac{e^{\beta\mu_\alpha}}{\Lambda_\alpha^3} e^{-\beta eq_\alpha\varphi} + \rho_f \right) \quad (9)$$

which turns out to be the most common route used to determine the average electrostatic potential φ.

3. Modelling protein-DNA interactions

Many DNA binding proteins happen to have a concave shape matching that of DNA. This is believed to optimize the recognition at the target site. In addition, these proteins need to be positively charged otherwise they would be repelled by DNA's high negative charge. In a previous work, we have suggested a toy model to study the relevance of the geometry in DNA-protein non specific interactions.[9,10] In this model, depicted in Fig. 1 (a), the model DNA (MDNA) is a uniformly charged cylinder and the model protein (MP) is a cylinder of larger radius but with an indentation of cylindrical shape that matches exactly the DNA shape, and is positively charged at the interface. They are also immersed in a RPM of a symmetric

1:1 electrolyte as described in Eq. (3). In the terminology of section (2), the charge density on these macromolecules – for a given distance $\mathcal{L}$ between them – corresponds to $\rho_f(\mathcal{L})$, the whole system is in a domain Σ and $\Omega(\mathcal{L})$ is the accessible region to ions i.e. anywhere in Σ except inside the macromolecules. The + and − ion bulk concentrations are assumed to be the same and denoted $n_b \equiv e^{\beta\mu_\pm}/\Lambda_\pm^3$. The grand potential $\mathcal{G}_{[\beta,\mathcal{M},V]}$

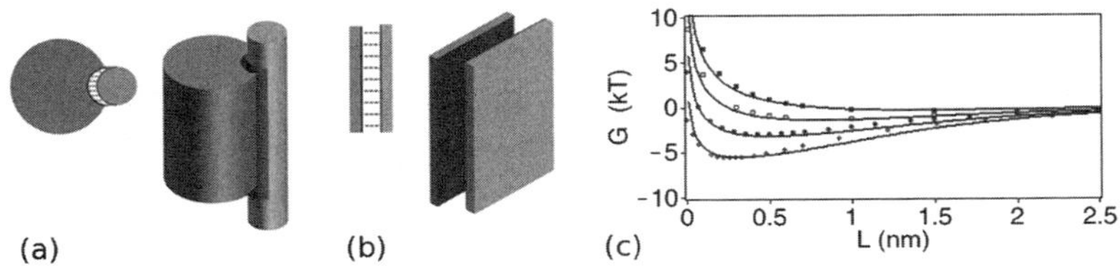

Fig. 1. **(a)** Toy model for a protein interacting with a DNA segment in solution. **(b)** The toy model can be mapped onto two oppositely but non-symmetrically charged plates in solution. **(c)** Free energy profiles obtained by integrating the PB equation for the plate plate system with (physiological) monovalent salt concentration of 0.1 mol/L, and for 4 different protein charge densities in the range from 0.06 to 0.3 times the absolute value of the DNA charge density.

of this whole system is implicitly dependent on the domain $\Omega(\mathcal{L})$ and the distribution $\rho_f(\mathcal{L})$. In fact, if we were to consider $\mathcal{L}$ as a random variable subject to thermal fluctuations, the grand potential $\mathcal{G}_{[\beta,\mathcal{M},V]}(\mathcal{L})$ would act as an *effective interaction energy*[16] between the MDNA and the MP such that each value l of $\mathcal{L}$ appears with a probability weight $e^{-\beta\mathcal{G}_{[\beta,\mathcal{M},V]}(l)}$.

Monte Carlo (MC) simulations of the system described above have been performed to compute exactly $\mathcal{G}_{[\beta,\mathcal{M},V]}(\mathcal{L})$ via a so called *thermodynamic integration.*[17] It was found[9,10] that $\mathcal{G}_{[\beta,\mathcal{M},V]}(l)$ is an increasing function for $l > l^*$ and a decreasing function if $l \in [0, l^*[$. In physical terms, l^* corresponds to an equilibrium distance between the MP and the MDNA segment.[18] It was also shown that the profile $\mathcal{G}_{[\beta,\mathcal{M},V]}(l)$ gotten from MC simulations could be matched with a PB theory for two plates (Fig. 1 (b)) by solving the PB equation (9) for φ and evaluating the expression (8) for every l. This "mapping" between the MC implementation of the toy model of Fig. 1 (a) and a PB treatment of a two plate system is valid provided that effective charge densities (related to that of the MDNA and MP) are used for the plates.[9,10] The actual values of these parameters depend on the particular modelling of ρ_f used in the MC simulations and therefore do not provide at the moment any more insights about what is happening in the system.

Overall, the intermittent behaviour observed for the DNA-protein system can be rationalized by considering the random nature of $\mathcal{L}$ and treating properly the physics of the ions: in a non-specific DNA-protein bound state, sliding is possible at the equilibrium distance l^*, while thermal fluctuations can still make the protein escape from DNA, in which case it would perform a jump. Let us finally note that, in practice, the density ρ_f cannot be measured experimentally, and is often inferred from structural data. Interestingly, the simple planar PB description that we have introduced is not only able to capture this physics, but also provides analytical expressions – as a function of effective charge densities ρ_f – for both l^* and $\mathcal{G}_{[\beta,\mathcal{M},V]}(l^*)$. Since these quantities directly determine the kinetic behaviour of the protein, a comparison with independent structural and dynamical data from experiments may be used as a simple alternative to estimate coarse grained surface densities at the protein-DNA interface.

References

1. A. D. Riggs, S. Bourgeois and M. Cohn, *J. Mol. Biol.* **53**, 401 (1970).
2. P. H. Richter and M. Eigen, *Biophysical Chemistry* **2**, 255 (1974).
3. P. von Hippel and O. Berg, *J. Biol. Chem.* **264**, 675 (1989).
4. M. Coppey, O. Bénichou, R. Voituriez and M. Moreau, *Biophys. J.* **87**, 1640 (2004).
5. N. Shimamoto, *J. Biol. Chem.* **274**, 15293 (1999).
6. P. Blainey, A. van Oijen, A. Banerjee, G. Verdine and X. Xie, *Proc. Natl. Acad. Sci. U.S.A.* **103**, 5752 (2006).
7. J. Elf, G. Li and X. Xie, *Science* **316**, 1191 (2007).
8. I. Bonnet, A. Biebricher, P.-L. Porté, C. Loverdo, O. Bénichou, R. Voituriez, C. Escud, W. Wende, A. Pingoud and P. Desbiolles, *Nucleic Acids Research* **36**, 4118 (2008).
9. V. Dahirel, F. Paillusson, M. Jardat, M. Barbi and J.-M. Victor, *Phys. Rev. Lett.* **102**, p. 228101 (2009).
10. F. Paillusson, V. Dahirel, M. Jardat, J.-M. Victor and M. Barbi, *Phys. Chem. Chem. Phys.* **13**, 12603 (2011).
11. D. McQuarrie, *Statistical mechanics* (University science books, 2000).
12. J.-M. Caillol, *J.Stat.Phys.* **115**, p. 1461 (2004).
13. D. Brydges and P. Martin, *J.Stat.Phys.* **96**, p. 1163 (1999).
14. R. Netz and H. Orland, *Eur. Phys.J. E* **1**, p. 203 (1999).
15. A. Maggs, *Eur. Phys.Lett.* **98**, p. 16012 (2012).
16. G. Voth, *Coarse-Graining of Condensed Phase and Biomolecular Systems* (CRC press, 2008).
17. D. Frenkel and S. B., *Understanding molecular simulation (second edition)* (Academic Press, 2002).
18. F. Paillusson, M. Barbi and J.-M. Victor, *Mol. Phys.* **107**, 1379 (2009).

HOW TO IDENTIFY LATERALLY TRANSFERRED GENES?

PIOTR ŚLIWKA

Fac. of Mathematics and Natural Sciences, St. Cardinal Wyszyński University,
Wóycickiego 1-3, 01-938 Warsaw, Poland
** E-mail: p.sliwka@uksw.edu.pl*

MAŁGORZATA DUDKIEWICZ

Dep. of Experimental Design and Bioinformatics, Warsaw University of Life Sciences
Nowoursynowska 159, 02-776 Warsaw, Poland
E-mail: malgorzata.dudkiewicz@sggw.waw.pl

The existence of the phenomenon of Horizontal Gene Transfer (HGT) is one of the important processes by means of which the progress of the evolution is explained. The purpose of this article was to propose a comprehensive and combined mathematical-statistical approach to the identification of HGT events, which is not found in the biological literature. In this article the database consisting of CLCA sequences and 13 whole bacterial genomes was used to test known methods of HGT events' identification. Moreover some new proposals were presented. One of them is the application of new measures, i.e. the angle and the area between the vectors identified with the codon genes. The statistical ranking of all used methods has been developed on the basis of the obtained results.

Keywords: *Horizontal Gene Transfer mechanism, CLCA, measure of similarity, statistcal ranking.*

1. Introduction

Horizontal gene transfer (HGT) is believed to be important mechanism for the attainment of genetic plasticity in many species of bacteria and is defined to be the movement of genetic material between cells other than by descent in which information travels through the generations as the cell divides. Horizontal gene transfer events were reported not only in bacterial species but, through bacterial intermediation, also in animals and plants cells. It is highly desirable to distinguish genes acquired by HGT from the whole pool of genes of the given species. With the progress of genomics and bioinformatics a vast amount of sequence data is now available which makes possible analyzing the general characteristics such as Codon Usage or GC content on the genomic scale. To study heterogeneity of those characteristic one various classical statistical or/and mathematics methods were used (e.g. factor analysis,[1] correspondence analysis,[1] correlation coefficients,[2] standard deviation,[2] L1 or L2 distance,[3] Kullback-Leibler divergence,[3,4] multivariate analysis,[5] etc.) and provided different results without leading method.

The purpose of this article was a comprehensive the combined mathematical-statistical approach to the identification of HGT events, which is not found in the biological literature. In this article the database consisting of CLCA sequences (calcium-activated chloride channel - one of the group of genes suspected to be horizontally transferred) and 13 whole bacterial genomes (abbreviation of the genome given in square brackets): *Afulgidus* [Af], *Alkalimnicola* [Al], *Apernix* [Ap], *Chloroflexus* [Ch], *Escherica coli* [Ec], *Gsulfurreducens* [Gs], *Herpetosiphon* [He], *Helicobacter pylori* [Hp], *Roseiflexus* [Ro], *Staphyloccocus aureus* [Sa], *Shewanella baltica* [Sh], *Shewanella MR1* [Sw], *Thermococcus* [Th]) was used to test known methods of HGT events identification. Moreover some new proposals were presented. One of them is the application of new measures, i.e. the angle and the area between the vectors identified with the codon genes. The statistical ranking of all used methods has been developed on the basis of the obtained results.

2. Material and methods

Let X^i be the n-dimensional vector of numbers of codons (or amino acids) identified with the i-th gene as the random variable, p_{X_i} - probability distribution of X_i, and X=$[X_i]_{i=1,\ldots,n}$ - the database genes of the genome (details on the construction of the r.v. X - see Appendix). The purpose of this analysis is to verify whether the genes suspected to originate in another genome (e.g. Y) actually does not come from the genome X.

In the first step classical methods for HGT events detection such as: discriminant analysis, exploratory and confirmatory factor analysis, correspondence analysis, and so on were verified. The obtained results did not allow for a clear way to assess whether a particular gene did not come from other than the considered genome. The combination of statistical tools (the Hypothesis Testing) and analytical methods (measure of the distance) was used. In the next step the comparative statistical analysis using 10 measures (abbreviation of these measures given in brackets) like: L2 norm (L2), Kolmogorov-Smirnov statistics (KS), measure of entropy (En), Pearson correlation coefficient (Co), Gini index (Gi), Brownian distance (Br),

Kullback-Leibler divergence (KL) and some new propositions (1) based on:

1. the angle ϕ between vectors (x,y) and the length of the vector (An, LoV),
2. the area between normed vectors x and y (Ar) were applied.

$$Ar_{x,y} = \frac{||x||\,||y||\sin\phi}{2} \tag{1}$$

To test the significance of differences between CLCA (suspected) and the gene from a genome the following six-step procedure separately at two levels was used:

1. first level - all amino acids (61 codons),
2. second level - individual amino acids (20 amino acids).

Step 1: Compute expected value (arithmetic or/and weighted average based on the length of the gene - Appendix) of the database (genome).

Step 2: Compute difference between expected value of the genome and each gene from database using described above measures: L2, KS, En, Co, Gi, Br, KL, An, LoV, Ar.

Step 3: Compute difference between CLCA (or suspected) gene and each gene from database (genome) using measures L2, KS ,..., Ar.

Step 4: Compute average of step 2 and average of step 3.

Step 5: Compare the averages of step 4 using appropriate statistical test (checking first the appropriate assumptions used for the test) and return p-value (the probability of obtaining a test statistic at least as extreme as the one that was actually observed, assuming that the H0 is true).

Step 6.1: Build the table with percentage of the p-value $\leqslant 0.05$ of the test for all amino acids (61 codons).

Step 6.2: Build the table with a mean of the p-value of the distribution for the individual amino acids (20 amino acids).

The consistence of the relevant comparisons of the probability distributions or its lack was verified on the basis of statistical tests[a]. No significant differences between CLCA (or suspected of external origin) gene and each gene from the database (suspected gene from the genome) if the p-value level exceeded 5% were found.

On this basis the percentage rate of the cases that did not meet the above condition for p-value as a new measure R_p was set.

[a]The following tests have been used: Levene test (the equality of variances in different samples), Welch t-test (robust for mean equality, intended for use with two samples having possibly unequal variances) and Brown-Forsythe test for mean (or median).

3. Results

As mentioned above the percentage ratio results (measure R_p) are included in Table 1 and Table 2 (G/M in the header of the table is Genome/Measure). The first column contains the abbreviation of the name of the genome used for the analysis. The first row contains the abbreviation of distance measure used for the analysis. The table contains the values of the R_p based on codons (Table 1) and based on amino acids (Table 2). R_p closer to 100 means that a greater percentage of codons, what indicates the lack of links between the CLCA and the genome. R_p closer to 0 confirms the origin of the suspected gene from the genome.

Table 1. R_p measure: the percentage of the p-value $\leqslant 0.05$ of W test - the all amino acids case (as a vector with 61 coordinates)

G/M	L2	KS	En	Co	Gi	Br	KL	An	LoV	Ar
Af	58%	58%	58%	0%	33%	0%	33%	50%	0%	58%
Al	100%	100%	100%	0%	0%	0%	100%	100%	100%	100%
Ap	50%	40%	50%	0%	40%	0%	40%	30%	0%	50%
Ch	0%	0%	0%	0%	0%	0%	0%	0%	0%	0%
Ec	88%	88%	88%	0%	75%	0%	50%	88%	25%	88%
Gs	78%	67%	78%	0%	0%	11%	33%	78%	11%	56%
He	0%	0%	0%	0%	0%	0%	0%	0%	0%	0%
Hp	100%	100%	100%	0%	25%	0%	25%	50%	0%	100%
Ro	0%	0%	0%	0%	0%	0%	0%	0%	0%	0%
Sa	44%	44%	67%	0%	11%	0%	11%	44%	22%	56%
Sh	100%	100%	0%	0%	100%	0%	0%	100%	0%	100%
Sw	100%	100%	100%	0%	0%	0%	100%	100%	0%	100%
Th	0%	0%	0%	0%	0%	0%	0%	0%	0%	0%

The better discrimination between the genes from the genome and the gene suspected of external origin was indicated by higher values in the Table 1 and Table 2. In Table 1 only L2 norm, Kolmogorov-Smirnov statistics, measure of entropy and area between normed vectors provide satisfactory results. In the case of the individual amino acids (Table 2) only L2 norm, Kolmogorov-Smirnov statistics, measure of entropy, Kullback-Leibler and area between normed vectors provide satisfactory results. Therefore a simple ranking of the measures ("the best discrimination" $> \ldots >$ "the worst discrimination") was done on the basis of the achieved results of R_p, :

Ar > L2 > KS - En - An > KL > LoV > Co > Br > Gi .

Table 2. R_p measure: the percentage of the p-value $\leqslant 0.05$ of W test - the individual amino acids case (20 amino acids)

G/M	L2	KS	En	Co	Gi	Br	KL	An	LoV	Ar
Af	78%	78%	79%	50%	3%	35%	80%	79%	75%	80%
Al	85%	80%	85%	65%	0%	35%	75%	75%	75%	80%
Ap	77%	77%	74%	49%	5%	42%	83%	74%	74%	77%
Ch	70%	70%	55%	60%	0%	25%	70%	70%	60%	70%
Ec	82%	82%	81%	64%	3%	44%	81%	82%	78%	82%
Gs	79%	79%	81%	58%	5%	36%	78%	79%	76%	80%
He	65%	60%	60%	45%	0%	35%	70%	65%	65%	65%
Hp	80%	80%	84%	53%	0%	31%	74%	80%	69%	80%
Ro	75%	75%	60%	45%	0%	30%	75%	75%	60%	75%
Sa	79%	78%	83%	47%	2%	36%	78%	79%	72%	79%
Sh	70%	70%	65%	50%	0%	40%	70%	70%	70%	70%
Sw	70%	70%	70%	55%	0%	45%	70%	70%	65%	70%
Th	70%	70%	65%	35%	0%	10%	55%	65%	75%	65%

4. Conclusions

The following conclusions can be made on the basis of the performed analyses and the obtained results, :

1. Classical methods like factor analysis, discriminant analysis,. . . ,etc. do not clearly differentiate gene suspected of Horizontal Gene Transfer (in the context of whole genome).

2. Positive results using discriminant analysis and proposed measures (61 coordinates) were obtained for 3 out of 7 suspected bacterial CLCA's sequences.

3. Positive results using proposed measures (20 amino acids) were obtained for 10 out of 13 suspected bacterial sequences from another genome.

4. From statistical point of view the proposed measures allowed to determine the probability that the gene from the tested genome was transferred from another genome.

The methodology of the analyses suggests that in the future the following methods should be used:

- Bayesian approach will be used to find the "a priori" distribution of the gene (a posteriori distribution based on "suspected" gene)

- Gibbs sampler and MCMC to the parameter estimation of the multivariate distribution of the genes.

5. Appendix

Representation of the database fragment containing the numbers of the codons (general 61 codons: Phe/TTT, Phe/TTC, Leu/TTA, Leu/TTG, Leu/CTT,. . . ., Gly/GGA, Gly/GGG, etc.) or amino acids (20 amino acids:

Phe, Leu, Ile,..., Gly, etc.) and the probability distribution of the amino acids given in Table 3 and Table 4 respectively. The first column in Table 3 and Table 4 indicates the length of the gene used to determine the weighted average (step 1 in section "Material and methods").

Table 3. Part of the distribution of the numbers of codons *(Shewanella baltica)*

Length of the gene	Phe TTT	Phe TTC	...	Gly GGT	Gly GGC	Gly GGA	Gly GGG
1389	12	11	...	14	5	3	3
1101	6	4	...	12	7	1	0
1083	16	1	...	10	7	4	3
...	...	...	...	...	...	...	...
138	1	1	...	1	3	0	0

Table 4. Part of the probability distribution of the amino acids *(Shewanella baltica)*

Length of the gene	Phe TTT	Phe TTC	...	Gly GGT	Gly GGC	Gly GGA	Gly GGG
1389	0.52	0.48	...	0.56	0.2	0.12	0.12
1101	0.6	0.4	...	0.6	0.35	0.05	0
1083	0.94	0.06	...	0.42	0.29	0.17	0.12
...	...	...	...	...	...	...	...
138	0.5	0.5	...	0.25	0.75	0.0	0.0

References

1. S. Garcia-Valve, E. Guzman, M. Montero, A. Romeu, HGT-DB: a database of putative horizontally transferred genes *Nucleic Acids Res.* **31(1)**, p.187 (2003).
2. S. Garcia-Valve,J. Palau, A. Romeu, Horizontal gene transfer in glycosyl hydrolases *Mol. Biol. Evol.* **16(9)**, p.1125 (1999).
3. I. Bahir, M. Fromer, Y. Prat, M. Linial, Viral adaptation to host: a proteome-based analysis of codon usage and amino acid preferences *Mol. Syst. Biol.* **5**, p.311 (2009).
4. RK. Azad, JG. Lawrence, Detecting laterally transferred genes: use of entropic clustering methods and genome position in *Nucleic Acids Res.* **35(14)**, p. 4629 (2007).
5. S. Kanaya, Y. Yamada, T. Ikemura, Studies of codon usage and tRNA genes of 18 unicellular organisms *Gene* **238**, p.143 (1999).

PATCH SIR MODELS ON THE K-REGULAR GRAPH

D. I. WALLACE *, E. HU-WANG and M. CHEN
Department of Mathematics, Dartmouth College, Hanover, NH 03755, U.S.A.
**E-mail: dwallace@math.dartmouth.edu*

The epidemiology of person-to-person communicable disease in large, homogeneous populations is modeled by three ordinary differential equations. These represent the susceptible, infectious and recovered individuals. For smaller populations an agent-based model can capture the inhomogeneous nature of human interactions by modeling each individual as a node in a network, or vertex in a graph. In this model we look at a set of large homogeneous populations, each modeled as a classic SIR epidemic (the patches), but which have immigration between them (the graph). We use this approach to simulate the spread of influenza during an outbreak and compare results of the simulation with available data.

It is a mathematically interesting and epidemiologically useful question to ask which aspects of disease transmission are controlled by the local properties of each SIR model, and which are controlled by the global connectivity of the graph. We prove that, for a general class of patch SIR models, the stability of the disease free equilibrium is a local property.

Keywords: SIR models, patch models, regular graphs, epidemiology.

1. Introduction and Models

The classic model of an infectious disease is given by three coupled ordinary differential equations for susceptible, infectious and recovered individuals (SIR models).[1] This model is appropriate for large homogeneous well-mixed populations. Models for diseases that spread from city to city consider populations that are separated into patches, each of which is described by an SIR model, with migration between patches given as a linear term. Equations 1-3 below give the general form for patch SIR models.

$$S_i' = f(S_i) - \beta S_i I_i - \Sigma_{j\sim i}(i,j)m_{(i,j)}S_i + \Sigma_{j\sim i}m_{(j,i)}S_j \tag{1}$$

$$I_i' = \beta S_i I_i - (v + b + \alpha) I_i - \Sigma_{j \sim i} m_{(i,j)} I_i + \Sigma_{j \sim i} m_{(j,i)} I_j \quad (2)$$

$$R_i' = v I_i - b R_i - \Sigma_{j \sim i} m_{(i,j)} R_i + \Sigma_{j \sim i} m_{(j,i)} R_j \quad (3)$$

Here $f(S_i)$ is the birth rate of the susceptible population, as any function of S_i only, b and v are birth and death rates, α is the death rate of the disease only, β is the transmission coefficient, $m_{(i,j)}$ is the relative rate of travel from vertex i to vertex j, assumed to be equal for susceptible, infectious and recovered populations at vertex i. Sums denoted by $j \sim i$ indicate vertices adjacent to the ith vertex. Note the $f(S_i)$ is not the most general form that could be taken, but it is good enough for many applications.

Such models can be considered hybrids between spacial models incorporating partial differential equations and single agent models evolving on graphs whose vertices represent individual members of a population. Patch models enable one to delineate local properties depending on the dynamics inside a patch from global properties depending on properties of the graph. For example, some graphs, such as Ramanujan graphs,[2] are known to be good propagators of information for their size. These properties should play a role in propagation of disease in patch models.

2. A Simulation of the U.S. influenza outbreak in 2009-2010

Six major U.S. areas connected by large airports (CA, CO, TX, GA, IL, NY) were used as a basis for parametrizing the model in equations 1-3. For this simulation, $f(S_i)$ was set to $a(1 - S_i/k_i) - dS_i$. The birth rate, a, and death rate, d, were taken as the U.S. average[34] and k_i was chosen to give an equilibrium population at the disease free equilibrium corresponding to the actual population of each city.[5] The transmission and death rates from this particular strain can be estimated from commonly known statistics[6] and the recovery rate can be estimated from data.[7] The graph between patches was the complete graph on six vertices, with migration rates estimated from airport data.[8] The resulting simulation was then compared with the patient data from clinics in each area (Regions 2,4,5,6,8,9) that report to the Center for Disease Control. The data gives the percent of those patients who, having presented with flu-like symptoms, tested positive for influenza.[9] Early cases were detected in California and Texas.[10] These were taken as initial conditions in the model, with disease incidence at all other locations taken to be zero. The resulting projections for all cities are shown in Figure 1, as well as the percent positive disease data for the U.S. as a whole. Figure

2 shows the projection for California plotted against the percent positive disease data for California.

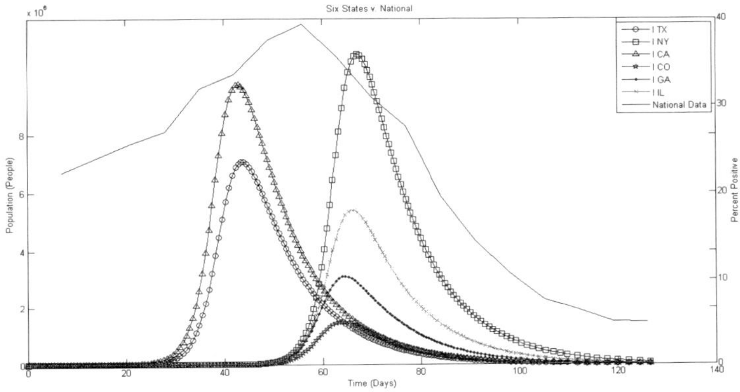

Fig. 1. Model predictions of all six cities, scale on left. National disease incidence (percent positive) data, scale on right.

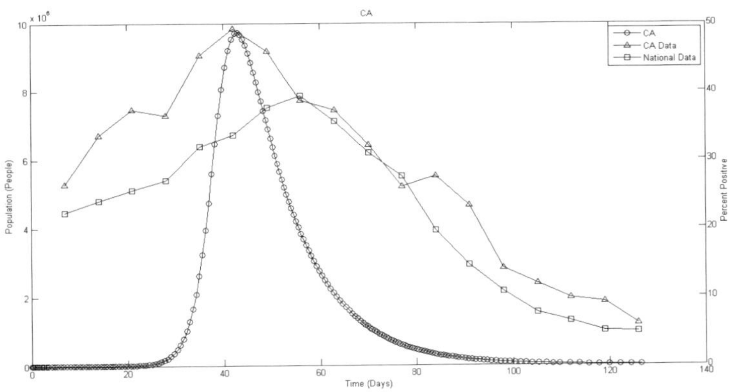

Fig. 2. Model prediction for California, scale on left. National and California disease incidence (percent positive) data, scale on right.

Data is not available for, nor is it easy to estimate, the actual number of influenza cases in a geographic region. However we might hope for a reasonable representation of timing of an outbreak, as measured by percent positive data. In Figure 1 the national average for the data sits right between the two projected peaks. The early peaks are for California and Texas, where the disease was presumed to start. The later peaks are for the other states, after migration brings them the disease. In Figure 2 the peak of the simulation for California lines up well with the peak of the data, as did the simulation for the New York region. The other four simulations (CO, GA, IL, TX) all peaked a week or two later than the data for the respective region.

3. A General Theorem for Homogeneous Systems on the k-regular Graph

Theorem 3.1. *Let a patch model on a graph be given by equations 1-3. Assume that* $m_{(i,j)} = m$ *for all edges. Assume that the graph is regular, with* n *vertices and* k *edges at each vertex. Assume that, at the disease free equilibrium all* S_i^* *are equal to* S^*. *Let* $f'(S^*)$ *denote the partial derivative of* f *with respect to* S_i *at the disease free equilibrium. Let* A *denote the adjacency matrix for the graph and let* $diag(c_i)$ *denote the* n *by* n *diagonal matrix with diagonal elements* c_i. *Assume further that*

(1) the eigenvalues of $diag(\beta S^* - (v+b+\alpha) - km) + mA)$ *are distinct from those of* $diag(f'(S^*) - \beta I_i - km) + mA$, *and*
(2) the eigenvalues of $diag(\beta S^* - (v+b+\alpha) - km) + mA)$ *are distinct from those of* $diag(-b - km) + mA$

Then the stability of the disease free equilibrium does not depend on n, k, m *or any other properties of the graph. The stability of the DFE depends only on* β, v, b, α, *and* $f'(S_i*)$.

Proof.

With assumptions and notation as above, the Jacobian, J, at the disease free equilibrium is given by:

$$\begin{pmatrix} \text{diag}((f'(S^*) - \beta I_i - km) + mA & \text{diag}(-\beta S_i) & 0 \\ 0 & \text{diag}(\beta S_i - (v+b+\alpha) - km) + mA) & 0 \\ 0 & \text{diag}(v) & \text{diag}(-b-km) + mA \end{pmatrix}. \tag{4}$$

For simplicity we will assign labels to each n by n submatrix as below.

$$\begin{pmatrix} M & P & 0 \\ 0 & Q & 0 \\ 0 & R & N \end{pmatrix}. \tag{5}$$

It is clear that

(1) vectors of the form $(V, 0, 0)^t$, where V is an eigenvector of M, and
(2) vectors of the form $(0, 0, W)^t$, where W is an eigenvector of N

are eigenvectors of J.

Let U^t be an eigenvector of Q with eigenvalue λ. Then $(A, U, B)^t$ is an eigenvector of J provided

(1) $MA^t + PU^t = \lambda A^t$, or equivalently $A^t = (\lambda I - M)^{-1}PU^t$
(2) $RU^t + NB^t = \lambda B^t$, or equivalently $B^t = (\lambda I - N)^{-1}RU^t$

Vectors A and B may be constructed because the matrices $(\lambda I - M)$ and $(\lambda I - N)$ are invertible under the hypotheses of the theorem. Thus we have that the eigenvalues of J are given by the eigenvalues of M, Q and N. For stability of the disease free equilibrium it suffices to show that the largest eigenvalue of any of these matrices is negative. In each case, the largest eigenvalue is bounded by the maximum of $(diag(f'(S^*) - km) + m\Lambda, diag(\beta S^* - (v + b + \alpha) - km) + m\Lambda, diag(-b - km) + m\Lambda)$, where Λ is the largest eigenvalue of A. For k-regular graphs this is known to be k, so the largest eigenvalue of J is therefore bounded by the maximum of $(diag(f'(S^*)), diag(\beta S^* - (v + b + \alpha)), diag(-b))$.

This gives the result. □

4. Summary

In this paper we give a general form for a patch SIR model with migration between connected patches represented as a graph with adjacency matrix A. The population growth term for each patch is given as any general function of the susceptible population. Often this is taken as a constant recruitment rate that is offset by a relative death rate. An even more general setting would allow population growth to be a function of the total population. However the assumption in the models here fits well with diseases that pass relatively quickly through a population, such as influenza.

A version of the general form is used as the basis of a numerical simulation of the influenza outbreak of 2010 using airport traffic data for six regions to estimate migration rates. The simulation shows a very good match in timing of outbreaks in two of six cities, and a somewhat late timing (compared to percent positive clinical data) in the other four. It suggests that simulations based on airport traffic data may be useful in bracketing the timing of outbreaks within a given period. Further, the peak of the outbreak for the U.S. as a whole sits neatly between the earliest and latest peaks of the simulation.

In a special case of the general model we take the patches to be identical, all migration rates to be identical, and the graph to be regular. For this situation we prove that the stability of the disease free equilibrium is a local property. It depends only on the parameters associated to birth, death, transmission and recovery for a single patch.

References

1. R. M. Anderson and R. M. and May, *Nature* **280**, 361 (1979).
2. A. Lubotzky, R. Phillips and P. Sarnak, *Combinatorica*8 **3**, 261 (1988).
3. Center for Disease Control and Prevention (Atlanta 2009), [Data File]. Retrieved May 3, 2012 from http://www.cdc.gov/nchs/births.htm
4. Center for Disease Control and Prevention (Atlanta 2009), [Data File]. Retrieved May 3, 2012 from `http://www.cdc.gov/nchs/deaths.htm`
5. Center for Disease Control and Prevention (Atlanta 2010). Retrieved May 3, 2012 from `http://www.cdc.gov/tb/statistics/reports/2010/pdf/report2010.pdf`
6. A. Gardner (ABC News, New York 2009). Retrieved May 3, 2012 from `http://abcnews.go.com/Health/Healthday/story?id=8439860`
7. Center for Disease Control and Prevention (Atlanta 2010) Retrieved May 3, 2012 from `http://www.cdc.gov/h1n1flu/qa.htm`
8. Airports Council International, North America (Montreal 2009), [Data File] Retrieved May 3, 2012 from `http://aci-na.org/content/airport-traffic-reports`
9. Center for Disease Control and Prevention (Atlanta, 2010). [Data Set]. Retrieved Oct. 3, 2012 from `http://gis.cdc.gov/grasp/fluview/fluportaldashboard.html`
10. Center for Disease Control and Prevention (Atlanta 2009). *MMWR* **58** **15**, 400-402. Retrieved Oct. 10, 2012 from `http://www.cdc.gov/mmwr/preview/mmwrhtml/mm5815a5.htm`

PARALLEL SESSION
Quantum Information and Representation Theory

20 & 21 & 22 AUGUST 2012

CONFUSABILITY GRAPHS FOR SYMMETRIC SETS OF QUANTUM STATES

GIULIO CHIRIBELLA

Center for Quantum Information, Institute for Interdisciplinary Information Sciences, Tsinghua University, Beijing, 100084, China
Email: gchiribella@mail.tinsghua.edu.cn

YUXIANG YANG

Department of Physics, Tsinghua University, Beijing, 100084, China
Email: yangyx09@mails.tsinghua.edu.cn

For a set of quantum states generated by the action of a group, we consider the graph obtained by considering two group elements adjacent whenever the corresponding states are non-orthogonal. We analyze the structure of the connected components of the graph and show two applications to the optimal estimation of an unknown group action and to the search for decoherence free subspaces of quantum channels with symmetry.

1. Introduction

A fundamental feature of quantum theory is the impossibility to reliably distinguish between non-orthogonal pure states.[1] This impossibility is at the root of the no-cloning theorem[2] and of the information-disturbance trade-offs that guarantee the security of quantum cryptography.[3] Non-orthogonality can be used to partition a set of pure states into subsets exhibiting genuine quantum features. Consider a map φ that encodes elements of a set X into unit vectors a Hilbert space $\mathscr{H}$

$$\varphi : \mathsf{X} \to \mathscr{H} \qquad x \in \mathsf{X} \mapsto |\varphi_x\rangle \in \mathscr{H} \tag{1}$$

The map φ endows X with a graph structure: The *confusability graph* associated to φ, denoted by $\Gamma(\mathsf{X}, \mathscr{H}, \varphi)$, is the graph where X is the set of vertices and two vertices $x, y \in \mathsf{X}$ are connected by an edge whenever $\langle \varphi_x | \varphi_y \rangle \neq 0$.

By this definition, two vertices $x, y \in \mathsf{X}$ are adjacent if and only if the corresponding states $|\varphi_x\rangle, |\varphi_y\rangle \in \mathscr{H}$ are confusable (i.e. non-orthogonal). A vertex x is connected to a vertex y if and only if we can go from the state

$|\varphi_x\rangle$ to the state $|\varphi_y\rangle$ through a path of confusable states.

Denoting by Γ_k, $k \in \mathsf{K}$ the connected components of $\Gamma(\mathsf{X}, \mathscr{H}, \varphi)$ and by X_k the subset of vertices belonging to Γ_k, we can partition the set of states $\varphi(\mathsf{X}) = \{|\varphi_x\rangle \mid x \in \mathsf{X}\}$ into mutually orthogonal subsets $\varphi(\mathsf{X}_k) = \{|\varphi_x\rangle \mid x \in \mathsf{X}_k\}$. Intuitively, the subsets $\varphi(\mathsf{X}_k)$ identify sectors of the Hilbert space $\mathscr{H}$ where the encoding φ exhibits genuinely quantum features. More precisely, one can define the mutually orthogonal subspaces

$$S_k := \mathsf{Span}\{|\varphi_x\rangle \mid x \in \mathsf{X}_k\} \qquad k \in \mathsf{K} \tag{2}$$

and the corresponding orthogonal projectors $P_k, k \in \mathsf{K}$. With this definition, the linear map $\mathcal{D}$ on trace-class operators defined by

$$\mathcal{D}(\rho) = \sum_{k \in \mathsf{K}} P_k \rho P_k \tag{3}$$

does not disturb states in the set $\varphi(\mathsf{X})$, that is, $\mathcal{D}(|\varphi_x\rangle\langle\varphi_x|) = |\varphi_x\rangle\langle\varphi_x|$ for every $x \in \mathsf{X}$. The map $\mathcal{D}$ represents a partial decoherence process that preserves the off-diagonal elements of a density matrix only within the subspaces S_k. Equivalently, we can interpret $\mathcal{D}$ as the result of a Lüders measurement that extracts information about the orthogonal subspaces S_k without disturbing states that have support inside these subspaces.

In this paper, we consider confusability graphs generated by the action of a group G. The map φ will be of the form

$$\varphi : \mathsf{G} \to \mathscr{H} \qquad g \in \mathsf{G} \mapsto |\varphi_g\rangle := U_g|\varphi\rangle, \tag{4}$$

where $|\varphi\rangle \in \mathscr{H}$ is a unit vector and $U : \mathsf{G} \to \mathcal{B}(\mathscr{H}), g \mapsto U_g$ is a unitary projective representation of G. In this special case, the confusability graph will be denoted by $\Gamma(\mathsf{G}, \mathscr{H}, U, |\varphi\rangle)$, to stress that the map φ is now specified by the choice of a representation U and of an input state $|\varphi\rangle$. In section 2, we will show that the component of the graph connected to the identity is a subgroup of G and that the other components are left-cosets. We then show the implications of this structure for the problem of quantum estimation of an unknown group action (section 3) and for the problem of finding decoherence free subspaces of quantum channels with symmetry (section 4). In particular, we will show that a covariant channel that preserves a single pure state $|\varphi\rangle \in \mathscr{H}$ will automatically preserve any state with support in one of the subspaces S_k. In other words, all the subspaces S_k will be automatically *decoherence free*.[4]

Before presenting our results, we would like to mention their connection to previous related works. An early appearance of a confusability graph is the *orthonormal representation of a graph* introduced by Lovász.[5] This notion, which is central to the study of zero error communication (see e.g.[6]), was recently generalized to the quantum case by Duan, Severini, and Winter.[7] In zero-error communication, the problem is to find maximal sets of disconnected (that is, non-confusable) points in the graph. Our discussion will take a complementary point of view, focussing instead on the connected components and on their role in quantum estimation and in the study of decoherence free subspaces.

2. Group theoretic structure of the confusability graph

Consider a confusability graph $\Gamma(\mathsf{G}, \mathscr{H}, U, |\psi\rangle)$. By definition, two group elements $g \in \mathsf{G}$ and $h \in \mathsf{G}$ are connected, denoted by $\sim$, if and only if there exists a finite path $\{g_i \in \mathsf{G}\}_{i=1}^{N}$ such that $g_1 = g$, $g_N = h$, and $\prod_{i=1}^{N-1} \langle \varphi_{g_i} | \varphi_{g_{i+1}} \rangle \neq 0$. It is then easy to prove the following

Proposition 2.1. *Let $\Gamma(\mathsf{G}, \mathscr{H}, U, |\varphi\rangle)$ be a confusability graph. Then,*

(1) the connected component $\mathsf{H} := \{h \in \mathsf{G} \mid h \sim e\}$ is a subgroup of G
(2) the connected components of $\Gamma(\mathsf{G}, \mathscr{H}, U, |\varphi\rangle)$ are the left cosets of G with respect to H.

Proof. If $g \sim g'$, then $hg \sim hg'$ for every $h \in \mathsf{G}$. Using this fact it is immediate to see that H is a group. It is also clear that $g \sim g'$ if and only if $g^{-1}g' \sim e$, that is, if and only if $g^{-1}g'$ belongs to H. Hence, the connected component $\mathsf{H}_g : \{g' \in \mathsf{G} \mid g' \sim g\}$ is the left coset $\mathsf{H}_g = g\mathsf{H}$. ■
Note that for a continuous representation of a topological group, the subgroup H always contains the component of the group topologically connected to the identity. However, the notion of connectedness that we use here makes sense even in cases where no topology is given a priori.

An important family of input states, containing the best possible input states for the estimation of an unknown group element $g \in \mathsf{G}$, is the family of the *class states*:[8] a class state $|\varphi\rangle \in \mathscr{H}$ can be concisely defined as a vector such that the characteristic function[9] $f(g) := \langle \varphi | U_g | \varphi \rangle$ is a class function, namely a function such that $f(h^{-1}gh) = f(g)$ for every $g, h \in \mathsf{G}$. For this particular type of states we have the condition

Proposition 2.2. *Let $\Gamma(\mathsf{G}, \mathscr{H}, U, |\varphi\rangle)$ be the confusability graph of a class state $|\varphi\rangle \in \mathscr{H}$. Then, the connected component H is a* normal *subgroup of G.*

Proof. For a class state, $g \sim g'$ implies $gh \sim g'h$ for every $h \in \mathsf{G}$. By the same proof of proposition 2.1, we then obtain that the connected components are right cosets of G. Hence, we must have $g\mathsf{H} = \mathsf{H}g$ for every

$g \in \mathsf{G}$, because $g\mathsf{H}$ and $\mathsf{H}g$ are two intersecting connected components. ■

In the case of class states, the decoherence map in Eq. (3) can be interpreted as the result of an projective measurement with outcome in the quotient group G/H.

3. Application to quantum estimation

The group-theoretic structure of the confusability graph has an immediate application to the problem of group parameter estimation.[1,8,10,11] Suppose that a quantum system with Hilbert space $\mathscr{H}$, initialized in a pure state $|\psi\rangle \in \mathscr{H}$, undergoes an unknown unitary transformation U_g. The problem is to find the best quantum measurement, represented by *positive operator valued measure (POVM)* $P(d\hat{g})$ in order to estimate g with maximum precision. The figure of merit is the minimization of a cost functional

$$c(\rho, P) := \sup_{g \in \mathsf{G}} \left\{ \int_{\mathsf{G}} c(\hat{g}, g)\ \langle \varphi_g | P(d\hat{g} | \varphi_g \rangle \right\}, \tag{5}$$

where $c(\hat{g}, g)$ is a cost function satisfying the property $c(h\hat{g}, hg) = c(\hat{g}, g), \forall \hat{g}, g, h \in \mathsf{G}$. In this problem, it is well known that the optimization can be restricted to the set of *covariant POVMs*,[1] making the problem significantly simpler. Inspecting the structure of the confusability graph, we can further simplify the problem, reducing the search of the optimal POVM to the search of an optimal POVM for the lower-dimensional subspace S_{H}, the subspace in Eq. (2) generated by the group elements in H. Indeed, we can write the unknown group element g as $g = \tilde{g}h$, where h belongs to H and $\tilde{g}$ is a fixed representative of the coset $g\mathsf{H}$. To estimate g, we can first identify $\tilde{g}$ by performing a projective measurement on the subspaces S_k, $k \in \mathsf{G}/\mathsf{H}$ defined in Eq. (2) (this measurement does not disturb the quantum state $|\varphi_g\rangle$). Then we can apply the unitary $U_{\tilde{g}}^\dagger$ to the state $|\varphi_g\rangle$, turning it into the state $|\varphi_h\rangle \in S_{\mathsf{H}}$. Besides the reduction of the dimensionality, this procedure has the additional bonus that the estimation problem inside S_{H} still has the form of Eq. (5), with the group G now replaced by its subgroup H. Hence, the problem of finding the optimal G-covariant POVM on $\mathscr{H}$ is reduced to the problem of finding the optimal H-covariant POVM on S_{H}.

4. Decoherence free subspaces of covariant channels

The confusability graph has a remarkable application in the search for decoherence free subspaces of quantum channels with group symmetry. Consider a quantum channel $\mathcal{C}$, namely a completely positive trace-preserving normal map transforming trace-class operators on $\mathscr{H}$. The channel $\mathcal{C}$ is said to be *covariant* if it satisfies the relation $\mathcal{C} \circ \mathcal{U}_g = \mathcal{U}_g \circ \mathcal{C}, \forall g \in \mathsf{G}$, where $\mathcal{U}_g$ is the unitary channel defined by $\mathcal{U}_g(\rho) := U_g \rho U_g^\dagger$. The following proposition can be used to simplify the search for decoherence free subspaces of covariant channels:

Proposition 4.1. ***(Decoherence free subspaces of covariant channels)*** *Suppose that the covariant channel $\mathcal{C}$ has a pure fixed point, i.e. that there is a pure state $|\psi\rangle \in \mathscr{H}$ such that $\mathcal{C}(|\psi\rangle\langle\psi|) = |\psi\rangle\langle\psi|$. Then, each subspace S_k, corresponding to a connected component of the confusability graph $\Gamma(\mathsf{G}, \mathscr{H}, U, |\varphi\rangle)$ is decoherence free.*

The proof is based on the observation that, due to covariance, one has $\mathcal{C}(|\psi_g\rangle\langle\psi_g|) = |\psi_g\rangle\langle\psi_g|$ for every $g \in \mathsf{G}$. The thesis follows by combining this fact with a simple lemma, which holds for arbitrary confusability graphs:

Lemma 4.1. *Suppose that the projectors on the pure states $\{\varphi_x \mid x \in \mathsf{X}\}$ are fixed points of the channel $\mathcal{C}$. Then, each subspace S_k associated to a connected component of the confusability graph $\Gamma(\mathsf{X}, \mathscr{H}, \varphi)$ is decoherence free.*

The lemma can be proved as a corollary of a more general result about the algebraic structure of the fixed points of quantum channels.[12] However, the lemma has also a very elementary proof, similar to the proof of the no-cloning theorem:[2]

Proof of lemma 4.1 Consider a unitary dilation of the channel $\mathcal{C}$, given by $\mathcal{C}(\rho) = \mathrm{Tr}_{\mathscr{H}_E}[U(\rho \otimes |\eta\rangle\langle\eta|)U^\dagger]$ for some Hilbert space $\mathscr{H}_E$, some unitary operator $U \in \mathcal{B}(\mathscr{H}_E)$, and some unit vector $|\eta\rangle \in \mathscr{H}_E$. Since the pure state $|\varphi_x\rangle$ is a fixed point of $\mathcal{C}$ we must have $U|\varphi_x\rangle|\eta\rangle = |\varphi_x\rangle|\eta_x\rangle$, where $|\eta_x\rangle \in \mathscr{H}_E$ is a unit vector possibly depending on x. For two points $x, y \in \mathsf{X}$, we then have $\langle\varphi_x|\varphi_y\rangle = \langle\varphi_x|\varphi_y\rangle\langle\eta_x|\eta_y\rangle$. Now, if $\langle\varphi_x|\varphi_y\rangle \neq 0$, we must have $|\eta_x\rangle = |\eta_y\rangle$. Hence, the vector $|\eta_x\rangle$ is constant on the connected components of the confusability graph $\Gamma(\mathsf{X}, \mathscr{H}, \varphi)$: $|\eta_x\rangle = |\eta_{x'} \equiv |\eta_k\rangle$ for every $x, x' \in \mathsf{X}_k$. Taking a generic linear combination $|\psi\rangle = \sum_{x \in \mathsf{X}_k} c_i |\varphi_x\rangle \in S_k$ we have $U|\psi\rangle|\eta\rangle = |\psi\rangle|\eta_k\rangle$, and, therefore $\mathcal{C}(|\psi\rangle\langle\psi|) = |\psi\rangle\langle\psi|$. ■

In addition to simplifying the search for decoherence free subspaces, proposition 4.1 places a constraint on the covariant channels that can achieve a desired task—such as quantum teleportation[13]—that involves the preservation of a pure state. For example, a covariant channel that teleports a single state $|\varphi\rangle$ must be able teleport all the states in the subspaces S_k corresponding to connected components of the confusability graph $\Gamma(\mathsf{G}, \mathscr{H}, U, |\varphi\rangle)$. The connected component S_{H} can easily be the whole Hilbert space: for a teleportation protocol that is covariant under time-translations,[14] assuming a non-degenerate Hamiltonian, the ability to teleport a quantum state $|\varphi\rangle = \sum_{n=0}^{\infty} c_n|n\rangle$, $c_n \neq 0, \forall n$ (and hence the whole trajectory $|\varphi_t\rangle = \sum_{n=0}^{\infty} c_n e^{i\omega_n t}|n\rangle$, with $\omega_n \neq \omega_{n'}$ for $n \neq n'$) is equivalent to the ability to teleport an arbitrary quantum state $|\psi\rangle \in \mathscr{H}$.

Acknowledgement

This work is supported the National Basic Research Program of China (973) 2011CBA00300 (2011CBA00301) and by the National Natural Science Foundation of China through Grants 61033001 and 61061130540.

References

1. C. Helstrom, *Quantum detection and estimation theory* (Academic Press, New York, 1976); A. S. Holevo, *Probabilistic and Statistical Aspects of Quantum Theory* (North Holland, Amsterdam, 1982).
2. W. Wootters and W. Zurek, *Nature* **299**, 802 (1982).
3. N. Gisin, G. Ribordy, W. Tittel, and H. Zbinden, *Rev. Mod. Phys.* **74**, 145 (2002).
4. G. Palma, K. Suominen, and A. Ekert, *Proc. Roy. Soc. Lond. A* **452**, 567 (1996); P. Zanardi and M. Rasetti, *Phys. Rev. Lett.* **79**, 3306 (1997); L. Duan and G. Guo, *Phys. Rev. A* **58**, 3491 (1998); D. Lidar, I. Chuang, and K. Whaley, *Phys. Rev. Lett.* **81**, 2594 (1998).
5. L. Lovász, *IEEE Trans. Infom. Th.* **IT-25**, 1 (1979).
6. S. Beigi, *Phys. Rev. A* **82**, 010303 (2010); T. Cubitt, D. Leung, W. Matthews, and A. Winter, *Phys. Rev. Lett.* **104**, 230503(2010).
7. R. Duan, S. Severini and A. Winter, Zero-error communication via quantum channels and a quantum Lovász θ-function, in *Info. Th. Proc., 2011 IEEE Int. Symposium on*, 64(2011).
8. G. Chiribella, *J. Phys. Conf. Ser.* **284**, 012001(2011).
9. I. Marvian and R. Spekkens, arXiv:1105.1816 (2011).
10. G. Chiribella, G. M. D'Ariano, and M. F. Sacchi, *Phys. Rev. A* **72**, 042338 (2005).
11. M. Hayashi, arXiv:1209.3463 (2012).
12. R. Blume-Kohout, H. K. Ng, D. Poulin, and L. Viola, *Phys. Rev. Lett.* **100**, 30501(2008); *Phys. Rev. A* **82**, 062306 (2010).
13. C. Bennett, G. Brassard, C. Crepeau, R. Jozsa, A. Peres, and W. Wootters, *Phys. Rev. Lett.* **70**, 1895 (1993).
14. G. Chiribella, V. Giovannetti, L. Maccone, and P. Perinotti, *Phys. Rev. A* **86**, 010304(R) (2012).

FOURIER ANALYTIC APPROACH TO ESTIMATION OF UNKNOWN ACTION OF NON-COMMUTATIVE GROUP

MASAHITO HAYASHI

Graduate School of Mathematics, Nagoya University, Furocho, Chikusaku, Nagoya, 464-860, Japan
Centre for Quantum Technologies, National University of Singapore, 3 Science Drive 2, Singapore 117542
**E-mail: masahito@math.nagoya-u.ac.jp*

This article proposes a unified method to estimation of group action by using Fourier analysis.

Keywords: Group representation; estimation; Fourier transform; dual space.

1. Introduction

In quantum theory, the reversible dynamics of a system is described by an element in a projective unitary representation of a group. In this case, the unitary acting on the real quantum system reflects important physical parameters. Therefore, we can estimate these physical parameters by estimating a unitary in a given projective representation of a group. In this setting, it is known that estimation of unitary has a quadratic speed up over the state estimation in quantum case.[4,5,7,9] The same quadratic speed-up can be found by optimizing the Fisher information, although we should remark that the limit of Fisher information does not give the attainable bound of precision of the estimation of unitary, as is pointed in the reference.[12]

The first studies[4] treated the phase estimation, which is essentially the estimation of the representation of U(1). Next, the estimation of SU(2) was studied.[5,7,9] Chiribella et al[6] established a general theory of estimation of unitary representation of a compact group. On the other hand, Imai et al[10] treated phase estimation by using Fourier analysis. This method can be extended to the case with energy constraint.[11]

In this paper, we focus on the Fourier analytic method as a unified approach to treat the non-compact case as well as the compact case. As a main result, we extend this method to the non-commutative case. Due to the page limit, we shall not describe the proofs, the application to the case of Heisenberg representation with energy constraint, and the relation with the results by Chiribella in the finite-dimensional case.[13] These parts are available in the full paper version.[14]

2. Formulation

We focus on a group G acting on the Hilbert space $\mathcal{H}$ of our interest. That is, we treat a projective unitary representation f of G over $\mathcal{H}$. Our aim is estimating the unknown unitary $f(g)$ under the assumption that $g \in G$. For this purpose, we can choose the input state ρ and the output measurement, which is described by the POVM M over the Hilbert space $\mathcal{H}$. Since the aim of the measurement is the estimation of the element of $g \in G$, the POVM M takes values in the group G. We describe the set of the above kinds of POVMs by $\mathcal{M}(G)$. Hence, our estimator is given as a pair of an input state ρ and a POVM M. This formulation covers the case when we choose the input state ρ_i with the probability p_i for $i = 1, \ldots, l$, and choose the output POVM M_i depending on the input state ρ_i. When the original input system is $\mathcal{H}'$, the above case can be described by the case when the input is $\sum_i p_i \rho_i \otimes |i\rangle\langle i|$ on $\mathcal{H}' \otimes \mathbb{C}^l$ and the POVM $M(\hat{g}) := \sum_i M_i(\hat{g}) \otimes |i\rangle\langle i|$ on $\mathcal{H}' \otimes \mathbb{C}^l$.

In order to treat this problem, we focus on the risk function R depending on the true value g and the estimate $\hat{g}$. Then, when the true value is g, the average error is given as

$$\mathcal{D}_{R,g}(\rho, M) := \int_G R(g, \hat{g}) \mathrm{Tr}\, f(g) \rho f(g)^\dagger M(d\hat{g}). \tag{1}$$

Given a prior distribution ν for g over G, we can define the Bayesian error:

$$\mathcal{D}_{R,\nu}(M) := \int_G \mathcal{D}_{R,g}(M) \nu(dg). \tag{2}$$

Hence, our aim is finding a pair of the input state ρ and POVM $M \in \mathcal{M}(G)$ minimizing $\mathcal{D}_{R,\nu}(\rho, M)$.

As an alternative criterion, we optimize the worst case as

$$\mathcal{D}_R(M) := \max_G \mathcal{D}_{R,g}(M), \tag{3}$$

which is called the mini-max criterion.

Since the difference between g and $\hat{g}$ is thought to be the same as that between $g'g$ and $g'\hat{g}$, we assume the left invariant condition in the following:

$$R(g, \hat{g}) = R(g'g, g'\hat{g}), \quad \forall g, \hat{g}, g' \in G. \tag{4}$$

According to Holevo,[1] as an important class of POVMs, we introduce a covariant POVM. A POVM M taking values in G is called covariant concerning the projective representation f when

$$f(g)M(B)f(g)^{\dagger} = M(gB). \tag{5}$$

Holevo[1] defined the concept for a general homogeneous space. The group G can be regarded as a special case of homogeneous spaces. We describe the set of covariant POVMs by $\mathcal{M}_{\mathrm{cov}}(G)$. For any covariant POVM $M \in \mathcal{M}_{\mathrm{cov}}(G)$, the average error $\mathcal{D}_{R,g}(\rho, M)$ does not depend on the true value g. Hence, we obtain

$$\mathcal{D}_{R,g}(\rho, M) = \mathcal{D}_{R,\nu}(\rho, M) = \mathcal{D}_{R}(\rho, M). \tag{6}$$

When G is compact, there exists an invariant probability measure μ_G. Then, we obtain the following theorem, which is called quantum Hunt-Stein theorem.[1]

Theorem 1. *When the risk function R is invariant and G is compact and Hausdorff, we obtain*

$$\begin{aligned} &\min_{M \in \mathcal{M}(G)} \mathcal{D}_{R,\mu_G}(\rho, M) = \min_{M \in \mathcal{M}(G)} \mathcal{D}_{R}(\rho, M) \\ &= \min_{M \in \mathcal{M}_{\mathrm{cov}}(G)} \mathcal{D}_{R,\mu_G}(\rho, M) = \min_{M \in \mathcal{M}_{\mathrm{cov}}(G)} \mathcal{D}_{R}(\rho, M). \end{aligned}$$

However, when G is not compact, it has no invariant probability measure. In this case, the above theorem can be generalized to the following way.[2,3]

Theorem 2. *When the risk function R is left invariant and G is locally compact and Hausdorff, we obtain*

$$\min_{M \in \mathcal{M}(G)} \mathcal{D}_{R}(\rho, M) = \min_{M \in \mathcal{M}_{\mathrm{cov}}(G)} \mathcal{D}_{R}(\rho, M). \tag{7}$$

Hence, in the following, in order to treat our problem without the compactness condition, we treat the minimization

$$\min_{\rho} \min_{M \in \mathcal{M}_{\mathrm{cov}}(G)} \mathcal{D}_{R}(\rho, M). \tag{8}$$

That is, we can restrict our measurement into covariant measurements without loss of generality. Given an input mixed state $\rho = \sum_i p_i |\phi_i\rangle\langle\phi_i|$, any measurement M satisfies

$$\mathcal{D}_{R,g}(\rho, M) = \sum_i p_i \mathcal{D}_{R,g}(|\phi_i\rangle\langle\phi_i|, M). \tag{9}$$

Hence, any covariant measurement M satisfies

$$\mathcal{D}_R(\rho, M) = \sum_i p_i \mathcal{D}_R(|\phi_i\rangle\langle\phi_i|, M). \tag{10}$$

Thus, we can restrict our input state into pure states without loss of generality.

Next, we characterize covariant POVMs. For this purpose, we consider the left invariant measure μ_G for the group G. Here, when G is compact, μ_G is chosen to be the probability measure. In this case, it is known that any covariant measurement M can be described by using a positive semi-definite operator T such that

$$M(B) = \int_B f(g) T f(g)^\dagger \mu_G(dg). \tag{11}$$

Conversely, the above kind of operator T satisfies

$$I = \int_G f(g) T f(g)^\dagger \mu_G(dg). \tag{12}$$

When a positive semi-definite T satisfies (12), it gives a covariant measurement by (11), which is denoted by M_T.

In order to characterize a positive semi-definite T satisfying (12), we make the irreducible decomposition of $\mathcal{H}$ concerning the projective representation f: $\mathcal{H} = \oplus_{\lambda \in S} \mathcal{U}_\lambda \otimes \mathcal{V}_\lambda$, where $\hat{G}[\mathcal{L}]$ is the set of symbols of irreducible projective representation of G with the factor system $\mathcal{L}$, S is a subset of $\hat{G}[\mathcal{L}]$, $\mathcal{U}_\lambda$ is the irreducible space corresponding to $\lambda \in \hat{G}[\mathcal{L}]$, and $\mathcal{V}_\lambda$ is the space describing the multiplicity of the irreducible space $\mathcal{U}_\lambda$. That is, the group G acts only on $\mathcal{U}_\lambda$ but not on $\mathcal{V}_\lambda$.

When G is compact, all of irreducible space $\mathcal{U}_\lambda$ is finite-dimensional. When G is not compact, there are finite-dimensional irreducible spaces $\mathcal{U}_\lambda$. In this case, we define the generalized dimension $\dim_G \mathcal{U}_\lambda$ by

$$I = \dim_G \mathcal{U}_\lambda \int_G f(g) \rho f(g)^\dagger \mu_G(dg), \tag{13}$$

where ρ is an arbitrary state on $\mathcal{U}_\lambda$. In the following, we assume that the generalized dimension $\dim_G \mathcal{U}_\lambda$ is finite for any $\lambda \in S$. For a pure state $|\phi\rangle\langle\phi|$, the family of output states $\{f(g)|\phi\rangle\langle\phi|f(g)^\dagger\}_{g \in G}$ belongs to a subspace $\oplus_{\lambda \in S} \mathcal{U}_\lambda \otimes \mathcal{V}'_\lambda$, where the dimension $\mathcal{V}'_\lambda$ is $\min\{\dim \mathcal{V}_\lambda, \dim \mathcal{U}_\lambda\}$. Hence, in the following, we assume that $\dim \mathcal{V}_\lambda \le \dim \mathcal{U}_\lambda$. In order to give a typical covariant POVM, we define a vector $|\mathcal{I}_\lambda\rangle := \sum_i |e_i\rangle\langle e_i| \in \mathcal{U}_\lambda \otimes \mathcal{U}_\lambda^*$ for CONS $\{e_i\}$ of $\mathcal{U}_\lambda$. Then, we define $|\mathcal{I}\rangle := \sum_\lambda \sqrt{d_\lambda} |\mathcal{I}_\lambda\rangle$. Here, the group G acts only on the first space $\mathcal{U}_\lambda$ under the representation space $\mathcal{U}_\lambda \otimes \mathcal{U}_\lambda^*$. Then, we obtain the following theorem.

Theorem 3. *Assume all of the above conditions. Given an inclusion $\mathcal{V}_\lambda \subset \mathcal{U}_\lambda^*$, for any input pure state $|\phi\rangle\langle\phi|$ on $\mathcal{H}$ and any positive semi-definite T satisfying (12), there exists a mixed input state ρ on $\mathcal{H}$ such that*

$$\mathcal{D}_R(|\phi\rangle\langle\phi|, M_T) = \mathcal{D}_R(\rho, M_{|\mathcal{I}\rangle\langle\mathcal{I}|}). \tag{14}$$

Since any mixed input state ρ on $\mathcal{H}$ can be written as a convex combination of pure states on $\mathcal{H}$,

$$\min_\rho \mathcal{D}_R(\rho, M_{|\mathcal{I}\rangle\langle\mathcal{I}|}) = \min_{|\phi\rangle\langle\phi|} \mathcal{D}_R(|\phi\rangle\langle\phi|, M_{|\mathcal{I}\rangle\langle\mathcal{I}|}). \tag{15}$$

That is, we obtain

$$\begin{aligned} &\min_\rho \min_{M\in\mathcal{M}_{\mathrm{cov}}(G)} \mathcal{D}_R(\rho, M) = \min_{|\phi\rangle\langle\phi|} \min_T \mathcal{D}_R(|\phi\rangle\langle\phi|, M_T) \\ &= \min_{|\phi\rangle\langle\phi|} \mathcal{D}_R(|\phi\rangle\langle\phi|, M_{|\mathcal{I}\rangle\langle\mathcal{I}|}). \end{aligned} \tag{16}$$

3. Relation to Fourier transform

In this section, in order to treat the relation to Fourier transform, we prepare notations. In the Fourier analysis for a general group, depending on the factor system $\mathcal{L}$, we define the space $L^2(\hat{G}[\mathcal{L}]) := \sum_{\lambda\in\hat{G}[\mathcal{L}]} \mathcal{U}_\lambda \otimes \mathcal{U}_\lambda^*$. Here, we can identify the space $\mathcal{U}_\lambda \otimes \mathcal{U}_\lambda^*$ with the space of the Hilbert Schmidt operators on $\mathcal{U}_\lambda$.

The Fourier transform with the factor system $\mathcal{L}$ is defined as a unitary map $\mathcal{F}_\mathcal{L}$ from $L^2(G)$ to $L^2(\hat{G}[\mathcal{L}])$ as follows. Given $\phi \in L^2(G)$, we define

$$(\mathcal{F}_\mathcal{L}[\phi])_\lambda := \sqrt{d_\lambda} \int_G f_\lambda(g)^\dagger \phi(g) \mu_G(dg). \tag{17}$$

The converse map $\mathcal{F}_\mathcal{L}^{-1}$ from $L^2(\hat{G}[\mathcal{L}])$ to $L^2(G)$ is given as

$$\mathcal{F}_\mathcal{L}^{-1}[A](g) := \sum_{\lambda\in\hat{G}[\mathcal{L}]} \sqrt{d_\lambda} \operatorname{Tr} f_\lambda(g) A_\lambda \tag{18}$$

for $A = (A_\lambda) \in L^2(\hat{G}[\mathcal{L}])$ when G is a compact group. When G is not compact group, by choosing a suitable measure $\mu_{\hat{G}[\mathcal{L}]}$ on $\hat{G}[\mathcal{L}]$, the converse map $\mathcal{F}_\mathcal{L}^{-1}$ is characterized as

$$\mathcal{F}_\mathcal{L}^{-1}[A](g) := \int_{\hat{G}[\mathcal{L}]} \sqrt{d_\lambda} \operatorname{Tr} f_\lambda(g) A_\lambda \mu_{\hat{G}[\mathcal{L}]}(d\lambda). \tag{19}$$

Since the group G acts on the first space $\mathcal{U}_\lambda$ in the space $\mathcal{U}_\lambda \otimes \mathcal{U}_\lambda^*$, the above representation space $\mathcal{H}$ can be regarded as a subspace of $L^2(\hat{G}[\mathcal{L}])$.

Now, when the factor system of the projective representation f is $\mathcal{L}$, we consider the case when the input state is $|\phi\rangle \in \mathcal{H} \subset L^2(\hat{G}[\mathcal{L}])$ and the measurement is $M_{|\mathcal{I}\rangle\langle\mathcal{I}|}$. Since

$$\langle \mathcal{I}|f(g)|\phi\rangle = \sum_{\lambda \in S} \mathrm{Tr}\,\sqrt{d_\lambda}\mathrm{Tr}\, f_\lambda(g)\phi_\lambda = \mathcal{F}_{\mathcal{L}}^{-1}[\phi](g), \tag{20}$$

the output $\hat{g}$ obeys the probability density function

$$\langle\phi|f(\hat{g})^\dagger|\mathcal{I}\rangle\langle\mathcal{I}|f(\hat{g})|\phi\rangle\mu_G(d\hat{g}) = |\mathcal{F}_{\mathcal{L}}^{-1}[\phi](\hat{g})|^2\mu_G(d\hat{g}). \tag{21}$$

Therefore, our optimization problem can be described by the terms of converse Fourier transform.

Acknowledgments

The author is partially supported by a MEXT Grant-in-Aid for Young Scientists (A) No. 20686026 and Grant-in-Aid for Scientific Research (A) No. 23246071. The Centre for Quantum Technologies is funded by the Singapore Ministry of Education and the National Research Foundation as part of the Research Centres of Excellence programme.

References

1. A. S. Holevo, Rep. Math. Phys., 16, 385 (1979)
2. N.A. Bogomolov, Theor. Prob. Appl., 26, 787 (1982)
3. M. Ozawa, Research Reports on Inf. Sci. (1980)
4. V. Buzek, et al, Phys. Rev. Lett. 82 2207 (1999)
5. G. Chiribella, et al, Phys. Rev. Lett. 93, 180503 (2004)
6. G. Chiribella, G. M. D'Ariano, M. F. Sacchi, Phys. Rev. A 72 042338 (2005)
7. E. Bagan, M. Baig, R. Munoz-Tapia, Phys. Rev. A 70, 030301(R) (2004)
8. A. Hayashi et al, Phys. Rev. A 71, 012326 (2005)
9. M. Hayashi, Phys. Lett., A 354, 183 (2006)
10. H. Imai and M. Hayashi, New J. Phys. 11, 043034 (2009)
11. M. Hayashi, Progress of Informatics, 8, 81-87 (2011)
12. M. Hayashi, Com. Math. Phys., **304**, No. 3, 689-709 (2011)
13. G. Chiribella, J. Phys.: Conf. Ser., **284**, 012001 (2011)
14. M. Hayashi, arXiv 1209.3463

PARALLEL SESSION
Finite Quantum Systems and Combinatorial Physics

20 & 21 & 23 & 24 AUGUST 2012

GENERALIZED BINOMIAL DISTRIBUTIONS

H. BERGERON

*ISMO, UMR 8214 CNRS, Univ Paris-Sud, France *E-mail: herve.bergeron@u-psud.fr*

J.-P. GAZEAU

APC, UMR 7164 CNRS, Univ Paris Diderot, France E-mail: gazeau@apc.univ-paris7.fr

E.M.F. CURADO and LIGIA M.C.S. RODRIGUES

CBPF Rio de Janeiro, Brazil E-mail: evaldo@cbpf.br, ligia@cbpf.br

We present a generalization of the binomial distribution associated with a sequence of positive numbers. It involves asymmetric and symmetric expressions of probabilities for *win-loss* sequences of trials. Our approach is based on generating functions and produces, in the symmetric case, polynomials of the binomial type. Poisson-like limits, Leibniz triangle rules and related entropy(ies) are considered. Our generalizations are illustrated by various analytical and numerical examples.

Keywords: Deformed binomial and Poisson distribution, generating functions, correlations.

1. Deformed binomial distribution and non-linear coherent states

In most of the realistic models in Physics one must take correlations into account: events which are usually presented as independent, like in a binomial Bernoulli process, are actually submitted to correlative perturbations. These perturbations lead to deformations of the mathematical independent laws. For instance, the deformation of the Poisson distribution upon which is based the construction of Glauber coherent states in quantum optics leads to the so-called nonlinear coherent states.[4,5] The realization of a special class of these states, adapted to this deformation, has been proposed in the quantized motion of a trapped atom in a Paul trap.[6,7] From the expansion of coherent states over the number states, $|\alpha\rangle = e^{-|\alpha|^2/2}\sum_{n=0}^{\infty}\alpha^n/\sqrt{n!}\,|n\rangle$, the probability $p_n(\eta)$ to detect n-photons using a ideal photodetector is Poisson: $p_n = e^{-|\alpha|^2}|\alpha|^{2n}/n! = e^{-\bar{n}}\bar{n}^n/n!$ where $\bar{n}$ is the expected number of photons in the coherent state $|\alpha\rangle$. Probability $p_n(\eta)$ to detect n-photons using a non-ideal photodetector with efficiency $(\eta < 1)$ is given in terms of the probability $p_m(\eta = 1)$ (using an ideal one):

$p_n(\eta) = \sum_{m=n}^{\infty} \binom{m}{n} \eta^n \, (1-\eta)^{m-n} \, p_m(\eta = 1)$. For coherent states this is a combination of binomial and Poisson distributions and implies that sub-unity quantum efficiency is equivalent to an ideal detector masked by a beam-splitter with transmission coefficient, $\eta \leq 1$: $p_n(\eta) = \frac{\eta^n |\alpha|^{2n}}{n!} \, e^{-\eta|\alpha|^2}$, which amounts to replace α by $\sqrt{\eta}\alpha$ in the expression of coherent states.

Associated to sequences $\mathcal{X} = (x_n)_{n\in\mathbb{N}}$ of positive numbers x_n for $n > 0$ and $x_0 = 0$ are families of normalized states for one-mode e.m. quantum field which have the following form in the corresponding Fock space: $|\alpha\rangle = \sum_{n=0}^{\infty} \frac{1}{\sqrt{\mathcal{N}(|\alpha|^2)}} \frac{\alpha^n}{\sqrt{x_n!}} |n\rangle$, $x_n! = x_1 x_2 \cdots x_n$, $x_0! \stackrel{\text{def}}{=} 1$. The function $\mathcal{N}(\mathrm{t}) = \sum_{n=0}^{\infty} \frac{t^n}{x_n!}$ reads as the "exponential" associated to the "factorials" $x_n!$. Discrete probability distribution $n \mapsto |\langle n|\alpha\rangle|^2 = \frac{1}{\mathcal{N}(|\alpha|^2)} \frac{|\alpha|^{2n}}{x_n!}$, is a deformation of the Poisson distribution with parameter $|\alpha|^2$. To a sequence $\mathcal{X}$ corresponds the formal distribution

$$\mathfrak{p}_k^{(n)}(\eta) = \frac{x_n!}{x_{n-k}! \, x_k!} \, \eta^k \, p_{n-k}(\eta) \, , \qquad \sum_{k=0}^{n} \mathfrak{p}_k^{(n)}(\eta) = 1 \, . \tag{1}$$

The polynomials $p_s(\eta)$ satisfy $p_s(0) = 1$, $p_s(1) = 0$, for all s, and the sequence $(p_s)_{n\in\mathbb{N}}$ is determined by recurrence: $p_0(\eta) = 1$, $p_1(\eta) = 1 - \eta, \ldots$ If $p_m(\eta) \geq 0$ for all m and $\eta \in [0,1]$, then probabilistic interpretation is valid, and $\eta = \langle x_k \rangle_n / x_n$. A remarkable result is that

$$G_{\mathcal{N},\eta}(t) := \frac{\mathcal{N}(t)}{\mathcal{N}(\eta t)} = \sum_{s=0}^{\infty} p_s(\eta) \frac{t^s}{x_s!} \tag{2}$$

is a generating functional for the polynomials $p_s(\eta)$. This is a crucial point for proving (and finding explicitly!) nontrivial sequences $\mathcal{X}$ for which Bernouilli-like distribution **<u>IS</u>** a probability distribution (positiveness condition). Through the deformed binomial, the probability $p_n(t;\eta)$ to detect n-photons with $(\eta < 1)$ is given in terms of the ideal probability $p_m(t;\eta = 1) = t^m/(\mathcal{N}(t) x_m!)$, $t \equiv |\alpha|^2$, by the simple formula:

$$p_n(t;\eta) = \sum_{m=n}^{\infty} \frac{x_m!}{x_{m-n}! \, x_n!} \, \eta^n \, p_{m-n}(\eta) \, p_m(t;\eta = 1) = \frac{(\eta \, t)^n}{\mathcal{N}(\eta t) \, x_n!} \, .$$

To this discrete probability distribution there corresponds the same family of normalized non-linear coherent states $|\alpha;\eta\rangle = |\sqrt{\eta}\alpha\rangle$. The above generalization can be interpreted as modifying the probabilities of k clicks (wins) in n events. With $y_0 = 0$, $y_n = x_n/n$ for $n > 0$: $\mathfrak{p}_k^{(n)}(\eta) = \binom{n}{k} \, \pi_k^{(n)}(\eta)$ is the probability to have k clicks and $n-k$ failures, whereas $\pi_k^{(n)}(\eta) \stackrel{\text{def}}{=} \frac{y_n!}{y_{n-k}! \, y_k!} \, \eta^k \, p_{n-k}(\eta)$ is the probability of one sequence word with k clicks and $n-k$ failures. This deformation of the binomial law is due to correlations among the events. This distribution is called asymmetric in the sense that $\mathfrak{p}_k^{(n)}(\eta)$ is not invariant under $k \to n-k$ and $\eta \to 1-\eta$, as it happens in the binomial case.

2. Solution of the positiveness problem

In Eq. 1 we have associated to the sequence $\mathcal{X} = (x_n)_{n\in\mathbb{N}}$ the polynomials $p_k^{(n)}(\eta)$. They possess a probabilistic interpretation as long as $\mathfrak{p}_k^{(n)}$ are themselves probabilities. This latter condition holds only if the $p_n(\eta)$ are nonnegative: non-negativeness of the p_n for $\eta \in [0,1]$ is a key-point to preserve the statistical interpretation. In this case, the quantity $\mathfrak{p}_k^{(n)}(\eta)$ can still be interpreted as the probability of having k wins and $n-k$ losses in a sequence of *correlated* n trials. But, due to the asymmetry, there can be bias favoring either win or loss, even in the case $\eta = 1/2$. In order to analyze positiveness constraint and characterize the solutions, an analysis based on generating functions has been developed in Ref.[3] First $\mathcal{N}(t)$ gives the generating function of $p_k(\eta)$ thanks to (2). Next define Σ as the set of entire series $\mathcal{N}(t) = \sum_{n=0}^{\infty} a_n t^n$ possessing a non-vanishing radius of convergence and verifying $a_0 = 1$ and $\forall n \geq 1,\ a_n > 0$ and associate to $\mathcal{N}$ the sequence $x_n = a_{n-1}/a_n = n\mathcal{N}^{(n-1)}(0)/\mathcal{N}^{(n)}(0)$ so that $a_n = 1/(x_n!)$. Then define Σ_0 as the set of entire series $f(z) = \sum_{n=0}^{\infty} a_n z^n$ possessing a non-vanishing radius of convergence and verifying the conditions $a_0 = 0,\ a_1 > 0$ and $\forall n \geq 2,\ a_n \geq 0$. Finally, we proved in[3] that $\Sigma_+ := \{\mathcal{N} \in \Sigma \,|\, \forall \eta \in [0,1),\ p_n(\eta) > 0\} = \{e^F \,|\, F \in \Sigma_0\}$ is the set of deformed exponentials such that the generating functions $G_{\mathcal{N},\eta}(t)$ solve the non-negativeness problem. Sequences $\mathcal{X}$ having this probabilistic content are named *sequences of complete statistical type* (CST)

An interesting property reenforcing the probabilistic interpretation is the Poisson-like limit of the asymmetric deformed binomial distribution. If the sequence $\{x_n\}_{n\in\mathbb{N}}$ is such that, at fixed finite m, $\lim\limits_{n\to\infty} \frac{x_{n-m}}{x_n} = 1$, then the limit when $n \to \infty$ of $\mathfrak{p}_k^{(n)}(\eta)$ with $\eta = t/x_n$ is equal to $\frac{1}{\mathcal{N}(t)} \frac{t^k}{x_k!}$.

Our result provides a fair estimate on the manner in which a CST sequence behaves in comparison with natural integers. More precisely, let us pick a $\mathcal{N} \in \Sigma_+$; its associated CST sequence $x_n = n\mathcal{N}^{(n-1)}(0)/\mathcal{N}^{(n)}(0)$ then verifies $0 \leq x_n \leq n x_1$. We give now two examples (see[3] for more):

- For any sequence $\{a_n\}_{n\in\mathbb{N}}$ of non-negative real numbers such that $\sum_{n=0}^{\infty} a_n < \infty$, and for any $\alpha \in [0,1]$, the function $\mathcal{N}$ defined as $\mathcal{N}(t) = \prod_{k=0}^{\infty}(1+\alpha a_k t)/(1-a_k t)$ belongs to Σ_+ (the assumption $\sum_{k=0}^{\infty} a_k < \infty$ is needed to obtain a convergent product). For instance, with $\mathcal{N}(t) = (1-at)^{-m} = \sum_{k=0}^{\infty}(t^k)/(x_k^{(m)}!)$, whose radius of convergence is $t < 1/a$, for $a > 0$,

$x_k^{(m)}! = k!\,(m-1)!/a^k(m-1+k)!$, and $x_k^{(m)} = k/(a(m-1+k)$. Notice that when $k \to \infty$ the coefficients $x_k^{(m)}$ approach the finite value $x_\infty^{(m)} = 1/a$. The corresponding polynomial probabilities $p_k(\eta)$, $0 \le \eta \le 1$, are hypergeometrical $p_k(m;\eta) = {}_2F_1(-m,-k;1-k-m;\eta)$. For instance, for $n=2$, the probability to have two wins in two trials is the same as in the binomial case, but the probability to have one win and one loss is smaller and the probability to have two losses is higher than the corresponding binomial one (the total sum is equal to one). The study in[3] of the joint probability distribution for two binary variables x_1 and x_2: $x_i = +1$ (win) and $x_i = -1$ (loss), $i = 1, 2$, is illuminating for better understanding correlations in the case $n = 2$.

• The simplest case with infinite convergence radius other than $\mathcal{N}(t) = e^t$, is $\mathcal{N}(t) = \exp\left(t + \frac{a}{2}t^2\right)$, with $a > 0$. The generating function $\exp\left((1-\eta)t + \frac{a}{2}(1-\eta^2)t^2\right)$ allows to express the $p_n(\eta)$'s and the x_n's in terms of Hermite polynomials and to derive the asymptotic behavior $x_n \sim \sqrt{n/a} + \cdots$.

With the more general deformed exponential $\mathcal{N}(t) = \exp(\sum_{n=1}^m a_n t^n)$, with $a_1 = 1$ and $a_n > 0$ for all $n \ge 2$, if written as $\mathcal{N}(t) = \sum_{n=1}^\infty t^n/x_n!$, the x_n coefficients verify $x_{n+1} = (n+1)/[1 + a_2 x_n + a_3 x_n x_{n-1} + a_4 x_n x_{n-1} x_{n-2} + \cdots + a_m x_n x_{n-1} \cdots x_{n-m+2}]$ leading to the asymptotic behavior $x_n \sim (n/a_m)^{1/m}$ when $n \to \infty$. This conjecture was recently proved by Hugh Denoncourt (private communication) and by C. Vignat and O. Lévêque in.[8]

3. Symmetric deformation of binomial distribution

The symmetric option is under study:[9]

$$\mathfrak{p}_k^{(n)}(\eta) = \frac{x_n!}{x_{n-k}!x_k!} q_k(\eta) q_{n-k}(1-\eta),$$

where the $q_k(\eta)$ are polynomials of degree k and where the $\mathfrak{p}_k^{(n)}(\eta)$ are constrained by $\forall n \in \mathbb{N}$, $\forall \eta \in [0,1]$, $\sum_{k=0}^n \mathfrak{p}_k^{(n)}(\eta) = 1$, (normalization) and non-negativeness condition $\forall n, k \in \mathbb{N}$, $\forall \eta \in [0,1]$, $\mathfrak{p}_k^{(n)}(\eta) \ge 0$. The normalization implies $\forall \eta \in [0,1]$, $\mathfrak{p}_0^{(0)}(\eta) = q_0(\eta) q_0(1-\eta) = 1 \Rightarrow q_0(\eta) = \pm 1$. Assuming $q_0(\eta) = 1$ implies $\forall n \in \mathbb{N}$, $\forall \eta \in [0,1]$, $\mathfrak{p}_0^{(n)}(\eta) = q_n(1-\eta)$. Therefore the non-negativeness condition is equivalent to the non-negativeness of the polynomials q_n on the interval $[0,1]$. Here, like in the asymmetric case, the quantity $\mathfrak{p}_k^{(n)}(\eta)$ can be interpreted as the probability of having k wins and $n-k$ losses in a sequence of *correlated* n trials. Besides, as we recover the invariance under $k \to n-k$ and $\eta \to 1-\eta$ of the binomial distribution, no bias (in the case $\eta = 1/2$) can exist favoring either win or loss.

Among possible generating functions, the simplest ones give rise to polynomials of binomial type. Starting from a x_n-generating function $\mathcal{N}(t) \in \Sigma$:

$$\forall \eta \in [0,1], \quad \tilde{G}_{\mathcal{N},\eta}(t) := \mathcal{N}(t)^{\eta} = \sum_{n=0}^{\infty} \frac{q_n(\eta)}{x_n!} t^n. \tag{3}$$

Then $\ln(\mathcal{N}(t))$ is analytical around $t = 0$, and so $\mathcal{N}(t)^{\eta} = \exp(\eta \ln(\mathcal{N}(t)))$ possesses a convergent series expansion around $t = 0$ (for all $\eta \in \mathbb{C}$).

The q_n's obey the following conditions.

(a) $q_0(\eta) = 1$, $q_1(\eta) = \eta$ and more generally

$$\forall n \in \mathbb{N},\, \forall \eta \in [0,1],\, q_{n+1}(\eta) = \eta \, \frac{x_{n+1}}{n+1} \sum_{k=0}^{n} \binom{x_n}{x_k} \frac{n-k+1}{x_{n-k+1}} \, q_k(\eta-1)\,.$$

(b) The q_n's are polynomials of degree n obeying $\forall n \in \mathbb{N}, q_n(1) = 1$, and $\forall n \neq 0, q_n(0) = 0$, and they fulfill the normalization condition.

(c) The q_n's fulfill the functional relation

$$\forall z_1, z_2 \in \mathbb{C},\, \forall n \in \mathbb{N},\, \sum_{k=0}^{n} \binom{x_n}{x_k} q_k(z_1) q_{n-k}(z_2) = q_n(z_1 + z_2)\,.$$

We note that these polynomials, suitably normalized, are of binomial type.

Since we already know that $q_0(\eta) = 1$ and $\forall n \neq 0, q_n(0) = 0$, the non-negativeness condition is equivalent to specify that for any $\eta \in]0,1]$, $q_n(\eta) > 0$ and then the function $t \mapsto \mathcal{N}(t)^{\eta}$ belongs to Σ. As for the asymmetric case, the set Σ_+ is proved to be also the set of generating functions solving the non-negativeness problem for the symmetric case. Note the simple expression for the expectation value of the k variable: $\langle k \rangle_n = \eta n$, which is the same value as for the ordinary binomial case. An example of such symmetric deformations is with $\mathcal{N}(t) = (1 - t/\alpha)^{-\alpha}$, $\alpha > 0$. This gives the sequence $x_n = n\alpha/(n + \alpha - 1)$ with $\lim_{n\to\infty} x_n = \alpha$. The corresponding polynomials are $q_n(\eta) = (\alpha\eta)_n/(\alpha)_n$, where $(x)_n = \Gamma(x+n)/\Gamma(x)$ (Pochhammer). Besides the mean value $\langle k \rangle_n = \eta n$ we have the expression for the square root of the variance: $(\sigma_k)_n = n\sqrt{\eta(1-\eta)}\sqrt{(1+\alpha/n)/(1+\alpha)}$, which becomes proportional to the mean value at large values of n. The Leibniz triangle rule

$$\varpi_k^{n-1} = \varpi_k^n + \varpi_{k+1}^n \tag{4}$$

is verified for the quantities $\varpi_k^n := \binom{y_n}{y_k} q_k(\eta) q_{n-k}(1-\eta) = \mathfrak{p}_k^{(n)}(\eta)/\binom{n}{k}$, with $y_n := x_n/n$.

A natural definition of the Boltzmann-Gibbs entropy for the above distribution is $S_{\rm BG}^{(n)} = -\sum_{k=0}^{n} \binom{n}{k} \varpi_k^n(\eta) \ln \varpi_k^n(\eta)$, the presence of the binomial coefficient means a counting of the degeneracy.

Following[10] the above result (4) implies that the entropy which is extensive when $n \to \infty$ is, like in the ordinary binomial case, Boltzmann-Gibbs, as is shown in Figure 1 where it is compared with the Tsallis entropy $S_q^{(n)} \stackrel{\rm def}{=} (1 - \sum_{k=0}^{n} \binom{n}{k} (\omega_k^n)^q)/(q-1)$.

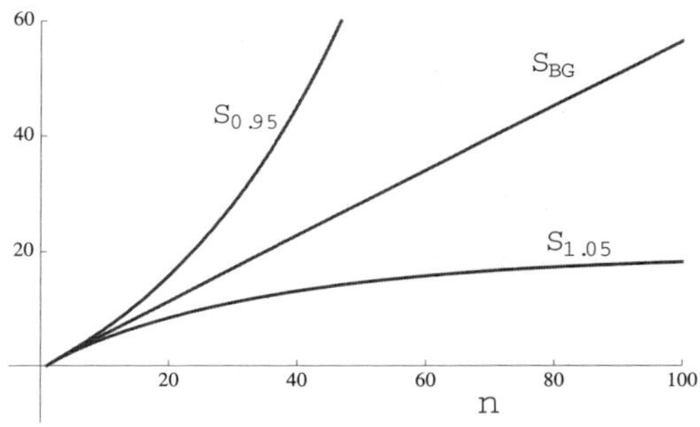

Fig. 1. n-dependence of entropies for the "q-gaussian" case ($\mathcal{N}(t) = (1 - t/\alpha)^{-\alpha}$). The Boltzmann-Gibbs (BG) and Tsallis (S_q) entropies of the distribution are shown for $\eta = 1/2$ and $\alpha = 3$. Upper curve is for $q = 0.95$. Bottom curve is for $q = 1.05$.

References

1. E. M. F. Curado, J.-P. Gazeau, and L. M. C. S. Rodrigues, Non-linear coherent states for optimizing Quantum Information, *Phys. Scr.* **82** 038108 (2010).
2. E.M.F. Curado, J.-P. Gazeau, L. M. C. S. Rodrigues, On a Generalization of the Binomial Distribution and Its Poisson-like Limit, *J. Stat. Phys.* **146** 264 (2012); arXiv:1105.3889 math-ph
3. H. Bergeron, E.M.F. Curado, J.-P. Gazeau, L. M. C. S. Rodrigues, Generating functions for generalized binomial distributions, *J. Math. Phys.* **53** 103304-1-21 (2012); arXiv:1203.3936 math-ph
4. V. V. Dodonov, *J. Opt. B: Quantum Semiclass. Opt.* **4** 1 (2002), and references therein.
5. J.-P. Gazeau, *Coherent States in Quantum Physics*, Wiley, 2009.
6. R. L. de Matos Filho and W. Vogel, *Phys Rev. A* **54** 4560 (1996).
7. Z. Kis, W. Vogel, and L. Davidovich, *Phys Rev. A* **64** 0033401 (2001).
8. C. Vignat and O. Lévêque, *"Proof of a conjecture by Gazeau et al using Gould Hopper polynomials*, arXiv:1203.5418v1[math-ph].
9. H. Bergeron, E.M.F. Curado, J.-P. Gazeau, and Ligia M.C.S. Rodrigues, *Symmetric generalized binomial distributions*, in preparation.
10. R. Hanel, S. Thurner, and C. Tsallis, *Eur. Phys. J. B* **72** 263 (2009).

QUANTUM DYNAMICS OF FINITE LEVEL SYSTEMS – MARKOVIANITY CRITERIA

DARIUSZ CHRUŚCIŃSKI

Institute of Physics, Nicolaus Copernicus University, Toruń, 87-100, Poland

We analyze the dynamics of a finite level quantum system using local in time master equation. Markovian dynamics is characterized in terms of divisible dynamical map. Moreover we provide a family of criteria which can distinguish Markovian and non-Markovian quantum evolution. They are based on a simple observation that Markovian dynamics implies monotonic behavior of several well known quantities like distinguishability of states, fidelity and relative entropy. Violation of any of these criteria implies that the corresponding dynamics is non-Markovian.

Keywords: Quantum open systems, quantum dynamics, Markovian evolution.

1. Introduction

The dynamics of finite level quantum systems attracts nowadays increasing attention.[1–3] It is relevant not only for the better understanding of quantum theory but it is fundamental in various modern applications of quantum mechanics. Since the system-environment interaction causes dissipation, decay and decoherence it is clear that dynamic of open systems is fundamental in modern quantum technologies, such as quantum communication, cryptography and computation.

The usual approach to the dynamics of an open quantum system with finite numbers of levels consists in applying the Markovian approximation, that leads to the following local master equation for the dynamical map Λ_t $(t \geq 0)$:

$$\dot{\Lambda}_t = L\,\Lambda_t\,, \qquad \Lambda_0 = \mathbb{1}\,, \tag{1}$$

where $\mathbb{1}$ denotes an identity map. Let us recall that a dynamical map $t \rightarrow \Lambda_t$ is a family of completely positive trace preserving maps in $\mathcal{B}(\mathcal{H})$.[4] In the case of finite dimensional systems a linear map $\Lambda : \mathcal{B}(\mathcal{H}) \rightarrow \mathcal{B}(\mathcal{H})$ is completely positive iff it admits the following Kraus representation[4]

$$\Lambda\, a = \sum_\alpha K_\alpha\, a\, K_\alpha^\dagger\,, \tag{2}$$

where $V_\alpha : \mathcal{H} \to \mathcal{H}$ are arbitrary (Kraus) operators. It is trace preserving iff $\sum_\alpha K_\alpha^\dagger K_\alpha = \mathbb{I}_{\mathcal{H}}$.

The corresponding evolution of the density operator is defined by $\rho_t = \Lambda_t \rho$. It was shown that the solution $\Lambda_t = e^{tL}$ defines completely positive trace preserving Λ_t iff L has the following well known GKSL form[5,6]

$$L\rho = -i[H, \rho] + \sum_k \left(V_k \rho V_k^\dagger - \frac{1}{2}\{V_k^\dagger V_k, \rho\} \right) , \tag{3}$$

where $\{a, b\} = ab + ba$, and V_k are arbitrary (noise) operators. The characteristic feature of Markovian semigroup $\Lambda_t = e^{tL}$ is that it neglects all memory effects caused by the nontrivial interaction of the system with an external world. However, recent theoretical studies and technological progress call for more refine approach based on non-Markovian evolution (see e.g. recent papers[7–18]).

The general legitimate quantum dynamics is described by a dynamical map Λ_t which satisfies local in time master equation

$$\dot{\Lambda}_t = L_t \Lambda_t , \quad \Lambda_0 = \mathbb{1} , \tag{4}$$

with time-dependent generator L_t. The formal solution to (4) reads as follows

$$\Lambda_t = \mathrm{T} \exp\left(\int_0^t L_u du \right) , \tag{5}$$

where T stands for the chronological operator. Interestingly, the problem of characterizing the properties of a local generator L_t which guarantee that the corresponding solution (5) gives rise to the legitimate dynamical map Λ_t is still open.

In this paper we characterize the classes of local generators for which one can control when the corresponding solution defines a legitimate dynamical map. Moreover, we discuss the class of Markovian evolution. Interestingly, there are several concepts of Markovianity available recently in the literature. We follow here the definition based on the mathematical notion of a divisible map. We provide simple necessary criteria for Markovianity which are based on monotonicity of several well known quantities like distinguishability of states, relative entropy or fidelity.

2. Divisible maps — Markovian dynamics

In this section we characterize important class of dynamical maps representing Markovian evolution. It is important to clarify this issues since there are few definitions used in the literature recently. We call a dynamical map Λ_t *divisible* if for any $t \geq s \geq 0$ one has the following decomposition

$$\Lambda_t = V_{t,s} \Lambda_s , \tag{6}$$

with completely positive propagator $V_{t,s}$. Note, that if Λ_t is invertible, then $V_{t,s} = \Lambda_t \Lambda_s^{-1}$, and hence the following composition law easily follows

$$V_{t,s} V_{s,u} = V_{t,u} \,, \tag{7}$$

for any $t \geq s \geq u$.

Definition 2.1. Dynamical map Λ_t corresponds to Markovian evolution if and only if it is divisible.

Interestingly, the property of being Markovian (or divisible) is fully characterized in terms of the local generator L_t. Note, that if Λ_t satisfies $\dot{\Lambda}_t = L_t \Lambda_t$, then $V_{t,s}$ satisfies

$$\partial_t V_{t,s} = L_t V_{t,s} \,, \quad V_{s,s} = \mathbb{1} \,, \tag{8}$$

and the corresponding solution reads $V_{t,s} = \mathrm{T} \exp\left(\int_s^t L_u du\right)$. The central result consists in the following

Theorem 1. *The map Λ_t is divisible if and only if L_t has the GKSL form for all t.*

Corollary 2.1. *Let $L_1, \ldots, L_N$ denote legitimate GKSL generators and consider*

$$L_t = \alpha_1(t) L_1 + \ldots + \alpha_N(t) L_N \,. \tag{9}$$

Note, that L_t a legitimate GKSL generator for all $t \geq 0$ if $\alpha_i(t) \geq 0$ for $t \geq 0$.

3. Commutative class of dynamical maps

As we already stressed the general solution to the local in time master equation (4) has only a formal meaning and in general we do not control the properties of L_t which guarantee that T-product exponential formula $\mathrm{T}\exp(\int_0^t L_u du)$ defines dynamical map. Note, however, that if L_t defines a commutative family of generators, that is

$$[L_t, L_u] = 0 \,, \tag{10}$$

for any $t, u \geq 0$, the formula (5) considerably simplifies: the 'T' product drops out and the solution is fully controlled by the integral $\int_0^t L_u du$.

Theorem 2. *If L_t defines a commutative family, then L_t is a legitimate generator of a quantum dynamical map if and only if $\int_0^t L_u du$ defines GKSL generator for all $t \geq 0$.*

Consider for example a set of legitimate GKSL generators $L_0, L_1, \ldots, L_N$ such that $L_0\rho = -i[H,\rho]$ with Hermitian H and

$$[L_i, L_j] = 0\,, \quad i,j = 0,1,\ldots,N\,. \tag{11}$$

Define a time-dependent generator by

$$L_t = \alpha_0(t)L_0 + \alpha_1(t)L_1 + \ldots + \alpha_N(t)L_N\,. \tag{12}$$

It is clear that L_t defines a commutative family. Note, that $\int_0^t L_u du$ defines a legitimate GKSL generator iff $\int_0^t \alpha_i(u)du \geq 0$, for $i = 1,\ldots,N$.

Example 3.1. As an example of commutative dynamics consider the following evolution of a qubit

$$L_t\rho = -\frac{i\omega_t}{2}[\sigma_z,\rho] + \frac{\gamma_t}{2}(\sigma_z\rho\sigma_z - \rho)\,, \tag{13}$$

where $\omega_t,\ \gamma_t : \mathbb{R}^+ \to \mathbb{R}$. The corresponding density matrix evolves as follows

$$\rho_t = \begin{pmatrix} \rho_{00} & \rho_{01}e^{-i\Omega_t - \Gamma_t} \\ \rho_{10}e^{i\Omega_t - \Gamma_t} & \rho_{11} \end{pmatrix}, \tag{14}$$

where $\Omega_t = \int_0^t \omega_u du$ and $\Gamma_t = \int_0^t \gamma_u du$. L_t gives rise to legitimate quantum evolution if and only if $\Gamma_t \geq 0$ for all $t \geq 0$ and hence the evolution corresponds to simple decoherence. One has therefore

- Λ_t defines a dynamical map iff $\Gamma_t \geq 0$,
- Λ_t is Markovian iff $\gamma_t \geq 0$
- Λ_t defines Markovian semigroup iff $\gamma_t = \gamma > 0$.

It is clear that $\gamma_t \geq 0$ implies $\Gamma_t \geq 0$. Whenever $\Gamma_t \geq 0$ but $\gamma_t \ngeq 0$, then the corresponding dynamics is non-Markovian.

4. Criteria for Markovianity

In this section we provide a series of necessary criteria for Markovian dynamics.[18] It turns out that dynamics represented by a divisible map displays characteristic monotonic behavior for several interesting quantities and hence violation of any of Markovianity criterium provides a clear sign for non-Markovian behavior.

Trace norm defines a natural distance between quantum states represented by density operators: given two density operators ρ and σ one defines

$$D[\rho,\sigma] = \frac{1}{2}||\rho - \sigma||_1\,, \tag{15}$$

where $||a||_1 := \mathrm{Tr}\,\sqrt{aa^\dagger}$. The quantity $D[\rho,\sigma]$ is usually interpreted as a measure of distinguishability of quantum states ρ and σ. It is well known that if Φ is a positive trace-preserving map, then $D[\Phi\rho, \Phi\sigma] \le D[\rho, \sigma]$, and hence if Λ_t is a dynamical map, then $D[\Lambda_t\rho, \Lambda_t\sigma] \le D[\rho, \sigma]$ for all $t \ge 0$.

Corollary 4.1. *If Λ_t is a divisible map, then*

$$\frac{d}{dt} D[\Lambda_t\rho, \Lambda_t\sigma] \le 0 , \tag{16}$$

that is, for the Markovian evolution distinguishability of any pair of initial states monotonically decreases.

Given two density operators ρ and σ one defines the corresponding fidelity by the following formula

$$F(\rho, \sigma) = \left(\mathrm{Tr}\left[\sqrt{\sqrt{\rho}\,\sigma\,\sqrt{\rho}}\right]\right)^2 . \tag{17}$$

Note, that if ρ is pure, that is, $\rho = |\psi\rangle\langle\psi|$, then $F(\rho, \sigma) = \langle\psi|\sigma|\psi\rangle$. If $\psi = |\phi\rangle\langle\phi|$ is pure as well, then $F(\rho, \sigma) = |\langle\psi|\phi\rangle|^2$ measures an overlap of $|\psi\rangle$ and $|\phi\rangle$. One proves

$$1 - F(\rho, \sigma) \le D[\rho, \sigma] \le \sqrt{1 - F(\rho, \sigma)^2} . \tag{18}$$

Moreover, if Φ is a quantum channel, then $F(\Phi\rho, \Phi\sigma) \ge F(\rho, \sigma)$. Hence if Λ_t is a dynamical map one has $F(\Lambda_t\rho, \Lambda_t\sigma) \ge F(\rho, \sigma)$ for all $t \ge 0$.

Corollary 4.2. *If Λ_t is a divisible map, then*

$$\frac{d}{dt} F(\Lambda_t\rho, \Lambda_t\sigma) \ge 0 , \tag{19}$$

that is, for the Markovian evolution fidelity of any pair of initial states monotonically increases.

A similar property holds for the relative entropy

$$S(\rho \,||\, \sigma) = \mathrm{Tr}(\rho[\log\rho - \log\sigma]) . \tag{20}$$

If Φ is a quantum channel, then $S(\Phi\rho \,||\, \Phi\sigma) \le S(\rho \,||\, \sigma)$ and hence if Λ_t is a dynamical map one has $S(\Lambda_t\rho \,||\, \Lambda_t\sigma) \le S(\rho \,||\, \sigma)$ for all $t \ge 0$.

Corollary 4.3. *If Λ_t is a divisible map, then*

$$\frac{d}{dt} S(\Lambda_t\rho \,||\, \Lambda_t\sigma) \le 0 , \tag{21}$$

that is, for the Markovian evolution relative entropy of any pair of initial states monotonically decreases.

5. Conclusions

We characterized a class of Markovian dynamics using the concept of divisible dynamical map. Interestingly Markovian dynamics is fully controlled in the local approach via the properties of the corresponding local generator. Moreover we provided a family of criteria which can distinguish Markovian and non-Markovian dynamics. These Markovianity criteria are based on a simple observation that Markovian dynamics implies monotonic behavior of several well known quantities like distinguishability of states, fidelity and relative entropy.

Acknowledgements

This work was partially supported by the National Science Centre project DEC-2011/03/B/ST2/00136.

References

1. H.-P. Breuer and F. Petruccione, *The Theory of Open Quantum Systems* (Oxford Univ. Press, Oxford, 2007).
2. U. Weiss, *Quantum Dissipative Systems*, (World Scientific, Singapore, 2000).
3. R. Alicki and K. Lendi, *Quantum Dynamical Semigroups and Applications*, (Springer, Berlin, 1987).
4. V. Paulsen, *Completely Bounded Maps and Operator Algebras*, Cambridge University Press, 2003.
5. V. Gorini, A. Kossakowski, and E. C. G. Sudarshan, J. Math. Phys. **17**, 821 (1976).
6. G. Lindblad, Comm. Math. Phys. **48**, 119 (1976).
7. J. Wilkie, Phys. Rev. E **62**, 8808 (2000); J. Wilkie and Yin Mei Wong, J. Phys. A: Math. Theor. **42**, 015006 (2009).
8. A. A. Budini, Phys. Rev. A **69**, 042107 (2004).
9. S. Maniscalco, Phys. Rev. A **72**, 024103 (2005); S. Maniscalco and F. Petruccione, Phys. Rev. A **73**, 012111 (2006); J. Piilo, S. Maniscalco, K. Härkönen and K.-A. Suominen, Phys. Rev. Lett. **100**, 180402 (2008)
10. H.-P. Breuer and B. Vacchini, Phys. Rev. Lett. **101** (2008) 140402; Phys. Rev. E **79**, 041147 (2009).
11. D. Chruściński and A. Kossakowski, Phys. Rev. Lett. **104**, 070406 (2010); Europhys. Lett. **97**, 20005 (2012).
12. H.-P. Breuer, E.-M. Laine, J. Piilo, Phys. Rev. Lett. **103**, 210401 (2009).
13. Á. Rivas, S.F. Huelga, and M.B. Plenio, Phys. Rev. Lett. **105**, 050403 (2010).
14. P. Haikka, J. D. Cresser, and S. Maniscalco, Phys. Rev. A **83**, 012112 (2011)
15. B. Vacchini, A. Smirne, E. Laine, J. Piilo, and H.-P. Breuer, New J. Phys. **13**, 093004 (2011) .
16. D. Chruściński, A. Kossakowski and Á. Rivas, Phys. Rev. A **83**, 052128 (2011).
17. D. Chruściński, A. Kossakowski, and S. Pascazio, Phys. Rev. A **81**, 032101 (2010).
18. D. Chruściński and A. Kossakowski, J. Phys. B: At. Mol. Opt. Phys. **45** 154002 (2012).

A MEASURE FREE APPROACH TO COHERENT STATES REFINED

A. HORZELA[a] and F. H. SZAFRANIEC[b]

[a] *Instytut Fizyki Jądrowej Polskiej Akademii Nauk,*
ul. Radzikowskiego 152, 31342 Kraków, Poland
E-mail: andrzej.horzela@ifj.edu.pl

[b] *Instytut Matematyki, Wydział Matematyki i Informatyki, Uniwersytet Jagielloński*
ul. Łojasiewicza 6, 30348 Kraków, Poland
E-mail: umszafra@cyf-kr.edu.pl

This is a follow up of our recent paper[1], which reorganizes the exposition and adds some comments made from a slightly different point of view. More precisely this is a purely reproducing kernel approach which allows to eliminate any use of measure in the standard constructions.

Keywords: Generalized coherent states, reproducing kernel Hilbert spaces.

1. Introduction

J.R.Klauder's approach[2–4] to coherent states comprises three postulates: (a) *normability*, (b) *continuity* and (c) *resolution of the identity*. J.-P. Gazeau's push on[5] extends substantially the area of applicability of Klauder's definition however it contains a rather restrictive version of (a): normability is replaced by the requirement the coherent state to be normalized - this seems to be too restrictive, cf. footnote b, p. 279. Our recent paper[1] provides the community with a challenging step further, though being still in flavor of Gazeau's generalized coherent states, makes coherent states free from any backing of $\mathcal{L}^2$ spaces. The present note is a kind of polish up of the previous study.

2. General formalism

The scheme presented in the paper[1] depends solely on two parameters, the given data:

1° an orthonormal basis [a] $(e_n)_{n=0}^d$, d being finite or infinite, in a given Hilbert space $\mathcal{H}$;

2° a sequence $(\Phi_n)_{n=0}^d$ of complex valued functions on a fixed set X such that

$$\sum_{n=0}^{d} |\Phi_n(x)|^2 < \infty \quad x \in X \tag{1}$$

and

$$(\alpha_n)_{n=0}^d \in \ell_d^2 \text{ and } \sum_{n=0}^{d} \alpha_n \Phi_n(x) = 0 \text{ for all } x \text{ imply all } \alpha_n\text{'s are } 0. \tag{2}$$

The *coherent states* $\{c_x\}_{x\in X}$ are now at hand:

$$c_x \stackrel{\text{def}}{=} \sum_{n=0}^{d} \Phi_n(x) e_n, \quad x \in X. \tag{3}$$

They are vectors in $\mathcal{H}$.

3. The RKHS couple

The condition Eq. (1) gives a rise to the following definition

$$K(x,y) \stackrel{\text{def}}{=} \sum_{n=0}^{d} \Phi_n(x)\overline{\Phi_n(y)}, \quad x, y \in X, \tag{4}$$

which is a positive definite kernel on X. It uniquely determines a Hilbert space $\mathcal{H}_K$ of complex valued functions on X, with the inner product $\langle \cdot | - \rangle_{\mathcal{H}_K}$, such that

the set of functions $\{K_x\}_{x\in X}$, $K_x \stackrel{\text{def}}{=} K(\cdot, x)$, is complete in $\mathcal{H}_K$;
the following formula, customarily called the *reproducing kernel property*,

$$f(x) = \langle f | K_x \rangle_{\mathcal{H}_K}, \quad f \in \mathcal{H}_K,\ x \in X \tag{5}$$

holds.

Due to Eq. (1) and Eq. (2) the sequence $(\Phi_n)_{n=0}^d$ is a basis in $\mathcal{H}_K$; cf. Fact B in[6] and in,[7] pp. 12-15. An introductory exposition of the matters can be found in the seminal paper[8].

---

[a] Recall a basis in a Hilbert space is a set of mutually orthogonal, norm 1 vectors, which is complete; a set $S \subset \mathcal{H}$ is called complete (sometimes called total) if $S^\perp = \{0\}$.

4. Unitarity of the Segal Bargmann transform

The operator $B : \mathcal{H} \ni h \mapsto Bh \in \mathcal{H}_{\mathcal{K}}$

$$Bh \stackrel{\text{def}}{=} \sum_{n=0}^{d} \Phi_n \langle e_n | h \rangle_{\mathcal{H}} \tag{6}$$

is well defined with the series converging in the $\mathcal{H}_{\mathcal{K}}$-topology. This is due to the fact that the sequence $(\langle e_n|h\rangle_{\mathcal{H}})_{n=0}^{d}$ of Fourier coefficients of $h \in \mathcal{H}$ is in ℓ_d^2. Now with $g, h \in \mathcal{H}$ we have

$$\langle Bg|Bh\rangle_{\mathcal{H}_{\mathcal{K}}} = \sum_{k,l=0}^{d} \langle \Phi_k|\Phi_l\rangle_{\mathcal{H}_{\mathcal{K}}} \langle e_l|f\rangle_{\mathcal{H}} \overline{\langle e_k|g\rangle_{\mathcal{H}}} = \sum_{n=0}^{d} \langle e_n|f\rangle_{\mathcal{H}} \overline{\langle e_n|g\rangle_{\mathcal{H}}} = \langle g|h\rangle_{\mathcal{H}}.$$

which means that B is an isometry. Because $Be_k = \Phi_k$ for all k's, B is surjective, hence unitary. The recapitulation of the above is as follows.

Theorem 4.1. *The Segal Bargmann tranform B as defined by Eq. (6) is a unitary operator between $\mathcal{H}$ and $\mathcal{H}_{\mathcal{K}}$. This means in particular that*

$$\boxed{\langle Bg|Bh\rangle_{\mathcal{H}_{\mathcal{K}}} = \langle g|h\rangle_{\mathcal{H}}} \tag{$\spadesuit$}$$

for $g, h \in \mathcal{H}$.

5. Back to Klauder's postulates

Let us look after Klauder's originating postulates under the present circumstances.

(a) c_x's to be normalizable[b] is due to the fact that each sequence $(\Phi_n(x))_{n=0}^{d}$ is that of Fourier coefficients of c_x with respect to the basis $(e_n)_{n=0}^{d}$ and by Eq. (1) it is in ℓ_d^2.

(b) If X is a topological space and the function

$$X \ni x \mapsto K(x,x) \in \mathbb{C} \tag{7}$$

is continuous then the mapping

$$X \ni x \mapsto c_x \in \mathcal{H}$$

[b] The reason while we get rid of explicite normalization in the definition Eq. (2) of coherent states and do replace it by a weaker condition of normalizability is the following: supposing that all the functions Φ_n's are holomorhic then $K(x,x)$, as a function of x, no longer is such, and therefore normalized coherent states, customarily used in the standard approach, ceased to be holomorphic too. The hidden beauty of the Bargmann construction[9] (which essence is to find a map between the Hilbert space of quantum states and the Hilbert space of analytic functions) is using unnormalized coherent states to find such a transformation.

is continuous in the norm of $\mathcal{H}$.

Indeed, because

$$\begin{aligned}\|c_x - c_y\|^2 &= \sum_{n=0}^{d} |\Phi_n(x) - \Phi_n(y)|^2 \\ &= \sum_{n=0}^{d} \left(K(x.x) - 2\,\mathfrak{Re}\,K(x,y) + K(y,y)\right)\end{aligned}$$

and the fact that continuity of the function of Eq. (7) implies joint continuity of the kernel the conclusion follows; cf. formula (1.3) in[7].

(c) Resolution of the identity of the coherent states $\{c_x\}_{x\in X}$ defined by Eq. (3) is precisely what formula (♠) in Theorem 4.1 says[c]. Even more, comparing to the by now classical considerations of Gazeau, it establishes, like in Bargmann[9], unitarity of the transform in question.

6. The reproducing kernel property now

A direct application of the reproducing kernel test (cf. condition (η) in[6] or condition (γ''), p. 8 in[7]) ensures us that the functions ($h \in \mathcal{H}$)

$$F_h(x) \stackrel{\text{def}}{=} \langle c_x | h \rangle_{\mathcal{H}}, \quad x \in X$$

belong to $\mathcal{H}_K$. Therefore the reproducing kernel property for them

$$F_h(x) = \langle F_h | K_x \rangle_{\mathcal{H}_K}$$

becomes relevant here as well. In the literature it is claimed to be a consequence of the resolution of the identity formula while in our approach it is in fact <u>independent</u> of that, following its own lines of arguing.

It is worthy to mention that, due to Eq. (3),

$$K(x,y) = \langle c_x | c_y \rangle_{\mathcal{H}}, \quad x, y \in X$$

which is another formula for the kernel K being in accordance with the construction on p. 5 of[6] and, with more details, with that labeled as (B_2) in[7], pp. 15-16.

---

[c] Subsection 6.4 in Hall's survey[10] is an in-depth discussion of this point of view. See also Section 7.

7. Resolution of the identity for malcontents

For those who have been reluctant so far to accept legitimacy of our approach let us make the following experiment. Revive for a moment the traditional Dirac notation $|x\rangle$ for c_x and suppose $\mathcal{H}_\mathcal{K}$ is isometrically imbedded in $\mathcal{L}^2(\mu)$ with some measure μ on X as it happens in Klauder's and Gazeau's. Then for $g, h \in \mathcal{H}$ we get from (B)

$$\begin{aligned}\langle h| \int_X |x\rangle\langle x| \, \mu(\mathrm{d}x) |g\rangle_{\mathcal{L}^2(\mu)} &= \int_X \langle h|x\rangle\langle x|g\rangle_{\mathcal{L}^2(\mu)} \, \mu(\mathrm{d}x) \\ &= \int_X \overline{(Bh)(x)}(Bg)(x) \, \mu(\mathrm{d}x) \\ &= \langle Bh|Bg\rangle_{\mathcal{L}^2(\mu)} = \langle Bh|Bg\rangle_{\mathcal{H}_\mathcal{K}} = \langle h|g\rangle_\mathcal{H}\end{aligned} \tag{8}$$

which concludes as [d]

$$\langle h| \int_X |x\rangle\langle x| \, \mu(\mathrm{d}x) |g\rangle_{\mathcal{L}^2(\mu)} = \langle h|g\rangle_\mathcal{H}$$

This is noting but the orthodox resolution of the identity formula whereas the last equality in Eq. (8) is exactly our (or Bargman's in the particular case) resolution of the identity.

8. To be integrable or not to be

Call a reproducing kernel Hilbert space $\mathcal{H}_\mathcal{K}$ *integrable* if there is a measure μ on X such that $\mathcal{H}_\mathcal{K}$ is isometrically imbedded in $\mathcal{L}^2(\mu)$. The following dichotomy becomes apparent

(α) $\mathcal{H}_\mathcal{K}$ is not integrable; a simple example of this is in the master paper[1], it is a kind of weighted ℓ^2 space.
(β) $\mathcal{H}_\mathcal{K}$ is integrable. Here the situation becomes more complex though quite interesting.

The case (α) is the foremost one which takes benefits of our approach. The other, (β) is intimately related (cf. the references[11,12]) to the complex moment problem (for the latter see[13]). Our guess is it might be its indeterminancy which would shed more light on the behavior of the Gazeau resolution of the identity formula; because this is the current subject of our investigation the out-coming results will appear elsewhere.

---

[d] Let us notice that the only legal way of acknowledging integrals with respect to operator valued measures, and such is the resolution of the identity formula, is to consider it as a week ones.

9. Conclusions

Let us emphasise that starting exclusively from the initial data 1° and 2° we reach the same goals as Klauder and Gazeau, without any use of measure. However in the presence of measure some new phenomena may be observed.

In our situation Klauder's requirement (a) turns out to be redundant. Continuity requirement (b) is preserved. The resolution of the identity formula (♠) becomes the strongest possible: unitarity of the Segal Bargmann like transform B instead of isometricity of Gazeau's. The reproducing kernel property attached to that lives its own live.

Aknowledgment

Research of the second author was supported by the MNiSzW grant NN201 1546438.

References

1. A. Horzela and F. H. Szafraniec, *J. Phys. A: Math. Theor.* **45** 244018 (2012).
2. J. R. Klauder *Annals of Physics* **11**, 123-168 (1960).
3. J. R. Klauder, *J. Math. Phys.* **4**, 1055–1058 (1963).
4. J. R. Klauder and B-S. Skagerstam, *Coherent States. Application in Physics and Mathematical Physics*, Ch.1, (World Scientific, Singapore, 1985).
5. J-P. Gazeau, *Coherent states in quantum physics*, Ch.5, (Wiley-VCH, Weinheim, 2009).
6. F. H. Szafraniec, *Oper. Theory: Adv. Appl.* **114**, 253–263 (2000).
7. F. H. Szafraniec, *Reproducing Kernel Hilbert Spaces* (Wydawnictwo Uniwersytetu Jagiellońskiego, Kraków, 2004, in Polish).
8. N. Aronszajn, *Trans. Amer. Math. Soc.* **68**, 337–404 (1950).
9. V. Bargmann, *Comm. Pure and Applied Mathematics* **14**, 187–214 (1961).
10. B. C. Hall, in *Contemp. Math.* **260**, 1-59 (Amer. Math. Soc., Providence, RI 2000).
11. F. H. Szafraniec, Integrability of RKHS, talk given at QTS5, Valladolid (Spain), 22-27, July, 2007 (slides available from the author upon request).
12. F. H. Szafraniec, *J. Comput. Appl Math.* **179**, 343-353 (2005).
13. J. Stochel and F. H. Szafraniec, *J. Funct. Anal.* **159**, 432-491 (1998).

THE Z_3-GENERALIZATION OF PAULI'S PRINCIPLE AND A CUBIC DIRAC EQUATION FOR QUARKS

R. KERNER*

LPTMC CNRS-URA 7600, Université Pierre et Marie Curie,
4, Place Jussieu, 75005 Paris, France
** E-mail: richard.kerner@upmc.fr*

Pauli's principle is generalized for the case of the Z_3 grading replacing the usual Z_2 grading, leading to ternary commutation relations and ternary algebras. Invariant cubic forms on such algebras are introduced, with the $SL(2, C)$ arising naturally as their invariance group in the case of lowest dimension, with two generators only. The wave equation generalizing the Dirac operator for the Z_3-graded case is introduced, whose diagonalization leads to a third-order equation. The solutions of this equation cannot propagate because their exponents contain a non-oscillating real damping factor. We show how certain cubic products can propagate nevertheless. The model suggests the origin of the color $SU(3)$ symmetry.

Keywords: Z_3 symmetry; Cubic Algebras; Pauli's principle, Quarks.

1. Z_3 generalization of Pauli's principle

The fundamental law ensuring the existence of electron shells and stable atoms forming Mendeleev's Periodic Table of Elements is the Pauli principle: fermionic creation operators must satisfy the anti-commutation relations $\Psi^a \Psi^b = -\Psi^b \Psi^a$, so that the wave functions of two-electron states are automatically anti-symmetric, which means that two electrons cannot coexist in the same state. The *Spin and Statistic Theorem* stipulates that fields describing particles which obey the Fermi-Dirac statistics, called *fermions*, transform under the half-integer representations of the Lorentz group, whereas fields describing particles which obey the Bose-Einstein statistics, called *bosons*, must transform under the integer representations of the Lorentz group.[1]

The strong interactions keeping binding protons and neutrons together inside atomic nuclei are apparently due to more elementary entities called *quarks*, behaving like almost point-like particles inside a nucleon, and obeying particular composition rules. Three quarks of three *different colors* are found in *two states*, "up" and "down", to form a proton (uud) or a neutron (udd). The corresponding anti-particles, or *anti-quarks*, belonging to three different *anti-colors*, compose the anti-proton and the anti-neutron. Finally, a combination of one quark with one anti-quark, of opposite colors produces bosons called *gluons*, carrying forces between quarks.

We claim that as long as there are three quarks composing a nucleon, we must probe the action of the S_3 symmetry groups which describes permutations of three different items. As there are only two species of quarks inside a nucleon, uud or udd, only the action of the Z_3 subgroup of cyclic permutations survives. The generalization of Pauli's principle in this case becomes quite obvious: the -1 element of the Z_2 group $(1, -1)$ has to be replaced by the cubic root of unity, $j = e^{\frac{2\pi i}{3}}$, satisfying $j + j^2 + 1 = 0$ in the following *cubic commutation relations*:[2]

$$\theta^A \theta^B \theta^C = j\, \theta^B \theta^C \theta^A = j^2\, \theta^C \theta^A \theta^B, \tag{1}$$

We shall also introduce a similar set of *conjugate* generators, $\bar{\theta}^{\dot{A}}$, $\dot{A}, \dot{B}, ... = 1, 2, ..., N$, satisfying similar condition with j^2 replacing j:

$$\bar{\theta}^{\dot{A}} \bar{\theta}^{\dot{B}} \bar{\theta}^{\dot{C}} = j^2\, \bar{\theta}^{\dot{B}} \bar{\theta}^{\dot{C}} \bar{\theta}^{\dot{A}} = j\, \bar{\theta}^{\dot{C}} \bar{\theta}^{\dot{A}} \bar{\theta}^{\dot{B}}, \tag{2}$$

A direct consequence of these constitutive relations is the impossibility of forming products of more than three quark or anti-quark operators. The proof is straightforward, using the associativity:

$$\theta^A \theta^B \theta^C \theta^D = j\, \theta^B \theta^C \theta^A \theta^D = j^2\, \theta^B \theta^A \theta^D \theta^C = j^3\, \theta^A \theta^D \theta^B \theta^C = j^4\, \theta^A \theta^B \theta^C \theta^D,$$

and because $j^4 = j \neq 1$, the only solution is $\theta^A \theta^B \theta^C \theta^D = 0$.

In order to make the constitutive relations for the set of operators θ^A and $\bar{\theta}^{\dot{B}}$ complete, we have to impose commutation relations on the binary products $\theta^A \bar{\theta}^{\dot{B}}$ and $\bar{\theta}^{\dot{B}} \theta^A$. We impose the following commutation relations:

$$\theta^A \bar{\theta}^{\dot{B}} = -j\, \bar{\theta}^{\dot{B}} \theta^A, \qquad \bar{\theta}^{\dot{B}} \theta^A = -j^2\, \theta^A \bar{\theta}^{\dot{B}}, \tag{3}$$

2. Invariant forms and the Lorentz symmetry

Consider a tri-linear form ρ^{α}_{ABC}, with values in some vector space ($\alpha = 1, 2, ..., n$). This 3-form will be called invariant under permutations if

$$\rho^{\alpha}_{ABC}\, \theta^A \theta^B \theta^C = \rho^{\alpha}_{BCA}\, \theta^B \theta^C \theta^A = \rho^{\alpha}_{CAB}\, \theta^C \theta^A \theta^B$$

$$= \frac{1}{3} \left[\rho^{\alpha}_{ABC}\, \theta^A \theta^B \theta^C + \rho^{\alpha}_{BCA}\, (j^2\, \theta^A \theta^B \theta^C) + \rho^{\alpha}_{CAB}\, j\, (\theta^A \theta^B \theta^C) \right],$$

From this it follows that we should have

$$\rho^{\alpha}_{ABC}\, \theta^A\theta^B\theta^C = \frac{1}{3}\left[\rho^{\alpha}_{ABC} + j^2\, \rho^{\alpha}_{BCA} + j\, \rho^{\alpha}_{CAB}\right]\theta^A\theta^B\theta^C, \tag{4}$$

from which we get the following properties of the ρ-cubic matrices:[3]

$$\rho^{\alpha}_{ABC} = j^2\, \rho^{\alpha}_{BCA} = j\, \rho^{\alpha}_{CAB}. \tag{5}$$

The conjugate forms $\bar{\rho}^{\dot{\alpha}}_{\dot{C}\dot{B}\dot{A}}$ satisfying the relations similar to (5) with j replaced be its conjugate, j^2:

$$\bar{\rho}^{\dot{\alpha}}_{\dot{A}\dot{B}\dot{C}} = j\, \bar{\rho}^{\dot{\alpha}}_{\dot{B}\dot{C}\dot{A}} = j^2\, \bar{\rho}^{\dot{\alpha}}_{\dot{C}\dot{A}\dot{B}} \tag{6}$$

In the simplest case of two generators, the j-skew-invariant forms have only two independent components, so here $\alpha, \beta = 1, 2$:

$$\rho^1_{121} = j\, \rho^1_{211} = j^2\, \rho^1_{112}, \qquad \rho^2_{212} = j\, \rho^2_{122} = j^2\, \rho^2_{221},$$

and we can set

$$\rho^1_{121} = 1,\ \rho^1_{211} = j^2,\ \rho^1_{112} = j, \qquad \rho^2_{212} = 1,\ \rho^2_{122} = j^2,\ \rho^2_{221} = j.$$

The constitutive cubic relations between the generators of the Z_3 graded algebra can be considered as intrinsic if they are conserved after linear transformations with commuting (pure number) coefficients, i.e. if they are basis-independent. Let $U^{A'}_{A}$ denote a non-singular $N \times N$ matrix, transforming the generators θ^A into another set of generators, $\theta^{B'} = U^{B'}_{B}\, \theta^B$.

We are looking for the solution of the covariance condition for the ρ-matrices:

$$S^{\alpha'}_{\beta}\, \rho^{\beta}_{ABC} = U^{A'}_{A}\, U^{B'}_{B}\, U^{C'}_{C}\, \rho^{\alpha'}_{A'B'C'}. \tag{7}$$

Now, $\rho^1_{121} = 1$, and we have two equations corresponding to the choice of values of the index α' equal to 1 or 2. For $\alpha' = 1'$ the ρ-matrix on the right-hand side is $\rho^{1'}_{A'B'C'}$, which has only three components,

$$\rho^{1'}_{1'2'1'} = 1, \quad \rho^{1'}_{2'1'1'} = j^2, \quad \rho^{1'}_{1'1'2'} = j,$$

which leads to the following equation:

$$S^{1'}_{1} = U^{1'}_{1}\, U^{2'}_{2}\, U^{1'}_{1} + j^2\, U^{2'}_{1}\, U^{1'}_{2}\, U^{1'}_{1} + j\, U^{1'}_{1}\, U^{1'}_{2}\, U^{2'}_{1} = U^{1'}_{1}\, (U^{2'}_{2}\, U^{1'}_{1} - U^{2'}_{1}\, U^{1'}_{2}),$$

because $j^2 + j = -1$, so that $S^{1'}_{1} = U^{1'}_{1}\, \det U$. The three remaining equations are obtained in a similar manner, with other choices of values for indeces, resulting in the following:

$$S^{2'}_{1} = -U^{2'}_{1}\, [\det(U)], \quad S^{1'}_{2} = -U^{1'}_{2}\, [\det(U)], \qquad S^{2'}_{2} = U^{2'}_{2}\, [\det(U)]. \tag{8}$$

The determinant of the 2×2 complex matrix $U^{A'}_{B}$ appears everywhere on the right-hand side. Taking the determinant of the matrix $S^{\alpha'}_{\beta}$ one gets immediately

$$\det(S) = [\det(U)]^3. \tag{9}$$

However, the U-matrices on the right-hand side are defined only up to the phase, which due to the cubic character of the covariance relations and they can take on three different values: 1, j or j^2, i.e. the matrices $j\, U^{A'}_{B}$ or $j^2\, U^{A'}_{B}$ satisfy the same relations as the matrices $U^{A'}_{B}$ defined above. The determinant of U can take on the values 1, j or j^2 if $det(S) = 1$ But for the time being, we have no reason yet to impose the unitarity condition. It can be derived from the conditions imposed on the invariance and duality.

In the Hilbert space of spinors the $SL(2, \mathbf{C})$ action conserved naturally two anti-symmetric tensors, $\varepsilon_{\alpha\beta}$ and $\varepsilon_{\dot{\alpha}\dot{\beta}}$, and their duals, $\varepsilon^{\alpha\beta}$ and $\varepsilon^{\dot{\alpha}\dot{\beta}}$.

Spinorial indeces thus can be raised or lowered using these fundamental $SL(2, \mathbf{C})$ tensors:

$$\psi_{\beta} = \epsilon_{\alpha\beta}\, \psi^{\alpha}, \quad \psi^{\dot{\delta}} = \varepsilon^{\dot{\delta}\dot{\beta}}\, \psi_{\dot{\beta}}.$$

In the space of quark states similar invariant form can be introduced, too. Theere is only one alternative: either the Kronecker delta, or the anti-symmetric 2-form ε. Supposing that our cubic combinations of quark states behave like fermions, there is no choice left: if we want to define the duals of cubic forms ρ^{α}_{ABC} displaying the same symmetry properties, we must impose the covariance principle as follows:

$$\epsilon_{\alpha\beta}\, \rho^{\alpha}_{ABC} = \varepsilon_{AD}\varepsilon_{BE}\varepsilon_{CG}\, \rho^{DEG}_{\beta}.$$

The requirement of the invariance of tensor ε_{AB}, $A, B = 1, 2$ with respect to the change of basis of quark states leads to the condition $\det U = 1$, i.e. again to the $SL(2, \mathbf{C})$ group.

3. The Z_3-generalization of Dirac's equation

Let us first underline the Z_2 symmetry of Maxwell and Dirac equations, implying their hyperbolic character which makes the propagation possible. Maxwell's equations *in vacuo* can be written as follows:

$$\frac{1}{c}\frac{\partial \mathbf{E}}{\partial t} = \nabla \wedge \mathbf{B}, \quad -\frac{1}{c}\frac{\partial \mathbf{B}}{\partial t} = \nabla \wedge \mathbf{E}. \tag{10}$$

These equations can be decoupled by applying the time derivation twice, which in vacuum, where $\operatorname{div}\mathbf{E} = 0$ and $\operatorname{div}\mathbf{B} = 0$ leads to the d'Alembert equation for both components separately:

$$\frac{1}{c^2}\frac{\partial^2 \mathbf{E}}{\partial t^2} - \nabla^2 \mathbf{E} = 0, \qquad \frac{1}{c^2}\frac{\partial^2 \mathbf{B}}{\partial t^2} - \nabla^2 \mathbf{B} = 0.$$

Nevertheless, neither of the components of the Maxwell tensor, be it **E** or **B**, can propagate separately alone. The Dirac equation for the electron displays a similar Z_2 symmetry, with two coupled equations which can be put in the following form:[4]

$$i\hbar\frac{\partial}{\partial t}\psi_+ - mc^2\psi_+ = i\hbar\boldsymbol{\sigma}\cdot\boldsymbol{\nabla}\psi_-, \quad -i\hbar\frac{\partial}{\partial t}\psi_- - mc^2\psi_- = -i\hbar\boldsymbol{\sigma}\cdot\boldsymbol{\nabla}\psi_+, \tag{11}$$

where ψ_+ and ψ_- are the positive and negative energy components of the Dirac equation; this is visible even better in the momentum representation:

$$\left[E - mc^2\right]\psi_+ = c\boldsymbol{\sigma}\cdot\mathbf{p}\psi_-, \qquad \left[-E - mc^2\right]\psi_- = -c\boldsymbol{\sigma}\cdot\mathbf{p}\psi_+. \tag{12}$$

The same effect (negative energy states) can be obtained by changing the direction of time, and putting the minus sign in front of the time derivative.

$$\left[\frac{1}{c^2}\frac{\partial^2}{\partial t^2} - \nabla^2 - m^2\right]\psi_\pm = 0$$

As in the electromagnetic case, neither of the components of this complex entity can propagate by itself; only all the components can.[5]

Now, if we want to include the action of Z_3 symmetry on three different fields, keeping also the Z_2-symmetry due to the existence of anti-particles as well as the additional Z_2-symmetry due to the existence of two components for each color field (u and d, for example), we should introduce three two-component fields, ψ, φ and χ. which may represent three colors, two quark states (e.g. "up" and "down"), and two anti-quark states (with anti-colors, respectively). Finally, in order to be able to implement the action of the $SL(2, \mathbf{C})$ group via its 2×2 matrix representation defined in the previous section, we choose the Hamiltonian $\hat{H}$ equal to the operator $\sigma \cdot \nabla$: The final symmetry is $Z_3 \times Z_2 \times Z_2$, resulting in *twelve-component* wave functions. The differential system that satisfies all these assumptions is as follows:

$$-i\hbar\frac{\partial}{\partial t}\psi_+ - mc^2\psi_+ = -i\hbar c\sigma\cdot\nabla\varphi_-, \quad i\hbar\frac{\partial}{\partial t}\varphi_- - jmc^2\varphi_- = -ji\hbar c\sigma\cdot\nabla\chi_+$$

$$-i\hbar\frac{\partial}{\partial t}\chi_+ - j^2mc^2\chi_+ = -j^2i\hbar c\sigma\cdot\nabla\psi_-, \quad i\hbar\frac{\partial}{\partial t}\psi_- - mc^2\psi_- = -i\hbar c\sigma\cdot\nabla\varphi_+,$$

$$-i\hbar\frac{\partial}{\partial t}\varphi_+ - jmc^2\varphi_+ = -ji\hbar c\sigma\cdot\nabla\chi_-, \quad i\hbar\frac{\partial}{\partial t}\chi_- - j^2mc^2\bar{\chi} = -j^2i\hbar c\sigma\cdot\nabla\psi_+,$$

leading to the common equation of sixth order that is satisfied by each of the components:

$$-\hbar^6 \frac{\partial^6}{\partial t^6}\psi - m^6 c^{12}\psi = -\hbar^6 \Delta^3 \psi. \tag{13}$$

Identifying energy and momentum operators, $-i\hbar\frac{\partial}{\partial t} \to E, \quad -i\hbar\nabla \to \mathbf{p}$, we can write (13) simply as $E^6 - m^6c^{12} = \mid \mathbf{p} \mid^6 c^6$. The equation (13) can be solved by separation of variables; the time-dependent and the space-dependent factors have the same structure:

$$A_1 \, e^{\omega \, t} + A_2 \, e^{j\, \omega \, t} + A_3 e^{j^2 \, \omega \, t}, \qquad B_1 \, e^{\mathbf{k}.\mathbf{r}} + B_2 \, e^{j\, \mathbf{k}.\mathbf{r}} + B_3 \, e^{j^2 \, \mathbf{k}.\mathbf{r}}$$

with ω and $\mathbf{k}$ satisfying the following dispersion relation:

$$\frac{\omega^6}{c^6} = \frac{m^6 c^6}{\hbar^6} + \mid \mathbf{k} \mid^6 . \tag{14}$$

The relation (14) is invariant under the action of $Z_2 \times Z_3$ symmetry, because to any solution with given real ω and $\mathbf{k}$ one can add solutions with ω replaced by $j\omega$ or $j^2\omega$, $j\mathbf{k}$ or $j^2\mathbf{k}$, as well as $-\omega$ for anti-particles. This yields nine solutions which all can be made real by appropriate linear comboinations. They are displayed in the 3×3 matrix below:

$$\begin{pmatrix} e^{\omega\, t - \mathbf{k}\cdot\mathbf{r}} & e^{\omega\, t + \frac{\mathbf{k}\cdot\mathbf{r}}{2}} \cos(\mathbf{k}\cdot\xi) & e^{\omega\, t + \frac{\mathbf{k}\cdot\mathbf{r}}{2}} \sin(\mathbf{k}\cdot\xi) \\ e^{-\frac{\omega\, t}{2} - \mathbf{k}\cdot\mathbf{r}} \cos\omega\tau & e^{-\frac{\omega\, t}{2} + \frac{\mathbf{k}\cdot\mathbf{r}}{2}} \cos(\omega\tau - \mathbf{k}\cdot\xi) & e^{-\frac{\omega\, t}{2} + \frac{\mathbf{k}\cdot\mathbf{r}}{2}} \cos(\omega\tau + \mathbf{k}\cdot\xi) \\ e^{-\frac{\omega\, t}{2} - \mathbf{k}\cdot\mathbf{r}} \sin\omega\tau & e^{-\frac{\omega\, t}{2} + \frac{\mathbf{k}\cdot\mathbf{r}}{2}} \sin(\omega\tau + \mathbf{k}\cdot\xi) & e^{-\frac{\omega\, t}{2} + \frac{\mathbf{k}\cdot\mathbf{r}}{2}} \sin(\omega\tau - \mathbf{k}\cdot\xi) \end{pmatrix}$$

where $\tau = \frac{\sqrt{3}}{2}\, t$ and $\xi = \frac{\sqrt{3}}{2}\mathbf{k}\cdot\mathbf{r}$; the same can be done with the conjugate solutions (with $-\omega$ instead of ω). The functions displayed in the matrix do not represent a wave; however, one can produce a propagating solution by forming certain cubic combinations, e.g. the product of diagonal elements of the above matrix gives the following expression:

$$e^{\omega\, t - \mathbf{k}\cdot\mathbf{r}} \, e^{-\frac{\omega\, t}{2} + \frac{\mathbf{k}\cdot\mathbf{r}}{2}} \cos(\omega\tau - \mathbf{k}\cdot\xi) \, e^{-\frac{\omega\, t}{2} + \frac{\mathbf{k}\cdot\mathbf{r}}{2}} \sin(\omega\tau - \mathbf{k}\cdot\xi) = \frac{1}{2}\sin(2\omega\tau - 2\mathbf{k}\cdot\xi).$$

Therefore, some special triple products of non-propagating solutions can yield propagating ones, but under the condition that the factors belong to three distinct subsets (later on identified as "colors").

References

1. W. Pauli, *Phys. Rev.* **58**, pp. 716-722 (1940)
2. R. Kerner, *C.R. Acad. Sci. Paris*, **312** (série II), pp. 191-196, (1991)
3. R. Kerner, *Journ. Math. Phys.*, **33**, pp. 403-411 (1992)
4. P.A.M. Dirac, *The Principles of Quantim Mechanics*, *Oxford Univ. Press*, (1958)
5. J.D. Bjorken and S.D. Drell, *Relativistic Quantum Mechanics*, *McGraw-Hill*, London-New York (1964).

PHASE STATES AND COHERENT STATES FOR GENERALIZED WEYL-HEISENBERG ALGEBRAS

M. R. KIBLER* and M. DAOUD

Institut de Physique Nucléaire, Université de Lyon (Lyon 1), Villeurbanne, F-69622, France
** E-mail: kibler@ipnl.in2p3.fr*

This report constitutes an introduction to three papers published by the authors in J. Phys. A [**43** (2010) 115303 and **45** (2012) 244036] and J. Math. Phys. [**52** (2011) 082101]. See these three papers for the relevant references.

Keywords: phase states; coherent states; mutually unbiased bases.

1. General outline

This paper is concerned with the construction of phase operators, phase states, vector phase states, and coherent states for a generalized Weyl-Heisenberg algebra. This polynomial algebra (that depends on real parameters) is briefly described. The various states are defined on a finite- or infinite-dimensional space depending on the parameters.

2. Generalized Weyl-Heisenberg Algebra

2.1. *The $\mathcal{A}_{\{\kappa\}}(1)$ algebra*

Let $\mathcal{A}_{\{\kappa\}}(1)$ be the generalized Weyl-Heisenberg algebra spanned by a^-, a^+ and N satisfying

$$[a^-, a^+] = F(N+I) - F(N), \quad [N, a^\pm] = \pm a^\pm$$
$$N = N^\dagger, \quad a^+ = (a^-)^\dagger, \quad F(N) = N\prod_{i=1}^{r}[I + \kappa_i(N-I)], \quad \kappa_i \in \mathbb{R}.$$

This definition yields a multi-parameter polynomial Weyl-Heisenberg algebra depending on r parameters $\{\kappa\} \equiv \{\kappa_1, \kappa_2, \ldots, \kappa_r\}$. The harmonic oscillator algebra corresponds to $\kappa_i = 0$ ($F(N) = N \Rightarrow F(N+I) - F(N) = I$).

2.2. *Hilbertian representation of $\mathcal{A}_{\{\kappa\}}(1)$*

We suppose that the operators $a^{\pm}$ and N act on a (finite or not) Hilbert space $\mathcal{F}_{\kappa}$ with an orthonormal basis $\{|n\rangle : n \text{ ranging}\}$ arising from the eigenvectors of N. The following actions

$$N|n\rangle = n|n\rangle, \quad a^{+}|n\rangle = \sqrt{F(n+1)}e^{-i[F(n+1)-F(n)]\varphi}|n+1\rangle$$
$$a^{-}|n\rangle = \sqrt{F(n)}e^{+i[F(n)-F(n-1)]\varphi}|n-1\rangle, \quad a^{-}|0\rangle = 0$$

(where $\varphi \in \mathbb{R}$) define an Hilbertian representation of $\mathcal{A}_{\{\kappa\}}(1)$.

The significance of the F structure function is clear. The operator

$$H = a^{+}a^{-} = F(N)$$

can be considered as the Hamiltonian for a quantum system.

2.3. *Dimension of $\mathcal{F}_{\kappa}$*

The dimension of the $\mathcal{F}_{\kappa}$ representation space of $\mathcal{A}_{\{\kappa\}}(1)$ is controlled by $F(n) \geq 0$. Two interesting cases deserve to be considered: (i) $\kappa_i \geq 0$ $(i = 1, 2, \cdots, r) \Rightarrow$ infinite-dimensional representation and (ii) $\kappa_1 < 0$, $\kappa_i \geq 0$ $(i = 2, 3, \cdots, r) \Rightarrow$ finite representation of dimension

$$d = 1 - \frac{1}{\kappa_1} \quad \text{for} \quad -\frac{1}{\kappa_1} \in \mathbb{N}^{*}.$$

The Fock-Hilbert space is spanned by $\{|n\rangle : n \in \mathbb{N}\}$ in the infinite-dimensional case and by $\{|n\rangle : n = 0, 1, \cdots, d-1\}$ in the finite-dimensional case. In finite dimension, two further conditions are fulfilled, viz, $a^{+}|d-1\rangle = 0$ and $(a^{-})^{d} = (a^{+})^{d} = 0$, which generalize the conditions for k-fermions which are objects, introduced by the authors, interpolating between fermions ($k = d = 2$) and bosons ($k = d \to \infty$).

2.4. *Example: The $\mathcal{A}_{\kappa}(1)$ algebra*

This example corresponds to $r = 1$, $\kappa_1 = \kappa$, and $\mathcal{A}_{\{\kappa\}}(1) \equiv \mathcal{A}_{\kappa}(1)$. The $\mathcal{A}_{\kappa}(1)$ one-parameter algebra is then defined by

$$[a^{-}, a^{+}] = I + 2\kappa N, \quad [N, a^{\pm}] = \pm a^{\pm}, \quad \left(a^{-}\right)^{\dagger} = a^{+}, \quad N^{\dagger} = N$$

and gives rise to three cases: (i) $\kappa = 0 \to \mathcal{A}_{\kappa}(1) = h_4$ (usual Weyl-Heisenberg algebra); (ii) $\kappa > 0 \to \mathcal{A}_{\kappa}(1) = su_{1,1}$; and (iii) $\kappa < 0 \to \mathcal{A}_{\kappa}(1) = su_2$. For $su_{1,1}$, $\kappa > 0$ corresponds to $2k\kappa = 1$, where k is the Bargmann index of the positive discrete series. For su_2, $\kappa < 0$ corresponds to $2j\kappa = -1$, where j is an angular momentum label associated with the $(2j)$ irreducible representation of su_2.

The H Hamiltonian associated with $\mathcal{A}_\kappa(1)$ can be redefined as $H_0 = X_+X_-$ with

$$a^{\pm} = \frac{1}{\sqrt{b}}X_{\pm}, \quad 2\kappa = \frac{a}{b}, \quad a \in \mathbb{R}, \quad b \in \mathbb{R}_+^* \Rightarrow H_0 = \frac{1}{2}aN(N-I) + bN$$

and gives rise to three particular cases: (i) $a = 0$, $b = 1$, $\kappa = 0$, h_4 case, $\dim \mathcal{F}_\kappa = \infty$: oscillator system; (ii) $a = 1$, $2b = u + v + 1$, $u > 1$, $v > 1$, $\kappa > 0$, $su_{1,1}$ case, $\dim \mathcal{F}_\kappa = \infty$: Pöschl-Teller system; and (iii) $a = -1$, $2b = \ell - 1$, $\ell \in \mathbb{N} \setminus \{0, 1\}$, $\kappa < 0$, su_2 case, $\dim \mathcal{F}_\kappa = \ell$: Morse system.

3. Phase operators and phase states for $\mathcal{A}_\kappa(1)$

We now continue with the case where $r = 1$. We define the E operator via

$$a^- = E\sqrt{F(N)} \quad \text{with} \quad E \equiv E_\infty \quad \text{or} \quad E \equiv E_d$$

for the infinite- or finite-dimensional case, respectively.

3.1. *Case $\kappa \geq 0$: The E_∞ operator*

The E_∞ (phase) operator is not unitary ($E_\infty E_\infty^\dagger = E_\infty^\dagger E_\infty + |0\rangle\langle 0| = I$). Its eigenstates are phase states given by

$$E_\infty|\varphi, \theta\rangle = \mathrm{e}^{\mathrm{i}\theta}|\varphi, \theta\rangle, \quad \theta \in [-\pi, \pi], \quad |\varphi, \theta\rangle = \sum_{n=0}^{\infty} \mathrm{e}^{-\mathrm{i}F(n)\varphi}\mathrm{e}^{\mathrm{i}n\theta}|n\rangle$$

on $\{\rho \mathrm{e}^{\mathrm{i}\theta} : \rho < 1\}$. The $|\varphi, \theta\rangle$ phase states depend on two continuous parameters φ and θ, are neither normalized nor orthogonal, are temporally stable with respect to H, and satisfy a closure property ($\int_{-\pi}^{+\pi} d\theta |\varphi, \theta\rangle\langle\varphi, \theta| = 2\pi I$).

The nonunitarity of E_∞ requires a truncation procedure *à la* Pegg and Barnett for the algebra $A_\kappa(1)$ – also useful for perturbation theory purposes. This leads to a truncated algebra $\mathcal{A}_{\kappa,s}(1)$ and a new representation space $\mathcal{F}_{\kappa,s}$ where s is the order of truncation (the dimension of $\mathcal{F}_{\kappa,s}$). This situation is similar to the finite case corresponding to $\kappa < 0$ ($\dim \mathcal{F}_\kappa = d$) to be considered below.

3.2. *Case $\kappa < 0$: The $\mathbf{E}_d$ and $\mathbf{G}_d$ operators*

3.2.1. *The E_d operator*

The E_d (phase) operator is unitary. Its eigenstates are given by

$$E_d|\varphi, m\rangle = \mathrm{e}^{2\pi \mathrm{i} m/d}|\varphi, m\rangle, \quad |\varphi, m\rangle = \frac{1}{\sqrt{d}}\sum_{n=0}^{d-1} \mathrm{e}^{\mathrm{i}[-F(n)\varphi + 2\pi mn/d]}|n\rangle$$

with $m \in \mathbb{Z}/d\mathbb{Z}$. The $|\varphi, m\rangle$ phase states depend on a discrete parameter m and a continuous parameter φ; they are temporally stable with respect to H; and, for fixed φ, they are orthonormal ($\langle\varphi, m|\varphi, m'\rangle = \delta_{m,m'}$, the overlap $\langle\varphi, m|\varphi', m'\rangle$ does not vanish for $\varphi' \neq \varphi$), they satisfy a closure relation ($\sum_{m=0}^{d-1} |\varphi, m\rangle\langle\varphi, m| = I$) and an equiprobability relation ($|\langle n|\varphi, m\rangle| = 1/\sqrt{d}$).

3.2.2. *The G_d operator*

For $\kappa < 0$, we can define another type of phase operator, namely,

$$G_d = a^- + \frac{(a^+)^{d-1}}{F(d-1)!}, \quad F(n)! = F(1)F(2)\ldots F(n), \quad F(0) = 1.$$

The eigenstates of G_d are phase states given by

$$G_d|\varphi, \mu\rangle = \mathrm{e}^{2\pi \mathrm{i}\mu/d}|\varphi, \mu\rangle, \quad |\varphi, \mu\rangle = c_0 \sum_{n=0}^{d-1} \frac{1}{\sqrt{F(n)!}} \mathrm{e}^{\mathrm{i}[-F(n)\varphi + 2\pi\mu n/d]}|n\rangle$$

with $\mu \in \mathbb{Z}/d\mathbb{Z}$. The $|\varphi, \mu\rangle$ phase states depend on a discrete parameter μ and a continuous parameter φ, are temporally stable with respect to H, not normalizable, not orthogonal, and do not satisfy a closure property. The $|\varphi, \mu\rangle$ phase states are similar to the Gazeau and Klauder coherent states except that: (i) they are eigenstates of a polynomial in a^- and a^+, (ii) their labeling includes an integer, and (iii) they correspond to a finite spectrum (finite sum).

4. Application to mutually unbiased bases

By quantizing φ according to $\varphi = -\pi(d-1)a/d$ with $a = 0, 1, \ldots, d-1$, the state vector $|\varphi, m\rangle$ becomes

$$|\varphi, m\rangle \equiv |am\rangle = \frac{1}{\sqrt{d}}\sum_{n=0}^{d-1} q^{n(d-n)a/2+nm}|n\rangle, \quad q = \exp\left(\frac{2\pi \mathrm{i}}{d}\right)$$

Note that $|am\rangle$ is a quadratic discrete Fourier transform (the case $a = 0$ corresponds to an ordinary discrete Fourier transform). This quantization leads to the following result. For d prime, the d bases $\{|am\rangle : m = 0, 1, \ldots, d-1\}$ for $a = 0, 1, \ldots, d-1$ and the canonical basis $\{|n\rangle : n = 0, 1, \ldots, d-1\}$ form a complete set of $d+1$ mutually unbiased bases (in the sense of quantum information).

5. Passing from $\mathcal{A}_\kappa(1)$ to $\mathcal{A}_\kappa(2)$

As an extension of $\mathcal{A}_\kappa(1)$ (with one degree of freedom) we can define a one-parameter generalized Weyl-Heisenberg algebra $\mathcal{A}_\kappa(2)$ involving two degrees of freedom. The $\mathcal{A}_\kappa(2)$ algebra is spanned by a_i^-, a_i^+ and N_i satisfying

$$[a_i^-, a_i^+] = I + \kappa(N_1 + N_2 + N_i), \quad [a_i^\pm, a_j^\pm] = 0, \quad [N_i, a_j^\pm] = \pm\delta_{i,j} a_i^\pm$$
$$N_i = N_i^\dagger, \qquad a_i^+ = (a_i^-)^\dagger, \qquad [a_i^\pm, [a_i^\pm, a_j^\mp]] = 0 \quad \text{for} \quad i \neq j$$

where $\kappa \in \mathbb{R}$ and $i, j \in \{1, 2\}$. This yields three cases: (i) $\kappa = 0 \to \mathcal{A}_\kappa(2) = h_4 \otimes h_4$; (ii) $\kappa > 0 \to \mathcal{A}_\kappa(2) = su_{2,1}$; and (iii) $\kappa < 0 \to \mathcal{A}_\kappa(2) = su_3$.

In the infinite case ($\kappa \geq 0$), we can define two types of nonunitary phase operators (associated with a_1^- and a_2^-). The corresponding phase states (neither normalized nor orthogonal) satisfy temporal stability and a closure property. A truncation *à la* Peggg and Barnett is necessary in view of the absence of unitarity.

In the finite case ($\kappa < 0$), we can define four types of unitary phase operators. The phase states are here vector phase states (which are analogs of vector coherent states introduced by Hecht 1987 – Zhang, Feng, Gilmore 1990 – Thirulogasanthar, Ali 2003 – Ali, Engliš, Gazeau 2004 – Ali, Bagarello 2005). They satisfy temporal stability, orthonormality, closure property, and equiprobability relation. They can be used for constructing mutually unbiased bases.

6. Coherent states for $\mathcal{A}_{\{\kappa\}}(1)$

We come back to the case of a multi-parameter generalized Weyl-Heisenberg algebra $\mathcal{A}_{\{\kappa\}}(1)$. We shall deal in turn with coherent states of type I (*à la* Klauder and Perelomov) and of type II (*à la* Barut and Girardello).

6.1. *Coherent states of type I*

The strategy is to look for states of the form $|z, \varphi\rangle = \sum_n \overline{a_n} z^n |n\rangle$ ($a_n \in \mathbb{C}$, $z \in \mathbb{C}$) with the replacements $|n\rangle \to a_n z^n$ and $a^+ \to d/dz$. The derivation of the a_n coefficients yields two results.

In infinite dimension, the states

$$|z,\varphi\rangle = \sum_{n=0}^{\infty} \frac{\sqrt{F(n)!}}{n!} z^n e^{-iF(n)\varphi}|n\rangle$$

exist only for $r = 1$. They satisfy $|z,\varphi\rangle = \exp(za^+)|0\rangle$ so that they are thus coherent states in the Klauder-Perelomov sense.

In finite dimension ($dim = d$ or s), we have the same formal expression for $|z,\varphi\rangle$ except that $\sum_{n=0}^{\infty} \to \sum_{n=0}^{dim-1}$ and $r = 1 \to r$ arbitrary. We thus get coherent states in finite dimension in the Klauder-Perelomov sense.

6.2. *Coherent states of type II*

The strategy here is to look for states of the form $|z,\varphi\rangle$ with the replacements $a_n \to b_n$, $|n\rangle \to b_n z^n$, and $a^+ \to z$. This leads to the two following results.

In infinite dimension, the states

$$|z,\varphi\rangle = \sum_{n=0}^{\infty} \frac{1}{\sqrt{F(n)!}} z^n e^{-iF(n)\varphi}|n\rangle$$

exist for any value of r. They satisfy $a^-|z,\varphi\rangle = z|z,\varphi\rangle$ and are thus coherent states in the Barut-Girardello sense.

Note that in infinite dimension, the coherent states of types I and II are identical solely in the case where $\kappa_i = 0$ ($i = 1, 2, \cdots, r$). The corresponding states are the well-known Glauber states.

In finite dimension ($dim = d$ or s), there are no Barut-Girardello coherent states for $z \in \mathbb{C}$. However, Barut-Girardello coherent states exist for $z \to \theta =$ Grassmann variable with $\theta^{dim} = 0$. They are given by

$$|\theta,\varphi\rangle = \sum_{n=0}^{dim-1} \frac{1}{\sqrt{F(n)!}} \theta^n e^{-iF(n)\varphi}|n\rangle$$

and satisfy $a^-|\theta,\varphi\rangle = \theta|\theta,\varphi\rangle$.

The properties of the coherent states of type I and II are described in the papers mentioned in the abstract.

7. Closing remarks

The developments in this paper should give an unified view of the three corresponding papers published by the authors. As open questions, let us mention the probabilistic interpretation and Bargmann functions associated with the coherent states of type I and II.

QUANTUM STATES ARISING FROM THE PAULI GROUPS: SYMMETRIES AND PARADOXES

M. PLANAT

Institut FEMTO-ST, CNRS
32 Avenue de l'Observatoire, F-25044 Besançon, France
E-mail: michel.planat@femto-st.fr

We investigate multiple qubit Pauli groups and the quantum states/rays arising from their maximal bases. Remarkably, the real rays are carried by a Barnes-Wall lattice BW_n ($n = 2^m$). We focus on the smallest subsets of rays allowing a state proof of the Bell-Kochen-Specker theorem (BKS). BKS theorem rules out realistic *non-contextual* theories by resorting to impossible assignments of rays among a selected set of maximal orthogonal bases. We investigate the geometrical structure of small BKS-proofs $v - l$ involving v rays and l $2n$-dimensional bases of n-qubits. Specifically, we look at the classes of parity proofs $18 - 9$ with two qubits (A. Cabello, 1996), $36 - 11$ with three qubits (M. Kernaghan & A. Peres, 1995) and related classes. One finds characteristic signatures of the distances among the bases, that carry various symmetries in their graphs.

Keywords: multiple qubits, Barnes-Wall lattices, Bell-Kochen-Specker theorem.

1. Real rays from the multiple qubit Pauli group and Barnes-Wall lattices

An n-dimensional Euclidean lattice $\mathcal{L}$ is a discrete additive subgroup of the real vector space $\mathbb{R}^n$, endowed with the standard Euclidean product, and spanned by a generator matrix M with rows in $\mathbb{R}^n$. The automorphism group $\text{Aut}(\mathcal{L})$ is the set of orthogonal matrices B such that under the conjugation action $U = MBM^{-1}$ by the generating matrix M, one has (i) $\det U = \pm 1$ and (ii) U is an integer matrix.[1] One is specifically interested by the family of Barnes-Wall lattices BW_n ($n = 2^m$ and $m > 1$) that generalize the root lattices $BW_4 \cong D_4$ and $BW_8 \cong E_8$ with BW_{16} (the densest known lattices) and higher.[2] Their automorphismgroup group is the so-called real Clifford group.[3,4]

Let us study the relationship between the Barnes-Wall lattices and the real rays arising from the n-qubit Pauli group $\mathcal{P}_n$.[5] The total number of states/rays appearing as eigenstates shared by the maximal commuting sets mcs of operators in $\mathcal{P}_m$ is nL, where $n = 2^m$ and the number of maximal commuting sets is $L = \prod_{i=1}^{m}(1 + 2^i)$, as shown in column 3 of table 1; the corresponding number of real rays is shown in column 4. The rays form maximal orthogonal bases whose orthogonality graph is defined in columns 5 and 6. One gets the following striking observations that

- The number of real rays identifies to the kissing number $k_n = (2^n + 2)k_{n-1}$ (with $k_1 = 1$) of a Barnes-Wall lattice BW_n ($n = 2^m$ and $m > 1$), that is isomorphic to $D_4, E_8, \Lambda_{16}, BW_{32} \cdots$.
- The automorphism group of the orthogonality graph attached to the real rays is that of BW_n, that is the real Clifford group $C_n^+ = 2_+^{1+2m}.\Omega_{2m}^+(2)$, of order $2^{m^2+m+1}(2^m - 1)\prod_{j=1}^{m-1}(4^j - 1)$, with 2_+^{1+2m} the extraspecial group of order 2^{2m+1} and $\Omega_{2m}^+(2)$ the orthogonal group [the derived subgroup of the general orthogonal group $O_{2m}^{\pm}(2)$].

One concludes that a natural frame for the real rays arising from the multiple qubit Pauli group $\mathcal{P}_m$ is the Barnes-Wall lattice BW_{2^m}.

Table 1. Cardinalities and symmetries of the real rays arising from the m-qubit Pauli group $\mathcal{P}_m$.

$n = 2^m$	$\#mcs$	#rays	#real rays	aut group	#aut group
2	3	6	4	D_4	8
4	15	60	24	$\mathbb{Z}_2^4 \rtimes (\mathbb{Z}_3^2 \rtimes D_4)$	1152
8	135	1080	240	$\mathbb{Z}_2^6 \rtimes S_8$	2580480
16	2295	36720	4320	$\cong aut(BW_{16})$	89181388800
32	75735	2423520	146880	$\cong aut(BW_{32})$	$\approx 4.8 \times 10^{15}$

2. The Bell-Kochen-Specker theorem for multiple qubits

The Bell-Kochen-Specker proof demonstrates the impossibility of Einstein's assumption, made in the famous Einstein-Podolsky-Rosen paper[6] that quantum mechanical observables represent 'elements of physical reality'. More specifically, the theorem excludes hidden variable theories that require elements of physical reality to be non-contextual (i.e. independent of the measurement arrangement).[7]

A non-bicoloring BKS proof consists of a finite set of rays/vectors that cannot be assigned truth values (1 for true, 0 for false) in such a way that

(i) one member of each complete orthonormal basis is true and (ii) no two orthogonal (that is, mutually compatible) projectors are both true.

A parity proof of BKS theorem is a set of v rays that form l bases (l odd) such that each ray occurs an even number of times over these bases. A proof of BKS theorem is ray critical (resp. basis critical) if it cannot be further simplified by deleting even a single ray (resp. a single basis).

The smallest state-independent proofs in three dimensions are of the $31-17$ type (31 rays located on 17 orthogonal triads).[8] The smallest BKS proof in dimension 4 (resp. 8) is a parity proof and corresponds to arrangements of real states arising from the two-qubit (resp. three-qubit) Pauli group, more specifically as eigenstates of operators forming Mermin's square (resp. Mermin's pentagram).[9]

Apart from the use of standard graph theoretical tools for characterizing the ray/base symmetries, we shall employ a useful signature of the proofs in terms of Bengtsson's distance D_{ab} between two orthonormal bases a and b defined as[10] $D_{ab}^2 = 1 - \frac{1}{d-1}\sum_{i,j}^{d}\left(|\langle a_i|b_j\rangle|^2 - \frac{1}{d}\right)^2$. The distance vanishes when the bases are the same and is maximal (equal to unity) when the two bases a and b are mutually unbiased, $|\langle a_i|b_j\rangle|^2 = 1/d$, and only then. We shall see that the bases of a BKS proof employ a selected set of distances which seems to be a universal feature of the proof.

2.1. *Small BKS parity proofs for two qubits from the Mermin's square*

$$\begin{array}{ccc} | & | & \| \\ -Z_1- & Z_2- & ZZ- \\ | & | & \| \\ -X_2- & X_1- & XX- \\ | & | & \| \\ -ZX- & XZ- & YY- \\ | & | & \| \end{array} \qquad (1)$$

The simplification of arguments in favour of a contextual view of quantum measurements started with Peres' work[8] and Mermin's report.[11] Observe that in (1), the three operators in each row and each column mutually commute and their product is the identity matrix, except for the right hand side column whose product is minus the identity matrix. There is no way of assigning multiplicative properties to the eigenvalues ± 1 of the nine operators while still keeping the same multiplicative properties for the operators.

The next step to be able to see behind the scene, and to reveal the simplest paradoxical/contextual set of rays and bases, was achieved by A. Cabello.[9] It is a $18-9$ BKS parity proof that can be given a remarkable diagrammatic illustration fitting the structure of a 24-cell.[12] More generally, it is already known that there exist four main types of parity proofs arising from 24 Peres rays,[8] that are of the type $18-9$, $20-11$, $22-13$ and $24-15$. Types $20-11$ and $22-13$ subdivide into two non-isomorphic ones A and B as shown in Table 2, fore more details see 13. The histogram of distances for various parity proofs $v-l$ obtained from Mermin's square is shown in table 2. The distances involved are $D=\{a_1,a_2,a_3,a_4,a_5\}=\{\frac{1}{\sqrt{3}},\frac{\sqrt{7}}{\sqrt{12}},\frac{\sqrt{2}}{\sqrt{3}},\frac{\sqrt{5}}{\sqrt{6}},1\}$. One can check the expected equality $2\sum a_i = l(l-1)$ in each proof.

Table 2.

proof $v-l$	#proofs	a_1	a_2	a_3	a_4	a_5
$24-15$	16	18	18	9	54	6
$22-13A$	96	12	18	3	42	3
$22-13B$	144	12	18	4	42	2
$20-11A$	96	6	18	0	30	1
$20-11B$	144	6	18	1	30	0
$18-9$	16	0	18	0	18	0

$$
\begin{array}{ccccccc}
\begin{pmatrix} 1 & 2 \\ 15 & 16 \end{pmatrix} & -1- & \begin{pmatrix} 1 & 3 \\ 17 & 18 \end{pmatrix} & -3- & \begin{pmatrix} 2 & 3 \\ 21 & 22 \end{pmatrix} & -2 \\
|16 & & |18 & & |22 & \\
\begin{pmatrix} 5 & 6 \\ 14 & 16 \end{pmatrix} & -5- & \begin{pmatrix} 5 & 7 \\ 18 & 20 \end{pmatrix} & -7- & \begin{pmatrix} 6 & 7 \\ 21 & 24 \end{pmatrix} & -6 \\
|14 & & |20 & & |24 & \\
\begin{pmatrix} 11 & 12 \\ 14 & 15 \end{pmatrix} & -12- & \begin{pmatrix} 10 & 12 \\ 17 & 20 \end{pmatrix} & -10- & \begin{pmatrix} 10 & 11 \\ 21 & 24 \end{pmatrix} & -11 \\
|15 & & |17 & & |21 &
\end{array} \qquad (2)
$$

The diagram for a $18-9$ proof is a 3×3 square. The 9 vertices of the graph are the 9 bases of the proof, the one-point crossing graph between the bases is the graph (2), with aut $= G_{72} = \mathbb{Z}_3^2 \rtimes D_4$. There are 18 (distinct) edges that encode the 18 rays, a selected vertex/base of the graph is encoded by the union of the four edges/rays that are adjacent to it.

As for the distances between the bases, two bases located in the same row (or the same column) have distance a_2, while two bases not in the same row (or column) have distance a_4, as readily discernible from Table 2. Indeed, any proof of the $18-9$ type has the same diagram as (2). Similar diagrams can be drawn to reflect the histogram of distances in proofs of a larger size.

2.2. *Small BKS parity proofs for three qubits from the Mermin's pentagram*

Quantum contextuality of a three-qubit system is also predicted in Mermin's report[11] in terms of its famous pentagram. Below we display it in a sligthly different as a magic" rectangle.

$$\begin{array}{ccccccccc}
 & | & & | & & | & & | & \\
 & Z_1 & & Z_1 & & X_1 & & X_1 & \\
 & | & & | & & | & & | & \\
 & Z_2 & & X_2 & & Z_2 & & X_2 & \\
 & | & & | & & | & & | & \\
 & Z_3 & & X_3 & & X_3 & & Z_3 & \\
 & | & & | & & | & & | & \\
= & ZZZ & = & ZXX & = & XZX & = & XXZ & = \\
 & | & & | & & | & & | &
\end{array} \tag{3}$$

Following 11, (3) is a parity proof of the BKS theorem because mutually commuting operators in the four columns multiply to the identity matrix while operators in the single row multiply to minus the identity matrix. Since each operator appears twice in this reasoning, it is impossible to assign truth values ± 1 to the eigenvalues while keeping the multiplicative properties of the operators.

The histogram of distances for various parity proofs $v - l$ obtained from Mermin's pentagram is given in table 3. The finite set of distances involved is $D = \{a_1, a_2, a_3\} = \{\frac{\sqrt{3}}{\sqrt{7}}, \frac{\sqrt{9}}{\sqrt{14}}, \frac{\sqrt{6}}{\sqrt{7}}\}$.

Table 3.

proof $v - l$	# proofs	a_1	a_2	a_3
$40 - 15$	64	20	30	55
$38 - 13$	640	12	30	36
$36 - 11$	320	4	30	21

$$\begin{array}{ccccccccc}
 & & & 2 & & & & & \\
6 - & & 7 - 8 & & & - 9 & == & 1 & \\
 & 10 & & & 11 & & & & \\
 & & & 3 & & & & & \\
 & 4 & & & & 5 & & &
\end{array} \tag{4}$$

As previously, a simple diagram illustrates how distances between the bases are distributed. Let us look at the $36-11$ parity proof. The 11 bases are displayed as a pentagram (4) plus the isolated reference base 1.

Two adjacent bases of the pentagram have two rays in common. The reference base has with each of the bases on the horizontal line of the pentagram four rays in common and is disjoint from any other base. The maximal distance a_3 is that between two disjoint bases; the intermediate distance a_2 occurs between two bases located in any line of the pentagram; and the shortest distance a_1 is that between the reference base and each of the four bases on the horizontal line of the pentagram.

3. Conclusion

We have performed a systematic investigation of small state parity proofs of the BKS theorem involving real rays of two and three qubits. Small non-parity proofs for four and five qubits are investigated in 13,14. Another ongoing work is the relation of the BKS operator proofs for three qubits and the geometric hyperplanes of the generalized hexagon $G_2(2)$.[15] Further work is also necessary to understand the link between the real rays coming from the Pauli group, Barnes-Wall lattices and the BKS state proofs.

References

1. J. H. Conway and N. J. Sloane, *Sphere Packings, Lattices and Groups* (Springer, New York, 1999, p. 90).
2. M. Planat, *Phys. Scr.* **T147** 014025 (2011).
3. G. Nebe, E. M. Rains and N. J. A Sloane, *Self-Dual Codes and Invariant Theory* (Springer, Berlin, 2006).
4. M. Planat, *Gen. Math. Notes* **2** 96-113 (2011).
5. M. Planat, *J. Phys. A: Math. Theor.* **44** 045301 (2011).
6. A. Einstein, B. Podolsky and N. Rosen, *Phys. Rev.* **47** 777-780 (1935).
7. S. Kochen and E. P. Specker, *J. Math. Mech.* **17** 59-87 (1967).
8. A. Peres, *Quantum Theory; Concepts and Methods* (Kluwer, Dordrecht, 1993).
9. A. Cabello, J. M. Estebaranz and G. García-Alcaine, *Phys. Lett. A* **212** 183-187 (1996).
10. P. Raynal, X. Lü and B. G. Englert, *Phys. Rev. A* **83** 062303 (2011).
11. N. D. Mermin, *Rev. Mod. Phys.* **65** 803 (1993).
12. M. Waegell and P. K. Aravind, *Found. Phys.* **41** 1786 (2011).
13. M. Planat, *Eur. Phys. J. Plus* **127** 86 (2012).
14. M. Planat, Preprint 1206.0105.
15. M. Saniga, M. Planat, P. Pracna and P. Levay, Preprint 1206.3436.

FINITE OSCILLATOR MODELS DESCRIBED BY THE LIE SUPERALGEBRA $\mathfrak{sl}(2|1)$

J. VAN DER JEUGT

Department of Applied Mathematics, Ghent University,
Krijgslaan 281-S9, 9000 Gent, Belgium
E-mail: Joris.VanderJeugt@UGent.be

We investigate new models for a finite quantum oscillator based upon the Lie superalgebra $\mathfrak{sl}(2|1)$, where the position and momentum operators are represented as odd elements of the algebra. We discuss properties of the spectrum of the position operator, and of the (discrete) position wavefunctions given by (alternating) Krawtchouk polynomials.

Keywords: Algebraic oscillator models, discrete wavefunctions, Lie superalgebra $\mathfrak{sl}(2|1)$, Krawtchouk polynomials.

1. Introduction and the Lie superalgebra $\mathfrak{sl}(2|1)$

The literature on quantum mechanics in a finite-dimensional Hilbert space is substantial. This paper is devoted to an algebraic model for a quantum oscillator allowing finite-dimensional representations, thus leading to a finite oscillator model.

The canonical one-dimensional quantum oscillator (in the convention $m = \omega = \hbar = 1$) is described by a position operator $\hat{q}$, a momentum operator $\hat{p}$ and a Hamiltonian $\hat{H}$ given by

$$H = \frac{\hat{p}^2}{2} + \frac{\hat{q}^2}{2}. \tag{1}$$

The *oscillator Lie algebra* is usually defined to be the Lie algebra generated by $\hat{q}, \hat{p}, \hat{H}$ and the identity operator 1, subject to the relations

$$[\hat{H}, \hat{q}] = -\mathrm{i}\, \hat{p}, \qquad [\hat{H}, \hat{p}] = \mathrm{i}\, \hat{q}, \tag{2}$$

$$[\hat{q}, \hat{p}] = \mathrm{i}. \tag{3}$$

Equation (3) is the canonical commutation relation (CCR); (2) is known as the Hamilton-Lie equations. In fact, (3) together with (1) imply the

Hamilton-Lie equations (2). The converse is not true. If one drops the CCR (3) but keeps (1) and (2), the resulting system is the *Wigner quantum oscillator*, also known as the *paraboson oscillator*.[1,2] The underlying algebra (generated by $\hat{q}$ and $\hat{p}$) is the Lie superalgebra $\mathfrak{osp}(1|2)$.

Most algebraic models for the quantum oscillator go even one step further, and also drop relation (1), thus keeping only the "dynamics" of the oscillator described by (2). In order to speak of *an algebraic oscillator model*, one requires[3]

- there should be three operators $\hat{q}$, $\hat{p}$, $\hat{H}$ belonging to some (Lie) algebra (or superalgebra) $\mathcal{A}$ such that (2) is satisfied,
- the spectrum of $\hat{H}$ in representations of $\mathcal{A}$ is equidistant.

The best known algebraic model for the quantum oscillator is based on the Lie algebra $\mathcal{A} = \mathfrak{su}(2)$.[3] $\hat{H}$, $\hat{q}$ and $\hat{p}$ are certain self-adjoint elements in $\mathfrak{su}(2)$, and the corresponding unitary representations are the common finite dimensional representations of dimension $2j+1$, with $2j$ a positive integer. The spectrum of $\hat{H}$ is of the form $n+\frac{1}{2}$ (with $n = 0, 1, \ldots, 2j$), and the spectrum of $\hat{q}$ is discrete and equidistant.The corresponding discrete position wavefunctions are expressed in terms of normalized Krawtchouk polynomials, tending to normalized Hermite polynomials when $j \to \infty$. Some interesting deformations of the $\mathfrak{su}(2)$ oscillator model were studied recently, in which case the discrete wavefunctions turn out to be normalized Hahn polynomials.[4,5]

Inspired by Wigner's work, where the paraboson oscillator is described by $\mathfrak{osp}(1|2)$, we have initiated the study of oscillator models based on a Lie superalgebra $\mathcal{A}$. We shall consider here an algebraic oscillator model based on $\mathcal{A} = \mathfrak{sl}(2|1)$. In order to do so, we need to consider an appropriate class of representations of $\mathfrak{sl}(2|1)$. Furthermore, we shall make a proper choice for $\hat{H}$, $\hat{q}$ and $\hat{p}$. In particular, $\hat{q}$ (and $\hat{p}$) should be odd elements of the Lie superalgebra. The main task is then to determine the spectrum of these operators and to construct the corresponding wavefunctions. In the current paper, we highlight just some of the main results for the $\mathfrak{sl}(2|1)$ oscillator. More detailed results are found in Ref. 6.

In terms of 3×3 Weyl matrices e_{ij} $(i, j = 1, 2, 3)$, one can choose the following basis for $\mathfrak{sl}(2|1)$[7]

$$F^+ = e_{32},\ G^+ = e_{13},\ F^- = e_{31},\ G^- = e_{23}, \tag{4}$$

$$H = \frac{1}{2}(e_{11} - e_{22}),\ E^+ = e_{12},\ E^- = e_{21},\ Z = \frac{1}{2}(e_{11} + e_{22}) + e_{33}, \tag{5}$$

where the first four elements are odd, and the last four even. The even subalgebra of $\mathfrak{sl}(2|1)$ is just $\mathfrak{sl}(2) \oplus \mathfrak{u}(1)$, where $\mathfrak{sl}(2)$ is spanned by H, E^+, E^- and $\mathfrak{u}(1)$ by Z. This follows from the commutators

$$[H, E^\pm] = \pm E^\pm, \quad [E^+, E^-] = 2H, \quad [Z, H] = [Z, E^\pm] = 0. \qquad (6)$$

For the odd basis elements, the mutual anti-commutators are

$$\begin{aligned} &\{F^\pm, G^\pm\} = E^\pm, \quad \{F^\pm, G^\mp\} = Z \mp H, \\ &\{F^\pm, F^\pm\} = \{G^\pm, G^\pm\} = \{F^\pm, F^\mp\} = \{G^\pm, G^\mp\} = 0. \end{aligned} \qquad (7)$$

The "mixed" commutators are given by

$$\begin{aligned} &[H, F^\pm] = \pm\frac{1}{2}F^\pm, \ [Z, F^\pm] = \frac{1}{2}F^\pm, \ [E^\pm, F^\pm] = 0, \ [E^\mp, F^\pm] = -F^\mp; \\ &[H, G^\pm] = \pm\frac{1}{2}G^\pm, \ [Z, G^\pm] = -\frac{1}{2}G^\pm, \ [E^\pm, G^\pm] = 0, \ [E^\mp, G^\pm] = G^\mp. \end{aligned}$$

We shall consider here a known class of representations of $\mathfrak{sl}(2|1)$, but with basis and action rewritten in a nicer and more appropriate form. These are *atypical* irreducible representations labelled by a non-negative integer j, with representation space W_j and $\dim(W_j) = 2j+1$. The (orthonormal) basis of W_j is denoted by $|j,m\rangle$ ($m = -j, -j+1, \ldots, +j$). Using the practical notation $\mathcal{E}(n) = 1$ if n is even and 0 otherwise; and $\mathcal{O}(n) = 1$ if n is odd and 0 otherwise, the actions of the odd basis elements is easily written as:

$$\begin{aligned} &F^\pm|j,m\rangle = \pm\mathcal{O}(j-m)\sqrt{\frac{j\pm m+1}{2}}\ |j, m\pm 1\rangle, \\ &G^\pm|j,m\rangle = \pm\mathcal{E}(j-m)\sqrt{\frac{j\mp m}{2}}\ |j, m\pm 1\rangle. \end{aligned} \qquad (8)$$

Those of the even basis elements read:

$$\begin{aligned} Z|j,m\rangle &= -\mathcal{E}(j-m)\frac{j}{2}\ |j,m\rangle - \mathcal{O}(j-m)(\frac{j+1}{2})\ |j,m\rangle, \\ H|j,m\rangle &= \frac{m}{2}\ |j,m\rangle, \\ E^\pm|j,m\rangle &= \frac{1}{2}\mathcal{E}(j-m)\sqrt{(j\mp m)(j\pm m+2)}\ |j,m\pm 2\rangle \\ &\quad +\frac{1}{2}\mathcal{O}(j-m)\sqrt{(j\mp m-1)(j\pm m+1)}\ |j,m\pm 2\rangle. \end{aligned} \qquad (9)$$

These are the known "dispin" representations:[8,9] W_j decomposes as $(\frac{j}{2}; -\frac{j}{2}) \oplus (\frac{j-1}{2}; -\frac{j+1}{2})$, where $(l;b)$ denotes the $\mathfrak{su}(2) \oplus u(1)$ representation "with isospin l and hypercharge b". Note that W_j is a *unitary* representation, for the adjoint operation

$$Z^\dagger = Z, \ H^\dagger = H, \ (E^\pm)^\dagger = E^\mp, \ (F^\pm)^\dagger = -G^\mp, \ (G^\pm)^\dagger = -F^\mp. \qquad (10)$$

2. The $\mathfrak{sl}(2|1)$ oscillator model

Let us now make a proper choice for position, momentum and Hamiltonian operator in $\mathfrak{sl}(2|1)$. Following the second requirement, we should take:

$$\hat{H} = 2H + j + \frac{1}{2}, \tag{11}$$

having spectrum $n + \frac{1}{2}$ $(n = 0, 1, \ldots, 2j)$. As concerns $\hat{q}$, an arbitrary self-adjoint odd element is of the form $\hat{q} = AF^+ + BG^+ - BF^- - AG^-$, with A and B arbitrary real numbers. An overall factor does not play a crucial role, however, so we can take $A^2 + B^2 = 1$. Assuming that A and B have the same sign (the other case is analogous), it follows that the most general form for $\hat{q}$ is

$$\hat{q} = \sqrt{p}\, F^+ + \sqrt{1-p}\, G^+ - \sqrt{1-p}\, F^- - \sqrt{p}\, G^-, \qquad (0 \le p \le 1). \tag{12}$$

Herein, p is a free parameter, and we shall treat only the generic case $0 < p < 1$. For the momentum operator, we have

$$\hat{p} = \mathrm{i}(\sqrt{p}\, F^+ + \sqrt{1-p}\, G^+ + \sqrt{1-p}\, F^- + \sqrt{p}\, G^-). \tag{13}$$

The expressions (11), (12) and (13) do indeed satisfy the Hamilton-Lie equations (2).

Next, we need to determine the eigenvalues and eigenvectors of the operators $\hat{q}$ and $\hat{p}$ in the representation W_j (we will consider only $\hat{q}$ here). From the actions of $F^\pm$ and $G^\pm$ in the (ordered) basis $\{|j,j\rangle, |j,j-1\rangle, \ldots, |j,-j\rangle\}$ of W_j, it follows that the matrix M_q of $\hat{q}$ is a symmetric tridiagonal matrix, with zero diagonal and off diagonal elements given by

$$R_1, S_1, R_2, S_2, \ldots, R_j, S_j,$$

where

$$R_k = \sqrt{p}\sqrt{j+1-k}, \qquad S_k = \sqrt{1-p}\sqrt{k} \qquad (k = 1, 2, \ldots, j).$$

The eigenvectors of this matrix M_q are described in terms of Krawtchouk polynomials.[10] These polynomials $K_n(x; p, N)$, of degree n $(n = 0, 1, \ldots, N)$ in the variable x, with parameter p, satisfy a discrete orthogonality relation:

$$\sum_{x=0}^{N} w(x; p, N) K_n(x; p, N) K_{n'}(x; p, N) = h(n; p, N)\, \delta_{nn'},$$

where $w(x; p, N) = \binom{N}{x} p^x (1-p)^{N-x}$ and $h(n; p, N) = \frac{n!(N-n)!}{N!} \left(\frac{1-p}{p}\right)^n$. The corresponding orthonormal Krawtchouk polynomials are denoted by

$$\tilde{K}_n(x; p, N) \equiv \frac{\sqrt{w(x; p, N)}}{\sqrt{h(n; p, N)}}\, K_n(x; p, N).$$

In terms of these, let U be the $(2j+1)\times(2j+1)$-matrix:

$$U_{2n,j} = (-1)^n \tilde{K}_0(n;p,j), \quad U_{2m+1,j} = 0,$$
$$U_{2n,j-k} = U_{2n,j+k} = \frac{(-1)^n}{\sqrt{2}} \tilde{K}_k(n;p,j),$$
$$U_{2n'+1,j-k} = -U_{2n'+1,j+k} = -\frac{(-1)^{n'}}{\sqrt{2}} \tilde{K}_{k-1}(n';p,j-1),$$

where $n \in \{0,1,\ldots,j\}$, $n' \in \{0,\ldots,j-1\}$ and $k \in \{1,\ldots,j\}$. Then the orthogonality of Krawtchouk polynomials implies that U is an orthogonal matrix: $UU^T = U^TU = I$. Furthermore, from forward and backward shift operator formulas[10] for Krawtchouk polynomials one can deduce that $M_qU = UD$, with D a diagonal matrix:

$$D = \mathrm{diag}(-\sqrt{j}, -\sqrt{j-1}, \ldots, -\sqrt{2}, -1, 0, 1, \sqrt{2}, \ldots, \sqrt{j-1}, \sqrt{j}).$$

In other words, the columns of U are the eigenvectors of M_q, and the eigenvalues of M_q (and hence of $\hat{q}$ in the representation W_j) are of the form $q_{\pm k} = \pm\sqrt{k}$ $(k = 0,1,\ldots,j)$. Note that in the columns of U, Krawtchouk polynomials with parameter j and with parameter $j-1$ alternate.

Let us denote the eigenvector for the eigenvalue q_k by $|j,q_k)$:

$$|j,q_k) = \sum_{m=-j}^{j} U_{j+m,j+k}|j,-m\rangle = \sum_{m=-j}^{j} \phi^{(p)}_{j+m}(q_k)|j,-m\rangle. \tag{14}$$

The reason to denote this coefficient by $\phi^{(p)}_{j+m}(q_k)$ is because overlaps between the normalized eigenstates of the position operator and the eigenstates of the Hamiltonian have an interpretation as position wavefunctions. Hence position wavefunctions are given in terms of (alternating) Krawtchouk polynomials, e.g. when $q_k > 0$:

$$\phi^{(p)}_{2n}(q_k) = \frac{(-1)^n}{\sqrt{2}} \tilde{K}_k(n;p,j), \qquad \phi^{(p)}_{2n+1}(q_k) = \frac{(-1)^n}{\sqrt{2}} \tilde{K}_{k-1}(n;p,j-1).$$

In Ref. 6 we have given some plots of these position wavefunctions, for various values of p, j and n. Their behaviour shows similarities with canonical oscillator wavefunctions, but of course they are a discrete version. When the dimension parameter j is large, the $\mathfrak{sl}(2|1)$ wavefunctions tend to paraboson wavefunctions (which is also confirmed by a limit calculation).

3. Further results

The determination of the eigenvalues and eigenvectors of $\hat{p}$ is very similar. The spectrum of $\hat{p}$ is the same as that of $\hat{q}$: $p_k = \pm\sqrt{k}$. The momentum wavefunctions are, up to a power of i, the same as the position wavefunctions:

$$\psi_{j+m}^{(p)}(p_k) = (\mathrm{i})^{j+m-1}\phi_{j+m}^{(p)}(p_k).$$

So one can determine the matrix F that transforms position wavefunctions into momentum wavefunctions; this is the $\mathfrak{sl}(2|1)$ discrete Fourier transform F. Since the wavefunctions have such a simple form, the matrix elements of F can be determined explicitly using the known bilinear generating function for Krawtchouk polynomials. This $\mathfrak{sl}(2|1)$ discrete Fourier transform has many properties similar to the standard Discrete Fourier Transform.

Choosing a different adjoint operation for $\mathfrak{sl}(2|1)$, the corresponding unitary representations are infinite dimensional. These discrete series representations of $\mathfrak{sl}(2|1)$, labelled by a positive number $\beta > 0$, have been determined recently.[11] In these representations, $\hat{H}$ has the same spectrum as the canonical oscillator. General forms for $\hat{q}$ (and $\hat{p}$) involve one parameter $\gamma > 0$ (like our p in the finite dimensional case). The spectrum of $\hat{q}$ is continuous ($= \mathbb{R}$) for $\gamma = 1$ and infinite discrete ($\pm\sqrt{\gamma^2 - 1}\sqrt{k}$, $k \in \mathbb{Z}_+$) for $\gamma > 1$ (similar for $0 < \gamma < 1$). The position wavefunctions $\Phi_n^{(\beta,\gamma)}(x)$ coincide with paraboson wavefunctions for $\gamma = 1$, and are related to (discrete) *Meixner polynomials* for $\gamma \neq 1$; the canonical oscillator is recovered for $\beta = 1/2$.

References

1. E. P. Wigner, *Phys. Rev.* **77**, 711 (1950).
2. T. D. Palev, *Czech J. Phys.* **B29**, 91 (1979).
3. N. M. Atakishiyev, G. S. Pogosyan, L. E. Vicent and K. B. Wolf, *J. Phys. A* **34**, 9381 (2001).
4. E. I. Jafarov, N. I. Stoilova and J. Van der Jeugt, *J. Phys. A* **44**, 265203 (2011).
5. E. I. Jafarov, N. I. Stoilova and J. Van der Jeugt, *J. Phys. A* **44**, 355205 (2011).
6. E. I. Jafarov and J. Van der Jeugt, *J. Phys. A* **45**, 275301 (2012).
7. L. Frappat, A. Sciarrino and P. Sorba, *Dictionary on Lie Algebras and Superalgebras* (Academic Press, London, 2000).
8. M. Marcu, *J. Math. Phys.* **21**, 1277 (1980).
9. M. Scheunert, W. Nahm and V. Rittenberg, *J. Math. Phys.* **18**, 155 (1977).
10. R. Koekoek, P. A. Lesky and R. F. Swarttouw, *Hypergeometric orthogonal polynomials and their q-analogues* (Springer-Verlag, Berlin, 2010).
11. E. I. Jafarov and J. Van der Jeugt, *Discrete series representations for* $\mathfrak{sl}(2|1)$, *Meixner polynomials and oscillator models*, arXiv:1208.2536 (2012).

ON THE CONTRACTION FROM DISCRETE FINITE TO FREE HAMILTONIAN SYSTEMS

KURT BERNARDO WOLF

Instituto de Ciencias Físicas, Universidad Nacional Autónoma de México, Av. Universidad s/n, Cuernavaca, Morelos 62251, México
E-mail: bwolf@fis.unam.mx

Finite discrete Hamiltonian systems are mothered by the Lie algebra su(2) and have wavefunctions (*signals*) of unit-spaced positions $q|_{-j}^{j}$. These wavefunctions are N-vectors ($N = 2j + 1$) that can be subject to the group U(2) of unitary *linear* transformations and overall phases. But moreover, they can *also* be subject to the much larger group of all unitary $N \times N$ matrices in U(N); those outside U(2) are '*nonlinear*' and correspond to *aberrations* in geometric optics. Here we examine in particular the contraction $j \to \infty$ of su(2) to iso(2), where the latter mothers infinite discrete free Hamiltonian systems. We find that the aberrations can be applied only to signals that are of appropriate decrease or bounded.

Keywords: Discrete Hamiltonian systems, Lie algebra contraction, Aberrations.

1. Discrete Hamiltonian systems

We characterize a D-dimensional Hamiltonian system by an N-parameter Lie algebra $\mathcal{L}$ whose generator basis $\{X_i\}_{i=1}^{N}$ satisfies three postulates:

Geometric: Among the N generators of $\mathcal{L}$ we identify D *position* operators $\{Q_i\}_{i=1}^{D}$, and a *Hamiltonian* operator H, such that the Lie brackets $[H, Q_i] = -\mathrm{i}P_i$, with $i|_1^D$, return *momentum* operators $\{P_i\}_{i=1}^{D} \in \mathcal{L}$ that are independent and elements of the same algebra but distinct from the position operators and Hamiltonian.

Dynamic: The Lie brackets of the Hamiltonian with the momentum operators return operators in the associative span of the position generators, $[H, P_i] = \mathrm{i}V_i'(\{Q_j\}) \in \overline{\mathcal{L}}$, the associative algebra of generator products.

Algebraic: Additionally we propose[1] a third postulate: The Lie brackets between the position and momentum operators, will return $[Q_i, P_j] = \mathrm{i}F_{i,j}(H, \{X_k\}, \mathit{1}) \in \overline{\mathcal{L}}$, in the span of the Hamiltonian and/or other generators $\{X_k\}$ of the algebra which are neither position nor momentum, and the central unit operator *1*.

A Hamiltonian system will be *discrete* if the position generators $\{Q_i\}_{i=1}^D$ are *compact*, so their *spectra* —the observable positions of the system— will be the point set of their discrete eigenvalues. These three postulates include mother q-algebras[2] as well as superalgebras.[3–5] Here we are concerned with one-dimensional ($D = 1$) *quadratic* Hamiltonian systems, where $V'(Q) = \sigma Q$ is linear in the argument, and where $[Q, P] = iH$ with $\sigma = +1$ describes the finite harmonic oscillator,[6,7] $\sigma = 0$ a free system,[8] and $\sigma = -1$ corresponds to a repulsive oscillator.[9] Among 3-dimensional Lie algebras,[10,11] this focuses our attention on $\mathsf{su}(2)$, $\mathsf{iso}(2)$ and $\mathsf{su}(1,1)$, which harbor a compact position operator.

The *contraction* of a mother Lie algebra $\mathcal{L}$ to $\mathcal{L}'$, such as $\mathsf{su}(2)$ to $\mathsf{iso}(2)$, where the position generator Q is invariant, describes the limit of the finite oscillator to a discrete but infinite free system, as recollected in Section 2. In Section 3 we note that N-point signals can be subject to a $\mathsf{U}(N)$ group of $N \times N$ matrices that we call *aberrations*, which include transformations beyond the 'linear' $\mathsf{SU}(2)$ ones that are domestic to the model,[12] and which can be classified as their geometric-optical counterparts.[13] In Section 3 we follow the fate of these transformations under the contraction of $\mathsf{su}(2)$ to $\mathsf{iso}(2)$. In the concluding Section 4 we recapitulate the restrictions on the signals to which these aberrations can be applied, and add some remarks on the Hamiltonian evolution of discrete functions that resemble their classical dependence on initial positions and velocities.

2. Contraction of su(2) to iso(2)

A contraction is a linear transformation of the generators of an algebra, $\mathcal{C}_\gamma : \mathcal{L} \mapsto \mathcal{L}$, $\gamma \in [1, \infty)$, which becomes singular when $\gamma \to \infty$, where some of the Lie brackets may vanish and determine a contracted algebra $\mathcal{L}'$. The irreducible unitary representations of $\mathsf{su}(2)$ and of $\mathsf{iso}(2)$ are determined by the invariant Casimir operator, whose representation is a multiple γ of the unit matrix,

$$C := \sigma Q^2 + P^2 + H^2 = \gamma 1. \tag{1}$$

For $\mathsf{su}(2)$, $\sigma = +1$, $\gamma = j(j+1)$ with $2j \in \mathsf{Z}_0^+$, and for $\mathsf{iso}(2)$, $\sigma = 0$, $\gamma = \tau^2$ with $\tau \in \mathsf{R}$. In $\mathsf{su}(2)$, the irreducible representation j has rows and columns $q|_{-j}^{j}$, and is of dimension $N = 2j+1$; the limit we examine is $N \to \infty$.

The Lie brackets of $\mathsf{su}(2)$, written as the two Hamilton equations and the algebraic postulate, are

$$[H,Q] = -\mathrm{i}P, \quad [H,P] = \mathrm{i}Q, \quad [Q,P] = -\mathrm{i}H. \tag{2}$$

We introduce a linear transformation to primed generators defined by the Casimir eigenvalue γ and a coefficient $\tau \neq 0$,

$$Q'_\gamma := Q, \quad P'_\gamma := \tau P/\sqrt{\gamma}, \quad H'_\gamma := \tau H/\sqrt{\gamma}. \tag{3}$$

When $\gamma \to \infty$ ($N \to \infty$), the Lie brackets (2) between these operators, become those which characterize the Euclidean Lie algebra $\mathsf{iso}(2)$ (indicated by primes), where the dynamical Hamilton equation becomes that of the free system $V'(q) = 0$,

$$[H',Q'] = -\mathrm{i}P', \quad [H',P'] = 0, \quad [Q',P'] = -\mathrm{i}H', \tag{4}$$

while the Casimir operator (1) becomes $C' = P'^2 + H'^2 = \tau^2 \mathit{1}$.

The self-adjoint representation matrices of position $\mathbf{Q} = \|Q_{q,q'}\|$ are real and diagonal, those of momentum $\mathbf{P} = \|P_{q,q'}\|$ are pure-imaginary and skew-symmetric (acting as second-neighbor difference), and the Hamiltonian $\mathbf{H} = \|H_{q,q'}\|$ is real and symmetric (second-neighbor average),

$$Q_{q,q'} = q\,\delta_{q,q'}, \tag{5}$$

$$P_{q,q'} = \mathrm{i}\tfrac{1}{2}\, C^{\gamma,\sigma}_{q,+}\,\delta_{q+1,q'} - \mathrm{i}\tfrac{1}{2}\, C^{\gamma,\sigma}_{q,-}\,\delta_{q-1,q'}, \tag{6}$$

$$H_{q,q'} = \tfrac{1}{2}\, C^{\gamma,\sigma}_{q,+}\,\delta_{q+1,q'} + \tfrac{1}{2}\, C^{\gamma,\sigma}_{q,-}\,\delta_{q-1,q'}, \tag{7}$$

where in $\mathsf{su}(2)$ $C^{j(j+1),+1}_{q,\pm} = \sqrt{(j\mp q)(j\pm q+1)}$, with $q|^{j}_{-j}$, and under the contraction become $C'^{\,\tau^2,0}_{q,\pm} = \tau$, with $q \in \mathsf{Z}$. When the latter act on an infinite signal $\mathbf{f} = \{f_q\}$ that is taken to be the set of Fourier series coefficients of a function on the circle $\widetilde{\mathbf{f}} = \{f(\theta)\}$, we have the following two realizations of the $\mathsf{iso}(2)$ algebra,

$$(Q'\,\mathbf{f})_q = q\,f(q), \;\; (Q'\,\widetilde{\mathbf{f}})(\theta) = \mathrm{i}\frac{\mathrm{d}}{\mathrm{d}\theta}\widetilde{f}(\theta), \tag{8}$$

$$(P'\,\mathbf{f})_q = \frac{\tau}{2\mathrm{i}}(f_{q+1} - f_{q-1}), \;\; (P'\,\widetilde{\mathbf{f}})(\theta) = \tau\sin\theta\,\widetilde{f}(\theta), \tag{9}$$

$$(H'\,\mathbf{f})_q = -\frac{\tau}{2}(f_{q+1} + f_{q-1}), \;\; (H'\,\widetilde{\mathbf{f}})(\theta) = -\tau\cos\theta\,\widetilde{f}(\theta). \tag{10}$$

The left-hand generators provide a model for free propagation in a discrete system with $q \in \mathsf{Z}$ positions, while the right-hand generators, the Fourier synthesis, model a vibrating ring. We recall also that in geometric optics, (10) corresponds to the classical Hamiltonian $h = -\sqrt{(\tau^2 - p^2)} = -\tau\cos\theta$ for rays at an angle θ from the optical axis, and whose momentum is $p = \tau\sin\theta$ in a medium of refractive index τ. Of course, the geometric Hamilton equation in Poisson brackets is there $\{h, q\} = -p/h$; only in the paraxial limit $h \approx -\tau$ can one 'wavize' to $\mathsf{iso}(2)$.[8]

It is interesting to note that while in $\mathsf{su}(2)$ the spectrum $\eta|_{-j}^{j}$ of H is discrete and has eigenfunctions of alternating parity, under the contraction (3) it becomes continuous, $\nu|_{-\tau}^{\tau}$, and doubly degenerate with even and odd eigenfunctions for $-\tau < \nu < \tau$. This can be traced back to the loss of the boundary conditions at $q = \pm j$ when $j \to \infty$.

3. Contraction of $\mathsf{U}(N)$ aberrations

In Ref.[12] we introduced *aberrations* on finite N-point signals in the context of the $\mathsf{su}(2)$ optical model for harmonic waveguides. These are unitary $N \times N$ self-adjoint matrices that generate the group $\mathsf{U}(N)$. This includes the $\mathsf{U}(1)$ overall phase generated by $\mathbf{1}$, and all independent products of the original set of $\mathsf{su}(2)$ generator matrices [13, Part IV]. Since the Casimir operator (1) restricts the powers of $\mathbf{H}$ to 0 or 1, we have two sets of generator matrices, for integer and half-integer *ranks* k and integer-spaced *weights* m,

$$\mathbf{M}^{0}_{k,m} := \left\{\mathbf{Q}^{k-m}\mathbf{P}^{k+m}\right\}_{\text{Weyl}}, \qquad m|_{-k}^{k},\ k \in \{0, \tfrac{1}{2}, 1, \ldots, j\} \tag{11}$$

$$\mathbf{M}^{1}_{k,m} := \left\{\mathbf{Q}^{k-m-1/2}\mathbf{P}^{k+m-1/2}\mathbf{H}\right\}_{\text{Weyl}}, \quad m|_{-k+1/2}^{k-1/2},\ k \in \{\tfrac{1}{2}, 1, \ldots, j\}, \tag{12}$$

where the *Weyl* ordering of the factors is demanded to ensure that the $\mathbf{M}^{0\,\text{or}\,1}_{k,m}$ matrices are self-adjoint. The matrices $\mathbf{M}^{0\,\text{or}\,1}_{k,m}$ of the same *rank* k form $(2k+1)$-dimensional ideals under $\mathsf{su}(2)$, and will transform among themselves under $\mathsf{SU}(2)$ corresponding to aberration order $A = 2k$.[12] In particular for $k = 2$, the $M^{0}_{2,m}$ multiplet consists of $D = 1$ spherical aberration, coma, curvature of field, distorsion and defocus, for $m = 2, 1, \ldots, -2$.

Under the contraction (4) from $\mathsf{su}(2)$ to $\mathsf{iso}(2)$, (11)–(12) become infinite matrices $\mathbf{M}'^{0\,\text{or}\,1}_{k,m}$. Their $\mathbf{P}'$ and $\mathbf{H}'$ factors, powers, and exponentiation present no problem since they only contribute with corresponding powers of $C'^{\tau^2,0}_{q,\pm} = \tau$; but powers of $\mathbf{Q}'$ do, because its matrix elements $q\delta_{q,q'}$ are unbounded. The right-hand realization in (8)–(10) indicates that a given $\mathbf{M}'^{\rho}_{k,m}$ $(\rho \in \{0, 1\})$ will contain a leading $(k-m-\frac{1}{2}\rho)^{\text{th}}$ derivative of the signal Fourier synthesis $\widetilde{f}(\theta)$. To keep the norm of a signal finite, $\widetilde{f}(\theta)$ should be thus at least $k-m-\frac{1}{2}\rho$ times differentiable, which implies that its Fourier series coefficients $\{f_q\}_{q=-\infty}^{\infty}$ must decrease faster than $|q|^{k-m-\rho/2}$.

4. Conclusions

Since the weight index ranges over $m|_{-k+\rho/2}^{k-\rho/2}$, this excludes any power of $\mathbf{Q}'$ above 1; because while $(e^{\mathrm{i}\alpha Q'}\widetilde{\mathbf{f}})(\theta) = \widetilde{f}(\theta-\alpha)$ and for differentiable functions $F(\theta)$, $\exp(\mathrm{i}\alpha F(\theta)Q')$ only produce a change the argument $\theta'(\theta,\alpha)$ and a multiplier phase, and thus will remain bounded, higher powers of Q' in the exponential will not; only spherical aberration and coma can be thus translated to $\mathsf{iso}(2)$. On the other hand, if the discrete signal $\mathbf{f} = \{f_q\}$ decreases faster than an exponential (such as $f_q \leq e^{-\alpha q^2}$ for $|q| \gg 1$) or is bounded (finite), then it can be subject to arbitrarily high aberrations. In doing so, finite signals will 'spill over' all $q \in \mathsf{Z}$.

Although we cannot give closed expressions even for the case for the elements of the $\mathsf{su}(2)$ $N \times N$ matrices $\mathbf{M}^{\rho}_{k,m}$ out of the elements of the matrix factors (5)–(7), that problem was evaded in Ref.[12] through numerical computation. In $\mathsf{iso}(2)$ we are at a greater disadvantage unless we remain with only a finite part of the signal when applying the aberration exponentials $\exp(\mathrm{i}\alpha\mathbf{M}'^{\rho}_{k,m})$ on $\{f_q\}$ or on $\{\widetilde{f}(\theta)\}$. Furthermore, since the commutator of two $\mathbf{M}^{\rho}_{k,m}$'s with $k > 1$ is generally a generator of rank higher than both, the contraction limit of the whole aberration group $\mathsf{U}(N)$ for $N \to \infty$ is not well defined. Only the subgroup generated by $\mathbf{M}^{0}_{k,k}$ (spherical aberration) and $\mathbf{M}^{0}_{k,k-1}$ (coma) and the two $\mathbf{M}^{1}_{k,m}$'s of highest weight m are retained. Yet, within the limits imposed by numerical computation, it is simpler than in the $\mathsf{su}(2)$ case discussed in Ref.[12] because the nonzero matrix diagonals have only powers of the constant τ.

The title "From finite unitarity to classical canonicity" was announced for the presentation at the ICGTMP conference. This is the name of a project rather than the sketch of results reported here. Under development are the implications contained in the form of the matrix elements (5)–(7) of the generators of the groups $\mathsf{SU}(2)$, $\mathsf{ISO}(2)$ and $\mathsf{SU}(1,1)$ that mother discrete quadratic systems, because these relate *second*-neighbor points, f_q with $f_{q\pm 2}$. The conjecture is that in this way initial velocities and values will evolve as their classical counterparts do driven by the Green function and its time derivative when applied separately to the odd and even points respectively. The structure of this classical limit is still under investigation.

Acknowledgements

This work is supported by the *Óptica Matemática* projects PAPIIT-UNAM 101011 and SEP-CONACyT 79899.

References

1. M. Arık, N.M. Atakishiyev and K.B. Wolf, Quantum algebraic structures compatible with the harmonic oscillator Newton equation, (*J. Phys. A* **32**, L371–L376 (1999)).
2. N.M. Atakishiyev, A.U. Klimyk and K.B. Wolf, Finite q-oscillator, *J. Phys. A* **37**, 5569–5587 (2004).
3. E.I. Jafarov, N.I. Stoilova, and J. Van der Jeugt, Finite oscillator models: the Hahn oscillator, *J. Phys. A* **44**, art. 265203 (2011).
4. E.I. Jafarov, N.I. Stoilova, and J. Van der Jeugt, The $su(2)_\alpha$ Hahn oscillator and a discrete Hahn-Fourier transform, *J. Phys. A* **44**, art. 355205 (2011).
5. E.I. Jafarov and J. Van der Jeugt, A finite oscillator model related to $sl(2|1)$, *J. Phys. A* **45**, art. 275301 (2012).
6. N.M. Atakishiyev and K.B. Wolf, Fractional Fourier-Kravchuk transform, *J. Opt. Soc. Am. A* **14**, 1467–1477 (1997).
7. N.M. Atakishiyev, G.S. Pogosyan, and K.B. Wolf, Finite models of the oscillator, *Phys. Part. Nuclei, Suppl. 3,* **36**, 521–555 (2005).
8. L.M. Nieto, N.M. Atakishiyev, S.M. Chumakov, and K.B. Wolf, Wigner distribution function for Euclidean systems, *J. Phys. A* **31**, 3875–3895 (1998).
9. C. Muñoz, J. Rueda-Paz, and K.B. Wolf, Discrete repulsive oscillator wavefunctions, *J. Phys. A* **42**, art. 485210 (2009).
10. C.W. Conatser, Contractions of the low-dimensional real Lie algebras, *J. Math. Phys.* **13**, 198–203 (1972).
11. M. Nestorenko and R. Popovych, Contractions of low-dimensional Lie algebras, *J. Math. Phys.* **47**, art. 123515 (2006).
12. K.B. Wolf, Linear transformations and aberrations in continuous and in finite systems, *J. Phys. A* **41**, art. 304026 (2008).
13. K.B. Wolf, *Geometric Optics on Phase Space* (Springer-Verlag, Heidelberg, 2004).

TRANSLATIONS AND INVERSIONS IN FINITE PHASE PLANE

J. ZAK

Department of Physics, Technion–Israel Institute of Technology
Haifa, Israel
E-mail: zak@physics.technion.ac.il

The inversion operators on a lattice in finite phase plane are used for building a complete set of mutually orthogonal Hermitian operators. The lattice is given by tc in the x direction and by $s\hbar\frac{2\pi}{Mc}$ in the p-direction; c is an arbitrary length constant and M is the dimension of the space; s and t assume the values from 0 to $M-1$. For M odd the M^2 inversion operators on the lattice form a complete set of mutually orthogonal operators. For M even we assign a sum of 4 inversion operators (a quartet) to each site of the lattice (t, s). We prove that these quartets for $t, s = 0, 1, \ldots, M-1$ form a mutually orthogonal set of M^2 Hermitian operators.

Keywords: Even and odd-dimensional spaces; Complete set of inversion operators.

1. Introduction

Quantum mechanics in finite phase plane goes all the way back to the works of Weyl[1] and Schwinger.[2] It later finds applications in a variety of areas such as the Wigner-function which was first introduced in infinite phase plane in 1932;[3] then considered by a number of authors in finite phase plane.[4–9] Quantum mechanics in finite phase plane in other areas one can find in the following review articles.[10–12]

It is well known that the Wigner function at the (x,p)-point in infinite phase plane equals up to a constant to the expectation value of the inversion operator around the point (x,p).[13,14] In a recent publication,[9] it was shown that this result also holds in finite phase plane of dimension M. However, one has to distinguish between M odd and M even. For M odd it suffices to define the inversion operator $I\left(tc, s\hbar\frac{2\pi}{Mc}\right)$ for the discrete values of

x and p

$$x = tc,\ p = s\hbar\frac{2\pi}{Mc},\ \ t, s = 0, 1, \ldots, M-1 \tag{1}$$

These points form a $M \times M$ lattice in the finite phase plane. Here M is the dimension of the vector space in the finite phase plane and c is an arbitrary unit length in the x-direction; correspondingly, Mc is the whole length of the phase plane in the x-direction and $\hbar\frac{2\pi}{c}$ is the length in the p-direction. For defining the Wigner function for M even one needs also to know the inversion operator at the middle points $\frac{x}{2}$, $\frac{p}{2}$ of the lattice in Eq. (1), $I\left(t\frac{c}{2}, s\hbar\frac{\pi}{Mc}\right)$. The lattice containing the middle points has $(2M)\times(2M)$ points and is given by

$$x = t\frac{c}{2},\ p = s\hbar\frac{\pi}{Mc},\ \ t, s = 0, 1, \ldots, 2M-1 \tag{2}$$

As was previously shown[9,15]

$$I\left(t\frac{c}{2}, s\hbar\frac{\pi}{Mc}\right) = D\left(tc, s\hbar\frac{2\pi}{Mc}\right) I \tag{3}$$

where t and s run from 0 to $2M-1$, $D\left(tc, s\hbar\frac{2\pi}{Mc}\right)$ is the shift operator in phase plane[16]

$$D\left(tc, s\hbar\frac{2\pi}{Mc}\right) = e^{-\frac{i}{\hbar}Ptc + i\frac{2\pi}{Mc}Xs} = e^{-\frac{i}{\hbar}Ptc}e^{i\frac{2\pi}{Mc}Xs}e^{i\pi st/M} \tag{4}$$

and I is the inversion operator around the zero in the phase plane. P and X in Eq. (4) are the momentum and coordinate operators respectively. In Eq. (2) there are $(2M)^2$ inversion operators, which means that only a $\frac{1}{4}$ of them, M^2, can be independent. This can explicitly be seen from the following relations:[9]

$$I\left[(t+M)\frac{c}{2}, s\hbar\frac{\pi}{Mc}\right] = (-1)^s I\left(t\frac{c}{2}, s\hbar\frac{\pi}{Mc}\right) \tag{5}$$

$$I\left[t\frac{c}{2}, (s+M)\hbar\frac{\pi}{Mc}\right] = (-1)^t I\left(t\frac{c}{2}, s\hbar\frac{\pi}{Mc}\right) \tag{6}$$

From Relations (5) and (6), one should expect that if t and s are limited by the requirement $0 \leq t, s < M$, then this separates out M^2 independent operators from Eq. (3). One can show that the following orthogonality relations holds for them[8] (it is convenient to carry out the calculation in the x-representation, $x = rc,\ r = 0, 1, \ldots, M-1$):

$$Tr\left[I\left(t\frac{c}{2}, s\hbar\frac{\pi}{Mc}\right) I\left(t'\frac{c}{2}, s'\hbar\frac{\pi}{Mc}\right)\right.$$

$$= \sum_{r=0}^{M-1} \langle rc| e^{-\frac{i}{\hbar}P(t-t')} e^{i\frac{2\pi}{Mc}X(s-s')} e^{i\frac{\pi}{M}(st+s't'-2st')} |rc\rangle$$

$$= M\Delta(t-t')\Delta(s-s') \tag{7}$$

which holds for $0 \le t, s, t', s' < M$. Here $\Delta(n) = 1$, for $n = 0$ modulo M and zero otherwise. Eq. (7) with $0 \le t, s, t', s' < M$ holds for both $M \equiv$ odd and $M \equiv$ even. This means that the M^2 inversion operators in Eq. (3) for $0 \le t, s < M$ are mutually orthogonal and therefore independent. Another obvious choice of t, s, t', s' for the operators in Eq. (7) to be orthogonal is [See Eqs. (5) and (6)]

$$M \le ts, t's' < 2M \tag{8}$$

It can be proven similar to the proof of Eq. (7) that for $M \equiv$ odd also another orthogonality relation is satisfied (see Eq. (10) below):

$$Tr\left[I\left(tc, s\hbar\frac{2\pi}{Mc}\right) I\left(t'c, s'\hbar\frac{2\pi}{Mc}\right)\right] = M\Delta(t-t')\Delta(s-s') \tag{9}$$

where $0 \le t, s, t', s' < M$

In an infinite phase plane, the operators in Eq. (3) were originally introduced by Groenewold[17] and Moyal[18] where they were called phase-space eigenfunctions. In Refs. 4 and 8 these operators are called phase-point operators. They were identified by Grossman and Royer[13,14] as having the meaning of inversion operators $I(x,p)$ in phase plane.

In this talk, we construct a set of M^2 orthogonal operators on the lattice in Eq. (1) out of the $(2M)^2$ inversion operators in Eq. (3).

2. Complete set of orthogonal lattice operators

In constructing the complete set we shall distinguish between M odd and M even.

a. $M \equiv$ odd

For $M \equiv$ odd we shall use the inversion operators on the lattice (1) (see Eq. (3))

$$I\left(tc, s\hbar\frac{2\pi}{Mc}\right) = D\left(2tc, 2s\hbar\frac{2\pi}{Mc}\right) I. \tag{10}$$

For t and s running from 0 to $M-1$, Eq. (10) contains M^2 operators $I\left(tc, s\hbar\frac{2\pi}{Mc}\right)$. According to Eq. (9), these operators are orthogonal and they therefore form a complete set of M^2 orthogonal operators in the M-dimensional phase plane.

b. $M \equiv$ even

For M even the orthogonality relation in Eq. (9) does not hold, because the inversion operators $I\left(tc, s\hbar\frac{2\pi}{Mc}\right)$ are not all independent, when M is even. Indeed, for $M \equiv$ even we have the following dependence relation for the operators $I\left(tc, s\hbar\frac{2\pi}{Mc}\right)$ (see Eqs. (3) and (10))

$$I\left[\left(t+\frac{M}{2}\right)c, \left(s+\frac{M}{2}\right)\hbar\frac{2\pi}{Mc}\right] = I\left(tc, s\hbar\frac{2\pi}{Mc}\right). \tag{11}$$

This relation follows from Eqs. (5) and (6). It shows that in Eq. (10) there are only $\left(\frac{M}{2}\right)^2$ independent operators. However, in Eq. (3) we have M^2 independent operators. We define the following M^2 operators $E\left(tc, s\hbar\frac{2\pi}{Mc}\right)$ for $t, s = 0, 1, \ldots, M-1$ on the lattice of Eq. (1).

$$E\left(tc, s\hbar\frac{2\pi}{Mc}\right) = \frac{1}{2}\sum_{m,n=0}^{1} I\left[(2t+n)\frac{c}{2}, (2s+m)\hbar\frac{\pi}{Mc}\right] \tag{12}$$

In what follows we show that the M^2 lattice operators $E\left(tc, s\hbar\frac{2\pi}{Mc}\right)$ in Eq. (12) for $t, s = 0, 1, \ldots, M-1$ are orthogonal for different t and s, namely:

$$Tr\left[E\left(tc, s\hbar\frac{2\pi}{Mc}\right) E\left(t'c, s'\hbar\frac{2\pi}{Mc}\right)\right] = M\Delta(t-t')\Delta(s-s') \tag{13}$$

as before $\Delta(n) = 1$, for $n = 0$ modulo M and zero otherwise. Each operator $E\left(tc, s\hbar\frac{2\pi}{Mc}\right)$ in Eq. (12) is a sum of 4 inversion operators as defined in Eq. (3). When t and s are in the range

$$0 \le t, s \le \frac{M}{2} - 1, \tag{14}$$

all the inversion operators on the right hand side of Eq. (12) are different and according to Eq. (7) they are orthogonal for different s and t. When t and s do not satisfy Eq. (14), something interesting happens. Each quartet on the right hand side of Eq. (12) for given t and s outside the range of Eq. (14) can be brought into a new quartet with t and s satisfying Eq. (14). This can be done by using Eqs. (5) and (6). We can therefore turn all the quartets for t and s outside Eq. (14) into quartets with t and s inside Eq. (14). But by doing so each transformed quartet will have 2 members with minus signs. As an example, consider the case of $M = 4$ and write E

of Eq. (12) for $t = 2, s = 1$, which do not satisfy Eq. (14) (s does satisfy Eq. (14), but t does not):

$$\begin{aligned} E\left(2c, \hbar\tfrac{2\pi}{4c}\right) = \tfrac{1}{2}\{&I\left(2c, \hbar\tfrac{2\pi}{4c}\right) + I\left(2c, 3\hbar\tfrac{\pi}{4c}\right) \\ &+I\left(5\tfrac{c}{2}, \hbar\tfrac{2\pi}{4c}\right) + I\left(5\tfrac{c}{2}, 3\hbar\tfrac{\pi}{4c}\right)\} \\ = \tfrac{1}{2}\{&I\left(0, \hbar\tfrac{2\pi}{4c}\right) - I\left(0, 3\hbar\tfrac{\pi}{4c}\right) \\ &+I\left(\tfrac{c}{2}, \hbar\tfrac{2\pi}{4c}\right) - I\left(\tfrac{c}{2}, 3\hbar\tfrac{\pi}{4c}\right)\}, \end{aligned} \tag{15}$$

where the second equality was obtained by using relation (5). $E\left(2c, \hbar\frac{2\pi}{4c}\right)$ is to be compared with $E\left(0, \hbar\frac{2\pi}{4c}\right)$ for $t = 0,\ s = 1$ which satisfy Eq. (14).

$$\begin{aligned} E\left(0, \hbar\tfrac{2\pi}{4c}\right) = \tfrac{1}{2}\{&I\left(0, \hbar\tfrac{2\pi}{4c}\right) + I\left(0, 3\hbar\tfrac{\pi}{4c}\right) \\ &+I\left(\tfrac{c}{2}, \hbar\tfrac{2\pi}{4c}\right) + I\left(\tfrac{c}{2}, 3\hbar\tfrac{\pi}{4c}\right)\} \end{aligned} \tag{16}$$

$E\left(2c, \hbar\frac{2\pi}{4c}\right)$ and $E\left(0, \hbar\frac{2\pi}{4c}\right)$ in Eqs. (15) and (16) have the same inversion operators, but in $E\left(2c, \hbar\frac{2\pi}{4c}\right)$ 2 inversion operators have a minus sign. As a consequence of this minus sign, the operators $E\left(2c, \hbar\frac{2\pi}{4c}\right)$ and $E\left(0, \hbar\frac{2\pi}{4c}\right)$ in Eqs. (15) and (16) respectively are orthogonal according to the orthogonality relation (7). With this example in mind, one can prove that all the 16 $E\left(tc, s\hbar\frac{2\pi}{4c}\right)$ for $M =$ are orthogonal to one another and they satisfy the orthogonality relation (13). This proof can be extended to any even M and we conclude[19] that the M^2 lattice operators $E\left(tc, s\hbar\frac{2\pi}{Mc}\right)$ in Eq. (12) satisfy the orthogonality relation in Eq. (13) and they form therefore a complete set of orthogonal operators in the finite phase plane for any even M.

3. Conclusions

We have constructed in this talk a complete and orthogonal set of M^2 operators in the finite phase plane of dimension M, by using the inversion operators. The construction is different for M odd and M even. For M odd the inversion operators $I\left(tc, s\hbar\frac{2\pi}{Mc}\right)$ themselves on the lattice in Eq. (1) (see Eq. (10)) form a complete and orthogonal set of M^2 operators. For M even we assign a sum of 4 inversion operators to each site on the lattice in Eq. (1). We denote this sum by $E\left(tc, s\hbar\frac{2\pi}{Mc}\right)$ (see Eq. (12)) and prove that for $t, s = 0, \ldots, M-1$ we obtain a complete and orthogonal set of M^2 operators. It is interesting that for $M = 2$ the 4 operators $E\left(tc, s\hbar\frac{2\pi}{2c}\right)$ coincide for $t, s = 0, 1$ with the phase-point operators in Eq. (10) of Ref. 4. Thus

$$E(0,0) = \frac{1}{2}(1 + \sigma_x + \sigma_y + \sigma_z) \tag{17}$$

where σ_x, σ_y and σ_z are the Pauli matrices. But unlike in Ref. 4, in the present paper we distinguish between the cases of M odd and M even and for each case a complete and orthogonal set of operators is constructed based on the inversion operators. The reason we have to distinguish between M odd and M even lies in Eq. (10). This equation shows a doubling feature between the coordinates of the sites in the inversion operators $I\left(tc, s\hbar\frac{2\pi}{Mc}\right)$ and the shift operators $D\left(2tc, 2s\hbar\frac{2\pi}{Mc}\right)$ (see also Eq. (3)). This doubling feature is crucial for the understanding of the difference between M odd and M even. For M odd, because of the periodicity of the finite phase plane, when t and s run from 0 to $M-1$, $2t$ and $2s$ cover the same range of sites from 0 to $M-1$. Thus, for $M=3$, when (s,t) run over 0,1,2, $(2s,2t)$ run over 0,2,1, because 4=1 Modulo 3. This was already pointed out in Ref. 2 and gives a very elementary explanation why there is a need to distinguish between M odd and M even. A similar doubling feature exists between the Wigner function and the wave functions.[9]

References

1. H. Weyl, *Theory of Groups and Quantum Mechanics* (New York, Dover, 1950).
2. J. Schwinger, *Proc. Nat. Acad. Sci. USA* **46**, 570 (1960).
3. E. P. Wigner, *Phys. Rev.* **40**, 749 (1932).
4. W. K. Wootters, *Ann. Phys.* **176**, 1 (1987).
5. D. Cohendet, Ph. Combe, M. Sirugue and M. Sirugue-Collin, *J. Phys. A: Math. Gen.* 1, 2875 (1988).
6. M. Horibe, A. Takami, F. Hashimoto and A. Hayashi, *Phys. Rev.* **A65**, 032105 (2002).
7. U. Leonhardt, *Phys. Rev.* **A53**, 2998 (1996).
8. C. Miquel, J. P. Paz and M. Saraceno, *Phys. Rev.* **A65**, 062309 (2002).
9. J. Zak, *J. Phys. A: Math. Theor.* **44**, 345305 (2011).
10. A. Vourdas, *Rev. Prog. Phys.* **67**, 1 (2004).
11. G. Bjork, A. B. Klimov and L. L. Sanchez-Soto, *Prog. Opt.* **51**, 469 (2008).
12. T. Durt, B. G. Englert, I. Bengtsson and K. Zyczkowski, *Int. J. Quantum Comput.* **8**, 535 (2010).
13. A. Grossmann, *Commun. Math. Physics* **48**, 191 (1976).
14. A. Royer, *Phys. Rev.* **A15**, 449 (1977).
15. A. M. F. Rivas and A. M. Ozorio de Almeida, *Annals of Physics* **276**, 223 (1999).
16. R. J. Glauber, *Phys. Rev.* **131**, 2766 (1963).
17. H. J. Groenewold, *Physica* **12**, 405 (1946).
18. J. E. Moyal, *Proc. Cambridgel Philos. Soc.* **41**, 99 (1949).
19. J. Zak, *J. Math. Phys.* **53**, 103514 (2012).

PARALLEL SESSION
Symmetries in String Theory and Supergravity

20 & 22 AUGUST 2012

BIPARTITA: PHYSICS, GEOMETRY & NUMBER THEORY

YANG-HUI HE*

Department of Mathematics, City University, London,
Northampton Square, London EC1V 0HB, UK;
School of Physics, NanKai University, Tianjin, 300071, P.R. China;
Merton College, University of Oxford, OX14JD, UK
** E-mail: hey@maths.ox.ac.uk*

Bipartite graphs, especially drawn on Riemann surfaces, have of late assumed an active rôle in theoretical physics, ranging from MHV scattering amplitudes to brane tilings, from dimer models and topological strings to toric AdS/CFT, from matrix models to dessins d'enfants in gauge theory. Here, we take a brief and casual promenade in the realm of brane tilings, quiver SUSY gauge theories and dessins, serving as a rapid introduction to the reader.

Keywords: Supersymmetric Gauge Theory, AdS/CFT, Toric Geometry, Calabi-Yau Manifolds, Dimer Models, dessins d'enfants.

1. Observatio Curiosa

We beg the readers to first indulge us with a seeming mere curiosity, and point out a peculiarity on a few theories well familiar to them. Take perhaps the most famous supersymmetric field theory: $\mathcal{N} = 4$ super-Yang-Mills theory in four dimensions. It is well known that this theory has three adjoint fields, say X, Y, Z, charged under a $U(N)$ gauge group, interacting via a superpotential which in $\mathcal{N} = 1$ language can be written as $W = \text{Tr}(X[Y, Z])$. We could represent this data in terms of a so-called "clover quiver":

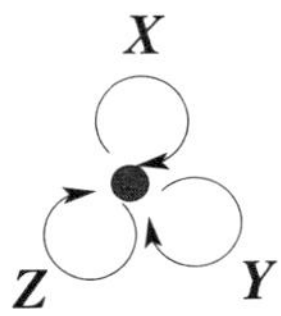

$$W = \text{Tr}(XYZ - XZY) \ . \tag{1}$$

Here, as is customary, the nodes correspond to factors in the gauge group which generically will take the form $\prod_i SU(N_i)$ in the low energy and arrows correspond to fields charged appropriately as adjoints or bi-fundamentals.

We see that the number N_G of gauge factors is one, the number of fields N_F, three and the number N_W of terms in the superpotential, two. It is immediately evident that $1 - 3 + 2 = 0$.

To see that this is not a mere exception, let us move on to another famous super-conformal field theory in four dimensions, the so-called "Klebanov-Witten conifold theory". This is an $SU(N) \times SU(N)$ theory with four fields $A_{i=1,2}$ and $B_{i=1,2}$ bi-fundamentally charged under the two gauge factors, interacting according to a quartic superpotential $W = \mathrm{Tr}(\epsilon_{i\ell}\epsilon_{jk}A_iB_jA_\ell B_k)$ where ϵ_{ij} is the Levi-Civita symbol. We can once again summarize this into a quiver, with the two nodes denoting the two $SU(N)$ gauge groups and the four arrows, the bi-fundamentals:

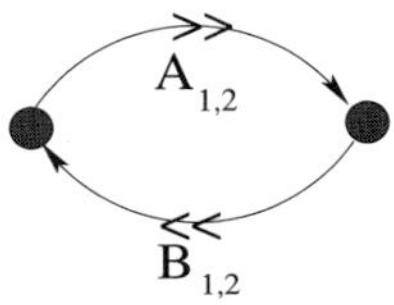

	$SU(N)$	$SU(N)$
$A_{i=1,2}$	$\square$	$\overline{\square}$
$B_{j=1,2}$	$\overline{\square}$	$\square$

$$W = \mathrm{Tr}(\epsilon_{il}\epsilon_{jk}A_iB_jA_lB_k) \tag{2}$$

Once more we find that the number of gauge groups, here $N_g = 2$, the number of fields, here $N_f = 4$ and the number of monomial terms in the superpotential, here $N_W = 2$, satisfy the constraint that $2 - 4 + 2 = 0$.

Of the plenitude of $\mathcal{N} = 1$ supersymmetric gauge theories in four dimensions, especially the wealth thereof being engineered by the AdS/CFT correspondence as well as others geometrically attainable as low-energy limits in string theory, we can readily verify that the charming relation

$$N_g - N_f + N_W = 0 \tag{3}$$

persists for the majority of those in the literature. This is not a coincidence: whenever the moduli space of the gauge theory is a toric variety, we shall see that (3) holds. In the context of string theory, almost all known gauge theories in the AdS/CFT correspondence have dual geometries which are toric because there are enough isometries for such spaces to allow for the explicit writing of (non-compact) Calabi-Yau metrics.[5] Indeed, our two examples above have moduli space of vacua being respectively $\mathbb{C}^3$ and the conifold (a quadric hypersurface in $\mathbb{C}^4$) which are canonical examples of affine toric varieties. To further proceed with this point, it is expedient to briefly remind the reader of the concept of the vacuum moduli space (VMS) for a supersymmetric gauge theory, both within and independent of the string theory context.

1.1. *The Geometry of Moduli Space of Vacua*

Consider a generic supersymmetric gauge theory in four dimension with action, written in $\mathcal{N}=1$ superspace notation, as given:

$$\int d^4x\ [\int d^4\theta\ \Phi_i^\dagger e^V \Phi_i + \frac{1}{4g^2}\int d^2\theta\ \mathrm{Tr}\,\mathcal{W}_\alpha \mathcal{W}^\alpha + \int d^2\theta\ W(\Phi) + \mathrm{h.c.}]\ . \quad (4)$$

Here Φ_i are chiral superfields, charged under some appropriate gauge group (not necessarily semi-simple), to which is associated some vector superfield V, field strength $\mathcal{W}$ and coupling g; the superpotential, W, is some (holomorphic) polynomial in Φ_i. In terms of the scalar component ϕ_i of Φ_i, the effective potential is $\sum_i \left|\frac{\partial W}{\partial \phi_i}\right|^2 + \frac{g^2}{4}(\sum_i q_i |\phi_i|^2)^2$, with q_i being the appropriate charges. Thus, the VMS is the minimum of this potential, which, being non-negative, compels it to be the simultaneous solutions of

$$\text{F-Terms: } \frac{\partial W}{\partial \phi_i} = 0\ ; \qquad \text{D-Terms: } \sum_i q_i |\phi_i|^2 = 0\ . \quad (5)$$

In general, these equations can become rather complicated, and the VMS, a non-trivial affine algebraic variety. Utilization of algorithmic and computational geometry to systematically study these equations was recently initiated.[4]

In string theory, when the gauge theory is engineered as the world-volume physics of a stack of D3-branes, the VMS is by construction the transverse six-dimensions which the branes probe. This is the key to the algebraic geometry of the $\mathrm{AdS}_5/\mathrm{CFT}_4$ correspondence: the VMS of world-volume theories is some affine Calabi-Yau threefold, existing as a real cone over some Sasaki-Einstein 5-manifold. Our two above examples, (1) and (2), are both embeddable into string theory, and we can readily check that the VMS are, respectively, (symmetrized products of) $\mathbb{C}^3$ as a cone over S^5 and the conifold as a cone over a space known as $T^{1,1}$.

By far the largest class of geometries which has dominated the literature, especially in AdS/CFT, is toric varieties and on these we shall henceforth focus. Moreover, for simplicity, we will take all the nodes of our quivers to be labeled by $N=1$ so that our gauge theories are Abelian; one could easily promote to arbitrary N, resulting in the VMS being an N-th symmetrized product. It shall be too great a digression for us to review toric geometry here, we simply mention that when the VMS is toric, one could use the gauge linear sigma-model[1] to study how the brane resolves singularities.[2] The algorithm of engineering the gauge theory, given an arbitrary toric Calabi-Yau diagram, was traditionally done by partial resolutions.[3]

2. Toric Moduli Space, Gauge Theory and Bipartiteness

The astute reader would recognize (3) as the Euler relation for a torus and we shall see how the aforementioned gauge theories can be written as tilings thereon.[6] The limitation of space does not permit us to delve into the vast web of correspondences which has emerged over the past decade and one is referred to marvelous reviews[7,8] and the myriad of works in both the physics and mathematics communities from which we draw a small sample representative of the various directions.[9–21] What we will do is to walk through the ensuing Figure, which illustrates the case for the conifold theory in some detail, and hope an appreciation, in a rather concrete manner, could be inspired in the reader:

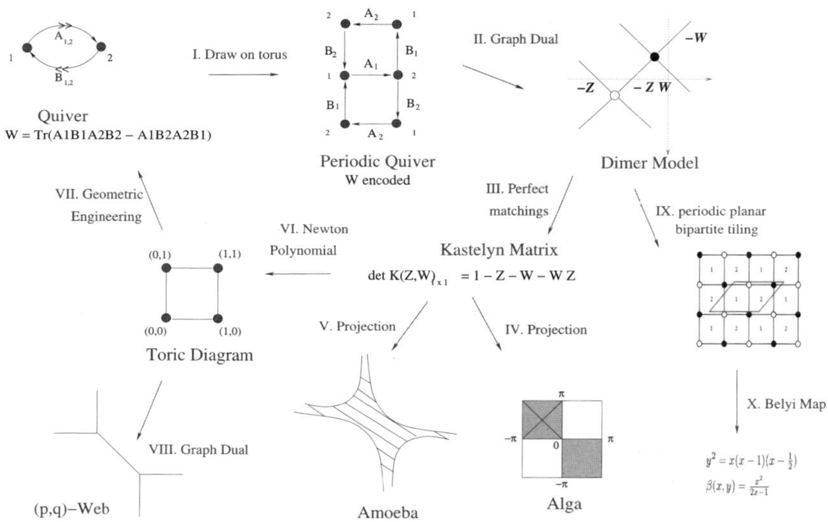

Arrow I: We begin with the theory, already presented in (2), having the quiver with four arrows and the quartic superpotential W. Following Arrow I, we recognize that if we drew the quiver periodically on the plane (with appropriately identified nodes and arrow explicitly marked), then, in the fundamental region, there will be two loops, one counter-clockwise, and the other, clockwise; these are precisely the two terms in W.

Arrows II: The graph dual - by replacing a counter-clockwise loop with a black node and clockwise, white - is then a bipartite graph periodic on a plane. For the conifold, this is a rectangular tiling with two squares in the fundamental domain $\mathcal{F}$. Equivalently, we can draw this as a *dimer model* on a torus – the origin of (3). Marking the two cycles of the T^2 with formal variables z and w and the (signed) crossings of the dimer relative to these cycles in $\mathcal{F}$ produce the indicated monomials.

Arrow III: In statistical mechanics and chemistry, an important problem is the enumeration of the number of *perfect-matchings* (where each pair of white-black is linked by a bi-molecular bond). This is achieved by a weighted adjacency, called the *Kasteleyn matrix* (here, there is only one pair of black-white nodes, thus it is just 1×1), whose determinant K gives a bi-variate polynomial in (z, w).

Arrows IV & V: Treating $K(z, w)$ as an algebraic variety (complex curve), we can perform two projections: (1) the *amoeba* projection $(z, w) \to (\log|z|, \log|w|)$ and (2) the *alga* (or co-amoeba) projection $(z, w) \to (\arg(z), \arg(w))$. These are real algebraic sets which constitute the study of "tropical geometry". One can see that the alga projection, being itself doubly periodic, very nicely contracts to the dimer.

Arrows VI, VII & VIII: The curve $K(z, w)$ is the *Newton polynomial* for the *toric diagram* of the conifold. Indeed, non-compact toric Calabi-Yau threefolds have the property that their toric diagram is a set of co-planar lattice points. The complementary amoeba projection, contracts nicely to the graph-dual of the (resolution) of the toric diagram. This dual is called the (p, q)*-web* and is where charged branes can wrap collapsing cycles. In summary, the (p, q)-web is the spine of the amoeba of $K(z, w)$ while the dimer is the contraction of the alga of $K(z, w)$. Given a geometry, such as specified by the toric diagram, the inverse algorithm of extracting the physics of the quiver theory is the process of geometrical engineering.

Arrow IX: We re-draw the dimer model by zooming out a little to make the periodic tiling of the plane apparent. The fundamental domain of the torus is marked in red.

Arrow X: When delving into bipartite graphs drawn on Riemann surfaces, one cannot resist but enter the subject of *dessins d'enfants*, which, as outlined in his "Esquisse d'un Programme", impressed Grothendieck more than any other topic in mathematics. For any such graph, one could find a Riemann surface Σ defined over the algebraic numbers $\overline{\mathbb{Q}}$ such that a rational map β takes Σ to $\mathbb{P}^1$ ramified only at three points: $0, 1, \infty$. The data (Σ, β) is called a *Belyi pair*. Here Σ is the torus realized concretely as a Weierstraß equation of an elliptic curve, which, remarkably, can be defined over $\overline{\mathbb{Q}}$. The pre-images of 0 (respectively 1) are the black (resp. white) nodes whose valency is equal to the ramification index of β, and pre-images of ∞ are the faces. There are many interesting manifestations of number theoretic properties in the physics, e.g., the degree of extension over $\mathbb{Q}$ of the R-charges is a Seiberg duality invariant.

Acknowledgments

Catharinae Sanctae Alexandriae et Ad Majorem Dei Gloriam.

I would like to thank the organizers of "The XXIX International Colloquium on Group-Theoretical Methods in Physics" for inviting me to the wonderful conference and take this opportunity to express my gratitude to my collaborators N. Benishti, S. Benvenuti, B. Feng, D. Forcella, S. Franco, D. Galloni, A. Hanany, V. Jejjala, K. Kennaway, J. McKay, J. Pasukonis, S. Ramgoolam, D. Rodriguez-Gomez, J. Sparks, A. Uranga, C. Vafa and A. Zaffaroni for our many adventures in *terra bipartita.* To the STFC, UK, for an AF and for grant ST/J00037X/1, to the Chinese Ministry of Education, for a Chang-Jiang Chair Professorship at NanKai University, and to the US NSF for grant CCF-1048082 I am most indebted.

References

1. E. Witten, "Phases of N=2 theories in two-dimensions," Nucl. Phys. B **403**, 159 (1993) [hep-th/9301042].
2. M. R. Douglas, B. R. Greene and D. R. Morrison, "Orbifold resolution by D-branes," Nucl. Phys. B **506**, 84 (1997) [hep-th/9704151].
3. B. Feng, A. Hanany and Y. -H. He, "D-brane gauge theories from toric singularities and toric duality," Nucl. Phys. B **595**, 165 (2001) [hep-th/0003085].
4. J. Gray, Y. -H. He, V. Jejjala and B. D. Nelson, "Exploring the vacuum geometry of N=1 gauge theories," Nucl. Phys. B **750**, 1 (2006) [hep-th/0604208].
5. J. P. Gauntlett, D. Martelli, J. F. Sparks and D. Waldram, "A New infinite class of Sasaki-Einstein manifolds," Adv. Theor. Math. Phys. **8**, 987 (2006) [hep-th/0403038].
6. A. Hanany and K. D. Kennaway, "Dimer models and toric diagrams," hep-th/0503149.
7. K. D. Kennaway, "Brane Tilings," Int. J. Mod. Phys. A **22**, 2977 (2007) [arXiv:0706.1660 [hep-th]].
8. M. Yamazaki, "Brane Tilings and Their Applications," Fortsch. Phys. **56**, 555 (2008) [arXiv:0803.4474 [hep-th]].
9. S. Franco, A. Hanany, K. D. Kennaway, D. Vegh and B. Wecht, "Brane dimers and quiver gauge theories," JHEP **0601**, 096 (2006) [hep-th/0504110].
10. I. Garcia-Etxebarria, F. Saad and A. M. Uranga, "Quiver gauge theories at resolved and deformed singularities using dimers," JHEP **0606**, 055 (2006) [hep-th/0603108].
11. A. Butti, D. Forcella and A. Zaffaroni, JHEP **0706**, 069 (2007) [hep-th/0611229].
12. B. Feng, Y. -H. He, K. D. Kennaway and C. Vafa, "Dimer models from mirror symmetry and quivering amoebae," Adv. Theor. Math. Phys. **12**, 489 (2008) [hep-th/0511287].
13. Jan Stienstra, "Hypergeometric Systems in two Variables, Quivers, Dimers and Dessins d'Enfants", arXiv:0711.0464
14. S. K. Ashok, F. Cachazo and E. Dell'Aquila, "Children's drawings from Seiberg-Witten curves," hep-th/0611082.
15. R. d. M. Koch and S. Ramgoolam, "From Matrix Models and Quantum Fields to Hurwitz Space and the absolute Galois Group," arXiv:1002.1634 [hep-th].
16. V. Jejjala, S. Ramgoolam, D. Rodriguez-Gomez, "Toric CFTs, Permutation Triples and Belyi Pairs," JHEP **1103**, 065 (2011) [arXiv:1012.2351 [hep-th]].
17. A. Hanany, Y. -H. He, V. Jejjala, J. Pasukonis, S. Ramgoolam and D. Rodriguez-Gomez, "The Beta Ansatz: A Tale of Two Complex Structures," JHEP **1106**, 056 (2011) [arXiv:1104.5490 [hep-th]].
    – "Invariants of Toric Seiberg Duality," Int. J. Mod. Phys. A **27**, 1250002 (2012) [arXiv:1107.4101 [hep-th]].
18. Y. -H. He, V. Jejjala and D. Rodriguez-Gomez, "Brane Geometry and Dimer Models," JHEP **1206**, 143 (2012) [arXiv:1204.1065 [hep-th]].
19. Nathan Broomhead, "Dimer models and Calabi-Yau algebras", arXiv:0901.4662.
20. S. Franco, "Bipartite Field Theories: from D-Brane Probes to Scattering Amplitudes," arXiv:1207.0807 [hep-th].
21. Y. -H. He and J. McKay, "N=2 Gauge Theories: Congruence Subgroups, Coset Graphs and Modular Surfaces," arXiv:1201.3633 [hep-th].

THREE-DIMENSIONAL (p, q) ANTI-DE SITTER SUPERSYMMETRY

SERGEI M. KUZENKO
School of Physics M013, The University of Western Australia
35 Stirling Highway, Crawley W.A. 6009, Australia
E-mail: sergei.kuzenko@uwa.edu.au

This article provides a mini-introduction to three-dimensional field theories with rigid (p, q) anti-de Sitter supersymmetry.

Keywords: Supersymmetry; conformal supergravity; AdS superspaces; nonlinear sigma models; hyperkähler geometry.

1. Introduction

In three dimensions, the anti-de Sitter (AdS) group is reducible,

$$\mathrm{SO}_0(2,2) \cong \Big(\mathrm{SL}(2,\mathbb{R}) \times \mathrm{SL}(2,\mathbb{R})\Big)/\mathbb{Z}_2 \ , \tag{1}$$

and so are its supersymmetric generalizations such as

$$\mathrm{OSp}(p|2;\mathbb{R}) \times \mathrm{OSp}(q|2;\mathbb{R}) \ . \tag{2}$$

As a result, $\mathcal{N}$-extended AdS supergravity exists in several incarnations known as the (p, q) AdS supergravity theories,[1] and so do their maximally symmetric solutions – the (p, q) AdS superspaces[2] which contain the usual AdS space, AdS_3, as their bosonic body. Here p and q are non-negative integers such that $\mathcal{N} = p + q$ and $p \geq q \geq 0$.

From the point of view of the superspace geometry for $\mathcal{N}$-extended conformal supergravity in three dimensions,[3,4] the (p, q) AdS superspaces naturally originate as maximally symmetric backgrounds with non-zero covariantly constant torsion and curvature.[2] These superspaces are necessarily conformally flat in the following cases: (i) $\mathcal{N} < 4$; (ii) $\mathcal{N} \geq 4$ and $p \geq q > 0$. The corresponding isometry groups belong to family (2). If $\mathcal{N} \geq 4$, the geometry of the $(\mathcal{N}, 0)$ AdS superspace is determined by two parameters. One of these parameters is the AdS curvature and the other an expectation value of the so-called supersymmetric Cotton tensor X^{IJKL}, which is completely antisymmetric in its SO($\mathcal{N}$) indices. In the case that $X^{IJKL} \neq 0$, the superspace is not conformally flat[2] and its isometry group does not belong to family (2).

By "three-dimensional (p, q) AdS supersymmetry" we mean the invariance of a field theory in AdS_3 under the isometry group of the (p, q) AdS superspace. Such theories can be realized in terms of superfields defined on the superspace under consideration, at least on the mass shell. For $p+q \leq 4$, off-shell formulations for these theories have been constructed[2,5,6] using the supergravity formalism developed in Ref. [4]. The remarkable result is that different choices of p and q, with $\mathcal{N} = p+q > 1$ fixed, lead to supersymmetric field theories with drastically different properties.[2,5,6] Such differences emerge even in the simplest extended case $\mathcal{N} = 2$.[5] For any sigma model with (1,1) AdS supersymmetry, its target space must be a non-compact Kähler manifold with an exact Kähler two-form. On the other hand, the (2,0) AdS supersymmetry allows compact Kähler manifolds as the target spaces of nonlinear sigma models.

This conference paper is organized as follows. In section 2 we give a brief review of the superspace formulation for $\mathcal{N}$-extended conformal supergravity in three dimensions. Section 3 describes the differential geometry of the (p, q) AdS superspaces. In section 4 we list the restrictions on the target space geometry of $\mathcal{N} = 3$ and $\mathcal{N} = 4$ supersymmetric nonlinear sigma models which are imposed by the (p, q) AdS supersymmetry.

2. Extended conformal supergravity

For any positive integer $\mathcal{N}$, three-dimensional $\mathcal{N}$-extended conformal supergravity possesses an off-shell formulation.[3] It can be described using a curved superspace which is parametrized by real bosonic (x^m) and real fermionic ($\theta^\mu_{\rm I}$) coordinates,

$$z^M = (x^m, \theta^\mu_{\rm I}) , \qquad m = 0, 1, 2 , \quad \mu = 1, 2 , \quad {\rm I} = 1, \cdots, \mathcal{N} . \qquad (3)$$

The superspace structure group can be chosen to be $SL(2, \mathbb{R}) \times SO(\mathcal{N})$, where $SL(2, \mathbb{R})$ is the Lorentz group and SO($\mathcal{N}$) the $\mathcal{N}$-extended R-symmetry group. The superspace differential geometry is encoded in covariant derivatives of the form

$$\mathcal{D}_A \equiv (\mathcal{D}_a, \mathcal{D}^I_\alpha) = E_A + \Omega_A + \Phi_A , \qquad (4)$$

where the tangent space indices take the values $\alpha = 1, 2$, $a = 0, 1, 2$, and $I = 1, \cdots, \mathcal{N}$. The first term on the right of (4) denotes the supervielbein,

$E_A = E_A{}^M \partial/\partial z^M$; the second term stands for the Lorentz connection,

$$\Omega_A = \frac{1}{2}\Omega_A{}^{bc}\mathcal{M}_{bc} = \frac{1}{2}\Omega_A{}^{\beta\gamma}\mathcal{M}_{\beta\gamma} , \quad \mathcal{M}_{ab} = -\mathcal{M}_{ba} , \quad \mathcal{M}_{\alpha\beta} = \mathcal{M}_{\beta\alpha} ; \quad (5)$$

finally, the third term is the SO($\mathcal{N}$)-connection,

$$\Phi_A = \frac{1}{2}\Phi_A{}^{KL}\mathcal{N}_{KL} , \qquad \mathcal{N}_{KL} = -\mathcal{N}_{LK} . \qquad (6)$$

The Lorentz generators with two vector indices ($\mathcal{M}_{ab}$) and two spinor indices ($\mathcal{M}_{\alpha\beta}$) are related to each other by the standard rule given in, e.g., Ref. [4]. The generators of the group SO($\mathcal{N}$) are denoted by $\mathcal{N}_{IJ}$.

In order to describe the $\mathcal{N}$-extended Weyl supermultiplet, the covariant derivatives have to obey certain constraints on the torsion tensor.[3] The Bianchi identities for these constraints were explicitly solved in Ref. [4]. In particular, the spinor-spinor anti-commutators reads as follows:

$$\begin{aligned}\{\mathcal{D}_\alpha^I, \mathcal{D}_\beta^J\} = {} & 2\mathrm{i}\delta^{IJ}\mathcal{D}_{\alpha\beta} - 2\mathrm{i}\varepsilon_{\alpha\beta}C^{\gamma\delta IJ}\mathcal{M}_{\gamma\delta} - 4\mathrm{i}S^{IJ}\mathcal{M}_{\alpha\beta} \\ & -4\mathrm{i}\Big(\varepsilon_{\alpha\beta}S^{K[I}\delta^{J]L} - C_{\alpha\beta}{}^{KL}\delta^{IJ} + 4C_{\alpha\beta}{}^{K(I}\delta^{J)L}\Big)\mathcal{N}_{KL} \\ & +\mathrm{i}\varepsilon_{\alpha\beta}X^{IJKL}\mathcal{N}_{KL} . \end{aligned} \qquad (7)$$

It turns out that all components of the torsion and curvature tensors are expressed in terms of three real dimension-one superfields,

$$X^{IJKL} = X^{[IJKL]} , \qquad S^{IJ} = S^{(IJ)} , \qquad C_a{}^{IJ} = C_a{}^{[IJ]} , \qquad (8)$$

and their covariant derivatives.

The supergeometry obtained is indeed suitable to describe conformal supergravity due to the fact that the underlying constraints are invariant under the super-Weyl transformations[4]

$$\delta\mathcal{D}_\alpha^I = \frac{1}{2}\sigma\mathcal{D}_\alpha^I + (\mathcal{D}^{\beta I}\sigma)\mathcal{M}_{\alpha\beta} + (\mathcal{D}_{\alpha J}\sigma)\mathcal{N}^{IJ} , \qquad (9a)$$

$$\begin{aligned}\delta\mathcal{D}_a = {} & \sigma\mathcal{D}_a + \frac{\mathrm{i}}{2}(\gamma_a)^{\alpha\beta}(\mathcal{D}^K_{(\alpha}\sigma)\mathcal{D}_{\beta)K} + \varepsilon_{abc}(\mathcal{D}^b\sigma)\mathcal{M}^c \\ & +\frac{\mathrm{i}}{16}(\gamma_a)^{\alpha\beta}([\mathcal{D}^{[K}_{(\alpha}, \mathcal{D}^{L]}_{\beta)}]\sigma)\mathcal{N}_{KL} . \end{aligned} \qquad (9b)$$

The super-Weyl transformations of the torsion and curvature tensors are

$$\delta S^{IJ} = \sigma S^{IJ} - \frac{\mathrm{i}}{8}([\mathcal{D}^{\rho(I}, \mathcal{D}^{J)}_\rho]\sigma) , \qquad (10a)$$

$$\delta C_a{}^{IJ} = \sigma C_a{}^{IJ} - \frac{\mathrm{i}}{8}(\gamma_a)^{\alpha\beta}([\mathcal{D}^{[I}_{(\alpha}, \mathcal{D}^{J]}_{\beta)}]\sigma) , \qquad (10b)$$

$$\delta X^{IJKL} = \sigma X^{IJKL} . \qquad (10c)$$

As is seen, the tensor X^{IJKL} transforms homogeneously. It proves to contain the Cotton tensor among its components.

3. (p, q) AdS superspaces

The $\mathcal{N}$-extended AdS superspaces correspond to those conformal supergravity backgrounds which satisfy the following requirements: (i) the torsion and curvature tensors are Lorentz invariant; and (ii) the torsion and curvature tensors are covariantly constant. The condition (i) implies that $C_a{}^{IJ} = 0$, while the condition (ii) requires the remaining components of the torsion and curvature tensors to be covariantly constant,

$$\mathcal{D}^I_\alpha S^{JK} = \mathcal{D}_a S^{JK} = 0 \ , \qquad \mathcal{D}^I_\alpha X^{JKLM} = \mathcal{D}_a X^{JKLM} = 0 \ . \tag{11}$$

The complete algebra of covariant derivatives takes the form:

$$\begin{aligned} \{\mathcal{D}^I_\alpha, \mathcal{D}^J_\beta\} &= 2\mathrm{i}\delta^{IJ}\mathcal{D}_{\alpha\beta} - 4\mathrm{i}S^{IJ}\mathcal{M}_{\alpha\beta} \\ &\quad + \mathrm{i}\varepsilon_{\alpha\beta}\Big(X^{IJKL} - 4S^{K[I}\delta^{J]L}\Big)\mathcal{N}_{KL} \ , \end{aligned} \tag{12a}$$

$$[\mathcal{D}_a, \mathcal{D}^J_\beta] = S^J{}_K(\gamma_a)_\beta{}^\gamma \mathcal{D}^K_\gamma \ , \tag{12b}$$

$$[\mathcal{D}_a, \mathcal{D}_b] = 4\, S^2\, \varepsilon_{abc}\mathcal{M}^c = -4\, S^2\, \mathcal{M}_{ab} \ . \tag{12c}$$

The two integrability conditions

$$\{\mathcal{D}^I_\alpha, \mathcal{D}^J_\beta\} S^{KL} = 0 \ , \qquad \{\mathcal{D}^I_\alpha, \mathcal{D}^J_\beta\} X^{KLMN} = 0 \tag{13}$$

are equivalent to the following algebraic constraints:

$$X^{IJN(K}S_N{}^{L)} = S^{IM}S_M{}^{(K}\delta^{L)J} - S^{JM}S_M{}^{(K}\delta^{L)I} \ , \tag{14a}$$

$$X^{IJN[K}X_N{}^{LPQ]} = 2S^{M[I}\delta^{J]N}\Big\{\delta^{[K}_N X_M{}^{LPQ]} - \delta^{[K}_M X_N{}^{LPQ]}\Big\} \ . \tag{14b}$$

The Bianchi identities

$$\sum_{[ABC)} [\mathcal{D}_A, [\mathcal{D}_B, \mathcal{D}_C\}\} = 0 \tag{15}$$

imply

$$\hat{S}^2 = S^2 \mathbb{1} \ , \qquad \hat{S} := (S^{IJ}) = \hat{S}^{\mathrm{T}} \ , \qquad S^2 := \frac{1}{\mathcal{N}}\,\mathrm{tr}(\hat{S}^2) \geq 0 \ . \tag{16}$$

In what follows the choice $S^2 > 0$ will be assumed. Then the relation (12c) tells us that the body of the superspace is an AdS space. Since $S^2 > 0$, it follows from (16) that $\hat{S}$ is a nonsingular symmetric $\mathcal{N} \times \mathcal{N}$ matrix, and $\hat{S}/S$ is an orthogonal matrix.

Due to the properties of $\hat{S}$, eq. (16), a local SO($\mathcal{N}$) transformation can be performed to diagonalize it:

$$S^{IJ} = S\,\mathrm{diag}(\overbrace{+1, \cdots, +1}^{p}, \overbrace{-1, \cdots, -1}^{q=\mathcal{N}-p}) \ , \qquad S > 0 \ . \tag{17}$$

In this gauge, there remains an unbroken local group $SO(p) \times SO(q)$. Now, the constraint (14a) implies that

$$q > 0 \quad \longrightarrow \quad X^{IJKL} = 0 \ . \tag{18}$$

Therefore X^{IJKL} may be non-vanishing only in the $(\mathcal{N}, 0)$ case for which the constraint (14b) reduces to

$$X_N{}^{IJ[K} X^{LPQ]N} = 0 \ . \tag{19}$$

For $\mathcal{N} = 4$, the general solution of this constraint is $X^{IJKL} = X\varepsilon^{IJKL}$, where ε^{IJKL} the Levi-Civita tensor and X is an arbitrary real parameter.

In the $\mathcal{N} = 4$ case, the algebra (12) simplifies if we switch from SO(4) isovector indices to pairs of $SU(2)_L \times SU(2)_R$ isospinor indices by making use of the isomorphism $SO(4) \cong \big(SU(2)_L \times SU(2)_R\big)/\mathbb{Z}_2$.

$$\begin{aligned} \{\mathcal{D}_\alpha^{i\bar{i}}, \mathcal{D}_\beta^{j\bar{j}}\} = \;& 2\mathrm{i}\varepsilon^{ij}\varepsilon^{\bar{i}\bar{j}}\mathcal{D}_{\alpha\beta} - 4\mathrm{i}\, S\, \varepsilon^{ij}\varepsilon^{\bar{i}\bar{j}}\mathcal{M}_{\alpha\beta} \\ & +2\mathrm{i}\varepsilon_{\alpha\beta}\varepsilon^{\bar{i}\bar{j}}(2S + X)\mathbf{L}^{ij} + 2\mathrm{i}\varepsilon_{\alpha\beta}\varepsilon^{ij}(2S - X)\mathbf{R}^{\bar{i}\bar{j}} \ , \end{aligned} \tag{20a}$$

$$[\mathcal{D}_a, \mathcal{D}_\beta^{j\bar{j}}] = S\,(\gamma_a)_\beta{}^\gamma \mathcal{D}_\gamma^{j\bar{j}} \ , \qquad [\mathcal{D}_a, \mathcal{D}_b] = -\,4\,S^2\,\mathcal{M}_{ab} \ . \tag{20b}$$

Here $\mathbf{L}^{ij}$ and $\mathbf{R}^{\bar{i}\bar{j}}$ are the generator of the groups $SU(2)_L$ and $SU(2)_R$ respectively. It is seen from (20) that there are two critical cases

$$X = \pm 2S \ . \tag{21}$$

4. Target space geometries

In Ref. [6] we classified all possible types of hyperkähler target spaces for the cases $\mathcal{N} = 3$ and $\mathcal{N} = 4$ by making use of two different realizations for the most general (p, q) supersymmetric sigma models: (i) off-shell formulations in terms of $\mathcal{N} = 3$ and $\mathcal{N} = 4$ projective supermultiplets; and (ii) on-shell formulations in terms of covariantly chiral scalar superfields in (2,0) AdS superspace. There is no space to discuss these realizations in the present conference paper, and we refer the interested reader to the original publication.[6] Here we only summarize the results of the classification.

With the exception of two *critical* (4,0) AdS supersymmetries, for which $X = \pm 2S$, all σ-model target spaces belong to the following large families of *non-compact* hyperkähler manifolds: (i) hyperkähler cones; and (ii) hyperkähler spaces with a U(1) isometry group which acts non-trivially on the two-sphere of complex structures (and necessarily leaves one complex structure invariant). Of course, all hyperkähler cones belong to family (ii). The target spaces of arbitrary nonlinear σ-models with (3,0), (3,1) and non-critical (4,0) supersymmetries in AdS_3 are hyperkähler cones. The main reason for this is the property that the R-symmetry group of such supersymmetric σ-models includes SU(2) as a subgroup.[2] The target spaces of arbitrary nonlinear σ-models with (2,1) and (2,2) supersymmetries in AdS_3 are hyperkähler manifolds belonging to the family (ii).

It was demonstrated in [6] that *compact* target spaces are allowed only for those nonlinear σ-models in AdS_3 which possess critical (4,0) supersymmetry such that $X = \pm 2S$. For concreteness, let us choose $X = 2S$. Then the target space of any supersymmetric σ-model has the form

$$\mathcal{M}_{\rm L} \times \mathcal{M}_{\rm R} \ , \tag{22}$$

where $\mathcal{M}_{\rm L}$ is a hyperkähler cone, while $\mathcal{M}_{\rm R}$ is an arbitrary hyperkähler manifold.

The target spaces of superconformal σ-models with six and eight supercharges are known to be hyperkähler cones. All nonlinear σ-models in AdS_3 with (3,0) and (3,1) supersymmetries are actually $\mathcal{N} = 4$ superconformal. As concerns the nonlinear σ-models in AdS_3 with (4,0) supersymmetry, they are superconformal only in the case $X = 0$.

Acknowledgments

I am happy to acknowledge Daniel Butter, Ulf Lindström and Gabriele Tartaglino-Mazzucchelli for collaboration on the projects this mini-review is based on. I would like to thank the organizers of *GROUP29* for the ideally organized conference. This was supported in part by the Australian Research Council under Grant No. DP1096372.

References

1. A. Achúcarro and P. K. Townsend, *Phys. Lett.* B **180**, 89 (1986).
2. S. M. Kuzenko, U. Lindström and G. Tartaglino-Mazzucchelli, *JHEP* **1208**, 024 (2012) [arXiv:1205.4622].
3. P. S. Howe, J. M. Izquierdo, G. Papadopoulos and P. K. Townsend, *Nucl. Phys. B* **467**, 183 (1996) [arXiv:hep-th/9505032].
4. S. M. Kuzenko, U. Lindström and G. Tartaglino-Mazzucchelli, *JHEP* **1103**, 120 (2011) [arXiv:1101.4013].
5. S. M. Kuzenko and G. Tartaglino-Mazzucchelli, *JHEP* **1112**, 052 (2011) [arXiv:1109.0496].
6. D. Butter, S. M. Kuzenko and G. Tartaglino-Mazzucchelli, "Nonlinear sigma models with AdS supersymmetry in three dimensions," arXiv:1210.5906.

PARTITION FUNCTION OF ABJM MATRIX MODEL

SANEFUMI MORIYAMA*

Kobayashi Maskawa Institute, Nagoya University, Nagoya, Japan, 464-0812
**E-mail: moriyama@math.nagoya-u.ac.jp*

We have studied the fermi gas system associated to the ABJM matrix model, aiming at understanding some hidden structures of M-theory. In trying to reduce the number of integrations in computing the partition functions $Z_k(N)$, we have found an identity relating two density matrices of the opposite parities. Using this relation, we have computed exact values of the partition functions for $k = 1$ up to $N = 9$.

Keywords: M-theory, Chern-Simons theory, Airy function.

1. Motivation

Here we would like to summarize our recent result in.[1] First, let us explain our previous work[2] as a motivation for the current study.

In our previous work[2] we have studied the ABJM matrix model ($g_s = 2\pi i/k$)

$$Z_k(N) = \frac{1}{(N!)^2}\int\prod_{i=1}^{N}\frac{d\mu_i}{2\pi}\prod_{k=1}^{N}\frac{d\nu_k}{2\pi}e^{-(\sum\mu_i^2-\sum\nu_k^2)/2g_s} \times\frac{\prod_{i<j}\Big(2\sinh\frac{\mu_i-\mu_j}{2}\Big)^2\prod_{k<l}\Big(2\sinh\frac{\nu_k-\nu_l}{2}\Big)^2}{\prod_{i,k}\Big(2\cosh\frac{\mu_i-\nu_k}{2}\Big)^2}. \tag{1}$$

We have shown that all the perturbative terms in the 't Hooft expansion are summed up into the Airy function (up to an N-independent overall factor),

$$Z_k(N) = \mathrm{Ai}\left(\left[\frac{\pi k^2}{\sqrt{2}}\right]^{2/3}\lambda_{\mathrm{ren}}\right), \tag{2}$$

where the Airy function is defined by

$$\mathrm{Ai}\,(z) = \frac{1}{2\pi i}\int_C \exp\left(-zt + \frac{1}{3}t^3\right), \tag{3}$$

with a certain contour C and the 't Hooft coupling is renormalized as

$$\lambda_{\mathrm{ren}} = \frac{N}{k} - \frac{1}{24} - \frac{1}{3k^2}. \tag{4}$$

This matrix model stems from the ABJM theory,[3] which proposes that N multiple M2-branes on the geometry $\mathbb{C}^4/\mathbb{Z}_k$ is described by a supersymmetric generalization of the Chern-Simons theory. It was known[4] from the AdS/CFT correspondence that N multiple M2-branes have the degree of freedom $N^{3/2}$. Hence, it was a challenge to derive the degree of freedom from the partition function of the ABJM theory.

Using the localization technique, the partition function of the ABJM theory, originally given by an infinite-dimensional path integral, was reduced[5] to a finite-dimensional matrix model (1).

Aside from its physical meaning in M-theory, the ABJM matrix model can be regarded as a natural generalization of the most fundamental $U(N)$ Gaussian matrix model.[6] There are two important generalizations for it. On one hand, if we replace the Vandermonde measure of the Gaussian matrix model by its q-deformed hyperbolic functions, we end up with the Chern-Simons matrix model, which is the partition function of the $U(N)$ Chern-Simons theory on S^3. If, on the other hand, we replace the measure by that of the supergroup $U(N|N)$, we obtain the supermatrix model. The ABJM matrix model is nothing but the model with both of these generalizations.

After various studies of the ABJM matrix model, finally in[7] the $N^{3/2}$ behavior was reproduced. This result, however, has infinitely many perturbative corrections. In[2] we have found that all of these perturbative corrections can be summed up compactly into the Airy function, as we mention above. Namely, we can regard our result of the Airy function as the perturbative completion of the $N^{3/2}$ behavior.

As we see in (3), the Airy function has an integral representation of the exponentiated cubic term reminiscent of the partition function of the Chern-Simons theory and appears also in other contexts of M-theory.[8] Therefore, it is expected that there is a deep relation among M-theory, the Chern-Simons theory and the Airy function.

In the history, string theory originates from the Veneziano amplitude which has nothing to do with the string worldsheet or the string conformal symmetry at the first sight. It is only after the works of Nambu and Virasoro that the worldsheet conformal symmetry was read off. The Airy function we have found might play a similar role as the Veneziano amplitude. It may give us a clue in discovering hidden structures of M-theory. Therefore we would like to study this matrix model further beyond the perturbation theory.

One of our main results in[1] is the computation of the partition functions up to $N = 9$ at $k = 1$, where there is no orbifold action and the supersymmetry is enhanced to the maximal case. (See also.[9]) The results are given by

$$\begin{aligned}
&Z(1) = \frac{1}{4}, \quad Z(2) = \frac{1}{16\pi}, \quad Z(3) = \frac{-3+\pi}{64\pi}, \quad Z(4) = \frac{-\pi^2+10}{1024\pi^2},\\
&Z(5) = \frac{26+20\pi-9\pi^2}{4096\pi^2}, \quad Z(6) = \frac{78-121\pi^2+36\pi^3}{147456\pi^3},\\
&Z(7) = \frac{-126+174\pi+193\pi^2-75\pi^3}{196608\pi^3},\\
&Z(8) = \frac{876-4148\pi^2-2016\pi^3+1053\pi^4}{18874368\pi^4},\\
&Z(9) = \frac{4140+8880\pi-15348\pi^2-13480\pi^3+5517\pi^4}{75497472\pi^4}.
\end{aligned} \tag{5}$$

2. Fermi Gas Formalism

After our summation by the Airy function,[2] an important hidden structure of this matrix model was realized beautifully in.[10] There, it was found that the ABJM partition function can be rewritten into the partition function of a fermi gas system

$$Z(N) = \frac{1}{N!}\sum_{\sigma}(-1)^{\sigma}\int\prod_i dq_i \langle q_i|\rho|q_{\sigma(i)}\rangle, \tag{6}$$

with the density matrix $\rho = e^{-H}$ defined by

$$\rho = \frac{1}{\sqrt{2\cosh\frac{q}{2}}}\frac{1}{2\cosh\frac{p}{2}}\frac{1}{\sqrt{2\cosh\frac{q}{2}}}. \tag{7}$$

The sum over permutations can be reexpressed as the sum over conjugacy classes. As is standard in the statistical mechanics, it is useful to consider the grand partition function

$$\Xi(z) = 1 + \sum_{N=1}^{\infty} Z(N) z^N = \det(1+z\rho), \tag{8}$$

by introducing the chemical potential. This relation implies that, to compute $Z(N)$, all we have to do is to compute $\operatorname{Tr}\rho^n$ for $n = 1, 2, \cdots, N$. In the following section, we would like to explain how to compute these quantities.

3. Anomaly

The difficulty in computing $\operatorname{Tr}\rho^N$ consists in the $2N$-dimensional integration. Let us try to reduce the dimension of the integrations. Using the canonical commutation relation $[q, p] = 2\pi i$, it is directly found from the Baker-Campbell-Hausdorff formula that

$$[2\cosh\tfrac{q}{2}] \cdot [2\cosh\tfrac{p}{2}] = i \cdot [2\sinh\tfrac{p}{2}] \cdot [2\sinh\tfrac{q}{2}]. \tag{9}$$

However, unfortunately, the naive reciprocal $[2\cosh\frac{p}{2}]^{-1} \cdot [2\cosh\frac{q}{2}]^{-1} = -i \cdot [2\sinh\frac{q}{2}]^{-1} \cdot [2\sinh\frac{p}{2}]^{-1}$, which would reduce the dimension of integrations, is not the case. Instead, comparing the matrix elements of both sides, we find that the anomaly term is controlled well and can be expressed as

$$\frac{1}{2\cosh\frac{p}{2}}\frac{1}{2\cosh\frac{q}{2}} = -i\frac{1}{2\sinh\frac{q}{2}}\Pi\frac{1}{2\sinh\frac{p}{2}}, \tag{10}$$

if we introduce a projection operator

$$\Pi = 1 - |p_{=0}\rangle\langle q_{=0}|. \tag{11}$$

Similarly, we find

$$\frac{1}{2\cosh\frac{q}{2}}\frac{1}{2\cosh\frac{p}{2}} = i\frac{1}{2\sinh\frac{p}{2}}\Pi^\dagger\frac{1}{2\sinh\frac{q}{2}}, \tag{12}$$

if we define $\Pi^\dagger = 1 - |q_{=0}\rangle\langle p_{=0}|$.

Though our original strategy of reducing the dimension of integrations fails, we find an important implication here. Using (12) and (10), we immediately find

$$\operatorname{Tr}\left(\Pi\frac{1}{2\cosh\frac{q}{2}}\frac{1}{2\cosh\frac{p}{2}}\right)^N = i^{2N}\operatorname{Tr}\left(\frac{1}{2\cosh\frac{p}{2}}\frac{1}{2\cosh\frac{q}{2}}\Pi^\dagger\right)^N. \tag{13}$$

Since the quantity $\operatorname{Tr}(\Pi[2\cosh\frac{q}{2}]^{-1}[2\cosh\frac{p}{2}]^{-1})^N$ is real, this is nothing but its hermitian conjugate $\operatorname{Tr}([2\cosh\frac{p}{2}]^{-1}[2\cosh\frac{q}{2}]^{-1}\Pi^\dagger)^N$. This observation implies that

$$\operatorname{Tr}\left(\Pi\frac{1}{2\cosh\frac{q}{2}}\frac{1}{2\cosh\frac{p}{2}}\right)^N = 0, \tag{14}$$

for any odd integer N. After expanding the projection operator (11), we find $\operatorname{Tr}\rho^N = \operatorname{Tr}([2\cosh\frac{q}{2}]^{-1}[2\cosh\frac{p}{2}]^{-1})^N$ is expressed by a sum of products of $\langle q_{=0}|([2\cosh\frac{q}{2}]^{-1}[2\cosh\frac{p}{2}]^{-1})^n|p_{=0}\rangle$.

More precisely, after taking the combinatorial factor into account carefully, we find that

$$\log \Xi(z)\Big|_{\rm odd} = \log G(z)\Big|_{\rm odd}, \tag{15}$$

where $G(z)$ is defined by

$$G(z) = \sum_{n=0}^{\infty} g_n z^n, \quad g_n = \langle q{=}0| \left(\frac{1}{2\cosh\frac{q}{2}} \frac{1}{2\cosh\frac{p}{2}} \right)^n |p{=}0\rangle. \tag{16}$$

Note that (15) only holds for odd powers of z.

What happens to the even powers? In fact, we have found an identity

$$\frac{\det(1+z\rho_-)}{\det(1-z\rho_+)} = G(z), \tag{17}$$

relating two parities of the density matrix $\rho_\pm$. Note that the density matrix $\rho(q_1,q_2) = \langle q_1|\rho|q_2\rangle$, preserving the wave-function parity $\rho(-q_1,-q_2) = \rho(q_1,q_2)$, can be decomposed into two parities as

$$\rho(q_1,q_2) = \rho_+(q_1,q_2) + \rho_-(q_1,q_2), \quad \rho_\pm(q_1,q_2) = \frac{\rho(q_1,q_2) \pm \rho(q_1,-q_2)}{2}. \tag{18}$$

The proof of the relation (17) is given in.[1]

Another structure, which is helpful in computing the partition functions, is a lemma from.[11] The lemma states that, if the density matrix ρ takes the form

$$\rho(q_1,q_2) = \frac{E(q_1)E(q_2)}{M(q_1)+M(q_2)}, \tag{19}$$

ρ^n is simply given by

$$\rho^n(q_1,q_2) = \frac{E(q_1)E(q_2)}{M(q_1)+(-1)^{n-1}M(q_2)} \sum_{m=0}^{\infty} (-1)^m \phi^m(q_1)\phi^{n-1-m}(q_2), \tag{20}$$

where $\phi^m(q)$ is recursively determined by

$$\phi^m(q) = \frac{1}{E(q)} \int \frac{dq}{2\pi} \rho(q,q')E(q')\phi^{m-1}(q'), \quad \phi^0(q) = 1. \tag{21}$$

This relation means that instead of multiplying the density matrices directly we can compute it by convoluting the density matrix to $E(q)$ one by one. After we decompose our density matrix into two parities $\rho_\pm$, the lemma is applicable to both of them.

4. Discussions and Further progress

We have found an interesting identity relating two grand partition functions of the opposite parities (17) and computed the partition functions of the ABJM theory at $k = 1$ up to $N = 9$ (5).

After obtaining the exact values of the partition functions, we can now proceed to studying the non-perturbative effects by subtracting the perturbative expression (2). After separating the known worldsheet instanton effects, we are able to propose an analytical expression for the D2-instanton. See[12] for details.

Also, in[1] we have found a Hankel matrix representation for the density matrix. The identity relating two parities (17) are extended to an identity for a general Hankel matrix in.[12]

Our original project is to compute $Z(N)$ for all N and find the exact energy spectrum of the fermi gas system. Also, as explained in the motivation section, we would like to find some hidden structures of M-theory. We hope to report progress in these directions in the near future.

References

1. Y. Hatsuda, S. Moriyama and K. Okuyama, arXiv:1207.4283 [hep-th].
2. H. Fuji, S. Hirano and S. Moriyama, JHEP **1108**, 001 (2011) [arXiv:1106.4631 [hep-th]].
3. O. Aharony, O. Bergman, D. L. Jafferis and J. Maldacena, JHEP **0810**, 091 (2008) [arXiv:0806.1218 [hep-th]].
4. I. R. Klebanov and A. A. Tseytlin, Nucl. Phys. B **475**, 164 (1996) [hep-th/9604089].
5. A. Kapustin, B. Willett and I. Yaakov, JHEP **1003**, 089 (2010) [arXiv:0909.4559 [hep-th]].
6. N. Drukker and D. Trancanelli, JHEP **1002**, 058 (2010) [arXiv:0912.3006 [hep-th]].
7. N. Drukker, M. Marino and P. Putrov, Commun. Math. Phys. **306**, 511 (2011) [arXiv:1007.3837 [hep-th]].
8. H. Ooguri, C. Vafa and E. P. Verlinde, Lett. Math. Phys. **74**, 311 (2005) [hep-th/0502211].
9. P. Putrov and M. Yamazaki, Mod. Phys. Lett. A **27**, 1250200 (2012) [arXiv:1207.5066 [hep-th]].
10. M. Marino and P. Putrov, J. Stat. Mech. **1203**, P03001 (2012) [arXiv:1110.4066 [hep-th]].
11. C. A. Tracy and H. Widom, Commun. Math. Phys. **179**, 667 (1996) [solv-int/9509003].
12. Y. Hatsuda, S. Moriyama and K. Okuyama, arXiv:1211.1251 [hep-th].

A DOUBLE-WELL SUSY MATRIX MODEL FOR 2D TYPE IIA SUPERSTRINGS*

FUMIHIKO SUGINO

Okayama Institute for Quantum Physics,
Kyoyama 1-9-1, Okayama 700-0015, Japan
E-mail: fumihiko_sugino@pref.okayama.lg.jp

We discuss an interpretation of a simple supersymmetric matrix model with a double-well potential as two-dimensional type IIA superstrings on a nontrivial Ramond-Ramond background. We find direct correspondence between single-trace operators in the matrix model and integrated vertex operators in type IIA theory.

1. Introduction

Solvable matrix models for two-dimensional quantum gravity or noncritical string theory were vigorously investigated around 1990, where a main motivation was to understand nonperturbative effects in string theory. While this approach has been succeeded for bosonic string theory, little has been known for superstring theory, in particular with target-space supersymmetry (SUSY). We would like to consider (solvable) matrix models describing superstring theory possessing target-space SUSY. We hope our analysis helpful to understand nonperturbative dynamics of matrix models of super Yang-Mills type for critical superstring theory.

2. Double-well SUSY matrix model

Ref. 3 discussed a following simple matrix model:

$$S = N\mathrm{tr}\left[\frac{1}{2}B^2 + iB(\phi^2 - \mu^2) + \bar{\psi}(\phi\psi + \psi\phi)\right], \qquad (1)$$

*This is based on the work with Tsunehide Kuroki.[1,2]

where B and ϕ are $N \times N$ hermitian matrices, and ψ and $\bar{\psi}$ are $N \times N$ Grassmann-odd matrices. The action S is invariant under SUSY transformations generated by Q and $\bar{Q}$:

$$Q\phi = \psi, \quad Q\psi = 0, \quad Q\bar{\psi} = -iB, \quad QB = 0, \tag{2}$$

$$\bar{Q}\phi = -\bar{\psi}, \quad \bar{Q}\bar{\psi} = 0, \quad \bar{Q}\psi = -iB, \quad \bar{Q}B = 0, \tag{3}$$

which leads to the nilpotency: $Q^2 = \bar{Q}^2 = 0$. After integrating out B, we have a scalar potential of a double-well shape: $\frac{1}{2}(\phi^2 - \mu^2)^2$. A large-$N$ saddle point equation for the eigenvalue distribution of the matrix ϕ: $\rho(x) \equiv \frac{1}{N}\mathrm{tr}\,\delta(x - \phi)$ reads

$$\int dy\, \rho(y)\, \mathrm{P}\frac{1}{x-y} + \int dy\, \rho(y)\, \mathrm{P}\frac{1}{x+y} = x^3 - \mu^2 x. \tag{4}$$

Its solution with filling fraction (ν_+, ν_-) is given by

$$\rho(x) = \begin{cases} \frac{\nu_+}{\pi}\, x\, \sqrt{(x^2-a^2)(b^2-x^2)} & (a < x < b) \\ \frac{\nu_-}{\pi}\, |x|\, \sqrt{(x^2-a^2)(b^2-x^2)} & (-b < x < -a) \end{cases} \tag{5}$$

with $a = \sqrt{\mu^2 - 2}$ and $b = \sqrt{\mu^2 + 2}$. The filling fractions satisfying $\nu_+ + \nu_- = 1$ indicate that $\nu_+ N$ ($\nu_- N$) eigenvalues are around the right (left) minimum of the double-well. The solution exists for $\mu^2 > 2$. The large-N free energy and $\left\langle \frac{1}{N}\mathrm{tr}\, B^n \right\rangle$ ($n = 1, 2, \cdots$) evaluated at the solution turn out to all vanish. This strongly suggests that the solution preserves SUSY. Thus, we conclude that the SUSY minima are infinitely degenerate and parametrized by (ν_+, ν_-). There exists a solution having the support of a single interval for $\mu^2 < 2$ that breaks SUSY.[4]

In what follows we will compute correlation functions at the saddle point (5) and find new logarithmic critical behavior as $\mu^2 \to 2 + 0$. Based on the result, we will discuss correspondence between the matrix model and two-dimensional type IIA superstring theory on a nontrivial Ramond-Ramond (RR) background. The logarithmic critical behavior is somewhat reminiscent of the $c = 1$ matrix model which is a matrix quantum mechanics of a single matrix variable.[5] The Penner model is known as a zero-dimensional matrix model exhibiting the same critical behavior as the matrix quantum mechanics.[6] It describes noncritical string theory whose target space is (Liouville direction) $\times$ (S^1 with self-dual radius). So, it is expected that our matrix model can be regarded as a SUSY version of the Penner model and describes two-dimensional superstring theory with SUSY on the target space (Liouville direction) $\times$ (S^1 with self-dual radius). Indeed, two-dimensional type II superstring theory with the identical target space is constructed,[7–10] where target space SUSY exist only at the self-dual radius of the circle.

Our matrix model is interpreted as the $O(n)$ model on a random surface with $n = -2$, whose critical behavior is described by the $c = -2$ topological gravity.[11] The partition function after B, ψ and $\bar{\psi}$ are integrated out is expressed as a Gaussian one-matrix model by the Nicolai mapping $H = \phi^2$, where H is integrated over the *positive definite* hermitian matrices, not over all the hermitian matrices. Ref. 12 discusses that the difference of the integration region has only effects which are nonperturbative in $1/N$, and the model can be regarded as the standard Gaussian matrix model at each order of genus expansion.

The Nicolai mapping changes the operators $\frac{1}{N}\mathrm{tr}\,\phi^{2n}$ ($n = 1, 2, \cdots$) to regular operators $\frac{1}{N}\mathrm{tr}\,H^n$. Hence, the behavior of their correlators is expected to be described by the Gaussian one-matrix (the $c = -2$ topological gravity) at least perturbatively in $1/N$. However, the operators $\frac{1}{N}\mathrm{tr}\,\phi^{2n+1}$ ($n = 0, 1, 2, \cdots$) are mapped to $\pm\frac{1}{N}\mathrm{tr}\,H^{n+1/2}$ that are singular at the origin. They are not observables in the $c = -2$ topological gravity, while they are natural observables as well as $\frac{1}{N}\mathrm{tr}\,\phi^{2n}$ in the original setting (1). In the next section, we will see that correlation functions among operators

$$\frac{1}{N}\mathrm{tr}\,\phi^{2n+1}, \qquad \frac{1}{N}\mathrm{tr}\,\psi^{2n+1}, \qquad \frac{1}{N}\mathrm{tr}\,\bar{\psi}^{2n+1} \qquad (n = 0, 1, 2, \cdots) \tag{6}$$

exhibit logarithmic singular behavior of powers of $\ln(\mu^2 - 2)$ at the planar topology.

In considering correspondence of the matrix model to superstring theory, the following observation will be helpful. Suppose ψ and $\bar{\psi}$ are regarded as target-space fermions in the corresponding superstring theory. Namely, ψ is interpreted as an operator in the (NS, R) sector and $\bar{\psi}$ in the (R, NS) sector in the RNS formalism. Then, under the so-called $(-1)^{\mathbf{F}_L}$ and $(-1)^{\mathbf{F}_R}$ transformations which change the signs of operators in the left-moving Ramond sector and those in the right-moving Ramond sector respectively, they transform as

$$(-1)^{\mathbf{F}_L}: \ \psi \to \psi, \qquad \bar{\psi} \to -\bar{\psi}, \tag{7}$$

$$(-1)^{\mathbf{F}_R}: \ \psi \to -\psi, \qquad \bar{\psi} \to \bar{\psi}. \tag{8}$$

In order for the matrix model action (1) to be invariant under the transformations, B and ϕ should transform as

$$(-1)^{\mathbf{F}_L}: \ B \to B, \qquad \phi \to -\phi, \tag{9}$$

$$(-1)^{\mathbf{F}_R}: \ B \to B, \qquad \phi \to -\phi. \tag{10}$$

This indicates that B corresponds to an operator in the (NS, NS) sector, and ϕ in the (R, R) sector.

3. Correlation functions

The planar one-point function $\left\langle \frac{1}{N}\mathrm{tr}\,\phi^n \right\rangle_0$ ($n = 1, 2, \cdots$) are computed as

$$\left\langle \frac{1}{N}\mathrm{tr}\,\phi^n \right\rangle_0 = \int dx\, x^n \rho(x)$$
$$= (\nu_+ + (-1)^n \nu_-)(2+\mu^2)^{n/2} F\left(-\frac{n}{2}, \frac{3}{2}, 3; \frac{4}{2+\mu^2}\right), \quad (11)$$

where the suffix "0" in the left hand side indicates the planar contribution. For n even, the expression is reduced to a polynomial of μ^2 giving nonsingular behavior as expected from the $c = -2$ topological gravity. On the other hand, when n is odd, it exhibits logarithmic singular behavior as $\mu^2 \to 2$:

$$\left\langle \frac{1}{N}\mathrm{tr}\,\phi^{2k+1} \right\rangle_0 \sim (\text{const.})\,(\nu_+ - \nu_-)\,\omega^{k+2} \ln \omega \quad (12)$$

with $\omega \equiv \frac{1}{4}(\mu^2 - 2)$. The symbol "$\sim$" denotes equality up to additive less singular terms.

We also computed two- and three-point functions at the planar level. The result suggests the behavior of general n-point function among operators $\Phi_{2k+1} = \frac{1}{N}\mathrm{tr}\,\phi^{2k+1} + (\text{mixing})$ ($k = 0, 1, 2, \cdots$), where "(mixing)" represents operators introduced to subtract nonuniversal singular parts which are cutoff-dependent and unphysical. The behavior is

$$\left\langle \prod_{i=1}^{n} \Phi_{2k_i+1} \right\rangle_{C,0} \sim (\text{const.})\,(\nu_+ - \nu_-)^n \omega^{2-\gamma+\sum_{i=1}^{n}(k_i-1)} (\ln \omega)^n. \quad (13)$$

The suffix "C" means taking the connected correlator. Here, we have $\gamma = -1$ which is the same as the string susceptibility of the $c = -2$ topological gravity. The gravitational scaling dimension of Φ_{2k+1} is seen to be k besides the logarithmic factor $(\ln \omega)^n$.

Finally, we computed planar two-point functions of fermionic operators

$$\Psi_{2k+1} = \frac{1}{N}\mathrm{tr}\,\psi^{2k+1} + (\text{mixing}), \quad \bar{\Psi}_{2k+1} = \frac{1}{N}\mathrm{tr}\,\bar{\psi}^{2k+1} + (\text{mixing}). \quad (14)$$

The result

$$\left\langle \Psi_{2k+1} \bar{\Psi}_{2\ell+1} \right\rangle_{C,0} \sim \delta_{k,\ell}(\text{const.})\,(\nu_+ - \nu_-)^{2k+1} \omega^{2k+1} \ln \omega \quad (15)$$

tells us that Ψ_{2k+1} and $\bar{\Psi}_{2k+1}$ have the gravitational scaling dimension k besides the logarithmic factor, which is the same as Φ_{2k+1} .

4. 2D type IIA superstring

The type II superstring theory discussed in Refs. 7–9 has the target space $(\varphi, x) \in$ (Liouville direction) $\times$ (S^1 with self-dual radius). The holomorphic energy-momentum tensor on the string world-sheet is

$$T = -\frac{1}{2}(\partial x)^2 - \frac{1}{2}\psi_x \partial \psi_x - \frac{1}{2}(\partial \varphi)^2 + \frac{Q}{2}\partial^2 \varphi - \frac{1}{2}\psi_\ell \partial \psi_\ell \tag{16}$$

with $Q = 2$, excepting ghosts' part. ψ_x and ψ_ℓ are superpartners of x and φ, respectively. Target-space supercurrents in the type IIA theory

$$q_+(z) = e^{-\frac{1}{2}\phi(z) - \frac{i}{2}H(z) - ix(z)}, \qquad \bar{q}_-(\bar{z}) = e^{-\frac{1}{2}\bar{\phi}(\bar{z}) + \frac{i}{2}\bar{H}(\bar{z}) + i\bar{x}(\bar{z})} \tag{17}$$

exist only for the S^1 target space of the self-dual radius. ϕ ($\bar{\phi}$) is the holomorphic (anti-holomorphic) bosonized superconformal ghost, and the fermions are bosonized as $\psi_\ell \pm i\psi_x = \sqrt{2}\, e^{\mp iH}$, $\bar{\psi}_\ell \pm i\bar{\psi}_x = \sqrt{2}\, e^{\mp i\bar{H}}$. Then the supercharges

$$Q_+ = \oint \frac{dz}{2\pi i}\, q_+(z), \qquad \bar{Q}_- = \oint \frac{d\bar{z}}{2\pi i}\, \bar{q}_-(\bar{z}) \tag{18}$$

are nilpotent: $Q_+^2 = \bar{Q}_-^2 = 0$, which indeed matches the supercharges Q and $\bar{Q}$ in the matrix model.

The NS sector vertex operator in the (-1)-picture is $T_k = e^{-\phi + ikx + p_\ell \varphi}$, and the R sector one in the $\left(-\frac{1}{2}\right)$-picture is $V_{k,\epsilon} = e^{-\frac{1}{2}\phi + \frac{i}{2}\epsilon H + ikx + p_\ell \varphi}$ with $\epsilon = \pm 1$. Locality with supercurrents, mutual locality, superconformal invariance and level matching condition determine physical vertex operators. The physical spectrum in the "winding background"[9] is given by

$$\begin{aligned}
&(\text{NS, NS}): && T_k \bar{T}_{-k} && (k \in \mathbf{Z} + \tfrac{1}{2}),\\
&(\text{R+, R}-): && V_{k,+1} \bar{V}_{-k,-1} && (k = \tfrac{1}{2}, \tfrac{3}{2}, \cdots),\\
&(\text{R}-, \text{R+}): && V_{-k,-1} \bar{V}_{k,+1} && (k = 0, 1, 2, \cdots),\\
&(\text{NS, R}-): && T_{-k} \bar{V}_{-k,-1} && (k = \tfrac{1}{2}, \tfrac{3}{2}, \cdots),\\
&(\text{R+, NS}): && V_{k,+1} \bar{T}_k && (k = \tfrac{1}{2}, \tfrac{3}{2}, \cdots),
\end{aligned} \tag{19}$$

where we take a branch of $p_\ell = 1 - |k|$ satisfying the locality bound.[13]

5. Correspondence between the matrix model and the type IIA theory

If we identify supercharges between the matrix model and the type IIA theory as $(Q, \bar{Q}) \Leftrightarrow (Q_+, \bar{Q}_-)$, single-trace operators in the matrix model naturally correspond to integrated vertex operators in the type IIA theory:

$$\Phi_{2k+1} \Leftrightarrow \int d^2z\, V_{k+\frac{1}{2},+1}\,\bar{V}_{-k-\frac{1}{2},-1},$$
$$\Psi_{2k+1} \Leftrightarrow \int d^2z\, T_{-k-\frac{1}{2}}\,\bar{V}_{-k-\frac{1}{2},-1},$$
$$\bar{\Psi}_{2k+1} \Leftrightarrow \int d^2z\, V_{k+\frac{1}{2},+1}\,\bar{T}_{k+\frac{1}{2}}. \tag{20}$$

Also, $\int d^2z\, T_{-k-\frac{1}{2}}\,\bar{T}_{k+\frac{1}{2}}$ are expected to be mapped to $\frac{1}{N}\mathrm{tr}\,(-iB)^{k+1}$ ($k = 0, 1, 2, \cdots$) perhaps with some mixing terms. It is consistent to the discussion in (7)-(10).

Note that (R−, R+) operators are singlets under the target-space SUSYs $Q_+, \bar{Q}_-$, and appear to have no counterpart in the matrix model side. Since the expectation value of operators carrying the nonzero RR charge $\langle\Phi_{2k+1}\rangle_0$ does not vanish as seen in (12), the matrix model is considered to correspond to the type IIA theory on a background of the (R−, R+) fields. We may introduce the (R−, R+) background in the form of vertex operators, when the strength of the background $(\nu_+ - \nu_-)$ is small. It turns out that computation of amplitudes in the type IIA theory gives consistent result to the correspondence (20).

Acknowledgments

The author is grateful to the organizers, especially Chengming Bai, for the invitation to the wonderful meeting and to the members of Chern Institute of Mathematics for warm hospitality.

References

1. T. Kuroki and F. Sugino, to appear in *Nucl. Phys. B*, arXiv:1208.3263 [hep-th].
2. T. Kuroki and F. Sugino, in preparation.
3. T. Kuroki and F. Sugino, *Nucl. Phys. B* **830**, 434 (2010).
4. T. Kuroki and F. Sugino, *Nucl. Phys. B* **844**, 409 (2011).
5. For a review, see I. R. Klebanov, arXiv:hep-th/9108019.
6. J. Distler and C. Vafa, *Mod. Phys. Lett. A* **6** 259 (1991).
7. D. Kutasov and N. Seiberg, *Phys. Lett. B* **251**, 67 (1990).
8. S. Murthy, *JHEP* **0311**, 056 (2003).
9. H. Ita, H. Nieder and Y. Oz, *JHEP* **0506**, 055 (2005).
10. P. A. Grassi and Y. Oz, arXiv:hep-th/0507168.
11. I. K. Kostov and M. Staudacher, *Nucl. Phys. B* **384**, 459 (1992).
12. D. Gaiotto, L. Rastelli and T. Takayanagi, *JHEP* **0505**, 029 (2005).
13. N. Seiberg, *Prog. Theor. Phys. Suppl.* **102**, 319 (1990).

PARALLEL SESSION
Loop Quantum Gravity

20 & 22 & 23 AUGUST 2012

LQG COSMOLOGY FROM THE FULL LQG

EMANUELE ALESCI*

Institute for Quantum Gravity, FAU Erlangen-Nürnberg, Staudtstr. 7, D-91058 Erlangen, Germany, EU.
Instytut Fizyki Teoretycznej, Uniwersytet Warszawski,ul. Hoża 69, 00-681 Warszawa, Poland, EU

FRANCESCO CIANFRANI

Institute for Quantum Gravity, FAU Erlangen-Nürnberg, Staudtstr. 7, D-91058 Erlangen, Germany, EU.

We present a new perspective on early cosmology based on Loop Quantum Gravity. We use projected spinnetworks, coherent states and spinfoam tecniques, to implement a quantum reduction of the full Kinematical Hilbert space of LQG, suitable to descrive inhomogeneous cosmological models. Some preliminar results on the solutions of the Scalar constraint of the reduced theory are also presented.

Keywords: Loop Quantum Gravity; Quantum Cosmology.

1. Introduction

The cosmological implementation of Loop Quantum Gravity (LQG) has been realized on a canonical level in Loop Quantum Cosmology (LQC)[1,2] and in a covariant way in Spin-Foam Cosmology[3] (see also Ref.[4]). In particular, the former is based on minisuperspace quantization, in which at first a phase-space reduction occurs thanks to the symmetry of a cosmological space-time and, then, the resulting system is quantized. This way, a complete description of the early Universe evolution is achieved, with a bounce generically replacing the initial singularity.

In this work, we discuss an alternative procedure to infer a quantum model for the early Universe, by reversing the ordering of "reduction" and "quantization". The reason for this choice is that it will provide us a description of a quantum geometrical model for the Universe. Within this scheme, all the techniques adopted in LQG can be applied and their consequences tested in a simplified framework. Among them, the most significant income from the full theory is the expression of the superHamiltonian operator, which can be self-consistently regularized as in Ref.[5]

In this respect, we consider an inhomogeneous extension of a Bianchi I model, in which reduced variables retain a dependence from spatial coordinates. This model can be seen on a classical level as an arbitrary reparametrization of the homogeneous phase or as an approximate description of the generic cosmological solution at each fixed Kasner epoch in the Mixmaster phase.[6] We will define a projection of the kinematical Hilbert space of LQG to a certain space in which: 1) all edges are parallel to the fiducial vectors associated with the Bianchi I model (cubical cell structure), 2) a gauge fixing of the internal SU(2) symmetry takes place. The condition 1) will break the invariance under full diffeomorphisms down to the transformations reflecting the cubical cell structure (reduced diffeomorphisms), while 2) will be implemented mimicking the procedure to solve the simplicity constraints in Spin-Foam models.[7] The latter represents the key point of our derivation, since by using projected[8] $U(1)$-networks all SU(2) group elements along an edge will be restricted to the $U(1)$ subgroup stabilizing some internal directions and reduced (coherent[9]) intertwiners will be predicted at nodes. At the end, geometrical operators (in particular the volume!) are diagonal and the expression of the superHamiltonian operator, adapted to the reduced cubical cell structure, can be explicitly computed. Furthermore, the regulator can be removed in the resulting expression by considering the action on s-knots[10] associated with reduced diffeomorphisms.

2. Inhomogeneous Bianchi I model

Let us consider the following form for the densitized triads E^a_i and Ashtekar-Barbero connections A^i_a in gravity phase space

$$E^a_i = p_i(t,x)\delta^a_i, \qquad A^i_a(t,x) = c_i(t,x)\delta^i_a. \tag{1}$$

These expressions can be obtained from the inhomogeneous extension of the Bianchi line element, *i.e.*

$$ds^2 = N^2(t,x)dt^2 - a_1^2(t,x)(dx^1)^2 - a_2^2(t,x)(dx^2)^2 - a_3^2(t,x)(dx^3)^2, \tag{2}$$

x^i being Cartesian coordinates on the (flat) spatial manifold, if each a_i is a function of the corresponding coordinate x^i only (case A), or can neglect the spatial gradients of the metric components with respect to the time derivatives (case B). The first case corresponds to a reparametrized Bianchi I homogeneous model, while the second can be realized approximatively during each Kasner describing the Mixmaster phase of the generic cosmological solution. After the restriction to the cases A) or B) the form (1) is uniquely determined only if one imposes the SU(2) gauge-fixing[11]

$$\chi_i = \epsilon_{ij}{}^{k} E^a_k \delta^j_a = 0. \tag{3}$$

We want to infer from the kinematical Hilbert space of the full theory $\mathcal{H}^{kin}$ a proper subspace describing the quantization of the reduced phase space (1). This will be done by a two step procedure.

Reduced diffeomorphisms - At first we implement the restriction to the graphs with edges e_i parallel to fiducial vectors $\omega_i = \partial_i$. This way, one singles out only those holonomies capturing the degrees of freedom associated with reduced variables $\{c_i, p^j\}$. The mathematical way to implement such a restriction is by defining a projector P which is nonvanishing only when acting on the holonomies h_{e_i} along the edges e_i, *i.e.*

$$Ph_e = \begin{cases} h_e & e = e_i \\ 0 & \text{otherwise} \end{cases}. \tag{4}$$

The image of P is a subspace $\mathcal{H}_P$ of the whole $\mathcal{H}^{kin}$, in which only a restricted class of diffeomorphisms are implemented (reduced diffeomorphisms). In fact, by projecting down the action $U(\varphi_\xi)$ of a generic diffeomorphisms φ_ξ, ξ being the infinitesimal parameters, one finds

$$^{red}U(\varphi_\xi) = PU(\varphi_\xi)P = \begin{cases} U(\varphi_\xi) & \xi^i = \xi^i(x^i) \\ 0 & \text{otherwise} \end{cases} \tag{5}$$

The set of reduced diffeomorphisms is thus made of all the transformations obtained by a generic diffeomorphisms along a given direction i followed by rigid translations along the remaing directions $j \neq i$.

SU(2) gauge fixing - In order to implement the gauge fixing (3), we are now going to adopt several techniques from Spin-Foam models. This is the case because we need to realize a condition which makes the whole system of constraints second-class, just like the simplicity constraints in BF theories.[7]

Let us define some reduced holonomies $\tilde{h}_{e_i}$ which belong to the $U(1)_i$ subgroups ($i = 1, 2, 3$) obtained by stabilizing the $SU(2)$ group around the internal directions $(1, 0, 0)$, $(0, 1, 0)$ and $(0, 0, 1)$, respectively. These objects are the classical holonomies obtained from the phase space restriction (1) and they can be seen as SU(2) group elements completely determined by their restrictions to the $U(1)_i$ subgroups. Hence, we define the associated wave functions $\tilde{\psi}$ by lifting the functions $\psi(h_i)$ of (abstract) $U(1)_i$ group elements h_i to those of the $SU(2)$ elements g via the Dupuis-Livine map:[8]

$$\tilde{\psi}(g) = \int_{U(1)} dh_i \; K(g,h_i)\psi(h_i), \qquad g \in SU(2), \tag{6}$$

in which the Kernel $K(g,h_i)$ is defined in terms $SU(2)$ and $U(1)_i$ characters, $\chi^{j(n)}(g)$ and $\chi^n(h_i)$, respectively, as follows

$$K(g,h_i) = \sum_n \int_{U(1)_i} dk_i \; \chi^{j(n)}(gk_i)\chi^n(k_i h_i), \qquad k_i \in U(1)_i. \tag{7}$$

The expression (6) can be evaluated by expanding ψ in the Peter-Weyl base for $U(1)_i$, so finding

$$\tilde{\psi}(g)_{e_i} = \sum_{n_i} {}^iD^{j(n_i)}_{m=n_i\, r=n_i}(g)\psi^{n_i}_{e_i}, \tag{8}$$

ψ^{n_i} being the coefficients of the expansion, while ${}^iD^{j(n_i)}_{mr}$ denote the Wigner matrices in the spin base $|j,m\rangle_i$ that diagonalize the operators J^2 and J_i.

We can now look for a solution of the condition (3). Let us define the fluxes $E_i(S^j)$ across the surfaces S^j dual to e_j and consider the master constraint $\chi^2 = \sum_i(\chi_i)^2 = 0$. By imposing the vanishing of the operator χ^2 one selects the degree of the representation $j(n_i)$. In fact in the limit $j \to +\infty$

$$\chi^2\tilde{\psi}(g)_{e_i} = 0 \to j(n_i) = |n_i|, \tag{9}$$

and the associated wave functions are given by

$$\tilde{\psi}(g)_{e_i} = \sum_{n_i} {}^iD^{j=|n_i|}_{n_i\, n_i}(g)\psi^{n_i}_{e_i} = \sum_j {}^iD^{|j|}_{jj}(g)\psi^j_{e_i}. \tag{10}$$

We denotes $\mathcal{H}^R$ the subspace of $\mathcal{H}_P$ containing the elements (10). In $\mathcal{H}^R$ the condition (3) holds weakly in view of the following relation

$$E_k(S^l)\tilde{\psi}(g)_{e_i} = 8\pi\gamma l_P^2 \delta^l_k \delta^k_i \sum_j j\, {}^iD^j_{jj}(g)\psi^j_{e_i}. \tag{11}$$

For later convenience, we also define the projector P_χ from $\mathcal{H}_P$ to $\mathcal{H}^R$.

3. Reduced recoupling theory

By combining P and P_χ one can restrict $\mathcal{H}^{kin}$ to the subaspace $\mathcal{H}^R$ describing the quantum system associated with the reduced phase space (1). This total projection allow us to discuss the implications of the original SU(2) gauge invariance in $\mathcal{H}^R$. This can be done by considering in $\mathcal{H}^{kin}$

the projector $P_{\mathcal{G}}$ to the SU(2)-invariant subspace ${}^{\mathcal{G}}\mathcal{H}^{kin}$ and by projecting it down, so getting the following projector in H^R

$$P_{\mathcal{G},\chi} = P^{\dagger}_{\chi}\, P_{\mathcal{G}} P_{\chi}. \tag{12}$$

This object acts at nodes and it selects the gauge invariant reduced Hilbert space ${}^{\mathcal{G}}\mathcal{H}^R$ whose basis states read

$$\langle h|\Gamma, j_e, x_v\rangle_R = \prod_{v\in\Gamma}\prod_{e\in\Gamma} \langle \mathbf{j_i}, \mathbf{x}|\mathbf{j_i}, \vec{\mathbf{e}}_i\rangle \cdot {}^{i}D^{j_{e_i}}_{j_{e_i} j_{e_i}}(h_{e_i}), \tag{13}$$

where $|\mathbf{j_i}, \mathbf{x}\rangle$ is the usual intetwiner base, while $|\mathbf{j_i}, \vec{\mathbf{e}}_i\rangle$ denotes Livine-Speziale coherent intertwiners[9] with normals $\vec{\mathbf{e}}_i$ tangent to the cubical cell. The coefficients are the intertwiners of the reduced model (reduced intertwiners) and they constitute the main new feature of our approach.

4. SuperHamiltonian

The superHamiltonian operator can be regularized directly in ${}^{\mathcal{G}}\mathcal{H}^R$, by repeating Thiemann construction.[5] In this respect, let us consider a cubulation C of the spatial manifold adapted to the cubical graph Γ on which wave functions are defined. For each node v and each pair of links e_i and e_j incident at v let us take a semi-analytic arcs a_{ij} whose end points s_{e_i}, s_{e_j} are interior points of e_i, e_j, respectively, and $a_{ij} \cap \Gamma = \{s_{e_i}, s_{e_j}\}$. Given the segments s_i (s_j) of e_i (e_j) from v to $s_i(s_j)$, a_{ij} is chosen such that $\alpha_{ij} := s_i \circ a_{ij} \circ s_j^{-1}$ is a rectangle. C is defined by repeating this construction for all e_i and e_j at v and for all $v \in \Gamma$. Finally, the super-Hamiltonian operator can be inferred from the expression of the full theory[5] by replacing holonomies and the volume operator with the reduced one $\tilde{h}^{(m)}$ and ${}^{R}V$:

$${}^{R}H^{m}_{\square}[N] := \frac{N(v)}{N_m^2}\, \epsilon^{ijk}\, \mathrm{Tr}\Big[\tilde{h}^{(m)}_{\alpha_{ij}} \tilde{h}^{(m)}_{s_k} \{\tilde{h}^{(m)-1}_{s_k}, {}^{R}V\}\Big], \tag{14}$$

in which the holonomies are taken in an irreducible representation determined by the index (m), while N_m is a normalizing factor. Now as in the full theory, the action of the superHamiltonian operator can be defined on the (reduced) diffeomorphisms invariant class of knots, where the regulator can be safely removed, but in this case the matrix element of (14) can be explicitly computed since the reduced volume, which is defined from reduced fluxes $E_i(S^i)$ (11), turns out to be diagonal on reduced basis elements.

5. Conclusions

We presented an innovative description for the cosmological sector of LQG based on a reduction of the Hilbert space of the full theory down to a proper space describing a certain inhomogeneous extension of the Bianchi I model. The resulting quantum space has been described in terms of a cell structure with cubic topology and of U(1) quantum numbers associated with quanta of area. A key-point of the procedure we defined was the emergence of non trivial intertwiners associated with the original SU(2) invariance. Having at our disposal a complete model for a quantum cosmological space-time we could apply all the techniques of LQG and, in particular, the regularization of the superHamiltonian could be accomplished. We also pointed out the possibility to infer an explicit expression for the matrix elements of the superHamiltonian operator. This scheme opens the perspective to furnish a self-consistent picture for the cosmological sector of LQG and to link the polymer parameter responsible for the bounce in LQC with a concrete realization of a discrete quantum geometry. Moreover, there is space left in our model to discuss the behavior of inhomogeneities in the limit of the approximation proper of the Belinski-Lipschitz-Kalatnikov conjecture.

Acknowledgment

F.C. thanks "Angelo Della Riccia" foundation for its financial support. This work was partially supported by the grant of Polish Narodowe Centrum Nauki nr 2011/02/A/ST2/00300.

References

1. Martin Bojowald, "Quantum cosmology: A fundamental description of the universe", *Lecture notes in physics*, Vol. 835 (Springer, New York, 2011).
2. A. Ashtekar, P. Singh, *Class. Quant. Grav.*, **28**, 213001 (2011).
3. Eugenio Bianchi, Thomas Krajewski, Carlo Rovelli, Francesca Vidotto, *Phys. Rev. D*, **83**, 104015 (2011).
4. E.F. Borja, J. Diaz-Polo, I. Garay and E. R. Livine, *Class. Quant. Grav.*, **27**, 235010 (2010).
5. T. Thiemann, *Class. Quant. Grav.*, **15**, (1998) 839.
6. V.A. Belinski, I.M. Khalatnikov, E.M. Lifshitz, *Adv. Phys.*, **19**, 525(1970); *Adv. Phys.*, **31**, 639(1982).
7. J. Engle, E. Livine, R. Pereira and C. Rovelli, *Nucl. Phys. B*, **799**, 136 (2008).
8. Mait Dupuis, Etera R. Livine, *Phys. Rev. D*, **82**, (2010) 064044.
9. E.R. Livine, S. Speziale, *Phys. Rev. D*, **76** (2007) 084028.
10. A. Ashtekar, J. Lewandowski, D. Marolf, J. Mourao and T. Thiemann, J. Math. Phys. **36**, 6456 (1995).
11. F. Cianfrani, G. Montani, *Phys. Rev. D*, **85**, (2012) 024027; F. Cianfrani, A. Marchini, G. Montani, *Europhys. Lett.*, **99**, (2012) 10003.

SPINFOAMS: SIMPLICITY CONSTRAINTS AND CORRELATION FUNCTIONS

YOU DING*

Department of Physics, Beijing Jiaotong University
Beijing, 100044, China
**E-mail: dingyou@bjtu.edu.cn*

The topic of this proceeding is some themes of spinfoam formalism, covariant formalism of loop quantum gravity. The focus is on the imposing of simplicity constraint and the computation of the Lorentzian correlation functions. The former is to find the way to connect different aspects related to the Engle-Pereira-Rovelli-Livine spinfoam model, and the latter is to test the resulting model and try to extract physics from that.

Keywords: Spinfoam formalism; Simplicity constraints; Correlation functions.

1. Introduction

Searching for a quantum theory of gravity is the major challenge in today's fundamental physics, which will merge the conceptual novelties of general relativity (GR) and quantum mechanics (QM). To find this consistent picture, if one takes the central lesson of GR seriously: gravity is geometry, it leads to the direction of research investigated by background independent approaches to quantum gravity. The *conceptual* framework of background-independent quantum gravity is as follows. On the kinematical level, we need a space of diffeomorphism-invariant states $\Psi[\phi^* q_{ab}] = \Psi[q_{ab}]$[a] (where q_{ab} is the space metric), and the scalar product of this space is diffeomorphism-invariant as well. On the dynamical level, according to the Lagrangian and the Hamiltonian versions in classical GR, we need two corresponding quantizations: Wheeler-DeWitt equation $\widehat{H}\Psi[q_{ab}] = 0$ in canonical version, Misner-Hawking path integral $Z = \int \delta g_{\mu\nu} \exp(\frac{i}{\hbar} S_{\text{gravity}[g_{\mu\nu}]})$

[a]This equation, as well as the next Wheeler-Dewitt equation, and also the Misner-Hawking path integral, are only formal, since the scalar product and the integral measure are ill-defined.

in covariant version, and the relation between the two versions. This is the main conceptual framework for background-independent quantum gravity.

Loop quantum gravity provides a mathematical description for this conceptual framework. In the canonical quantization approach, the Hilbert space of kinematical states and the field operators that act on it are obtained from classical GR following a rather standard canonical quantization strategy, which provide a mathematically well-defined and physically compelling description of the kinematics of quantum gravity. However, the Hilbert space of dynamical states, where the Hamiltonian solves, is not easy to obtain because of the non-polynomial structure of the current Hamiltonian operator.

This problem can be solved by spin-foam formalism, which circumvents the complexity of the canonical LQG Hamiltonian operator: the dynamics can be expressed by a vertex amplitude of a spin-foam. The topic of this proceeding is some themes of spinfoam formalism. The focus is on the imposing of simplicity constraint and the computation of the Lorentzian correlation function. The former is to find the way to connect different aspects related to the Engle-Pereira-Rovelli-Livine (EPRL) spinfoam model, and the latter is to test the resulting model and try to extract physics from that.

2. Simplicity constraints

There are several approaches to derive spinfoam formalism; here we will consider it as a constrained BF theory. In classical level, the Platini action of GR can be considered as the constrained BF action: i.e. $S_{\mathrm{BF}} = \int \mathrm{Tr}(B \wedge F)$ plus the simplicity constraints ${}^*B = e \wedge e$, where F is the curvature of a connection, and B is a Lie algebra valued 2-form. Motivated by this result, on the quantum level, one can derive a spinfoam model by imposing the simplicity constraints on the boundary states for the BF vertex amplitude.

To give sense to the "path-integral", we discretize it on an oriented 2-complex $\mathcal{K} := (V(\mathcal{K}), E(\mathcal{K}), F(\mathcal{K}))$ which consists of sets of *vertices* $v \in V(\mathcal{K})$, *edges* $e \in E(\mathcal{K})$ and *faces* $f \in F(\mathcal{K})$, equipped with a *boundary relation* ∂ associating an ordered pair of vertices $(s(e), t(e))$ ("source" and "target") to each edge e and a finite sequence of edges $\{e_k^{\epsilon_{e_k f}}\}_{k=1,\dots,n}$ to each face f. The boundary graph γ_v of a single vertex v is the graph whose nodes are the edges e in ∂v and whose links are the faces f in ∂v. We now express the BF partition function on the 2-complex:

$$Z_{BF}(\mathcal{K}) = \int dg_{fv} \prod_{f} \delta(\prod_{v\in\partial f} g_{fv}) \prod_{v} A_v(g_{fv}) \,, \tag{1}$$

where g_{fv} is the group element associated to the couple (f, v) and the vertex amplitude $A_v(g_f)$ is defined by

$$A_v(g_f) = \int \prod_{e\in\partial v} dg_e \prod_{f\in\partial v} \delta(g_{e_f} g_f g_{e'_f}^{-1}) \,. \tag{2}$$

We now introduce the simplicity constraints. In classical discretized form, the simplicity constraints can be expressed as: (i) the linearized simplicity constraint: there exists a unit vector n_I for each node e, s.t., $\forall f \in \partial e$ $C_f^J = n_I {}^*\Sigma_f^{IJ} = 0$ and (ii) the closure constraint: $G^{IJ} = \sum_{f\in\partial e} \Sigma_f^{IJ} = 0$, with $B_f = {}^*\Sigma + \frac{1}{\gamma}\Sigma_f$.

The "time" gauge where $n_I = (0, 0, 0, 1)$ gives the fixed constraints

$$C_f^i = \gamma\, L_f^i - sK_f^i = 0\,, \quad G_L^i = \sum_{f\in\partial e} L_f^i = 0\,, \quad G_K^i = \sum_{f\in\partial e} K_f^i = 0\,, \tag{3}$$

with signature $s = +1$ for Euclidean case and $s = -1$ for Lorentzian case, $L_f^j := \frac{1}{2}\epsilon^j{}_{kl} B_f^{kl}$, $K_f^j := B_f^{0j}$. Thus the simplicity constraints can be quantized by replacing the variables L and K with rotation and boost generators of Lie algebra respectively.

To solve the simplicity constraint operators, one uses *weak* solution in the sense of matrix elements. γ-simple subspace solves weakly the linearized simplicity constraint in the sense that

$$\langle\widetilde{\Phi}|C^i|\widetilde{\Psi}\rangle = 0 \quad \forall\widetilde{\Phi}, \widetilde{\Psi} \in \widetilde{\mathcal{H}}_{\gamma_v} \,. \tag{4}$$

One can find details and construction of γ-simple subspace in.[1–3] Gauge-invariant γ-simple subspace solves the closure constraint

$$\langle\Phi|G_L^i|\Psi\rangle = \langle\Phi|G_K^i|\Psi\rangle = 0 \quad \forall\Phi, \Psi \in \mathcal{H}_{\gamma_v} \equiv \mathrm{Inv}_{SU(2)}\Big[\widetilde{\mathcal{H}}_{\gamma_v}\Big] \,. \tag{5}$$

The projection from BF boundary to the constrained boundary results in the EPRL vertex.[2,3]

3. Lorentzian correlation functions

To test the semiclassical behavior of the EPRL spinfoam amplitude,[4] we compute the two-point correlation functions for the Penrose metric operator in the Lorentzian signature.[5–14] In particular, the correlation functions are calculated on a *time-oriented* semiclassical state,[15] which is peaked on the space-like boundary geometry of a Lorentzian 4-simplex.

The boundary of a Lorentzian 4-simplex consists of five space-like tetrahedra a which meet at ten triangles (ab) $(a, b = 1...5$ and $a < b)$. Suppose all the time-like normals to the tetrahedra are outward-pointing, wedges are then classified into two types: it is called in[16,17] *thick wedge* if the incident tetrahedra are of same pointing type, which means both future-pointing or both past-pointing, otherwise called *thin wedge*. Let us assign a quantity to the triangle (ab) to denote this classification:

$$\Pi_{ab} = \begin{cases} 0, \text{ thick wedge} \\ \pi, \text{ thin wedge} \end{cases} . \tag{6}$$

Consider the complete graph Γ dual to the boundary of the 4-simplex, the basis of $\mathcal{H}_\Gamma$ can be chosen as *Lorentzian coherent spin network states*[15] with nodes labeled by *Lorentzian Livine-Speziale coherent intertwiners*:[15]

$$|\Gamma, j_{ab}, \Upsilon_a(\vec{n})\rangle = \exp\big(-i \sum_{(a<b)} \Pi_{ab} j_{ab}\big) |\Gamma, j_{ab}, \Phi_a(\vec{n})\rangle \, . \tag{7}$$

Now let us come to the Lorentzian semiclassical state superposed by (to ???) the Lorentzian coherent spin network states. Let ϕ_o label the simplicial extrinsic curvature, which is an angle associated to the triangle shared by the tetrahedra; a Lorentzian semiclassical state peaked both on intrinsic and extrinsic geometry can be given by a superposition of Lorentzian coherent spin network states: $|\Psi_o\rangle = \sum_{j_{ab}} \psi_{j_o,\phi_o}(j) |j, \Upsilon(\vec{n})\rangle$, with coefficients $\psi_{j_o,\phi_o}(j)$ given by a gaussian times a phase.[18]

The connected two-point correlation function G^{abcd}_{nm} on a semiclassical boundary state $|\Psi_o\rangle$ is defined as

$$G^{abcd}_{nm} = \langle E^a_n \cdot E^b_n \; E^c_m \cdot E^d_m \rangle - \langle E^a_n \cdot E^b_n \rangle \, \langle E^c_m \cdot E^d_m \rangle \, , \tag{8}$$

where $(E^a_n)_i$ is a flux operator through a surface f_{an} dual to the triangle between the tetrahedra a and n, parallel transported in the tetrahedron n. Here the dynamical expectation value of an operator $\mathcal{O}$ on the semiclassical state $|\Psi_o\rangle$ is defined via $\langle \mathcal{O} \rangle = \frac{\langle W|\mathcal{O}|\Psi_o\rangle}{\langle W|\Psi_o\rangle}$, where W is the EPRL spinfoam amplitude.[4] The correlation function (8) can be reexpressed in terms of group integral:[15]

$$G^{abcd}_{nm} = \frac{\sum_j \psi_j \int \mathrm{d}^4 g \, \mathrm{d}^{10} z \; q^{ab}_n q^{cd}_m e^S}{\sum_j \psi_j \int \mathrm{d}^4 g \, \mathrm{d}^{10} z \; e^S} - \frac{\sum_j \psi_j \int \mathrm{d}^4 g \, \mathrm{d}^{10} z \; q^{ab}_n e^S}{\sum_j \psi_j \int \mathrm{d}^4 g \, \mathrm{d}^{10} z \; e^S} \, \frac{\sum_j \psi_j \int \mathrm{d}^4 g \, \mathrm{d}^{10} z \; q^{cd}_m e^S}{\sum_j \psi_j \int \mathrm{d}^4 g \, \mathrm{d}^{10} z \; e^S} . \tag{9}$$

Notation is as follows. The coefficient ψ_j stands for $\psi_{j_o,\phi_o}(j)$. d^4g is the Haar measure of $SL(2,\mathbb{C})^4$, one per each tetrahedron, and one redundant integral is removed to obtain the finite integral.[19] We consider an irreducible representation of the Lorentz group $SL(2,\mathbb{C})$ on the space $\mathcal{H}_{(k,p)}$ of homogeneous functions $f^j_m(z)^{(k,p)}$ on *complex projective line* $\mathcal{P}$, and the integral $d^{10}z$ is over one $\mathcal{P}$ per each triangle, up to a factor $-(\langle Z_{ab}, Z_{ab}\rangle\langle Z_{ba}, Z_{ba}\rangle)^{-1}$, with $Z_{ab} = g_a^\dagger z$ and $Z_{ba} = g_b^\dagger z$. The "action" S is given by

$$S(g,z) = \sum_{(a<b)} \left(j_{ab} \log \frac{\langle J\xi_{ab}, Z_{ab}\rangle^2 \langle Z_{ba}, \xi_{ba}\rangle^2}{\langle Z_{ab}, Z_{ab}\rangle\langle Z_{ba}, Z_{ba}\rangle} + i\gamma j_{ab} \log \frac{\langle Z_{ba}, Z_{ba}\rangle}{\langle Z_{ab}, Z_{ab}\rangle} - i\Pi_{ab} j_{ab} \right),$$

with γ the Barbero-Immirzi parameter; the spinor $\xi \in \mathbb{C}^2$ is related to the coherent state $|j, \vec{n}\rangle$ in the sense that the corresponding $SU(2)$ group element $n(\xi)$ rotate the direction $(0,0,1)$ in sphere $\mathbb{S}^2$ into the $\vec{n}$ direction. The antipodal vector $-\vec{n}(\xi)$ can be associated to $J\xi$, i.e. $-\vec{n}(\xi) = \vec{n}(J\xi)$. the insertion q_n^{ab} is given by

$$q_n^{ab} = \gamma^2 j_{an} j_{bn} \frac{\langle \vec{\sigma} Z_{an}, \xi_{an}\rangle}{\langle Z_{an}, \xi_{an}\rangle} \cdot \frac{\langle \vec{\sigma} Z_{bn}, \xi_{bn}\rangle}{\langle Z_{bn}, \xi_{bn}\rangle}. \tag{10}$$

The large-spin asymptotics of the correlation function (9) can be obtained via stationary phase approximation,[13,17] which is given by (for details, see[15]):

$$G^{abcd}_{nm}(\alpha) = (\gamma j_o)^3 (R^{abcd}_{nm}(\alpha) + O(\gamma)) \tag{11}$$

with

$$R^{abcd}_{nm} = \frac{1}{\gamma^3 j_o^3} \sum_{p<q,r<s} Q^{-1}_{(pq)(rs)} \frac{\partial q_n^{ab}}{\partial j_{pq}} \frac{\partial q_m^{cd}}{\partial j_{rs}}. \tag{12}$$

and

$$Q_{(ab)(cd)} = -\frac{\gamma \alpha^{(ab)(cd)}}{\sqrt{(j_o)_{ab}}\sqrt{(j_o)_{cd}}} + S''_{\text{Regge}}. \tag{13}$$

If we take the classical limit, introduced in,[13,15] where the Barbero-Immirzi parameter is taken to zero, $\gamma \to 0$, and the spin of the boundary state is taken to infinity, $j \to \infty$, keeping the size of the quantum geometry $A \sim \gamma j$ finite and fixed, the two-point function (11) we obtain exactly matches the one obtained from Lorentzian Regge calculus.[20]

4. Conclusion and Discussion

In this proceeding we studied two themes of spinfoam formalism, the imposing of simplicity constraint and the computation of the Lorentzian correlation functions. Since the simplicity constraint is secondary class, we use Gupta-Bleuler procedure to impose the simplicity constraint *weakly*, namely asking the matrix elements of the simplicity constraints to vanish on physical boundary states. We compute the two-point function for the Penrose metric operator from the simplicial EPRL spinfoam model, with a single vertex, in Lorentzian theory. In some reasonable limit, the obtained two-point function exactly matches the one obtained from Lorentzian Regge calculus.

Acknowledgement

This work is supported by the Fundamental Research Funds for the Central Universities.

References

1. Y. Ding and C. Rovelli, *Class. Quant. Grav.* **27**, p. 165003 (2010).
2. Y. Ding and C. Rovelli, *Class. Quant. Grav.* **27**, p. 205003 (2010).
3. Y. Ding, M. Han and C. Rovelli, *Phys. Rev.* **D83**, p. 124020 (2011).
4. J. Engle, E. Livine, R. Pereira and C. Rovelli, *Nucl. Phys.* **B799**, 136 (2008).
5. E. Bianchi, L. Modesto, C. Rovelli and S. Speziale, *Class. Quant. Grav.* **23**, 6989 (2006).
6. E. R. Livine and S. Speziale, *JHEP* **11**, p. 092 (2006).
7. E. Bianchi and L. Modesto, *Nucl. Phys.* **B796**, 581 (2008).
8. J. D. Christensen, E. R. Livine and S. Speziale, *Phys. Lett.* **B670**, 403 (2009).
9. E. Alesci and C. Rovelli, *Phys. Rev.* **D76**, p. 104012 (2007).
10. E. Alesci and C. Rovelli, *Phys. Rev.* **D77**, p. 044024 (2008).
11. E. Alesci, E. Bianchi and C. Rovelli, *Class. Quant. Grav.* **26**, p. 215001 (2009).
12. S. Speziale, *Adv. Sci. Lett.* **2**, 280 (2009).
13. E. Bianchi, E. Magliaro and C. Perini, *Nucl. Phys.* **B822**, 245 (2009).
14. Y. Neiman (2011), * Temporary entry *.
15. E. Bianchi and Y. Ding, *J.Phys.Conf.Ser.* **360**, p. 012044 (2012).
16. J. W. Barrett and T. J. Foxon, *Class. Quant. Grav.* **11**, 543 (1994).
17. J. W. Barrett, R. J. Dowdall, W. J. Fairbairn, F. Hellmann and R. Pereira, *Class. Quant. Grav.* **27**, p. 165009 (2010).
18. C. Rovelli, *Phys. Rev. Lett.* **97**, p. 151301 (2006).
19. J. Engle and R. Pereira, *Phys. Rev.* **D79**, p. 084034 (2009).
20. T. Regge, *Nuovo Cim.* **19**, 558 (1961).

SPINFOAMS CONTRIBUTING IN FIRST ORDER OF VERTEX EXPANSION TO THE DIPOLE COSMOLOGY TRANSITION AMPLITUDE

MARCIN KISIELOWSKI[1,2]

[1] Instytut Fizyki Teoretycznej, Uniwersytet Warszawski, ul. Hoża 69, 00-681 Warszawa (Warsaw), Polska (Poland)

[2] St. Petersburg Department of Steklov Mathematical Institute RAS, 27, Fontanka, 191023 St. Petersburg, Russia

E-mail: Marcin.Kisielowski@fuw.edu.pl

In this article we present a method for finding all foams with given boundary, one internal vertex and no edges connecting this vertex with itself. We apply the method to the Dipole Cosmology model and find all spinfoams contributing to the transition amplitude in first order of vertex expansion.

Keywords: Spinfoam models, Loop Quantum Gravity.

1. Introduction

Spinfoams may be thought to be histories of spinnetwork states, the states of quantum gravity according to Loop Quantum Gravity (LQG). The states, called spinnetwork states, are associated with colored graphs and the histories of those graphs are associated with colored two-complexes — the two-complexes are called foams, and together with a coloring form spinfoams. There have been a few proposals for a spinfoam models of quantum gravity, however only one so far was successfully defined for all spinnetwork states — the Engle–Pereira–Rovelli–Livine (EPRL) model.[1,2] The extended model is therefore a candidate for dynamics of Loop Quantum Gravity — it may provide transition amplitudes between spinnetwork states. Basing on this model Bianchi, Rovelli and Vidotto introduced a model (called Dipole Cosmology model) which is a candidate for a spinfoam model of quantum cosmology.[3] They calculated a transition amplitude between coherent states peaked on homogeneous, isotropic geometries using certain approximations, and showed that the resulting amplitude satisfies a quantum Friedmann constraint. One of the approximations was a truncation of the LQG state space to states associated with a *dipole graph* — a graph which has two nodes connected with four links. Second was that of large volume of the universe, and third was that the transition amplitude is calculated using a certain foam which has one internal vertex — this is a restriction to a first order of vertex expansion. We searched for other possible foams in the first order of vertex expansion. Strictly speaking we searched for foams which have one internal vertex, no edges connecting this vertex with itself, and whose boundary graph is a disjoint sum of two dipole graphs (one associated with the initial state, and one associated with the final state).

In order to solve the problem we used graph diagrams.[4] First of all because they define a class of two-complexes which is just broad enough to be compatible with the extension of the EPRL model, but not too broad.[2,4] Second reason was that the graph diagrams are useful in classifying foams, in particular they proved to be useful in finding foams with given boundary graph, and given order of the vertex expansion.[5]

2. Graph diagrams

A graph diagram is a pair $(\mathcal{G}, \mathcal{R})$, where $\mathcal{G} = \{\Gamma_I : I \in \mathcal{I}\}$ is a set of oriented, connected and closed graphs indexed by elements of an index set $\mathcal{I}$ and $\mathcal{R}$ is a family of relations:

- $\mathcal{R}_{\text{node}}$ — a symmetric relation on a set of nodes of the graphs Γ_I, $I \in \mathcal{I}$. A node may be in relation with only one different node or may be unrelated. Two nodes can be in relation, if the number of incoming/outgoing links at one node is equal to the number of outgoing/incoming links at the other node. A link, which intersects a node twice, is counted twice, once as incoming, and once as outgoing. The unrelated nodes are called *boundary nodes*. The relation $\mathcal{R}_{\text{node}}$ is called a *node relation.*
- $\mathcal{R}_{\text{link}} = \{\mathcal{R}_{\text{link}}^{(n,n')}\}_{(n,n')\in\mathcal{R}_{\text{node}}}$ — a family of symmetric relations on the set of links of the graphs Γ_I, $I \in \mathcal{I}$, collectively called a *link relation.* A link relation is defined by a family of bijective maps $\{\phi_{(n,n')}\}_{(n,n')\in\mathcal{R}_{\text{node}}}$, such that: (i) $\phi_{(n,n')}$ is a map from the set of links at the node n' onto the set of links at the node n mapping each incoming/outgoing link at a node n' into a unique outgoing/incoming link at the related node n, (ii) $\phi_{(n',n)} = \phi^{-1}_{(n,n')}$. The family of bijections $\phi_{(n,n')}$ induces the family of symmetric relations $\{\mathcal{R}_{\text{link}}^{(n,n')}\}_{(n,n')\in\mathcal{R}_{\text{node}}}$ in the following way: $(\ell, \ell') \in \mathcal{R}_{\text{link}}^{(n,n')}$ iff (i) ℓ intersects n and ℓ' intersects n', (ii) $\ell = \phi_{(n,n')}(\ell')$.

Each graph gives rise to an elementary building block — a foam, which is an image of a homotopy of this graph into a point. The node and the link relation code the way one glues those blocks and obtains a foam corresponding to the graph diagram. As a result, to each graph diagram corresponds a unique foam.[4] Moreover, there is a natural coloring of the graph diagram, which induces a coloring of the corresponding foam and turns the foam into a spinfoam. A graph diagram together with such a coloring is called an operator spinnetwork diagram.[4] As a result, each operator spinnetwork diagram can be mapped into a unique spinfoam. We consider the class of foams defined by graph diagrams to be optimal for spinfoam models,[4] and use the operator spinnetwork diagram formalism instead of standard spinfoam approach. The diagrammatic approach turned out to be a useful tool in finding all foams with given properties. It is reflected by considerations in next section, where we construct all foams with the boundary given by two dipole graphs, possessing one internal vertex and no edges connecting this vertex with itself.

3. The algorithm

In order to solve the problem of finding all foams with given structure of the boundary graph, we introduce a notion of a static diagram corresponding to a graph Γ. Static diagrams change the problem of finding all foams with given boundary graph into a problem of finding all foams whose boundary nodes have a given structure. A structure of a node is encoded in a squid, and a structure of all boundary nodes is encoded in a squid set.

A *squid* is a graph with the following vertices: a distinguished one, called a head of the squid, and the remaining ones called leg-nodes. Links, called legs, are connecting each leg-node with the head of the squid. A *squid set* is a set of oriented squids. Given a graph Γ, one may construct a squid set $\mathcal{S}_\Gamma$ corresponding to this graph: each node n of the graph Γ defines uniquely a squid $\mathcal{X}_n$ ($\mathcal{X}_n$ is a unique squid satisfying a condition that the number of incoming/outgoing links at its head is equal to the number of incoming/outgoing links at the node n), and therefore a graph Γ defines a squid set $\mathcal{S}_\Gamma = \{\mathcal{X}_n : n \in \Gamma^{(0)}\}$, where $\Gamma^{(0)}$ denotes the set of nodes of the graph Γ. We therefore have a natural mapping from a set of all oriented graphs $\mathfrak{G}$ to a set of all squid sets $\mathfrak{S}$, $\pi : \mathfrak{G} \to \mathfrak{S}$, $\pi(\Gamma) := \mathcal{S}_\Gamma$. The map π is not injective, and therefore not invertible. The pre-image $\pi^{-1}(\mathcal{S}_\Gamma)$ is the set of all graphs, which have the same structure of nodes as the graph Γ.

A *static diagram* corresponding to a graph Γ is a graph diagram $\mathcal{D}_\Gamma = (\{\widetilde{\theta}_{\mathcal{X}_n} : \mathcal{X}_n \in \mathcal{S}_\Gamma, n \in \Gamma^{(0)}\}, \mathcal{G}_{\text{static}})$, defined as follows:

- Denote by $\overline{\mathcal{X}}_n$ a unique graph obtained from $\mathcal{X}_n$ by reversing an orientation of each link of $\mathcal{X}_n$. Define $\theta_{\mathcal{X}_n}$ as a unique graph in $\pi^{-1}(\{\mathcal{X}_n, \overline{\mathcal{X}}_n\})$

which does not have link-loops, i.e. such that every link of the graph is connecting the two different nodes. Then each graph $\widetilde{\theta}_{\mathcal{X}_n}$ is obtained from $\theta_{\mathcal{X}_n}$ by splitting each link ℓ of the graph $\theta_{\mathcal{X}_n}$ into two links by introducing an extra node n_ℓ on ℓ.

- The node and link relations $\mathcal{G}_{\text{static}}$ are defined by a structure of the graph Γ in the following way. One can identify links of each squid $\mathcal{X}_n$ with the links of the graph Γ which are intersecting n. Denote by $s(\ell)$ the starting point of ℓ, by $t(\ell)$ the endpoint of ℓ and by $\gamma^{(1)}$ a set of links of a graph γ. Then to a link ℓ there corresponds a unique link in $\mathcal{X}^{(1)}_{s(\ell)}$ as well as in $\mathcal{X}^{(1)}_{t(\ell)}$. This identification leads to symmetric relation in $\bigcup_{n\in\Gamma^{(0)}} \mathcal{X}^{(1)}_n$. It induces a symmetric relation in the set of nodes $\{n_\ell \in \widetilde{\theta}^{(0)}_{\mathcal{X}_n} : \ell \in \theta^{(1)}_{\mathcal{X}_n},\ \mathcal{X}_n \in \mathcal{S}_\Gamma\}$. This is the node relation (the nodes in $\widetilde{\theta}^{(0)}_{\mathcal{X}_n}$ corresponding to heads of the squids $\mathcal{X}_n$ and $\overline{\mathcal{X}}_n$ remain unrelated). Each node n_ℓ is two-valent, one link is incoming and one is outgoing at n_ℓ — therefore there is a unique family of link relations $\mathcal{R}^{(n_\ell, n'_\ell)}_{\text{link}}$.

The name 'static' comes from the fact, that the graph diagram $\mathcal{D}_\Gamma$ describes a static (trivial) evolution of the graph Γ into itself. The static graph diagram $\mathcal{D}_\Gamma$ may be equipped with a natural coloring which turns it into an operator spinnetwork diagram.[4,5] The resulting operator spinnetwork diagram describes a trivial evolution of a spinnetwork into itself.

A static diagram $\mathcal{D}_\Gamma$ can be glued with an interaction diagram $\mathcal{D}_{\text{int}}$ (defined below) — a result of this procedure will be a graph diagram $\mathcal{D}_\Gamma \# \mathcal{D}_{\text{int}}$ which has the given boundary Γ. [a] An *interaction diagram* $\mathcal{D}_{\text{int}}$ corresponding to a graph Γ is a graph diagram, such that $\mathcal{S}_{\partial\mathcal{D}_{\text{int}}} = \mathcal{S}_\Gamma$, where $\partial\mathcal{D}_{\text{int}}$ denotes a boundary graph of $\mathcal{D}_{\text{int}}$. The diagram $\mathcal{D}_\Gamma \# \mathcal{D}_{\text{int}}$ will be constructed in two steps. First, consider a graph diagram $\mathcal{D}_\Gamma \sqcup \mathcal{D}_{\text{int}}$ defined in the following way: (i) a set of graphs of $\mathcal{D}_\Gamma \sqcup \mathcal{D}_{\text{int}}$ is a sum of the set of graphs of $\mathcal{D}_\Gamma$ and the set of graphs of $\mathcal{D}_{\text{int}}$, (ii) node and link relations are induced from the node and link relations in $\mathcal{D}_\Gamma$ and $\mathcal{D}_{\text{int}}$. Second, extend the node relation and the family of link relations of $\mathcal{D}_\Gamma \sqcup \mathcal{D}_{\text{int}}$, such that the following two conditions are satisfied (the extension is not unique): (i) each boundary node of $\mathcal{D}_{\text{int}}$ is in relation with a boundary node of $\mathcal{D}_\Gamma$ corresponding to a head of a squid $\overline{\chi}_n$, $n \in \Gamma^{(0)}$, (ii) boundary nodes of $\mathcal{D}_\Gamma$ corresponding to heads of the squids χ_n, $n \in \Gamma^{(0)}$ remain unrelated. A resulting diagram is $\mathcal{D}_\Gamma \# \mathcal{D}_{\text{int}}$. Its boundary graph is the graph Γ.

In order to find all graph diagrams with a given boundary it remains

[a] A boundary of a graph diagram is a boundary graph of the corresponding foam.[4]

to find all possible interaction diagrams $\mathcal{D}_{\rm int}$. A general algorithm may be found in our recent article.[5] In the case of foams with one internal vertex and no edges connecting this vertex with itself the interaction diagram is simply a graph. The set of all possible interaction graphs is $\pi^{-1}(\mathcal{S}_\Gamma)$.

Let us apply now the procedure to Dipole Cosmology (we restrict to first order of vertex expansion). In this case the boundary (unoriented) graph is a disjoint union of two (unoriented) dipole graphs $|\Gamma| = |\Gamma_{4\theta}| \cup |\Gamma_{4\theta}|$. The algorithm is the following:

(1) $\Gamma = \Gamma_{4\theta} \cup \Gamma_{4\theta}$: choose an orientation of each link of each of the two graphs $|\Gamma_{4\theta}|$ (Fig. 1 a).

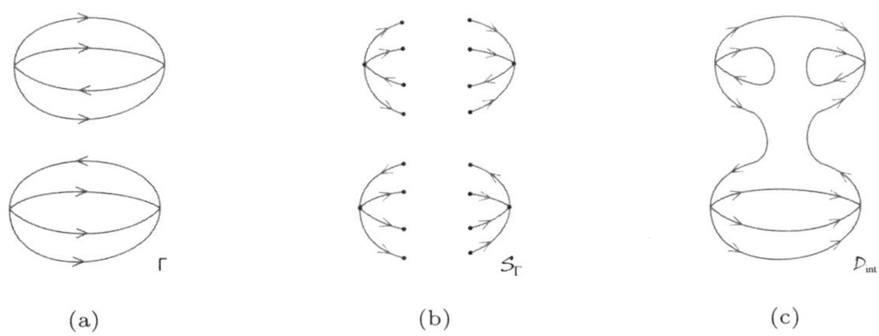

Fig. 1. Construction of the graph diagram. (a) Step (*i*) – choose an orientation of each link. (b) Step (*ii*) – construct the squid set $\mathcal{S}_\Gamma$. (c) Step (*iii*) – construct an interaction graph $\mathrm{D}_{\rm int} = \Gamma_{\rm int}$; an example is depicted.

(2) $\mathcal{S}_\Gamma$: Construct a squid set $\mathcal{S}_\Gamma$ of the graph Γ (Fig. 1 b).
(3) $\mathcal{D}_{\rm int}$: Find a graph diagram $\mathcal{D}_{\rm int}$ such that $\mathcal{S}_{\partial\mathcal{D}_{\rm int}} = \mathcal{S}_\Gamma$. This is the interaction diagram $\mathcal{D}_{\rm int}$. In the first order of vertex expansion $\mathcal{D}_{\rm int}$ is just a graph ($\partial\mathcal{D}_{\rm int} \cong \mathcal{D}_{\rm int}$), an element of the set $\pi^{-1}(\mathcal{S}_\Gamma)$ (Fig. 1 c).
(4) $\mathcal{D}_\Gamma\#\mathcal{D}_{\rm int}$: Use the static graph diagram $\mathcal{D}_\Gamma$ of the graph Γ (Fig. 2 a) and construct a diagram $\mathcal{D}_\Gamma\#\mathcal{D}_{\rm int}$ (Fig. 2 b).
(5) Coloring: Define arbitrary coloring of the diagram $\mathcal{D}_\Gamma\#\mathcal{D}_{\rm int}$ which turns it into an operator spinnetwork diagram.
(6) Consider all possible: orientations of the graph Γ, interaction graphs $\mathcal{D}_{\rm int}$ (i.e. elements of the set $\pi^{-1}(\mathcal{S}_\Gamma)$), extensions of the graph diagram $\mathcal{D}_\Gamma \sqcup \mathcal{D}_{\rm int}$ to $\mathcal{D}_\Gamma\#\mathcal{D}_{\rm int}$, and colorings.

In the Dipole Cosmology (in first order of vertex expansion) each interaction graph is obtained by assigning an orientation to a graph having precisely **4 nodes**, each of which is **four-valent**. It comes out, that there

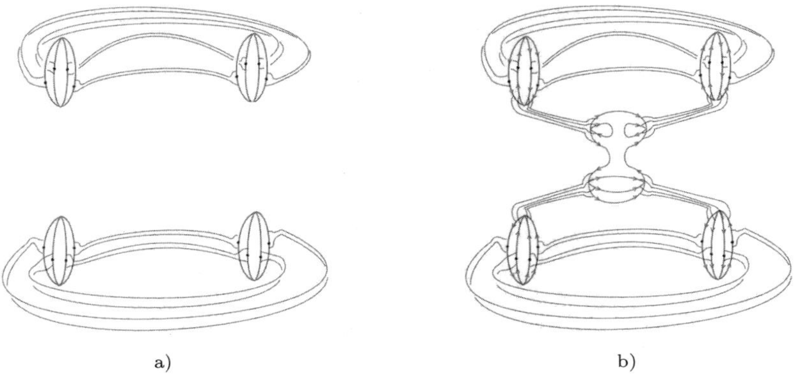

Fig. 2. Construction of the graph diagram. Step (*iv*) – the static graph diagram $\mathcal{D}_\Gamma$ ((a) - the dotted lines denote the link relations) is attached to the diagram $\mathcal{D}_{\text{int}}$ (b), and the final diagram $\mathcal{D}_\Gamma \# \mathcal{D}_{\text{int}}$ is obtained.

are 20 such graphs.[5] However, the number of possible graph diagrams is in general different from this number. First, note that given an orientation of the boundary graph it is not always possible to assign a compatible orientation to every of the 20 unoriented interaction graphs — in other words the number of elements of the set $\pi^{-1}(\mathcal{S}_{\Gamma_{4\theta} \cup \Gamma_{4\theta}})$ is less than or equal to 20. Second, an interaction graph does not determine uniquely a graph diagram $\mathcal{D}_\Gamma \# \mathcal{D}_{\text{int}}$ — there is still an ambiguity in node and link relations.

Acknowledgements

Marcin Kisielowski acknowledges financial support from the project 'International PhD studies at the Faculty of Physics University of Warsaw' of Foundation for Polish Science, cofinanced from the programme IE OP 2007-2013 within European Regional Development Fund. The work was also partially supported by the grants N N202 104838, and 182/N-QGG/2008/0 (PMN) of Polish Ministerstwo Nauki i Szkolnictwa Wyższego.

References

1. J. Engle, E. Livine, R. Pereira and C. Rovelli, *Nucl.Phys.* **B799**, 136 (2008).
2. W. Kaminski, M. Kisielowski and J. Lewandowski, *Class.Quant.Grav.* **27**, p. 095006 (2010).
3. E. Bianchi, C. Rovelli and F. Vidotto, *Phys.Rev.* **D82**, p. 084035 (2010).
4. M. Kisielowski, J. Lewandowski and J. Puchta, *Class.Quant.Grav.* **29**, p. 015009 (2012).
5. M. Kisielowski, J. Lewandowski and J. Puchta, *arXiv:1203.1530 [gr-qc]* (2012).

PARALLEL SESSION
Tensor and Group Field Theories for Quantum Gravity

21 & 24 AUGUST 2012

ASYMPTOTIC FREEDOM OF RANK 4 TENSOR GROUP FIELD THEORY

J. BEN GELOUN

Perimeter Institute, 31 Caroline Str N, Waterloo, ON N2L 2Y5, Canada
E-mail: jbengeloun@perimeterinstitute.ca

Recently, a rank four tensor group field theory has been proved renormalizable. We provide here the key points on the renormalizability of this model and its UV asymptotic freedom.

Keywords: Random tensors; Renormalization; β-function; Asymptotic freedom.

1. Introduction

Random tensor models[1–5] have been introduced in the early 90's as a natural generalization of matrix models.[6] Random matrices defined a very successful framework for addressing the quantization of gravity in 2D via random triangulations. Until recently, the essential tool used for achieving all analytical results in matrix models, namely the t'Hooft 1/N expansion, was crucially missing for higher rank extension of these models. Indeed, Gurau for a particular class of models called colored[7–9] succeeded in defining such a 1/N expansion.[5,10–13] Important results follow rapidly this breakthrough. To mention a few of these, the statistical analysis of the colored models shows a phase transition with computable critical exponent,[14] long-standing mathematical physics issues[15,16] and universal behavior of these tensors find a sense,[17] and, finally, one defines the first renormalizable tensor models of rank higher than 2.[18–22] We refer such a category of quantum field theory to as Tensor Group Field Theories (TGFT's).[1,4]

The present work concerns the first of these renormalizable models addressing the generation of 4D simplicial pseudo-manifold via a quantum field theoretical formalism. We provide here an overview of the renormalizability[18] and UV asymptotic freedom[20] of this model. Despite the fact that a contact with a quantization of gravity is not yet understood at this stage, such a framework is definitely useful for addressing future investigations on emergent geometries.[1,4]

2. Renormalizability of the model

Consider a complex rank four tensor over the group $U(1)$, $\varphi : U(1)^4 \to \mathbb{C}$ which admits the decomposition in Fourier modes as

$$\varphi(h_1, h_2, h_3, h_4) = \sum_{p_j \in \mathbb{Z}} \varphi_{[p_1,p_2,p_3,p_4]} e^{ip_1\theta_1} e^{ip_2\theta_2} e^{ip_3\theta_3} e^{ip_4\theta_4} \tag{1}$$

where $h_i \in U(1)$, $\theta_i \in [0, 2\pi)$ and p_j are momentum indices. We denote $\varphi_{[p_1,p_2,p_3,p_4]} = \varphi_{1,2,3,4}$ and assume no symmetry under permutation of the arguments for this tensor.

The kinetic term of the model, written in momentum space, reads

$$S^{\text{kin},0} = \sum_{p_j} \bar{\varphi}_{1,2,3,4} \Big[\sum_{s=1}^{4} (p_s)^2 + m^2 \Big] \varphi_{1,2,3,4} \tag{2}$$

where the sum is performed over all momentum values p_j. Clearly, such a kinetic term is inferred from a Laplacian dynamics which can be motivated in different ways.[1,23] To this kinetic term, we associate a Gaussian measure $d\mu_C[\varphi]$ with covariance $C = (\sum_s p_s^2 + m^2)^{-1}$.

Trace invariant objects[16] define the interaction part of the model:

$$S_{6;1} = \sum_{p_j} \varphi_{1,2,3,4} \bar{\varphi}_{1',2,3,4} \varphi_{1',2',3',4'} \bar{\varphi}_{1'',2',3',4'} \varphi_{1'',2'',3'',4''} \bar{\varphi}_{1,2'',3'',4''} + \text{perm.}$$

$$S_{6;2} = \sum_{p_j} \varphi_{1,2,3,4} \bar{\varphi}_{1',2',3',4} \varphi_{1',2',3',4'} \bar{\varphi}_{1'',2,3,4'} \varphi_{1'',2'',3'',4''} \bar{\varphi}_{1,2'',3'',4''} + \text{perm.} \tag{3}$$

$$S_{4;1} = \sum_{p_j} \varphi_{1,2,3,4} \, \bar{\varphi}_{1',2,3,4} \, \varphi_{1',2',3',4'} \, \bar{\varphi}_{1,2',3',4'} + \text{perm.} \tag{4}$$

$$S_{4;2} = \Big[\sum_{p_j} \bar{\varphi}_{1,2,3,4} \, \varphi_{1,2,3,4} \Big] \Big[\sum_{p'_j} \bar{\varphi}_{1',2',3',4'} \, \varphi_{1',2',3',4'} \Big] \tag{5}$$

where the sum (+perm.) is over all 24 permutations of the four color indices. The flow of $S_{6;2}$ generates a ϕ^4-type interaction that we will refer to as *anomalous term* (5) and so should be part of our action.

Feynman graphs have a tensor structure. Lines are four-stranded and represent propagators (Figure 1) whereas vertices are nonlocal objects (as depicted in Figure 2 and Figure 3).

Introducing an ultraviolet cutoff Λ and counterterms, one proves[18] that the model described by (2) and (3)-(5) is renormalizable at all orders of

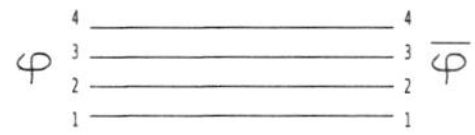

Fig. 1. The propagator.

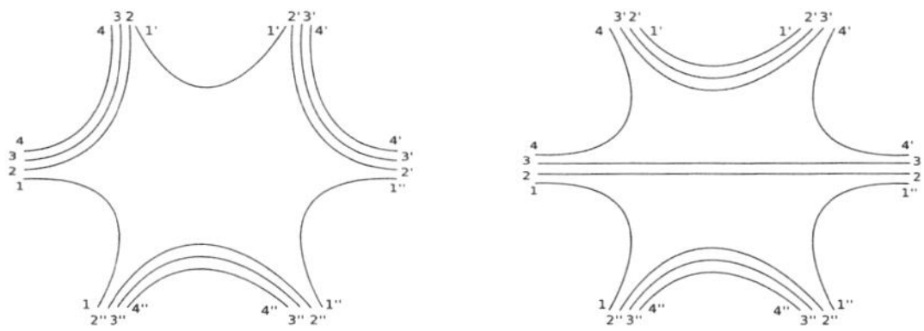

Fig. 2. Vertices of the type $\phi^6_{(1)}$ from $S_{6;1}$ (left) and of the type $\phi^6_{(2)}$ from $S_{6;1}$ (right).

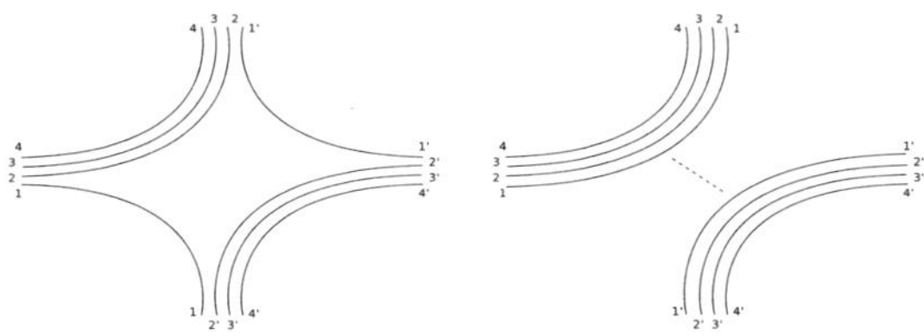

Fig. 3. Vertices of the type $\phi^4_{(1)}$ from $S_{4;1}$ (left) and the anomalous term $\phi^4_{(2)} = (\phi^2)^2$ from $S_{4;2}$ (right).

perturbation theory. This statement is proved by multiscale analysis[24] and the study of the graph topology. We review now the main ingredients of this proof.

• *A slice decomposition*: Use the Schwinger representation of the propagator $C = \int_0^\infty e^{-\alpha(\sum_s p_s^2 + m^2)} d\alpha$, to slice this quantity as $C = \sum_{i=0}^\infty C_i$ where $C_i = \int_{M^{-2i}}^{M^{-2i+2}} e^{-\alpha(\sum_s p_s^2+m^2)} d\alpha \leq K M^{-2i} e^{M^{-2i}(\sum_s p^2 + m^2)}$, for $M > 0$.

• *The multi-scale expansion* of a graph amplitude using the above bound yields the *crude power-counting* at a momentum attribution μ, $A_{\mathcal{G};\mu} \leq K^n \prod_{(i,k)} M^{\omega_d(G^i_k)}$, where K, n are some constants, G^i_k are called the quasi-local subgraphs[24] of $\mathcal{G}$, $\omega_d(G) = -2L(G) + F_{\rm int}(G)$ is the degree of divergence of the graph G, $L(G)$ the number of internal lines and $F_{\rm int}(G)$ the number of closed loops or faces of the subgraph G. Quasi-local subgraphs define the *generalized notion of locality* here.

• *A refined power-counting theorem*: A graph $\mathcal{G}$ in this theory admits a five-color extension $\mathcal{G}_{\text{color}}$ which is itself a rank four tensor graph. A jacket J of $\mathcal{G}_{\text{color}}$ is a ribbon subgraph of $\mathcal{G}_{\text{color}}$ defined by a color cycle $(0abcd)$ up to a cyclic permutation. The jacket $\widetilde{J}$ is the closed jacket obtained from J after closing all external legs present in J. The boundary $\partial\mathcal{G}$ of the graph $\mathcal{G}$ is the closed graph defined by vertices corresponding to external legs and by lines corresponding to external strands of $\mathcal{G}$.[8] It is, presently, a vacuum rank 3 colored graph. A boundary jacket J_∂ is a jacket of $\partial\mathcal{G}$.

The anomalous vertex $(\phi^2)^2$ is disconnected from the point of view of its strands. We define *connected* component graphs by reducing such vertices to twice ϕ^2-vertices.

The divergence degree of any connected graph $\mathcal{G}$ is given by

$$\omega_d(\mathcal{G}) = -\tfrac{1}{3}\Big[\sum_{J} g_{\widetilde{J}} - \sum_{J_\partial} g_{J_\partial}\Big] - [C_{\partial\mathcal{G}} - 1] - V_4 - 2[V_2 + V_2''] - \tfrac{1}{2}\,[N_{\text{ext}} - 6] \quad (6)$$

where $g_{\widetilde{J}}$ and g_{J_∂} are the genus of $\widetilde{J}$ and J_∂, respectively, $C_{\partial\mathcal{G}}$ is the number of connected components of the boundary graph $\partial\mathcal{G}$; the first sum is performed on all closed jackets $\widetilde{J}$ of $\mathcal{G}_{\text{color}}$ and the second sum is performed on all boundary jackets J_∂ of $\partial\mathcal{G}$; V_4 is the number of $\phi^4_{(1)}$ vertices, V_2 the number of vertices of the type ϕ^2 (mass counterterms), $V_2'' = 2V_4'$ is twice the number of vertices of type $\phi^4_{(2)} = (\phi^2)^2$, N_{ext} its number of external legs.

One notices that $\omega_d(\mathcal{G})$ does not depend on vertices of the type ϕ^6. Finally, the following table lists all primitively divergent graphs:

N_{ext}	$V_2 + V_2''$	V_4	$\sum_{J_\partial} g_{J_\partial}$	$C_{\partial\mathcal{G}} - 1$	$\sum_{\widetilde{J}} g_{\widetilde{J}}$	$\omega_d(\mathcal{G})$
6	0	0	0	0	0	0
4	0	0	0	0	0	1
4	0	1	0	0	0	0
4	0	0	0	1	0	0
2	0	0	0	0	0	2
2	0	1	0	0	0	1
2	0	2	0	0	0	0
2	0	0	0	0	6	0
2	1	0	0	0	0	0

Table 1: List of primitively divergent graphs

Call graphs satisfying $\sum_{\widetilde{J}} g_{\widetilde{J}} = 0$ "melonic" graphs or simply "melons".[14] Thus, most of graphs in Table 1 are melons with melonic boundary.

3. β-functions

We will discuss the β-functions of the model given by the interaction

$$S = \tfrac{1}{3}\lambda_{6;1}\, S_{6;1} + \lambda_{6;2}\, S_{6;2} + \tfrac{1}{2}\lambda_{4;1}\, S_{4;1} + \tfrac{1}{2}\lambda_{4;2}\, S_{4;2} \quad (7)$$

where some symmetry factors are introduced. It turns out that the behavior of the renormalized coupling constants $\lambda^{\rm ren}_{6;1/2}$ in the UV are those significant for this model.

The β-functions of the $\phi^6_{(1/2)}$ are evaluated with two ingredients: (1) the truncated and amputated one particle irreducible (1PI) 6-point functions $\Gamma_{6;1/2}(\cdot)$ the external data of which are designed in the form of the initial (bare) interaction; (2) the wave function renormalization given by $Z = 1 - \partial_{b_1^2}\Sigma|_{b_{1,2,3,4}=0}$, where $\Sigma(b_1, b_2, b_3, b_4) = \langle \varphi_{1,2,3,4}\bar{\varphi}_{1,2,3,4}\rangle^t_{1PI}$ is the so-called self-energy or sum of all amputated 1PI two-point functions. We identify all relevant graphs related to the six- and two-point functions using Table 1.

The following ratios encode the β-functions of the coupling constants:

$$\lambda^{\rm ren}_{6;1/2} = -\frac{\Gamma_{6;1/2}(0,0,0,0,0,0,0,0,0,0,0,0)}{Z^3} \qquad (8)$$

At two loops (for the first) and four loops (for the second), one finds that the renormalized coupling constants satisfy the equations

$$\lambda^{\rm ren}_{6;2} = \lambda_{6;2} + 2\lambda_{6;2}\lambda_{6;1}S^1 + 3\lambda^2_{6;2}[S^1 + S^{12}] + O(\lambda^3) \qquad (9)$$

$$\lambda^{\rm ren}_{6;1} = \lambda_{6;1} + 8\lambda^3_{6;1}\mathcal{S} + O(\lambda^4_{6;1}) \qquad (10)$$

where $S^{1,12}$ and $\mathcal{S}$ are formal log-divergent sums.[19] The β-functions of this model at this order of perturbation are given by

$$\beta_{6;2;\,(2)} = 3 \qquad \beta_{6;2;\,(12)} = 2 \qquad \beta_{6;1} = 8 \qquad (11)$$

Assuming positive coupling constants, (9)-(11) show that the overall model is asymptotically free in UV.

In fact, the complete study of all renormalized coupling equations shows that[19] there exists a UV fixed manifold for the model given by

$$\lambda_{6;1/2} = 0 \qquad \forall\lambda_{4;1} \qquad \lambda_{4;2} = 0 \qquad (12)$$

Perturbing the system around this fixed manifold by adding small quantities to the bare couplings, the renormalized coupling constants increase in the infrared (IR). This fact generally hints at a phase transition, a scenario very promising for finding new models in the continuum.

Acknowledgements

The organizers of the XXIXth International Colloquium on Group-Theoretical Methods in Physics, Nankai, China, are warmly thank for their welcome and hospitality. Discussions with R. Gurau and V. Rivasseau are gratefully acknowledged. Research at Perimeter Institute is supported by the Government of Canada through Industry Canada and by the Province of Ontario through the Ministry of Research and Innovation.

References

1. V. Rivasseau, *The Tensor Track: an Update*, (a contribution to this proceeding) arXiv:1209.5284 [hep-th].
2. V. Rivasseau, *PoS C NCFG2010*, **004** (2010).
3. V. Rivasseau, *Quantum Gravity and Renormalization: The Tensor Track*, arXiv:1112.5104 [hep-th].
4. D. Oriti, *The group field theory approach to quantum gravity*, arXiv:gr-qc/0607032.
5. R. Gurau, *A review of the large N limit of tensor models*, (a contribution to this proceeding) arXiv:1209.4295 [math-ph].
6. P. Di Francesco, P. H. Ginsparg and J. Zinn-Justin, *Phys. Rept.* **254**, 1 (1995).
7. R. Gurau, *Commun. Math. Phys.* **304**, 69 (2011).
8. R. Gurau, *Annales Henri Poincare* **11**, 565 (2010).
9. R. Gurau, *Class. Quant. Grav.* **27**, 235023 (2010).
10. R. Gurau, *Annales Henri Poincare* **12**, 829 (2011).
11. R. Gurau and V. Rivasseau, *Europhys. Lett.* **95**, 50004 (2011).
12. R. Gurau, *Annales Henri Poincare* **13**, 399 (2012).
13. R. Gurau and J. P. Ryan, *SIGMA* **8**, 020 (2012).
14. V. Bonzom, R. Gurau, A. Riello and V. Rivasseau, *Nucl. Phys. B* **853**, 174 (2011).
15. R. Gurau, *Nucl. Phys. B* **852**, 592 (2011).
16. R. Gurau, *Nucl. Phys. B* **865**, 133 (2012).
17. R. Gurau, *Universality for Random Tensors*, arXiv:1111.0519 [math.PR].
18. J. Ben Geloun and V. Rivasseau, *A Renormalizable 4-Dimensional Tensor Field Theory*, arXiv:1111.4997 [hep-th]; Commun. Math. Phys. (2012) DOI 10.1007/s00220-012-1549-1;
    J. Ben Geloun and V. Rivasseau, *Addendum to 'A Renormalizable 4-Dimensional Tensor Field Theory'* arXiv:1209.4606 [hep-th]; to appear *Commun. Math. Phys.* (2012).
19. J. Ben Geloun and D. O. Samary, *3D Tensor Field Theory: Renormalization and One-loop β-functions*, arXiv:1201.0176 [hep-th].
20. J. Ben Geloun, *Two and four-loop β-functions of rank 4 renormalizable tensor field theories*, to appear in *Class. Quant. Grav.* (2012), arXiv:1205.5513 [hep-th].
21. J. Ben Geloun and E. R. Livine, *Some classes of renormalizable tensor models*, arXiv:1207.0416 [hep-th].
22. S. Carrozza, D. Oriti and V. Rivasseau, *Renormalization of Tensorial Group Field Theories: Abelian U(1) Models in Four Dimensions*, arXiv:1207.6734 [hep-th].
23. J. Ben Geloun and V. Bonzom, *Int. J. Theor. Phys.* **50**, 2819 (2011).
24. V. Rivasseau, *From perturbative to constructive renormalization*, (Princeton Univ. Pr., Princeton, 1991).

SCHWINGER-DYSON EQUATIONS IN GROUP FIELD THEORIES OF QUANTUM GRAVITY

"The XXIX International Colloquium on Group-Theoretical Methods in Physics"
Chern Institute of Mathematics
Tianjin, China
August 20-26, 2012

THOMAS KRAJEWSKI

Centre de Physique Théorique
Campus de Luminy
F-13288 Marseille cedex 09 France
E-mail: Thomas.Krajewski@cpt.univ-mrs.fr

In this talk, we elaborate on the operation of graph contraction introduced by Gurau in his study of the Schwinger-Dyson equations. After a brief review of colored tensor models, we identify the Lie algebra appearing in the Schwinger-Dyson equations as a Lie algebra associated to a Hopf algebra of the Connes-Kreimer type. Then, we show how this operation also leads to an analogue of the Wilsonian flow for the effective action. Finally, we sketch how this formalism may be adapted to group field theories.

Keywords: Matrix models, tensor models, Schwinger-Dyson equations, spin networks, infinite dimensional Lie algebras.

1. A very short overview of colored tensor models

Colored models have been introduced by Gurau in the context of group field theory.[1] These models admit a perturbative expansion dominated by spherical topologies[2] and are at the roots of the first renormalizable group field theory.[3] Here, we follow closely the formalism introduced by Gurau in the study of the Schwinger-Dyson equations.[4] In dimension D, the basic dynamical variables of a colored tensor model are two complex conjugate rank D tensors $M_{i_1 \dots i_D}$ and $\overline{M}_{i_1 \dots i_D}$, with $i_k \in \{1, \dots, N\}$. To any bipartite graph with black and white vertices of valence D and a proper coloring with D colors, we associate a $\big(\mathrm{U}(N)\big)^D$ invariant denoted $\mathrm{Tr}_\Gamma(M, \overline{M})$ constructed as follows. To any white (resp. black) vertex we associate a tensor $M_{i_1 \cdots D}$

(resp. $\overline{M}_{i_1 \ldots i_D}$) and contract the nth indices of the tensors following the edges of color n in Γ. For instance,

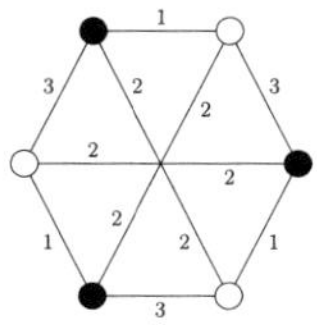

$$\rightarrow \sum_{i_a,\, j_b,\, k_c} \overline{M}_{i_1 i_2 i_3} \overline{M}_{j_1 j_2 j_3} \overline{M}_{k_1 k_2 k_3} M_{i_1 k_2 j_3} M_{j_1 i_2 k_3} M_{k_1 j_2 i_3} \quad (1)$$

The main object of interest is the generating function

$$\mathcal{W}[\lambda_\Gamma] = \log \int [DMD\overline{M}] \exp\left\{ \sum_\Gamma \frac{\lambda_\Gamma}{C_\Gamma} \mathrm{Tr}_\Gamma(M, \overline{M}) \right\}, \quad (2)$$

where $\int [DMD\overline{M}]$ is the Lebesgue measure on the real and imaginary part of the tensors and C_Γ is the cardinal of the automorphism group of Γ. The integral $\mathcal{W}[\lambda_\Gamma]$ is defined as a formal power series as follows. We single out the contribution of the dipole graph

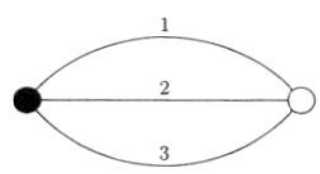

$$\rightarrow \sum_{i_a} \overline{M}_{i_1, i_2, i_3} M_{i_1 i_2 i_3} \quad (3)$$

whose exponential defines a Gaußian measure on tensors. Then, we use Wick's theorem to expand $\mathcal{W}[\lambda_\Gamma]$ as a formal power series in the coefficients λ_Γ, where the number of vertices Γ is strictly greater than 2. From a geometric perspective, each graph Γ represents a triangulation of a space of dimension D, with D-simplexes on its vertices glued together along their boundary $(D-1)$-simplexes following the edges of Γ. The Feynman graphs appearing in the expansion of $\mathcal{W}[\lambda_\Gamma]$ are associated with cellular decompositions of $(D+1)$-dimensional spaces.

2. Graph contraction and Schwinger-Dyson equations

In order to derive the Schwinger-Dyson equations for tensor models,[4] it is convenient to define the contraction of a pair of vertices $v, \overline{v}$ of different colors in a non necessarily connected, bipartite colored graph Γ of degree D. It is the graph $\Gamma/\overline{v}v$ obtained by removing the vertices v and $\overline{v}$ and attaching the resulting free half-edges of identical colors. For example, in $D = 3$,

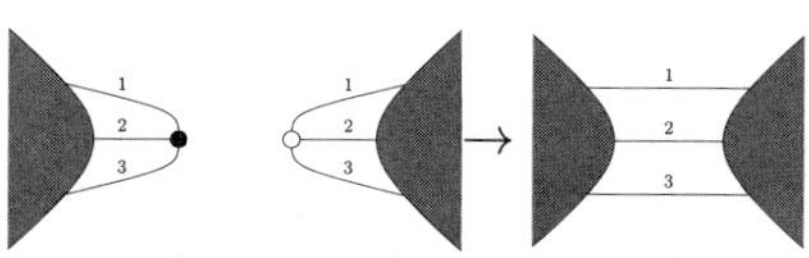

vertices not sharing an edge

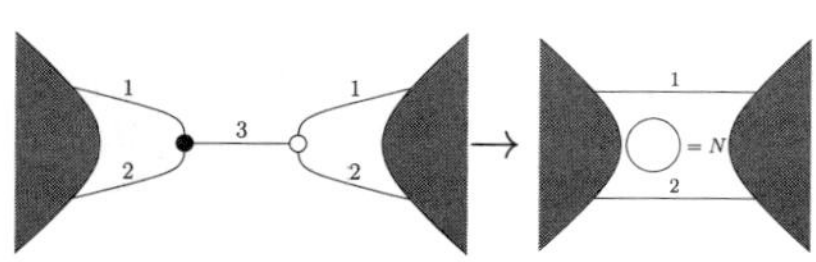

vertices sharing an edge

In the second illustration, v and $\overline{v}$ share an edge, which results in a extra loop in the contracted graph $\Gamma/\overline{v}v$. This extra loop simply yields a factor of N and can be omitted, In all case, $\Gamma/\overline{v}v$ is also a colored bipartite graph of degree D. However, $\Gamma/\overline{v}v$ may not be connected even if Γ is.

Then the Schwinger-Dyson equations for colored tensor models[4] read

$$\Big(\sum_{v=\circ\in\Gamma_0} \prod_{\substack{k \text{ connected} \\ \text{components}}} \frac{\partial}{\partial\lambda_{[\Gamma_0/\overline{v}_0 v]_k}} + \sum_{\Gamma \text{ connected},\, v=\circ\in\Gamma} \lambda_\Gamma \frac{\partial}{\partial\lambda_{(\Gamma_0\Gamma)/\overline{v}_0 v}} \Big) \mathcal{W}[\lambda_\Gamma] = 0 \tag{4}$$

with the the convention that $\frac{\partial}{\partial\lambda_{[\Gamma_0/\overline{v}_0 v]_k}} = N$ if $[\Gamma_0/\overline{v}_0 v]_k$ is a loop without vertex. To derive these equation we choose a graph Γ and a black vertex $\overline{v}_0 \in \Gamma_0$ and perform the infinitesimal change of variable $M \to M + \delta_{(\Gamma,\overline{v}_0)} M$ in the integral defining $\mathcal{W}[\lambda_\Gamma]$, with $\delta_{(\Gamma,\overline{v}_0)} M$ equal to $\mathrm{Tr}_\Gamma(M, \overline{M})$ with the contribution of vertex $\overline{v}_0$ removed. The first term in (4) arises from the Jacobian whereas the second one comes from the variation of the integrand. It is worthwhile to notice that Schwinger-Dyson equations may be used to determine the large N limit.[5]

Geometrically, Schwinger-Dyson equations can be represented as

$\sum_{\substack{v\in\Gamma_0 \\ v\neq v_0}} \Big\langle$ 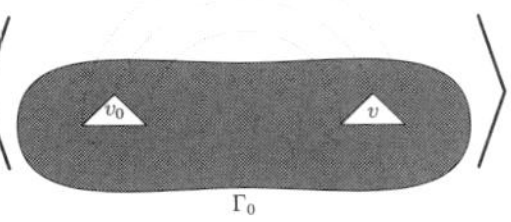 $\Big\rangle + \sum_{\substack{\Gamma \\ v\in\Gamma}} \Big\langle$ 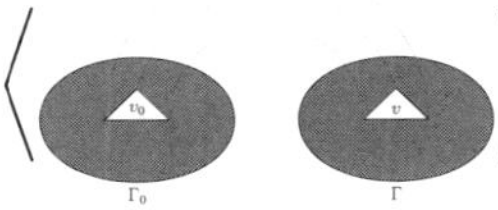 $\Big\rangle = 0.$ (5)

From an algebraic point of view, Schwinger-Dyson equations are constraints on the generating function $\mathcal{W}[\lambda_\Gamma]$ which can be written as

$$\mathcal{L}_{(\Gamma_0,\overline{v}_0)} \mathcal{W}[\lambda_\Gamma] = 0. \tag{6}$$

The differential operators form a Lie algebra representing the Lie algebra of infinitesimal change of variables. For $D = 2$, tensor models reduce to non hermitian matrix models, graphs Γ are necklaces and the operators $\mathcal{L}_n$ define the Witt algebra $[\mathcal{L}_m, \mathcal{L}_n] = \mathcal{L}_{m-n}$. In the $D > 2$, this algebra can be associated with a Hopf algebra of graphs of the Connes-Kreimer type.

3. Lie and Hopf algebras of graphs

A Hopf algebra $\mathcal{H}$ is an associative algebra together with a counit $\epsilon : \mathcal{H} \to \mathbb{R}$, a coproduct $\Delta : \mathcal{H} \to \mathcal{H} \otimes \mathcal{H}$ and an antipode $S : \mathcal{H} \to \mathcal{H}$ obeying some axioms (see for instance the book[6]). Let us define $\mathcal{H}_D$ as the commutative algebra generated by pairs (Γ, v) where Γ is a connected bipartite graph

Γ with vertices of order D and a proper edge coloring with D colors and $v \in \Gamma$ a vertex. The coproduct is of the Connes-Kreimer type[7]

$$\Delta(\Gamma, v) = (\Gamma, v)\otimes 1 + 1\otimes(\Gamma, v) + \sum_{\substack{\Gamma_n \text{ disjoint subgraphs, } v\notin\Gamma_n \\ \text{with } D \text{ legs with different colors}}} \left(\prod_n (\widehat{\Gamma}_n, v_n)\right)\otimes\left(\Gamma\Big/\prod_n \Gamma_n\,, v\right) \tag{7}$$

Γ_n is assumed to be connected and necessarily has all external legs attached to either black or white vertices. We define $\widehat{\Gamma}_n$ by attaching an extra vertex v_n of the opposite color to the external legs of Γ_n. The reduced graph $\Gamma/\prod_n \Gamma_n$ is obtained by shrinking each Γ_n to a single vertex in Γ.

For any commutative Hopf algebra, the set of characters (i.e. linear and multiplicative maps from $\mathcal{H}$ to $\mathbb{C}$) form a group for the convolution product $\alpha * \beta = (\alpha \otimes \beta) \circ \Delta$ with unit ϵ and inverse $\alpha^{-1} = \alpha \circ S$. Thus, characters of $\mathcal{H}_D$ define a group whose elements are fully described by their values on the generators (Γ, v). The Lie algebra of this group is the Lie algebra of constraints of tensor models. Its generators are indexed by pairs (Γ, v) and its bracket is the linearized form of the bracket derived from the convolution product.

4. Effective action in colored tensor models

In the context of tensor models, the analogue of the Wilsonian effective action is defined as

$$S_t[T, \overline{T}] = -\log \int \underbrace{[D\overline{Q}DQ]_{K_t}}_{\text{Gaußian measure}} \exp -S_0[T + Q, \overline{T} + \overline{Q}] \tag{8}$$

with T a fixed tensor and Q a fluctuating one. The latter is integrated out with a Gaußian weight K_t that depends on a parameter t analogous to the floating infrared cut-off in quantum field theory. In the combinatorial model studied here, we focus on counting graphs instead of deriving scaling laws. Therefore, we choose the simplest Gaußian weight $K_t(M, \overline{M}) = t^{-1}\mathrm{Tr}_{\mathrm{dipole}}\big[T, \overline{T}\big]$. The effective action is expanded over all colored bipartite graphs (included disconnected ones) with vertices of valence D

$$S_t[T, \overline{T}] = \sum_{\Gamma} \frac{\lambda_\Gamma(t)}{C_\Gamma} \mathrm{Tr}_\Gamma(T, \overline{T})$$

For example, in the case of $2d$ gravity and matrix models, the expansion reads

$$S[M, M^\dagger] = \sum_{\{n_k\}} \lambda_{\{n_k\}} \prod_k \frac{1}{k^{n_k} n_k!} \left[\mathrm{Tr}(MM^\dagger)^k\right]^{n_k} \tag{9}$$

The effective action obeys a differential equation analogous to the Wilsonian flow equation in quantum field theory

$$\frac{d\lambda_\Gamma}{dt} = \sum_{0 \leq k \leq D} \sum_{\substack{\text{edge cuts} \\ \{e_1,\dots,e_k\}}} \left\{ N^{D-k} \lambda_{\Gamma - \{e_1,\dots,e_k\}} + \sum_{\Gamma_1 \cup \Gamma_2 = \Gamma - \{e_1,\dots,e_k\}} \lambda_{\Gamma_1} \lambda_{\Gamma_2} \right\} \tag{10}$$

The sum runs over all edge cuts of k edges with different colors. The graph $\Gamma - \{e_1, \dots, e_k\}$ is obtained by cutting k edges with different colors and attaching two new vertices to the resulting free half-edges. For instance,

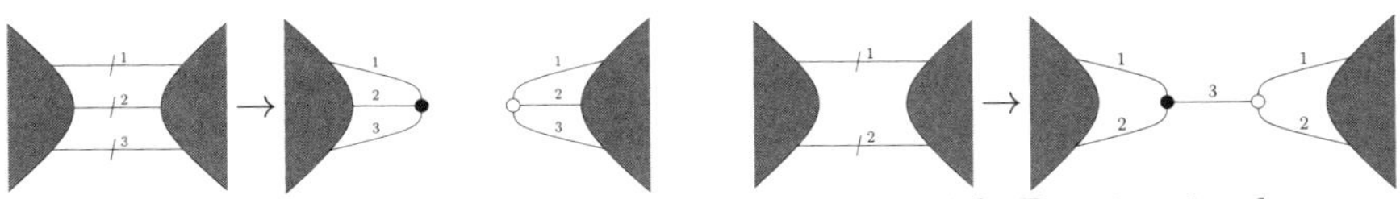

cut with $D = 3$ colors   cut with $D - 1 = 2$ colors

Note that these operations play a important role in the proof of the renormalizabilty of some colored models.[3]

From a geometrical perspective, this equation can be depicted as

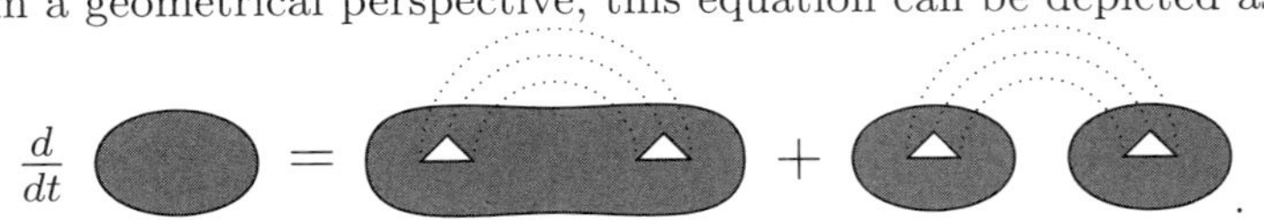

In this equation, the effective action is a sum over connected Feynman graph of the tensor model. Therefore, it may be understood as a sum over connected $(D+1)$-dimensional triangulations, whose boundary may be disconnected. Then, the evolution of such a triangulation results either from an identification of two D-simplexes in the boundary of the same connected $(D+1)$-dimensional triangulation or in different $(D+1)$-dimensional triangulations.

5. Extension to group field theory

Group field theories[8] are particular quantum field theories defined on several copies of a group G which may be SU(2), SU(2) $\times$ SU(2) or SL(2, $\mathbb{C}$). In the sequel, we restrict our attention to SU(2). Their perturbative expansion yields a sum over $(D+1)$-dimensional triangulations weighted by the corresponding spin foam amplitudes appearing in quantum gravity.[9] The graphs Γ are at the basis of the loop quantum gravity boundary Hilbert spaces $\mathcal{H}_\Gamma = \mathrm{L}^2(\mathrm{SU}(2)^{E/V})$ of gauge invariant functions $\Psi(h_e) = \Psi(g_{s(e)} h_e g_{t(e)}^{-1})$ with $s(e)$ and $t(e)$ source and target of the edge e. The group field theory generating function analogous to (2) reads

$$\mathcal{W}[\Psi_\Gamma] = \log \int [D\Phi] \exp\Big\{ \sum_\Gamma \int dh_e dg_{v,e} \\ \frac{\Psi_\Gamma(h_e)}{C_\Gamma} \prod_e \delta\big(h_e\, g_{s(e),e}^{-1} g_{t(e),e}\big) \prod_v \Phi(g_{v,e_1}, \cdots, g_{v,e_D})\Big\} \quad (11)$$

A convenient basis of $\mathcal{H}_\Gamma$ is provided by spin networks $s = (\Gamma, j_e, i_v)$ constructed by associating a spin to every edge and to every vertex an intertwiner between all the spins associated to the edges incident to that vertex. Therefore, all the previous formalism can be extended mutatis mutandis to group field theory replacing graphs by spin networks.

Acknowledgements

The author warmly thanks Vincent Rivasseau for his invitation to this conference and Yongge Ma for his hospitality at Beijing Normal University.

References

1. R. Gurau, *Colored Group Field Theory'* Commun. Math. Phys. **304** (2011) 69-93. arXiv:0907.2582 [hep-th]
2. R. Gurau, *The complete 1/N expansion of colored tensor models in arbitrary dimensions* Annales Henri Poincare **13** (2012) 399 arXiv:1102.5759 [gr-qc]
3. J. Ben Geloun and V. Rivasseau, *A Renormalizable 4-Dimensional Tensor Field Theory* arXiv:1111.4997 [hep-th]
4. R. Gurau, Nucl. Phys. B **865** (2012) 133 *The Schwinger Dyson equations and the algebra of constraints of random tensor models at all orders'* arXiv:1203.4965 [hep-th]
5. V. Bonzom, *Revisiting random tensor models at large N via the Schwinger-Dyson equations* arXiv:1208.6216 [hep-th]
6. V. Chari and A. Pressley, *A guide to quantum groups* Cambridge University Press 1994
7. A. Connes and D. Kreimer, *Renormalization in quantum field theory and the Riemann-Hilbert problem. 1. The Hopf algebra structure of graphs and the main theorem* Commun. Math. Phys. **210** (2000) 249 [hep-th/9912092]
8. L. Freidel, *Group field theory: An Overview* Int. J. Theor. Phys. **44** (2005) 1769-1783. arXiv:hep-th/0505016
9. C. Rovelli, *Quantum gravity* Cambridge University Press 2007

THE QUANTUM GEOMETRY OF TENSORIAL GROUP FIELD THEORIES

D. ORITI*

Albert Einstein Institute
Am Mühlenberg 1, 14476 Potsdam-Golm, Germany, EU
**E-mail: daniele.oriti@aei.mpg.de*

We remark the importance of adding suitable pre-geometric content to tensor models, obtaining what has recently been called tensorial group field theories, to have a formalism that could describe the structure and dynamics of quantum spacetime. We also review briefly some recent results concerning the definition of such pre-geometric content, and of models incorporating it.

Introduction

The last few years have witnessed a revival of tensor models,[2] as a way to generalize to higher dimensions the successes of matrix models[1] in describing 2d quantum gravity as a theory of random surfaces. Historically, this revival started[10] in the area of spin foam models,[4] a covariant version of the dynamics of loop quantum gravity (LQG).[3] A complete definition of such dynamics was indeed proposed in the form of *group field theories* (GFTs),[7] combinatorially non-local field theories on group manifolds whose Feynman diagrams are given by d-dimensional simplicial complexes, and whose Feynman amplitudes are given by the same spin foam amplitudes encoding the quantum dynamics of spin networks states in LQG. This formulation also suggested[14] a change of perspective on the same dynamics. The spin foam approach has developed to a great extent, with the construction of new models and an increased understanding of their quantum geometric aspects.[4,12] To go beyond the truncation of degrees of freedom represented by any single simplicial complex, towards an approximately continuum physics, remains however a pressing issue. This is basically a problem in renormalization and of extracting effective dynamics from the fundamental one. One strategy is suggested by a lattice gauge theory perspective on spin foam amplitudes and involves background independent coarse graining.[18] The other strategy uses the GFT implementation of the spin foam dynamics and standard QFT renormalization[11] and mean field theory[16] tools. This second strategy rests on the new developments of tensor models,[5,6] in particular the discovery of a large-N expansion,[17] which have led to improved analytic control. This also opens the possibility of a better analytic control over GFTs, and thus the full dynamics of spin foam models and LQG, thanks to these tensorial tools. In order to stress this possibility we refer to them in the following as *Tensorial Group field Theories.* Rather than calling for a wider application of TGFTs in spin foam and LQG research, the main point we want to make in this paper is that, if the goal is to solve the problem of Quantum Gravity, we need to study interesting tensorial group field theories with their richer quantum geometric structure rather than the simpler tensor models, for which most of the analytic results have been obtained up to now.

1. Basics of tensorial field theories

The type of models that have been studied up to now fall into two categories. The first is *tensor models*, an uncolored simplicial $d = 3$ example being:

$$S(T) = \frac{1}{2}\sum_{i,j,k} T_{ijk}T_{kji} - \frac{\lambda}{4!\sqrt{N^3}} \sum_{ijklmn} T_{ijk}T_{klm}T_{mjn}T_{nli} \tag{1}$$

where the complex tensor over $(\mathbb{Z}_N)^3$ can be graphically associated to a triangle with edges labelled by $i = 1,..,N$, and the interaction has the combinatorics of the gluing of four triangles along edges to form a tetrahedron. The coloring of the same model with additional labels on the tensors could be used to define interactions with a $U(N)^d$ invariance, corresponding to more general polytopes.[6] The second is *Tensorial Group Field theories*, an uncolored, simplicial 3d example[a] being the Boulatov model:

$$S_{3d}[\varphi,\bar{\varphi}] = \frac{1}{2}\int [dg]^3 \bar{\varphi}(g_1,g_2,g_3)\varphi(g_3,g_2,g_1) -$$
$$- \frac{\lambda}{4!}\int [dg]^6 \varphi(g_1,g_2,g_3)\varphi(g_3,g_4,g_5)\varphi(g_5,g_2,g_6)\varphi(g_6,g_4,g_1) \quad + \quad c.c.$$

where the basic variables is a (complex) field over G^3, with G a Lie group ($SU(2)$ for the 3d quantum gravity model), assumed to possess the invariance: $\varphi(g_i) = \varphi(hg_i)$, and the same combinatorics as the simple tensor model above. The invariance property is a simple example of additional feature imposed on a tensorial field, motivated, as the choice $G = SU(2)$, by quantum geometric considerations.[7]

All the above models are then defined, at the quantum level, by the perturbative expansion of the partition function in Feynman diagrams Γ,

[a]Clearly, if neither coloring nor tensor invariances are used, what we have is really just an ordinary GFT, which we however treat as a special case of TGFTs.

which correspond to arbitrary gluings of d-simplices (or other polytopes) along their (d-1)-faces. What physics one seeks to encode in the formalism depends on the exact choices of action (interaction kernel and propagator) and data associated to the basic tensorial field (domain space). This area of research has witnessed an impressive growth in recent years, with results on many aspects of the formalism: the construction of models for quantum gravity as well as statistical systems, studies on classical and quantum symmetries of the same models, analyses on the topology of the cellular complexes generated in perturbative expansion, the large-N limit, studies on perturbative TGFT renormalization, works on summability and critical behaviour, and the extraction of effective physics. For all this, we refer to the literature. Now, we focus on the definition of quantum gravity models.

2. Pre-geometric data: phase space, quantization maps, flux representation

The phase space underlying quantum gravity TGFT models is the cotangent bundle over a Lie group: $\mathcal{T}^*G \simeq G \times \mathfrak{g}$, with, in particular, $G = SU(2)$. This is the building block of the phase space of both simplicial gravity and LQG.[3,4,7] The group elements are interpreted as parallel transports of an $SU(2)$ connection along elementary links of a graph or of a (dual of a) simplicial complex, and the conjugate Lie algebra elements as fluxes of a dual (densitized) co-triad across (d-2)-faces dual to the same links. We refer to the literature for more details on the quantization. Here we report on some recent work[13] on a new representation for the quantum theory.[?]

The fundamental Poisson brackets are

$$\{\zeta^i(g), \zeta^j(g)\} = 0 \ , \ \{X_i, \zeta^j(g)\} = \tilde{\mathcal{L}}_i \zeta^j(g) \ , \ \{X_i, X_j\} = \kappa \epsilon_{ij}{}^k X_k \qquad (2)$$

where ζ are coordinates on the group manifold and $\mathcal{L}$ is a Lie derivative. Any choice of *quantization map* $\mathcal{Q} : C^\infty(\mathcal{T}^* SU(2)) \to \mathcal{A}$ will give the corresponding algebra of operators acting on some Hilbert space $\mathcal{H}$. Given the commutativity of functions on the group, a standard basis is given by group-labelled states, which provide a realization of the Hilbert space as $L^2(G)$. One can also look for a dual realization in terms of non-commutative functions of Lie algebra elements $X \in \mathfrak{su}(2) \simeq \mathbb{R}^3$, endowed with a $\star$-product following uniquely from the quantization map:

$$f \star h \equiv \mathcal{Q}^{-1}(\mathcal{Q}(f)\mathcal{Q}(h)) \quad . \qquad (3)$$

This allows to define the (flux) representation

$$\hat{\zeta}^i \tilde{\psi}(X) = -i\hbar \frac{\partial}{\partial X_i} \tilde{\psi}(X) \quad , \quad \hat{X}_i \tilde{\psi}(X) = \tilde{\psi}(X) \star X_i \quad . \qquad (4)$$

The coefficients of the change of basis from group-labelled states to Lie algebra-labelled states are non-commutative plane waves:

$$E_g(X) := e_\star^{\frac{i}{\hbar\kappa}k_g\cdot X} = \mu(g)e^{\frac{i}{\hbar}\zeta(g)\cdot X} \tag{5}$$

k_g are the coordinates obtained by inverse exponential map, and the $\star$-exponential is defined by the series expansion in $\star$-monomials of Lie algebra elements. The plane waves are thus generically $\star$-exponentials for the $\star$-product defined from the quantization map, and can be written as standard exponentials for some choice of coordinates on the group which also follows uniquely from the quantization map. They satisfy: $(E_{g_1} \star E_{g_2})(X) = E_{g_1g_2}(X)$. Using them, one can then define a unitary intertwining map between group and Lie algebra representations: a non-commutative Fourier transform. The flux representations and non-commutative Fourier transforms that follow from various quantization maps have been studied in .[13] The one based on the plane waves $E_g(x) = e^{i\zeta_g\cdot x}$, with $\zeta_g^i = \sin\theta n^i$, with $g = \cos\theta I + i\sin\theta n\cdot\sigma$ in the fundamental representation (σ^i are the Pauli matrices), has already found several applications in quantum gravity.[?,8]

3. A TGFT model for 4d quantum gravity

We now give an example of a TGFT model for 4d gravity.[9] We aim at a description of quantum spacetime as the result of the interaction of fundamental building blocks represented by quantum tetrahedra, the quanta of our TGFT field, and at encoding appropriately their quantum geometry. A geometric tetrahedron in $\mathbb{R}^4$ can be described by four bivectors $B_i \in \wedge^2\mathbb{R}^4 \simeq \mathfrak{so}(4) \simeq \mathfrak{su}(2)\oplus\mathfrak{su}(2)$, associated to its four triangles, and a vector $k \in S^3$, interpreted as its (unit) normal, satisfying:

$$N_I\left(*B_i^{IJ}\right) = 0 \qquad (simplicity) \qquad \sum_i B_i^{IJ} = 0 \qquad (closure) \tag{6}$$

So the classical phase space for a tetrahedron, before the imposition of constraints, is $[\mathcal{T}^*SO(4)]^4 \simeq [\mathcal{T}^*SU(2)\times\mathcal{T}^*SU(2)]^4$. In selfdual and anti-selfdual components, the first condition becomes $B_+^i + kB_-^i k^{-1} = 0$, where $\bar{k} := (\bar{k}_-, \bar{k}_+) \in SO(4)$ maps the vector $N^I = (1,0,0,0)$ to k^I, then $k = \bar{k}_+\bar{k}_-^{-1} \in SU(2)$. One can also change variables to $\bar{B} = B + \frac{1}{\gamma}*B$ so that $\beta\bar{B}_+^i + k\bar{B}_-^i k^{-1} = 0$ with $\beta = \frac{\gamma-1}{\gamma+1}$. We define a TGFT field representing a quantum tetrahedron $\varphi_k(g_1,\cdots g_4) \leftrightarrow \varphi_k(x_1,\cdots x_4)$, with $g_i \in SO(4)$ and $x_i \in \mathfrak{so}(4)$ (representing triangle bivectors). On this, we impose:

$$(C \triangleright \varphi)_k = \int dh\, E_h \cdots E_h \star \varphi_{h^{-1}\triangleright k} \quad . \tag{7}$$

Upon integration over k, this gives the closure condition (this is a generalization of the invariance of the Boulatov TGFT field). Then, we impose the simplicity condition using the function:

$$S_k^\beta(x) := \delta_{-kx^-k^{-1}}(\beta x^+) = \int_{SU(2)} du\, e^{itr[k^{-1}ukx^-]} e^{i\beta tr[ux^+]} \tag{8}$$

where $\delta_{-a}(b) := \delta(a+b)$ is the $\mathfrak{su}(2)$ non-commutative delta function, as:

$$(S^\beta \triangleright \varphi)_k(x_1, \cdots, x_4) = \prod_{j=1}^{4} S_k^\beta(x_j) \star \varphi_k(x_1, \ldots x_4) \quad . \tag{9}$$

Defining $\widehat{\Psi}_k := S \triangleright C \triangleright \varphi_k = C \triangleright S \triangleright \varphi_k$, the action for the TGFT model imposing all the geometric conditions is

$$S = \frac{1}{2} \int [d^6 x_i]^4\, dk\, \varphi_{k1234} \star \varphi_{k1234} \; +$$
$$+ \frac{\lambda}{5!} \int [d^6 x_i][dk_a]\, \widehat{\Psi}_{1234k_a} \star \widehat{\Psi}_{4567k_b} \star \widehat{\Psi}_{7389k_c} \star \widehat{\Psi}_{962\,10k_d} \star \widehat{\Psi}_{10\,851k_e} \tag{10}$$

where the star product pairs repeated indices.

The Feynman amplitudes of this TGFT take the form of a non-commutative simplicial path integral for the Holst-Plebanski action (with Immirzi parameter γ), and a quantum measure including geometric constraints on both bivectors and discrete connection. The expansion of the same amplitudes in group representations gives a spin foam model, encoding these constraints as conditions on the embedding of $SU(2)$ representations into $SO(4)$ representations.

4. Why adding pre-geometric information

Let us now summarize some motivations for enriching tensor models with pre-geometric data to give TGFTs. The first reason is exemplified by the TGFT presented above: due to the additional data, the Feynman amplitudes of TGFTs can be given by simplicial gravity path integrals and, dually, by spin foam models. Thus, TGFTs are a 2nd quantized formulation of the spin network dynamics of LQG and connect directly with other quantum gravity approaches. From them, they can import techniques and physical insights, which in turn can guide both TGFT model building and the analysis of the resulting models. Compared to tensor models, TGFTs can be endowed with new symmetries, e.g. the analogue of 3d simplicial diffeomorphisms,[15] possibly leading to different critical behaviour and phase structure. A true renormalization flow for TGFTs can be defined and the issue of renormalizability addressed.[11] Last, the additional pre-geometric information is an important guide for 'reading out'possible geometric and physical meaning from the TGFT field and action and states, using the insights coming from LQG, spin foam models and simplicial gravity. The possibility of such physically motivated guesses and approximations will be crucial in the analysis of any TGFT model for gravity, and in any attempt to relate it to effective continuum physics (e.g. via mean field methods[16]).

References

1. P. Di Francesco, P. Ginsparg, J. Zinn-Justin, Phys. Rept. 254 (1995) 1-133
2. J. Ambjorn, B. Durhuus, T. Jonsson, Mod. Phys. Lett. **A6**, 1133-1146, (1991); N. Sasakura, Mod.Phys.Lett. A6 (1991) 2613-2624
3. T. Thiemann, *Modern canonical quantum General Relativity*, Cambridge University Press, Cambridge (2007)
4. A. Perez, to appear in Living Reviews, arXiv:1205.2019 [gr-qc];
5. R. Gurau, J. Ryan, SIGMA 8 (2012) 020, arXiv:1109.4812 [hep-th]
6. V. Bonzom, R. Gurau, V. Rivasseau, Phys.Rev. D85 (2012) 084037, arXiv:1202.3637 [hep-th]
7. D. Oriti, in G. Ellis et al. (eds.), *Foundations of space and time*, Cambridge University Press, Cambridge (2012), arXiv:1110.5606 [hep-th]; D. Oriti, in: D. Oriti (ed.), *Approaches to quantum gravity*, Cambridge University Press, Cambridge (2009), arXiv: gr-qc/0607032
8. A.Baratin, D.Oriti, PRL 105 (2010) 221302, arXiv:1002.4723[hep-th]
9. A. Baratin, D. Oriti, Phys.Rev. D85 (2012) 044003, arXiv:1111.5842 [hep-th]
10. R. De Pietri, L. Freidel, K. Krasnov, C. Rovelli, Nucl. Phys. B **574**, 785 (2000), [arXiv: hep-th/9907154]
11. S. Carrozza, D. Oriti, V. Rivasseau, arXiv:1207.6734 [hep-th]; S. Carrozza, D. Oriti, V. Rivasseau, to appear; V. Rivasseau, PoS CNCFG2010 (2010) 004, arXiv:1103.1900 [gr-qc]; J. Ben Geloun, arXiv:1205.5513 [hep-th];J. Ben Geloun, V. Rivasseau, Comm. Math Phys. 318, pp 69-109 (2013)
12. M.Dupuis, J.Ryan, S.Speziale, SIGMA 8 (2012) 052, arXiv:1204.5394[gr-qc]
13. C. Guedes, D. Oriti, M. Raasakka, to appear
14. D. Oriti, PoS(QG-Ph)030, [arXiv:0710.3276]; L. Sindoni, arXiv:1105.5687 [gr-qc]; D. Oriti, J.Phys.Conf.Ser. 67 (2007) 012052, arXiv: hep-th/0612301
15. A. Baratin, F. Girelli and D. Oriti, Phys. Rev. D **83**, 104051 (2011) [arXiv:1101.0590 [hep-th]]
16. D. Oriti, L. Sindoni, New J.Phys. 13 (2011) 025006, arXiv:1010.5149 [gr-qc]; E. Livine, D. Oriti, J. Ryan, Class.Quant.Grav. 28 (2011) 245010, arXiv:1104.5509 [gr-qc]; W. Fairbairn, E. Livine, Class.Quant.Grav. 24 (2007) 5277-5297, arXiv: gr-qc/0702125
17. R. Gurau and V. Rivasseau, Europhys. Lett. **95**, 50004 (2011) [arXiv:1101.4182 [gr-qc]]; R. Gurau, arXiv:1102.5759 [gr-qc]
18. B. Bahr, B. Dittrich, F. Hellmann, W. Kaminski, arXiv:1208.3388 [gr-qc]; B. Dittrich, F. Eckert, M. Martin-Benito, arXiv:1109.4927 [gr-qc]

CANONICAL TENSOR MODEL WITH LOCAL TIME AND ITS UNIQUENESS

N. SASAKURA

Yukawa Institute for Theoretical Physics, Kyoto University,
Kyoto, 606-8502, Japan
E-mail: sasakura@yukawa.kyoto-u.ac.jp

A canonical formalism of the rank-three tensor model with the notion of local time is proposed. The consistency of the local time evolution is guaranteed by imposing that local Hamiltonians and the $so(N)$ kinematical symmetry of the tensor model should form a first class constraint algebra. By imposing some physically reasonable assumptions, it is shown that there exist only two such local Hamiltonians with a slight difference in index contraction. The first class constraint algebra is shown to approach the DeWitt constraint algebra of the general relativity in a certain locality limit. Quantization of the system is briefly discussed.

Keywords: Tensor model; Canonical formalism; DeWitt constraint algebra; General relativity; Quantum gravity.

1. Introduction

Formulation of quantum gravity is remaining as one of the major difficulties in physics. An interesting direction is to pursue the possibility that the space-time is an emergent phenomenon. The idea of emergent space-time and gravity has attracted much attention in recent years.[1]

A class of models with such an idea is given by the matrix models. Feynman diagrams generated from a matrix model can be interpreted as dual diagrams of two-dimensional simplicial manifolds. From various aspects, it is established that the matrix models correctly describe the two-dimensional gravity.[2]

A natural extension of the matrix models to higher dimensions was proposed as tensor models.[3–5] They have tensors in place of matrices as their degrees of freedom. The tensor models generate Feynman diagrams which may be regarded as dual diagrams of higher dimensional simplicial manifolds. However, the original proposal with symmetric tensors has turned out to have various difficulties. Recently, a more successful class of tensor models with unsymmetric tensors, called colored tensor models, has been proposed and triggered a number of successive studies.[6]

In the above description, however, there exists a fundamental issue in regarding the tensor models as theory of emergent space-time and gravity; the ranks of tensors depend on dimensions of simplicial manifolds. This is serious because dimensions should be a dynamical quantity rather than an input parameter in the idea of emergent space-time.

A way to circumvent this issue is to interpret rank-three tensor models in another way than above. A fuzzy space defines a space in terms of an algebra of functions f_a $(a = 1, 2, \ldots, N)$ on it as $f_a f_b = C_{ab}{}^c f_c$, where $C_{ab}{}^c$ is the structure constant. By identifying $C_{ab}{}^c$ with a dynamical variable of a rank-three tensor model, it may be regarded as a theory of dynamical fuzzy spaces.[7] In this interpretation, a rank-three tensor model can in principle deal with any dimensional spaces, since fuzzy spaces can approximate any continuous spaces. This interpretation will be justified, only when its dynamics is shown to reproduce a notion of space-time with properties required by the nature, such as classicality, smoothness, three-dimensionality, locality, causality, symmetries, containing gravity and matters. This is a long-range purpose of the study of rank-three tensor models.

The main topic here is how to introduce time in a rank-three tensor model by constructing local Hamiltonians. A non-trivial issue in the introduction is the mutual consistency of local time evolutions. As in the general relativity, this will be guaranteed by a first class constraint algebra formed by local Hamiltonians and a kinematical symmetry of a rank-three tensor model.

2. The rank-three tensor model

The rank-three tensor model in this paper has a rank-three tensor which satisfies the generalized Hermiticity condition,

$$M_{abc} = M_{bca} = M_{cab} = M^*_{bac} = M^*_{acb} = M^*_{cba}, \tag{1}$$

where $a, b, c = 1, 2, \cdots, N$. Because of the generalized Hermiticity condition, the symmetry of the tensor model is the orthogonal group,

$$M'_{abc} = L_a{}^{a'} L_b{}^{b'} L_c{}^{c'} M_{a'b'c'}, \quad L \in O(N). \tag{2}$$

I also introduce the canonical conjugate to M_{abc}, which satisfies the

Poisson bracket, $\{M_{abc}, P_{def}\} = \delta_{abc,def}$, where

$$\delta_{abc,def} \equiv \delta_{ad}\delta_{be}\delta_{cf} + \delta_{ae}\delta_{bf}\delta_{cd} + \delta_{af}\delta_{bd}\delta_{ce}, \tag{3}$$

$$P_{abc} = P_{bca} = P_{cab} = P^*_{bac} = P^*_{acb} = P^*_{cba}. \tag{4}$$

The generators of the symmetry (2) are given by

$$\mathcal{D}_{[ab]} = \frac{1}{2}\left(P_{acd}M_{bcd} - M_{acd}P_{bcd}\right), \tag{5}$$

where $[ab]$ denotes anti-symmetric indices.

3. The interpretation in terms of fuzzy spaces

A fuzzy space is defined by an algebra of functions f_a $(a = 1, 2, \ldots, N)$ on it,

$$f_a f_b = M_{abc} f_c. \tag{6}$$

Because of the generalized Hermiticity condition (1), a fuzzy space associated to a rank-three tensor model is equipped with an inner product which has a trace-like property as

$$\langle f_a | f_b \rangle = \delta_{ab}, \tag{7}$$

$$\langle f_a | f_b f_c \rangle = \langle f_a f_b | f_c \rangle = \langle f_c f_a | f_b \rangle, \tag{8}$$

$$f_a^* = f_a, \tag{9}$$

$$(f_a f_b)^* = f_b f_a. \tag{10}$$

Note that associativity is not assumed in general. A non-associative algebra with such a trace-like inner product may be called a non-associative Frobenius algebra. An algebra without associativity does not often tend to show physically interesting general properties. On the other hand, this trace-like feature of the inner product plays key roles in showing that the kind of fuzzy spaces have various quantum mechanical properties and that the symmetry transformation (2) can be expressed by n-ary transformations.[7]

I will give here two simple examples of the above kind of fuzzy spaces. The function algebra of a usual space is given by

$$f_{z_1} f_{z_2} = \delta^D(z_1 - z_2) f_{z_1}, \tag{11}$$

where z_i are D-dimensional coordinates. This can be understood by considering functions of x which are labeled by z, $f_z(x) \equiv \delta^D(x - z)$ and that $f_{z_1}(x) f_{z_2}(x) = \delta^D(x - z_1)\delta^D(x - z_2) = \delta^D(z_1 - z_2) f_{z_1}(x)$. In fact, all the functions of x can be expressed as a linear combination of f_z. Thus, with M_{abc}, a usual space can be represented by $M_{z_1 z_2 z_3} = \delta^D(z_1 - z_2)\delta^D(z_2 - z_3)$.

One may consider a fuzzy space by regularizing the delta functions by Gaussians as

$$M_{z_1 z_2 z_3} = B \exp\left(-\beta\left((z_1 - z_2)^2 + (z_2 - z_3)^2 + (z_3 - z_1)^2\right)\right), \tag{12}$$

where B, β are positive numbers. Since this respects the D-dimensional Poincaré symmetry, the fuzzy space should be regarded as a fuzzy D-dimensional flat space.

4. Symmetry of tensor models and diffeomorphism on fuzzy spaces

From the relation (6), the symmetry transformation (2) of the tensor model is mapped to a transformation of a fuzzy space,

$$f'_a = L_a{}^{a'} f_{a'}, \quad L \in O(N). \tag{13}$$

Since the functions f_a can intuitively be considered to represent fuzzy "points", the transformation (13) can be regarded as a fuzzy analogue of coordinate transformation.

One can discuss this more precisely. The generators (5) satisfy the $so(N)$ Lie algebra,

$$\{D(V_1), D(V_2)\} = D([V_1, V_2]), \tag{14}$$

$$D(V) \equiv V^{[ab]} \mathcal{D}_{[ab]}, \tag{15}$$

where $V^{[ab]}$ denotes an anti-symmetric matrix, and $[V_1, V_2]$ is the matrix commutator. Now let me formally replace the indices with D-dimensional coordinates, and consider

$$V^{[xy]} \equiv \frac{1}{2}\left(v^\mu(x) + v^\mu(y)\right) \delta^D_\mu(x - y), \tag{16}$$

which is parameterized by a vector field $v^\mu(x)$. Then one can find that $V_3 = [V_1, V_2]$ leads to

$$v_3^\mu(x) = v_1^\nu(x)\partial_\nu v_2^\mu(x) - v_2^\nu(x)\partial_\nu v_1^\mu(x) = \mathcal{L}_{v_1} v_2. \tag{17}$$

This is indeed the algebra of diffeomorphism.

5. Construction of local Hamiltonians

Mutual consistency of local time evolutions requires that commutation of local Hamiltonians should vanish up to symmetries, namely,

$$\{\mathcal{H}_a, \mathcal{H}_b\} = \text{symmetry generators.} \tag{18}$$

In the case of tensor models, there actually exists a large symmetry (2), where N is intuitively the number of "points".

The search for such consistent local Hamiltonians is a complicated process. The final result[8] is that there exist only two such Hamiltonians,

$$\mathcal{H}_a = P_{a(bc)}P_{bde}M_{cde} \text{ or } \mathcal{H}_a = P_{a(bc)}P_{bde}M_{ced}, \tag{19}$$

with (ab) denoting symmetrization of the indices, which have a slight difference in index contraction, on the following assumptions: (A1) Local Hamiltonian has one index. (A2) Form a consistent first class constraint algebra. (A3) Invariant under time reversal. (A4) At most cubic polynomial function. (A5) Terms are connected.

6. First class constraint algebra and DeWitt algebra

The local Hamiltonians (19) and the $so(N)$ generators (5) satisfy

$$\{H(T_1), H(T_2)\} = D([\tilde{T}_1, \tilde{T}_2]), \tag{20}$$

$$\{D(V), H(T)\} = H(VT), \tag{21}$$

where $H(T) \equiv T^a\mathcal{H}_a,\ \tilde{T}_{ab} \equiv T^c P_{c(ab)}$. With (14), they form a first class constraint algebra. Note that (20) has structure functions rather than constants, because of the dependence of $\tilde{T}$ on P_{abc}. This means that the algebra is in fact an open algebra, which closes only on the constraint surfaces.

This situation is very similar to the general relativity. In fact, one can show that, by formally replacing the indices with coordinates, substituting (12) into P_{abc} in the right-hand side of (20), and taking $\beta \to \infty$ limit, one reproduces the DeWitt constraint algebra of general relativity.[9]

7. Quantization

A way of quantization is to promote the canonical variables to the operators satisfying $[\hat{M}_{abc}, \hat{P}_{def}] = i\delta_{abc,def}$. Then the local Hamiltonians are generally corrected by operator ordering as

$$\hat{\mathcal{H}}_a = \hat{P}_{a(bc)}\hat{P}_{bde}\hat{M}_{cde} + i\lambda\hat{P}_{abb}. \tag{22}$$

This is similar to the other Hamiltonian. One can see that the first class constraint algebra is not deformed by the quantization. Therefore the algebra gives a consistent quantization of the system.

Another way of quantization is to apply the path integral quantization of constraint systems.[10] It can formally be written as

$$Z = \int \mathcal{D}M\mathcal{D}P\, \delta(\mathcal{H}_a)\delta(\mathcal{D}_{ab})\delta(\chi_a)\delta(\chi_{ab})\mathrm{Det}(\{\mathcal{H}\,\mathcal{D}, \chi\})e^{i\int PdM}, \tag{23}$$

where χ_a, χ_{ab} are gauge-fixing conditions.

8. Summary and future prospects

A canonical tensor model with the notion of local time has been proposed. The consistency of local time evolutions is guaranteed by a first class constraint algebra among local Hamiltonians and symmetry generators. The two-fold uniqueness of local Hamiltonians has been proved on some physically reasonable assumptions. It would be highly interesting to study whether the quantization of the canonical tensor model leads to the emergence of space-time with the features required by the nature.

It would also be highly interesting to seek for connections to the CDT.[11]

Acknowledgments

I would like to thank the organizers for this nice opportunity to present this work at the workshop.

References

1. S. Carlip, "Challenges for Emergent Gravity," arXiv:1207.2504 [gr-qc].
2. P. Di Francesco, P. H. Ginsparg and J. Zinn-Justin, "2-D Gravity and random matrices," Phys. Rept. **254**, 1 (1995) [hep-th/9306153].
3. J. Ambjorn, B. Durhuus and T. Jonsson, "Three-Dimensional Simplicial Quantum Gravity And Generalized Matrix Models," Mod. Phys. Lett. A **6**, 1133 (1991).
4. N. Sasakura, "Tensor Model For Gravity And Orientability Of Manifold," Mod. Phys. Lett. A **6**, 2613 (1991).
5. N. Godfrey and M. Gross, "Simplicial Quantum Gravity In More Than Two-Dimensions," Phys. Rev. D **43**, 1749 (1991).
6. R. Gurau and J. P. Ryan, "Colored Tensor Models - a review," SIGMA **8**, 020 (2012) [arXiv:1109.4812 [hep-th]].
7. N. Sasakura, "Tensor models and 3-ary algebras," J. Math. Phys. **52**, 103510 (2011) [arXiv:1104.1463 [hep-th]].
8. N. Sasakura, "Uniqueness of canonical tensor model with local time," Int. J. Mod. Phys. A **27**, 1250096 (2012) [arXiv:1203.0421 [hep-th]].
9. N. Sasakura, "Canonical tensor models with local time," Int. J. Mod. Phys. A **27**, 1250020 (2012) [arXiv:1111.2790 [hep-th]].
10. L. D. Faddeev, "Feynman integral for singular Lagrangians," Theor. Math. Phys. **1**, 1 (1969) [Teor. Mat. Fiz. **1**, 3 (1969)].
11. J. Ambjorn, A. Goerlich, J. Jurkiewicz and R. Loll, "Nonperturbative Quantum Gravity," arXiv:1203.3591 [hep-th].

CONSTRUCTIVE RENORMALIZATION OF THE TWO DIMENSIONAL GROSSE-WULKENHAAR MODEL

ZHITUO WANG

Dipartimento di Matematica, Universita di Roma Tre, 00146, Roma, Italy
**E-mail: zhituo@mat.uniroma3.it*

In this talk we prove that the perturbation series of the 2-dimensional Grosse-Wulkenhaar model is Borel summable. The interested reader could look at[1] for more details.

1. Introduction

Quantum field theory defined on noncommutative manifold[2,3] is a promising way to study the physics at the Planck scale, at which the space-time becomes fuzzy or quantized. The simplest noncommutative space is the Moyal space, which is a symplectic manifold generated by the coordinates operators x_i obeying the commutation relation $[x_i, x_j] = i\theta_{ij}$, where θ_{ij} is a constant. The first perturbative renormalizable noncommutative quantum field theory in 4 dimensional Moyal space is the Grosse-Wulkenhaar model.[4,5] It is is defined by changing the product of fields in the commutative scalar ϕ^4 theory into the noncommutative Moyal product as well as adding an harmonic oscillator term, so that the action obeys a symmetry called the Langmann-Szabo duality.[6] The β function of this model is vanishing when we are approaching the fixed point of this model.[7–9] In this paper we consider the 2 dimensional Grosse-Wulkenhaar mode at the self-dual point. We prove that this model is constructive renormalizable, namely the perturbation series of the connected Schwinger's function is Borel summable, with the method of loop vertex expansion (LVE for short).[1,10–13] The LVE is a combination of the intermediate fields techniques with the BKAR forest formula. It is a new method in constructive physics and is suitable for dealing with the models with nonlocal interactions, for which the traditional methods like the cluster/Mayer expansions break down.

2. Moyal space and Grosse-Wulkenhaar Model

The 2-dimensional Grosse-Wulkenhaar model in the matrix basis is defined by (see[1] for more details):

$$S[\phi] = 2\pi\theta \mathrm{Tr}\Big[\frac{1}{2}\phi\Delta\phi + :\frac{\lambda}{4}\phi_\star^4:\Big]$$

$$= 2\pi\theta \sum_{m,n,k,l} \Big[\frac{1}{2}\phi_{mn}\Delta_{mn;kl}\phi_{kl} + \frac{\lambda}{4} : \phi_{mn}\phi_{nk}\phi_{kl}\phi_{lm} : \Big]. \tag{1}$$

where the scalar field ϕ is a Hermitian matrix. We have put the ultraviolet cutoff Λ to the matrix indices. Namely, $m,\ n = 1, \cdots, \Lambda$. We have taken the Wick ordering of the interaction terms so as to make the renormalization more explicite:

$$\phi_{mn}\phi_{nk}\phi_{kl}\phi_{lm} := \phi_{mn}\phi_{nk}\phi_{kl}\phi_{lm} - 8\phi_{mp}\phi_{pm}T_m^\Lambda + 6(T^\Lambda)^2_m, \tag{2}$$

where $T_m^\Lambda = \log\frac{\Lambda+m}{m} \sim \log\Lambda$, for $1 \leqslant m \ll \Lambda$, $\mathrm{Tr}(T_m^\Lambda)^2 \sim (2\ln^2 2 + \frac{\pi^2}{6})\Lambda$.

The elements of the covariance matrix $[C_{mn,kl}]$ reads:

$$C_{mn,kl} = \frac{1}{\mu^2 + \frac{4}{\theta}(m+n+1)}\delta_{ml}\delta_{nk}. \tag{3}$$

It is a $\Lambda^2 \times \Lambda^2$ dimensional diagonal matrix in the double indices.

3. The intermediate field representation and the Loop vertex expansion

3.1. *The intermediate field representation*

The partition function for the matrix model reads:

$$Z(\lambda) = \int d\mu(\phi_{mn}) e^{-S[\phi_{mn}]}, \tag{4}$$

where

$$d\mu(\phi_{mn}) = \pi^{-N^2/2} e^{-1/2[\sum_{mn}\phi_{mn}\Delta_{mn}\phi_{nm}]} \prod_{mn} d\mathrm{Re}(\phi_{mn}) d\mathrm{Im}(\phi_{mn}). \tag{5}$$

is the normalized Gaussian measure of the matrix fields ϕ with covariance C and $S[\phi_{mn}]$ is the Wick ordered interaction term.

We introduce the Hermitian matrix σ as an intermediate field and integrate out the original matrix fields ϕ. The result reads:

$$Z(\lambda) = \int d\mu(\sigma) e^{2i\sqrt{2\lambda}\mathrm{Tr}\, T^\Lambda\sigma - \frac{1}{2}\mathrm{Tr}\log[1 + i\sqrt{2\lambda}\ C^{1/2}(I\otimes\sigma + \sigma\otimes I)C^{1/2}] + \frac{5}{2}\lambda T_\Lambda^2}, \tag{6}$$

where $\mathrm{Tr}T^\Lambda\sigma = \sum_m T^\Lambda_m \sigma_{mm}$, I is the $\Lambda\times\Lambda$ dimensional identity matrix. $C^{1/2}$ is the square root of the covariance C. So it is still a diagonal matrix.

3.2. *The BKAR Tree formula*

The most interesting quantities in our model are the connected Schwinger's functions. We shall use the BKAR tree formula to calculate them.

Define the interaction terms in (6) as $\hat{V}$. We expand the exponential as $\sum_n \frac{V(\hat{\sigma})^n}{n!}$ using the replica trick, then the vacuum Schwinger's function is given by:

Theorem 3.1 (Loop Vertex Expansion[10]).

$$\log Z(\lambda,\Lambda,\mathcal{V}) = \sum_{n=1}^{\infty}\frac{1}{n!}\sum_{\mathcal{T}\text{ with } n \text{ vertices}} G_{\mathcal{T}} \tag{7}$$

$$G_{\mathcal{T}} = \Big\{\prod_{\ell\in\mathcal{T}}\sum_{m_\ell,n_\ell,p_\ell,q_\ell}[\int_0^1 dw_\ell]\Big\}\int d\nu_{\mathcal{T}}(\{\sigma^v\},\{w\})$$

$$\Big\{\prod_{\ell\in\mathcal{T}}[\delta_{n_\ell p_\ell}\delta_{m_\ell q_\ell}\frac{\delta}{\delta\sigma^{v(\ell)}_{m_\ell,n_\ell}}\frac{\delta}{\delta\sigma^{v'(\ell)}_{p_\ell,q_\ell}}]\Big\}\prod_{v=1}^{n}\hat{V}_v,$$

where

- *each line ℓ of the tree joins two different loop vertices $V^{v(\ell)}$ and $V^{v'(\ell)}$ which are identified through the function $\delta_{m_\ell q'_\ell}\delta_{n_\ell p'_\ell}$.*
- *the sum is over spanning trees joining all n loop vertices.*
- *the normalized Gaussian measure $d\nu_{\mathcal{T}}(\{\sigma_v\},\{w\})$ over the fields σ_v has now a covariance*

$$<\sigma^v_{mn},\sigma^{v'}_{kl}>=\delta_{nk}\delta_{ml}w^{\mathcal{T}}(v,v',\{w\}), \tag{8}$$

which depend on the "fictitious" indices. Here $w^{\mathcal{T}}(v,v',\{w\})$ equals to 1 if $v=v'$, and equals to the infimum of the w_ℓ for ℓ running over the unique path from v to v' in $\mathcal{T}$ if $v\neq v'$.

We define the full resolvent $\hat{R}(\lambda,\sigma)$ as:

$$\hat{R}_{mn}(\sigma,\lambda) = -\frac{2}{i\sqrt{2\lambda}}\frac{\partial}{\partial\sigma_{mn}}[-\frac{1}{2}\mathrm{Tr}\log(1+i\sqrt{2\lambda}C^{1/2}\sigma C^{1/2})]$$

$$= \big[C^{1/2}\frac{1}{1+i\sqrt{2\lambda}C^{1/2}\sigma C^{1/2}}C^{1/2}\big]_{mn}. \tag{9}$$

Define also the resolvent as $R=\frac{1}{1+i\sqrt{2\lambda}C^{1/2}\sigma C^{1/2}}$ and we have $\hat{R} = C^{1/2}\,R\,C^{1/2}$.

3.3. *The graph representation*

A typical graph of the LVE is a tree of loop vertices with the decoration of counter terms. There exist also the dual representation of LVE, which is a single big disc. Different loop vertices of the direct representation is drawn as different regions in the big disc.

We shall work in the dual representation for the rest of this paper. The

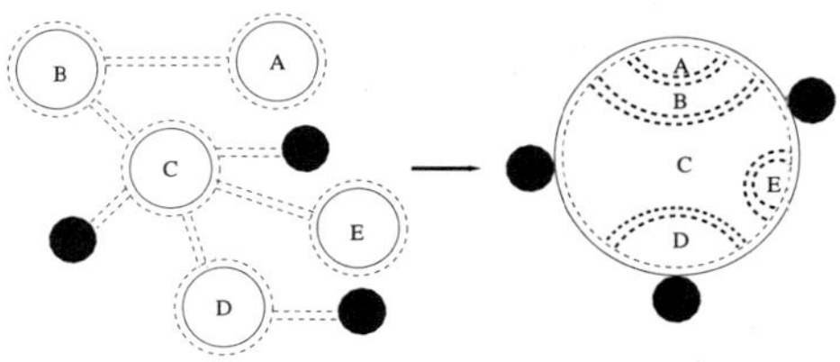

Fig. 1. The graphs of the direct (left) and the dual (right) representations.

amplitude for a tree of loop vertices reads:

$$G_\Lambda = \sum_{\mathcal{T}\text{ lines, } |\mathcal{T}|=n-1} (-\lambda)^{n-1} \int d\nu(\sigma, w) \mathrm{Tr}\Big\{ \vec{\prod}_{u\in\mathcal{O}} P_u \Big\}, \tag{10}$$

where $\mathcal{O}$ is the set of the resolvents and the counterterms. P_u means either the resolvent $\hat{R}$ or the counter term T_Λ, by following the cyclic ordering in the dual graph. By using the fact that the resolvent is bounded in norm by one and the Cayley's theorem which states that the number of labeled trees of n vertices is bounded by $n^{n-2} \sim n!$, we find easily that the amplitude of an arbitrary graph of order n is bounded by:

$$|G_\Lambda| \leq \sum_n \frac{n^{n-2}}{n!} \lambda^n (\ln\Lambda)^n \leq \sum (\lambda K)^n (\ln\Lambda)^n, \tag{11}$$

which is divergent when $\Lambda \to \infty$. We should introduce a renormalization process, which is called the cleaning expansion, to cancel the divergent counter terms. We shall work in the multiscale representation of the amplitudes for the cleaning expansions.

4. The Cleaning Expansion

The idea of the cleaning expansion is to expand and then contract the σ fields hidden in the resolvent R to compensate the counter terms. We shall use the integration by parts. In this process we would generate either an inner tadpole, a crossing line or a nesting line. The amplitude of an inner tadpole is exactly the same as the counter term but with a minus sign, so the compensation is exact. The amplitude of a crossing or nesting line of scale j is proportional to M^{-2j}. We impose also the stopping rule for the cleaning expansion so that we generate enough convergent terms but don't expand forever (otherwise this would generate the full perturbation series which is divergent due to the combinatorial factors). We stop the expansion until the number of crossings or nesting lines reaches $n = aj_{max}^2$, so that we could gain a convergent factor $e^{-aj_{max}^3}$ after we stop the cleaning expansion. The interested readers could look at[1,13] for more details.

5. Nelson's argument and the bound of the connected function

There might be still arbitrarily number of counter terms in the uncleaned part after we stop the cleaning expansion and they are divergent. We shall resum them by using the integration by parts.

After the resummation we shall generate a divergent factor $e^{\lambda(\ln\Lambda)^3}$. But this factor is not dangerous, as we could bound it by the following formula:

$$e^{-aj_{max}^3} \cdot (j_{max}^2)! \cdot e^{\lambda \ln^3 \Lambda} \sim e^{-aj_{max}^3 + 2j_{max}^2 \ln j_{max} + \lambda j_{max}^3} < 1 \qquad (12)$$

as long as a is chosen properly, for example $a = 1.4\lambda$. Here we have use the fact that fact that $\ln \Lambda \sim j_{max}$ and $(aj_{max}^2)! \sim e^{2j_{max}^2 \ln j_{max}}$.

6. The Borel summability

We could easily prove the Borel summability of the perturbation series by using the fact that each resolvent is still bounded by finite number in the Borel plane, the complex analytic domain for the Schwinger's function. The interested reader could consult[1] and the reference therein for the details.

7. Conclusions and prospectives

In this paper we have constructed the 2-dimensional Grosse Wulkenhaar model with the method of loop vertex expansion (LVE). We expect that this method can be used to construct the 4 dimensional Grosse-Wulkenhaar model. The LVE could also be employed to study the random tensor field theories.

References

1. Zhituo Wang, "Construction of 2-dimensional Grosse-Wulkenhaar Model," arXiv:1104.3750 [math-ph]. submitted.

2. M. R. Douglas and N. A. Nekrasov, "Noncommutative field theory," Rev. Mod. Phys. **73** (2001) 977 [hep-th/0106048].
3. R. J. Szabo, "Quantum field theory on noncommutative spaces," Phys. Rept. **378** (2003) 207 [hep-th/0109162].
4. H. Grosse and R. Wulkenhaar, "Power-counting theorem for non-local matrix models and renormalisation," Commun. Math. Phys. **254** (2005) 91 [arXiv:hep-th/0305066].
5. H. Grosse and R. Wulkenhaar, "Renormalisation of phi**4 theory on noncommutative R**4 in the matrix base," Commun. Math. Phys. **256**, 305 (2005) [arXiv:hep-th/0401128].
6. E. Langmann and R. J. Szabo, Duality in scalar field theory on noncommutative phase spaces, *Phys. Lett.* **B533** (2002) 168–177,
7. H. Grosse and R. Wulkenhaar, "The beta-function in duality-covariant noncommutative phi**4 theory," Eur. Phys. J. C **35** (2004) 277 [arXiv:hep-th/0402093].
8. M. Disertori and V. Rivasseau, "Two and three loops beta function of non commutative phi(4)**4 theory," Eur. Phys. J. C **50** (2007) 661 [arXiv:hep-th/0610224].
9. M. Disertori, R. Gurau, J. Magnen and V. Rivasseau, "Vanishing of beta function of non commutative phi(4)**4 theory to all orders," Phys. Lett. B **649** (2007) 95 [arXiv:hep-th/0612251].
10. V. Rivasseau, "Constructive Matrix Theory," JHEP **0709** (2007) 008 [arXiv:0706.1224 [hep-th]].
11. J. Magnen and V. Rivasseau, Constructive field theory without tears, Ann. Henri Poincaré **9** 403-424 (2008), arXiv:0706.2457[math-ph].
12. V. Rivasseau and Zhituo Wang, "How are Feynman graphs resumed by the Loop Vertex Expansion?", arXiv:1006.4617 [math-ph].
13. V. Rivasseau and Zhituo Wang, "Constructive Renormalization for Φ_2^4 Theory with Loop Vertex Expansion", arXiv:1104.3443 [math-ph].
14. D. Brydges and T. Kennedy, Mayer expansions and the Hamilton-Jacobi equation, Journal of Statistical Physics, **48**, 19 (1987).
15. A. Abdesselam and V. Rivasseau, "Trees, forests and jungles: A botanical garden for cluster expansions," in Constructive Physics, Lecture Notes in Physics 446, Springer Verlag, 1995, arXiv:hep-th/9409094.

PARALLEL SESSION
Conformal Field Theories

20 & 21 & 23 AUGUST 2012

A BRAIDED MONOIDAL CATEGORY FOR SYMPLECTIC FERMIONS

A. DAVYDOV[1], I. RUNKEL[2*]

[1] *Dept. Math., Ohio Univ., Athens, Ohio 45701, USA*
[2] *Dept. Math., Hamburg Univ., Bundesstr. 55, 20146 Hamburg, Germany*

We describe a class of examples of braided monoidal categories which are built from Hopf algebras in symmetric categories. The construction is motivated by a calculation in two-dimensional conformal field theory and is tailored to contain the braided monoidal categories occurring in the study of the Ising model, their generalisation to Tamabara-Yamagami categories, and categories occurring for symplectic fermions.

1. Introduction

In this short note we summarise some of the results in Refs. 1 and 2, where also more extensive references can be found.

We are interested in a particular type of $\mathbb{Z}/2\mathbb{Z}$-graded braided monoidal categories. The grade 0 component is the monoidal category $\mathrm{Rep}_{\mathcal{S}}(H)$ of modules over a Hopf algebra H in a symmetric monoidal category $\mathcal{S}$, and the grade 1 component is the category $\mathcal{S}$ itself. We will write $\mathcal{C} = \mathcal{C}_0 + \mathcal{C}_1$ with $\mathcal{C}_0 = \mathrm{Rep}_{\mathcal{S}}(H)$ and $\mathcal{C}_1 = \mathcal{S}$. The tensor product functor $*$ on the various components is defined as:

A	B	$A * B$		comments
$\mathcal{C}_0$	$\mathcal{C}_0$	$A \otimes B$	$\in \mathcal{C}_0$	the H-action is via the coproduct of H
$\mathcal{C}_0$	$\mathcal{C}_1$	$F(A) \otimes B$	$\in \mathcal{C}_1$	$F : \mathrm{Rep}_{\mathcal{S}}(H) \to \mathcal{S}$ is the forgetful functor
$\mathcal{C}_1$	$\mathcal{C}_0$	$A \otimes F(B)$	$\in \mathcal{C}_1$	
$\mathcal{C}_1$	$\mathcal{C}_1$	$H \otimes A \otimes B$	$\in \mathcal{C}_0$	the H-action is by multiplication

This somewhat ad-hoc looking definition of the tensor product is actually quite natural. The mixed tensor products are the natural left and right action of the monoidal category $\mathrm{Rep}_{\mathcal{S}}(H)$ on $\mathcal{S}$. To obtain the last line in the table, assume that $\mathcal{C}$ can be made rigid. Writing T for the tensor unit of $\mathcal{S}$ considered as an object in $\mathcal{C}_1$, we have for all H-modules M

$$\mathrm{Hom}_{\mathcal{C}_0}(T^* * T, M) \;\cong\; \mathrm{Hom}_{\mathcal{C}_1}(T, T * M) \;\cong\; \mathrm{Hom}_{\mathcal{S}}(1, F(M)) \;. \qquad (1)$$

This means that $T^* * T$ is a representing object for the functor $M \mapsto \mathrm{Hom}_{\mathcal{S}}(1, F(M))$, and so $T^* * T \cong H$ as an H-module. If we in addition

*IR thanks the organisers of the *XXIX Internat. Colloq. on Group-Theoretical Methods in Physics* (August 20–26, 2012, Chern Inst., Tianjin) for the opportunity to speak.

demand that $T^* \cong T$, the last line in the above table follows.

Given the above form of the tensor product functor $*$ on $\mathcal{C}$, one can ask if it is possible to describe associativity and braiding isomorphisms for $*$ in terms of Hopf algebraic data on H. Our results for this question are given in Section 4. But before getting there, in Sections 2 and 3 we would like to give the two examples of such $\mathbb{Z}/2\mathbb{Z}$-graded braided monoidal categories which were our main motivation when setting up the formalism.

2. Tambara-Yamagami categories

For simplicity, we will work over the field $\mathbb{C}$. Consider a fusion category $\mathcal{C}$ over $\mathbb{C}$ whose simple objects are labelled by $G \cup \{m\}$ where G is a finite group and m is an extra label. Suppose that the tensor product $*$ is of the form, for $a, b \in G$,

$$a * b \cong ab \quad , \quad m * a \cong m \cong a * m \quad , \quad m * m \cong \bigoplus_{g \in G} g \; . \tag{2}$$

This tensor product is a special case of the one in the above table: the underlying symmetric category $\mathcal{S}$ is $vect(\mathbb{C})$, the category of finite dimensional $\mathbb{C}$ vector spaces. The component $\mathcal{C}_0$ is spanned by the simple objects $g \in G$; the component $\mathcal{C}_1$ is spanned by m alone, so that $\mathcal{C}_1 \cong vect(k)$. The Hopf algebra $H \in \mathcal{S}$ is the function algebra $\mathrm{Fun}(G, \mathbb{C})$.

For any such fusion category $\mathcal{C}$, the group G is necessarily abelian and $\mathcal{C}$ is monoidally equivalent to $\mathcal{C}(\chi, \tau)$, which is defined as follows [3, Thm. 3.2]. $\mathcal{C}(\chi, \tau)$ has simple objects and fusion rules as in (2), and its associator is determined by a symmetric non-degenerate bicharacter $\chi : G \times G \to \mathbb{C}^\times$ and a choice of $\tau \in \mathbb{C}^\times$ such that $\tau^2 = |G|^{-1}$. The associator is a bit lengthy and we refer to Ref. 3.

The category $\mathcal{C}(\chi, \tau)$ allows for a braiding if and only if G is an elementary 2-group (i.e. $gg = e$ for all $g \in G$).[4] The braiding isomorphisms are determined by a quadratic form σ associated to the bicharacter[a] χ and a number β such that $\beta \neq 0$ and $\beta^2 = \tau \sum_{a \in G} \sigma(a)$. Explicitly, under the identifications (1) the braiding is $(a, b \in G)$

$$c_{a,b} = \chi(a,b)\, id_{ab} \; , \; c_{a,m} = \sigma(a)\, id_m = c_{m,a} \; , \; c_{m,m} = \bigoplus_{g \in G} \beta\, \sigma(g)^{-1}\, id_g \; . \tag{3}$$

An important example of a braided monoidal category of the above type is provided by the two-dimensional critical Ising model. There, one considers

[a] This means that $\sigma : G \to \mathbb{C}^\times$ satisfies $\sigma(a) = \sigma(a^{-1})$, $\sigma(e) = 1$, and that $\chi(a,b)\sigma(a)\sigma(b) = \sigma(ab)$ for all $a, b \in G$.

the three irreducible representations $\hat{\mathbf{1}}$, $\hat{\varepsilon}$, $\hat{\sigma}$ of the Virasoro algebra which have central charge $c = \frac{1}{2}$ and lowest L_0-weights $h_{\hat{\mathbf{1}}} = 0$, $h_{\hat{\varepsilon}} = \frac{1}{2}$ and $h_{\hat{\sigma}} = \frac{1}{16}$. The fusion rules are of the form (1) where $\hat{\mathbf{1}}, \hat{\varepsilon}$ generate the group $G = \mathbb{Z}/2\mathbb{Z}$ and $m = \hat{\sigma}$ has fusion rule $\hat{\sigma} * \hat{\sigma} \cong \hat{\mathbf{1}} \oplus \hat{\varepsilon}$. The braiding isomorphism $c_{r,s}$ – projected to the simple object $t \in r*s$ – is multiplication by $\exp(\pi i(h_r + h_s - h_t))$. Comparing to (3) shows that the braided monoidal structure is determined by $\sigma(\hat{\varepsilon}) = \exp(\pi i/2)$, $\beta = \exp(\pi i/8)$ and thus $\chi(\hat{\varepsilon}, \hat{\varepsilon}) = -1$, $\tau = 1/\sqrt{2}$.

3. Symplectic fermions

Continuing with examples from two-dimensional conformal field theory, we now consider symplectic fermions.[5] The mode algebra of n pairs of symplectic fermsions is determined by a $2n$-dimensional symplectic vector space $\mathfrak{h}$. It is convenient to think of $\mathfrak{h}$ as a purely odd abelian Lie super-algebra with non-degenerate super-symmetric pairing $(-,-)$; we will use this language in the following. The symplectic fermion mode algebra is the affinisation $\hat{\mathfrak{h}}$ of $\mathfrak{h}$ with central element K and graded bracket $[a_m, b_n] = m\,(a,b)\,\delta_{m+n,0}\,K$, where $m, n \in \mathbb{Z}$ for untwisted (Neveu-Schwarz) representations and $m, n \in \mathbb{Z} + \frac{1}{2}$ for twisted (Ramond) representations.

Denote by $S(\mathfrak{h})$ the symmetric algebra of $\mathfrak{h}$ in $svect(\mathbb{C})$, the category of finite-dimensional complex super-vector spaces. Note that as a vector space, $S(\mathfrak{h})$ is simply the exterior algebra of the vector space underlying $\mathfrak{h}$; in particular, $S(\mathfrak{h})$ is finite-dimensional. The categories of untwisted and twisted representations of $\hat{\mathfrak{h}}$ (of a certain type) are equivalent to [2, Thms. 2.4 & 2.8]:

$$\text{(untwisted)} \;\; \mathcal{C}_0 := \mathrm{Rep}_{svect}\, S(\mathfrak{h}) \qquad \text{(twisted)} \;\; \mathcal{C}_1 := svect(\mathbb{C})\;. \tag{4}$$

We would like to stress that for $\dim \mathfrak{h} = 2n > 0$, $\mathcal{C}_0$ is not semi-simple.

A conformal field theory calculation endows the category $\mathcal{C} = \mathcal{C}_0 + \mathcal{C}_1$ with a $\mathbb{Z}/2\mathbb{Z}$-graded tensor product [2, Thm. 3.13]. This tensor product is of the form stated in Section 1 with symmetric category $\mathcal{S} = svect(\mathbb{C})$ and Hopf algebra $H = S(\mathfrak{h})$.

The associativity isomorphism is determined by the copairing $C \in \mathfrak{h} \otimes \mathfrak{h}$ dual to the super-symmetric pairing $(-,-)$ on $\mathfrak{h}$, and by a top-form λ on $S(\mathfrak{h})$ such that $(\lambda \otimes \lambda)(e^{-C}) = 1$ [2, Thm. 6.2]. To be more specific, pick a basis $\{e_i\}_{i=1,\dots,2n}$ of $\mathfrak{h}$ such that the pairing takes the standard form $(e_{2k-1}, e_{2k}) = 1 = -(e_{2k}, e_{2k-1})$ for $k = 1, \dots, n$. Then $C = \sum_{k=1}^{n} (e_{2k} \otimes e_{2k-1} - e_{2k-1} \otimes e_{2k})$ and, if we set $\hat{C} = -2 \sum_{k=1}^{n} e_{2k} \otimes e_{2k-1}$, the top-form λ is determined by $\lambda(\hat{C}^n) = n!(-2i)^n$. The explicit form of the associativity isomorphisms will be given as a special case of Theorem 4.1 below.

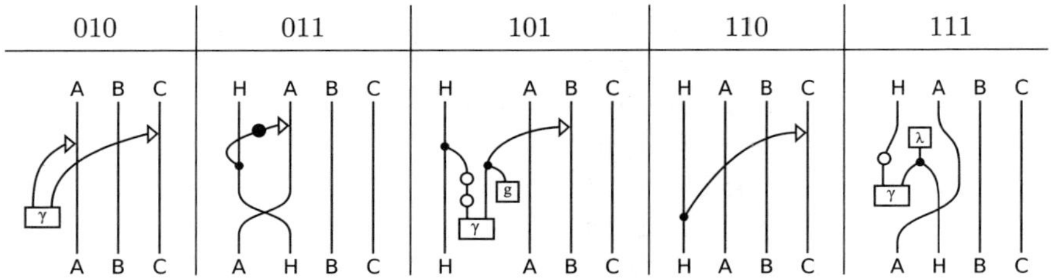

Figure 1. Associativity isomorphism $\alpha_{A,B,C} : A*(B*C) \to (A*B)*C$. The label *abc* means that $A \in \mathcal{C}_a$, $B \in \mathcal{C}_b$, $C \in \mathcal{C}_c$. In the three non-listed cases 000, 001, 100, $\alpha_{A,B,C}$ is the identity (or rather the associator of the underlying category $\mathcal{S}$). The diagrams are read from bottom to top, the empty and solid dot denote S and S^{-1}, respectively, and the three-valent vertices are the product and coproduct. The arrowhead depicts the action of H on a module.

To describe the braiding, denote by ω_V the parity involution on a super-vector space V, and by $s_{V,W} : V \otimes W \to W \otimes V$ the symmetric structure on $svect(\mathbb{C})$. Then [2, Thm. 6.4]:

$$\begin{array}{lll} A & B & c_{A,B} \;:\; A*B \to B*A \\ \hline \mathcal{C}_0 & \mathcal{C}_0 & s_{A,B} \circ \exp(-C) \\ \mathcal{C}_0 & \mathcal{C}_1 & s_{A,B} \circ \big(\exp(\tfrac{1}{2}\hat{C}) \otimes id_B\big) \\ \mathcal{C}_1 & \mathcal{C}_0 & s_{A,B} \circ \big(id_A \otimes \exp(\tfrac{1}{2}\hat{C})\big) \circ (id_A \otimes \omega_B) \\ \mathcal{C}_1 & \mathcal{C}_1 & e^{-i\pi\frac{n}{4}} \cdot \big(id_{S(\mathfrak{h})} \otimes s_{A,B}\big) \circ \big(\exp(-\tfrac{1}{2}\hat{C}) \otimes id_A \otimes \omega_B\big) \end{array} \tag{5}$$

In the last line, note that for $A, B \in \mathcal{C}_1$ we have $A * B = S(\mathfrak{h}) \otimes A \otimes B$.

4. A unified framework

The braiding isomorphisms (3) and (5) in the two examples just discussed may look quite different at first glance, but – just as was the case for the tensor product $*$ itself – they are actually two instances of the same structure.

Namely, let $\mathcal{S}$ be a pivotal symmetric monoidal category (i.e. a ribbon category with symmetric braiding $s_{A,B} : A \otimes B \to B \otimes A$) and let H be a Hopf algebra in $\mathcal{S}$ with invertible antipode. Denote by $\mathrm{Rep}_{\mathcal{S}}(H)$ the monoidal category of left H-modules in $\mathcal{S}$ and set

$$\mathcal{C} = \mathcal{C}_0 + \mathcal{C}_1 \qquad \text{with} \qquad \mathcal{C}_0 = \mathrm{Rep}_{\mathcal{S}}(H) \quad , \quad \mathcal{C}_1 = \mathcal{S} \; . \tag{6}$$

On $\mathcal{C}$ we fix the $\mathbb{Z}/2\mathbb{Z}$-graded tensor product functor $*$ from Section 1. Denote by μ the multiplication of H, by Δ the coproduct, and by S the antipode. For a morphisms $x : \mathbf{1} \to H$ in $\mathcal{S}$, write ${}_xM$ and M_x for the endomorphism of H given by left- and right-multiplication with x, respectively, and

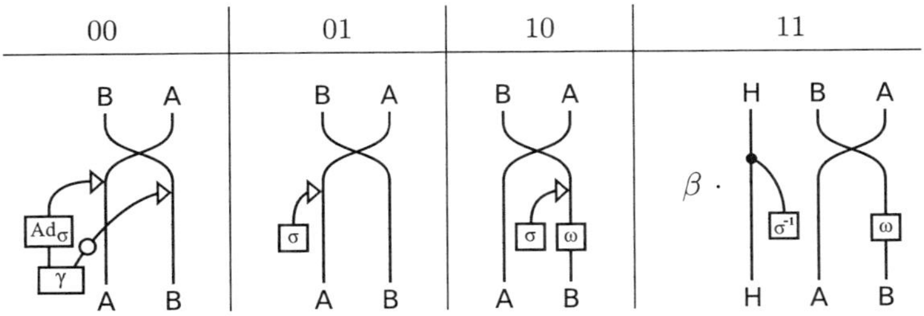

Figure 2. Braiding isomorphisms $c_{A,B} : A * B \to B * A$. The notation is as in Figure 1.

Ad_x for the endomorphism given by $x(-)x^{-1}$. Associativity isomorphisms for $*$ can be obtained from Hopf-algebraic data as follows [1, Cor. 3.17]:

Theorem 4.1. *Let $\gamma : \mathbf{1} \to H \otimes H$ and $\lambda : H \to \mathbf{1}$ be two morphisms in $\mathcal{S}$ such that*

(1) γ is a non-degenerate Hopf-copairing,
(2) λ is a right cointegral for H, such that there exits $g : \mathbf{1} \to H$ with $(id \otimes \lambda) \circ \Delta = g \circ \lambda$, and such that $(\lambda \otimes \lambda) \circ (id \otimes S) \circ \gamma = id_{\mathbf{1}}$,
(3) γ satisfies the symmetry condition $s_{H,H} \circ \gamma = (id \otimes (S^2 \circ \mathrm{Ad}_g^{-1})) \circ \gamma$.

Then the natural isomorphisms in Figure 1 define associativity isomorphisms for $*$.

Note that it is not claimed that the above description gives *all* associativity isomorphisms for $*$; outside of $\mathcal{S} = vect(k)$ this would require extra assumptions. In Ref. 1, the above theorem is actually proved in the more general setting of $\mathcal{S}$ being ribbon but not necessarily symmetric. The relevant Hopf algebra notions are reviewed in [1, Sec. 2].

Example 4.1. 1. The Tambara-Yamagami categories are recovered for $\mathcal{S} = vect(\mathbb{C})$ and $H = \mathrm{Fun}(G, \mathbb{C})$. The cointegral and copairing are $\lambda = \tau \sum_{a \in G} \delta_a$ and $\gamma = \sum_{a,b \in G} \chi(a,b)\, \delta_a \otimes \delta_b$.

2. For symplectic fermions take $\mathcal{S} = svect(\mathbb{C})$ and $H = S(\mathfrak{h})$. The copairing is $\gamma = e^C$ and the cointegral λ is as given in Section 3.

3. Another example for $\mathcal{S} = vect(\mathbb{C})$ is provided by Sweedler's four-dimensional Hopf algebra, which is not semi-simple [1, Sec. 3.8.3]. This illustrates that Theorem 4.1 is more general than Tambara-Yamagami categories even in the vector space case.

For the braiding isomorphisms we need to fix an involutive monoidal automorphism ω of the identity functor on $\mathcal{S}$. For $\mathcal{S} = vect(\mathbb{C})$, ω is necessarily the identity, but for $\mathcal{S} = svect(\mathbb{C})$ there are already two choices: the identity and parity involution. We have [1, Thm. 1.2 & Rem. 4.11]:

Theorem 4.2. *Let H, γ, λ and g be as in Theorem 4.1 and let $\sigma : \mathbf{1} \to H$ and $\beta : \mathbf{1} \to \mathbf{1}$ be invertible. Suppose that*

(1) γ is determined through σ by $\gamma = ({}_{\sigma^{-1}}M \otimes M_{\sigma^{-1}}) \circ \Delta \circ \sigma$.
(2) λ satisfies $\lambda \circ S = \lambda \circ \mathrm{Ad}_\sigma$ and $\lambda \circ \sigma = \beta \circ \beta$.
(3) Ad_σ is a Hopf-algebra isomorphism $H \to H_{\mathrm{cop}}$ (the opposite coalgebra).
(4) $S \circ \sigma = {}_gM \circ \sigma = M_{g^{-1}} \circ \sigma$.
(5) ω evaluated on H satisfies $(id \otimes \omega_H) \circ \gamma = \big(\mathrm{Ad}_\sigma \otimes (\mathrm{Ad}_\sigma^{-1} \circ S)\big) \circ \gamma$.

Then the natural isomorphisms in Figure 2 define a braiding on $\mathcal{C}$.

If σ^2 is central in H, then $\mathcal{C}$ can be made into a ribbon category with twist isomorphisms $\theta_A = \sigma^{-2}.(-)$ for $A \in \mathcal{C}_0$ (i.e. the left action of σ^{-2} on the H-module A), and $\theta_A = \beta^{-1}\,\omega_A$ for $A \in \mathcal{C}_1$, see [1, Prop. 4.18].

Example 4.2. 1. In the Tambara-Yamagami case, σ and β are as in Section 2. Comparing (3) and Figure 2 shows that the braiding isomorphisms match ($S = id_H$ for an elementary 2-group, and $\mathrm{Ad}_\sigma = id_H$ since H is commutative).

2. For symplectic fermions choose $\sigma = \exp(\frac{1}{2}\hat{C})$ and $\beta = e^{-\pi i n/4}$. Then (5) agrees with Figure 2.

3. Sweedler's Hopf algebra is quasi-triangular, but the resulting braiding on $\mathcal{C}_0$ does not extend to all of $\mathcal{C}$ (at least via the above construction). However, one can find a 16-dimensional semi-simple Hopf algebra in $vect(\mathbb{C})$ which is neither commutative nor co-commutative for which Theorems 4.1 and 4.2 apply [1, Sec. 4.7.4]. This is another instance where our setting is more general than the Tambara-Yamagami case even for $\mathcal{S} = vect(\mathbb{C})$.

Bibliography

1. A. Davydov and I. Runkel, *$\mathbb{Z}/2\mathbb{Z}$-extensions of Hopf algebra module categories by their base categories*, 1207.3611 [math.QA].
2. I. Runkel, *A braided monoidal category for free super-bosons*, 1209.5554 [math.QA].
3. D. Tambara and S. Yamagami, *Tensor categories with fusion rules of self-duality for finite abelian groups*, J. Alg. **209** (1998) 692–707.
4. J. Siehler, *Braided near-group categories*, math.QA/0011037.
5. H.G. Kausch, *Curiosities at c = -2*, hep-th/9510149.

PARTITION FUNCTIONS, MAPPING CLASS GROUPS AND DRINFELD DOUBLES

JENS FJELSTAD

Department of Physics, Nanjing University, Nanjing, China

JÜRGEN FUCHS and CARL STIGNER

Teoretisk fysik, Karlstad University, Karlstad, Sweden

CHRISTOPH SCHWEIGERT

Algebra und Zahlentheorie, Hamburg University, Hamburg, Germany

Higher genus partition functions of two-dimensional conformal field theories have to be invariants under linear actions of mapping class groups. We illustrate recent results [1,3] on the construction of such invariants by concrete expressions obtained for the case of Drinfeld doubles of finite groups. The results for doubles are independent of the characteristic of the underlying field, and the general results do not require any assumptions of semisimplicity.

Keywords: Mapping class group, factorizable Hopf algebra, Drinfeld double, conformal field theory.

1. General results

To start, let us sketch the incentive for our work from two-dimensional conformal field theory (CFT). The building blocks for CFT correlators with given chiral data are the conformal blocks, which form vector bundles with projectively flat connection over the moduli space of complex curves with marked points. It is fairly well understood that essential properties of the monodromies of these connections are encoded in the structure of a ribbon category $\mathcal{C}$. In many applications to physical systems, it is unnatural to require $\mathcal{C}$ to be semisimple. Indeed, non-semisimple ribbon categories arise in logarithmic conformal field theories, a topic of much recent activity.

In this note we consider linear representations of the mapping class groups of Riemann surfaces of genus g with m holes, for arbitrary values of g and m, on complex vector spaces that are specific morphism spaces of a finite ribbon category $\mathcal{C}$. Recall that a *finite tensor category* is a rigid monoidal category with finite-dimensional morphism spaces and finitely many isomorphism classes of simple objects, each of which has a projective cover, and with every object having finite length; a *ribbon category* is a rigid monoidal category endowed with a compatible braiding and twist.

A central idea in quantum field theory is the one of "summing over all intermediate states". Beyond the semisimple setting, one cannot give a naive mathematical meaning to this by restricting the summation by hand to, say, simple or indecomposable or projective objects. The idea is, however, appropriately implemented by the categorical notion of a *coend* $\int^{X\in\mathcal{C}} G(X,X)$ of a certain functor $G\colon \mathcal{C}^{\mathrm{op}}\times\mathcal{C}\to\mathcal{D}$. We have[4,5]

Theorem 1. *Let $\mathcal{C}$ be a finite ribbon category.*
(i) *The coend*

$$L(\mathcal{C}) := \int^{U\in\mathcal{C}} U^\vee \otimes U \tag{1}$$

of the functor $\mathcal{C}^{\mathrm{op}}\times\mathcal{C}\to\mathcal{C}$ that acts on objects as $(U,V)\mapsto U^\vee\otimes V$ exists.
(ii) *The coend $L(\mathcal{C})$ carries a natural structure of a Hopf algebra in $\mathcal{C}$. It is endowed with an integral and with a Hopf pairing $\varpi_{L(\mathcal{C})} : L(\mathcal{C})\otimes L(\mathcal{C})\to \mathbf{1}$.*
(iii) *Via the rigidity of $\mathcal{C}$, every object of $\mathcal{C}$ carries a natural structure of right comodule over $L(\mathcal{C})$.*

Proposition 2. [3, Rem. 2.3] *For $\mathcal{C}$ a finite ribbon category, every object of $\mathcal{C}$ carries a natural structure of left-right Yetter-Drinfeld module over $L(\mathcal{C})$, with the right comodule structure being the one of Theorem* 1(iii) *and the left module structure obtained via the braiding of $\mathcal{C}$.*
This provides a fully faithful embedding of $\mathcal{C}$ into the category of left-right Yetter-Drinfeld modules over $L(\mathcal{C})$.

A *factorizable* finite ribbon category is a finite ribbon category $\mathcal{C}$ for which the Hopf pairing $\varpi_{L(\mathcal{C})}$ is non-degenerate; for such a category the integral of the Hopf algebra $L(\mathcal{C})$ is two-sided and $L(\mathcal{C})$ also has a two-sided cointegral. In the sequel we restrict our attention to a specific subclass of such categories. We denote by $\Bbbk$ an algebraically closed field of characteristic zero and by $H\equiv(H,m,\eta,\Delta,\varepsilon,\mathrm{s},R,v)$ a finite-dimensional ribbon Hopf $\Bbbk$-algebra. Here m, η, Δ, ε and s are the product, unit, coproduct, counit and antipode of H, $R\in H\otimes_\Bbbk H$ is the R-matrix and $v\in H$ the ribbon element of H. Such a Hopf algebra has, uniquely up to scalars, a non-zero left integral $\Lambda\in H$ and a right cointegral $\lambda\in H^*$; the antipode of H is invertible. We denote by $Q:=R_{21}\cdot R\in H\otimes_\Bbbk H$ the monodromy matrix and by $f_Q=(d_H\otimes id_H)\circ(id_{H^*}\otimes Q)\colon H^*\to H$ the *Drinfeld map*.

We also assume that the Hopf algebra H is *factorizable*, meaning that the Drinfeld map f_Q is invertible. In this case the integral Λ is two-sided, and the Drinfeld map sends the cointegral λ to a non-zero multiple of the integral. We can (and do) normalize them such that $\lambda \circ \Lambda = 1$ and $f_Q(\lambda) = \Lambda$.

Now denote by H-Mod the category of (finite-dimensional, left) H-modules. The following is an immediate consequence of well-known results:

Proposition 3. *The category H-Mod carries a natural structure of a factorizable finite ribbon category.*

The category H-Bimod of finite-dimensional H-bimodules can be treated very much in the same vein, giving (see Ref. 1)

Corollary 4. *The category H-*Bimod *of finite-dimensional H-bimodules has a natural structure of a factorizable finite ribbon category.*

Remark 5. (i) It is worth pointing out that the relevant monoidal structure is *not* the one for which the tensor product is taken over H (and thus treats the Hopf algebra H just as an algebra).
(ii) We can use the ribbon structure on H to endow this tensor category with the structure of a ribbon category. To this end, we can use the simultaneous left action and right action of either the R-matrix and the ribbon element, or else the inverse R-matrix and inverse ribbon element. Altogether this results in four structures of ribbon category on H-Bimod; these are pairwise isomorphic. For our purpose it is crucial to take one of the two ribbon structures in which mutually inverse elements are used on the left and on the right; for concrete expressions see the formulas (3.3) and (4.19) of Ref. 1. This is precisely the ribbon structure that makes H-Bimod, for factorizable H, braided equivalent to the Drinfeld center of H-Mod.

As shown in Appendix A.2 of Ref. 1, there is an equivalence

$$H\text{-Mod}^{\mathrm{rev}} \boxtimes H\text{-Mod} \xrightarrow{\simeq} H\text{-Bimod} \tag{2}$$

of ribbon categories. We use this equivalence to tacitly identify these two categories. In particular, we think of the coend $L(H\text{-Mod}^{\mathrm{rev}} \boxtimes H\text{-Mod})$, as introduced in (1), as an H-bimodule. We denote this bimodule by K; its structure as an H-bimodule and as a Hopf algebra in H-Bimod are described in detail in Appendix A.3 of Ref. 1.

The simplest partition function in the semisimple case is the charge conjugation modular invariant; this is of Peter-Weyl form and thus a direct sum of terms in bijection with the isomorphism classes of simple H-modules. In the non-semisimple case, the direct sum gets again generalized to a coend, this time of a functor $G^H\colon H\text{-Mod}^{\rm op}\times H\text{-Mod}\to H\text{-Bimod}$. The functor G^H in question is the one obtained by composing the functor $H\text{-Mod}^{\rm op}\times H\text{-Mod}\to H\text{-Mod}^{\rm rev}\boxtimes H\text{-Mod}$ that acts on objects as $U\times V\mapsto U^\vee\boxtimes V$ with the equivalence (2).

Proposition 6. [1, Propos. A.3] *The coend F of the functor $G^H\colon H\text{-}Mod^{\rm op}\times H\text{-}Mod\to H\text{-Bimod}$ exists. Its underlying object – also denoted by F – is the* coregular bimodule, *i.e. the vector space H^* dual to H endowed with the duals of the left and right regular actions of H on itself.*

Similarly as the coend K, also the coend F turns out to carry important additional algebraic structure internal to H-Bimod:

Theorem 7. [2, Theorem 2] *The coend F carries a natural structure of a commutative symmetric Frobenius algebra with trivial twist in the ribbon category H-Bimod. The algebra and coalgebra structural maps are given by*

$$\begin{aligned} &m_F:=\Delta^*\,,\qquad \eta_F:=\varepsilon^*\,,\qquad \varepsilon_F:=\Lambda^*\qquad\text{and}\\ &\Delta_F:=[(id_H\otimes(\lambda\circ m))\circ(id_H\otimes \mathrm{s}\otimes id_H)\circ(\Delta\otimes id_H)]^*, \end{aligned}\tag{3}$$

with Λ and λ the integral and cointegral of H, respectively.

We now turn to the action of mapping class groups. The results in Ref. 5 imply that for any triple $(g,p,q)\in(\mathbb{Z}_{\geq 0})^3$ the morphism space $V_{g;p,q} = \mathrm{Hom}_{H|H}(K^{\otimes g}\otimes F^{\otimes q},F^{\otimes p})$ naturally carries a projective representation $\pi_{g;p,q}$ of the subgroup $\mathrm{Map}_{g;p,q}$ of the mapping class group of Riemann surfaces of genus g with $p+q$ holes that leaves two selected subsets of sizes p and q separately invariant. For details see Propos. 2.4 and Rem. 2.6 of Ref. 3. $\pi_{g;p,q}$ is in fact a genuine linear representation (Rem. 5.5 of Ref. 1).

In applications to CFT, F is a candidate for the bulk state space, and the spaces $V_{g;p,q}$ play the role of genus-g conformal blocks with p incoming and q outgoing insertions of the bulk state space. The corresponding correlation functions are specific elements in those spaces that have to be invariant under the mapping class group action and be compatible with sewing.

We now present elements $\mathrm{Cor}_{g;p,q}\in V_{g;p,q}$ that are candidates for these correlation functions. Denote by $m_F^{(r)}$ and $\Delta_F^{(r)}$ multiple products and coproducts of F, respectively, and by ρ_F^K the natural action (see Propos. 2) of the Hopf algebra $K\in H$-Bimod on the H-bimodule F, and set

$$\begin{aligned} &\mathrm{Cor}_{0;1,1}:=id_F\,,\qquad \mathrm{Cor}_{1;1,1}:=m_F\circ(\rho_F^K\otimes id_F)\circ(id_K\otimes\Delta_F)\,,\\ &\mathrm{Cor}_{g;1,1}:=\mathrm{Cor}_{1;1,1}\circ(id_K\otimes\mathrm{Cor}_{g-1;1,1})\quad\text{for}\quad g>1\,,\\ &\mathrm{Cor}_{g;p,q}:=\Delta_F^{(p)}\circ\mathrm{Cor}_{g;1,1}\circ\big(id_{K^{\otimes g}}\otimes m_F^{(q)}\big)\,. \end{aligned}\tag{4}$$

Theorem 8. *The element* $\mathrm{Cor}_{g;p,q}$ *of* $V_{g;p,q}$ *is invariant under the action* $\pi_{g;p,q}$ *of the group* $Map_{g;p,q}$.

This statement is the main result of Ref. 3. It provides yet another instance of a surprising conspiracy of algebraic structures – in the case at hand, finite-dimensional factorizable Hopf algebras – and geometric ones – here, the fundamental groups of moduli spaces of complex curves.

2. Drinfeld doubles

Associated to any finite group G there is the quasitriangular Hopf algebra $\mathcal{D}G$, called the (Drinfeld) double of G. As a vector space, $\mathcal{D}G = \Bbbk(G)\otimes_{\Bbbk}\Bbbk[G]$ with $\Bbbk(G)$ the algebra of functions on G and $\Bbbk[G]$ the group algebra over an (algebraically closed) field $\Bbbk$. From here on we do not assume any longer that $\Bbbk$ has characteristic zero, and thus do not require $\mathcal{D}G$ to be semisimple. A natural basis of $\mathcal{D}G$ is given by $b_{g|x} := \delta_g\otimes b_x$ with $g,x\in G$, where δ_g and b_x are the usual natural bases of $\Bbbk(G)$ and $\Bbbk[G]$. In terms of this basis the Hopf algebra structure of $\mathcal{D}G$ reads

$$b_{g|x}\cdot b_{h|y} = \delta_{g,xhx^{-1}}\, b_{g|xy}\,, \qquad \eta(1) = b_{1|e} = \sum\nolimits_{g\in G} b_{g|e}\,, \qquad \varepsilon(b_{g|x}) = \delta_{g,e}\,,$$

$$\Delta(b_{g|x}) = \sum\nolimits_{h\in G} b_{h|x}\otimes b_{h^{-1}g|x}\,, \qquad \mathrm{s}(b_{g|x}) = b_{x^{-1}g^{-1}x|x^{-1}}\,. \tag{5}$$

Note that $\mathrm{s}^2 = id_{\mathcal{D}G}$. The Hopf algebra $\mathcal{D}G$ is factorizable quasitriangular, with R-matrix and associated monodromy matrix given by

$$R = \sum\nolimits_{g\in G} b_{g|e}\otimes b_{1|g} \qquad \text{and} \qquad Q = \sum\nolimits_{g,h\in G} b_{h|g}\otimes b_{g|g^{-1}hg}\,. \tag{6}$$

$\mathcal{D}G$ is also ribbon, with ribbon element $\nu = \sum_{g\in G} b_{g|g^{-1}}$, and it has a two-sided integral and a two-sided cointegral, given (upon choice of normalizations) by $\Lambda = \sum_g b_{e|g}$ and by $\lambda = \sum_g \beta_{g|e}\in\mathcal{D}G^*$, where $\{\beta_{g|x}\}$ is the basis of the dual space $\mathcal{D}G^*$ dual to $\{b_{g|x}\}$. The integral satisfies $\varepsilon\circ\Lambda = |G|$; accordingly, by Maschke's theorem the category $\mathcal{D}G$-Mod of left $\mathcal{D}G$-modules is semisimple iff the characteristic of $\Bbbk$ does not divide the order $|G|$.

3. Invariants from Drinfeld doubles

Let us apply some of the general results from Section 1 to the factorizable Hopf algebra $H = \mathcal{D}G$. First, the Hopf algebra K in $\mathcal{D}G$-Bimod is $(\mathcal{D}G^*)\otimes_{\Bbbk}(\mathcal{D}G^*)$ as a vector space, with the coadjoint left and right $\mathcal{D}G$-actions on the first and second tensor factor, respectively, and its Hopf algebra structure is easily expressed in terms of the dual basis $\{\beta_{g|x}\}$. The

Frobenius algebra F in $\mathcal{D}G$-Bimod is $\mathcal{D}G^*$ as a vector space, with the coregular left and right $\mathcal{D}G$-actions. The Frobenius algebra structure is obtained by specializing formula (3) to the present situation; we find

$$\begin{aligned} \eta_F &= \sum\nolimits_{x\in G} \beta_{e|x} , \quad & m_F(\beta_{g|x}, \beta_{h|y}) &= \delta_{x,y}\, \beta_{hg|x} , \\ \varepsilon_F(\beta_{g|x}) &= \delta_{g,e} , & \Delta_F(\beta_{g|x}) &= \sum\nolimits_{h\in G} \beta_{h^{-1}g|x} \otimes \beta_{h|x} . \end{aligned} \tag{7}$$

Now consider the action ρ_F^K of the Hopf algebra K on the Frobenius algebra F, which is a linear map from $(\mathcal{D}G^*)^{\otimes_{\Bbbk} 3}$ to $\mathcal{D}G^*$. We obtain

$$\rho_F^K(\beta_{g|x}{\otimes}\beta_{h|y}{\otimes}\beta_{k|z}) = \delta_{z,xzy}\, \delta_{k,x}\, \beta_{g^{-1}xg|g^{-1}zy^{-1}hy} . \tag{8}$$

We can now insert the specific expressions (7) and (8) into the general formula (4). We restrict our attention to the case $p{=}1{=}q$; the extension to general values of p and q is straightforward. At genus 1 we have

$$\mathrm{Cor}_{1;1,1}(\beta_{g|x}{\otimes}\beta_{h|y}{\otimes}\beta_{k|z}) = \delta_{y^{-1},z^{-1}xz}\, \delta_{y^{-1}hy,z^{-1}gz}\, \beta_{k[x,g]|z} , \tag{9}$$

with $[x,g]{=}x^{-1}g^{-1}xg$ the group commutator. By iteration we arrive at

$$\mathrm{Cor}_{n;1,1}(\beta^{(n)}_{gh|xy}{\otimes}\beta_{k|z}) = \prod\nolimits_{i=1}^{n} \delta_{y_i^{-1},z^{-1}x_iz}\, \delta_{y_i^{-1}h_iy_i,z^{-1}g_iz}\, \beta_{k[x_n,g_n]\cdots[x_1,g_1]|z} , \tag{10}$$

where we introduced the short-hand notation

$$\beta^{(n)}_{gh|xy} := \beta_{g_1|x_1} \otimes \beta_{h_1|y_1} \otimes \beta_{g_2|x_2} \otimes \beta_{h_2|y_2} \otimes \cdots \otimes \beta_{g_n|x_n} \otimes \beta_{h_n|y_n} . \tag{11}$$

According to Theorem 8 the morphisms (10) are invariant under the action of $\mathrm{Map}_{g;p,q}$ if $\Bbbk$ has characteristic zero. But we actually expect that this remains true for general characteristic, including the case that the characteristic divides the order of G so that $\mathcal{D}G$ is non-semisimple.

Indeed, we have verified that the torus partition function $\mathrm{Cor}_{1;0,0}$ is modular invariant irrespective of the characteristic of $\Bbbk$. Checking mapping class group invariance of correlators with field insertions and at higher genus amounts to specializing various expressions from [3, Sect. 5]; this would be straightforward, but some of the computations involved tend to be lengthy.

References

1. J. Fuchs, C. Schweigert, and C. Stigner, J. of Algebra 363 (2012) 29.
2. J. Fuchs, C. Schweigert, and C. Stigner, in: *Strings, Gauge Fields, and the Geometry Behind. The Legacy of Maximilian Kreuzer* (A. Rebhan et al, eds.; World Scientific, Singapore 2012), p. 289.
3. J. Fuchs, C. Schweigert, and C. Stigner, e-print math.QA/1207.6863.
4. T. Kerler and V.V. Lyubashenko, *Non-Semisimple Topological Quantum Field Theories for 3-Manifolds with Corners* (Springer Verlag, New York 2001).
5. V.V. Lyubashenko, Commun. Math. Phys. 172 (1995) 467.

GAUGE DEFECT NETWORKS IN TWO-DIMENSIONAL CFT

RAFAŁ R. SUSZEK

Katedra Metod Matematycznych Fizyki, Wydział Fizyki, Uniwersytet Warszawski
ul. Hoża 74, PL-00-682 Warszawa, Poland
E-mail: suszek@fuw.edu.pl

An interpretation of the gauge anomaly of the two-dimensional multi-phase σ-model is presented in terms of an obstruction to the existence of a topological defect network implementing a local trivialisation of the gauged σ-model.

Keywords: Sigma models, defects; Gauge anomaly; Gerbe theory.

Introduction. The Symmetry Principle and the Gauge Principle figure among the most fundamental ideas of local QFT. In particular, they serve to distinguish theories with a non-anomalous realisation of classical symmetries, admitting descent to the space of orbits of an action of the symmetry group on the space of states. This note gives an account of recent progress in realising the Principles in 2d CFT in the mixed geometro-cohomological framework developed by Gawędzki *et al.* and adapted[1,2] to the study of CFT defects. As we shall argue, it is through application of the ensuing correspondence between defects and CFT dualities that the categorial construction[3] of the gauged CFT is justified and brought to completion.[4] Our discussion emphasises the naturalness of Gawędzki's approach and paves the way towards a rigorous description of T-duality and T-folds.

The multi-phase σ-model. The mono-phase 2d non-linear σ-model is a theory of maps $X \in C^1(\Sigma, M)$ from a compact closed 2d metric manifold (Σ, η) to a metric manifold (M, g) supporting a bundle gerbe $\mathcal{G}$. The gerbe is a geometric realisation[5] of an integral class $[\frac{1}{2\pi}\mathrm{H}] \in H^3(M)$. It gives rise to a Cheeger–Simons differential character $\mathrm{Hol}_{\mathcal{G}}(X)$ defined as the image of $[X^*\mathcal{G}]$ under the isomorphism $\check{H}^2(\Sigma, \mathrm{U}(1)) \xrightarrow{\cong} \mathrm{U}(1)$. The theory is determined by the Principle of Least Action applied to the functional

$$S_\sigma[X;\eta] := -\tfrac{1}{2}\int_\Sigma \mathrm{g}(\mathsf{d}X \overset{\wedge}{,} \star_\eta \mathsf{d}X) - \mathsf{i}\,\log \mathrm{Hol}_{\mathcal{G}}(X)$$

in which the Hodge operator $\star_\eta$ acts on the first and the metric g contracts the second tensor factor in $\mathsf{d}X(\sigma) = \partial_a X^\mu(\sigma)\,\mathsf{d}\sigma^a \otimes \partial_\mu|_{X(\sigma)}$ (in local coordinates $\{\sigma^a\}^{a\in\{1,2\}}$ on Σ and $\{X^\mu\}^{\mu\in\overline{1,\dim M}}$ on M). The class $[\frac{1}{2\pi}\mathrm{H}]$, necessary for the cancellation of the Weyl anomaly, is represented by a 2-cocycle in the Deligne hypercohomology of M, and it is in this guise that gerbes first entered the description of the σ-model,[6,7] leading to a classification of field theories with fixed $(M, \mathrm{g}, \mathrm{H})$ by elements of $H^2(M, \mathrm{U}(1))$, and to a geometric quantisation of the σ-model through transgression.[7]

Drawing inspiration from models of spin lattices with lines of spin frustration and from orbifold string theory, we are led to consider[8] mappings from Σ to a disjoint union of manifolds M, with discontinuities along lines $\ell \subset \Sigma$ at which the component of the energy-momentum current generating diffeomorphisms of Σ that preserve ℓ is continuous, whence the name conformal defects. Amidst these, we find topological defects with the property that all components of the current are continuous at ℓ. A formal path-integral argument[1] shows that in a factorisable CFT with defect lines we are bound to consider intersections of the latter. Together with defect lines, they compose an oriented graph $\Gamma \subset \Sigma$ that splits Σ into patches. Thus, we arrive at the definition of the multi-phase σ-model:[1] It is a theory of C^1-smooth maps X that send the patches to a metric space (M, g), defect lines to a smooth space Q with smooth maps $\iota_1, \iota_2 : Q \to M$ and $\omega \in \Omega^2(Q)$, and defect junctions of valence $n \geq 3$ to a smooth space T_n with smooth maps $\pi_n^{k,k+1} : T_n \to Q$, $k \in \mathbb{Z}/n\mathbb{Z}$ subject to additional consistency constraints. The target space $\mathscr{F} := M \sqcup Q \sqcup T$, $T := \bigsqcup_{n\geq 3} T_n$ is the base of a geometric structure $\mathfrak{B}$ from the 2-category $\mathfrak{BGrb}^\nabla(\mathscr{F})$ of abelian bundle gerbes with connection[9] over $\mathscr{F}$. The structure consists of a 0-cell (gerbe) $\mathcal{G}$, a 1-cell (1-isomorphism) $\Phi : \iota_1^*\mathcal{G} \xrightarrow{\cong} \iota_2^*\mathcal{G} \otimes I_\omega$ (for I_ω the trivial gerbe of curving ω), and 2-cells (2-isomorphisms) $\varphi_n : \circ_{k\in\mathbb{Z}/n\mathbb{Z}}\, \pi_n^{k,k+1\,*}\left(\Phi^{\varepsilon_n^{k,k+1}} \otimes \mathrm{Id}\right) \xRightarrow{\cong} \mathrm{Id}$ (with $\varepsilon_n^{k,k+1} = \pm 1$ for an incoming/outgoing defect line with index k). The triple $(\mathcal{G}, \Phi, \varphi_n)$ is a geometric realisation of the 2π-integral class of the relative 3-form $\Theta := \mathrm{H} \oplus \omega \oplus 0$ in the cohomology of the complex of cochain groups[4] $C^p_{\mathrm{dR}}(M, Q, T_n | \Delta_Q, \Delta_{T_n}) := \Omega^p(M) \oplus \Omega^{p-1}(Q) \oplus \Omega^{p-2}(T_n)$, $p \in \mathbb{N}$ (here, $\Omega^{-1}(X) := \mathbb{R}^{\pi_0(X)}$ and $\Omega^{m<-1}(X) \equiv 0$) endowed with the differential $\widetilde{\mathsf{d}}(\omega_M \oplus \omega_Q \oplus \omega_{T_n}) := \mathsf{d}\omega_M \oplus (-\mathsf{d}\omega_Q - \Delta_Q\omega_M) \oplus (\mathsf{d}\omega_{T_n} - \Delta_{T_n}\omega_Q)$, written for $\Delta_Q := \iota_2^* - \iota_1^*$ and $\Delta_{T_n} := \sum_{k\in\mathbb{Z}/n\mathbb{Z}} \varepsilon_n^{k,k+1}\, \pi_n^{k,k+1\,*}$. The said theory is determined by the Principle of Least Action applied to the functional[1]

$$S_\sigma[(X|\Gamma);\gamma] := -\tfrac{1}{2}\int_\Sigma \mathrm{g}(\mathsf{d}X \overset{\wedge}{,} \star_\gamma \mathsf{d}X) - \mathrm{i}\log \mathrm{Hol}_{(\mathcal{G},\Phi,\varphi_n\,|\,n\in\mathbb{N}_{\geq 3})}(X|\Gamma)\,,$$

in which the last term is a differential character expressed in terms of Deligne data of the $(\mathcal{G}, \Phi, \varphi_n)$. The definition enables us to label conformal defects with fixed $(M, \mathrm{g}, \mathcal{G}, Q, \iota_\alpha, \omega)$ by classes in $H^1(Q, \mathrm{U}(1))$, and (in-equivalent) choices of the φ_n by elements of $\mathrm{U}(1)^{\pi_0(T_n)}$. It also determines a geometric quantisation of (the twisted sector of) the σ-model through a relative variant of transgression.[2,10]

Defects vs. dualities. A relation between topological defects and CFT dualities has been predicted[11] within the categorial approach to the quantisation of the CFT developed by Fröhlich, Fuchs, Schweigert, Runkel *et al.* Its realisation in the geometric framework outlined above was worked out in the author's recent paper,[2] preceded by a case study[1] of a class of defects in the Wess–Zumino–Witten–Gawędzki σ-model.[7,10] It stems from the simple world-sheet intuition: A (space-like) circular defect $\ell_\circ$ separating two patches of the σ-model establishes a correspondence between states from the respective phases, represented by the Cauchy data localised at the circular components of the boundary of the patches joining at $\ell_\circ$. A canonical analysis of the relation between the two sets of Cauchy data that follows from the gluing condition imposed at $\ell_\circ$, as part of the variational principle, upon classical (*i.e.*, minimal) field configurations X demonstrates that the subspace $\mathfrak{I}_\sigma \subset \mathsf{P}_\sigma^{\times 2}$ composed of pairs of states in cross-defect correspondence is isotropic with respect to $\mathrm{pr}_1^*\Omega_\sigma - \mathrm{pr}_2^*\Omega_\sigma$ (here, the $\mathrm{pr}_\alpha : \mathsf{P}_\sigma^{\times 2} \to \mathsf{P}_\sigma$ are the canonical projections). The defect data induce a full-fledged (pre)quantum CFT duality, that is $\mathfrak{I}_\sigma$ is a graph of a symplectomorphism preserving the Hamiltonian density and lifting to an isomorphism $\mathrm{pr}_1^*\mathcal{L}_\sigma|_{\mathfrak{I}_\sigma} \cong \mathrm{pr}_2^*\mathcal{L}_\sigma|_{\mathfrak{I}_\sigma}$ between pullbacks of the prequantum bundle $\mathcal{L}_\sigma$ of the mono-phase σ-model, iff the maps $\mathsf{L}Q \to \mathsf{L}M : \phi \mapsto \iota_\alpha \circ \phi$ are surjective submersions, the defect is topological and extra technical conditions are satisfied.[2] This is the case, in particular, when the defect carries the data of a 1-isomorphism $\Phi_F : \mathrm{pr}_1^*\mathcal{G} \xrightarrow{\cong} \mathrm{pr}_2^*\mathcal{G}$ over the graph $Q = (\mathrm{id}_M \times F)(M)$ of an isometric diffeomorphism F of (M, g). The associated duality relates mappings into a single component of M, and so it represents a global symmetry of the mono-phase σ-model. This brings us to the main point of our discussion.

The Gauge Principle through defect networks. Global symmetries of the σ-model form a subgroup $\mathrm{G}_\sigma \subset \mathrm{Diff}(\mathscr{F})$ acting on $\mathscr{F}$ as $\ell : \mathrm{G}_\sigma \times \mathscr{F} \to \mathscr{F}$, with the ι_α and $\pi_n^{k,k+1}$ assumed G_σ-equivariant. Elements of G_σ satisfy the defining relation $S_\sigma[(\ell_g(X)|\Gamma); \eta] = S_\sigma[(X|\Gamma); \eta]$. In order to understand the underlying geometry, it is instructive to examine infinitesimal symmetry transformations, induced by flows $\psi_\cdot :]-\epsilon, \epsilon[\times \mathscr{F} \to \mathscr{F}$, $\epsilon > 0$ of the G_σ-fundamental vector fields $\mathscr{K} \in \mathfrak{X}(\mathscr{F})$. Inspection of the identity $\frac{\mathsf{d}}{\mathsf{d}t}|_{t=0} S_\sigma[(\psi_t \circ X|\Gamma); \eta] = 0$ reveals that the symmetries correspond to those sections $\mathscr{K} \oplus \mathrm{K} \in \mathfrak{X}(\mathscr{F}) \oplus C^1_{\mathrm{dR}}(M, Q, T_n | \Delta_Q, \Delta_{T_n})$ of the generalised tangent bundles (here, $\Gamma\left((T_n \times \mathbb{R})_0\right) = \mathbb{R}^{\pi_0(T_n)}$)

$$\mathrm{E}\mathscr{F} := (\mathsf{T}M \oplus \mathsf{T}^*M) \sqcup (\mathsf{T}Q \oplus (Q \times \mathbb{R})) \sqcup \bigsqcup_{n \geq 3} (\mathsf{T}T_n \oplus (T_n \times \mathbb{R})_0) \to \mathscr{F}$$

which obey the relations $\mathscr{L}_{\mathscr{K}}\mathrm{g} = 0$ and $\widetilde{\iota}_{\mathscr{K}}\Theta = -\widetilde{\mathsf{d}}\mathrm{K}$, expressed in terms of the relative contraction[4] $\widetilde{\iota}_{\mathscr{V}}(\omega_M \oplus \omega_Q \oplus \omega_{T_n}) := \iota_{\mathscr{V}}\omega_M \oplus (-\iota_{\mathscr{V}}\omega_Q) \oplus \iota_{\mathscr{V}}\omega_{T_n}$. There is an essentially unique extension of the Lie bracket on $\mathfrak{X}(\mathscr{F})$ to $\Gamma(\mathrm{E}\mathscr{F})$ that closes on the set $\Gamma_\sigma(\mathrm{E}\mathscr{F})$ of these sections, to wit,

$$[\mathscr{V} \oplus \upsilon, \mathscr{W} \oplus \varpi]^{\Theta}_{\mathrm{C}} := [\mathscr{V}, \mathscr{W}] \oplus [\mathscr{L}_{\mathscr{V}}\varpi - \mathscr{L}_{\mathscr{W}}\upsilon - \tfrac{1}{2}\widetilde{\mathsf{d}}(\widetilde{\iota}_{\mathscr{V}}\varpi - \widetilde{\iota}_{\mathscr{W}}\upsilon) + \widetilde{\iota}_{\mathscr{V}}\widetilde{\iota}_{\mathscr{W}}\Theta],$$

and it is natural to think of it as a relative Θ-twisted Courant bracket. Indeed, it restricts to the H-twisted Courant bracket on sections of the standard generalised tangent bundle $\mathsf{T}M \oplus \mathsf{T}^*M$ whose rôle in the description of symmetries of the mono-phase σ-model has long been known.[12] The intrinsically gerbe-theoretic nature of the bracket is readily established with the help of Hitchin-type isomorphisms.[4] With the algebraic structure on $\Gamma_\sigma(\mathrm{E}\mathscr{F})$ thus determined, we are ready to discuss the Gauge Principle.

Geometrically, the Principle stipulates that the covariant configuration (c.c.) bundle $\mathscr{F}' := \Sigma \times \mathscr{F}$ of the σ-model be replaced by the bundle $\mathsf{P} \times_\ell \mathscr{F}$ associated, through ℓ, to an arbitrary principal G_σ-bundle $\mathsf{P} \to \Sigma$ and carrying a fibrewise action of the adjoint bundle $\mathsf{P} \times_{\mathrm{Ad}} \mathrm{G}_\sigma$. In this manner, the action of the symmetry group (encoded in the isotype G_σ of the fibre of $\mathsf{P} \times_{\mathrm{Ad}} \mathrm{G}_\sigma$) is rendered local while preserving the original field content, the latter being determined by the isotype $\mathscr{F}$ of the fibre of $\mathsf{P} \times_\ell \mathscr{F}$. Clearly, we must insist that the new c.c. bundle admit a *global* section, to be identified with the lagrangean field of the gauged σ-model under construction. The geometric descent from the extended c.c. bundle $\widetilde{\mathscr{F}} := \mathsf{P} \times \mathscr{F}$ to $\mathsf{P} \times_\ell \mathscr{F}$ is straightforward as the G_σ-action on the former is free. The nontrivial task is to lift this action equivariantly to an extension $\widetilde{\mathfrak{B}}_{\mathcal{A}}$ of the original $\mathfrak{B}$ to $\widetilde{\mathscr{F}}$ obtained by coupling its pullback to the gauge field given by the principal G_σ-connection 1-form $\mathcal{A} \in \Omega^1(\mathsf{P}) \otimes \mathfrak{g}_\sigma$, $\mathfrak{g}_\sigma := \mathrm{Lie}\,\mathrm{G}_\sigma$ on P. This is to be done in such a manner that $\widetilde{\mathfrak{B}}_{\mathcal{A}}$ descends to the quotient $\mathsf{P} \times_\ell \mathscr{F}$ in the sense captured by a variant[3] of Stevenson's Principle of Descent[9] (to be viewed as a (2-)categorification of pullback), and defines a CFT invariant under the action of the gauge group $\mathsf{P} \times_{\mathrm{Ad}} \mathrm{G}_\sigma$.

It is convenient to begin the analysis with the topologically trivial case $\mathsf{P} = \Sigma \times \mathrm{G}_\sigma$, in which we keep $\mathscr{F}'$ as the c.c. bundle and replace $\mathcal{A}$ by the connection 1-form $\mathrm{A} = \mathrm{A}^A \otimes t_A \in \Omega^1(\Sigma) \otimes \mathfrak{g}_\sigma$, written in terms of the generators t_A of $\mathfrak{g}_\sigma$ subject to the relations $[t_A, t_B] = f_{ABC}\, t_C$ in which the f_{ABC} are the structure constants of $\mathfrak{g}_\sigma$. Upon choosing a basis $\mathfrak{K}_A := \mathscr{K}_A \oplus (\kappa_A \oplus k_A \oplus 0)$, $A \in \overline{1, \dim \mathfrak{g}_\sigma}$ in $\Gamma_\sigma(\mathsf{E}\mathscr{F})$ projecting to the set of G_σ-fundamental vector fields $\mathscr{K}_A$ associated with the t_A, we obtain the following (minimal) extension[3] $\mathfrak{B}'_{\mathrm{A}}$ of $\mathfrak{B}$ to $\mathscr{F}'$ (subscripts 1,2 denote pullbacks along the canonical projections): the metric

$\mathrm{g_A} := \mathrm{g}_2 - \mathrm{g}(\mathscr{K}_A,\cdot)_2 \otimes \mathrm{A}_1^A - \mathrm{A}_1^A \otimes \mathrm{g}(\mathscr{K}_A,\cdot)_2 + \mathrm{g}(\mathscr{K}_A,\mathscr{K}_B)_2\, \mathrm{A}_1^A \otimes \mathrm{A}_1^B$ and the 0-cell $\mathcal{G}_\mathrm{A} := \mathcal{G}_2 \otimes I_{\rho_\mathrm{A}}$ with $\rho_\mathrm{A} := \kappa_{A\,2} \wedge \mathrm{A}_1^A - \frac{1}{2}\,(\iota_{\mathscr{K}_A}\kappa_B)_2\, \mathrm{A}_1^A \wedge \mathrm{A}_2^B$ over M'; the 1-cell $\Phi_\mathrm{A} := \Phi_2 \otimes J_{\lambda_\mathrm{A}}$ with the trivial 1-cell (trivial line bundle) J_{λ_A} with a connection 1-form $\lambda_\mathrm{A} := -k_{A\,2}\,\mathrm{A}_1^A$ over Q'; the 2-cells $\varphi_{n\,\mathrm{A}} := \varphi_{n\,2}$ over the T_n'. In terms of these, we define – for $\xi := (\mathrm{id}_\Sigma, X)$ with X as before – the gauged σ-model coupled to the topologically trivial gauge field

$$S_\sigma[(X|\Gamma);\mathrm{A},\gamma] := -\tfrac{1}{2}\int_\Sigma \mathrm{g_A}(\mathsf{d}\xi \overset{\wedge}{,} \star_\gamma \mathsf{d}\xi) - \mathsf{i}\,\log \mathrm{Hol}_{(\mathcal{G}_\mathrm{A},\Phi_\mathrm{A},\varphi_{n\,\mathrm{A}}\,|\,n\in\mathbb{N}_{\geq 3})}(\xi|\Gamma)$$

We may next impose the demand that the above be invariant under gauge transformations $(X,\mathrm{A}) \mapsto (\ell_\chi(X), \mathrm{Ad}_\chi \mathrm{A} - \mathsf{d}\chi\,\chi^{-1})$ defined for arbitrary $\chi \in C^\infty(\Sigma,\mathrm{G}_\sigma)$. For χ homotopic to the identity, this yields the condition that the triple $(\bigoplus_{A=1}^{\dim\mathfrak{g}_\sigma} C^\infty(\mathscr{F},\mathbb{R})\,\mathfrak{K}_A, [\cdot,\cdot]_\mathrm{C}^\Theta, \alpha_{\mathsf{T}\mathscr{F}} \equiv \mathrm{pr}_1)$ define a Lie algebroid, and the latter is then canonically isomorphic[3] with the action algebroid $\mathfrak{g}_\sigma \ltimes_\ell \mathscr{F}$. Invariance under large gauge transformations requires the existence of a 1-cell $\Upsilon : \ell^*\mathcal{G} \xrightarrow{\cong} \mathrm{pr}_2^*\mathcal{G} \otimes I_{\rho_{\theta_L}}$ (for θ_L the left-invariant Maurer–Cartan 1-form on G_σ) and a 2-cell $\Xi : \ell^*\Phi \overset{\cong}{\Longrightarrow} [((\mathrm{id}_{\mathrm{G}_\sigma} \times \iota_2)^*\Upsilon^{-1} \otimes \mathrm{Id}) \circ (\mathrm{pr}_2^*\Phi \otimes \mathrm{Id}) \circ (\mathrm{id}_{\mathrm{G}_\sigma} \times \iota_1)^*\Upsilon] \otimes J_{\lambda_{\theta_L}}$ (subject to additional constraints) from the 2-category $\mathfrak{BGrb}^\nabla(\mathrm{G}_\sigma \times \mathscr{F})$ based on the arrow manifold of the action groupoid $\mathrm{G}_\sigma \ltimes_\ell \mathscr{F}$. Upon recalling that $\mathfrak{g}_\sigma \ltimes_\ell \mathscr{F}$ is the tangent algebroid of $\mathrm{G}_\sigma \ltimes_\ell \mathscr{F}$, we are led to identify the latter as the algebro-differential structure underlying those symmetries of the σ-model that can be gauged. This observation is elucidated with the help of the equivalence[4] $\mathrm{G}_\sigma\text{-}\mathfrak{Bun}(\Sigma\|\mathscr{F}) \cong \mathrm{G}_\sigma \ltimes_\ell \mathscr{F}\text{-}\mathfrak{Bun}(\Sigma)$ between the groupoid of principal G_σ-bundles $\mathsf{P} \to \Sigma$ with the property that $\mathsf{P} \times_\ell \mathscr{F}$ admits a global section and the groupoid of principal $\mathrm{G}_\sigma \ltimes_\ell \mathscr{F}$-bundles[13] over Σ. The equivalence has a natural world-sheet interpretation, reminiscent of the categorial generalised-orbifold construction:[14] An object P of $\mathrm{G}_\sigma\text{-}\mathfrak{Bun}(\Sigma\|\mathscr{F})$ (data of the gauged σ-model, forgetting the connection) trivialising over an open cover $\mathcal{O} := \{\Sigma_i\}_{i\in\mathscr{I}}$ of Σ corresponds to a family of smooth mappings $X_i : \Sigma_i \to \mathscr{F}$ under which images of points $\sigma \in \Sigma_i \cap \Sigma_j$ are related as $X_i(\sigma) = \ell_{g_{ij}}(X_j)(\sigma)$ by arrows of $\mathrm{G}_\sigma \ltimes_\ell \mathscr{F}$ defined in terms of transition maps g_{ij} of P. This demonstrates the necessity of incorporating nontrivial principal G_σ-bundles in a complete description of the gauged σ-model as they are necessary to reproduce the twisted sector of the descended CFT. Their presence appears to ensure self-consistency of the quantised theory.[3]

The goal of extending the Gauge Principle to topologically nontrivial bundles P can be attained through local trivialisation of P over $\mathcal{O}$ followed by the gluing of local data (X_i, A_i) over the Σ_i and of the local phases $\mathfrak{B}'_{\mathrm{A}_i}$ of the gauged σ-model by means of the g_{ij} at the edges $\ell_{ij} \subset \Sigma_i \cap \Sigma_j$

of a trivalent graph $\Gamma_{\mathcal{O}}$ associated with $\mathcal{O}$, with $\Gamma_{\mathcal{O}} \cap \Gamma$ discrete and not containing vertices of Γ. We then check[4] that the pair (Υ, Ξ), required to exist by previous arguments, can be used to induce the desired structure of a topological defect at the ℓ_{ij}, alongside that of a (trans-)defect junction at $\Gamma_{\mathcal{O}} \cap \Gamma$. These jointly implement the gluing of local data in the presence of Γ, and the gluing itself emerges as a duality between the trivialised phases $\mathfrak{B}'_{\mathrm{A}_i}$. The hitherto assignments already fix[1] the structure to be pulled back to the vertices of $\Gamma_{\mathcal{O}}$, and so we are led to demand the existence of a 2-cell $\gamma : (d_2^{(0)\,*}\Upsilon \otimes \mathrm{Id}) \circ d_2^{(2)\,*}\Upsilon \overset{\cong}{\Longrightarrow} d_2^{(1)\,*}\Upsilon$ of $\mathfrak{BGrb}^\nabla(\mathrm{G}_\sigma^2 \times M)$, written in terms of the face maps $d_l^{(k)} : \mathrm{G}_\sigma^l \times M \to \mathrm{G}_\sigma^{l-1} \times M$, $k \in \overline{0,l}$ of the nerve of $\mathrm{G}_\sigma \ltimes_\ell \mathscr{F}$ and playing a rôle analogous to that of φ_3. At this stage, we still have to demand that the gauged σ-model thus sewn from the local phases $\mathfrak{B}'_{\mathrm{A}_i}$ be independent of the arbitrary choices made, *e.g.*, the choice of $\mathcal{O}$ and $\Gamma_{\mathcal{O}}$, of the trivialisation and of index assignments. This further constrains (Υ, γ, Ξ), so that ultimately we have to demand that the triple define a G_σ-equivariant structure[4] on $\mathfrak{B}$. By the Principle of Descent, the gauged σ-model then yields[3] a (manifestly gauge invariant) CFT on $\mathsf{P} \times_\ell \mathscr{F}$, equivalent to a σ-model on $\mathscr{F}/\mathrm{G}_\sigma$ whenever the latter exists as a manifold. Obstructions to the existence of the G_σ-equivariant structure (known as gauge anomalies), as well as its inequivalent realisations are classified[3] by a G_σ-equivariant extension of the relative Deligne hypercohomology.

References

1. I. Runkel & R.R. Suszek, *Adv. Theor. Math. Phys.* **13**, 1137 (2009).
2. R.R. Suszek, *Hamb. Beitr. Math.* **360** (2011) [arXiv: 1101.1126 [hep-th]].
3. K. Gawędzki, R.R. Suszek & K. Waldorf, *Adv. Theor. Math. Phys.* **15**, 621 (2011); *Commun. Math. Phys.* **302**, 513 (2010); arXiv: 1202.5808 [hep-th].
4. R.R. Suszek, *Hamb. Beitr. Math.* **361** (2011) [arXiv: 1209.2334 [hep-th]].
5. M.K. Murray, *J. Lond. Math. Soc.* **54**, 403 (1996); M.K. Murray & D. Stevenson, *J. Lond. Math. Soc.* **62**, 925 (2000).
6. O. Alvarez, *Commun. Math. Phys.* **100**, 279 (1985).
7. K. Gawędzki, In *Nonperturbative QFT*, 101, Plenum Press (1988).
8. J.A. Fuchs, C. Schweigert & K. Waldorf, *J. Geom. Phys.* **58**, 576 (2008).
9. D. Stevenson, "The Geometry of Bundle Gerbes", PhD thesis, Univ. of Adelaide (2000); K. Waldorf, *Theory Appl. Categories* **18**, 240 (2007).
10. K. Gawędzki & N. Reis, *Rev. Math. Phys.* **14**, 1281 (2002); *J. Geom. Phys.* **50**, 28 (2003); K. Gawędzki, *Commun. Math. Phys.* **258**, 23 (2005).
11. J. Fröhlich, J.A. Fuchs, I. Runkel & C. Schweigert, *Phys. Rev. Lett.* **93**, 070601 (2004); *Nucl. Phys.* **B763**, 354 (2007).
12. A.Yu. Alekseev & T. Strobl, *JHEP* **0503**, 035 (2005).
13. I. Moerdijk, *Ann. Inst. Fourier* **41**, 189 (1991).
14. J. Fröhlich, J.A. Fuchs, I. Runkel & C. Schweigert, In *Proceedings of the XVIth ICMP*, 608, World Scientific (2009).

CRITICAL D-MODULE REPS FOR FINITE SUPERCONFORMAL ALGEBRAS AND THEIR SUPERCONFORMAL MECHANICS

F. TOPPAN

Teo Dep., CBPF, Rua Dr. Xavier Sigaud 150 (Urca),
Rio de Janeiro (RJ), cep 22290-180, Brazil
E-mail: toppan@cbpf.br

The simple finite Lie superalgebras $D(2,1;\alpha)$, $G(3)$, $D(4,1)$, $D(2,2)$, $A(3,1)$ and $F(4)$ admit D-module representations, given by a set of differential operators of a single variable $t \in \mathbb{R}$, at a critical value of the scaling dimension λ. These superalgebras are one-dimensional $\mathcal{N}$-extended superconformal algebras with $\mathcal{N}=4$ ($D(2,1;\alpha)$), $\mathcal{N}=7$ ($G(3)$) and $\mathcal{N}=8$ (the remaining ones). The critical D-module reps induce invariant actions in the Lagrangian framework for superconformal mechanics in D target dimensions. The $\mathcal{N}=8$ critical scalings $\lambda = \frac{1}{D-4}$ are linked to the D-dimensional harmonic functions with $D = 1,2,3,5,6,7,8$. This talk is based on J. Math. Phys. **53** (2012) 043513 (arXiv:1112.0995), J. Math. Phys. **53** (2012) 103518 (arXiv:1208.3612) and some extra material.

1. Introduction

The finite one-dimensional superconformal algebras are the simple Lie superalgebras[1–3] $\mathcal{G}$ which admit a grading $\mathcal{G} = \mathcal{G}_{-1} \oplus \mathcal{G}_{-\frac{1}{2}} \oplus \mathcal{G}_0 \oplus \mathcal{G}_{\frac{1}{2}} \oplus \mathcal{G}_1$ and whose even sector is $\mathcal{G}_{even} = sl(2) \oplus R$ (the subalgebra R is known as R-symmetry). The odd sector $(\mathcal{G}_{\frac{1}{2}} \oplus \mathcal{G}_{-\frac{1}{2}})$ is spanned by $2\mathcal{N}$ generators. The positive sector $\mathcal{G}_{>0}$ is isomorphic to the algebra of the supersymmetric quantum mechanics[4] (at fixed $\mathcal{N}$). Let D, H, K be the $sl(2)$ generators; $\mathcal{G}_1$ $(\mathcal{G}_{-1})$ is spanned by H (K), while $\mathcal{G}_0 = D\mathbb{C} \oplus R$. Up to $\mathcal{N} \leq 8$ we have the following list of finite superconformal algebras over $\mathbb{C}$ (for $\mathcal{N}=0$ we recover the $sl(2)$ algebra):

$$
\begin{aligned}
&\mathcal{N}=0: A_1, \\
&\mathcal{N}=1: B(0,1), \\
&\mathcal{N}=2: A(1,0), \quad C(2), \\
&\mathcal{N}=3: B(1,1), \\
&\mathcal{N}=4: A(1,1), \quad \underline{D(2,1;\alpha)}, \\
&\mathcal{N}=5: B(2,1), \\
&\mathcal{N}=6: A(2,1), \quad B(1,2), \quad D(3,1), \\
&\mathcal{N}=7: B(3,1), \quad \underline{G(3)}, \\
&\mathcal{N}=8: D(4,1), \quad D(2,2), \quad A(3,1), \quad \underline{F(4)},
\end{aligned} \tag{1}
$$

(the exceptional simple Lie superalgebras have been underlined). Their associated real forms can be found in [5].

A D-module rep of $sl(2)$, non-critical since it closes for any $\mu \in \mathbb{C}$, is given by

$$H = \tfrac{d}{dt}, \quad D = -t\tfrac{d}{dt} - \mu, \quad K = -t^2\tfrac{d}{dt} - 2\mu t, \tag{2}$$

(with non-vanishing commutators $[D,H] = H, [D,K] = -K, [H,K] = 2D$).

In [6,7] the extensions of the (2) D-module rep to D-module reps (realized by a set of differential operators of a single variable $t \in \mathbb{R}$ acting on n even and n odd component fields) for the superconformal algebras with $\mathcal{N} > 0$ were constructed.

They are based on the following two requirements:
i) a D-module rep of the $\mathcal{G}_{>}0$ subalgebra must coincide with a corresponding D-module rep of the global, one dimensional, $\mathcal{N}$-extended supersymmetry (based on [8], the latters have been classified, for $\mathcal{N} \le 8$ and minimal number of bosonic and fermionic component fields, in [9]),
ii) the $sl(2)$ generators act diagonally (see [6]) on the component fields and μ is identified with their respective scaling dimensions.

For global $\mathcal{N} = 4, 8$ supersymmetries the minimal D-module reps are uniquely characterized by the assignment of the scaling dimensions of the component fields. They are specified by $(D, \mathcal{N}, \mathcal{N} - D)$, ($D$ fields of scaling dimension λ, $\mathcal{N}$ fields of scaling dimension $\lambda + \frac{1}{2}$ and $\mathcal{N} - D$ fields of scaling dimension $\lambda + 1$) (here $D = 0, 1, \ldots, \mathcal{N}$). For $\mathcal{N} = 7$ there exists a unique global supermultiplet which cannot be extended to a global $\mathcal{N} = 8$ supermultiplet. Its "field content" $(1, 7, 7, 1)$ means that the assignment of the scaling dimensions is as follows: 1 field of s.d. λ, 7 fields of s.d. $\lambda + \frac{1}{2}$, 7 fields of s.d. $\lambda + 1$, 1 field of s.d. $\lambda + \frac{3}{2}$.

For $\mathcal{N} = 4, 7, 8$ the (so far) arbitrary λ is the overall scaling dimension of the global supermultiplet.

The construction of a D-module rep for the one-dimensional superconformal algebras is based on two inputs: a given D-module rep for the global $\mathcal{N}$-extended supersymmetry algebra and the overall scaling dimension λ (entering the $sl(2)$ subalgebra generators) of the supermultiplet.

Once identified these differential operators, there is no guarantee that their (repeated) commutators close producing a D-module rep of a finite superconformal algebra (to be identified). This is the case if a "closure condition" is satisfied. The closure condition means, in particular, that the global supersymmetry generators of the $\mathcal{G}_{\frac{1}{2}}$ sector belong to a representation of the R-symmetry. In our investigation three possibilities were encountered:

i) the closure condition is satisfied for any λ. A D-module rep of a finite superconformal algebra is obtained for any λ,

ii) the closure condition is never satisfied,

iii) the closure condition is satisfied for a critical value $\lambda = \lambda_{cr}$.

2. The critical $\mathcal{N} = 4, 7, 8$ D-modules.

The minimal, global, $\mathcal{N} = 4, 7, 8$ supermultiplets produce critical D-module representations of the corresponding superconformal algebras with the following identifications:

$\mathcal{N} = 4$: D-module reps are obtained for the $D(2, 1; \alpha)$ superalgebras from the global $(D, 4, 4 - D)$ supermultiplets, for any λ, with the identification $\alpha = (2 - D)\lambda$,

$\mathcal{N} = 7$: a $G(3)$ D-module rep is obtained from $(1, 7, 7, 1)$ at the critical value $\lambda = -\frac{1}{4}$,

$\mathcal{N} = 8$: D-module reps are obtained from $(D, 8, 8 - D)$, $D \neq 4$, at the critical scalings $\lambda_D = \frac{1}{D-4}$, with the identifications: $D(4, 1)$ for $D = 0, 8$, $F(4)$ for $D = 1, 7$, $A(3, 1)$ for $D = 2, 6$ and $D(2, 2)$ for $D = 3, 5$.

Several comments are in order. We list a few of them.

For other values of $\mathcal{N}$ (in particular $\mathcal{N} = 2, 3$) D-module reps with no criticality (closing for any value λ, were encountered).

The $\mathcal{N} = 8$ global supermultiplet $(4, 8, 4)$ does not produce a D-module rep of an $\mathcal{N} = 8$ superconformal algebra.

The $\mathcal{N} = 7, 8$ critical scaling dimensions are (partially) explained from the decomposition of the supermultiplets into two $\mathcal{N} = 4$ supermultiplets, taking into account the relation between α and λ and the isomorphism $D(2, 1; \alpha) \equiv D(2, 1; \alpha')$ for α, α' related by an S_3-group transformation.

The D-module reps of the finite superconformal algebras allow to construct invariant actions (in Lagrangian framework) for superconformal mechanics (e.g., a uniquely defined $F(4)$-invariant action for the $(1,8,7)$ supermultiplet is produced,[7,9,10] etc.).

For $\mathcal{N}=4$ the relation between α and λ gives constraints on possible superconformal actions for interacting supermultiplets. In particular an admissible solution is found[7] for $\alpha=\varphi$, the golden ratio, for the $D(2,1;\alpha)$-invariance of the $(1,4,3)$ and $(3,4,1)$ supermultiplets.

3. Link between the $\mathcal{N}=8$ critical scalings and the harmonic functions

A global $\mathcal{N}=8$-invariant action depends on a D-dimensional harmonic function[11,12] Φ ($\Box_D\Phi=0$). Its bosonic sector is a one-dimensional sigma model $\mathcal{S}=\int dt g_{ij}\dot{x}^i\dot{x}^j$ whose D-dimensional target metric is conformally flat ($g_{ij}=\delta_{ij}\Phi$), with Φ the conformal factor.

A conformal factor $\Phi=r^b$, dependent on the radial coordinate $r=\sqrt{\sum_i(x^i)^2}$ alone, is harmonic if the equation $b+D=2$ is satisfied. On the other hand, the action $\mathcal{S}$ is scale invariant and contains no dimensional parameter if the scaling dimension $\lambda=[x^i]$ satisfies the condition $(b+2)\lambda-1+2=0$, where we have assumed, consistently with the conventions used in the previous Section, that $[t]=-1$. It is worth observing that, combining these two separate requirements, we recover the critical scaling dimensions $\lambda=\frac{1}{D-4}$ obtained from the closure condition of the D-module reps of the $\mathcal{N}=8$ superconformal algebras. Indeed, at the given target dimension D and associated critical scaling we recover a superconformal mechanics invariant under the associated superconformal algebra. The scalar curvature R of the target, computed from Φ, is $R=\frac{1}{4}r^{D-4}[(D-1)(D-2)^2(D-6)]$.

These results can be summarized in the table

D	Φ	R	λ_{cr}	$\mathcal{S}$	$\mathcal{G}$
0	$-$	$-$	$-\frac{1}{4}$	$-$	$D(4,1)$
1	r	0	$-\frac{1}{3}$	$+$	$F(4)$
2	1	0	$-\frac{1}{2}$	$+$	$A(3,1)$
3	r^{-1}	$-\frac{3}{2}r^{-1}$	-1	$+$	$D(2,2)$
4	r^{-2}	-6	$-$	$-$	$-$
5	r^{-3}	$-18r$	1	$+$	$D(2,2)$
6	r^{-4}	0	$\frac{1}{2}$	$+$	$A(3,1)$
7	r^{-5}	$\frac{75}{2}r^3$	$\frac{1}{3}$	$+$	$F(4)$
8	r^{-6}	$126r^4$	$\frac{1}{4}$	$+$	$D(4,1)$

(3)

Therefore superconformal mechanics can be constructed from D-module representations of superconformal algebras. See [13] and references therein for a recent review on superconformal mechanics and its applications (see also [14] for a recent paper).

4. Dual supermultiplets

The $\mathcal{N} = 8$ supermultiplets, both globally supersymmetric and superconformal, are dually related under the exchange $D \leftrightarrow 8 - D$. In the superconformal case the duality gets manifested with the recovering of the same superconformal algebra (the critical scalings are related via the transformation $\lambda_D \leftrightarrow -\lambda_{8-D}$).

Let $\Psi \equiv \Psi_D$ denotes a superconformal multiplet with 8 bosonic and 8 fermionic component fields, while $\widetilde{\Psi} \equiv \Psi_{8-D}$ denotes a dual counterpart. A pairing $< \cdot|\cdot >$ exists such that the integral $\int dt < \widetilde{\Psi}|\Psi >$ is superconformally invariant. For a given superalgebra a generalized superconformal action $\mathcal{S}_{\alpha,\beta,\gamma}(\Psi, \widetilde{\Psi})$ can be given by setting $\mathcal{S}_{\alpha,\beta,\gamma}(\Psi, \widetilde{\Psi}) = \alpha\mathcal{S}_D(\Psi) + \beta\mathcal{S}_{8-D}(\widetilde{\Psi}) + \gamma\int dt < \widetilde{\Psi}|\Psi >$, with α, β, γ arbitrary non-dimensional constants and the $\mathcal{S}_D$, $\mathcal{S}_{8-D}$ actions given in Section **3**.

This classical superconformal action induces, in the path integral formulation, a superconformally invariant theory with partition function $\mathcal{Z}(\alpha, \beta, \gamma; J, \widetilde{J})$ ($J, \widetilde{J}$ are superconformal currents), given by

$$\mathcal{Z}(\alpha, \beta, \gamma; J, \widetilde{J}) = \int\int \mathcal{D}\Psi\mathcal{D}\widetilde{\Psi} e^{-\mathcal{S}_{\alpha,\beta,\gamma}(\Psi,\widetilde{\Psi})+\int dt<\widetilde{J}|\Psi>+\int dt<\widetilde{\Psi}|J>}. \quad (4)$$

At least formally the measure $\mathcal{D}\Psi\mathcal{D}\widetilde{\Psi}$ is superconformally invariant. Both the action and the measure are dimensionless. Indeed, the dimension $[\mathcal{D}\Psi] = D\lambda_D - 8(\lambda_D + \frac{1}{2}) + (D - 8)(\lambda_D + 1) = 4 - D$ is compensated by $[\mathcal{D}\widetilde{\Psi}] = D - 4$. This construction naturally leads to superconformal models with double target manifolds.

5. Conclusions and open problems

The critical D-module representations induce invariant actions in the Lagrangian framework for classical superconformal mechanics. The quantization based on the functional approach via path integral is also naturally encoded. The superconformal invariance of the measure requires considering dually related supermultiplets leading to sigma-model actions with double target manifolds.

Several open questions, inspired by an operatorial quantization approach, can be raised. We are listing here a few of them.

A D-module rep induces a specific lowest weight representation (there exists a l. w. vector $|w>$ s.t. $H|w>=0$ and $Q_i|w>=0$ for $i=1,\ldots,\mathcal{N}$) at the critical scaling dimension λ. Are the l.w.r.'s induced by a D-module rep unitary for some given real form of the $d=1$ SCA? Furthermore, is there an intrinsic, representation-theoretical characterization, to single out a l.w.r. induced by a D-module, from a generic l.w.r. of the corresponding superalgebra?

It seems possible that unitary l.w.r.'s which are *not* induced by a D-module rep could define a quantum mechanics which does not admit a classical counterpart and a Lagrangian description.

Acknowledgments

This work was supported by CNPq. The material here reported is partly based on a collaboration with Z. Kuznetsova and S. Khodaee.

References

1. V. G. Kac, Comm. Math. Phys. **53** (1977) 31.
2. W. Nahm, V. Rittenberg and M. Scheunert, J. Math. Phys. **17** (1976) 1626 and J. Math. Phys. **17** (1976) 1640.
3. L. Frappat, A. Sciarrino and P. Sorba, *Dictionary on Lie algebras and superalgebras*, Academic Press, London (2000). Also available in arXiv:hep-th/9607161.
4. E. Witten, Nucl. Phys. **B 188** (1981) 513.
5. A. Van Proeyen, *Tools for supersymmetry.* Lectures in the spring school in Calimanesti, Romania, April 1998; arXiv:hep-th/9910030.
6. Z. Kuznetsova and F. Toppan, J. Math. Phys. **53** (2012) 043513; arXiv:1112.0995[hep-th].
7. S. Khodaee and F. Toppan, J. Math. Phys. **53** (2012) 103518; arXiv:1208.3612[hep-th].
8. A. Pashnev and F. Toppan, J. Math. Phys. **42** (2001) 5257; arXiv:hep-th/0010135.
9. Z. Kuznetsova, M. Rojas, F. Toppan, JHEP **0603** (2006) 098; arXiv:hep-th/0511274.
10. F. Delduc and E. Ivanov, Phys. Lett. **B 654** (2007) 200; arXiv:0706.2472[hep-th].
11. S. Bellucci, E. Ivanov, S. Krivonos and O. Lechtenfeld, Nucl. Phys. **B 699** (2004) 226; arXiv:hep-th/0406015.
12. M. Gonzales, M. Rojas and F. Toppan, Int. J. Mod. Phys. **A 24** (2009), 4317; arXiv:0812.3042[hep-th].
13. S. Fedoruk, E. Ivanov and O. Lechtenfeld, J. Phys. **A 45** (2012) 173001; arXiv:1112.1947[hep-th].
14. N. B. Copland, S. M. Ko and J.-H. Park, JHEP 07 (2012) 076; arXiv:1205.3869[hep-th]

EXTENDED GRIESS ALGEBRAS, CONFORMAL DESIGNS AND MATSUO-NORTON TRACE FORMULAE

HIROSHI YAMAUCHI

Department of Mathematics, Tokyo Woman's Christian University
2-6-1 Zempukuji, Suginami, Tokyo 167-8585 Japan
E-mail: `yamauchi@lab.twcu.ac.jp`

We introduce the extended Griess algebra for a vertex operator superalgebra and derive its Matsuo-Norton trace formula based on conformal design structure. This article is an exposition of the paper [Y12].

Keywords: Vertex operator algebras; Virasoro algebras; Finite simple groups.

1. Conformal design

The definition of vertex operator algebras (VOAs) is regarded as an algebraic formulation of the 2-dimensional conformal field theory. The conformal symmetry in 2-dimension is represented by the Virasoro algebra. Like symmetric groups and orthogonal groups play central role in block and spherical designs, a conformal design is formulated by the symmetry of the Virasoro algebra upon a vertex operator algebra.

Let $V = \oplus_{n\geq 0} V_n$ be a vertex operator algebra and M its module. Let ω be the conformal vector of V and denote $\langle\omega\rangle$ the subalgebra of V generated by ω. Here we suppose that $\langle\omega\rangle$ is a direct summand of V as a $\langle\omega\rangle$-module and we set the projection map $\pi : V \to \langle\omega\rangle$.

Definition 1.1. ([H08]) An $L(0)$-homogeneous subspace X of M is called a *conformal t-design* if it satisfies: For any $a \in \oplus_{0\leq n\leq t} V_n$ one has

$$\mathrm{tr}_M\, \mathrm{o}(a) = \mathrm{tr}_M\, \mathrm{o}(\pi(a)), \tag{1}$$

where $\mathrm{o}(a) = a_{(\mathrm{wt}(a)-1)}$ is the zero-mode action of a.

In the definition of conformal designs the vacuum module $\langle\omega\rangle = \pi(V)$ plays a role of the Virasoro invariants.

2. Matsuo-Norton trace formulae

A VOA V is called *of OZ-type*[a] if it has the $L(0)$-grading $V = \oplus_{n\geq 0} V_n$ such that $V_0 = \mathbb{C}\mathbb{1}$ and $V_1 = 0$. Then the weight two subspace V_2 possesses a structure of a commutative algebra with an invariant bilinear form. This commutative algebra structure on V_2 is called the *Griess algebra* of V. Matsuo [Ma01] derived formulae, called Matsuo-Norton trace formulae, which describes traces of adjoint actions in the Griess algebra based on its conformal design structure.

Theorem 2.1 ([Ma01]). *Let V be a VOA of OZ-type. If V_2 forms a conformal $2t$-design with $t \leq 5$ then the traces $\mathrm{tr}_{V_2}\, o(a^1)\cdots o(a^k)$ with $a^1, \cdots, a^k \in V_2$, $1 \leq k \leq t$, satisfy Matsuo-Norton formulae.*

We give formulae of Matsuo-Norton type for extended Griess algebras.

3. Extended Griess algebra

Let $V = V^0 \oplus V^1$ be a vertex operator superalgebra (SVOA).

Condition 3.1. We assume that V^0 is of OZ-type and V^1 have the $L(0)$-grading $V^1 = \oplus_{n\geq 0} V_{n+h}$, $V_h \neq 0$, $h \in \mathbb{Z} + 1/2$ and $h > 0$.

The subspace V_h and its weight h are called the *top level* and the *top weight* of V^1, respectively. Consider the subspace $V_2 \oplus V_h$ of $V^0 \oplus V^1$. We define a product and a bilinear form on it as follows. For $a, b \in V_2$ and $u, v \in V_h$,

$$\begin{aligned} &ab = a_{(1)}b, \quad au := a_{(1)}u, \quad ua := u_{(1)}a, \quad uv := u_{(2h-3)}v, \\ &(a|b)\mathbb{1} = a_{(3)}b, \quad (a|u) = (u|a) = 0, \quad (u|v)\mathbb{1} = u_{(2h-1)}v. \end{aligned} \tag{2}$$

Proposition 3.1 ([Y12]). *The product in (2) equips the space $V_2 \oplus V_h$ with a structure of a commutative algebra which extends the Griess algebra structure on V_2. Moreover, the bilinear form above is invariant with respect to the product.*

We call the commutative algebra structure on $V_2 \oplus V_h$ as the *extended* Griess algebra.

Remark 3.1. The space $V_2 \oplus V_h$ carries a $\mathbb{Z}_2$-grading. The product on this space in (2) is really commutative and not super-commutative. Namely, for $u, v \in V_h$, one has $uv = vu$.

---

[a] "OZ" stands for One-Zero.

The extended Griess algebra $V_2 \oplus V_h$ is a $\mathbb{Z}_2$-graded extension of the Griess algebra and one can consider square roots of idempotents of V_2.

Proposition 3.2 ([Y12],[Za85]). *Let e be an idempotent in V_2 and $x \in V_h$ is a square root of e in the extended Griess algebra $V_2 \oplus V_h$.*
(1) If $h = 1/2$ then $(x|x) = 1/4$, $(e|e) = 1/16$ and $[x_{(m)}, x_{(n)}]_+ = \delta_{m+n,-1}/4$. Therefore, the subalgebra $\langle x \rangle$ is isomorphic to a simple SVOA $L(1/2,0) \oplus L(1/2,1/2)$.
(2) If $h = 3/2$ then $3(x|x) = 4(e|e)$ and the subalgebra $\langle x \rangle$ is isomorphic to the $N = 1$ $c = 8(e|e)$ Virasoro SVOA.
(3) If $h = 5/2$ and $x_{(n)}x \in \langle e \rangle$ for $n \geq 0$, then the subalgebra $\langle x \rangle$ has a simple quotient $L(-13/14, 0) \oplus L(-13/14, 5/2)$.

4. Trace formulae for extended Griess algebras

Condition 4.1. We consider an SVOA $V = V^0 \oplus V^1$ satisfying the following conditions.
(1) V satisfies Condition 3.1.
(2) V has a non-degenerate invariant bilinear form.
(3) The subalgebra $\langle \omega \rangle$ is a direct summand of V as a $\langle \omega \rangle$-module.
(4) The central charge c of V is none of $0, -22/5, 1/2, -68/7, -46/3, -3/5, -232/11$.[b]

Let $\pi : V \to \langle \omega \rangle$ be the projection map, which is a $\langle \omega \rangle$-homomorphism. Then $V = \langle \omega \rangle \oplus \operatorname{Ker} \pi$ by (3) of Condition 4.1.

Lemma 4.1. *With respect to the invariant bilinear form, $\langle \omega \rangle \perp \operatorname{Ker} \pi$.*

Proof. Let $a \in \langle \omega \rangle$ and $b \in \operatorname{Ker} \pi$. Then $(a|b) = \operatorname{Res}_z(Y(a,z)\mathbb{1}|b) = \operatorname{Res}_z(\mathbb{1}|Y^*(a,z)b)$, where $Y^*(a,z)$ is the adjoint operator of $Y(a,z)$ (cf. [FLM88]). Since $\operatorname{Ker} \pi$ is a $\langle \omega \rangle$-submodule, one has $\operatorname{Res}_z Y^*(a,z)b \in \operatorname{Ker} \pi$. Since $\operatorname{Ker} \pi \cap \mathbb{C}\mathbb{1} = 0$, we obtain $(a|b) = 0$ as claimed. □

By the lemma above together with (2) of Condition 4.1, we have:

Proposition 4.1. *Under Condition 4.1, the invariant bilinear form restricted on $\langle \omega \rangle$ is non-degenerate.*

---

[b] We have to exclude these values by the structure theory of Verma modules over the Virasoro algebra. Also, these values appear as zeros of denominators of formulae.

Now set $d := \dim V_h$ and let $\{u^i\}$ a basis of V_h and $\{u_i\}$ its dual basis. Consider the following generalized Casimir vector.

$$\kappa_n := \sum_{i=1}^{d} u^i_{(2h-n-1)} u_i \in V^0_n. \tag{3}$$

Lemma 4.2. *The top level V_h forms a conformal n-design based on V^0 if and only if $\kappa_j \in \langle \omega \rangle$ for $0 \le j \le n$.*

Theorem 4.1. *Suppose V satisfies Condition 4.1 and the top level V_h forms a conformal $2n$-design based on V^0 with $1 \le n \le 5$. Then traces $\mathrm{tr}_{V_h}\, o(a^1)\cdots o(a^k)$ with $a^1, \cdots, a^k \in V_2$, $1 \le k \le n$, satisfy certain formulae of Matsuo-Norton type.*

Corollary 4.1. *Suppose V satisfies Condition 4.1 and the top level V_h forms a conformal $2n$-design based on V^0 with $1 \le n \le 5$. Let $e \in V_2$ be a $c = 2(e|e)$ Virasoro vector. Then traces $\mathrm{tr}_{V_h}\, o(e)^k$, $1 \le k \le n$, are given as follows.*

$$\mathrm{tr}_{V^h}\, o(e) = \frac{2hd}{c}(e|e) \quad (t=1)$$

$$\mathrm{tr}_{V^h}\, o(e)^2 = \frac{4hd(5h+1)}{c(5c+22)}(e|e)^2 + \frac{2hd(22h-c)}{c(5c+22)}(e|e) \quad (t=2)$$

$$\mathrm{tr}_{V^h}\, o(e)^k = D_{2k}(c)^{-1} \textstyle\sum_{j=1}^{t} E^{(k)}_j (e|e)^j \quad (k=3,4,5),$$

where $D_m(c) \in \mathbb{Z}[c]$ and $E^{\bullet}_j \in \mathbb{Z}[c,d,h]$ are explicitly given in [Y12].

5. Application

Here we consider the Baby-monster SVOA $V\!B^{\natural} = V\!B^{\natural,0} \oplus V\!B^{\natural,1}$ introduced by Höhn in [H95] which affords an action of the Baby-monster sporadic finite simple group $\mathbb{B}$. It is known that $V\!B^{\natural}$ satisfies Condition 4.1.

Remark 5.1. Denote $\theta = (-1)^{2L(0)}$ the canonical $\mathbb{Z}_2$-symmetry of the superalgebra. It is shown in [H02],[Y05] that $\mathrm{Aut}(V\!B^{\natural}) = \mathrm{Aut}(V\!B^{\natural,0}) \times \langle \theta \rangle$, $\mathrm{Aut}(V\!B^{\natural,0}) \simeq \mathbb{B}$ and the even part $V\!B^{\natural,0}$ has three irreducible modules which are all $\mathbb{B}$-stable.

The top level of $V\!B^{\natural,1}$ is of dimension $d = 4371$ and has the top weight $h = 3/2$. It is shown in Lemma 2.6 of [DLM96] (see also [H03],[H08]) that $V\!B^{\natural,0}$ is of class $\mathcal{S}^6$. Therefore the top level of $V\!B^{\natural,1}$ forms a conformal 6-design based on $V\!B^{\natural,0}$. (Actually, it is shown in [H03] and [H08] that $V\!B^{\natural,0}$ is of class $\mathcal{S}^7$ and $V\!B^{\natural,1}_{3/2}$ is a conformal 7-design, respectively.)

Let t be a 2A-involution of $\mathbb{B}$. Then $C_{\mathbb{B}}(t) \simeq 2 \cdot {}^2E_6(2){:}2$ (cf. [ATLAS]). The Griess algebra of $V\!B^{\natural,0}$ is of dimension 96256 and we have the following decompositions as a module over $\mathbb{B}$ and $C_{\mathbb{B}}(t)$ (cf. [HLY12]).

$$\begin{aligned} V\!B^{\natural,0}_2 &= \underline{\mathbf{1}} \oplus \underline{\mathbf{96255}} && \text{over } \mathbb{B}, \\ &= \underline{\mathbf{1}} \oplus \underline{\mathbf{1}} \oplus \underline{\mathbf{1938}} \oplus \underline{\mathbf{48620}} \oplus \underline{\mathbf{45696}} && \text{over } C_{\mathbb{B}}(t). \end{aligned} \tag{4}$$

The $\mathbb{B}$-invariant subalgebra of the Griess algebra is 1-dimensional spanned by the conformal vector ω of $V\!B^{\natural}$, but the $C_{\mathbb{B}}(t)$-invariant subalgebra of the Griess algebra forms a 2-dimensional commutative (and associative) subalgebra spanned by two mutually orthogonal Virasoro vectors. It is shown in [HLY12] that central charges of these Virasoro vectors are $7/10$ and $114/5$. (The sum is the conformal vector of $V\!B^{\natural}$.) Let e be the shorter one, the simple $c = 7/10$ Virasoro vector fixed by $C_{\mathbb{B}}(t)$. For the odd part we have the following decompositions.

$$\begin{aligned} V\!B^{\natural,1}_{3/2} &= \underline{\mathbf{4371}} && \text{over } \mathbb{B}, \\ &= \underline{\mathbf{1}} + \underline{\mathbf{1938}} + \underline{\mathbf{2432}} && \text{over } C_{\mathbb{B}}(t). \end{aligned} \tag{5}$$

Since e is fixed by $C_{\mathbb{B}}(t)$, the zero-mode $\mathrm{o}(e)$ acts as scalars on $C_{\mathbb{B}}(t)$-irreducible components. Applying Theorem 4.1 and Corollary 4.1, we obtain the following.

$$\begin{array}{rcccc} V\!B^{\natural,1}_{3/2} & = & \underline{\mathbf{1}} & + \ \underline{\mathbf{1938}} & + \ \underline{\mathbf{2432}} \\ \mathrm{o}(e) & : & \dfrac{3}{2} & \dfrac{1}{10} & 0 \end{array} \tag{6}$$

We can choose a non-zero vector x in the 1-dimensional $C_{\mathbb{B}}(t)$-invariant subspace of $V\!B^{\natural,1}_{3/2}$ which is a square root of e in the extended Griess algebra $V\!B^{\natural,0}_2 \oplus V\!B^{\natural,1}_{3/2}$. By Proposition 3.2 the subalgebra $\langle x \rangle$ is isomorphic to the $N = 1$ $c = 7/10$ Virasoro SVOA $L(7/10, 0) \oplus L(7/10, 3/2)$. It is known (cf. [Mi96],[HLY12]) that the map $\sigma_e := \exp(10\pi \mathrm{o}(e))$ is well-defined and one has $t = \theta\sigma_u$ on $V\!B^{\natural}$. The correspondence is known to one to one.

Theorem 5.1 ([HLY12]). *There is a one-to-one correspondence between the subalgebras of $V\!B^{\natural}$ isomorphic to the $N = 1$ $c = 7/10$ simple Virasoro SVOA $L(7/10, 0) \oplus L(7/10, 3/2)$ and the 2A-elements of the Baby-monster $\mathbb{B}$ given by the association $u \mapsto \sigma_e\theta$ where e is the conformal vector of the sub SVOA, $\sigma_e = \exp(10\pi \mathrm{o}(u))$ and $\theta = (-1)^{2L(0)}$.*

Remark 5.2. Let $V = V^0 \oplus V^1$ be an SVOA such that V^1 has the top weight $3/2$. Suppose the top level $V_{3/2}$ forms a conformal 6-design based on

V^0. It is shown in [H08] that if $\dim V_{3/2} > 1$ then the central charge c of V is either 16 or 47/2. Suppose further that there is a simple $c = 7/10$ Virasoro vector $e \in V_2$ which has a square root $x \in V_{3/2}$ in the extended Griess algebra. In this case one can show that $c = 47/2$ and $\dim V_{3/2} = 4371$, those of $VB^\natural$. The author naively expects that $VB^\natural$ is the unique object satisfying these conditions.

Acknowledgments

This work was supported by JSPS Grant-in-Aid for Young Scientists (Start-up) No. 19840025, (B) No. 21740011 and No. 24740027.

References

[ATLAS]. J.H. Conway, R.T. Curtis, S.P. Norton, R.A. Parker and R.A. Wilson, ATLAS of finite groups. Clarendon Press, Oxford, 1985.

[DLM96]. C. Dong, H. Li and G. Mason, Some twisted sectors for the moonshine module, Moonshine, the Monster, and related topics (South Hadley, MA, 1994), 25–43, Contemp. Math., 193, Amer. Math. Soc., Providence, RI, 1996.

[FLM88]. I.B. Frenkel, J. Lepowsky and A. Meurman, Vertex Operator Algebras and the Monster. Academic Press, New York, 1988.

[H95]. G. Höhn, Selbstduale Vertexoperatorsuperalgebren und das Babymonster. Ph.D. thesis, Bonn 1995, Bonner Mathematische Schriften **286**. `arXiv:0706.0236`

[H02]. G. Höhn, The group of symmetries of the shorter moonshine module. `arXiv:math/0210076`.

[H03]. G. Höhn, Generalized moonshine for the Baby Monster. Preprint. `http://www.math.ksu.edu/~gerald/papers/baby8.ps`.

[H08]. G. Höhn, Conformal designs based on vertex operator algebras. *Adv. Math.* **217** (2008), 2301–2335.

[HLY12]. G. Höhn, C.H. Lam and H. Yamauchi, McKay's E_7 observation on the Babymonster, *Internat. Math. Res. Notices* **2012** (2012), 166–212. doi:10.1093/imrn/rnr009

[Ma01]. A. Matsuo, Norton's trace formulae for the Griess algebra of a vertex operator algebra with larger symmetry. *Commun. Math. Phys.* **224** (2001), 565–591.

[Mi96]. M. Miyamoto, Griess algebras and conformal vectors in vertex operator algebras. *J. Algebra* **179** (1996), 528–548.

[Y05]. H. Yamauchi, 2A-orbifold construction and the baby-monster vertex operator superalgebra. *J. Algebra* **284** (2005), 645–668.

[Y12]. H. Yamauchi, Extended Griess algebras and Matsuo-Norton trace formulae, submitted, `arXiv:1206.3380`

[Za85]. A.B. Zamolodchikov, Infinite additional symmetries in two dimensional conformal quantum field theory, *Theor. Math. Phys.* **65** (1985) 1205.

PARALLEL SESSION
Supersymmetry and Quantum Groups

20 & 21 & 22 & 23 AUGUST 2012

GALILEAN SUPERCONFORMAL ALGEBRAS WITH CENTRAL EXTENSION AND THEIR REALIZATIONS

N. AIZAWA

Department of Mathematics and Information Sciences,
Graduate School of Science, Osaka Prefecture University,
Nakamozu Campus, Sakai, Osaka 599-8531, Japan
E-mail: aizawa@mi.s.osakafu-u.ac.jp

The Galilean conformal algebra (GCA) is a class of non-semisimple Lie algebra specified by spin ℓ and dimension of space d. We investigate $\mathcal{N} = 2$ supersymmetric extensions of GCA and introduce a novel super-GCA for $d = 2$ but arbitrary ℓ. We also give a realization of the novel superalgebra in terms of bosons and fermions.

Keywords: Galilean conformal algebra; Supersymmetry; Bose-Fermi realization.

1. Introduction

Recent discovery of the non-relativistic AdS/CFT correspondence[1–3] attracts renewed interest in a class of Lie algebra which we call the Galilean conformal algebra (GCA) in this paper. This is a non-semisimple Lie algebra whose structure or representations have not been thoroughly studied. Each GCA is specified by two parameters $d \in \mathbb{N}$ of dimension of space and $\ell \in \frac{1}{2}\mathbb{N}$ where $\mathbb{N}$ denotes the set of positive integers. The parameter d also specifies the maximal semisimple subalgebra $sl(2,\mathbb{R}) \oplus so(d)$ of GCA. The parameter ℓ specifies the ℓd dimensional abelian ideal of GCA which carries the spin ℓ representation of $sl(2,\mathbb{R})$ subalgebra. The abelian ideal has central extensions for certain values of (d,ℓ).[3–7] The $\ell = 1/2$ member, called Schrödinger algebra, is the best known example of this class of algebra.

In this paper we investigate $\mathcal{N} = 2$ supersymmetric extensions of GCA and their central extensions. Our main result is the introduction of a novel super-GCA with central extension in two dimensional space but arbitrary ℓ. We also give a realization of the novel super-GCA by bosons and fermions (see also Ref. 8). To clarify our motivation, we briefly review the bosonic

GCA and previous works on supersymmetric extensions in the next section. Indeed, the previous works suggest that there exist a unknown supersymmetric extension of the GCA for arbitrary ℓ and the extension is possible only in two dimensional space. We give a definition of the novel super-GCA in §3 and study its central extensions. It is shown in §4 that the superalgebra is realized in terms of uncoupled bosons and fermions. This realization leads an interesting relation of the super-GCA and a certain bose-fermi oscillator Hamiltonian.

2. Bosonic Galilean conformal algebra and its supersymmetric extensions

Galilean conformal algebra for a given pair of (d, ℓ) has generators

$$D,\ H,\ C,\ M_{ij},\ P_i^{(n)},$$

where $i, j = 1, 2, \cdots, d$ and $n = 0, 1, \cdots, 2\ell$. The sets of generators $\langle D, H, C\rangle$ and $\langle M_{ij}\rangle$ span $sl(2,\mathbb{R})$ and $so(d)$ subalgebras, respectively. $\langle P_i^{(n)}\rangle$ is a basis of abelian ideal of the GCA. The nonvanishing commutation relations are given by[9]

$$\begin{aligned}
&[D, H] = 2H, \qquad [D, C] = -2C, \qquad [C, H] = D,\\
&[M_{ij}, M_{k\ell}] = -\delta_{ik}M_{j\ell} - \delta_{j\ell}M_{ik} + \delta_{i\ell}M_{jk} + \delta_{jk}M_{i\ell},\\
&[H, P_i^{(n)}] = -nP_i^{(n-1)}, \qquad [D, P_i^{(n)}] = 2(\ell - n)P_i^{(n)},\\
&[C, P_i^{(n)}] = (2\ell - n)P_i^{(n+1)}, \qquad [M_{ij}, P_k^{(n)}] = -\delta_{ik}P_j^{(n)} + \delta_{jk}P_i^{(n)}.
\end{aligned} \tag{1}$$

It is known that the algebra has the two types of central extensions:[3]
(i) mass extension existing for any d and half-integer ℓ

$$[P_i^{(m)}, P_j^{(n)}] = \delta_{ij}\,\delta_{m+n,2\ell}\,I_m M, \qquad I_m = (-1)^{m+\ell+\frac{1}{2}}(2\ell - m)!\,m!. \tag{2}$$

(ii) exotic extension existing only for $d = 2$ and integer ℓ

$$[P_i^{(m)}, P_j^{(n)}] = \epsilon_{ij}\,\delta_{m+n,2\ell}\,\check{I}_m\Theta, \qquad \check{I}_m = (-1)^m(2\ell - m)!\,m!, \tag{3}$$

where ϵ_{ij} is the antisymmetric tensor with $\epsilon_{12} = 1$. The structure constants are determined, up to an overall factor, from the Jacobi identity.

Next we briefly summarize the supersymmetric extensions of GCA. Supersymmetric extension of $\ell = 1/2$ GCA has been completed by Duval and Horváthy.[10] They provided a systematic method to construct $\mathcal{N}$-supersymmetric extension of Schrödinger algebra in d dimensional space with central extension. It was shown that there exist two types of supersymmetric extensions. They called them *standard* and *exotic* extensions.

One may has the standard extension in any dimension of space, while the exotic extension only in $d = 2$.

There are some attempts to extend $\ell \geq 1$ GCA to superalgebra, however, no systematic method to obtain $\mathcal{N}$-supersymmetric extension has been established yet. Such an attempt was initiated by de Azcárraga and Lukierski[11] for $\ell = 1$ and $\mathcal{N} = 2k$ (even) in $(3+1)$ dimensional spacetime. It was followed by the work of Sakaguchi,[13] then the work of Bagchi and Mandal.[12] These works introduced $\ell = 1$ super-GCA's in $d = 2$ or $d = 3$ with some values of $\mathcal{N}$. A possible general structure of $\ell = 1$ super-GCA was discussed by Fedoruk and Lukierski.[14] Very recently, Masterov introduced $\mathcal{N} = 2$ super-GCA for any d and ℓ.[15] We remark that central extension is not considered in any works on $\ell \geq 1$ super-GCA mentioned above.

Msterov's $\mathcal{N} = 2$ super-GCA has the following generatores: the generators of bosonic GCA, supercharges $Q, \bar{Q}$, conformal supercharges $S, \bar{S}$ and fermionic partner of $P_i^{(n)}$ denoted by $X_i^{(n)}, \bar{X}_i^{(n)}$ $(n = 0, 1, \cdots, 2\ell - 1)$ and additional bosonic generators R, $J_i^{(n)}$ $(n = 0, 1, \cdots, 2\ell - 2)$. The suffix $i = 1, 2, \cdots, d$ denotes the space coordinate. It has the superalgebra $sl(2/1)$, generated by $\langle Q, \bar{Q}, S, \bar{S}, D, H, C, R\rangle$, as a subalgebra. The subset $\langle P_i^{(n)}, X_i^{(n)}, \bar{X}_i^{(n)}, J_i^{(n)}\rangle$ forms an abelian ideal of the whole superalgebra. If we set $\ell = 1/2$, then the *standard* super Schrödinger algebra (without central extension) of Duval and Horváthy is recovered. Thus the Masterov's superalgebra is a generalization of the standard super Schrödinger algebra to higher ℓ. It is easy to verify that the superalgebra for higher ℓ has the mass and exotic central extensions depending on the parity of 2ℓ.[8]

From this observation, one may ask the following questions. (i) is it possible to generalize the $\mathcal{N} = 2$ *exotic* super Schrödinger algebra to higher ℓ ? (ii) if it is the case, are the mass and the exotic central extension generalized to the superalgebra with higher ℓ ? We give an affirmative answer to these questions in the next section.

3. Exotic supersymmetric extension of GCA

We would like to introduce a Lie superalgebra which is a generalization of the $\mathcal{N} = 2$ exotic super Schrödinger algebra to higher ℓ. One may be able to this based on the definition of Lie superalgebra. Namely, we set up a set of (anti)commutation relations, then verify that they satisfy the super Jacobi identity.

As the exotic supersymmetric extension for $\ell = 1/2$ exists only in two dimensional space, we set $d = 2$. The space rotation is generated by one element M_{12}. We introduce fermionic generators: supercharges Q, Q^*, conformal supercharges S, S^*, and fermionic partner of $P_i^{(n)}$ denoted by $X_i^{(n)}$ ($n = 0, 1, \cdots, 2\ell - 1,\ i = 1, 2$). We also introduce an additional bosonic generator R. We then set up the following nonvanishing commutators as well as (1): fermionic sector:

$$\begin{aligned}
&\{Q, Q\} = \{Q^*, Q^*\} = 2H, \ \{S, S\} = \{S^*, S^*\} = 2C, \\
&\{Q, S\} = \{Q^*, S^*\} = D, \\
&\{Q, S^*\} = -M_{12} + R, \qquad \{Q^*, S\} = M_{12} - R, \\
&\{Q, X_i^{(n)}\} = -P_i^{(n)}, \qquad \{Q^*, X_i^{(n)}\} = \sum_k \epsilon_{ik} P_k^{(n)}, \\
&\{S, X_i^{(n)}\} = -P_i^{(n+1)}, \qquad \{S^*, X_i^{(n)}\} = \sum_k \epsilon_{ik} P_k^{(n+1)},
\end{aligned} \tag{4}$$

bosonic-fermionic sector:

$$\begin{aligned}
&[H, S] = -Q, \ \ [H, S^*] = -Q^*, \ \ [C, Q] = S, \qquad [C, Q^*] = S^*, \\
&[D, S] = -S, \ \ [D, S^*] = -S^*, \ \ [D, Q] = Q, \qquad [D, Q^*] = Q^*, \\
&[H, X_i^{(n)}] = -nX_i^{(n-1)}, \qquad [C, X_i^{(n)}] = (2\ell - n - 1)X_i^{(n+1)}, \\
&[D, X_i^{(n)}] = (2\ell - 2n - 1)X_i^{(n)}, \ [R, X_i^{(n)}] = (2\ell + 1)\sum_k \epsilon_{ik} X_k^{(n)}, \\
&[M_{12}, S] = S^*, \ [M_{12}, S^*] = -S, \ [M_{12}, Q] = Q^*, \ [M_{12}, Q^*] = -Q, \\
&[R, S] = 2S^*, \ \ [R, S^*] = -2S, \ \ [R, Q] = 2Q^*, \ \ [R, Q^*] = -2Q, \\
&[Q, P_i^{(n)}] = nX_i^{(n-1)}, \qquad [Q^*, P_i^{(n)}] = n\sum_k \epsilon_{ik} X_k^{(n-1)}, \\
&[S, P_i^{(n)}] = -(2\ell - n)X_i^{(n)}, \qquad [S^*, P_i^{(n)}] = -(2\ell - n)\sum_k \epsilon_{ik} X_k^{(n)}.
\end{aligned} \tag{5}$$

additional bosonic commutator:

$$[R, P_i^{(n)}] = (2\ell - 1)\sum_k \epsilon_{ik} P_k^{(n)}. \tag{6}$$

One may verify by the straightforward computation that these (anti) commutation relations satisfy the super Jacobi identity. It is also verified that by setting $\ell = 1/2$ we recover the exotic super Schrödinge algebra of Duval and Horváthy.[10] Thus we have defined the $\mathcal{N} = 2$ exotic super-GCA without central extension. As seen from the (anti)commutation relations the exotic super-GCA has the $osp(2/2)$ subalgebra generated by $\langle Q, Q^*, S, S^*, D, H, C, R\rangle$. The set of generators $\langle P_i^{(n)}, X_i^{(n)}\rangle$ forms an abelian ideal of the exotic superalgebra. This exotic super-GCA also differs from the standard one on the number of the X-type generator. It is the half of the standard superalgebra. Furthermore, the exotic super-GCA does not have J generators.

Now we consider central extensions of the exotic super-GCA. For half-integer ℓ we introduce the mass central extension defined by (2) and

$$\{X_i^{(m)}, X_j^{(n)}\} = \delta_{ij}\delta_{m+n,2\ell-1}\alpha_m M, \quad \alpha_m = (-1)^{m+\ell-\frac{1}{2}}(2\ell-1-m)!\,m!. \tag{7}$$

For integer ℓ we introduce the exotic central extension by (3) and

$$\{X_i^{(m)}, X_j^{(n)}\} = \epsilon_{ij}\delta_{m+n,2\ell-1}\check{\alpha}_m \Theta, \quad \check{\alpha}_m = (-1)^{m+1}(2\ell-1-m)!\,m!. \tag{8}$$

One may verify by the straightforward computation that these commutation relations satisfy the super Jacobi identity. Thus we have obtained the $\mathcal{N}=2$ exotic super-GCA with mass or exotic central extensions.

4. Realization of exotic super-GCA

In this section we give a realization of the $\mathcal{N}=2$ exotic super-GCA with central extension in terms of bosons and fermions. We introduce $2\ell+1$ independent bosons $b_m, b_m^\dagger$ $(m=1,\cdots,2\ell+1)$ and 2ℓ independent fermions $a_m, a_m^\dagger$ $(m=1,\cdots,2\ell)$. They satisfy the (anti)commutation relations

$$[b_m, b_n^\dagger] = \{a_m, a_n^\dagger\} = \delta_{mn},$$

and all other (anti)commutators vanish.

It turns out that the realization of the $osp(2/2)$ sector is common for the mass and the exotic central extensions:

$$H = \sum_{m=1}^{2\ell}(m-2\ell-1)b_{m+1}^\dagger b_m + \sum_{m=1}^{2\ell-1}(m-2\ell)a_{m+1}^\dagger a_m,$$

$$D = \sum_{m=1}^{2\ell+1} 2(m-\ell-1)b_m^\dagger b_m + \sum_{m=1}^{2\ell}(2m-2\ell-1)a_m^\dagger a_m,$$

$$C = \sum_{m=1}^{2\ell} m b_m^\dagger b_{m+1} + \sum_{m=1}^{2\ell-1} m a_m^\dagger a_{m+1}, \qquad M_{12} = -i\sum_{m=1}^{2\ell+1} b_m^\dagger b_m,$$

$$R = (2\ell-1)i\sum_{m=1}^{2\ell+1} b_m^\dagger b_m + (2\ell+1)i\sum_{m=1}^{2\ell} a_m^\dagger a_m, \tag{9}$$

$$Q = \sum_{m=1}^{2\ell}\left((2\ell+1-m)b_m a_m^\dagger + b_{m+1}^\dagger a_m\right), \qquad Q^* = i\sum_{m=1}^{2\ell}\left((2\ell+1-m)b_m a_m^\dagger - b_{m+1}^\dagger a_m\right),$$

$$S = -\sum_{m=1}^{2\ell}\left(m b_{m+1} a_m^\dagger + b_m^\dagger a_m\right), \qquad S^* = -i\sum_{m=1}^{2\ell}\left(m b_{m+1} a_m^\dagger - b_m^\dagger a_m\right).$$

The realization of the $\langle P_i^{(n)}, X_i^{(n)} \rangle$ sector, of course, depends on the central extension. In the case of mass central extension, the $\langle P_i^{(n)}, X_i^{(n)} \rangle$ sector is realized as follows:

$$P_1^{(m)} = \sqrt{\frac{\mu}{2}}(I_m b_{m+1} + b^\dagger_{2\ell+1-m}), \quad P_2^{(m)} = \frac{1}{i}\sqrt{\frac{\mu}{2}}(I_m b_{m+1} - b^\dagger_{2\ell+1-m}),$$

$$X_1^{(m)} = \sqrt{\frac{\mu}{2}}(\alpha_m a_{m+1} + a^\dagger_{2\ell-m}), \quad X_2^{(m)} = \frac{1}{i}\sqrt{\frac{\mu}{2}}(\alpha_m a_{m+1} - a^\dagger_{2\ell-m}).$$

While the case of exotic central extension it is realized by the equations:

$$P_1^{(m)} = i\sqrt{\frac{i\theta}{2}}(\check{I}_m b_{m+1} + b^\dagger_{2\ell+1-m}), \quad P_2^{(m)} = \sqrt{\frac{i\theta}{2}}(\check{I}_m b_{m+1} - b^\dagger_{2\ell+1-m}),$$

$$X_1^{(m)} = i\sqrt{\frac{i\theta}{2}}(\check{\alpha}_m a_{m+1} + a^\dagger_{2\ell-m}), \quad X_2^{(m)} = \sqrt{\frac{i\theta}{2}}(\check{\alpha}_m a_{m+1} - a^\dagger_{2\ell-m}).$$

It is seen from (9) that the generator D is a linear combination of bosonic and fermionic oscillator. This suggests that D may be interpreted as a Hamiltonian. Then each generator of the exotic super-GCA is either a raising operator or a lowering operator or a symmetry operator of the eigenvalue of D. This interpretation is different from the ordinary supersymmetric theories where the generator H plays the role of Hamiltonian. However, we think that this is an interesting observation resulting from the boson-fermion realization.

References

1. D. T. Son, *Phys. Rev.* **D78** 046003, (2008).
2. K. Balasubramanian and J. McGreevy, *Phys. Rev. Lett.* **101** 061601, (2008).
3. D. Martelli and Y. Tachikawa, *JHEP* **1005** 091, (2010).
4. M. Henkel, *J. Stat. Phys.* **75** 1023, (1994).
5. P. C. Stichel and W. J. Zakrzewski, *Ann. Phys.* **310** 158, (2004).
6. J. Lukierski, P. C. Stichel and W. J. Zakrzewski, *Phys. Lett.* **A357** 1, (2006).
7. J. Lukierski, P. C. Stichel and W. J. Zakrzewski, *Phys. Lett.* **B650** 203, (2007).
8. N. Aizawa, *Preprint*, arXiv:1206.2708.
9. J. Negro, M. A. del Olmo and A. Rodriguez-Marco, *J. Math. Phys.* **38** 3786, (1997).
10. C. Duval and P. A. Horváthy, *J. Math. Phys.* **35** 2516, (1994).
11. J. A. de Azcárraga and J. Lukierski, *Phys. Lett.* **B678** 411, (2009).
12. A. Bagchi and I. Mandal, *Phys. Rev.* **D80** 086011, (2009).
13. M. Sakaguchi, *J. Math. Phys.* **51** 042301, (2010).
14. S. Fedoruk and J. Lukierski, *Phys. Rev* **D84** 065002, (2011).
15. I. Masterov, *J. Math. Phys.* **53** 072904, (2012).

BOSONIC SINGLETONS AND THEIR SYMMETRY ALGEBRAS

XAVIER BEKAERT

Laboratoire de Mathématiques et Physique Théorique
Unité Mixte de Recherche 7350 *du CNRS*
Fédération de Recherche 2964 *Denis Poisson*
Université François Rabelais, Parc de Grandmount
37200 Tours, France
xavier.bekaert@lmpt.univ-tours.fr

Singletons are those unitary irreducible modules of the Poincaré or (anti) de Sitter group that can be lifted to unitary modules of the conformal group. Higher-spin algebras are the corresponding realizations of the universal enveloping algebra of the conformal algebra on these modules. The ambient formulation of the maximal symmetry algebra of tensorial singletons is reviewed.

Keywords: Singleton, Conformal symmetry, Higher spins.

1. Introduction

The celebrated singletons are rather "remarkable representations", as coined by Dirac in his seminal paper[1] on the subject. Indeed, these representations of the anti de Sitter spacetime isometry group possess several surprising properties which are so exceptional that they distinguish singletons from all other such representations.[a] The many faces of singletons provides some flavor of their ubiquitous appearances in such seemingly unrelated areas of mathematical physics as the AdS/CFT correspondence, the hydrogen atom spectrum, the electric-magnetic duality, the infinite-component Majorana equation, etc. The goal of this short proceedings is to briefly review the ambient formulation of the symmetry algebras for singletons of any integer spin and in any dimension.

---

[a]Several of these properties have been reviewed elsewhere[2] as an elementary introduction to singletons through a list, presumably inexhaustive, of their distinct but equivalent definitions.

2. Anti de Sitter spacetime

The most transparent realization of AdS_{n+1} ($n \geqslant 1$) is via a global isometric embedding in a flat ambient space:

- The **ambient space** $\mathbb{R}^{n,2}$ is endowed with the
  - Cartesian coordinates X^A (where $A = 0, 0', 1, 2, \ldots, n$) and
  - ("mostly plus") metric $\eta_{AB} = \text{diag}(-1, -1, +1, +1, \ldots, +1)$.
- The **anti de Sitter spacetime** AdS_{n+1} is the codimension one quadric (more precisely, the one-sheeted hyperboloid)

$$\eta_{AB} X^A X^B = -R^2 \,,$$

where $R > 0$ is the curvature radius, endowed with the induced metric.

So the **isometry algebra** is manifestly the real Lie algebra

$$\mathfrak{o}(n, 2) = \text{span}_{\mathbb{R}}\{\mathrm{L}_{AB}\}$$

which is **linearly realized** on $\mathbb{R}^{n,2}$ through the generators (the ambient "orbital angular momenta")

$$\mathrm{L}_{AB} = \mathrm{X}_A \mathrm{P}_B - \mathrm{X}_B \mathrm{P}_A$$

where $\mathrm{P}_A = -i \frac{\partial}{\partial X^A}$.

3. Conformal boundary of anti de Sitter spacetime

In order to make the conformal symmetries of the Minkowski spacetime $\mathbb{R}^{n-1,1}$ well defined globally, it is necessary to complete by adding "points at infinity". The conformal compactification of $\mathbb{R}^{n-1,1}$ can be identified with the conformal boundary of AdS_{n+1}, whose most transparent realization is via its global (conformal isometric) embedding in the projectivization of the ambient space $\mathbb{R}^{n,2}$:

- The ambient space is now the projective space $\mathbb{P}(\mathbb{R}^{n,2}) \cong \mathbb{RP}^{n+1}$ endowed with the
  - **Homogeneous coordinates** X^A (where $A = 0, 0', 1, 2, \ldots, n$),
  - Equivalence relation $X^A \sim \lambda\, X^A$ (for any $\lambda \in \mathbb{R}_0$)
  - Conformal metric (i.e. the equivalence class of) η_{AB}

  As usual, points of the projective space $\mathbb{P}(\mathbb{R}^{n,2})$ are rays of $\mathbb{R}^{n,2}$.

- The **Dirac hypercone**[b] is the embedded codimension one quadric (more precisely, the null cone with the tip excluded)

$$\eta_{AB}X^AX^B = 0$$

quotiented by the equivalence relation, endowed with the induced conformal metric. This conformal space is the conformal boundary of the anti de Sitter spacetime AdS_{n+1}.

Geometrically, the points of ∂AdS_{n+1} are null rays of the ambient space $\mathbb{R}^{n,2}$. Heuristically, the boundary of the anti de Sitter spacetime is the asymptotic (i.e. located "at infinity") region of intersection between the hyperboloid and the hypercone. Actually, it may be identified with any of the three constant curvature spacetimes (supplemented by "points at infinity").

The **conformal isometry algebra** of the conformal boundary ∂AdS_{n+1} is manifestly the real Lie algebra $\mathfrak{o}(n,2)$ linearly realized on $\mathbb{P}(\mathbb{R}^{n,2})$ through the ("ambient orbital angular momentum") generators. More concretely, it is linearly realized on $\mathbb{R}^{n,2}$ through the generators $\mathrm{L}_{AB} = \mathrm{X}_{[A}\mathrm{P}_{B]}$, where the square bracket denotes the antisymmetrization.

4. Singletons as kernels of the Howe dual algebra

Integer-spin singletons can be defined as very specific conformal primary fields on ∂AdS_{n+1} (see[2] for a review). In order to make conformal symmetries manifest, they can lifted to ambient fields on $\mathbb{R}^{n,2}$. More explicitly, spin-s singletons can be obtained as distributions on an auxilliary superspace, subject to first-class constraints spanning the superalgebra $\mathfrak{osp}(2s|2)$ which is Howe dual to the conformal algebra $\mathfrak{o}(n,2)$, as reviewed now.

Let the letters α, β be superindices corresponding to a $(2|2s)$-dimensional symplectic superspace $T^*\mathbb{R}^{1|s} \cong \mathbb{R}^{2|2s}$ with symplectic form $\mathcal{J}_{\alpha\beta}$, manifestly preserved by the orthosymplectic algebra $\mathfrak{osp}\,(2s\,|\,2)$. **Ambient multiforms** are functions on the superspace

$$\mathbb{R}^{n,2} \oplus \Pi(\mathbb{R}^{n,2} \otimes \mathbb{R}^s) \cong \mathbb{R}^{n+2\,|\,s(n+2)} \qquad \text{with}$$

- $n+2$ even coordinates X^A on $\mathbb{R}^{n,2}$
- $s(n+2)$ odd coordinates ϑ_i^A on $\Pi(\mathbb{R}^{n,2} \otimes \mathbb{R}^s)$.

[b]This construction is the Lorentzian analogue of the "Möbius model" in the mathematical literature. It was introduced a long time ago in physics by Dirac in a paper[3] which still remains a splendid introduction to the ambient formulation.

Let $(P_A|\pi^i_B)$ be the conjugates of the supercoordinates $(X^A|\theta^B_i)$. The phase (super)space coordinates on the cotangent bundle $T^*\mathbb{R}^{n+2|s(n+2)}$ are collectively denoted by

$$Z^A_\alpha := (X^A, P_B|\, \theta^A_i, \pi^j_B)$$

where the superindex α takes $2+2s$ values. The graded Poisson bracket originating from the symplectic structure on the phase superspace is

$$\{Z^A_\alpha, Z^B_\beta\} = \eta^{AB}\mathcal{J}_{\alpha\beta}$$

The phase space coordinates Z^A_α are natural coordinates on the tensor product $\mathbb{R}^{n,2}\otimes\mathbb{R}^{2|2s}$. The algebra $\mathfrak{o}(n,2)$ is linearly realized on $\mathbb{R}^{n+2\,|\,s(n+2)}$ as

$$\mathrm{J}^{AB} = \mathrm{X}^A\mathrm{P}^B - \mathrm{X}^B\mathrm{P}^A \,-\, i\,\vartheta^A_i\frac{\partial}{\partial\vartheta^i_B} + i\,\vartheta^B_i\frac{\partial}{\partial\vartheta^i_A}$$

The Weyl symbols of these generators of $\mathfrak{o}(n,2)$ are the bilinears $J^{AB} = \mathcal{J}^{\alpha\beta}Z^A_\alpha Z^B_\beta$. The Lie superalgebra $\mathfrak{osp}(2s|2) = \mathrm{span}\{\mathrm{T}_{\alpha\beta}\}$ can be presented

- by its generators $\mathrm{T}_{\alpha\beta} = \mathrm{T}_{\beta\alpha}$
- modulo the graded commutation relations

$$[\mathrm{T}_{\alpha\beta},\mathrm{T}_{\gamma\delta}] \;=\; i\,\mathcal{J}_{\beta\gamma}\mathrm{T}_{\alpha\delta} \,+\, \text{(anti)symmetrizations}\,,$$

where the Weyl symbols of these constraints are the bilinears $T_{\alpha\beta} = \eta_{AB}Z^A_\alpha Z^B_\beta$. Their Moyal graded commutators reproduce the corresponding graded commutation relations. The respective realizations of $\mathfrak{o}(n,2)$ and $\mathfrak{osp}(2s|2)$ are maximal commutants in the algebra of quadratic Weyl symbols: they form a Howe dual pair.

Ambient construction of tensorial singletons:[4] *The spin-s ($\in\mathbb{N}$) singleton $\mathfrak{o}(n,2)$-module can be realized as a space of distributions on the superspace $\mathbb{R}^{n+2\,|\,s(n+2)}$ which are annihilated by the $\mathfrak{osp}(2s|2)$ superalgebra, which is Howe dual to $\mathfrak{o}(n,2)$ in the superalgebra of linear operators.*

4.1. *Higher-spin algebras as algebras of symmetries*

The symmetries of the previous spin-s singleton module can be captured in the following definitions:

- A **symmetry generator of a free spin s singleton** is an element of the centralizer of $\mathfrak{osp}(2s|2)\subset A_{n+2|s(n+2)}$ in the Weyl superalgebra $A_{n+2|s(n+2)}$ quotiented by the subspace $A_{n+2|s(n+2)}\mathfrak{osp}(2s|2)$, i.e. a differential operator that commutes with all operators $\mathrm{T}_{\alpha\beta}$ on their kernel.

- A **trivial symmetry generator of a spin** s **singleton** is a symmetry generator in the subspace $A_{n+2|s(n+2)}\mathfrak{osp}(2s|2)$, so it vanishes on the singleton module.

The equation $\mathrm{T}_{\alpha\beta}\Psi(X) = 0$ for the distribution Ψ on $\mathbb{R}^{n,2}$ describing the spin-s singleton is preserved by the transformation $\Psi \mapsto \mathrm{O}\Psi$ when O is a symmetry generator, and $\mathrm{O}\Psi = 0$ when the symmetry generator is trivial. The realisation of the universal enveloping algebra $\mathcal{U}\big(\mathfrak{o}(n,2)\big)$ on the singleton module is automatically an algebra of such symmetry generators. Indeed, one of the leitmotiv behind higher-spin symmetries is that, while super symmetries correspond to "square roots" of isometry generators, higher symmetries are "powers" of isometry generators.

With the help of two ingredients (the Howe duality and the Sergeev theory of classical Lie superalgebra invariants), one can determine the

Singleton maximal symmetry algebra:[4] *The following algebras are isomorphic:*

- *The maximal algebra of symmetry generators of a free spin-s singleton,*
- *The realization of the universal enveloping algebra* $\mathcal{U}\big(\mathfrak{o}(n+2,\mathbb{C})\big)$ *on the spin-s singleton module,*
- *The subalgebra of differential operators on the superspace* $\mathbb{R}^{n+2|s(n+2)}$ *whose Weyl symbols are* $\mathfrak{osp}(2s|2)$*-invariants modulo* $\mathfrak{o}(n+2)$*-traces, i.e. polynomials in the bilinears* $J^{AB} = \mathcal{J}^{\alpha\beta} Z^A_\alpha Z^B_\beta$ *modulo* $T_{\alpha\beta} = \eta_{AB} Z^A_\alpha Z^B_\beta$.

Moreover, the adjoint representation of these algebras is a completely reducible $\mathfrak{o}(n,2)$*-module which decomposes as the sum of all irreducible* $\mathfrak{o}(n,2)$*-modules labeled by Young diagrams such that*

- *(i) all columns are of even length,*
- *(ii) the sum of the lengths of any two columns is not greater than* $n+2$*, and*
- *(iii) any column on the right of the* $2s$*-th column is of length two,*

where each such irreducible module appears with multiplicity one.

Remarks: These results

- generalize the previous works on the particular cases: any $n \geqslant 3$ but[5,6] $s = 0$ or any $s \in \mathbb{N}/2$ but[7] $n = 4$.

- are formulated in terms of the above ambient definition of singleton symmetries, somewhat stronger than the local definition.[4,5]
- can be summarized in a slogan:
  All symmetry generators of a free singleton are polynomials in the isometry generators.

Acknowledgments

I am grateful to I. Bandos, G. Barnich, N. Boulanger, M. Grigoriev, C. Iazeolla, M. Rausch de Traubenberg, D. Sorokin, P. Sundell, M. Tsulaia and M. Valenzuela for various enjoyable discussions and collaborations, scattered over the years, involving singletons and/or higher-spin algebras.

I also acknowledge all the organisers of the " XXIX International Colloquium on Group-Theoretical Methods in Physics" for their kind invitation to present my talks and the opportunity to publish the present material in the present proceedings.

References

1. P. A. M. Dirac, "A remarkable representation of the 3+2 de Sitter group," *J. Math. Phys.* **4** (1963) 901.
2. X. Bekaert, "The many faces of singletons" Physics Ann. Univ. Craiova **21** (2011) 154; "Singletons and their maximal symmetry algebras" in the *Proceedings of the 6th mathematical physics meeting: summer school and conference on modern mathematical physics, Belgrade (Serbie) 14-23 September 2010,* `arXiv:1111.4554` [`math-ph`].
3. P. A. M. Dirac, "Wave equations in conformal space," Annals Math. **37** (1936) 429.
4. X. Bekaert and M. Grigoriev, "Manifestly conformal descriptions and higher symmetries of bosonic singletons," SIGMA **6** (2010) 038 [`arXiv:0907.3195` [`hep-th`]].
5. M. G. Eastwood, "Higher symmetries of the Laplacian," Annals Math. **161** (2005) 1645 [`arXiv:hep-th/0206233`].
6. M. A. Vasiliev, "Nonlinear equations for symmetric massless higher spin fields in (A)dS(d)," *Phys. Lett.* **B567** (2003) 139 [`hep-th/0304049`].
7. M. A. Vasiliev, "Conformal higher spin symmetries of 4D massless supermultiplets and $\mathfrak{osp}(L, 2M)$ invariant equations in generalized (super)space," Phys. Rev. **D66** (2002) 066006 [`hep-th/0106149`];
   S. C. Anco and J. Pohjanpelto, "Generalized symmetries of massless free fields on Minkowski space," *SIGMA* **4** (2008) 004[`arXiv:0801.1892` [`math-ph`].

NEW CLASS OF QUANTUM DEFORMATIONS OF $D = 4$ EUCLIDEAN SUPERSYMMETRIES

A. BOROWIEC, J. LUKIERSKI* and M. MOZRZYMAS

Institute for Theoretical Physics, Wroclaw University,
pl. Maxa Borna 9, 50-204 Wroclaw, Poland
**E-mail: lukier@ift.uni.wroc.pl*
www.ift.uni.wroc.pl

V. N. TOLSTOY

Lomonosov Moscow State University,
Skobeltsyn Institute of Nuclear Physics
Moscow 119991, Russian Federation
E-mail: tolstoy@nucl-th.sinp.msu.ru

We present the class of deformations of simple Euclidean superalgebra, which describe the supersymmetrization of some Lie algebraic noncommutativity of $D = 4$ Euclidean space-time. The presented deformations are generated by the supertwists. We provide new explicit formulae for a chosen twisted $D = 4$ Euclidean Hopf superalgebra and describe the corresponding quantum covariant deformation of chiral Euclidean superspace.

Keywords: Euclidean SUSY; Quantum Deformations; Drinfeld twist; Pseudo-conjugation; Chiral superspace.

1. Introduction

In the applications of quantum deformations to fundamental interaction theories it should be taken into considerations that supersymmetry is one of the main basic building blocks of the string and M-theory. We recall that the classification of deformations of $D = 4$ Poincaré symmetries is almost complete[1,2] (see also Ref. 3,4), but systematic studies of supersymmetric extensions of deformed Poincare algebras quite recently were considered.[5–7] In this note we shall deal with the superextensions of quantum $D = 4$ Euclidean symmetries providing the Lie-algebraic structure of corresponding Euclidean space-time commutators. We add that Euclidean theories are often very useful as intermediate step in the calculations in Minkowski space.

Historically first and the most known example of Lie-algebraic space-time deformations are the $\kappa-$deformations of Minkowski space-time[8–10]

$$[\widehat{x}_\mu, \widehat{x}_\nu] = \frac{i}{\kappa}(a_\mu \widehat{x}_\nu - a_\nu \widehat{x}_\mu) \tag{1}$$

where $a_\mu = (1,0,0,0)$ corresponds to the standard (time-like) κ-deformation[11] and if $a_\mu = (1,1,0,0)$ we get the light-cone κ-deformation.[12]

In this short paper we would like to consider an example of $D = 4$ Euclidean supersymmetries, which in its bosonic sector provides the Lie-algebraic deformation of $D = 4$ space-time sector. This choice of deformation follows from our recent paper[7] where we did present the Euclidean counterparts of Zakrzewski list of $D = 4$ Poincare classical r-matrices[2] and classified their supersymmetrizations in the case of simple chiral (holomorphic) Euclidean superalgebras [a].

Plan of our presentation is the following. In section 2 we deliver the list of Euclidean $D = 4$ classical r-matrices and their supersymmetrizations presented in Ref. 7. We show that the conjugation and pseudoconjugation of Euclidean superalgebras are related with quaternionic structure of fundamental $O(3)$ and $O(4)$ spinors. In section 3 we give explicit formulae for particular $D = 4$ Euclidean quantum twisted superalgebra, which leads to the Lie-algebraic deformation of space-time sector. Further we provide the deformation of corresponding Euclidean chiral $N = (\frac{1}{2}, 0)$ superspace and indicate the link with Seiberg $N = \frac{1}{2}$ deformation of superspace. In section 4 we present short outlook.

2. $D = 4$ Euclidean twisted deformations and their $N = \frac{1}{2}$ superextenstions

2.1. *Euclidean twisted deformations*

Let us recall $D = 4$ Euclidean Lie algebra $E(4) = o(4) \ltimes P_4$, where six Euclidean rotation generators $M_{\mu\nu} = -M_{\nu\mu} \in o(4)$ $(\mu, \nu = 0, 1, 2, 3)$ and

[a] Simple chiral (holomorphic) $D = 4$ Euclidean superalgebra with four complex supercharges is usually denoted as $N = (\frac{1}{2}, \frac{1}{2})$ Euclidean SUSY; by Hermitean conjugation of supercharges one obtains simple antichiral (antiholomorphic) $N = (\frac{1}{2}, \frac{1}{2})$ superalgebra. The lowest dimensional real Euclidean $N = (1,1)$ superalgebra is described by eight complex supercharges satisfying $SU(2)$-Majorana reality conditions and it contains $N = (\frac{1}{2}, \frac{1}{2})$ chiral and antichiral sectors.

Euclidean four-momenta $P_\mu \in P_4$ satisfy the relations

$$[M_{\mu\nu}, M_{\lambda\rho}] = i(\delta_{\nu\lambda}M_{\rho\mu} - \delta_{\nu\rho}M_{\lambda\mu} - \delta_{\mu\lambda}M_{\rho\nu} + \delta_{\mu\rho}M_{\lambda\nu}) \qquad (2)$$

$$[M_{\mu\nu}, P_\rho] = i(\delta_{\rho\mu}P_\nu - \delta_{\rho\nu}P_\mu), \qquad [P_\mu, P_\nu] = 0. \qquad (3)$$

The generators $(M_{\mu\nu}, P_\mu)$ are Hermitean. We recall that from relations (2)-(3) we can obtained real $D = 4$ Poincaré algebra $P(3,1) = o(3,1) \ltimes \widetilde{P}_4$ ($(\widetilde{M}_{\mu\nu} \in o(3,1)$, $\widetilde{P}_\mu \in P_4$ where $\mu, \nu = 0,1,2,3$) after the substitution $(i,j = 1,2,3)$

$$\widetilde{M}_{ij} = M_{ij}, \qquad \widetilde{M}_{i0} = -iM_{i0}, \qquad \widetilde{P}_i = P_i, \qquad \widetilde{P}_0 = -iP_0. \qquad (4)$$

Strictly speaking it means that one can introduce, on a complex algebra generated by the relations (2)-(3), two non-isomorphic real structures (i.e. $\star-$conjugation operations): the Euclidean one with respect to which the Euclidean generators $(M_{\mu\nu}, P_\rho)$ are Hermitean (i.e. selfconjugated) and the Poincaré one with $(\widetilde{M}_{\mu\nu}, \widetilde{P}_\rho)$ Hermitean. Both conjugations (real forms) coincide on the generators (M_{ij}, P_k).[7]

In Ref. 7 we observed that all the real Poincaré r-matrices introduced by S. Zakrzewski in Ref. 2 describe equally well the classical r-matrices for complex $D = 4$ Euclidean Lie algebra $o(4;\mathbb{C}) \ltimes P_4^{\mathbb{C}}$ which describes also $D = 4$ Poincaré algebra. It appears that out of the Zakrzewski list with 21 classical r-matrices only 8 admits the reality conditions leading to $D = 4$ Euclidean real classical r-matrices. The list of such matrices is the following (we use $M_i = \frac{1}{2}\epsilon_{ijk}M_{jk}$, $N_i = M_{0i}$, $P_\pm = P_0 \pm P_3$)

$$\begin{aligned}
&r_1 = i\xi N_3 \wedge M_3 + \zeta_1 P_+ \wedge P_- + \zeta_2 P_1 \wedge P_2 \\
&r_2 = \xi P_0 \wedge M_3 + r_8, \qquad r_3 = \xi P_3 \wedge N_3 + r_8, \qquad r_4 = \xi P_1 \wedge M_3 + r_8, \\
&r_5 = \xi P_+ \wedge M_3 + r_8, \qquad r_6 = \zeta_1 P_1 \wedge P_+, \\
&r_7 = \zeta_2 P_1 \wedge P_2, \qquad r_8 = \zeta_1 P_0 \wedge P_3 + \zeta_2 P_1 \wedge P_2,
\end{aligned} \qquad (5)$$

where the parameters ξ, ζ_1, ζ_2 are real.

It is interesting to deduce from (5) that the classical r-matrix describing standard light-cone κ-deformation does not have its Euclidean counterparts.

2.2. *Quaternionic structure of Euclidean spinors, Euclidean supercharges conjugation and pseudoconjugation*

The supercharges of Euclidean supersymmetry are described by fundamental $O(4)$ spinors, which are $O(4)$ Clifford algebra modules. Because $o(4) = o(3) \oplus o(3)$ let us consider firstly $o(3)$ spinors. Following the list of Clifford modules and fundamental spinors for orthogonal groups[13–16] they are described by single real quaternion $q = q_0 + q_r e_r$, where e_r describe the quaternionic units ($e_r e_s = -\delta_{rs} + \epsilon_{rst} e_t$) and q_0, q_r are real. The irreducible fundamental representation of spinorial covering $\overline{SO(3)} = Sp(1) \equiv U(1; H) = SU(2)$ is described by real unit quaternions $q\overline{q} = q_0^2 + q_r^2 = 1$, ($\overline{q} = q_0 - q_r e_r$). We recall that in physics one uses rather complex realization of quaternions. In order to define a complex realization $\mathbf{q}$ of real quaternions we should firstly introduce matrix realizations $e_r = -i\sigma_r$, where σ_r are 2×2 Pauli matrices and one obtains unit quaternions as decribing 2×2 $SU(2)$ matrices

$$U = \begin{pmatrix} q_0 - iq_3 & -q_2 - iq_1 \\ q_2 - iq_1 & q_0 + iq_3 \end{pmatrix} = \begin{pmatrix} z_1 & -z_2^* \\ z_2 & z_1^* \end{pmatrix} \tag{6}$$

where $z_1 = q_0 - iq_3$, $z_2 = q_2 - iq_1$ and $\det U \equiv |z_1|^2 + |z_2|^2 = 1$. Here z^* denotes a complex conjugation of $z \in \mathbb{C}$. The one-component quaternionic spinor q can be represented as two-component complex spinor, described by the first column of the relation (6)

$$\mathbf{q} = \begin{pmatrix} q_0 - iq_3 \\ q_2 - iq_1 \end{pmatrix} = \begin{pmatrix} z_1 \\ z_2 \end{pmatrix} \in \mathbb{C}^2 \tag{7}$$

The quaternionic nature of complex spinors (7) is exhibited by the presence of quaternionic pseudoconjugation ($\epsilon \equiv i\sigma_2$)[15,16]

$$(\mathbf{q}^H)_\alpha = \epsilon_{\alpha\beta} \mathbf{q}^*_\beta \qquad (\mathbf{q}^H)^H = -\mathbf{q} \tag{8}$$

which commutes with $SU(2) \simeq SO(3)$ rotations, i.e. $(U\mathbf{q})^H = U\mathbf{q}^H$. The condition (8) is a complex realization of the quaternionic map $q \to q^H = -qe_2$ [b], because the group $U(1; H) = SU(2)$ acts on q from the left, its commutativity with the mapping $q \to q^H$ is obvious. In order to interpret the relation (8) in a quaternionic framework one can deduce from (7) that complex conjugation $\mathbf{q} \to \mathbf{q}^*$ represents the quaternionic map $q \to -e_2 q e_2$, and one gets $q^H = -qe_2 = -e_2(-e_2 q e_2)$, where in complex realization (8) we inserted $e_2 = -i\sigma_2$.

The Euclidean supercharges for general $D = 4$ Euclidean (n, m) superalgebra (n, m are half-integers) are described by n left chiral $SU_L(2)$ spinors denoted by $Q^a_{L\alpha}$ ($a = 1, ..., n$, $\alpha = 1, 2$) and m right chiral $SU_R(2)$ spinors denoted by $Q^b_{R\alpha}$ ($b = 1, .., m$). In order to obtain the simplest $D = 4$ Euclidean chiral (holomorphic) $N = (\frac{1}{2}, \frac{1}{2})$ superalgebra we need a doublet of

[b]This quaternionic map as well as its complex counterpart given by (8) is not unique. One can replace $e_2 \to \alpha_i e_i$ (with $\Sigma|\alpha_i|^2 = 1$) in $q^H = -qe_2$.

complex supercharges $(Q_{L\alpha}, Q_{R\beta})$ which enters into the basic SUSY relation (see e.g. Ref. 17 where is denoted $Q_{L\alpha} \equiv Q_{\alpha;}, Q_{R\alpha} \equiv Q_{;\alpha}$)

$$\{Q_{L\alpha}, Q_{R\beta}\} = 2(\sigma^E_\mu)_{\alpha\beta}\mathcal{P}^\mu \tag{9}$$

$$\{Q_{L\alpha;}, Q_{L\beta}\} = \{Q_{R\alpha}, Q_{R\beta}\} = 0 \tag{10}$$

where $Q_{L\alpha}, Q_{R\beta}$ transforms respectively under $SU_L(2), SU_R(2)$ and $\mathcal{P}^\mu$ are complexified fourmomenta. Further

$$(\sigma^E_\mu) = (iI_2, \sigma_i) \tag{11}$$

describe $D = 4$ Euclidean Pauli matrices which transform under $D = 4$ Euclidean rotations $SO(4) = SU_L(2) \otimes SU_R(2)$ as Euclidean fourvector $(U_{L,R} \in SU_{L,R}(2))$

$$U_L \sigma^E_\mu U^T_R = A^\nu_\mu \sigma^E_\nu \tag{12}$$

where real 4×4 matrices A^ν_μ represent $SO(4)$ four-vector rotations. This is due to the fact that the Euclidean Pauli matrices (11) satisfy the constraints

$$(\sigma^E_\mu)^* = \epsilon\, \sigma^E_\mu\, \epsilon \tag{13}$$

while for $SU(2)$ matrices (6) one has $U^* = -\epsilon\, U \epsilon$ instead. This can be used to embed Euclidean space-time onto quaternions. We point out that the superalgebra (9-10) cannot be constrained by any reality condition, because the supercharges $Q_{L\alpha}$,$Q_{R\beta}$ transform under two independent $SU(2)$ groups.

From physical point of view it would be desirable to obtain in the superalgebra (9) the real (Hermitean) fourmomenta $\mathcal{P}_\mu$. For that purpose one can adopt the quaternionic pseudoconjugation (8) to the Euclidean supercharges in the following two ways:

i) One introduces an abstract inner $SU(2)-$invariant graded antiautomorphism of fourth order ($\tau^2 = -1 \Rightarrow \tau^4 = 1$) [c]

$$\tau(Q_{L\alpha}) = \epsilon_{\alpha\beta} Q_{L\beta} \qquad \tau(Q_{R\alpha}) = \epsilon_{\alpha\beta} Q_{R\beta} \tag{14}$$

which satisfies the antilinearity property: $\tau(zQ) = \bar{z}\tau(Q)$. The pseudoconiugation (14) was introduced by Nahm, Rittenberg, Scheinert,[18] Berezin, Tolstoy[19] and Manin[20] (see also[21,22]). Now one can check explicitly (taking into account (13)) that the relation (9-10) remains invariant under pseudoconjugation (14) if the Euclidean fourmomentum generators are real.

[c]It is well known that it does not exist $SU(2)-$invariant antiautomorphism of second order (conjugation) in two dimensions, i.e. matrices (6) cannot be made real at the same time by change of basis in $\mathbb{C}^2$.

ii) one can assume that the supercharges $Q_{L\alpha}, Q_{R\beta}$ admit well defined Hermitean conjugation which is an antilinear involutive outer map provided linearly independent conjugated antichiral generators $Q^+_{L\alpha} \equiv Q_{L\dot{\alpha}}, Q^+_{R\beta} \equiv Q_{R\dot{\beta}}$. The generators $Q^+_{L\dot{\alpha}}, Q^+_{R\dot{\beta}}$ form a basis of complex-conjugate antichiral (antiholomorphic) $N = (\frac{1}{2}, \frac{1}{2})$ superalgebra obtained by Hermitean conjugation of relations (9-10). One can introduce the following outer antilinear antiautomorphism of fourth order (i.e. pseudoconjugation) of $N = (\frac{1}{2}, \frac{1}{2})$[23,24]

$$\tau(Q_{L\alpha}) = \epsilon_{\alpha\beta} Q^+_{L\beta} \qquad \tau(Q_{R\alpha}) = \epsilon_{\alpha\beta} Q^+_{R\beta} \tag{15}$$

which maps the chiral $N = (\frac{1}{2}, \frac{1}{2})$ Euclidean superalgebra into antichiral one [d]. If we use the relation (13) one can show easily that both superalgebras will have the same algebraic form (9-10) if the fourmomenta generators $\mathcal{P}_\mu$ are Hermitean (real). The pseudoconjugation (15) has the form of quaternionic pseudoconjugation (8), applied to two-component complex spinorial supercharges.

2.3. *Twisted quantum deformations od $D = 4$ $N = (\frac{1}{2}, \frac{1}{2})$ holomorphic Euclidean superalgebra*

In this subsection we shall consider the Euclidean superalgebra (9-10), invariant under the pseudoconjugation (14) what implies that the fourmomenta $\mathcal{P}^\mu$ are real and Euclidean.

We shall study further the superextensions of the classical $D = 4$ Euclidean r-matrices r_A $(A = 1, .., 8)$ listed in formulae (5)

$$r_A \to r_A^{SUSY} = r_A + s_A \tag{16}$$

where s_A is the odd part of the supersymmetric r-matrix which is bilinear in $Q_{L\alpha}$ and $Q_{R\beta}$. The general complex formula for s_A is the following (we define $Q \vee Q' = Q \otimes Q' + Q' \otimes Q$)

$$s_A = s_A^{(1)\alpha\beta} Q_{L\alpha} \vee Q_{L\beta} + s_A^{(2)\alpha\beta} Q_{L\alpha} \vee Q_{R\beta} + s_A^{(3)\alpha\beta} Q_{R\alpha} \vee Q_{R\beta} \tag{17}$$

If we substitute (11) into the graded classical YB equation and use the

---

[d] The pseudoconjugation property $\tau^2 = -1$ follows if $(Q^*)^+ = (Q^+)^*$ what implies that the antichiral supercharges $Q^+_{L\alpha}$,$Q^+_{R\beta}$ are mapped into $Q_{L\alpha}$,$Q_{R\beta}$ by the formula with the same algebraic form (15).

$D = 4$ Euclidean superalgebra relations ($M_i^{\pm} = \frac{1}{2}(M_i \mp N_i)$

$$[M_i^+, Q_{L\alpha}] = -\frac{1}{2}(\sigma_i)_\alpha^\beta Q_{L\beta} \qquad [M_i^-, Q_{L\alpha}] = [M_i^+, Q_{R\alpha}] = 0$$
$$[M_i^-, Q_{R\alpha}] = \frac{1}{2}(\sigma_i)_\alpha^\beta Q_{R\beta}; \tag{18}$$

one gets that the classical $r-$matrices r_A (see (5)) for $A = 1, ..., 5$ do have a common superextension $s \equiv s_A$ given by the formula

$$s = \eta Q_{L2} \vee Q_{L1} \tag{19}$$

and for $A = 6, 7, 8$ we obtain another three-parameter superextension

$$\widetilde{s} = \eta^{\alpha\beta} Q_{L\alpha} \vee Q_{L\beta} \tag{20}$$

It should be added that the formulae (19-20) after the replacement $Q_{L\alpha} \to Q_{R\alpha}$ provide right-handed chiral extensions of $r-$matrices (5), which as well describe the supersymmetrization of (16) satisfying the supersymmetric YB equation.

The quantum deformations described by classical $r-$matrices r_6, r_7, r_8 describe the canonical twist deformations, which leads to $\mathbb{C}-$number commutator of quantum space-time coordinates and deforms Grassmann superspace variables into the generators of $o(4)$ Clifford algebra[7] . More complicated are the the deformations described by $r_1, ..., r_5$. The first, denoted by r_1, generates the quadratic deformation of Euclidean space-time algebra and the remaining four r_2, r_3, r_4,r_5 lead to Lie-algebraic formulae for non-commutative Euclidean space-time algebra. In next paragraph we shall describe in detail the deformation described by superymmetric classical r-matrix r_2^{SUSY} (see (5), (16) and (19)).

3. $D = 4$ Euclidean Hopf algebra providing Lie-algebraic deformation of the bosonic Euclidean space sector

We shall consider now in detail quantum twisted Euclidean supersymmetry generated by the classical $r-$matrix r_2 (see (5)), and supersymmetrically extended by the term (19). The algebraic sector of Euclidean twisted superalgebra remains undeformed (see (3) but the coproducts are modified by the similarity transformation[25]

$$\Delta = F^{-1} \Delta_0 F \tag{21}$$

where Δ_0 describes the undeformed primitive coproduct. The twist factor $F \in U(E_4) \otimes U(E_4)$ corresponding to r_2 looks as follows [e] . Here ξ, ζ_1, ζ_2

[e] In this section we shall drop the index L, i.e. $Q_{L\alpha} \to Q_\alpha$.

and η are real deformation parameters in order to assure invariance ("unitarity") under the Euclidean Hermitean conjugation in the bosonic sector and pseudoconjugation (14) in the fermionic one. We get

$$F = \exp(i\xi M_3 \wedge P_0)\exp(i\zeta_1 P_3 \wedge P_0 + i\zeta_2 P_2 \wedge P_1 + \eta Q_1 \vee Q_2) \qquad (22)$$

The fourmomentum coproducts are the following

$$\Delta(P_0) = \Delta_0(P_0) = \mathbf{1} \otimes P_0 + P_0 \otimes \mathbf{1}$$
$$\Delta(P_j) = P_j \vee \cos(\xi P_0) + \sin(\xi P_0) \wedge \epsilon_{3jk} P_k, \qquad j,k = 1,2,3 \qquad (23)$$

The coproducts for the "space" rotation generators are

$$\begin{aligned}\Delta(M_j) = {} & M_j \vee \cos(\xi P_0) + \sin(\xi P_0) \wedge \epsilon_{3jk} M_k + \delta_{j3}\{\Delta_0(M_3) - M_3 \vee \cos(\xi P_0)\} \\ & - \zeta_2\{\epsilon_{j1k} P_k \cos(\xi P_0) \wedge P_2 \cos(\xi P_0) - \epsilon_{j1k} P_k \sin(\xi P_0) \vee P_1 \cos(\xi P_0) \\ & + P_1 \cos(\xi P_0) \wedge \epsilon_{j2k} P_k \cos(\xi P_0) - P_2 \cos(\xi P_0) \vee \epsilon_{j2k} P_k \sin(\xi P_0)\} \\ & - \zeta_1\{\epsilon_{j3k} P_k \vee P_0 \cos(\xi P_0) + P_0 \sin(\xi P_0) \wedge P_j\} + \frac{\eta}{2} R_j \qquad (24)\end{aligned}$$

where R_j is a shortcut for the following expression

$$\begin{aligned} R_j = {} & (\sigma_j)_2^\beta\{Q_\beta \cos(\tfrac{\xi}{2}P_0) \vee Q_1 \cos(\tfrac{\xi}{2}P_0) - iQ_\beta \sin(\tfrac{\xi}{2}P_0) \wedge Q_1 \cos(\tfrac{\xi}{2}P_0)\} \\ & + (\delta_{j3}\delta_2^\beta - i\epsilon_{j3k}(\sigma_k)_2^\beta)\{iQ_\beta \cos(\tfrac{\xi}{2}P_0) \wedge Q_1 \sin(\tfrac{\xi}{2}P_0) \\ & + Q_\beta \sin(\tfrac{\xi}{2}P_0) \vee Q_1 \sin(\tfrac{\xi}{2}P_0)\} + (\sigma_j)_1^\beta\{Q_2 \cos(\tfrac{\xi}{2}P_0) \vee Q_\beta \cos(\tfrac{\xi}{2}P_0) \\ & + iQ_2 \cos(\tfrac{\xi}{2}P_0) \wedge Q_\beta \sin(\tfrac{\xi}{2}P_0)\} - (\delta_{j3}\delta_1^\beta - i\epsilon_{j3k}(\sigma_k)_1^\beta) \\ & \times \{iQ_2 \sin(\tfrac{\xi}{2}P_0) \wedge Q_\beta \cos(\tfrac{\xi}{2}P_0)\} + Q_2 \sin(\tfrac{\xi}{2}P_0) \vee Q_\beta \sin(\tfrac{\xi}{2}P_0)\}. \end{aligned}$$

The coproducts for Euclidean "boosts" are the following ($j = 1,2,3$)

$$\begin{aligned}\Delta(N_j) = {} & N_j \vee \cos(\xi P_0) + \sin(\xi P_0) \wedge \epsilon_{3jl} N_l - \xi\{\cos(\xi P_0) \wedge P_j \\ + \quad & \sin(\xi P_0) \wedge \epsilon_{3jk} P_k\} + \zeta_2 \delta_{2j}\{P_0 \cos(\xi P_0) \vee P_1 + P_0 \sin(\xi P_0) \wedge P_2\} \\ - \quad & \zeta_1\{\epsilon_{3jk} P_k \cos(\xi P_0) \wedge P_3 \sin(\xi P_0) + P_0 \cos(\xi P_0) P_3 \wedge P_j \cos(\xi P_0)\} \\ - \quad & \zeta_2 \delta_{1j}\{P_0 \cos(\xi P_0) \vee P_2 - P_0 \sin(\xi P_0) \wedge P_1\} + \frac{\eta}{2} R_j \qquad (25)\end{aligned}$$

One can observe that for the generators $M_j^{\pm}$ (see (18)) the R_j (supersymmetric) contribution appears only in the antichiral sector M_j^- while the chiral one remains bosonic. For the supercharges we obtain ($Q_{R\alpha} \to \tilde{Q}_\alpha$)

$$\Delta(Q_\alpha) = Q_\alpha \vee \cos(\tfrac{\xi}{2}P_0) + (\sigma_3)_\alpha^\beta Q_\beta \wedge \sin(\tfrac{\xi}{2}P_0) \qquad (26)$$

$$\Delta(\tilde{Q}_\alpha) = \cos(\tfrac{\xi}{2}P_0) \vee \tilde{Q}_\alpha + \sin(\tfrac{\xi}{2}P_0) \wedge (\sigma_3)_\alpha^\beta \tilde{Q}_\beta + 4\eta S_{\alpha(12)} \qquad (27)$$

where $S_{\alpha(12)} = \frac{1}{2}\left(S_{\alpha 12} + S_{\alpha 21}\right)$ denotes the symmetrization and

$$\begin{aligned}
S_{\alpha 12} &= Q_1 \wedge \cos(\frac{\xi}{2}P_0)\{(\sigma_0)_{2\alpha}P_0 + (\sigma_3)_{2\alpha}P_3\} \\
&+ \; i(\sigma_3)_1^{\beta}Q_\beta \vee \sin(\frac{\xi}{2}P_0)\{(\sigma_0)_{2\alpha}P_0 + (\sigma_3)_{2\alpha}P_3\} \\
&+ \; Q_1 \cos(\xi P_0) \wedge \cos(\frac{\xi}{2}P_0)\{\sum_{i=1,2}(\sigma_i)_{2\alpha}P_i\} \\
&+ \; i(\sigma_3)_1^{\beta}Q_\beta \cos(\xi P_0) \vee \sin(\frac{\xi}{2}P_0)\{\sum_{i=1,2}(\sigma_i)_{2\alpha}P_i\} \\
&+ \; Q_1 \sin(\xi P_0) \vee \cos(\frac{\xi}{2}P_0)\{\sum_{i=1,2}(\sigma_i)_{2\alpha}\epsilon_{3ik}P_k\} \\
&+ \; i(\sigma_3)_1^{\beta}Q_\beta \sin(\xi P_0) \wedge \sin(\frac{\xi}{2}P_0)(\sum_{i=1,2}(\sigma_i)_{2\alpha}\epsilon_{3ik}P_k)
\end{aligned}$$

We use the Mandelstam realization (see Ref. 7) of undeformed $N = (\frac{1}{2}, \frac{1}{2})$ Euclidean superalgebra on left chiral Euclidean superspace (x^μ, θ^α), where x^μ are real Euclidean space-time coordinates and $\theta_{L\alpha} \equiv \theta_\alpha$ describes (left) chiral Grassmannian spinors transforming under $SU_L(2)$ ($\partial_\mu \equiv \frac{\partial}{\partial x^\mu}$, $\partial_\alpha \equiv \frac{\partial}{\partial \theta^\alpha}$)

$$P_\mu = -i\partial_\mu, \qquad M_{\mu\nu} = ix_{[\mu}\partial_{\nu]} + \frac{1}{2}(\theta\sigma_{\mu\nu})^\alpha\partial_\alpha \tag{28}$$

$$Q_\alpha = i\partial_\alpha, \qquad \tilde{Q}_\beta = 2\theta^\alpha(\sigma^\mu)_{\alpha\beta}\partial_\mu \tag{29}$$

Further we insert (21) into the $\star-$product formula of chiral Euclidean fields $\Phi(x_\mu, \theta_\alpha)$

$$\Phi(x_\mu, \theta_\alpha) \star \Psi(x_\mu, \theta_\alpha) := m[F^{-1} \triangleright \Phi(x_\mu, \theta_\alpha) \otimes \Psi(x_\mu, \theta_\alpha)] \tag{30}$$

If $\triangleright$ describes in (24) the realization (23) of Euclidean superalgebra generators occurring in F^{-1} one obtains the formulae for deformed superspace relations

$$[x^\mu, x^\nu]_\star = 2i(\zeta_1\delta_3^{[\mu}\delta_0^{\nu]} + \zeta_2\delta_2^{[\mu}\delta_0^{\nu]}) + 2i\xi(\delta_2^{[\mu}\delta_0^{\nu]}x^1 + \delta_0^{[\mu}\delta_1^{\nu]}x^2) \tag{31}$$

$$[x^\mu, \theta^\alpha]_\star = \xi\delta_0^\mu(\sigma_3)_\beta^\alpha\theta^\beta \tag{32}$$

$$\{\theta^\alpha, \theta^\beta\}_\star = -2\eta\delta_1^{(\alpha}\delta_2^{\beta)} \tag{33}$$

We observe that all deformation parameters are dimensionfull. The first part of relation (31) obtained by putting $\xi = 0$, together with the relation (33), provides canonical type of deformation, while the second part of (31) together with (32) ($\xi \neq 0$) leads to space-time deformation of Lie-algebraic type. We add that the relations (32-33) are invariant under pseudoconjugation (14) if defined for the Grassmann coordinates θ^α, provided that the Euclidean spacetime coordinates x^μ are Hermitean.

We recall that first deformation of chiral $N = (\frac{1}{2}, \frac{1}{2})$ Euclidean supersymmetries was proposed by Seiberg[26] with primary deformation introduced a priori by the canonical modification (33) of the anti-commutativity of Grassmann variables θ^α. In such a framework the quantum-deformed supersymmetries do not appear, but the anticommutators of supercharges are modified and it appears that half of the considered Euclidean supersymmetries are broken explicitly [f].

4. Outlook

In this lecture we presented mainly the results of our paper Ref. 7 but we also provided new results in Section 3, where the explicit coproducts for twisted $D = 4$, $N = (\frac{1}{2}, \frac{1}{2})$ Euclidean chiral Hopf superalgebra are written down. The case of $N = (\frac{1}{2}, \frac{1}{2})$ Euclidean SUSY corresponds to $N = 1$ Poincaré supersymmetry, but as follows from Sect. 2.2, it can be obtained from Poincaré case after doubling of supercharges (in $N = 1$ Poincaré superalgebra we have four real supercharges, and in $N = (\frac{1}{2}, \frac{1}{2})$ Euclidean case four complex). This passage from Poincaré to Euclidean SUSY is consistent with Osterwalder-Schrader analytic continuation procedure[27] from Poincaré invariant theories in Minkowski space into Euclidean invariant theories in Euclidean $D = 4$ space, where the doubling of fermions is emphasized. It should be stressed that imposing the pseudoconjugation (14) or (15) cannot be used for the reduction of degrees of freedom of $N = (\frac{1}{2}, \frac{1}{2})$ supercharges.

The similarity between Euclidean and Poincaré case can be achieved however in the cases of $N = (1, 1)$ Euclidean and $N = 2$ Poincaré supersymmetries, in both cases described by four complex or eight real supercharges. This property follows from the introduction for $N = (1, 1)$ Euclidean SUSY the conjugation $\#$ ($\#^2 = 1$), which can be used for imposing the reality constraints. These reality constraints are called $SU(2)-$Majorana condition[15] and are the consequences of quaternionic structure of $D = 4$ Euclidean space, as we demonstrate below.

In order to introduce the reality condition consistent with the quaternionic structure one has to consider a pair of quaternions q^a $(a = 1, 2)$

[f]Such explicit breaking has been also called $N = \frac{1}{2}$ breaking.[26]

and define the quaternionic conjugation $(q^a)^\#$ $(((q^a)^\#)^\# = q^a)$ in terms of quaterionic pseudoconjugation (8) as follows

$$(q^a)^\# = -\epsilon^{ab}(q^b)^H \qquad \Leftrightarrow \qquad (\mathbf{q}^a_\alpha)^\# = \epsilon^{ab}\epsilon_{\alpha\beta}\bar{\mathbf{q}}^b_\beta \tag{34}$$

The quaternionic reality condition ($SU(2)$ Majorana reality condition[15]) takes the form

$$q^a = (q^a)^\# \qquad \Leftrightarrow \qquad \mathbf{q}^a_\alpha = \epsilon^{ab}\epsilon_{\alpha\beta}\bar{\mathbf{q}}^b_\beta \tag{35}$$

If we wish to obtain real Euclidean $N = (1,1)$ superalgebra with real four-momenta one should introduce the pairs of left and right chiral supercharges $(Q^a_{L\alpha}, Q^a_{R\alpha})$, $a = 1,2$ which satisfy the superalgebraic counterpart of the reality condition (35).

$$Q^a_{L\alpha} = \epsilon^{ab}\epsilon_{\alpha\beta}(Q^b_{L\beta})^\dagger \qquad Q^a_{R\alpha} = \epsilon^{ab}\epsilon_{\alpha\beta}(Q^b_{R\beta})^\dagger \tag{36}$$

In our next paper we shall consider the real $N = (1,1)$ Euclidean superalgebra and its quantum twist deformations. It should be added that by considering only the doublets of supercharges $Q^a_{L\alpha}$ ($Q^a_{R\alpha}$) we obtain $N = (2,0)$ chiral ($N = (0,2)$ antichiral) Euclidean superalgebras which can be mapped into each other by the $N = 2$ extensions of the pseudoconjugations (15).

Acnowledgements

Three of the authors (A.B., J.L. and V.N.T. would like to thank Prof. Cheng-Ming Bao and Prof. Mo-Lin Ge for warm hospitality at Nankai University in Tianjin. Two of us (A.B. and J.L.) were supported by the Polish NCN Grant 2011/51/B/ST2/03354 and V.N.T was partly supported by the RFBR grant No.11-01-00980-a and the grant No.12-09-0064 of the Academic Fund Program of the National Research University Higher School of Economics.

References

1. P.Podles, S.L. Woronowicz, *Comm. Math. Phys.* **178**, 61 (1996).
2. S. Zakrzewski, *Comm. Math. Phys.*, **187**, 285 (1997).
3. V.N. Tolstoy, *Bulg. J. Phys.* **35**, 441 (2008); arXiv: 0712.3962 [hep-th].
4. A. Borowiec, J. Lukierski, V.N. Tolstoy, *Eur. Phys. Journ. C* **48**, 633 (2006); hep-th/060146.
5. A. Borowiec, J. Lukierski, V.N. Tolstoy, *Eur. Phys. Journ.C* **44**, 139 (2005); hep-th/0412131.

6. A. Borowiec, J. Lukierski, V.N. Tolstoy, in *Proceedings of International Workoshop: Supersymmetries and Quantum Symetries (SQS'07)*, 205-215 (Dubna 2008), eds S.Fedoruk, E.Ivanov, ISBN 5-930-0184-3; arXiv:0803.4167.
7. A. Borowiec, J. Lukierski, M.Mozrzymas, V.N. Tolstoy, *JHEP* **1206**:154 (2012); arXiv:1112.1936.
8. S. Zakrzewski, *Journ. of Phys. A* **27**, 2075 (1994).
9. S. Majid and H. Ruegg, *Phys Lett. B* **334**, 348 (1994).
10. J. Lukierski, H. Ruegg and W. J. Zakrzewski, *Ann. Phys.* **243**, 90 (1995).
11. J. Lukierski, A. Nowicki, H. Ruegg and V.N. Tolstoy, *Phys. Lett. B* **264**, 331 (1991).
12. A. Ballesteros, F.J. Herranz, M.A. del Olmo, M. Santander, *Phys. Lett. B***351**, 137 (1995).
13. M.F. Atiyah, R. Bott, A. Shapiro, *Topology* Suppl.**3**, 1 (1964).
14. R. Coquereaux, Phys. Lett. 115B, 389 (1982).
15. T. Kugo, P. Townsend, *Nucl. Phys. B* **221**, 357 (1983).
16. J. Lukierski, in *Supersymmetry and its applications: superstrings, anomalies and supergravity*, ed. G.W. Gibbons at al. (Cambridge University Press 1986) p.463.
17. J. Lukierski, A. Nowicki, *J. Math. Phys.* **25**, 2545 (1984).
18. M. Scheunert, W. Nahm, V. Rittenberg, *J. Math. Phys.* **18**, 146 (1977), ibid. **18**, 155 (1977).
19. F.A. Berezin, V.N. Tolstoy, *Commun. Math. Phys.* **78**, 409 (1981).
20. Y.I. Manin, *Gauge field theory and complex geometry* (Springer, Berlin 1987) Sec. 3.6.
21. E. Ivanov, O. Lechtenfeld, B. Zupnik JHEP 02 (2004) 012
22. O. Azorkina, A. Banin, I. Buchbinder, N. Pletnov, *Phys.Lett. B* **633**, 389 (2006).
23. J. Lukierski, *Czech. Jour. Phys.* **37**, 359 (1987).
24. D.G.C. McKeon, T.N. Sherry, *Ann. Phys.* **285**, 221 (2000), ibid. **288**, 2 (2001).
25. V.G. Drinfeld, *Leningrad J. Math.* **1**, 1419 (1990).
26. N. Seiberg, *JEHP* **06**:010 (2003); hep-th/ 0308012.
27. K. Osterwalder, R. Schrader, *Helv. Phys. Acta* **46**, 277 (1973).

BACKGROUND FIELD METHOD IN $\mathcal{N} = 2$, $d = 3$ SUPERSPACE

I.L. BUCHBINDER

Department of Theoretical Physics, Tomsk State Pedagogical University, Tomsk 634061, Russia, E-mail: joseph@tspu.edu.ru

N.G. PLETNEV

Department of Theoretical Physics, Institute of Mathematics, 630090 Novosibirsk, Russia, Email: pletnev@math.nsc.ru

I.B. SAMSONOV *

INFN, Sezione di Padova, via F. Marzolo 8, 35131 Padova, Italy E-mail: samsonov@mph.phtd.tpu.ru

We review the features of the background field method for three-dimensional gauge theories in $\mathcal{N} = 2$ superspace. As an application, we compute one-loop low-energy effective actions in $\mathcal{N} = 2, 4, 8$, $d = 3$ SYM theories.

Keywords: Superspace, Effective action, Chern-Simons theory.

1. Introduction and conclusions

Quantum aspects of three-dimensional gauge theories with extended supersymmetry have been extensively studied during last several years. The interest to these models was re-inspired by the discovery of three-dimensional superconformal field theories wanted for the AdS_4/CFT_3 correspondence. These are familiar BLG and ABJM theories.[1,2] It is believed that these models capture the low-energy degrees of freedom of M2 branes, similarly as the $\mathcal{N} = 8$, $d = 3$ SYM theory describes a set of D2 branes. An interesting open problem is to study low-energy effective action in these field theories, to understand quantum properties of multiple M2 and D2 branes. This should be compared with the four-dimensional case for which there is good agreement between the $\mathcal{N} = 4$, $d = 4$ SYM low-energy effective action and the D3 brane dynamics.[3]

*On leave from Tomsk Polytechnic University, 634050 Tomsk, Russia.

Motivated by this problem, we develop the superspace methods for computing low-energy effective action in three-dimensional gauge theories with extended supersymmetry. We employ the $\mathcal{N}=2$ superfield approach which is mimic to the well known $\mathcal{N}=1$, $d=4$ superspace.[4] Our main result is the construction of the background field method for gauge theories in this superspace. The gauge-fixing procedure for thee-dimensional gauge theories has some features because of the presence of the topological Chern-Simons term. We construct the operators of quadratic fluctuations both for the SYM and Chern-Simons terms for arbitrary gauge superfield background. As an application, we compute one-loop low-energy effective actions in $\mathcal{N}=2,4,8$ SYM theories.

One of the possible further applications of these results could the the study of low-energy effective in the ABJM theory which should describe the dynamics of a set of M2 branes. However, this problem requires separate studies because of weak understanding of the stability of quadratic fluctuations above the vacuum in Chern-Simons-matter theories.

The present contribution is based on our previous publications.[5]

2. The background field method

The $\mathcal{N}=2$, $d=3$ superspace is parametrized by the coordinates $z^M=(x^m,\theta_\alpha,\bar\theta_\alpha)$ and is supplied by covariant spinor derivatives D_α, $\bar D_\alpha$, $\{D_\alpha,\bar D_\beta\}=-2i\partial_{\alpha\beta}$. In this superspace we consider the $\mathcal{N}=2$ supersymmetric Yang-Mills-Chern-Simons theory,

$$S=S_{\rm SYM}+S_{\rm CS}\,, \tag{1}$$

$$S_{\rm SYM}=\frac{1}{g^2}{\rm tr}\int d^7z\,G^2=-\frac{1}{2g^2}{\rm tr}\int d^5z\,W^\alpha W_\alpha\,, \tag{2}$$

$$S_{\rm CS}=\frac{ik}{8\pi}{\rm tr}\int_0^1 dt\int d^7z\,\bar D^\alpha(e^{-2tV}D_\alpha e^{2tV})e^{-2tV}\partial_t e^{2tV}\,, \tag{3}$$

where V is the $\mathcal{N}=2$ gauge superfield with the superfield strengths $G=\frac{i}{4}\bar D^\alpha(e^{-2V}D_\alpha e^{2V})$ and $W_\alpha=\bar D_\alpha G$; g is gauge coupling constant and k is the Chern-Simons level. The action is invariant under the following gauge transformations

$$e^{2V}\to e^{i\bar\lambda}e^{2V}e^{-i\lambda}\,, \tag{4}$$

where λ is a chiral parameter.

The gauge superfield V splits into the background V and quantum v parts,

$$e^{2V}\to e^{2V}e^{2gv}\,, \tag{5}$$

so that the background-gauge-covariant spinor derivatives are given by

$$\nabla_\alpha = e^{-2gv}\mathcal{D}_\alpha e^{2gv}\,, \qquad \bar{\nabla}_\alpha = \bar{\mathcal{D}}_\alpha\,, \tag{6}$$

where

$$\mathcal{D}_\alpha = e^{-2V} D_\alpha e^{2V}\,, \qquad \bar{\mathcal{D}}_\alpha = \bar{D}_\alpha\,. \tag{7}$$

The above decomposition of the gauge superfield V into the background V and quantum v superfields is analogous to the background-quantum splitting in $\mathcal{N}=1$, $d=4$ superfield Yang-Mills theory,[6] but differs from the splitting in conventional Yang-Mills theory[7] and in $\mathcal{N}=2$, $d=4$ super Yang-Mills theory in harmonic superspace formulation.[8]

The action (1) can be decomposed in a series with respect to the quantum gauge superfield v,

$$S[V,v] = S[V] + S_1[V,v] + S_2[V,v] + \dots\,, \tag{8}$$

$$S_1[V,v] = \mathrm{tr}\int d^7z\, v\left(\frac{i}{g}\mathcal{D}^\alpha\bar{\mathcal{D}}_\alpha G + \frac{gk}{\pi}G\right), \tag{9}$$

$$S_2[V,v] = \mathrm{tr}\int d^7z\, v\left(\frac{1}{8}\mathcal{D}^\alpha\bar{\mathcal{D}}^2\mathcal{D}_\alpha - iW^\alpha\mathcal{D}_\alpha + \frac{ig^2k}{4\pi}\bar{\mathcal{D}}^\alpha\mathcal{D}_\alpha\right)v\,. \tag{10}$$

Here we do not consider the terms with vertices for quantum superfields as we restrict ourself to one-loop computations only.

The operator of quadratic fluctuations in the action S_2 is degenerate and requires gauge fixing,

$$S_{\mathrm{gf}} = \mathrm{tr}\int d^7z\, v\left[-\frac{1}{16\alpha}\{\mathcal{D}^2,\bar{\mathcal{D}}^2\} + \frac{i}{\beta}\bar{\mathcal{D}}^2 + \frac{i}{\beta}\mathcal{D}^2\right]v\,. \tag{11}$$

Here α and β are gauge fixing parameters.

The ghost superfield action has the same structure as in standard $\mathcal{N}=1$, $d=4$ gauge theory,

$$S_{\mathrm{gh}} = \mathrm{tr}\int d^7z(\bar{b}c - b\bar{c} - \bar{\varphi}\varphi) + O(g)\,. \tag{12}$$

Anticommuting ghost superfields b, c, φ are covariantly chiral.

It is convenient to choose the gauge fixing parameters as $\alpha = 1$ and $\beta = \frac{8\pi}{kg^2}$. Then the sum of the actions S_2 and S_{gf} reads

$$S_2 + S_{\mathrm{gf}} = \mathrm{tr}\int d^7z\, v(-\Box_{\mathrm{v}} + H)v\,, \tag{13}$$

where

$$\Box_{\mathrm{v}} = \mathcal{D}^m\mathcal{D}_m + G^2 + iW^\alpha\mathcal{D}_\alpha - i\bar{W}^\alpha\bar{\mathcal{D}}_\alpha\,, \tag{14}$$

$$H = \frac{ig^2k}{8\pi}(\bar{\mathcal{D}}^\alpha\mathcal{D}_\alpha + \mathcal{D}^\alpha\bar{\mathcal{D}}_\alpha + \bar{\mathcal{D}}^2 + \mathcal{D}^2)\,. \tag{15}$$

Here $\Box_{\rm v}$ is the covariant d'Alembertian operator in the space of real superfields and H originates from the Chern-Simons part of the action.

As a result, the one-loop effective action in the $\mathcal{N}=2$ gauge superfield model can be schematically written as

$$\Gamma = \Gamma_{\rm v} + \Gamma_{\rm gh}\,, \qquad \Gamma_{\rm v} = \frac{i}{2}{\rm Tr}_{\rm v}\ln(\Box_{\rm v} - H)\,, \quad \Gamma_{\rm gh} = -\frac{3i}{2}{\rm Tr}_{+}\ln\Box_{+}\,. \quad (16)$$

The contribution $\Gamma_{\rm v}$ to the one-loop effective action comes from the quantum gauge superfield while $\Gamma_{\rm gh}$ is due to ghosts. Here ${\rm Tr}_{\rm v}$ and ${\rm Tr}_{+}$ are the functional traces of the operators acting in the spaces of real and chiral superfields, respectively. The operator $\Box_{+}$ is the covariant d'Alembertian operator acting in the space of covariantly chiral superfields which was introduced in ref. 5,

$$\Box_{+} = \frac{1}{16}\bar{\mathcal{D}}^2\mathcal{D}^2 = \mathcal{D}^m\mathcal{D}_m + G^2 + \frac{i}{2}(\mathcal{D}^\alpha W_\alpha) + iW^\alpha\mathcal{D}_\alpha\,. \quad (17)$$

The explicit expressions for the traces of these operators can be found after one specifies the gauge group and the background gauge superfield.

3. Superfield effective actions

3.1. $\mathcal{N}=2$ *SYM*

Consider pure $\mathcal{N}=2$ SYM model with classical action (2). In eq. (11) we send $\beta\to\infty$ to remove the part responsible for the gauge fixing in the Chern-Simons action. Then the expression for the one-loop effective action (16) slightly modifies,

$$\Gamma^{\mathcal{N}=2}_{\rm SYM} = \Gamma_{\rm v} + \Gamma_{\rm gh}\,, \qquad \Gamma_{\rm v} = \frac{i}{2}{\rm Tr}_{\rm v}\ln\Box_{\rm v}\,, \quad \Gamma_{\rm gh} = -\frac{3i}{2}{\rm Tr}_{+}\ln\Box_{+}\,, \quad (18)$$

where the operators $\Box_{\rm v}$ and $\Box_{+}$ are given in (14) and (17).

To compute the functional traces, we have to specify the gauge group and the background gauge superfield V. For simplicity, we consider the gauge group SU(2) spontaneously broken down to U(1). So, the background gauge superfield V is Abelian and obeys classical equations of motion, $\bar{D}^\alpha D_\alpha G = 0$. Moreover, we restrict it to be constant with respect to the space-time derivative, $\partial_m G = 0$. For such a background the one-loop effective action can be computed explicitly. Here we present the leading terms,

$$\Gamma^{\mathcal{N}=2}_{\rm SYM} = -\frac{3}{2\pi}\int d^7z\, G\ln G + \frac{9}{128\pi}\int d^7z\frac{W^2\bar{W}^2}{G^5} + \ldots\,. \quad (19)$$

In the component field decomposition, this action is responsible for the scale-invariant and supersymmetric F^2 and F^4 terms while the dots mean higher orders with respect to the Maxwell field.

3.2. $\mathcal{N} = 4$ *SYM*

Consider the $\mathcal{N} = 4$ SYM model with the classical action

$$S^{\mathcal{N}=4}_{\rm SYM} = \frac{1}{g^2}{\rm tr}\int d^7z\left(G^2 - \frac{1}{2}e^{-2V}\bar{\Phi}e^{2V}\Phi\right), \tag{20}$$

where Φ is a chiral superfield in the adjoint representation. This chiral superfield contributes to the one-loop effective action similarly as the chiral ghost superfields, but with opposite sign. Hence, the one-loop effective action can be written as

$$\Gamma^{\mathcal{N}=4}_{\rm SYM} = \frac{i}{2}{\rm Tr}_{\rm v}\ln(\Box_{\rm v} + \bar{\Phi}\Phi) - i{\rm Tr}_{+}\ln(\Box_{+} + \bar{\Phi}\Phi)\,. \tag{21}$$

These functional traces are computed in a usual way. As a result, we get leading terms in the effective action,

$$\Gamma^{\mathcal{N}=4}_{\rm SYM} = \frac{2}{\pi}\int d^7z\left[\sqrt{G^2+\bar{\Phi}\Phi} - G\ln(G+\sqrt{G^2+\bar{\Phi}\Phi}) + \frac{1}{32}\frac{W^2\bar{W}^2}{(G^2+\bar{\Phi}\Phi)^{5/4}} + \ldots\right]. \tag{22}$$

The first two terms in the rhs of this expression are responsible for $\mathcal{N} = 4$ supersymmetric (and superconformal) generalization of the Maxwell F^2 term while the third term gives F^4 among other components and the dots stand for higher-order terms.

3.3. $\mathcal{N} = 8$ *SYM*

The classical action of $\mathcal{N} = 8$ SYM theory has three chiral superfields Φ_i, $i = 1, 2, 3$ in the adjoint representation,

$$\begin{aligned} S^{\mathcal{N}=8}_{\rm SYM} &= \frac{1}{g^2}{\rm tr}\int d^7z\left(G^2 - \frac{1}{2}e^{-2V}\bar{\Phi}^i e^{2V}\Phi_i\right) \\ &\quad + \frac{1}{12g^2}\left({\rm tr}\int d^5z\,\varepsilon^{ijk}\Phi_i[\Phi_j, \Phi_k] + c.c.\right). \end{aligned} \tag{23}$$

These three chiral superfields completely cancel contributions from three ghost superfields in (16) and the one-loop effective action is simply

$$\Gamma^{\mathcal{N}=8}_{\rm SYM} = \frac{i}{2}{\rm Tr}_{\rm v}\ln(\Box_{\rm v} + \bar{\Phi}^i\Phi_i) = \frac{3}{32\pi}\int d^7z\frac{W^2\bar{W}^2}{(G^2 + \bar{\Phi}^i\Phi_i)^{5/2}} + \ldots \tag{24}$$

This expression shows that the low-energy effective action in the $\mathcal{N} = 8$ SYM theory does not contain F^2 terms. The non-renormalization of the F^2 term was pointed out by Dine and Seiberg.[9]

Acknowledgments

The authors are grateful to the RFBR grant Nr. 12-02-00121 and LRSS grant Nr. 224.2012.2 for partial support. I.L.B. and I.B.S. acknowledge the support from DFG grant LE 838/12/1. The work of I.B.S. was also supported by the Padova University Project CPDA 119349. N.G.P. acknowledges the support from RFBR grant Nr. 11-02-00242.

References

1. J. Bagger, N. Lambert, Phys. Rev. D75 (2007) 045020, hep-th/0611108; J. Bagger, N. Lambert, Phys. Rev. D77 (2008) 065008, arXiv:0711.0955; J. Bagger, N. Lambert, JHEP 02 (2008) 105, arXiv:0712.3738; J. Bagger, N. Lambert, Phys. Rev. D79 (2009) 025002, arXiv:0807.0163; A. Gustavsson, Nucl. Phys. B811 (2009) 66, arXiv:0709.1260; A. Gustavsson, JHEP 04 (2008) 083, arXiv:0802.3456.
2. O. Aharony, O. Bergman, D.L. Jafferis, J. Maldacena, JHEP 10 (2008) 091, arXiv:0806.1218.
3. I. Chepelev, A.A. Tseytlin, Nucl. Phys. B515 (1998) 73, hep-th/9709087; A.A. Tseytlin, hep-th/9908105; I.L. Buchbinder, S.M. Kuzenko, A.A. Tseytlin, Phys. Rev. D62 (2000) 045001, hep-th/9911221; S.M. Kuzenko, I.N. McArthur, Phys. Lett. B591 (2004) 304, hep-th/0403082; S.M. Kuzenko, I.N. McArthur, Nucl. Phys. B697 (2004) 89, hep-th/0403240; S.M. Kuzenko, JHEP 03 (2005) 008, hep-th/0410128.
4. I.L. Buchbinder, S.M. Kuzenko, *Ideas and Methods of Supersymmetry and Supergravity*, IOP Publishing, Bristol and Philadelphia (1998).
5. I.L. Buchbinder, N.G. Pletnev, I.B. Samsonov, JHEP 01 (2011) 121, arXiv:1010.4967; JHEP 04 (2010) 124, arXiv:1003.4806; arXiv:1206.5711; I.L. Buchbinder, N.G. Pletnev, JHEP 11 (2011) 085, arXiv:1108.2966.
6. M.T. Grisaru, W. Siegel, Nucl. Phys. B159 (1979) 429; Nucl. Phys. B187 (1981) 149.
7. B.S. DeWitt, Phys. Rev. 162 (1967) 1195.
8. I.L. Buchbinder, E.I. Buchbinder, S.M. Kuzenko, B.A. Ovrut, Phys. Lett. B417 (1998) 61, hep-th/9704214.
9. M. Dine, N. Seiberg, Phys. Lett. B409 (1997) 239, hep-th/9705057.

CONSERVATION LAWS FOR SO(p,q)

V.K. DOBREV

Institute for Nuclear Research and Nuclear Energy, Bulgarian Academy of Sciences, Tsarigradsko Chaussee 72, BG-1784 Sofia, Bulgaria

Using our previous results on the systematic construction of invariant differential operators for non-compact semisimple Lie groups we classify the conservation laws in the case of SO(p,q).

1. Introduction

In a recent paper[1] we started the systematic explicit construction of invariant differential operators. We gave an explicit description of the building blocks, namely, the *parabolic subgroups and subalgebras* from which the necessary representations are induced. Thus we have set the stage for study of different non-compact groups.

Since the study and description of detailed classification should be done group by group we had to decide which groups to study first. We decided to start with a subclass of the hermitian-symmetric algebras which share some special properties of the conformal algebra $so(n,2)$. That is why, in view of applications to physics, we called these algebras '*conformal Lie algebras*' (CLA), (or groups)[2]. Later we gave a natural way to go beyond this subclass using essentially the same results. For this we introduce the new notion of *parabolic relation* between two non-compact semisimple Lie algebras $\mathcal{G}$ and $\mathcal{G}'$ that have the same complexification and possess maximal parabolic subalgebras with the same complexification[3].

Thus, for example, using results for the conformal algebra $so(n,2)$ (for fixed n) we can obtain results for all pseudo-orthogonal algebras $so(p,q)$ such that $p+q=n+2$. In this way, in[3] (among other things) cases we gave the main and the reduced multiplets of indecomposable elementary representations for $so(p,q)$ including the necessary data for all relevant invariant differential operators.

We specially stressed that the classification of all invariant differential operators includes as special cases all possible *conservation laws* and *con-*

served currents, unitary or not. In the present paper we give explicitly the conservation laws in the case of $so(p,q)$.

This paper is a short sequel of[3]. Due to the lack of space we refer to[3] for motivations and extensive list of literature on the subject.

2. Preliminaries

Let $\mathcal{G} = so(p,q)$, $p \geq q$, $p+q > 4$. We choose a maximal parabolic $\mathcal{P} = \mathcal{M} \oplus \mathcal{A} \oplus \mathcal{N}$ such that:

$$\mathcal{M} = so(p-1, q-1), \quad \dim \mathcal{A} = 1 , \quad \dim \mathcal{N} = p+q-2 . \qquad (1)$$

With this choice we get for the conformal algebra $so(n,2)$ the Bruhat decomposition $\mathcal{G} = \mathcal{P} \oplus \tilde{\mathcal{N}}$ with direct physical meaning $(\tilde{\mathcal{N}} \cong \mathcal{N})$[3].

We label the signature of the representations of $\mathcal{G}$ as follows:

$$\begin{aligned} \chi &= \{ n_1 , \ldots , n_h ; c \} , \\ & n_j \in \mathbb{Z}/2 , \quad c = d - \tfrac{p+q-2}{2} , \quad h \equiv [\tfrac{p+q-2}{2}], \\ & |n_1| < n_2 < \cdots < n_h , \quad p+q \ \text{even} , \\ & 0 < n_1 < n_2 < \cdots < n_h , \quad p+q \ \text{odd} , \end{aligned} \qquad (2)$$

where the parameter c (related to the conformal weight d) labels the characters of $\mathcal{A}$, and the first h entries are labels of the finite-dimensional (nonunitary for $q \neq 1$) irreps μ of $\mathcal{M}$.

We call the above induced representations $\chi = \mathrm{Ind}_{\mathcal{P}}^{\mathcal{G}}(\mu \otimes \nu \otimes 1)$ *elementary representations* of $G = SO(p,q)$[4]. Their spaces of functions are:

$$\mathcal{C}_\chi = \{ \mathcal{F} \in C^\infty(G, V_\mu) \mid \mathcal{F}(gman) = e^{-\nu(H)} \cdot D^\mu(m^{-1}) \, \mathcal{F}(g) \}$$

where $a = \exp(H)$, $H \in \mathcal{A}$, $m \in M = SO(p-1, q-1)$, $n \in N = \exp \mathcal{N}$. The representation action is the *left regular action*:

$$(\mathcal{T}^\chi(g)\mathcal{F})(g') = \mathcal{F}(g^{-1}g') , \quad g, g' \in G . \qquad (3)$$

• An important ingredient in our considerations are the *highest/lowest weight representations* of $\mathcal{G}^{\mathbb{C}}$. These can be realized as (factor-modules of) Verma modules V^Λ over $\mathcal{G}^{\mathbb{C}}$, where $\Lambda \in (\mathcal{H}^{\mathbb{C}})^*$, $\mathcal{H}^{\mathbb{C}}$ is a Cartan subalgebra of $\mathcal{G}^{\mathbb{C}}$, weight $\Lambda = \Lambda(\chi)$ is determined uniquely from χ[5].

Actually, since our ERs are induced from finite-dimensional representations of $\mathcal{M}$ the Verma modules are always reducible. Thus, it is more convenient to use *generalized Verma modules* $\tilde{V}^\Lambda$ such that the role of the highest/lowest weight vector v_0 is taken by the (finite-dimensional) space $V_\mu \, v_0$. For the generalized Verma modules (GVMs) the reducibility is controlled only by the value of the conformal weight d, or the parameter c.

Relatedly, for the intertwining differential operators only the reducibility w.r.t. non-compact roots is essential.

• One main ingredient of our approach is as follows. We group the (reducible) ERs with the same Casimirs in sets called *multiplets*[6]. The multiplet corresponding to fixed values of the Casimirs may be depicted as a connected graph, the *vertices* of which correspond to the reducible ERs and the *lines (arrows)* between the vertices correspond to intertwining operators. The multiplets contain explicitly all the data necessary to construct the intertwining differential operators. Actually, the data for each intertwining differential operator consists of the pair (β, m), where β is a (non-compact) positive root of $\mathcal{G}^{\mathbb{C}}$, $m \in \mathbb{N}$, such that the *BGG Verma module reducibility condition*[7] (for highest weight modules) is fulfilled:

$$(\Lambda + \rho, \beta^\vee) \ = \ m \ , \quad \beta^\vee \equiv 2\beta/(\beta,\beta) \tag{4}$$

where ρ is half the sum of the positive roots of $\mathcal{G}^{\mathbb{C}}$. When the above holds then the Verma module with shifted weight $V^{\Lambda - m\beta}$ (or $\tilde{V}^{\Lambda - m\beta}$ for GVM and β non-compact) is embedded in the Verma module V^Λ (or $\tilde{V}^\Lambda$). This embedding is realized by a singular vector v_s expressed by a polynomial $\mathcal{P}_{m,\beta}(\mathcal{G}^-)$ in the universal enveloping algebra $(U(\mathcal{G}_-))\ v_0$, $\mathcal{G}^-$ is the subalgebra of $\mathcal{G}^{\mathbb{C}}$ generated by the negative root generators[8]. More explicitly,[5] $v^s_{m,\beta} = \mathcal{P}_{m,\beta}\, v_0$ (or $v^s_{m,\beta} = \mathcal{P}_{m,\beta}\, V_\mu\, v_0$ for GVMs).[a] Then there exists[5] an *intertwining differential operator* of order $m = m_\beta$:

$$\mathcal{D}_{m,\beta} \ : \ \mathcal{C}_{\chi(\Lambda)} \ \longrightarrow \ \mathcal{C}_{\chi(\Lambda - m\beta)} \tag{5}$$

given explicitly by:

$$\mathcal{D}_{m,\beta} \ = \ \mathcal{P}_{m,\beta}(\widehat{\mathcal{G}^-}) \tag{6}$$

where $\widehat{\mathcal{G}^-}$ denotes the *right action* on the functions $\mathcal{F}$.

Thus, in each such situation we have an *invariant differential equation* of order $m = m_\beta$:

$$\mathcal{D}_{m,\beta}\, f \ = \ f' \ , \qquad f \in \mathcal{C}_{\chi(\Lambda)} \ , \quad f' \in \mathcal{C}_{\chi(\Lambda - m\beta)} \ . \tag{7}$$

In most of these situations the invariant operator $\mathcal{D}_{m,\beta}$ has a non-trivial invariant kernel in which a subrepresentation of $\mathcal{G}$ is realized. Thus, studying the equations with trivial RHS:

$$\mathcal{D}_{m,\beta}\, f \ = \ 0 \ , \qquad f \in \mathcal{C}_{\chi(\Lambda)} \ , \tag{8}$$

[a]For explicit expressions for singular vectors we refer to[9].

is also very important. For example, in many physical applications in the case of first order differential operators, i.e., for $m = m_\beta = 1$, equations (8) are called *conservation laws*, and the elements $f \in \ker \mathcal{D}_{m,\beta}$ are called *conserved currents.* Below we give them explicitly for $so(p,q)$.

3. Classification of Conservation Laws for $so(p, q)$

Using results of[4,10–13] we present the main multiplets (which contain the maximal number of ERs with this parabolic) with the following explicit parametrization of the ERs in the multiplets (following[12]):

$$\begin{aligned}
\chi_1^\pm &= \{\epsilon\, n_1\,, \dots, n_h\,;\, \pm n_{h+1}\}\,, \quad n_h < n_{h+1}\,,\\
\chi_2^\pm &= \{\epsilon\, n_1\,, \dots, n_{h-1}\,,\, n_{h+1}\,;\, \pm n_h\}\\
\chi_3^\pm &= \{\epsilon\, n_1, \dots, n_{h-2}, n_h, n_{h+1}\,;\, \pm n_{h-1}\}\\
&\dots\\
\chi_{h-1}^\pm &= \{\epsilon\, n_1\,, n_2\,, n_4\,, \dots, n_h\,,\, n_{h+1}\,;\, \pm n_3\}\\
\chi_h^\pm &= \{\epsilon\, n_1\,, n_3\,, \dots, n_h\,,\, n_{h+1}\,;\, \pm n_2\}\\
\chi_{h+1}^\pm &= \{\epsilon\, n_2\,, n_3\,, \dots, n_h\,,\, n_{h+1}\,;\, \pm n_1\}\\
\epsilon &= \begin{cases} \pm, & p+q \ \ even \\ 1, & p+q \ \ odd \end{cases}
\end{aligned} \tag{9}$$

($\epsilon = \pm$ is correlated with $\chi^\pm$). Clearly, the multiplets correspond 1-to-1 to the finite-dimensional irreps of $so(p+q, \mathbb{C})$ with signature $\{n_1, \dots, n_h, n_{h+1}\}$ and we are able to use previous results due to the parabolic relation between the $so(p,q)$ algebras for $p+q$ -fixed. Note that the two representations in each pair $\chi^\pm$ are called *shadow fields.*

Further, we denote by $C_i^\pm$ the representation space with signature $\chi_i^\pm$.

The ERs in the multiplet are related by *intertwining integral and differential operators.*

The *integral operators* were introduced by Knapp and Stein[14]. Here these operators intertwine the pairs $C_i^\pm$ (cf. (9)):

$$G_i^\pm \ :\ C_i^\mp \longrightarrow C_i^\pm\,, \quad i = 1, \dots, 1+h\,. \tag{10}$$

The *intertwining differential operators* correspond to non-compact positive roots of the root system of $so(p+q, \mathbb{C})$, cf.[5]. In the current context, compact roots of $so(p+q, \mathbb{C})$ are those that are roots also of the subalgebra $\mathcal{M}^{\mathbb{C}}$, the rest of the roots are non-compact. We denote the differential operators by $d_i\,, d'_i$. The spaces from (9) they intertwine are:

$$\begin{aligned}
d_i \ &: \ \mathcal{C}_i^- \longrightarrow \mathcal{C}_{i+1}^- \,, \quad i = 1, \dots, h \ ; \\
d'_i \ &: \ \mathcal{C}_{i+1}^+ \longrightarrow \mathcal{C}_i^+ \,, \quad i = 1, \dots, h-1 \ ; \\
d_h \ &: \ \mathcal{C}_{h+1}^+ \longrightarrow \mathcal{C}_h^+ \,, \quad (p+q) - \text{even} \ ; \\
d'_h \ &: \ \mathcal{C}_h^- \longrightarrow \mathcal{C}_{h+1}^+ \,, \quad (p+q) - \text{even} \ ; \\
d'_h \ &: \ \mathcal{C}_{h+1}^- \longrightarrow \mathcal{C}_h^+ \,, \quad (p+q) - \text{even} \ ; \\
d'_h \ &: \ \mathcal{C}_{h+1}^+ \longrightarrow \mathcal{C}_h^+ \,, \quad (p+q) - \text{odd} \ ; \\
d_{h+1} \ &: \ \mathcal{C}_{h+1}^- \longrightarrow \mathcal{C}_{h+1}^+ \,, \quad (p+q) - \text{odd} \ .
\end{aligned} \tag{11}$$

The degrees of these intertwining differential operators are given just by the differences of the c entries[12]:

$$\begin{aligned}
&\deg d_i = \deg d'_i = n_{h+2-i} - n_{h+1-i} \,, \qquad i = 1, \dots, h \,, \qquad (12) \\
&\deg d'_h = n_2 + n_1 \,, \quad (p+q) - \text{even} \,, \\
&\deg d_{h+1} = 2n_1 \,, \quad (p+q) - \text{odd} \ .
\end{aligned}$$

where d'_h is omitted from the first line for $(p+q)$ even.

Thus, we are able to list all cases of **first order** intertwining differential operators or **conservation laws** (we give only the signatures):

$$\begin{aligned}
d_1 \ &: \ \{\epsilon\, n_1\,, n_2\,, \dots, n_h\,; -n_h - 1\} \longrightarrow \{\epsilon\, n_1\,, n_2\,, \dots, n_{h-1}\,, n_h + 1\,; -n_h\}\,; \\
d'_1 \ &: \ \{n_1\,, \dots, n_{h-1}\,, n_h + 1\,; n_h\} \longrightarrow \{n_1\,, \dots, n_h\,; n_h + 1\} \ ; \\
&\dots \qquad\qquad\qquad\qquad\qquad\qquad\qquad\qquad\qquad\qquad\qquad (13) \\
d_i \ &: \ \{\epsilon\, n_1\,, n_2\,, \dots, n_{h+1-i}\,, n_{h+3-i}\,, \dots, n_{h+1}\,; -n_{h+1-i} - 1\} \longrightarrow \\
&\quad \{\epsilon\, n_1\,, n_2\,, \dots, n_{h-i}\,, n_{h+1-i} + 1\,, n_{h+3-i}\,, \dots, n_{h+1}\,; -n_{h+1-i}\}\,, \\
&\qquad i = 2, \dots, h-1 \ ; \\
d'_i \ &: \ \{n_1\,, \dots, n_{h-i}\,, n_{h+1-i} + 1\,, n_{h+3-i}\,, \dots, n_{h+1}\,; n_{h+1-i}\} \longrightarrow \\
&\quad \{n_1\,, \dots, n_{h+1-i}\,, n_{h+3-i}\,, \dots, n_{h+1}\,; n_{h+1-i} + 1\}\,, \\
&\qquad i = 2, \dots, h-1 \ ; \\
&\dots \\
d_h \ &: \ \{\epsilon\,(n_2 - 1)\,, n_3\,, \dots, n_{h+1}\,; -n_2\} \longrightarrow \\
&\quad \{\epsilon n_2\,, n_3\,, \dots, n_{h+1}\,; 1 - n_2\}\,, \quad n_2 > \tfrac{1}{2} \ ; \\
d_h \ &: \ \{n_2\,, n_3\,, \dots, n_{h+1}\,; n_2 - 1\} \longrightarrow \{n_2 - 1\,, n_3\,, \dots, n_{h+1}\,; n_2\}\,, \\
&\qquad n_2 > \tfrac{1}{2}\,, \quad (p+q) - \text{even}; \\
d'_h \ &: \ \{n_2 - 1\,, n_3\,, \dots, n_{h+1}\,; -n_2\} \longrightarrow \{n_2\,, n_3\,, \dots, n_{h+1}\,; 1 - n_2\}\,, \\
&\qquad n_2 > \tfrac{1}{2}\,, \quad (p+q) - \text{even} \ ;
\end{aligned}$$

$$d'_h \ : \ \{-n_2\,,\, n_3\,,\dots,n_{h+1}\,;\, n_2-1\} \ \longrightarrow \ \{1-n_2\,,n_3\,,\dots,n_{h+1}\,;\, n_2\}\ ,$$
$$n_2 > \tfrac{1}{2}\ , \quad (p+q) - \text{even}\ ;$$
$$d'_h \ : \ \{n_2\,,\, n_3\,,\dots,n_{h+1}\,;\, n_2-1\} \ \longrightarrow \ \{n_2-1\,,n_3\,,\dots,n_{h+1}\,;\, n_2\}\ ,$$
$$n_2 > 1\ , \quad (p+q) - \text{odd}\ ;$$
$$d'_{h+1} \ : \ \{-n_2\,,\, n_3\,,\dots,n_{h+1}\,;\, -\tfrac{1}{2}\} \ \longrightarrow \ \{n_2\,,n_3\,,\dots,n_{h+1}\,;\, \tfrac{1}{2}\}\ ,$$
$$n_2 > 1\ , \quad (p+q) - \text{odd}\ ;$$
$$\epsilon = -(-1)^{p+q}\ .$$

Some of these conservation laws were given also in[15]. Although the operators are valid for arbitrary $so(p,q)$ $(p+q \geq 5)$ the contents of the ERs is very different. This analysis was done in detail in[3].

Besides the above cases there are other ERs (not related to finite-dimensional irreps of $so(p+q,\mathbb{C})$), and which also give rise to first order intertwining differential operators, resp. conservation laws:

$$d_\epsilon \ : \ \tilde{\chi}^-_\epsilon \ \longrightarrow \ \tilde{\chi}^+_\epsilon\ , \qquad (14)$$
$$\tilde{\chi}^\pm_\epsilon = \{\pm\epsilon\,\tfrac{1}{2}\,, n_3\,,\dots,\, n_{h+1}\,;\, \pm\tfrac{1}{2}\}\ ,$$

where ϵ is defined as in (9), but here it is not correlated with $\tilde{\chi}^\pm$. There is also a Knapp-Stein integral operator acting from $\tilde{\chi}^+_\epsilon$ to $\tilde{\chi}^-_\epsilon$. The representations $\tilde{\chi}^-_\epsilon$ are called *minimal representations.*

References

1. V.K. Dobrev, Rev. Math. Phys. **20** (2008) 407-449.
2. V.K. Dobrev, J. Phys. **A42** (2009) 285203.
3. V.K. Dobrev, arXiv:1208.0409 [hep-th], CERN-PH-TH/2012-215.
4. V.K. Dobrev, G. Mack, V.B. Petkova, S.G. Petrova and I.T. Todorov, Lecture Notes in Physics, Vol. 63 (Springer, Berlin, 1977).
5. V.K. Dobrev, Rept. Math. Phys. **25** (1988) 159-181; first as ICTP Trieste preprint IC/86/393 (1986).
6. V.K. Dobrev, Lett. Math. Phys. **9**, 205-211 (1985).
7. I.N. Bernstein, I.M. Gel'fand and S.I. Gel'fand, Funct. Anal. Appl. **5** 1–8 (1971).
8. J. Dixmier, *Enveloping Algebras*, (North Holland, New York, 1977).
9. V.K. Dobrev, Lett. Math. Phys. **22** (1991) 251-266; N. Chair, V.K. Dobrev and H. Kanno, Phys. Lett. **B283** (1992) 194-202.
10. V.K. Dobrev and V.B. Petkova, Rept. Math. Phys. **13** (1978) 233-277.
11. V.B. Petkova and G.M. Sotkov, Lett. Math. Phys. **8** (1984) 217-226, Erratum-ibid. 9 (1985) 83.
12. V.K. Dobrev, Suppl. Rendiconti Circolo Matematici di Palermo, Serie II, Numero 43 (1996) 15-56.
13. V.K. Dobrev, J. Phys. **A41** (2008) 425206; arXiv:0712.4375.
14. A.W. Knapp and E.M. Stein, Ann. Math. **93** (1971) 489-578.
15. R.R. Metsaev, Anomalous conformal currents, shadow fields and massive AdS fields, arXiv:1110.3749 [hep-th].

GAUGED $\mathcal{N} = 4$ SUPERSYMMETRIC MECHANICS WITH SPIN SUPERMULTIPLETS*

S. FEDORUK† and E. IVANOV

Bogoliubov Laboratory of Theoretical Physics, JINR, Joliot-Curie 6, 141980 Dubna, Moscow region, Russia
E-mails: fedoruk,eivanov@theor.jinr.ru

O. LECHTENFELD

Institut für Theoretische Physik, Leibniz Universität Hannover, Appelstraße 2, D-30167 Hannover, Germany
E-mail: lechtenf@itp.uni-hannover.de

We present elements of the superfield approach to constructing $\mathcal{N}=4$ supersymmetric quantum mechanics developed in arXiv:0812.4276, arXiv:0905.4951, arXiv:0912.3508, arXiv:1112.1947 and arXiv:1204.4474.

Keywords: Supersymmetric quantum mechanics; Harmonic superspace.

1. Introduction

Recently, a new type of models of $\mathcal{N}=4$ supersymmetric mechanics was discovered and studied. Their distinguishing feature is the coupling of three types of irreducible $\mathcal{N}=4$ multiplets, viz. dynamical, "semi-dynamical" and "topological" multiplets.

The dynamical multiplets produce normal kinetic terms for all components.

The spin (or "semi-dynamical") multiplets are characterized by the kinetic terms the order of which in time derivative is lower by one as compared with the standard sitution: the fermionic variables are auxiliary whereas the bosonic component fields appear through the $d=1$ Wess-Zumino (WZ) action which is of the first order. Their dynamics is an example of the

---

*Contribution to the XXIXth International Colloquium on Group-Theoretical Methods in Physics, Nankai, China, August 20-26, 2012.

† On leave of absence from V.N. Karazin Kharkov National University, Ukraine.

Chern-Simons mechanics[1–4]. After quantization, the semi-dynamical variables become spin degrees of freedom parametrizing fuzzy manifolds.

Gauge ("topological") multipltets are intended for gauging some isometries in the models. The relevant gauge transformations are capable to gauge away all components of gauge multiplets *locally*; however, in these multiplets there are components which cannot be gauged away *globally*[5].

So far there were considered gauged superfield models with semi-dynamical **(4,4,0)** multiplets coupled to dynamical **(1,4,3)** multiplets[6–9], **(4,4,0)** multiplets[10] or **(3,4,1)** multiplets,[11] as well as non-gauged model with dynamical **(1,4,3)** and semi-dynamical **(3,4,1)** multiplets[12].

A slightly different treatment of the semi-dynamical spin variables (without gauging) was employed in[13,14] .

It is important that the consistent superfield description of $\mathcal{N}{=}\,4$ spin supermultiplets is only possible within $\mathcal{N}{=}\,4$ harmonic superspace[15]. As opposed to the usual $\mathcal{N}{=}\,4$, $d{=}\,1$ superspace $(t, \theta_i, \bar\theta^k)$, $i,k = 1,2$, the harmonic superspace is parametrized by the coordinates $(t, \theta^\pm, \bar\theta^\pm, u^\pm_i)$, $\theta^\pm = \theta^i u^\pm_i$, $\bar\theta^\pm = \bar\theta^i u^\pm_i$, where commuting SU(2)-doublets $u^\pm_i$, $u^{+i}u^-_i = 1$, are harmonic coordinates. The harmonic superspace contains the harmonic analytic subspace with half of Grassmann coordinates: $(\zeta, u) = (t_A, \theta^+, \bar\theta^+, u^\pm_i)$, $t_A = t - i(\theta^+\bar\theta^- + \theta^-\bar\theta^+)$. The $\mathcal{N}{=}\,4$ superfields defined on the harmonic analytic subspace are called the harmonic analytic ones.

In the present talk we sketch the basic elements of the models with spin supermultiplets and dynamical **(1,4,3)** multiplets.

2. Dynamical (1,4,3) + semi-dynamical (4,4,0) + gauging

The supermultiplet **(1,4,3)** is described by the real superfield $\mathcal{X}(t, \theta^i, \bar\theta_i)$ subjected to the constraints $D^i D_i \,\mathcal{X} = \bar D_i \bar D^i \,\mathcal{X} = [D^i, \bar D_i]\,\mathcal{X} = 0$[16]. In harmonic superspace this supermultiplet is described by the harmonic superfield $\mathcal{X}(t, \theta^\pm, \bar\theta^\pm, u^\pm_i)$ satisfying the constraints $D^{++}\mathcal{X} = 0$, $D^+D^-\mathcal{X} = 0$, $(D^+\bar D^- + D^-\bar D^+)\mathcal{X} = 0$[15]. A different description of the supermultiplet **(1,4,3)** is provided by the analytic harmonic prepotential $\mathcal{V}(\zeta, u)$ which is defined up to gauge transformations $\delta\mathcal{V} = D^{++}\lambda^{--}(\zeta, u)$ and is related to the superfield $\mathcal{X}$ by the harmonic integral transform[5]

$$\mathcal{X}(t, \theta_i, \bar\theta^i) = \int du\, \mathcal{V}(t_A, \theta^+, \bar\theta^+, u)\Big|_{\theta^\pm = \theta^i u^\pm_i,\ \bar\theta^\pm = \bar\theta^i u^\pm_i}\,. \tag{1}$$

The supermultiplet **(4,4,0)** is described by the analytic superfields $\mathcal{Z}^+(\zeta, u)$ subjected to the constraints $D^{++}\mathcal{Z}^+ = 0$[15]. Its WZ action is an integral of general function of $\mathcal{Z}^+$ and $u^\pm$ over harmonic analytic subspace.

2.1. *Matrix model*

This model contains n^2 general superfields $\mathcal{X} = (\mathcal{X}_a^b)$, $a, b = 1, \ldots, n$ and n analytic superfields $\mathcal{Z}_a^+(\zeta, u)$. The gauge matrix connection $V^{++}(\zeta, u)$ covariantizes the harmonic derivative, $\mathcal{D}^{++} = D^{++} + i\,V^{++}$.

The model is described by the action of $\mathcal{N}=4$ $d=1$ supergauge theory[6]

$$S_n = -\tfrac{1}{4(1+\alpha)}\mathrm{Tr}\int \mu_H\, \mathcal{X}^{-1/\alpha} + \tfrac{1}{2}\int \mu_A \mathcal{V}_0\, \widetilde{\mathcal{Z}}^+\mathcal{Z}^+ + \tfrac{i}{2}\,c\int \mu_A\, \mathrm{Tr}\, V^{++}\,, \quad (2)$$

where μ_H, μ_A are the measures of integration over harmonic and analytic superspaces and $\mathcal{V}_0$ is the prepotential for $\mathrm{Tr}(\mathcal{X})$.

It is the Lagrange multiplier $\mathcal{V}_0$ in the second term of the action (2) which ensures the invariance under the $\mathcal{N}=4$ superconformal group $D(2,1;\alpha)$. The gauge freedom with respect to the local $\mathrm{U}(n)$ transformations
$\mathcal{X}' = e^{i\lambda}\mathcal{X}e^{-i\lambda}$, $\quad \mathcal{Z}^{+\prime} = e^{i\lambda}\mathcal{Z}^+$, $\quad V^{++\prime} = e^{i\lambda}\,V^{++}\,e^{-i\lambda} - i\,e^{i\lambda}(D^{++}e^{-i\lambda})$
allows us to choose the WZ gauge $V^{++} = -2i\,\theta^+\bar\theta^+ A(t_A)$. Then

$$\begin{aligned} S_n = &\int dt\Big[\mathrm{Tr}\left(\nabla X\nabla X + c\,A\right) + \tfrac{i}{2}X_0\big(\bar Z_k\nabla Z^k - \nabla\bar Z_k Z^k\big) + \tfrac{n}{8}(\bar Z^{(i}Z^{k)})(\bar Z_i Z_k)\Big] \\ &-i\mathrm{Tr}\int dt\left(\bar\Psi_k\nabla\Psi^k - \nabla\bar\Psi_k\Psi^k\right) - \int dt\,\frac{\Psi_0^{(i}\bar\Psi_0^{k)}(\bar Z_i Z_k)}{X_0}\,, \end{aligned} \quad (3)$$

where $X(t)$, $\Psi^i(t)$, $Z^i(t)$ are first components of $\mathcal{X}$, $\mathcal{Z}^+$ and $X_0 := \mathrm{Tr}(X)$, $\Psi_0^i := \mathrm{Tr}(\Psi^i)$, $\bar\Psi_0^i := \mathrm{Tr}(\bar\Psi^i)$.

Due to applying the gauging procedure, the fields Z_a^k are subject to the constraints $\bar Z_i^a Z_a^i = c$ for all $a = 1, \ldots, n$. So, Z_a^i play the role of target harmonics. Modulo center-of-mass part, the bosonic limit of the action (3)

$$S_n^{(b)} = \int dt\Big\{\sum_a \dot x_a \dot x_a + \sum_{a\neq b}\frac{\mathrm{Tr}(S_a S_b)}{4(x_a - x_b)^2}\Big\}$$

is none other than the integrable $\mathrm{U}(2)$-spin Calogero model in the formulation of Polychronakos,[20] where $(S_a)_i{}^j \equiv \bar Z_i^a Z_a^j$ form $u(2)$ algebras

$$[(S_a)_i{}^j, (S_b)_k{}^l]_D = i\delta_{ab}\left\{\delta_i^l (S_a)_k{}^j - \delta_k^j (S_a)_i{}^l\right\}.$$

for each a.

2.2. *Superconformal mechanics*

The action (2), for $n = 1$, yields the following component action[7,8]

$$\begin{aligned} S_1 = &\int dt\left[\dot x\dot x + \tfrac{i}{2}\left(\bar z_k \dot z^k - \dot{\bar z}_k z^k\right) - \frac{\alpha^2(\bar z_k z^k)^2}{4x^2} - A\left(\bar z_k z^k - c\right)\right] \quad (4) \\ &-i\int dt\left(\bar\psi_k\dot\psi^k - \dot{\bar\psi}_k\psi^k\right) + \tfrac{2}{3}\int dt\,\frac{3\alpha\psi^i\bar\psi^k z_{(i}\bar z_{k)} + (1+2\alpha)\psi^i\bar\psi^k\psi_{(i}\bar\psi_{k)}}{x^2}. \end{aligned}$$

Using the Noether procedure, we can find the odd generators

$$Q^i = p\,\psi^i + 2i\alpha\frac{z^{(i}\bar{z}^{k)}\psi_k}{x} + i(1+2\alpha)\frac{\psi_k\psi^k\bar{\psi}^i}{x}, \qquad S^i = -2\,x\psi^i + t\,Q^i$$

(and c.c.) which produce the whole superalgebra $D(2,1;\alpha)$ in their closure.

The angular variables z^i, $\bar{z}_i$ define the first Hopf map $S^3 \to S^2$. In particular, the quantum quantities $Y_a = \frac{1}{2}\,\hat{\bar{z}}_i(\sigma_a)^i{}_j\hat{z}^j$ satisfy the relations

$$[Y_a, Y_b] = i\,\epsilon_{abc}Y_c, \qquad Y_aY_a = \tfrac{c}{2}\left(\tfrac{c}{2}+1\right)$$

which are the defining relations for the fuzzy sphere coordinates[17–19].

Using the holomorphic realization of spinor variables, we find that the state vector is formed by the functions $A_{i'}(x,v^+)$, $B'_i(x,v^+)$, $B''_i(x,v^+)$ which carry the representations of the groups SU(1,1), $SU_L(2)$ and $SU_R(2)$:

	$r_0, \mathbf{T}^2 = r_0(r_0-1)$	$j, \mathbf{J}^2 = j(j+1)$	$i, \mathbf{I}^2 = i(i+1)$
$A_{k'}(x,v^+)$	$\frac{\lvert\alpha\rvert(c+1)+1}{2}$	$\frac{c}{2}$	$\frac{1}{2}$
$B'_k(x,v^+)$	$\frac{\lvert\alpha\rvert(c+1)+1}{2} - \frac{1}{2}\mathrm{sign}(\alpha)$	$\frac{c}{2}-\frac{1}{2}$	0
$B''_k(x,v^+)$	$\frac{\lvert\alpha\rvert(c+1)+1}{2} + \frac{1}{2}\mathrm{sign}(\alpha)$	$\frac{c}{2}+\frac{1}{2}$	0

These $D(2,1;\alpha)$ representations are just those which were re-obtained in a recent paper[21] by algebraic methods.

3. Dynamical (1,4,3) + semi-dynamical (3,4,1)

The supermultiplet **(3,4,1)** is described by the even analytic superfield $L^{++}(\zeta,u)$ subjected to the harmonic constraint $D^{++}L^{++} = 0$[15]. The general action for its interaction with the dynamical **(1,4,3)** multiplet reads[12]

$$S = \int \mu_H\,\mathcal{L}(\mathcal{X}) - \tfrac{i}{2}\int \mu_A\,\mathcal{L}^{(+2)}(L^{++},u) + \tfrac{i}{2}\int \mu_A\,\mathcal{V}(L^{++}+c^{++}), \quad (5)$$

where $c^{++} = c^{ik}u^+_iu^+_k$ and the constants c^{ik} are parameters of the model. In the case $\mathcal{L}(\mathcal{X}) = -\frac{1}{2}\mathcal{X}^2$ we obtain the following component action

$$S = \int dt\,\Big[\dot{x}\dot{x} - \tfrac{1}{4}(v_a+c_a)(v_a+c_a) - \mathcal{A}_a\dot{v}_a + B\,(x-\mathcal{U}) + i\left(\dot{\bar{\psi}}^k\psi_k - \bar{\psi}^k\dot{\psi}_k\right) - (\mathcal{R}_b\mathcal{R}_b)^{-1}\mathcal{R}_a\psi^i(\sigma_a)_i{}^k\bar{\psi}_k\Big]$$

where the potentials $\mathcal{U}$, $\mathcal{A}_a$, $\mathcal{R}_a$ depend on the triplet components v_a of the **(3,4,1)** supermultiplet and satisfy the constraints

$$\partial_a\partial_a\,\mathcal{U} = 0, \qquad \mathcal{R}_a = \partial_a\,\mathcal{U},$$
$$\partial_a\partial_a\,\mathcal{A}_b = 0, \quad \partial_a\mathcal{A}_a = 0, \quad \partial_a\mathcal{A}_b - \partial_b\mathcal{A}_a = -\epsilon_{abc}\partial_c\mathcal{U}.$$

So, the potentials $\mathcal{A}_a$, $\mathcal{R}_a$ are defined by the harmonic function $\mathcal{U}(v)$ which can be taken as a superposition of the multi-center solutions.

The auxiliary field B appears as a Lagrange multiplier for the constraint $x = \mathcal{U}(v)$. We treat x as an independent phase variable:

$$v_a = v_a(x, \ell_{\underline{a}}), \qquad a = 1,2,3, \qquad \underline{a} = 1,2,$$

where $\ell_{\underline{a}}$ are semi-dynamical (spin) variables which decouple from the dynamical degrees of freedom: $[\ell_{\underline{a}}, x]_D = [\ell_{\underline{a}}, p]_D = [\ell_{\underline{a}}, \psi_k]_D = 0$.

The constraints of the model guarantee the fulfillment of the equations

$$[p, v_a]_D = \tfrac{1}{2}\,\epsilon_{abc}\,[v_b, v_c]_D\,,$$

which coincide with the SDiff(Σ_2) Nahm equations[22]. Moreover, the fulfillment of the basic relations $\{\hat{Q}^i, \hat{\bar{Q}}_k\} = 2\hbar\,\delta^i_k \hat{H}$, $\{\hat{Q}^i, \hat{Q}^k\} = 0$ for the quantum supersymmetry generators

$$\hat{Q}^i = \hat{p}\,\hat{\psi}^i + i\,(\hat{v}_a + c_a)\,\sigma_a^{ik}\hat{\psi}_k\,, \qquad \hat{\bar{Q}}_i = \hat{p}\,\hat{\bar{\psi}}_i - i\,(\hat{v}_a + c_a)\,\sigma_{a\,ik}\hat{\bar{\psi}}^k$$

requires the validity of the operator Nahm equations[23]

$$\hbar\,\frac{\partial}{\partial\hat{x}}\,\hat{v}_a = \tfrac{i}{2}\,\epsilon_{abc}[\hat{v}_b, \hat{v}_c]\,.$$

So, the Nahm equations ensure the existence of the $\mathcal{N}=4$ supersymmetry, both at the classical and the quantum levels.

In the one-monopole case the operator Nahm equations are satisfied for $\hat{v}_a = (\hat{\ell}_a/\hat{x}) + k_a$, where $\hat{\ell}_a$, $\hat{\ell}_a\hat{\ell}_a = g^2$, are spin variables. Using the matrix realization for $\hat{\ell}_a$ we reproduce the spectrum of the previous section.

In the two-monopole case the spin variables $\ell_{\underline{a}} = (\ell_3, \varphi)$ are defined as

$$\ell_3 := \frac{g_1(v_3 - k_1)}{|\vec{v} - \vec{k}_1|} + \frac{g_2(v_3 - k_2)}{|\vec{v} - \vec{k}_2|}, \qquad \varphi := \arctan\left(\frac{v_2}{v_1}\right),$$

where $\vec{k}_{1,2}$ define the position of the poles. As distinct from the one-monopole case, the expansion for v_3 contains not only the linear terms in ℓ_3. Therefore we are led to modify the classical-quantum correspondence,

$$\hat{v}_a(\hat{x}, \hat{\ell}_{\underline{a}}) \quad\to\quad \hat{v}_a(\hat{x}, \hat{\ell}_{\underline{a}}; \hbar)$$

in order to save the operator Nahm equations. In[12] we presented the first terms (in $\hbar$) of the quantum supercharges in some limiting cases.

4. Summary

We outlined the universal approach to building new $\mathcal{N}=4$ $d=1$ models. Their distinguishing salient features are the gauging of some isometries of the models and the use of semi-dynamical multiplets in addition to the truly dynamical ones. Due to the spin variables, the quantum spectrum in general describes different irreducible representations of the symmetry group of the model. Using the "“topological” gauge supermultiplets picks up just one fixed irreducible representation. In addition, imposing different gauges (besides WZ gauge) allows us to obtain various equivalent formulations of the given system. Considering one type of these models, we have also discovered a surprising correspondence between the famous Nahm equations and the presence of $\mathcal{N}=4$ supersymmetry in the model.

The proposed approach can be used to build up new models with $\mathcal{N}\neq 4$ extended supersymmetry. It is also interesting to use new supersymmetric mechanics models for constructing various versions of the Dolbeault and Dirac complexes and calculating their indices, like this was done in[24,25].

Acknowledgments

This work was supported by RFBR grants 11-02-90445, 12-02-00517 (S.F. & E.I.) and a DFG grant, project No. 436 RUS/113/669 (E.I. & O.L.). S.F. would like to thank the Organizing Committee of the XXIXth International Colloquium on Group-Theoretical Methods in Physics for the warm hospitality at the Chern Institute of Mathematics, Nankai University.

References

1. L. Faddeev, R. Jackiw, *Phys. Rev. Lett.* **60**, 1692 (1988).
2. R. Floreanini, R. Percacci, E. Sezgin, *Nucl. Phys.* **B322**, 255 (1989).
3. P.S. Howe, P.K. Townsend, *Class. Quant. Grav.* **7**, 1655 (1990).
4. A.P. Polychronakos, *Phys. Lett.* **B266**, 29 (1991).
5. F. Delduc, E. Ivanov, *Nucl. Phys.* **B753**, 211 (2006), *ibid* **B770**, 179 (2007).
6. S. Fedoruk, E. Ivanov, O. Lechtenfeld, *Phys. Rev.* **D79**, 105015 (2009).
7. S. Fedoruk, E. Ivanov, O. Lechtenfeld, *JHEP* **0908**, 081 (2009).
8. S. Fedoruk, E. Ivanov, O. Lechtenfeld, *JHEP* **1004**, 129 (2010).
9. S. Fedoruk, E. Ivanov, O. Lechtenfeld, *J. Phys.* **A45**, 173001 (2012).
10. E.A. Ivanov, M.A. Konyushikhin, A.V. Smilga, *JHEP* **1005**, 003 (2010).
11. E. Ivanov, M. Konyushikhin, *Phys. Rev.* **D82**, 085014 (2010).
12. S. Fedoruk, E. Ivanov, O. Lechtenfeld, *JHEP* **1206**, 147 (2012).
13. S. Bellucci, S. Krivonos, *Phys. Rev.* **D80**, 065022 (2009).
14. S. Krivonos, O. Lechtenfeld, *Phys. Rev.* **D80**, 045019 (2009).
15. E. Ivanov, O. Lechtenfeld, *JHEP* **0309**, 073 (2003).
16. E.A. Ivanov, S.O. Krivonos, V.M. Leviant, *J. Phys.* **A22**, 4201 (1989).
17. J. Madore, *Phys. Lett.* **B263**, 245 (1991); *Class. Quant. Grav.* **9**, 69 (1992).
18. E. Ivanov, L. Mezincescu, P.K. Townsend, `arXiv:hep-th/0311159`.
19. K. Hasebe, *Phys. Rev. Lett.* **94**, 206802 (2005).
20. A.P. Polychronakos, *J. Phys.* **A39**, 12793 (2006).
21. K. Govil, M. Gunaydin, `arXiv:1209.0233 [hep-th]`.
22. R.S. Ward, *Phys.Lett.* **B234**, 81 (1990).
23. W. Nahm, *Lect. Notes Phys.* **180**, 456 (1983).
24. E. Ivanov, A. Smilga, *Intern. J. Mod. Phys.* **A27**, 1230024 (2012).
25. S. Fedoruk, E. Ivanov, A. Smilga, *Intern. J. Mod. Phys.* **A27**, 1250146 (2012).

TWISTING, 'FINITE' POINCARÉ TRANSFORMATIONS AND QFT ON NONCOMMUTATIVE SPACE

GAETANO FIORE*

Dip. di Matematica e Applicazioni, Università "Federico II"
V. Claudio 21, 80125 Napoli, Italy,
& I.N.F.N., Sez. di Napoli, Complesso MSA, V. Cintia, 80126 Napoli, Italy
**E-mail: gaetano.fiore@na.infn.it*

We briefly sketch our formulation[1] of finite twisted Poincaré transformations and of the corresponding covariant (scalar) Quantum Field Theory on Grönewold-Moyal-Weyl noncommutative spacetime.

Keywords: Noncommutative spaces; quantum groups; quantum field theory.

1. Introduction

The idea of noncommutative (nc) spacetime is rather old.[2] The Grönewold-Moyal-Weyl (GMW) nc spaces are the simplest; their coordinates fulfill

$$[\hat{x}^\mu, \hat{x}^\nu] = i\mathbf{1}\theta^{\mu\nu} \qquad \Leftrightarrow \qquad [x^\mu \overset{\star}{,} x^\nu] = i\mathbf{1}\theta^{\mu\nu}, \tag{1}$$

where θ is a constant real antisymmetric matrix. In the formulation at the lhs the nc coordinates $\hat{x}^\mu$ are introduced from the start; while at the rhs a nc, associative product $\star$ is introduced in the vector space of functions of the commutative coordinates x^μ so that $x^\mu \star x^\nu - x^\nu \star x^\mu = i\mathbf{1}\theta^{\mu\nu}$. Here $\mu, \nu = 0, 1, 2, 3$, so as to obtain a deformation of Minkowski space; indices are raised or lowered multiplying by the standard Minkowski metric $\eta_{\mu\nu}$.

Clearly, (1) are translation invariant, *not* Lorentz-covariant. At the end of Ref. 3 Oeckl briefly suggested a Hopf algebra H, deformation (with parameters θ) of that $Fun(G)$ of the functions on the Poincaré group G, such that (1) are covariant under the H-coaction; H is the Poisson-Lie group deformation[4] [based on the twist (5)] of $Fun(G)$. Ref. 3 has ever since received very little attention. In 2004 it was noted[5–7] that (1) are covariant under the action of the deformation H' of the Universal Enveloping Algebra (UEA) $U\mathbf{g}$ of G obtained by Drinfeld twisting.[4] By duality,[8,9] the latter statement is essentially equivalent (up to completion issues) to the former.

Poincaré covariance is a basic requirement of Quantum field Theory (QFT) on Minkowski spacetime. Wightman axiomatized it in the form

R1. (**Poincaré Covariance**) The (active) relativistic transformation law of the Hilbert space of states $\mathcal{H}$ is given by a strongly continuous unitary operator $(A,y)\in SL(2,\mathbb{C})\ltimes\mathbb{R}^4\mapsto U(y,A)=U(y)\,U(A)$. There is a unique invariant state Ψ_0 (the vacuum). The field operators transform as follows,

$$U(y,A)\,\varphi^\alpha(x)\,U^{-1}(y,A)=S^\alpha_\beta(A^{-1})\,\varphi^\beta\big[\Lambda(A)x+y\big],\qquad(2)$$

with S a finite-dimensional representation of $SL(2,\mathbb{C})$, $\Lambda(A)\in SO^+(3,1)$ the Lorentz matrix corresponding to $A\in SL(2,\mathbb{C})$. Denoting as $m_{\mu\nu}$ the representation of $M_{\mu\nu}\in sl(2,\mathbb{C})$ on 4-vectors and as $\omega(\Lambda)$ a matrix such that $\Lambda=e^{m_{\mu\nu}\omega^{\mu\nu}}$, the self-adjoint operators $\sigma(P_\mu),\sigma(M_{\mu\nu})$ such that

$$U(y)=e^{i\sigma(P)\cdot y},\qquad U(\Lambda)=e^{i\sigma(M_{\mu\nu})\omega^{\mu\nu}}.\qquad(3)$$

are resp. the energy-momentum and the Lorentz operators of the theory.

q) **What is the twisted Poincaré covariant analog of R1?** Here we sketch our answer[1] based on the coaction of H. This fills a hole we left in Ref. 10,11, where we adapted Wightman axioms to GMW spaces. The generators of H are the 'quantized coordinates' of G, i.e. the observables $\hat{y}^\mu,\Lambda^\mu_\nu$ parametrizing 'finite' changes of Reference Frames (RF); as they do not commute, RF are to be considered as 'quantum objects'. Previous answers[10–21] to *q)* were incomplete, based on the action of H'. Interpreting[17,20] the actions of $\mathbf{1}+g/n\in H'$ ($g\in\mathbf{g}$, $n\to\infty$) and $e^g=lim_{n\to\infty}(\mathbf{1}+g/n)^n$ as the trasformations resp. under an 'infinitesimal' and a 'finite' change of RF is no longer justified if $\theta\neq0$, since within H' there is no full deformed analog of the Lie algebra $\mathbf{g}$ (albeit there exist notions[22,23] that reproduce part of the properties of $\mathbf{g}$). For simplicity, we stick to a free scalar field and use $SO^+(3,1)$ instead of $SL(2,\mathbb{C})$ for the Lorentz group.

2. The Hopf $*$-algebras H, H' and their (co)modules

H is generated by real $\mathbf{1},\Lambda^\mu_\nu,\hat{y}^\mu$ fulfilling

$$\begin{aligned}&[\Lambda^\mu_\rho,\Lambda^\nu_\sigma]=0,\qquad \Lambda^\mu_\rho\Lambda^\nu_\sigma\eta^{\rho\sigma}=\eta^{\mu\nu}\mathbf{1},\qquad [\hat{y}^\mu,\Lambda^\nu_\sigma]=0,\\ &[\hat{y}^\mu,\hat{y}^\nu]=i(\theta^{\mu\nu}\mathbf{1}-\theta^{\rho\sigma}\Lambda^\mu_\rho\Lambda^\nu_\sigma),\qquad \det\Lambda=\mathbf{1},\qquad \Lambda^0_0>0.\end{aligned}\qquad(4)$$

The coproduct, counit, antipode Δ,ε,S are not deformed:

$$\begin{aligned}&\varepsilon(\Lambda^\mu_\nu)=\delta^\mu_\nu,\ \Delta(\Lambda^\mu_\nu)=\Lambda^\rho_\nu\otimes\Lambda^\mu_\rho,\qquad S(\Lambda^\mu_\nu)=(\eta\Lambda^T\eta)^\mu_\nu\equiv\Lambda^{-1}{}^\mu_\nu,\\ &\varepsilon(\hat{y}^\mu)=0,\ \ \Delta(\hat{y}^\mu)=\hat{y}^\nu\otimes\Lambda^\mu_\nu+\mathbf{1}\otimes\hat{y}^\mu,\quad S(\hat{y}^\mu)=-\Lambda^{-1}{}^\mu_\nu\hat{y}^\nu.\end{aligned}$$

H' has the same $*$-algebra structure, counit ε' and antipode S' as $U\mathbf{g}$, hence also the same irreducible representations: the classification of elementary particles is not deformed.[5] The coproducts Δ', Δ'_c of $H', U\mathbf{g}$ are related by the unitary transformation $\mathcal{F}$ (the *twist*)

$$\Delta'(g) = \mathcal{F}\Delta'_c(g)\mathcal{F}^{-1} \equiv \sum_I g^I_{1'} \otimes g^I_{2'}, \qquad \mathcal{F} \equiv \mathcal{F}^\alpha \otimes \mathcal{F}_\alpha := e^{\frac{i}{2}\theta^{\mu\nu} P_\mu \otimes P_\nu} \tag{5}$$

(sum over α). More explicitly, on the generators $M_{\mu\nu}, P_\mu, \mathbf{1}'$,

$$\Delta'(M_\omega) = M_\omega \otimes \mathbf{1}' + \mathbf{1}' \otimes M_\omega + P_\mu [\omega\eta, \theta\eta]^\mu_\nu \otimes P^\nu, \qquad \Delta'(P_\mu) = P_\mu \otimes \mathbf{1}' + \mathbf{1}' \otimes P_\mu,$$

where $M_\omega := \omega^{\mu\nu} M_{\mu\nu}$, $\omega^{\nu\mu} = -\omega^{\mu\nu}$, η is the Minkowski metric, and $\omega\eta$, etc., are row-by-column matrix products. H, H' are dual, namely there exists a nondegenerate bilinear map $\langle\, ,\rangle : H' \otimes H \to \mathbb{C}$ (pairing) such that

$$\begin{aligned}
&\langle g, \mathbf{1} \rangle = \varepsilon'(g), && \langle \mathbf{1}', b \rangle = \varepsilon(b), \\
&\langle g, bc \rangle = \textstyle\sum_I \langle g^I_{1'}, b \rangle \langle g^I_{2'}, c \rangle, && \langle gh, b \rangle = \textstyle\sum_I \langle g, b^I_1 \rangle \langle h, b^I_2 \rangle, \\
&\text{where } \Delta(b) = \textstyle\sum_I b^I_1 \otimes b^I_2, && \langle S'(g), b \rangle = \langle g, S(b), \rangle \\
&\langle g^{*'}, b \rangle = \overline{\langle g, S(b)^* \rangle}, && \langle g, b^* \rangle = \overline{\langle S'(g)^{*'}, b \rangle}.
\end{aligned} \tag{6}$$

$\langle\, ,\rangle : H' \otimes H \to \mathbb{C}$ is recursively determined from the relations

$$\begin{aligned}
&\langle \mathbf{1}', \mathbf{1} \rangle = 1, && \langle \mathbf{1}', \Lambda^\mu_\nu \rangle = \delta^\mu_\nu, && \langle \mathbf{1}', \hat{y}_\mu \rangle = 0, \\
&\langle M_\omega, \mathbf{1} \rangle = 0, && \langle M_\omega, \Lambda^\mu_\nu \rangle = 2i(\omega\eta)^\mu_\nu, && \langle M_\omega, \hat{y}^\nu \rangle = 0, \\
&\langle P_\mu, \mathbf{1} \rangle = 0, && \langle P_\mu, \Lambda^\mu_\nu \rangle = 0, && \langle P_\mu, \hat{y}^\nu \rangle = i\delta^\mu_\nu,
\end{aligned} \tag{7}$$

as in the undeformed case. The **Lorentz** Hopf subalgebra $H_L \subset H$ generated by $\mathbf{1}, \Lambda^\mu_\nu$ **is central and undeformed**: in a 'quantum' change of RF only the position of the origin of one RF w.r.t. another has a 'quantum' uncertainty. Whereas in H' the **translation** Hopf subalgebra $H'_T \subset H'$ generated by the P_μ is **undeformed**. Any (left) H'-module $(*)$-algebra $\hat{\mathcal{A}}$ is also a (right) H-comodule $(*)$-algebra, and conversely. The action $\triangleright : H' \otimes \hat{\mathcal{A}} \to \hat{\mathcal{A}}$ and coaction $\delta : \hat{\mathcal{A}} \to \hat{\mathcal{A}} \otimes H$ are related to each other by

$$\delta(a) = \textstyle\sum_I a^I_{\bar{1}} \otimes a^I_{\bar{2}} \qquad \Rightarrow \qquad g \triangleright a = \textstyle\sum_I a^I_{\bar{1}} \langle g, a^I_{\bar{2}} \rangle, \tag{8}$$

$$\delta(a) = \mathcal{T}(\triangleright \otimes \cdot)(a \otimes \mathbf{1}) = \sum_{J \in \mathcal{J}} (e^J \triangleright a) \otimes e_J, \tag{9}$$

where $\{e_J\}_{J \in \mathcal{J}}$ is any basis of the vector space $V(H)$ underlying H, $\{e^J\}_{J \in \mathcal{J}} \subset H'$ the dual basis and

$$\mathcal{T} = \sum_{J \in \mathcal{J}} e^J \otimes e_J \in H' \otimes H \qquad \text{(canonical element)}. \tag{10}$$

One can deform any left $U\mathbf{g}$-module $*$-algebra $\mathcal{A}$ into a left H'*-module $*$-algebra* $\mathcal{A}_\star$ endowing the vector space $V(\mathcal{A})$ with the product

$$a\star b := \left(\overline{\mathcal{F}}^{\alpha}\triangleright a\right)\left(\overline{\mathcal{F}}_{\alpha}\triangleright b\right), \qquad \overline{\mathcal{F}}^{\alpha}\otimes\overline{\mathcal{F}}_{\alpha} := \mathcal{F}^{-1} \tag{11}$$

(sum over α). We may choose $\mathcal{A}$ as the algebra $\mathcal{X}$ of polynomials in x^h, or of fast decaying smooth functions on $\mathbb{R}^4$ (as usual $g\in\mathbf{g}$ act as first order differential operators, e.g. $P_\mu \simeq i\partial_\mu$); or $\mathcal{A}=\mathcal{D}\equiv$algebra of smooth differential operators on $\mathcal{X}$; or their tensor products; etc. (11), (5) imply $x^h\star x^k = x^h x^k + \frac{i}{2}\theta^{hk}$, whence $(1)_2$. If a,b admit Fourier transforms $\tilde{a}$,$\tilde{b}$, then

$$a(x)\star b(x) = \int d^m p \int d^m q \; e^{i(p\cdot x + q\cdot x - p_h\theta^{hk}q_k/2)}\tilde{a}(p)\tilde{b}(q) \tag{12}$$

extends the definition (11) to larger domains [e.g. makes sense even if $a\in\mathcal{S}(\mathbb{R}^4)$ and $b\in\mathcal{S}'(\mathbb{R}^4)$, or $a,b\in\mathcal{L}^2(\mathbb{R}^4)$]. We shall denote $\mathcal{A}_\star$ as $\hat{\mathcal{A}}$ if we can define it also by generators and relations; e.g. $\hat{\mathcal{X}}$ is generated by $\mathbf{1},\hat{x}^\mu$ fulfilling $(1)_1$. Abbreviating $x_1^\mu \equiv x^\mu\otimes\mathbf{1}\otimes..\otimes\mathbf{1}$, $\quad x_2^\mu \equiv \mathbf{1}\otimes x^\mu\otimes..\otimes\mathbf{1}$,... one finds real variables $\hat{x}_i^\mu$ fulfilling

$$[\hat{x}_i^\mu,\hat{x}_j^\nu] = \mathbf{1}i\theta^{\mu\nu}, \qquad \Leftrightarrow \qquad [x_i^\mu \overset{\star}{,} x_j^\nu] = \mathbf{1}i\theta^{\mu\nu} \tag{13}$$

or, in Weyl form,

$$e^{ip\cdot\hat{x}_i}e^{iq\cdot\hat{x}_j} = e^{iq\cdot\hat{x}_j}e^{ip\cdot\hat{x}_i}e^{-ip\theta q}, \qquad p\theta q := p_\mu\theta^{\mu\nu}q_\nu \tag{14}$$

Note that coordinate differences $\hat{x}_i^\mu - \hat{x}_j^\mu$ are H-invariant and central; equivalently, $\xi_i^\mu := x_{i+1}^\mu - x_i^\mu$ are invariant, $\star$-central and $f(\xi)\star g = f(\xi)g = g\star f(\xi)$. Any x_i^μ is a combination of ξ_i^μ and a single $\star$-noncommutative X^μ fulfilling (1), e.g. $X^\mu = x_1^\mu$. **Integration over the space** and the **differential calculus are not deformed**. We use the notation $\partial_{x_i^\mu}\equiv\frac{\partial}{\partial x_i^\mu}$ for the partial derivatives. On $\hat{\mathcal{X}}$ the action is determined by

$$P_\mu\triangleright\hat{x}^\rho = i\delta_\mu^\rho, \qquad M_{\mu\nu}\triangleright\hat{x}^\rho = i(\delta_\nu^\rho\hat{x}_\mu - \delta_\mu^\rho\hat{x}_\nu), \tag{15}$$

the coaction by $\delta(\hat{x}^\mu) = \hat{x}^\nu\otimes\Lambda_\nu^\mu + \mathbf{1}_{\mathcal{X}}\otimes\hat{y}^\mu$, or, in equivalent 'Manin' form,

$$\hat{x}'^\mu \equiv \delta(\hat{x}^\mu) = \hat{x}^\nu\Lambda_\nu^\mu + \hat{y}^\mu = \Lambda_\nu^\mu\hat{x}^\nu + \hat{y}^\mu \qquad \Leftrightarrow \qquad \hat{x}' = \Lambda\hat{x} + \hat{y}, \tag{16}$$

after the redefinitions $\hat{x}\otimes\mathbf{1}, \mathbf{1}_{\mathcal{X}}\otimes\Lambda, \mathbf{1}_{\mathcal{X}}\otimes\hat{y} \mapsto \hat{x},\Lambda,\hat{y}$. We extend δ to space $\hat{\mathcal{X}}'$ of distributions using the generalized basis $\{e_p\}_{p\in\mathbb{R}^4}$, $e_p(\hat{x}) = e^{-ip\cdot\hat{x}}$. Applying δ to the identity $(2\pi)^4 e_p(\hat{x}) = \int d^4q \int d^4z e^{i(q-p)\cdot z} e_q(\hat{x})$, and changing integration variables $q\mapsto\Lambda q$ we find that $\delta(e_p)$ is the following combination of e_q:

$$\delta(e_p) = e^{-ip\cdot\hat{y}}\int d^4q \int \frac{d^4z}{(2\pi)^4}\, e^{i(\Lambda q - p)\cdot z}\; e_q \tag{17}$$

So we need to enlarge H so as to contain also 'functions' of $\hat{y},\Lambda$ like $e^{ip\cdot\hat{y}}$, $e^{i\Lambda q\cdot z}$. (17) holds also for $p\in\mathcal{P}^+\subset\mathbb{R}^4\equiv$the upper hyperboloid of eq. $p^2 = m^2$.

3. Finite Poincaré transformation of the free scalar field

On the GMW spaces we can define[10] a free hermitean scalar quantum field $\hat{\varphi}$ compatible with H'-covariant Wightman axioms by the decomposition

$$\hat{\varphi}(\hat{x}) = \int d\mu(p)\,[e^{-ip\cdot\hat{x}}\hat{a}^p + \hat{a}^\dagger_p e^{ip\cdot\hat{x}}], \tag{18}$$

where $d\mu(p) = \delta(p^2 - m^2)\vartheta(p^0)d^4p$, $p = (p^0, \mathbf{p})$, $\omega_\mathbf{p} = \sqrt{\mathbf{p}^2 + m^2}$, and the deformed creation, annihilation operators $\hat{a}^p, \hat{a}^\dagger_p$ fulfill eq. (44) of ref. 10:

$$\begin{aligned} &\hat{a}^\dagger_p\hat{a}^\dagger_q = e^{ip\theta q}\,\hat{a}^\dagger_q\hat{a}^\dagger_p, && \hat{a}^p\hat{a}^q = e^{ip\theta q}\,\hat{a}^q\hat{a}^p,\\ &\hat{a}^p\hat{a}^\dagger_q = e^{-ip\theta q}\,\hat{a}^\dagger_q\hat{a}^p + 2\omega_\mathbf{p}\delta^3(\mathbf{p}-\mathbf{q})\mathbf{1}, && p, q \in \mathcal{P}^+,\\ &[\hat{a}^p, f(\hat{x})] = [\hat{a}^\dagger_p, f(\hat{x})] = 0. \end{aligned} \tag{19}$$

The commutator of the field at two different spacetime points $\hat{x}_1, \hat{x}_2$ reads

$$[\hat{\varphi}(\hat{x}_1), \hat{\varphi}(\hat{x}_2)] = \int \frac{d\mu(p)}{(2\pi)^3}\left[e^{-ip\cdot(\hat{x}_1-\hat{x}_2)} - e^{ip\cdot(\hat{x}_1-\hat{x}_2)}\right] \tag{20}$$

and is practically undeformed. [We do not consider here the alternative definition of $\hat{\varphi}$ leading to (20) based on eq. (46) of ref. 10.]

Any monomial $\hat{\varphi}(\hat{x}_1)...\hat{\varphi}(\hat{x}_n)$ can be regarded as an element of the algebra $\hat{\Phi}^e := \hat{\mathcal{A}}^+ \otimes \widehat{\bigotimes}_{i=1}^{\infty}\hat{\mathcal{X}}'$, where $\hat{\mathcal{A}}^+$ is generated by $\hat{a}^p, \hat{a}^\dagger_p, \mathbf{1}$, and $\widehat{\bigotimes}_{i=1}^{\infty}\hat{\mathcal{X}}'$ is the space of distributions with an arbitrary number of arguments [the 1st, 2nd,... tensor factor $\hat{\mathcal{X}}'$ refers to the dependence of the distribution on $\hat{x}_1, \hat{x}_2, ...$ fulfilling (13-14); the dependence of the above monomial on $\hat{x}_h$ is trivial for $h > n$]. We extend $\triangleright, \delta$ to $\hat{\Phi}^e$ by requiring that $\hat{\mathcal{A}}^+$ is invariant.

As a $*$-algebra $\hat{\mathcal{A}}^+$ is isomorphic[24] to that $\mathcal{A}^+$ of creation, annihilation operators $a^p, a^\dagger_p$ and has the same vacuum and representation on Fock space $\mathcal{H}_F$ as $\mathcal{A}^+$. For free fields the $*$-algebra map $\sigma : U\mathbf{g} \mapsto \mathcal{A}^+$ appearing in (3) is the 'Jordan-Schwinger' realization[24] of $U\mathbf{g}$; in particular, $\sigma(P_\mu)$ reads

$$\sigma(P_\mu) = \int d\mu(p)\,p_\mu a^+_p a^p \simeq \int d\mu(p)\,p_\mu \hat{a}^+_p\hat{a}^p. \tag{21}$$

The expression after $\simeq$ is obtained from the preceding one expressing $a^p, a^\dagger_p$ as "functions" of $\hat{a}^p, \hat{a}^\dagger_p, \mathbf{1}$, by the isomorphism $\hat{\mathcal{A}}^+ \simeq \mathcal{A}^+$.

In ref. 1 we prove the **deformed analog of R1** with (2) replaced by

$$\mathcal{U}\,[\hat{\varphi}(\hat{x})\otimes\mathbf{1}]\,\mathcal{U}^{-1} = \delta[\hat{\varphi}(\hat{x})], \tag{22}$$

where $\mathcal{U} = \sigma(e^I)\otimes S(e_I) \in \hat{\mathcal{A}}^+\otimes H$, $\mathcal{U}^{*\otimes*} = \mathcal{U}^{-1} = \sigma(e^I)\otimes e_I$; more suggestively, this can be recast in the form

$$\begin{aligned} \mathcal{U}(\hat{y},\Lambda)\,\hat{\varphi}(\hat{x})\,\mathcal{U}^{-1}(\hat{y},\Lambda) &= \hat{\varphi}(\Lambda\hat{x}+\hat{y}),\\ \mathcal{U}(\hat{y},\Lambda) = U(\hat{y})\,U(\Lambda) &= e^{i\sigma(P)\cdot\hat{y}}e^{i\sigma(M_{\mu\nu})\omega^{\mu\nu}}, \end{aligned} \tag{23}$$

where $\sigma(g)$ is understood to be expressed in terms of $\hat{a}^p, \hat{a}^\dagger_p$. The unitary operator $\mathcal{U}$ (*active Poincaré transformation*) leaves $\widehat{\bigotimes}_{i=1}^{\infty}\hat{\mathcal{X}}'$ invariant and trasforms $\mathcal{H}_F\otimes\mathcal{H}_{RF}$, where $\mathcal{H}_{RF}$ is the Hilbert space of the RF. The coaction δ (*passive Poincaré transformation*) leaves $\mathcal{H}_F, \hat{\mathcal{A}}^+$ invariant and trasforms $\widehat{\bigotimes}_{i=1}^{\infty}\hat{\mathcal{X}}'\otimes\mathcal{H}_{RF}$. The appearence of $\mathcal{H}_{RF}$ is a new phenomenon. Eqs. (22-23) state that the result of the passive Poincaré transformation $x \mapsto \Lambda x + y$ at the rhs equals that of the active Poincaré transformation at the lhs. This has some resemblance with the dual approach of ref. 21. $\hat{\mathcal{A}}^+$ is also[1] a comodule algebra of H^o (the Hopf algebra H with opposite product).

We can postulate (22-23) also for an interacting field, if σ is a realization of H' as an algebra of operators on the corresponding Hilbert space.

References

1. G. Fiore, "Covariance of QFT on Grönewold-Moyal-Weyl spaces under finite deformed Poincaré transformations", in preparation.
2. W. Heisenberg, *Letter of Heisenberg to Peierls* (1930), in: Wolfgang Pauli, Scientific Correspondence, vol. II, 15, Ed. Karl von Meyenn, Springer (1985).
3. R. Oeckl, Nucl. Phys. **B581** (2000), 559-574.
4. V. G. Drinfeld, Sov. Math. Dokl. **28** (1983), 667.
5. M. Chaichian, P. Kulish, K. Nishijima, A. Tureanu, Phys.Lett.**B604** (2004), 98
6. J. Wess, "Deformed Coordinate Spaces; Derivatives", hep-th/0408080
7. F. Koch, E. Tsouchnika, Nucl.Phys. **B717** (2005), 387-403.
8. E. Abe, *Hopf Algebras*, Cambridge University Press (1980).
9. V. Chari, A. Pressley, *A Guide to Quantum Groups*, Cambridge Univ. Press (1994).
10. G. Fiore, J. Wess, Phys. Rev. **D75** (2007), 105022.
11. G.Fiore, in *Quantum Field Theory and Beyond*, p. 64. World Scientific (2008).
12. M. Chaichian, P. Presnajder, A. Tureanu, Phys. Rev. Lett. **94** (2005), 151602.
13. A. P. Balachandran, G. Mangano, A. Pinzul, S. Vaidya, Int. J. Mod. Phys. **A21** (2006), 3111.
14. F. Lizzi, S. Vaidya, P. Vitale, Phys. Rev. **D73** (2006), 125020.
15. J. Bu, H. Kim, Y. Lee, C. Vac, J. Yee, Phys. Rev. **D73** (2006), 125001.
16. Y. Abe, Int. J. Mod. Phys. **A22** (2007), 1181.
17. A. P. Balachandran, T. R. Govindarajan, G. Mangano, A. Pinzul, B. A. Qureshi, S. Vaidya, Phys. Rev. **D75** (2007), 045009.
18. J. Zahn, Phys. Rev. **D73** (2006), 105005.
19. A. P. Balachandran, A. Ibort, G. Marmo, M. Martone, JHEP**1103** (2011), 057.
20. G. Piacitelli, Commun. Math. Phys. **295** (2010), 701-729.
21. J. Lukierski, M. Woronowicz, Int. Jour. Mod. Phys. **A27** (2012), 1250084; electronic form in IJMP Conf. Series vol. **13** (2012) 127-140.
22. V. Lyubashenko, A. Sudbery, J. Math. Phys. **39** (1998), 3487-3504.
23. P.Aschieri, M.Dimitrijevic, F.Meyer, J.Wess, Class.Quant.Grav.**23** (2006), 1883.
24. G. Fiore, J. Phys. A: Math. Theor. **43** (2010), 155401.

HALF-QUANTUM LINEAR ALGEBRA

A. ISAEV

Bogoliubov Laboratory of Theoretical Physics, JINR
141980 Dubna, Moscow Region, Russia
E-mail: isaevap@theor.jinr.ru

O. OGIEVETSKY*

Center of Theoretical Physics
Aix Marseille Université, CNRS, UMR 7332, 13288 Marseille, France
Université de Toulon, CNRS, UMR 7332, 83957 La Garde, France
E-mail: oleg@cpt.univ-mrs.fr

The Cayley–Hamilton–Newton theorem for half-quantum matrices is proven.

Keywords: Yang–Baxter, quantum matrix, Cayley–Hamilton, Newton identities.

1. Introduction

The Cayley–Hamilton theorem in the linear algebra and the Newton identities in the theory of symmetric functions have a common predecessor, discovered in[3] and called there Cayley–Hamilton–Newton (CHN) identities. These are matrix identities, which naturally include the Cayley–Hamilton identity and whose trace reproduces the Newton identities. The CHN identities admit a q-deformation, see Ref. 3. Later,[4] the CHN identities were established for a larger class of quantum matrix algebras which includes both the algebra of functions on a quantum group of GL type and the deformed universal enveloping algebra of $\mathfrak{gl}$ type. This class of algebras, defined with the help of a compatible pair $(\hat{R}, \hat{F})$ of R-matrices (see the precise definition in Section 2), can be described differently, as braided Hopf algebras universally coacting (in the braided sense) on two, left and right, quantum spaces. When the R-matrix $\hat{F}$ is the flip, this is the usual universally coacting Hopf algebra.

---

*On leave of absence from P. N. Lebedev Phys. Inst., Leninsky Pr. 53, Moscow, Russia.

There is, nowadays, a certain interest in the algebras universally coacting on just one quantum space, see[1] and the literature cited there. We give the definition of the braided Hopf algebra $\mathcal{M}_{SA}(\hat{R}, \hat{F})$ universally coacting (again in the braided sense) on one quantum space. The definition appeals again to a compatible pair $(\hat{R}, \hat{F})$ of R-matrices. The matrix whose entries generate $\mathcal{M}_{SA}(\hat{R}, \hat{F})$ is called *half-quantum* (HQ-)matrix. An example of HQ-matrices arises naturally in the theory of quantum groups of GL type with spectral parameters, see Section 2.

The main result of the present article is the CHN theorem for the HQ-matrices defined by a pair $(\hat{R}, \hat{F})$ where $\hat{R}$ is of Hecke type, that is, the spectral decomposition of $\hat{R}$ has two projectors, S and A. The algebras $\mathcal{M}_{SA}(\hat{R}, \hat{F})$ coact on the "right" quantum space associated to S. The formal exchange of projectors leads to the statements about the algebras coacting on the "left" quantum space, associated to S, so it is enough to investigate the algebras $\mathcal{M}_{SA}(\hat{R}, \hat{F})$.

The coefficients of the CHN identities are analogues of the sets of elementary and complete symmetric functions. Contrary to the algebra generated by the full quantum matrix, none of these sets is commutative for HQ-matrices. As for full quantum matrices, the CHN theorem for HQ-matrices uses modified matrix powers. We define four, in general, different powers, $M^{\rightarrow k}$ and $M^{\leftarrow k}$, of an HQ-matrix M; if M is a full quantum matrix then $M^{\rightarrow k} = M^{\leftarrow k}$. As a direct consequence of the CHN theorem we obtain the Cayley–Hamilton theorem and Newton identities for HQ-matrices.

The HQ-matrices can be defined in a more general setting, when one knows which eigen-projectors of an R-matrix belong to the symmetric and anti-symmetric part (see Section 3 in[5] for a general discussion). This is the case, in particular, of the R-matrices for simple Lie groups (in any representation). The CHN theorem for full quantum matrices is known[6] in the simplest situation, for the orthogonal and symplectic groups in the defining representation and it will be interesting to generalize it for the HQ-matrices.

Notation. Let V be a finite-dimensional $\mathbb{C}$-vector space and $V^{\otimes M} := V \otimes V \otimes \dots$ (M times, $M = 0, 1, 2, \dots$; by convention, $V^{\otimes 0} := \mathbb{C}$). Let $\mathfrak{A}$ be an associative algebra and X an $\mathfrak{A}$-valued operator on V (that is, the matrix elements of X belong to $\mathfrak{A}$). We denote by X_j the operator which acts as X on the j-th copy of the space V in $V \otimes V \otimes \dots$ (and as the identity on other copies); for an $\mathfrak{A}$-valued operator Y on $V \otimes V$, we denote by $Y_{j,k}$ the operator which acts as Y on the j-th and k-th copies of the space V in $V \otimes V \otimes \dots$ (and as the identity on other copies) *etc.* We write sometimes $Y_j := Y_{j,j+1}$ for brevity. The identity operator on $V^{\otimes M}$ we denote by Id when the value of M is clear from the context. We denote by $P \in \mathrm{Aut}\,(V \otimes V)$ the permutation operator, $P(v \otimes w) := w \otimes v$. The operation of taking traces in copies of V with the numbers $i_1 \dots i_k$ is denoted by $\mathrm{tr}_{\,(i_1 \dots i_k)}$. For an $\mathfrak{A}$-valued operator Z on $V^{\otimes m}$ define the operator $Z^{\uparrow k}$, $k = 0, 1, 2, \dots$, in $V^{\otimes m+k}$ by $Z^{\uparrow k} := \mathrm{Id}^{\otimes k} \otimes Z$. We use symmetric q-numbers, $j_q := q^{j-1} + q^{j-3} + \cdots + q^{-j+1}$ for $j = 0, 1, \dots$ (by convention, the empty sum is 0). Given a numerical q, we shall say that a positive integer j is q-admissible if $k_q \neq 0$ for $k = 1, 2, \dots, j$.

2. Generalized half-quantum matrices

2.1. *Operators $\hat{R}$ and $\hat{F}$*

Generalized half-quantum matrix algebras are defined with the help of two operators $\hat{R}, \hat{F} \in \mathrm{Aut}\,(V \otimes V)$ which satisfy, depending on a property in question, to a combination of conditions **(i)**-**(iv)** below.

(i) $(\hat{R}, \hat{F})$ is a *compatible pair* of R-matrices:

$$\hat{R}_{12}\hat{R}_{23}\hat{R}_{12} = \hat{R}_{23}\hat{R}_{12}\hat{R}_{23}\,, \qquad \hat{F}_{12}\hat{F}_{23}\hat{F}_{12} = \hat{F}_{23}\hat{F}_{12}\hat{F}_{23}\,, \tag{1}$$

$$\hat{R}_{12}\hat{F}_{23}\hat{F}_{12} = \hat{F}_{23}\hat{F}_{12}\hat{R}_{23}\,, \qquad \hat{F}_{12}\hat{F}_{23}\hat{R}_{12} = \hat{R}_{23}\hat{F}_{12}\hat{F}_{23}\,. \tag{2}$$

(ii) The R-matrix $\hat{R}$ is of *Hecke type*: its projector decomposition is $\hat{R} = qS^{(2)} - q^{-1}A^{(2)}$ where $q \in \mathbb{C}^*$ is such that 2 is q-admissible. The projectors $S^{(2)}$ and $A^{(2)}$ (called, respectively, q-symmetrizer and q-antisymmetrizer) are complementary, $S^{(2)} + A^{(2)} = \mathrm{Id}$.

The higher q-symmetrizers and q-antisymmetrizers $S^{(k+1)}, A^{(k+1)} \in \mathrm{End}\,(V^{\otimes k+1})$, $k > 1$, can be defined, if $k+1$ is q-admissible, in several ways, *e.g.*, inductively, in any of the forms

$$A^{(k)\uparrow 1}\Big(q^k\,\mathrm{Id} - k_q\hat{R}_1\Big)A^{(k)\uparrow 1} =: (k+1)_q A^{(k+1)} := A^{(k)}\Big(q^k\,\mathrm{Id} - k_q\hat{R}_k\Big)A^{(k)}\,, \tag{3}$$

$$S^{(k)\uparrow 1}\Big(q^{-k}\,\mathrm{Id} + k_q\hat{R}_1\Big)S^{(k)\uparrow 1} =: (k+1)_q S^{(k+1)} := S^{(k)}\Big(q^{-k}\,\mathrm{Id} + k_q\hat{R}_k\Big)S^{(k)}\,. \tag{4}$$

By convention, $A^{(1)} := \mathrm{Id}$ and $S^{(1)} := \mathrm{Id}$; then (3) and (4) hold for $k = 1$ as well.

(iii) The Hecke R-matrix $\hat{R}$ is *even of height*[2] n: n is q-admissible, $\mathrm{rank}\,(A^{(n)}) = 1$ and $A^{(n)}(q^n\,\mathrm{Id} - n_q\hat{R}_n)A^{(n)} = 0$.

(iv) The R-matrix $\hat{F}$ is invertible and *skew-invertible*: there exists $\Psi \in \mathrm{Aut}\,(V \otimes V)$ satisfying $\mathrm{tr}_{\,2}(\Psi_{12}\hat{F}_{23}) = P_{13}$.

2.2 F-trace. The *quantum trace* of a matrix X with arbitrary entries is $\mathrm{tr}_{\hat{F}}(X) := \mathrm{tr}\,(DX)$, where $D \in \mathrm{End}\,(V)$ is defined by $D_1 := \mathrm{tr}\,_2(\Psi_{12})$. We have

$$\mathrm{tr}_{\hat{F}\,(2)}(\hat{F}_1) = \mathrm{Id}_1\ ,\ \hat{F}_1 D_1 D_2 = D_1 D_2 \hat{F}_1\ ,\ \mathrm{tr}_{\hat{F}\,(2)}(\hat{F}_1^{\epsilon}\, X_1\, \hat{F}_1^{-\epsilon}) = \mathrm{Id}_1\, \mathrm{tr}_{\hat{F}}(X)\ ,\ \epsilon = \pm 1 .$$

2.3 Let M be an operator on V with arbitrary entries. Define inductively

$$M_{\overline{1}} := M_1\ , \qquad M_{\overline{k+1}} := \hat{F}_{k,k+1} M_{\overline{k}} \hat{F}^{-1}_{k,k+1}\ . \tag{5}$$

Definition. *Let $\mathcal{M}_{SA}(\hat{R}, \hat{F})$ be the unital associative algebra generated by the entries of M with the defining relations*

$$S^{(2)}_{12}\, M_{\overline{1}}\, M_{\overline{2}}\, A^{(2)}_{12} = 0\ . \tag{6}$$

We call M the half-quantum SA*-matrix (*HQ*-matrix, for brevity).*

To explain the meaning of the definition, let $\mathcal{X}_S$ be the algebra generated by components $\{x^a\}$ of a vector from V with the defining quadratic relations $(S^{(2)})^{ab}_{cd} x^c x^d = 0$ (the Einstein summation convention is assumed throughout the text). The HQ-matrix coacts, $x^a \to M^a_b x^b$, on the algebra $\mathcal{X}_S$ and generates the braided Hopf algebra universally coacting on $\mathcal{X}_S$. The braiding between the entries of M and x is given by $X_1 M_2 = M_{\overline{2}} X_1$, where $X^a_b := c_b x^a$, c_b are independent commuting variables. As for the braided coproduct, let M and M' be HQ-matrices which satisfy

$$\hat{F}_{12} M_1 \hat{F}_{12}^{-1} M'_1 = M'_1 \hat{F}_{12} M_1 \hat{F}_{12}^{-1}\ .$$

Then MM' is again an HQ-matrix.

Notation. For natural l and k, $l \leqslant k$, set $M_{\overline{l\to k}} := M_{\overline{l}} \ldots M_{\overline{k}}$, $\hat{F}_{l\to k} := \hat{F}_l \ldots \hat{F}_{k-1}$, $\hat{R}_{l\to k} := \hat{R}_l \ldots \hat{R}_{k-1}$ and $\hat{R}_{k\leftarrow l} := \hat{R}_{k-1} \ldots \hat{R}_l$ (by convention, the empty product is 1).

2.4 Let $(\hat{R}, \hat{F})$ be a pair satisfying the conditions **(i)** and **(iv)**, Subsection **2.1**. The following Lemma, valid for a matrix M with arbitrary entries, was proved in Ref. 4.

Lemma 1. ***a)*** *We have*

$$\hat{F}_i M_{\overline{k}} = M_{\overline{k}} \hat{F}_i \quad \text{and} \quad \hat{R}_i M_{\overline{k}} = M_{\overline{k}} \hat{R}_i \quad \text{for } k \neq i, i+1\ , \tag{7}$$

$$\hat{F}_{i\to k+1} M_{\overline{i\to k}} = M_{\overline{i+1\to k+1}} \hat{F}_{i\to k+1} \quad \text{for } i \leq k\ . \tag{8}$$

b) *Let $\alpha(Y^{(k)}) := \mathrm{tr}_{\hat{F}\,(1,\ldots,k)}(Y^{(k)} M_{\overline{1\to k}})$ where $Y^{(k)}$ is a polynomial in $\hat{R}_1, \ldots, \hat{R}_{k-1}$. Then*

$$\mathrm{tr}_{\hat{F}\,(i+1,\ldots,i+k)}(Y^{(k)\uparrow i} M_{\overline{i+1\to i+k}}) = \mathrm{Id}_{1,\ldots,i}\, \alpha(Y^{(k)})\ . \tag{9}$$

2.5 Assume now that M is an HQ-matrix. Since $\hat{R}_1 = qS^{(2)} - q^{-1}A^{(2)}$ and $S^{(2)} + A^{(2)} = \mathrm{Id}$, the relations (6) can be rewritten in the four following equivalent (when $q + q^{-1} \neq 0$) forms

$$\begin{aligned} \hat{R}_1 M_{\overline{1}} M_{\overline{2}} A^{(2)} &= -q^{-1}\, M_{\overline{1}} M_{\overline{2}} A^{(2)} \,, \quad S^{(2)} M_{\overline{1}} M_{\overline{2}} \hat{R}_1 = q\, S^{(2)} M_{\overline{1}} M_{\overline{2}} \,, \\ A^{(2)} M_{\overline{1}} M_{\overline{2}} A^{(2)} &= M_{\overline{1}} M_{\overline{2}} A^{(2)} \,, \quad S^{(2)} M_{\overline{1}} M_{\overline{2}} S^{(2)} = S^{(2)} M_{\overline{1}} M_{\overline{2}} \,. \end{aligned} \tag{10}$$

By induction on $k = 1, 2, \ldots$ we obtain, using Lemma 1,

$$S^{(2)}_{j,j+1} M_{\overline{j}}\, M_{\overline{j+1}} A^{(2)}_{j,j+1} = 0 \,, \tag{11}$$

$$A^{(k)\uparrow i} M_{\overline{i+1 \to i+k}} A^{(k)\uparrow i} = M_{\overline{i+1 \to i+k}} A^{(k)\uparrow i} \,, \; i = 0, 1, 2, \ldots \tag{12}$$

$$S^{(k)\uparrow i} M_{\overline{i+1 \to i+k}} S^{(k)\uparrow i} = S^{(k)\uparrow i} M_{\overline{i+1 \to i+k}} \,, \; i = 0, 1, 2, \ldots \tag{13}$$

In eqs.(12)-(13), k is assumed to be q-admissible.

2.6 Example. Let $(\hat{R}, \hat{F})$ be a compatible pair with $\hat{R}$ of Hecke type. Let $\hat{R}(u) := \hat{R} + (q - q^{-1})/(u^2 - 1)$ be the Baxterizaton of $\hat{R}$ and let $T(u)$ be the matrix generating the algebra with the defining relations $\hat{R}_{12}(u/v) T_{\overline{1}}(u) T_{\overline{2}}(v) = T_{\overline{1}}(v) T_{\overline{2}}(u) \hat{R}_{12}(u/v)$ (in the simplest cases, this can be an algebra related to quantum or classical (super)-Yangians or affine algebras). Now, $\hat{R}(q)$ is proportional to the symmetrizer, so setting $u = qv$, multiplying from the right by $A^{(2)}_{12}$ and writing $T(qv) = q^{v\partial_v} T(v) q^{-v\partial_v}$ we find $S^{(2)}_{12} q^{v\partial_v} T_{\overline{1}}(v) q^{-v\partial_v} T_{\overline{2}}(v) A^{(2)}_{12} = 0$, which implies that $q^{-v\partial_v} T$ and $T q^{-v\partial_v}$ are HQ-matrices, $S^{(2)}_{12} q^{-v\partial_v} T_{\overline{1}}(v) q^{-v\partial_v} T_{\overline{2}}(v) A^{(2)}_{12} = 0$ and $S^{(2)}_{12} T_{\overline{1}}(v) q^{-v\partial_v} T_{\overline{2}}(v) q^{-v\partial_v} A^{(2)}_{12} = 0$.

3. Cayley–Hamilton–Newton theorem

3.1. *Symmetric functions*

Let M be an HQ-matrix. Define four sets, $\{s_k(M)\}$, $\{\overline{s}_k(M)\}$, $\{\sigma_k(M)\}$ and $\{\tau_k(M)\}$, $k = 0, 1, 2, \ldots$, of elements of the algebra $\mathcal{M}_{SA}(\hat{R}, \hat{F})$ by $s_0(M) = \overline{s}_0(M) = \sigma_0(M) = \tau_0(M) = 1$ and, for $k > 0$,

$$s_k(M) := \mathrm{tr}_{\hat{F}(1\ldots k)}(\hat{R}_{1\to k} M_{\overline{1\to k}}) \,, \quad \overline{s}_k(M) := \mathrm{tr}_{\hat{F}(1\ldots k)}(\hat{R}_{k\leftarrow 1} M_{\overline{1\to k}}) \,, \tag{14}$$

$$\sigma_k(M) := q^k\, \mathrm{tr}_{\hat{F}(1\ldots k)}(A^{(k)} M_{\overline{1\to k}}) \,, \quad \tau_k(M) := q^{-k}\, \mathrm{tr}_{\hat{F}(1\ldots k)}(S^{(k)} M_{\overline{1\to k}}) \,. \tag{15}$$

3.2. *Powers of* HQ*-matrices*

Define the matrices $M^{\to k}$, $\tilde{M}^{\to k}$, $M^{\leftarrow k}$ and $\tilde{M}^{\leftarrow k}$, *k-th powers* of M, by

$$(M^{\to k})_1 := \mathrm{tr}_{\hat{F}(2,\ldots,k)}(M_{\overline{1\to k}} \hat{R}_{1\to k}) \,, \quad (M^{\leftarrow k})_1 := \mathrm{tr}_{\hat{F}(2,\ldots,k)}(\hat{R}_{k\leftarrow 1} M_{\overline{1\to k}}) \,, \tag{16}$$

$$(\tilde{M}^{\rightarrow k})_1 := \mathrm{tr}_{\hat{F}(2,\ldots,k)}(\hat{R}_{1\rightarrow k} M_{\overline{1\rightarrow k}}), \quad (\tilde{M}^{\leftarrow k})_1 := \mathrm{tr}_{\hat{F}(2,\ldots,k)}(M_{\overline{1\rightarrow k}} \hat{R}_{k\leftarrow 1}). \tag{17}$$

Define the k-th wedge powers $M^{\wedge k}$, $\tilde{M}^{\wedge k}$ and the k-th symmetric powers M^{sk}, $\tilde{M}^{sk}$ of M by

$$(M^{\wedge k})_1 := \mathrm{tr}_{\hat{F}(2,\ldots,k)}(A^{(k)} M_{\overline{1\rightarrow k}}), \quad (M^{sk})_1 := \mathrm{tr}_{\hat{F}(2,\ldots,k)}(M_{\overline{1\rightarrow k}} S^{(k)}), \tag{18}$$

$$(\tilde{M}^{\wedge k})_1 := \mathrm{tr}_{\hat{F}(2,\ldots,k)}(M_{\overline{1\rightarrow k}} A^{(k)}), \quad (\tilde{M}^{sk})_1 := \mathrm{tr}_{\hat{F}(2,\ldots,k)}(S^{(k)} M_{\overline{1\rightarrow k}}). \tag{19}$$

3.3. *Cayley–Hamilton–Newton theorem for* **HQ**-*matrices* (Cf.[3])

Assume that the pair $(\hat{R}, \hat{F})$ satisfies conditions **(i)**, **(ii)** *and* **(iv)**, *Subsection* ***2.1***, *and assume that the positive integer j is q-admissible. The* HQ-*matrix satisfies the following matrix identities*

$$j_q\, M^{\wedge j} = \sum_{k=0}^{j-1} (-1)^{j-k+1} M^{\leftarrow j-k} \sigma_k(M)\ , \tag{20}$$

$$j_q M^{sj} = \sum_{k=0}^{j-1} M^{\rightarrow j-k}\ \tau_k(M)\,, \tag{21}$$

$$j_q\, \tilde{M}^{\wedge j} = \sum_{k=0}^{j-1} (-1)^{j-k+1} \tilde{M}^{\leftarrow j-k} \sigma_k(M)\ , \tag{22}$$

$$j_q \tilde{M}^{sj} = \sum_{k=0}^{j-1} \tilde{M}^{\rightarrow j-k}\ \tau_k(M)\,. \tag{23}$$

We sketch the *proof* of (20). For k, $1 \le k < j$, rewrite $(M^{\leftarrow j-k}\sigma_k(M))_1$ as

$$\begin{aligned} & q^k \mathrm{tr}_{\hat{F}(2\ldots j-k)}(\hat{R}_{j-k\leftarrow 1} M_{\overline{1\rightarrow j-k}}) \mathrm{tr}_{\hat{F}(1\ldots k)}(A^{(k)} M_{\overline{1\rightarrow k}}) \\ &= q^k \mathrm{tr}_{\hat{F}(2\ldots j)}(A^{(k)\uparrow j-k} \hat{R}_{j-k\leftarrow 1} M_{\overline{1\rightarrow j-k}} M_{\overline{j-k+1\rightarrow j}}) \\ &= q^k \mathrm{tr}_{\hat{F}(2\ldots j)}(A^{(k)\uparrow j-k} \hat{R}_{j-k\leftarrow 1} M_{\overline{1\rightarrow j}})\ . \end{aligned} \tag{24}$$

Here we used (9). We use the recurrence for the q-antisymmetrizers in the form $q^k A^{(k)\uparrow 1} = (k+1)_q A^{(k+1)} + k_q A^{(k)\uparrow 1} \hat{R}_1 A^{(k)\uparrow 1}$ to rewrite the last expression in (24) as

$$\begin{aligned} & (k+1)_q \mathrm{tr}_{\hat{F}(2\ldots j)}(A^{(k+1)\uparrow j-k-1} \hat{R}_{j-k\leftarrow 1} M_{\overline{1\rightarrow j}}) \\ & \quad + k_q \mathrm{tr}_{\hat{F}(2\ldots j)}(A^{(k)\uparrow j-k} \hat{R}_{j-k} A^{(k)\uparrow j-k} \hat{R}_{j-k\leftarrow 1} M_{\overline{1\rightarrow j}})\ . \end{aligned} \tag{25}$$

In the second term of (25), the right $A^{(k)\uparrow j-k}$ commutes with $\hat{R}_{j-k\leftarrow 1}$ and the left $A^{(k)\uparrow j-k}$ can be cyclically moved. So, the last term takes the form

$$\begin{aligned} & k_q \mathrm{tr}_{\hat{F}(2\ldots j)}(\hat{R}_{j-k+1\leftarrow 1} A^{(k)\uparrow j-k} M_{\overline{1\to j}} A^{(k)\uparrow j-k}) \\ &= k_q \mathrm{tr}_{\hat{F}(2\ldots j)}(\hat{R}_{j-k+1\leftarrow 1} M_{\overline{1\to j}} A^{(k)\uparrow j-k}) \\ &= k_q \mathrm{tr}_{\hat{F}(2\ldots j)}(A^{(k)\uparrow j-k} \hat{R}_{j-k+1\leftarrow 1} M_{\overline{1\to j}}) \ . \end{aligned}$$

We used (12) in the first equality; in the second equality, we moved $A^{(k)}$ to the left using the cyclic property of the trace. Finally, we obtain

$$\begin{aligned} (M^{\leftarrow j-k}\sigma_k(M))_1 = (k+1)_q \mathrm{tr}_{\hat{F}(2\ldots j)}(A^{(k+1)\uparrow j-k-1} \hat{R}_{j-k\leftarrow 1} M_{\overline{1\to j}}) \\ + k_q \mathrm{tr}_{\hat{F}(2\ldots j)}(A^{(k)\uparrow j-k} \hat{R}_{j-k+1\leftarrow 1} M_{\overline{1\to j}}) \ . \end{aligned} \tag{26}$$

Eq.(20) follows from the expressions (26) for $M^{\leftarrow j-k}\sigma_k(M)$, $k = 1, 2, \ldots, j-1$, and $M^{\leftarrow j}\sigma_0(M) = M^{\leftarrow j}$. ■

3.4. *Corollaries*

Newton relations. Taking the trace in (20), we find, for a q-admissible j,

$$q^{-j} j_q \sigma_j(M) = \sum_{k=0}^{j-1} (-1)^{j-k+1}\, \overline{s}_{j-k}(M)\sigma_k(M) \ , \tag{27}$$

$$q^j j_q \tau_j(M) = \sum_{k=0}^{j-1} s_{j-k}(M)\tau_k(M) \ . \tag{28}$$

Cayley-Hamilton theorem. Assume that $\hat{R}$ is even of height n. Then there exist tensors $\epsilon_{a_1 a_2 \ldots a_n}$ and $\epsilon^{a_1 a_2 \ldots a_n}$ such that

$$(A^{(n)})^{b_1\ldots b_n}_{a_1\ldots a_n} = \epsilon_{a_1\ldots a_n}\epsilon^{b_1\ldots b_n} \quad \text{and} \quad \epsilon_{a_1\ldots a_n}\epsilon^{a_1\ldots a_n} = 1 \ .$$

By (12) for $k = n$ and $i = 0$, we have

$$(M_{\overline{1\to n}})^{a_1\ldots a_n}_{b_1\ldots b_n}\epsilon^{b_1\ldots b_n} = \det_q(M)\epsilon^{a_1\ldots a_n} \ , \tag{29}$$

where the *quantum determinant* of the HQ-matrix M is defined, up to a normalization factor, by

$$\det_q(M) := \epsilon_{a_1\ldots a_n}(M_{\overline{1\to n}})^{a_1\ldots a_n}_{b_1\ldots b_n}\,\epsilon^{b_1\ldots b_n} \ .$$

Eq.(29) implies that

$$\tilde{M}^{\wedge n} = \det_q(M)\mathcal{D} \ ,$$

where the matrix $\mathcal{D}$ is defined by

$$\mathcal{D}_1 := \mathrm{tr}_{\hat{F}(2\ldots n)}(A^{(n)}) \ .$$

Taking $j = n$ in (22) we obtain the Cayley–Hamilton theorem for the half-quantum matrices:

$$\sum_{k=0}^{n-1}(-1)^{n-k}\tilde{M}^{\leftarrow n-k}\sigma_k(M) + n_q\mathrm{det}_q(M)\mathcal{D} = 0\ . \tag{30}$$

Standard $\hat{R}$. Let $(\hat{R}_{DJ}, P)$ be the compatible pair with the standard multi-parametric Drinfeld–Jimbo R-matrix in dimension n. In[8] , a version of the Cayley–Hamilton theorem, with usual matrix powers but diagonal matrices as coefficients, has been proven. Namely, the HQ-matrix M satisfies the matrix identity

$$M^n + \sum_{k=1}^{n}(-1)^k\Sigma_k M^{n-k} = 0\,, \tag{31}$$

with certain diagonal matrices Σ_m, $m = 1, \dots, n$. By the same technics as in[7] one can relate modified powers of M - in this situation, i.e., for the compatible pair $(\hat{R}_{DJ}, P)$ - with usual powers and deduce (31) from (30).

References

1. A. Chervov, G. Falqui and V. Rubtsov, Adv. Appl. Math. **43** (2009) 239–315.
2. D. I. Gurevich, Leningrad Math. J. **2** (1991) 801– 828.
3. A. P. Isaev, O. V. Ogievetsky and P. N. Pyatov, Czech. J. Phys. **48** (1998) 1369–1374.
4. A. P. Isaev, O. V. Ogievetsky and P. N. Pyatov, J. Phys. A: Math. Gen. **32** (1999) L115–L121.
5. O. V. Ogievetsky, Contemp. Math. **294**, Amer. Math. Soc., Providence, RI (2002) 161-232.
6. O. Ogievetsky and P. Pyatov, Orthogonal and symplectic quantum matrix algebras and Cayley-Hamilton theorem for them. ArXiv: math.QA/0511618
7. O. Ogievetsky, A. Vahlas, Lett. Math. Phys. **65** (2003) 49–57.
8. J. J. Zhang, J. Pure Appl. Algebra **129** (1998) 101–109.

INTEGRABLE SPIN CHAINS AND REPRESENTATION THEORY

P. P. KULISH*

St. Petersburg Department of Steklov Mathematical Institute, Russia
** E-mail: kulish@euclid.pdmi.ras.ru*

A series of integrable spin chains related to special tensor product representations of Temperley - Lieb algebra is constructed. The generator of the TL-algebra is a rank one projector acting on $\mathbb{C}^n \otimes \mathbb{C}^n$. The spectra of these spin chains with appropriate boundary conditions are the same and equivalent to the spin 1/2 XXZ-model while the multiplicities of the eigenvalues are different and depend on n. The generators of the TL-algebra with rankX_j bigger than 1 are also mentioned.

Keywords: QISM, integrable spin chains, Yang - Baxter equation, Temperley - Lieb algebra.

Quantum inverse scattering method (QISM) gives rise to the algebraic treatment of the famous isotropic Heisenberg spin $\frac{1}{2}$ chain (XXX-model), and to construction of higher spin chains $s = 1, \frac{3}{2}, 2, \ldots$ preserving integrability, and application of the algebraic Bethe Ansatz.[1–3] The method can be applied to new integrable models corresponding to simple Lie algebras and superalgebras, their deformations: quantum groups, and their representations. Essential role in the QISM is played by R-matrices which are solutions to the Yang - Baxter equation.[4,5] Already analysis of the $XXX_{1/2}$-model results in symmetry algebra $U(\mathfrak{sl}(2))$, and the group algebra $\mathbb{C}[\mathcal{S}_N]$ of the symmetric group $\mathcal{S}_N$ which interchanges local state spaces $\mathbb{C}^2_j$, $j = 1, 2, \ldots, N$:

$$\begin{aligned} H_{XXX} &= \sum_{m=1}^{N} \left(\sigma^x_m \sigma^x_{m+1} + \sigma^y_m \sigma^y_{m+1} + (\sigma^z_m \sigma^z_{m+1} - 1)\right) \\ &= 2 \sum_{m=1}^{N} \left(\mathcal{P}_{m\, m+1} - I_{m\, m+1}\right). \end{aligned}$$

$$\mathcal{H} = \bigotimes_{m=1}^{N} \mathbb{C}^2_m . \qquad H_{XXX} \in \mathbb{C}[\mathcal{S}_N].$$

The global action of $U(\mathfrak{sl}(2)$ on the space of states $\mathcal{H}$ is given by the generators:

$$S^z = \frac{1}{2}\sum_{m=1}^{N} \sigma^z_m, \qquad S^{\pm} \sum_{m=1}^{N} \sigma^{\pm}_m,$$

while the action of $\mathbb{C}[\mathcal{S}_N]$ represented by the permutation operators $\mathcal{P}(v \otimes w) = (w \otimes v), \quad v, w \in \mathbb{C}^2$. These two actions commute in between, and the global $U(\mathfrak{sl}(2))$ and $\mathbb{C}[\mathcal{S}_N]$ are mutual centralisers (the Shur - Weyl duality).[6] The higher integrals of motion are also $U(\mathfrak{sl}(2))$ invariant and they are elements of $\mathbb{C}[\mathcal{S}_N]$. Similar situation takes place for anisotropic XXZ-model. However, one has to impose a special boundary condition

$$H_{XXZ} = \sum_{k=1}^{N-1} \left(\sigma^x_k \sigma^x_{k+1} + \sigma^y_k \sigma^y_{k+1} + \cosh\eta\, (\sigma^z_k \sigma^z_{k+1} - 1) + \right.$$
$$\left. \sinh\,\eta\, (\sigma^z_k - \sigma^z_{k+1})\right) .$$

This Hamiltonian is $U_q(\mathfrak{sl}(2))$ invariant ($q := e^{\eta}$) while the centralizer of the $U_q(\mathfrak{sl}(2))$ action on $\mathcal{H}$ is a deformation of $\mathbb{C}[\mathcal{S}_N]$ — the Hecke algebra $\mathfrak{H}_N(q)$ or more precisely the Temperley - Lieb algebra $\mathrm{TL}_N(q)$.[7] The Hamiltonian densities are the generators of this centralizer. Generalization of these models to higher spin $s = 1, 3/2, \ldots$ results in corresponding generalizations of centralizer algebras. The Hamiltonian of the spin 1 XXX-model is

$$H^{(s)}_{XXX} = \sum_{m=1}^{N-1} (\vec{S}_m, \vec{S}_{m+1}) - (\vec{S}_m, \vec{S}_{m+1})^2$$

with the space of states $\mathcal{H} = \otimes_{m=1}^{N} \mathbb{C}^3_m$, and the centralizer is the Brauer algebra or the Birman - Wenzl - Murakami algebra in the q-deformed case. These two mutually commuting algebras: a symmetry algebra U_s and its centralizer $\mathcal{C}_N$, give rise to the multiplicity free decomposition of the space of states on irreducible representations V_l of U_s and W_l of $\mathcal{C}_N$:

$$\mathcal{H} = \bigotimes_{m=1}^{N} \mathbb{C}^k_m = \bigoplus_{l} V_l \otimes W_l .$$

In the case of the XXX-model $U_s = U(\mathfrak{sl}(2))$, $\mathcal{C}_N = \mathbb{C}[\mathcal{S}_N]$ and the Clebsch - Gordan decomposition is

$$\mathcal{H} = \bigotimes_{m=1}^{N} \mathbb{C}^2_m = \bigoplus_{k}^{N} V_k \otimes W_k, \quad \dim V_k = k+1, \quad \dim W_k = \binom{N}{k} - \binom{N}{k-1} .$$

The latter decomposition is a consequence of the ring structure of the finite dimensional representations of $U(\mathfrak{sl}(2))$ with defining relations for irreducible representations

$$V_1 \otimes V_k = V_{k+1} \oplus V_{k-1}.$$

Taking the trace of identity operator over this space one gets

$$2p_k = p_{k+1} + p_{k-1}, \qquad p_k = \dim V_k = k+1.$$

Constructing special representation of TL-algebra acting on $\mathcal{H} = \bigotimes_{m=1}^{N} \mathbb{C}_m^n$ with fixed n yields as the centralizer corresponding quantum algebra $U_q(n)$ with representation ring structure isomorphic to the ring of $U(\mathfrak{sl}(2))$. Howeveer, although the defining relations of $U_q(n)$ representations are the same the dimensions of V_k are given by recurrent relations of the second kind Chebyshev polynomials:

$$np_k = p_{k+1} + p_{k-1}, \qquad V_0 = \mathbb{C}, \quad V_1 = \mathbb{C}^n, V_k = p_k(n).$$

Let us remind the definition of Hecke and Temperley - Lieb algebras as quotients of the group algebra of the braid group $\mathcal{B}_N$. The latter one is generated by $(N-1)$ invertible elements $\check{R}_j$, $j = 1, 2, \ldots, N-1$ with defining relations

$$\check{R}_j \check{R}_k \check{R}_j = \check{R}_k \check{R}_j \check{R}_k, \text{ for } |j-k| = 1 \quad \text{and} \quad \check{R}_j \check{R}_k = \check{R}_k \check{R}_j, \text{ for } |j-k| > 1.$$

Additional quadratic characteristic equation

$$\left(\check{R}_j - q\right)\left(\check{R}_j + 1/q\right) = 0$$

yields the (finite dimensional) Hecke algebra $\mathfrak{H}_N(q)$. One can introduce generators X_j proportional to an idempotent

$$\check{R}_j = q\mathbb{I} + X_j, \quad X_j^2 = -\left(q + \frac{1}{q}\right) X_j,$$

and satisfying cubic-linear relations

$$X_j X_k X_j - X_j = X_k X_j X_k - X_k, \quad |j-k| = 1.$$

Require that each side of this relation is zero one gets the TL-algebra $TL_N(q)$ defined by the generators

$$X_j^2 = -(q + \frac{1}{q}) X_j, \quad X_j X_k = X_k X_j, |j-k| > 1, \tag{1}$$

$$X_j X_k X_j = X_j, |j-k| = 1. \tag{2}$$

The tensor product reducible representations of TL-algebra $TL_N(q)$ acting on $\mathcal{H} = \bigotimes_{m=1}^{N} \mathbb{C}_m^n$. can be parametrized by invertible $n \times n$ matrix b.[8] To construct the generators X_j we introduce notation $\bar{b} := b^{-1}$, and interpret these matrices as n^2 component vectors $b_{ab}, \bar{b}_{cd}$. The action of X_j on the space $\mathcal{H}$ being nontrivial on $\mathbb{C}_j^n \otimes \mathbb{C}_{j+1}^n$ is given by $(X_j)_{ab,cd} = b_{ab}\bar{b}_{cd}$ and identity matrix on other factors $\mathbb{C}_m^n$, $m \neq j, j+1$. It is easy to check that relations of X_j are valid with

$$X_j^2 = -\nu(q)X_j, \qquad \nu(q) = (q + \frac{1}{q}) = -trb^t\bar{b}.$$

Constant solution of the braid group relation is

$$\check{R}^{\pm 1} = q^{\pm 1}I + X \in \mathrm{End}_{U_q(n)}\mathbb{C}^n \otimes \mathbb{C}^n.$$

The spectral parameter R-matrix $\check{R}(u)$ is obtained by the Yang - Baxterization procedure

$$\check{R}(u) = u\check{R} - \frac{1}{u}\check{R}^{-1} = \omega(uq)\,(I - P) - \omega(u/q)P,$$

where $P = -(q + 1/q)^{-1}X$ is the rank one projector, and $\omega(q) = q - 1/q$. The degeneracy points $u = q^{\pm 1}$ give rise to fusion procedure, monodromy $T^{(k)}(u)$ and transfer matrices $t^{(k)}(u) = \mathrm{tr}T^{(k)}(u)$, $k = 1, 2, \ldots$ in irreducible representations V_k of quantum algebra $U_q(n)$. The L-operator is $L_{a,\,j}(u) = \mathcal{P}_{a,\,j}\check{R}_{a,\,j}$. One can find a reference state (bare vacuum) $\Omega = \otimes_1^N \omega$ and to prove by the analytic Bethe Ansatz that the spectrum of this integrable spin chain is equivalent to the spectrum of spin $1/2 - XXZ$-model.[8] In the case of spin chains with open boundary Sklyanin's formalism[9] has to be applied. Detailed study of physical characteristics of particular models of this type can be found in papers.[10,11]

Finite dimensional irreducible representations $V_{m\,n}$ of $sl(3)$ are parametrized by highestl weights $m\omega_1 + n\omega_2$. As in the case of $sl(2)$, according to the Clebsch-Gordan decomposition, the recurrent relations for $p_{mn} = \dim V_{mn}$ can be obtained

$$xp_{mn} = p_{m+1n} + p_{m-1n+1} + p_{mn-1},$$

$$yp_{mn} = p_{mn+1} + p_{m+1n-1} + p_{m-1n},$$

$$sl(3): \; x = y = 3, \quad p_{mn} = \frac{1}{2}(m+1)(n+1)(m+n+2), \quad p_{00} = 1.$$

These relations define generalized Chebyshev polynomials $p_{mn}(x, y)$.[12] We do not know R-matrices for ring of representations with these relations and dimentions of the fundamental representations V_{10}, V_{01} different from 3.

Let us point out that there are tensor product representations of $TL_N(q)$ with idempotents of higher rank (two, three, etc.)[13,14]

$$\sqrt{2}X = \begin{pmatrix} 1 & 0 & 0 & i \\ 0 & 1 & i & 0 \\ 0 & -i & 1 & 0 \\ -i & 0 & 0 & 1 \end{pmatrix},$$

$$\check{R}(q) = qP_+ - \frac{1}{q}P_-, \quad \text{rank} P_\pm = 2, \quad q = \frac{i-1}{\sqrt{2}},$$

$$\sqrt{3}X = \left(\begin{array}{ccc|ccc|ccc} 1 & & & & & m & & 1 & \\ & 1 & & m & & & & & m \\ & & 1 & & 1 & & 1 & & \\ \hline & m^2 & & 1 & & & & & 1 \\ & & 1 & & 1 & & 1 & & \\ m^2 & & & & & 1 & & m^2 & \\ \hline & & 1 & & 1 & & 1 & & \\ 1 & & & & & m & & 1 & \\ & m^2 & & 1 & & & & & 1 \end{array}\right), \quad m^3 = 1,$$

$$\check{R}(q) = qP_+ - \frac{1}{q}P_-, \quad \text{rank} P_- = 3, \quad q = \frac{i-\sqrt{3}}{2}.$$

These realizations of $TL_N(q)$ are used[13,14] in the quantum information theory. The paper[15] contains the representations of the quantum algebra defind by the first 4×4 matrix. Special features of these representations are that the parameter q is a root of unity, and corresponding R-matrices are not skew invertible[16]

Acknowledgments

The author thanks the organizers of The XXIX International Colloquium on Group-Theoretical Methods in Physics, August 20-26, 2012 at Chern Institute of Mathematics, Tianjin, China for invitation and warm hospitality. This work is supported by grant RFBR 12-01-00207.

References

1. L. A. Takhtajan and L. D. Faddeev, *Upekhi Mat. Nauk* **34**, No. 5, 13 (1979).
2. P. P. Kulish and E. K. Sklyanin, *Lect. Notes Phys.* **151**, 61 (Springer, 1982).
3. L. D. Faddeev, in *Quantum symetries*, Proceed. Les Houches, Session LXIV, 149 (North-Holland, 1998).
4. C. N. Yang, *Phys. Rev. Lett.* **19**, 1312 (1967); *Phys. Rev.* **168**, 1920 (1968).
5. R. J. Baxter, *Exactly solved models in statistical mechanics*, (Academic Press, San Diego, 1982).
6. L. A. Takhtajan and L. D. Faddeev, *J. Sov. Math.* **24**, 241 (1984).
7. P. P. Martin, *Potts models and related problems in statistical mechanics*, (World Scientific, 1991).
8. P. P. Kulish, *J. Phys. A* **36**, L489 (2003).
9. E. K. Sklyanin, *J. Phys. A* **21**, 2375 (1988).
10. B. Aufgebauern and A. Kluemper, *J. Stat. Mech.* P05018 (2010).
11. A. Doikou, *Nucl. Phys. B* **725**, 493 (2005).
12. T. H. Koornwinder, *Indag. Math.* **36**, 357 (1974).
13. J. L. Chen, K. Xue and M. L. Ge , *Phys. Rev. A* **76**, 042324 (2007).
14. C. Sun, G. Wang, T. Hu, C. Zhou, Q. Wang, K. Xue, *IJQI* **7**, 1285 (2009).
15. D. Arnaudon, A. Chakrabarti, V. Dobrev, S. Mikhov, *Int. J. Mod.Phys. A.* **18**, 4201 (2003).
16. O. V. Ogievetsky, *Contem. Math.* **294**, 161 (2002).

NON-LINEAR DEFORMATIONS OF DUALITY–SYMMETRIC THEORIES

P. PASTI, D. SOROKIN and M. TONIN *

Dipartimento di Fisica e Astronomia Galileo Galilei,
Universitá degli Studi di Padova
and
INFN, Sezione di Padova
Padova, 35131, Italy
** E-mail: pasti@pd.infn.it, sorokin@pd.infn.it, tonin@pd.infn.it*

We discuss general conditions which are imposed on the form of actions for duality-symmetric theories by the requirement of manifest duality symmetry and space-time invariance.

Keywords: duality symmetry; supergravity; quantum field theory.

1. Introduction

Duality symmetry plays an important role in many theoretical models of physical interest. One of such theories is $N = 8$ supergravity[1] which has faced a substantial revival of interest related to its miraculous finiteness up to three and four loops.[2–6] It is assumed that the cancelation of the divergencies in this theory happens not only due to the maximal supersymmetry but also because of the invariance of the theory under duality group $E_{7(7)}$. A radical assumption advocated in[7,8] is that the duality symmetry may be in charge of the perturbative finiteness of $N = 8$ supergravity. However, as was shown in[9–14] supersymmetric and $E_{7(7)}$–invariant higher–order counterterms of the form $\partial^{2k}\mathcal{F}^4$ (where $\mathcal{F}^i$ are Abelian field strengths in the 56–dimensional representation of $E_{7(7)}$) may appear at seven loops. It is, therefore, important to understand how the contribution of these terms to the effective action modify the linear (so–called twisted) self–duality condition obeyed by $\mathcal{F}$ in the classical theory

$$\mathcal{F}^i_{mn} = \frac{1}{2}\Omega^i{}_j\,\varepsilon_{mnpq}\mathcal{F}^{pq\,j} \equiv \Omega^i{}_j\tilde{\mathcal{F}}^i_{mn}\,, \tag{1}$$

where $\Omega^i{}_j$ is the $E_{7(7)}$–invariant symplectic form ($\Omega^2 = -1$). This condition establishes the on–shell relation between 28 'electric' fields F^A appearing in the action and their 28 'magnetic' duals $G_{\bar{A}}$ all together combined in $\mathcal{F}^i = (F^A, G_{\bar{A}})$. It thus restricts the number of independent physical gauge fields to 28.

One can formulate the problem in a more general setting. Namely, what conditions should be satisfied by non-linear terms in the action for the theory to remain duality-symmetric and whether a non–linear deformation of (1) is compatible with supersymmetry? There are two ways of tackling these problems.

In the first approach the action contains only 'electric' fields and is, therefore, not duality–invariant. Duality symmetry manifests itself only on the mass–shell as the linear symmetry of the duality relation similar to eq. (1). This imposes certain conditions on the form of the action and, in particular, on the possible form of non–linear terms, as was originally pointed out in.[15]

In the second approach the action depends on the double number of gauge fields, i.e. the 'electric' and 'magnetic' ones and is manifestly duality invariant. The duality relations between the 'electric' and 'magnetic' fields follow from their equations of motion. The advantage of this approach is that duality symmetry gets lifted to the action level, which is useful *e.g.* for studying its implications in quantum theory or couplings to gravity. The price to pay is subtleties which one should face with the Lorentz invariance of the duality–symmetric actions. In what follows we shall discuss these issues.

2. Duality-invariant action with non-manifest Lorentz invariance

One of the ways to construct duality–symmetric actions is to sacrifice manifest Lorentz covariance. The first action of this type for describing Maxwell theory coupled to monopoles and dions was constructed in.[16] An alternative formulation was considered in[17] and was further developed for duality–symmetric fields in other dimensions in[18] and.[19] For instance, the duality–symmetric Lagrangian for Maxwell theory, which we here discuss for simplicity but without the loss of generality, has the following form

$$L = \frac{1}{8}\mathcal{F}^i_{\mu\nu}\mathcal{F}^{\mu\nu i} + \frac{1}{4}(\mathcal{F}^i_{0a} - \varepsilon^{ij}\tilde{\mathcal{F}}^j_{0a})(\mathcal{F}^{i0a} - \varepsilon^{ij}\tilde{\mathcal{F}}^{j0a})\,, \tag{2}$$

where $\mathcal{F}^i = dA^i$ ($i = 1, 2$) are the field strengths of the two (off–shell independent) gauge fields A^i, electric and magnetic ones. The Lagrangian is manifestly invariant under $SO(2)$ duality rotations of A^i. Its first term is the sum of two Maxwell Lagrangians, while the second term breaks manifest Lorentz invariance, since the $4d$ indices $\mu = 0, 1, 2, 3$ are split into time and 3d space indices $a = 1, 2, 3$ in a non–covariant way. However, the Lagrangian is invariant under modified Lorentz transformations

$$\delta A^i_\mu = \delta_\Lambda A^i_\mu + x^a \Lambda_a{}^0 (\mathcal{F}^i_{0a} - \varepsilon^{ij} \tilde{\mathcal{F}}^j_{0a}) \,, \tag{3}$$

where the first term is the conventional Lorentz rotation with the parameter $\Lambda_\mu{}^\nu$.

The twisted self–duality condition, which relates the one field strength with the Poincaré dual of the second one, is obtained by integrating the second–order equation of motion obtained from the Lagrangian (3)

$$\frac{\delta L}{\delta A^i} = 0 \;\Rightarrow\; \mathcal{F}^i_{\mu\nu} - \varepsilon^{ij} \tilde{\mathcal{F}}^j_{\mu\nu} = 0 \;\Rightarrow\; F^1_{\mu\nu} = \tilde{F}^2_{\mu\nu} \,. \tag{4}$$

One may note that on the mass shell the conventional Lorentz covariance gets restored. In the non–covariant formulation the issue of the non–linear deformation of duality–symmetric theories was revised in.[20]

3. Lorentz covariant duality–symmetric actions

It is always useful to maintain covariance of the action, e.g. for studying consistent couplings of duality–symmetric fields to gravity, supersymmetrization or non–linear deformations of the theory. A covariant construction of the duality–symmetric actions was proposed in.[21,22] It uses a single auxiliary scalar field $a(x)$ which enters the Lagrangian only under a normalized derivative $v_\mu = \frac{\partial_\mu a}{\sqrt{(\partial a)^2}}$, $v_\mu v^\mu = 1$

$$L = \frac{1}{8} \mathcal{F}^i_{\mu\nu} \mathcal{F}^{\mu\nu i} + \frac{1}{4} v^\mu (\mathcal{F}^i_{\mu\nu} - \varepsilon^{ij} \tilde{\mathcal{F}}^j_{\mu\nu})(\mathcal{F}^{i\rho\nu} - \varepsilon^{ij} \tilde{\mathcal{F}}^{j\rho\nu}) v_\rho \,. \tag{5}$$

In addition to conventional Abelian gauge symmetries, this action is also invariant under the following local symmetries

$$\delta_I A^i_\mu = \Phi(x) v_\mu \,, \qquad \delta a(x) = 0 \,, \tag{6}$$

$$\delta_{II} a(x) = \phi(x) \,, \qquad \delta_{II} A^i_\mu = \frac{\phi}{\sqrt{(\partial a)^2}} v^\mu (\mathcal{F}^i_{\mu\nu} - \varepsilon^{ij} \tilde{\mathcal{F}}^j_{\mu\nu}) \,. \tag{7}$$

The symmetry (6), which implies that $A^i_\mu v^\mu$ components are pure gauge, is crucial for the derivation of the self–duality condition by integrating the second–order field equations. The second symmetry implies that $a(x)$ is completely auxiliary and can be gauge fixed, e.g. in such a way that

$v_\mu = \delta^0_\mu$. In this gauge the Lagrangian (5) reduces to the non–covariant one (2).

4. Non–linear generalization

We would now like to add to the Lagrangian (5) higher order terms in the gauge field strengths and their derivatives (which may also contain couplings to other matter fields of the theory e.g. scalars and fermions) in such a way that the symmetries (6) and (7) are preserved.[23] This will guarantee that the corresponding non–linearly deformed self–duality condition will be the consequence of the equations of motion which follow from this Lagrangian, and thus the non–linear theory will remain duality–symmetric. To find an ansatz for a possible form of the non–linear terms, let us first rewrite the Lagrangian (5) in the following form

$$L = \frac{1}{4}(v^\mu \mathcal{F}^i_{\mu\nu})(v_\rho \tilde{\mathcal{F}}^{i\rho\nu}) - \frac{1}{4}(v^\mu \tilde{\mathcal{F}}^i_{\mu\nu})(v_\rho \tilde{\mathcal{F}}^{i\rho\nu})\,. \tag{8}$$

Note that the pure gauge components $A^i_\mu v^\mu$ enter only the first term in this Lagrangian under a total derivative, while $v^\mu \tilde{\mathcal{F}}^i_{\mu\nu}$ do not contain $A^i_\mu v^\mu$. This is the origin of the symmetry (6). So to keep this symmetry we will assume that the higher order corrections to this Lagrangian only depend on $v^\mu \tilde{\mathcal{F}}^i_{\mu\nu}$ and their derivatives

$$L = \frac{1}{4}(v^\mu \mathcal{F}^i_{\mu\nu})(v_\rho \tilde{\mathcal{F}}^{i\rho\nu}) - \frac{1}{4}(v^\mu \tilde{\mathcal{F}}^i_{\mu\nu})(v_\rho \tilde{\mathcal{F}}^{i\rho\nu}) - \frac{1}{4}\mathcal{L}[v\tilde{\mathcal{F}}, \partial(v\tilde{\mathcal{F}})]\,. \tag{9}$$

To find how the second symmetry (7) gets modified, let us notice that the variation of A^i in (7) is proportional to the (left hand side of) twisted self–duality condition (4) projected along v_μ. So, one can assume that in the non–linear case this variation should be proportional to the deformed self–duality condition. One gets this condition by varying the Lagrangian with respect to A^i_μ and integrating the obtained equations of motion with the use of the symmetry (6). The result is[23]

$$v^\mu (\mathcal{F}^i_{\mu\nu} - \varepsilon^{ij} \tilde{\mathcal{F}}^j_{\mu\nu}) - \varepsilon^{ij} \frac{\delta \mathcal{L}}{\delta(v_\mu \tilde{\mathcal{F}}^{\mu\nu}_j)} = 0\,, \tag{10}$$

and the corresponding local transformations (7) under which (9) should be invariant are assumed to be

$$\delta_{II} a(x) = \phi(x)\,, \quad \delta_{II} A^i_\mu = \frac{\phi}{\sqrt{(\partial a)^2}}[v^\mu (\mathcal{F}^i_{\mu\nu} - \varepsilon^{ij} \tilde{\mathcal{F}}^j_{\mu\nu} - \varepsilon^{ij} \frac{\delta \mathcal{L}}{\delta(v_\mu \tilde{\mathcal{F}}^{\mu\nu}_j)}]\,. \tag{11}$$

The requirement of invariance under (11) imposes the condition on the non–linear terms $\mathcal{L}$ of (9), which in the differential form notation is

$$\delta_{II} L = 0 \;\Rightarrow\; \varepsilon^{ij} d \left[\frac{v}{\sqrt{(\partial a)^2}} \left(i_v \tilde{\mathcal{F}}^i + \frac{1}{2} \frac{\delta \mathcal{L}}{\delta i_v \tilde{\mathcal{F}}_i} \right) \frac{\delta \mathcal{L}}{\delta i_v \tilde{\mathcal{F}}_j} \right] = 0 \,. \tag{12}$$

Eq. (12) is the covariant counterpart of similar conditions on higher–order deformations found in[20,24] in the on–shell duality symmetric description of[15] and in the non manifestly covariant formulation.[20] The appearance of the quadratic term in $\mathcal{L}$ in (12) tells us that if we deform the quadratic Lagrangian with a single higher–order term, e.g. of the form $\partial^{2k}\mathcal{F}^4$, then to satisfy the condition (12) one should add to the Lagrangian an infinite series of even higher–order terms. The well know example of such a deformation is Born–Infeld (BI) theory, but there exist more complicated recent examples including higher derivatives (see e.g.[20,24,25]).

5. Supersymmetry

Once a non–linear deformation of the duality–symmetric action has been consistently done, one may ask the question about whether this deformation is consistent with supersymmetry. This is important e.g. for the analysis of consistent deformations of $N = 8$, $D = 4$ supergravity by counterterms $\partial^{2k}\mathcal{F}^4$. As we mentioned, such terms may appear at seven loops. These terms are supersymmetric and $E_{7(7)}$ invariant on the mass shell of the linear twisted self–duality condition (1). However, as we have seen, when included in the effective action (9) such terms give rise to an infinite series of higher–order terms which make the duality relation non–linear (10). All this does not guarantee that the whole construction preserves supersymmetry. Known examples of non–linear duality–symmetric and supersymmetric theories include $N = 1, 2$[24,26–29] and $N = 4$[30] (D3–brane) BI–like deformations of Abelian gauge theories with the duality group $U(1)$, and the case of the M5–brane,[31] *i.e.* a non–linear (2,0) self–dual $6d$ gauge theory with 16 supersymmetries (i.e. $N = 4$, from the $D = 4$ perspective). It should be mentioned that consistent couplings of *external* supersymmetric BI–like models to $N = 1$ and 2 supergravities are known,[32,33] however, an important issue which remains is whether non-linear deformations are possible for vector fields *inside* supergravity multiplets, in particular, in $N = 4, 8$ supergravities. In this respect let us also mention that non–linearities in field theories are often associated with spontaneous (super)symmetry breaking, while $N = 8$, $D = 4$ supergravity is maximally supersymmetric and there is no conventional field theories with more than 32 supercharges. So, if appear, the non–linearities in this theory should have a different nature, e.g. stringy corrections.

References

1. E. Cremmer and B. Julia, *Nucl.Phys.* **B159**, p. 141 (1979).
2. Z. Bern, L. J. Dixon and R. Roiban, *Phys.Lett.* **B644**, 265 (2007).
3. Z. Bern, J. Carrasco, L. J. Dixon, H. Johansson, D. Kosower *et al.*, *Phys.Rev.Lett.* **98**, p. 161303 (2007).
4. Z. Bern, J. Carrasco, L. J. Dixon *et al.* *Phys.Rev.* **D78**, p. 105019 (2008).
5. Z. Bern, J. Carrasco and H. Johansson (2009), arXiv:0902.3765.
6. Z. Bern, J. Carrasco, L. J. Dixon, H. Johansson and R. Roiban, *Phys.Rev.Lett.* **103**, p. 081301 (2009).
7. R. Kallosh, *JHEP* **1106**, p. 073 (2011).
8. R. Kallosh, *JHEP* **1203**, p. 083 (2012).
9. G. Bossard, P. Howe and K. Stelle, *JHEP* **1101**, p. 020 (2011).
10. N. Beisert, H. Elvang, D. Z. Freedman, M. Kiermaier, A. Morales *et al.*, *Phys.Lett.* **B694**, 265 (2010).
11. N. Berkovits, *Phys.Rev.Lett.* **98**, p. 211601 (2007).
12. M. B. Green, J. G. Russo and P. Vanhove, *Phys.Rev.Lett.* **98**, p. 131602 (2007).
13. M. B. Green, J. G. Russo and P. Vanhove, *JHEP* **1006**, p. 075 (2010).
14. G. Bossard, P. Howe, K. Stelle and P. Vanhove, *Class.Quant.Grav.* **28**, p. 215005 (2011).
15. M. K. Gaillard and B. Zumino, *Nucl.Phys.* **B193**, p. 221 (1981).
16. D. Zwanziger, *Phys. Rev.* **D3**, p. 880 (1971).
17. S. Deser and B. Zumino, *Phys. Lett.* **B62**, p. 335 (1976).
18. M. Henneaux and C. Teitelboim, *Phys. Lett.* **B206**, p. 650 (1988).
19. J. H. Schwarz and A. Sen, *Nucl. Phys.* **B411**, 35 (1994).
20. G. Bossard and H. Nicolai, *JHEP* **1108**, p. 074 (2011).
21. P. Pasti, D. P. Sorokin and M. Tonin, *Phys. Rev.* **D52**, 4277 (1995).
22. P. Pasti, D. P. Sorokin and M. Tonin, *Phys. Rev.* **D56**, 2473 (1997).
23. P. Pasti, D. Sorokin and M. Tonin, *Phys.Rev.* **D86**, p. 045013 (2012).
24. J. Carrasco, R. Kallosh and R. Roiban, *Phys.Rev.* **D85**, p. 025007 (2012).
25. W. Chemissany, R. Kallosh and T. Ortin, *Phys.Rev.* **D85**, p. 046002 (2012).
26. S. Cecotti and S. Ferrara, *Phys.Lett.* **B187**, p. 335 (1987).
27. S. V. Ketov, *Mod.Phys.Lett.* **A14**, 501 (1999).
28. S. M. Kuzenko and S. Theisen, *JHEP* **0003**, p. 034 (2000).
29. J. Broedel, J. Carrasco, S. Ferrara *et al.* *Phys.Rev.* **D85**, p. 125036 (2012).
30. M. Cederwall, A. von Gussich, B. E. Nilsson and A. Westerberg, *Nucl.Phys.* **B490**, 163 (1997).
31. I. A. Bandos, K. Lechner, A. Y. Nurmagambetov, P. Pasti, D. P. Sorokin and M. Tonin, *Phys. Rev. Lett.* **78**, 4332 (1997).
32. S. M. Kuzenko and S. A. McCarthy, *JHEP* **0302**, p. 038 (2003).
33. S. M. Kuzenko, *JHEP* **1206**, p. 012 (2012).

HIGHER ORDER ALGEBRAS AND GROUPS

M. RAUSCH de TRAUBENBERG*

Université de Strasbourg
IPHC-DRS-Physique Théorique des Hautes Énergies
23 rue du loess - BP28 - 67037 Strasbourg cedex 2, France
** E-mail: Michel.Rausch@IPHC.cnrs.fr*

Cubic extensions of the Poincaré algebra can be constructed in consistency with physical assumptions and provide new insights for the description of phenomena. In this paper we review some recent results associated to these structures and give the salient steps to construct their associated groups.

Keywords: Cubic extension of the Poincaré algebra; ternary algebras and groups.

1. Introduction

Lie algebras and Lie-superalgebras are central in the description of elementary particles. However, recently some non-quadratic structures emerged in different contexts in the description of physical phenomena. For instance, this generalised algebraic approach is given by the n-linear algebras in Quantum Mechanics,[1] ternary algebras in the description of multiple M_2-branes[2,3] or higher order extensions of the Poincaré algebra.[4–6]

2. Lie algebras of order F: definition and examples

A (real or complex) vector space $\mathfrak{g} = \mathfrak{g}_0 \oplus \mathfrak{g}_1$ with basis $\{X_i, i = 1, \cdots, \dim(\mathfrak{g}_0)\}, \{Y_a, a = 1, \cdots, \dim(\mathfrak{g}_1)\}$,[a] is called an elementary Lie algebra of order F if it satisfies the following requirements:

(1) $\mathfrak{g}_0$ is a Lie algebra with bracket $[X_i, X_j] = f_{ij}{}^k X_k$;
(2) $\mathfrak{g}_1$ is a representation of $\mathfrak{g}_0$, that is $[X_i, Y_a] = R_{ia}{}^b Y_b$;
(3) the composition of F elements in $\mathfrak{g}_1$ closes within $\mathfrak{g}_0$: $\{Y_{a_1}, \cdots, Y_{a_F}\} = Q_{a_1 \cdots a_F}{}^i X_i$, where $\{\cdots\}$ is the fully symmetric product of F elements ;

[a]From now on, generic element of $\mathfrak{g}_0$ are denoted X, of $\mathfrak{g}_1$ denoted Y and of $\mathfrak{g}$ denoted Z.

(4) the following fundamental identities are satisfied: $[\{Y_{a_2},\cdots,Y_{a_F+1}\},Y_{a_1}]+\cdots+[\{Y_{a_1},\cdots,Y_{a_F}\},Y_{a_F+1}]=0$.

There is basically two ways to obtain Lie algebras of order F: either by matrix representations or by an induction theorem. For the matrix representation,[7] as an example, consider $\mathfrak{gl}_{\text{el}}(m_1,m_2,m_3)$ the set of $(m_1+m_2+m_3)\times(m_1+m_2+m_3)$ matrices of the form

$$\mathfrak{gl}_{\text{el}}(m_1,m_2,m_3)=\left\{\begin{pmatrix} a_0 & b_1 & 0\\ 0 & a_1 & b_2\\ b_0 & 0 & a_2\end{pmatrix}\right\},$$

with $a_0\in\mathfrak{gl}(m_1), a_1\in\mathfrak{gl}(m_2), a_2\in\mathfrak{gl}(m_3)$, and $b_1\in\mathcal{M}_{m_1,m_2}(\mathbb{R}), b_2\in\mathcal{M}_{m_2,m_3}(\mathbb{R}), b_0\in\mathcal{M}_{m_3,m_1}(\mathbb{R})$. It is obvious to show that

$$\begin{aligned}\mathfrak{g}=\mathfrak{g}_0\oplus\mathfrak{g}_1= &\Big(\mathfrak{gl}(m_1)\oplus\mathfrak{gl}(m_2)\oplus\mathfrak{gl}(m_3)\Big)\oplus\\ &\Big(\mathcal{M}_{m_1,m_2}(\mathbb{R})\oplus\mathcal{M}_{m_2,m_3}(\mathbb{R})\oplus\mathcal{M}_{m_3,m_1}(\mathbb{R})\Big)\,,\end{aligned}$$

$\mathfrak{g}$ is a Lie algebra of order-three. The second types of algebras are based upon an induction theorem.[5] As an example consider $\mathfrak{g}_0$ a semi-simple Lie algebra and $\mathfrak{g}_1$ its adjoint representation. Let $\{J_a, a=1,\cdots,\dim\mathfrak{g}_0\}$ be a basis of $\mathfrak{g}_0$ and $\{A_a, a=1,\cdots,\dim\mathfrak{g}_0\}$ be the corresponding basis of $\mathfrak{g}_1$. Let $g_{ab}=\text{Tr}(A_aA_b)$ be the Killing form and $f_{ab}{}^c$ be the structure constants of $\mathfrak{g}_0$. Then one can endow $\mathfrak{g}=\mathfrak{g}_0\oplus\mathfrak{g}_1$ with a Lie algebra of order-three structure

$$[J_a,J_b]=f_{ab}{}^cJ_c,\quad [J_a,A_b]=f_{ab}{}^cA_c,\quad \{A_a,A_b,A_c\}=g_{ab}J_c+g_{ac}J_b+g_{bc}J_a. \tag{1}$$

It was then realised that within the framework of Lie algebras of order F higher order extensions of the Poincaré algebra can be defined. Historically the first higher order extensions of the Poincaré algebra were infinite dimensional and related to anyons in $(1+2)-$ dimensional space-time, (*i.e.*, discrete series of $\mathfrak{so}(1,2)$).[9–11] Later on, it was observed that the cubic algebra (1) with $\mathfrak{g}_0=\mathfrak{so}(1,4)$ leads, by means of an Inönü-Wigner contraction,[5,6,8] to the cubic extension of the Poincaré algebra $\mathfrak{g}=I\mathfrak{so}(1,3)\oplus\mathbb{R}^{1,3}$ with $I\mathfrak{so}(1,3)$ the Poincaré algebra in $(1+3)-$dimensions generated by $\{L_{\mu\nu},P_\mu\}$ and $\mathbb{R}^{1,3}$ the vector representation of $\mathfrak{so}(1,3)$ generated by $\{V_\mu\}$.

The algebra of order-three $\mathfrak{g}$ is defined by the following brackets

$$
\begin{aligned}
[L_{\mu\nu}, L_{\rho\sigma}] &= \eta_{\nu\sigma} L_{\rho\mu} - \eta_{\mu\sigma} L_{\rho\nu} + \eta_{\nu\rho} L_{\mu\sigma} - \eta_{\mu\rho} L_{\nu\sigma}, \\
[L_{\mu\nu}, P_\rho] &= \eta_{\nu\rho} P_\mu - \eta_{\mu\rho} P_\nu, \ \ [P_\mu, P_\nu] = 0, \\
[L_{\mu\nu}, V_\rho] &= \eta_{\nu\rho} V_\mu - \eta_{\mu\rho} V_\nu, \ \ [P_\mu \, V_\nu] = 0, \\
\{V_\mu, V_\nu, V_\rho\} &= \eta_{\mu\nu} P_\rho + \eta_{\nu\rho} P_\mu + \eta_{\rho\mu} P_\nu \ ,
\end{aligned}
$$

with $\eta_{\mu\nu} = \text{diag}(1,-1,-1,-1)$ the Minkowski metric. This algebra was intensively studied[12,13] and invariant actions were explicitly constructed. It was also noted that it admits a natural generalisation to any space-time dimensions. Subsequently higher dimensional invariant actions were constructed, and in this case it turns out that the higher dimensional cubic extensions of the Poincaré algebra induce a symmetry on generalised gauge fields or $p-$forms[14] from the natural operations on $p-$forms (wedge product, inner product, *etc*).

Next we have studied several other possibilities to extend the Poincaré algebra along the lines of cubic algebras: (i) we have studied in a systematic way all the possible cubic extensions of the Poincaré algebra in $(1+3)-$dimensions,[6] (ii) we have defined an adapted superspace associated to the cubic extensions of the Poincaré algebra,[15] (iii) we have studied kinematical algebras of order-three and shown that they are related by deformations and/or contractions.[8] It was also shown that quartic extensions of the Poincaré algebra can be defined: these extensions further turned out to be strongly related to $N = 2$ supersymmetry.[16]

3. Groups associated to Lie algebras of order F

A priori, the structure of group seems to be incompatible with the structure of ternary algebra. Indeed, for groups the product of two elements is always defined although for ternary algebras only the product of three elements is defined. However, groups associated to Lie algebras of order-three can be constructed using Hopf algebras.[6] We recall here the basic steps to obtain a group associated to a Lie algebra of order-three (see[6] for more details).

(1) Consider $\mathfrak{g} = \mathfrak{g}_0 \oplus \mathfrak{g}_1$ a real Lie algebra of order-three.
(2) The universal enveloping algebra $\mathcal{U}(\mathfrak{g})$ is the quotient of the tensor algebra $T(\mathfrak{g})$ by the two sided-ideal generated by

$$X_1 \otimes X_2 - X_2 \otimes X_1 - [X_1, X_2],$$
$$X_1 \otimes Y_2 - Y_2 \otimes X_1 - [X_1, Y_2],$$
$$Y_1 \otimes Y_2 \otimes Y_3 + \text{ perm. } - \{Y_1, Y_2, Y_3\} .$$

Then, as a vector space, we have the following isomorphism

$$\mathcal{U}(\mathfrak{g}) \cong \mathcal{U}(\mathfrak{g}_0) \oplus \Lambda^3(\mathfrak{g}_1) ,$$

with $\mathcal{U}(\mathfrak{g}_0)$ the usual universal enveloping algebra of the Lie algebra $\mathfrak{g}_0$, $\Lambda^3(\mathfrak{g}_1)$ be Roby (or three-exterior) algebra.[11,17] The three-exterior algebra $\Lambda_3(\mathbb{R}^m)$ is generated by the elements $\theta^i, i = 1, \cdots, m$ subjected to the relation $\{\theta^i, \theta^j, \theta^k\} = 0$. The three-exterior algebra is $\mathbb{Z}_3$−graded and has the following decomposition

$$\Lambda_3(\mathbb{R}^m) = \Lambda_3(\mathbb{R}^m)_0 \oplus \Lambda_3(\mathbb{R}^m)_1 \oplus \Lambda_3(\mathbb{R}^m)_2 ,$$

where $\Lambda_i \equiv \Lambda_3(\mathbb{R}^m)_i$ corresponds to the set of polynomials of degree i mod. 3. However, in contrast to the usual Grassmann algebra, the Roby algebra is infinitely generated. A basis of $\Lambda_3(\mathbb{R}^m)$ were identified in.[17] We call the elements of the basis of $\Lambda_3(\mathbb{R}^n)$ the Roby elements.

(3) A Poincaré-Birkhoff-Witt theorem is proven and a basis of $\mathcal{U}(\mathfrak{g})$ is identified.

(4) Endow $\mathcal{U}(\mathfrak{g}_\mathbb{C})$ ($\mathfrak{g}_\mathbb{C}$ is the complexified of $\mathfrak{g}$) with a Hopf algebra structure

$$\begin{array}{ll} \text{co-product} & \Delta 1 = 1 \hat{\otimes} 1, \ \ \Delta Z = Z \hat{\otimes} 1 + 1 \hat{\otimes} Z \ , Z \in \mathfrak{g} \ , \\ \text{antipode} & S(1) = 1, \ \ S(Z) = -Z \ , Z \in \mathfrak{g} \ , \\ \text{co-unity} & \epsilon(1) = 1, \ \ \epsilon(Z) = 0 \ , Z \in \mathfrak{g} \ , \end{array}$$

with $\hat{\otimes}$ the $\mathbb{Z}_3$−twisted tensor product (the elements of $\mathfrak{g}_0$ (resp. $\mathfrak{g}_1$) are of grade zero (resp. one)). Extend the definition of the co-product, antipode and co-unit to any elements of $\mathcal{U}(\mathfrak{g})$.

(5) Consider the dual of $\mathcal{U}(\mathfrak{g})$ and endow it with a Hopf algebra structure. The following vector space isomorphism can be proved

$$\mathcal{U}(\mathfrak{g})^* \cong \mathbb{C}[\mathfrak{g}_0] \oplus \Lambda_3(\mathfrak{g}_1) .$$

Furthermore, one can show that $g = \alpha X + \beta Y, \alpha \in \mathbb{C}, \beta \in \Lambda_1$ is group-like

$$\Delta e^g = e^g \ \hat{\otimes} \ e^g \ , \quad S(e^g) = e^{-g} \ , \quad \epsilon(e^g) = 1 \ .$$

(6) To define a group associated to $\mathfrak{g}$ some care is needed. Indeed, we have to a take care with (i) Baker-Campbell-Hausdorff formulæ since there

is no quadratic relations upon the Y's, (ii) the fact that $\Lambda_3(\mathbb{C}^k)$ is infinitely generated. One can show that

$$G(\Lambda) = \Big\{ \mathrm{e}^{\alpha^i X_i} \prod_{k=1}^{m} \mathrm{e}^{\beta^{k\,a}{}_b \theta^b Y_a}, \quad m \in \mathbb{N},\ \alpha^i, \beta^{k\,a}{}_b \in \mathbb{C}, \\ \text{with } \beta^{k\,a}{}_b \theta^b Y_a \text{ nilpotent matrices} \Big\}$$

is a group. If we apply this construction to the Lie algebra of order three $\mathfrak{g}_{\mathrm{el}}(m_1, m_2, m_3)$ we obtain a group that we denote $GL_{\mathrm{el}}(m_1, m_2, m_3, \Lambda)$.

Differently, and in a straight analogy with Lie super-groups, one can construct a group associate to Lie algebras of order three. This construction has the advantage not to invoke the heavy machinery of Hopf algebra. First of all we introduce some set of matrices with coefficients in $\Lambda_3(\mathbb{R}^m)$:

$$\begin{aligned} \mathcal{M}_k(\Lambda_0) &= \mathcal{M}_k(\mathbb{R}) \ \otimes \Lambda_0, \\ \mathcal{M}_{k,\ell}(\Lambda_1) &= \mathcal{M}_{k,\ell}(\mathbb{R}) \ \otimes \Lambda_1, \\ \mathcal{M}_{k,\ell}(\Lambda_2) &= \mathcal{M}_{k,\ell}(\mathbb{R}) \ \otimes \Lambda_2. \end{aligned}$$

A matrix, *e.g.* in $\mathcal{M}_k(\Lambda_0)$ is given by

$$A(\Lambda_0) = A_{(0)} + A_{(3)\,ijk}\theta^i\theta^j\theta^k + A_{(6)\,ijk\ell mn}\theta^i\theta^j\theta^k\theta^\ell\theta^m\theta^n + \cdots \tag{2}$$

where the sum above is finite and taken only over the Roby elements, *i.e.*, upon the basis vectors. We can show that the matrix given in (2) is invertible if and only if $A_{(0)}$ is invertible.[7]

We denote now by $GL(m_1, \Lambda_0)$ the set of invertible matrices of $\mathcal{M}_{m_1}(\mathbb{R}, \Lambda_0)$ such that the matrix $A(\Lambda_0)$ and its inverse $A(\Lambda_0)^{-1}$ have finite expansions. We then introduce $A_0 \in GL(m_1, \Lambda_0), A_1 \in GL(m_2, \Lambda_0), A_2 \in GL(m_3, \Lambda_0),\ B_0 \in \mathcal{M}_{m_3,m_1}(\Lambda_1), B_1 \in \mathcal{M}_{m_1,m_2}(\Lambda_1), B_2 \in \mathcal{M}_{m_2,m_3}(\Lambda_1)$ and $C_0 \in \mathcal{M}_{m_2,m_1}(\Lambda_2), C_1 \in \mathcal{M}_{m_3,m_2}(\Lambda_2), C_2 \in \mathcal{M}_{m_1,m_3}(\Lambda_2)$, and define the $(m_1 + m_2 + m_3) \times (m_1 + m_2 + m_3)$ matrices

$$GL(m_1, m_2, m_3) = \left\{ M = \begin{pmatrix} A_0 & B_1 & C_2 \\ C_0 & A_1 & B_2 \\ B_0 & C_1 & A_2 \end{pmatrix} \right\}.$$

Then, $GL(m_1, m_2, m_3)$ is a group. It is obvious to show that the product of two elements in $GL(m_1, m_2, m_3)$ is in $GL(m_1, m_2, m_3)$. The associativity is simply a consequence of the associativity of the matrix algebra and of the 3−exterior algebra. Finally one can obtain explicit formulæ for the inverse matrix.[7]

To end, we would like to notice that the two groups constructed so far are not equal and we have

$$GL_{\text{el}}(m_1, m_2.m_3, \Lambda) \subset GL(m_1, m_2, m_3) \ .$$

References

1. Y. Nambu *Phys. Rev.* **D7** 2405 (1973); T. Curtright and C. K. Zachos *Phys. Rev.* **D68** 085001 (2003) [arXiv:hep-th/0212267].
2. J. Bagger and N. Lambert *Phys. Rev.* **D7** 065008 (2008) [arXiv:0711.0955 [hep-th]]; J. Bagger and N. Lambert *Phys. Rev.* **D79** 025002 (2009) [arXiv:0807.0163 [hep-th]]; A. Gustavsson *Nucl. Phys.* **B811** 66 (2009) [arXiv:0709.1260 [hep-th]].
3. J. A. de Azcárraga and J. M. Izquierdo *J. Phys. A: Math. Theor.* **43** 293001 (2010) [arXiv:math-ph/1005.1028v1].
4. M. Rausch de Traubenberg and M. J. Slupinski *J. Math. Phys.* **41** 4556 (2000) [arXiv:hep-th/9904126].
5. M. Rausch de Traubenberg and M. J. Slupinski *J. Math. Phys.* **43** 5145 (2002) [arXiv:hep-th/0205113].
6. M. Goze, M. Rausch de Traubenberg and A. Tanasa *J. Math. Phys.* **48** 093507 (2007) [arXiv:math-ph/0603008].
7. M. Rausch de Traubenberg *J. Phys. Conf. Ser.* **128** 012060 (2008) [arXiv:0710.5368 [math-ph]]; M. Goze and M. Rausch de Traubenberg *J. Math. Phys.* **50** 063508 (2009) [arXiv:0809.4212 [math-ph]].
8. R. Campoamor-Stursberg and M. Rausch de Traubenberg, *J. Math. Phys.* **49** 063506 (2008) [arXiv:0801.2630 [hep-th]].
9. M. Rausch de Traubenberg and M. J. Slupinski, *Mod. Phys. Lett.* **A12** 3051 (1997) [hep-th/9609203].
10. M. Rausch de Traubenberg, *J. Phys. Conf. Ser.* **175** 012003 (2009) [arXiv:0811.1465 [hep-th]].
11. M. Rausch de Traubenberg, hep-th/9802141.
12. N. Mohammedi, G. Moultaka and M. Rausch de Traubenberg *Int. J. Mod. Phys.* **A19** 5585 (2004) [arXiv:hep-th/0305172].
13. G. Moultaka, M. Rausch de Traubenberg and A. Tanasa *Int. J. Mod. Phys.* **A20** 5779 (2005) [arXiv:hep-th/0411198].
14. G. Moultaka, M. Rausch de Traubenberg and A. Tanasa, in *Proceedings of the XIth International Conference Symmetry Methods in Physics, Prague 21-24 June 2004* [arXiv:hep-th/0407168].
15. R. Campoamor-Stursberg and M. Rausch de Traubenberg, *J. Phys. A; Math. Theor.* **42** 495202 (2009) [arXiv:0907.2149 [hep-th]].
16. R. Campoamor-Stursberg and M. Rausch de Traubenberg, *J. Phys.A: Math. Theor.* **43** 455201 (2010) [arXiv:1005.4994 [hep-th]].
17. N. Roby *Bull. Sc. Math.* **94** 49 (1970).

STRUCTURES OF GENERALIZED MAP VIRASORO ALGEBRAS

YUCAI SU*, HENAN WU, SONG WANG, XIAOQING YUE

Department of Mathematics, Tongji University,
Shanghai, China
**E-mail: ycsu@tongji.edu.cn*

In this paper, we study the structure theory of a class of generalized map Virasoro algebras. In particular, the derivation algebras, the automorphism groups and the second cohomology groups of generalized map centerless Virasoro algebras are determined.

Keywords: generalized map Virasoro algebras, derivations, automorphisms, 2-cocycles.

1. Introduction

Many generalizations of the Virasoro algebras were introduced in recent years, and structure theories of these algebras have been considered by some authors (see, e.g.,[1–7]). In this paper, we will consider the following generalization of the (centerless) Virasoro algebra.

Let Γ be an additive subgroup of a field $\mathbb{F}$ of characteristic 0, and $\mathbb{F}[\Gamma]$ the group algebra of Γ with basis $\{e^\alpha \,|\, \alpha \in \Gamma\}$. Denote $\mathscr{W}(\Gamma) = \mathbb{F}[\Gamma]\partial$ (where $\partial(e^\alpha) = \alpha e^\alpha$), which is usually referred to as the (*generalized*) *Witt algebra* or (*generalized centerless*) *Virasoro algebra*.[6] Since $\mathscr{W}(\Gamma) \cong \mathscr{W}(\Gamma')$ if $\Gamma \cong \Gamma'$,[2] and for any $0 \neq \alpha \in \Gamma$ we have $\Gamma \cong \frac{1}{\alpha}\Gamma$, by replacing Γ by $\frac{1}{\alpha}\Gamma$ if necessary, we always assume that $1 \in \Gamma$. Let A be a unital commutative associative algebra. We denote $\mathscr{W}_A(\Gamma) = \mathscr{W}(\Gamma) \otimes A$, which is referred to as a (*generalized*) *map Witt algebra* or (*generalized*) *map* (*centerless*) *Virasoro algebra*. In case $A = \mathbb{F}[t, t^{-1}]$, we get the (generalized) loop Virasoro algebras, which play important roles in the structure theory of Lie algebras, e.g., in affine Lie algebras.[8]

The structure theory and cohomology theory of Lie algebras play important roles in Lie theory, as it is known that central extensions (determined by second cohomology groups) are closely related to structures of Lie algebras (e.g.,[8]). Thus computations of second cohomology groups seem to be important and interesting as well. We will determine the derivation algebra, the automorphism group and the second cohomology group of $\mathscr{W}_A(\Gamma)$.

We can write $\mathscr{W}_A(\Gamma)$ as

$$\mathscr{W}_A(\Gamma)= \bigoplus_{\alpha\in\Gamma} \mathscr{W}_A(\Gamma)_\alpha, \quad \text{where } \mathscr{W}_A(\Gamma)_\alpha = \text{span}\{L_\alpha x \mid \alpha\in\Gamma, x\in A\}, \quad (1)$$

with relations

$$[L_\alpha x, L_\beta y] = (\beta-\alpha)L_{\alpha+\beta}xy, \quad \forall\, \alpha,\beta\in\Gamma,\ x,y\in A. \quad (2)$$

We wish to remark that the structure of $\mathscr{W}_A(\Gamma)$ depends heavily on the structure of A. For instance, the dimension of the second cohomology group of $\mathscr{W}_A(\Gamma)$ varies, depending on the dimension of A (Theorem 4.1): the algebra $\mathscr{W}(\Gamma)$ (i.e., $A=\mathbb{F}$) has only one universal central extension, while the loop centerless Virasoro algebra (i.e., $A=\mathbb{F}[t^{\pm1}]$) admits infinitely many central extensions.

The paper is organized as follows. In Section 2, we determine the derivation algebra of the (generalized) map Witt algebra $\mathscr{W}_A(\Gamma)$. Then in Section 3, the automorphism group for $\mathscr{W}_A(\Gamma)$ is determined. Finally in Section 4, we compute the second cohomology group for $\mathscr{W}_A(\Gamma)$. The main results of the paper are stated in Theorems 2.1, 3.1 and 4.1.

2. Derivation algebra of $\mathscr{W}_A(\Gamma)$

Denote by $(\text{Der}\,\mathscr{W}_A(\Gamma))_\gamma$ the space of derivations of degree γ. Note that every derivation $D\in \text{Der}\,\mathscr{W}_A(\Gamma)$ can be written as $D=\sum_{\gamma\in\Gamma} D_\gamma$ (e.g.,[3,5]), where $D_\gamma \in (\text{Der}\,\mathscr{W}_A(\Gamma))_\gamma$ and for every $x\in\mathscr{W}_A(\Gamma)$ only *finitely many* $D_\gamma(x)\neq 0$. In case $\gamma\neq 0$, we can easily verify that D_γ is simply the inner derivation $D_\gamma = \text{ad}_{-\gamma^{-1}D(L_0)}$. From this one can prove as in [5] that $\text{Der}\,\mathscr{W}_A(\Gamma) = \oplus_{\gamma\in\Gamma}(\text{Der}\,\mathscr{W}_A(\Gamma))_\gamma$. Now we fix $D\in(\text{Der}\,\mathscr{W}_A(\Gamma))_0$ and assume

$$D(L_\alpha x) = L_\alpha f(\alpha,x) \text{ for } \alpha\in\Gamma,\ x\in A \text{ and some } f(\alpha,x)\in A. \quad (3)$$

Applying D to (2), we have

$$(\beta-\alpha)f(\alpha+\beta,xy) = (\beta-\alpha)(yf(\alpha,x)+xf(\beta,y)). \quad (4)$$

Lemma 2.1. *We have* $f(\alpha+\beta,xy) = yf(\alpha,x)+xf(\beta,y)$ *for* $\alpha,\beta\in\Gamma,\ x,y\in A$.

Proof. Take any $\gamma \in \Gamma\backslash\{0, \alpha, -\beta\}$ satisfying $2\gamma \neq \beta - \alpha$. Then from (4) we can obtain $f(\alpha+\beta, xy) = yf(\alpha+\gamma, x) + xf(\beta-\gamma, y) = yf(\alpha, x) + xyf(\gamma, 1) + xf(\beta, y) + xyf(-\gamma, 1) = yf(\alpha, x) + xf(\beta, y)$. □

Denote by $\mathrm{Hom}(\Gamma, A)$ the space of group homomorphisms from Γ to A. Then for any $\phi \in \mathrm{Hom}(\Gamma, A)$ we can define a derivation D_ϕ as follows:

$$D_\phi(L_\alpha x) = L_\alpha x\phi(\alpha), \quad \forall\, \alpha \in \Gamma,\ x \in A. \tag{5}$$

We still use $\mathrm{Hom}(\Gamma, A)$ to denote the corresponding subspace of $\mathrm{Der}\,\mathscr{W}_A(\Gamma)$.

Denote by $\mathrm{Der}\,A$ the space of derivations of A. For any $\rho \in \mathrm{Der}\,A$, we can also define a derivation D^ρ as follows:

$$D^\rho(L_\alpha x) = L_\alpha\rho(x), \forall\, \alpha \in \Gamma,\ x \in A. \tag{6}$$

Similarly, we denote the corresponding subspace of $\mathrm{Der}\,\mathscr{W}_A(\Gamma)$ by $\mathrm{Der}\,A$.

Now we have the following main theorem in this section.

Theorem 2.1. *The derivation space of $\mathscr{W}_A(\Gamma)$ can be decomposed into a direct sum $\mathrm{Der}\,\mathscr{W}_A(\Gamma) = \oplus_{\gamma\in\Gamma}(\mathrm{Der}\,\mathscr{W}_A(\Gamma))_\gamma$, such that $(\mathrm{Der}\,\mathscr{W}_A(\Gamma))_\gamma = (\mathrm{Inn}\,\mathscr{W}_A(\Gamma))_\gamma$ if $\gamma \neq 0$ and $(\mathrm{Der}\,\mathscr{W}_A(\Gamma))_0 = \mathrm{Hom}(\Gamma, A) \oplus \mathrm{Der}\,A$.*

Proof. It suffices to prove $(\mathrm{Der}\,\mathscr{W}_A(\Gamma))_0 = \mathrm{Hom}(\Gamma, A) \oplus \mathrm{Der}\,A$. Take an element $D \in (\mathrm{Der}\,\mathscr{W}_A(\Gamma))_0$ and assume $D(L_\alpha x) = L_\alpha f(\alpha, x)$ as in (3). Denote by ϕ the map sending α to $f(\alpha, 1)$ and ρ the map sending x to $f(0, x)$. Then $\phi \in \mathrm{Hom}(\Gamma, A)$, $\rho \in \mathrm{Der}\,A$ and $D = D_\phi + D^\rho$ by Lemma 2.1. It is easy to see that $\mathrm{Hom}(\Gamma, A) \cap \mathrm{Der}\,A = 0$. This completes the proof. □

Corollary 2.1. *In case $\Gamma = \mathbb{Z}$, $\mathrm{Der}\,\mathscr{W}_A(\mathbb{Z}) = \mathrm{Inn}\,\mathscr{W}_A(\mathbb{Z}) \oplus \mathrm{Der}\,A$.*

Proof. Take a homorphism $\phi \in \mathrm{Hom}(\mathbb{Z}, A)$, one has $\phi(i) = i\phi(1), \forall i \in \mathbb{Z}$. Then we obtain that

$$D_\phi(L_i x) = L_i x\phi(i) = L_i x i\phi(1) = [L_0\phi(1), L_i x], \quad \forall\, i \in \mathbb{Z},\ x \in A.$$

Hence $\mathrm{Hom}(\mathbb{Z}, A) \subset \mathrm{Inn}\,\mathscr{W}_A(\mathbb{Z})$. It is obvious that $\mathrm{Inn}\,\mathscr{W}_A(\mathbb{Z}) \cap \mathrm{Der}\,A = 0$. Thus $\mathrm{Der}\,\mathscr{W}_A(\mathbb{Z}) = \mathrm{Inn}\,\mathscr{W}_A(\mathbb{Z}) \oplus \mathrm{Der}\,A$. □

3. Automorphism group of $\mathscr{W}_A(\Gamma)$

We require A to be an integral domain in this section. An element $x \in \mathscr{W}_A(\Gamma)$ is called *ad-locally finite* if for any $y \in \mathscr{W}_A(\Gamma)$, the space spanned by $\{\mathrm{ad}_x^i(y) \mid i \in \mathbb{Z}^+\}$ is of finite dimension.

Lemma 3.1. *The set of ad-locally finite elements of $\mathscr{W}_A(\Gamma)$ is contained in $\mathscr{W}_A(\Gamma)_0$.*

Proof. Assume $v = \sum_{\alpha\in\Gamma} L_\alpha x_\alpha$ (finite sum) is a locally finite element. Choose a well order on Γ compatible with its group structure. Let $\alpha_0 = \max\{\alpha \in \Gamma \,|\, x_\alpha \neq 0\}$. Suppose $\alpha_0 \neq 0$. Take $u = L_{2\alpha_0}$. For any $i \in \mathbb{Z}^+$, the highest term of $\mathrm{ad}_v^i(u)$ with respect to the given order on Γ is $i!\alpha_0^i L_{(i+2)\alpha_0} x_{\alpha_0}^i \neq 0$. Thus these elements are linearly independent, and the dimension of $\mathrm{span}\{\mathrm{ad}_v^i(u) \,|\, i \in \mathbb{Z}^+\}$ is infinite. Hence $\alpha_0 = 0$. Similarly, $\min\{\alpha \in \Gamma \,|\, x_\alpha \neq 0\} = 0$, i.e., v belongs to $\mathscr{W}_A(\Gamma)_0$. □

Now for any automorphism $\sigma \in \mathrm{Aut}\mathscr{W}_A(\Gamma)$, by Lemma 3.1 we can assume $\sigma(L_0) = L_0 a$ for some $a \in A$. For any $0 \neq \beta \in \Gamma$, assume $\sigma(L_\beta) = \sum_{\gamma\in\Gamma} L_\gamma z_\gamma$ for some $z_\gamma \in A$. Applying σ to (2) gives $\sum_{\gamma\in\Gamma} \gamma L_\gamma a z_\gamma = \beta(\sum_{\gamma\in\Gamma} L_\gamma z_\gamma)$. Comparing the coefficients implies that $z_\gamma = 0$ except that $a\gamma = \beta$. We deduce that $a \in \mathbb{F}^*$, $a\Gamma = \Gamma$, and

$$\sigma(L_\beta) = L_{a^{-1}\beta} z_{a^{-1}\beta}. \tag{7}$$

We take $T(\Gamma) = \{a \in \mathbb{F}^* \,|\, a\Gamma = \Gamma\}$. For $a \in T(\Gamma)$, we define

$$\sigma_a(L_\alpha x) = a L_{a^{-1}\alpha} x \text{ for } \alpha \in \Gamma,\ x \in A, \tag{8}$$

which is clearly an automorphism of $\mathscr{W}_A(\Gamma)$. We use the same symbol $T(\Gamma)$ to denote the set of such autoumorphisms. Thus we have the following.

Lemma 3.2. *The group $T(\Gamma)$ is an subgroup of* $\mathrm{Aut}\mathscr{W}_A(\Gamma)$.

Denote $\mathrm{Aut}^0\mathscr{W}_A(\Gamma) = \{\sigma \in \mathrm{Aut}\mathscr{W}_A(\Gamma) \,|\, \sigma(\mathscr{W}_A(\Gamma)_\alpha) \subset \mathscr{W}_A(\Gamma)_\alpha \text{ for } \alpha \in \Gamma\}$, a subgroup of $\mathrm{Aut}\mathscr{W}_A(\Gamma)$.

Lemma 3.3. *The group* $\mathrm{Aut}\mathscr{W}_A(\Gamma)$ *admits a semidirect product* $\mathrm{Aut}\mathscr{W}_\mathrm{A}(\Gamma) = T(\Gamma) \ltimes \mathrm{Aut}^0\mathscr{W}_A(\Gamma)$.

Proof. Let $\sigma \in \mathrm{Aut}\mathscr{W}_A(\Gamma)$. By (7) we may assume $\sigma(L_\alpha x) = L_{a^{-1}\alpha}\phi(\alpha, x)$ for some $\phi(\alpha, x) \in A$. Define $\sigma_0(L_\alpha x) = L_\alpha a^{-1}\phi(\alpha, x)$. It is easy to see that $\sigma_0 \in \mathrm{Aut}^0\mathscr{W}_A(\Gamma)$, $\sigma = \sigma_a\sigma_0$ and $T(\Gamma) \cap \mathrm{Aut}^0\mathscr{W}_A(\Gamma) = 1$. For $\tau \in \mathrm{Aut}^0\mathscr{W}_A(\Gamma)$, one can verify that $\sigma_a\tau\sigma_{a^{-1}} \in \mathrm{Aut}^0\mathscr{W}_A(\Gamma)$, i.e., it is a normal subgroup. The proof is completed. □

Now we will give an explicit description about $\mathrm{Aut}^0\mathscr{W}_A(\Gamma)$. Take any $\sigma \in \mathrm{Aut}^0\mathscr{W}_A(\Gamma)$ and assume $\sigma(L_\alpha x) = L_\alpha g(\alpha, x)$, where g is a map from $\Gamma \times A$ to A. Applying σ to (2) gives $(\beta - \alpha)g(\alpha, x)g(\beta, y) = (\beta - \alpha)g(\alpha + \beta, xy)$.

Similar to the proof of Lemma 2.1, we have $g(\alpha, x)g(\beta, y) = g(\alpha+\beta, xy)$. Define $\sigma_1(L_\alpha x) = L_\alpha x g(\alpha, 1)$ and $\sigma_2(L_\alpha x) = L_\alpha g(0, x)$. Then $\sigma = \sigma_1\sigma_2$.

Denote by Aut A the automorphism group of A. Then every element of this set corresponds to an element of $\mathrm{Aut}^0 \mathscr{W}_A(\Gamma)$. So we construct a group homorphism from Aut$A$ to $\mathrm{Aut}^0 \mathscr{W}_A(\Gamma)$.

Denote by $\mathrm{Hom}(\Gamma, \mathrm{A}^*)$ the space of homorphisms from Γ to A^*, which admits a group structure in an obvious way. For any element $\varphi \in \mathrm{Hom}(\Gamma, \mathrm{A}^*)$, we use the same symbol φ to define $\varphi(L_\alpha x) = L_\alpha x\varphi(\alpha)$. Then $\varphi \in \mathrm{Aut}^0 \mathscr{W}_A(\Gamma)$. So $\mathrm{Hom}(\Gamma, \mathrm{A}^*)$ also induces a subgroup of $\mathrm{Aut}^0 \mathscr{W}_A(\Gamma)$.

Lemma 3.4. *The group* $\mathrm{Hom}(\Gamma, \mathrm{A}^*)$ *is a normal subgroup of* $\mathrm{Aut}^0 \mathscr{W}_A(\Gamma)$.

Proof. Let $\varphi \in \mathrm{Hom}(\Gamma, \mathrm{A}^*)$ and $\sigma \in \mathrm{Aut}^0 \mathscr{W}_A(\Gamma)$. Assume $\sigma^{-1}(L_\alpha x) = L_\alpha g(\alpha, x)$. Define $\psi(\alpha) = g(0, \phi(\alpha))$. One can check that $\psi \in \mathrm{Hom}(\Gamma, \mathrm{A}^*)$ and $\sigma^{-1}\varphi\sigma = \psi$. Thus $\mathrm{Hom}(\Gamma, \mathrm{A}^*)$ is a normal subgroup. □

Lemma 3.5. *The group* $\mathrm{Aut}\mathscr{W}_A(\Gamma)$ *admits a semidirect product* $\mathrm{Aut}^0 \mathscr{W}_A(\Gamma) = \mathrm{Aut}\, A \ltimes \mathrm{Hom}(\Gamma, \mathrm{A}^*)$.

Proof. It suffices to prove $\mathrm{Aut}\, A \cap \mathrm{Hom}(\Gamma, \mathrm{A}^*) = 1$. Let $\sigma \in \mathrm{Aut}A \cap \mathrm{Hom}(\Gamma, \mathrm{A}^*)$. Then $\sigma(L_\alpha x) = L_\alpha \phi(x) = L_\alpha x \varphi(\alpha)$ for $\alpha \in \Gamma$, $x \in A$, and some $\phi \in \mathrm{Aut}A$, $\varphi \in \mathrm{Hom}(\Gamma, \mathrm{A}^*)$. Setting $\alpha = 0$ we obtain that $\phi(x) = x$ for all $x \in A$, i.e., $\sigma = 1$. □

Theorem 3.1. *We have* $\mathrm{Aut}\mathscr{W}_{\mathrm{A}}(\Gamma) = T(\Gamma) \ltimes (\mathrm{Aut}A \ltimes \mathrm{Hom}(\Gamma, \mathrm{A}^*))$.

Proof. It follows from Lemmas 3.3 and 3.5. □

4. Second cohomology group of $\mathscr{W}_A(\Gamma)$

Lemma 4.1. *Let ψ be a 2-cocycle on $\mathscr{W}_A(\Gamma)$. Then there exists a 2-cocycle ϕ in the cohomology class of ψ such that*

$$\phi(\mathscr{W}_A(\Gamma)_\alpha, \mathscr{W}_A(\Gamma)_\beta) = 0 \quad \text{if } \alpha + \beta \neq 0.$$

and $\phi(L_{-1}1, L_1 x) = 0$ for $x \in A$.

Proof. Let $f : \mathscr{W}_A(\Gamma) \to \mathbb{F}$ be a linear function on $\mathscr{W}_A(\Gamma)$ defined by

$$f(L_0 x) = \tfrac{1}{2}\psi(L_{-1}, L_1 x), \;\; f(L_\alpha x) = \alpha^{-1}\psi(L_0, L_\alpha x) \text{ if } \alpha \neq 0. \tag{9}$$

Set $\phi = \psi - \psi_f$, then one can verify ϕ satisfies the listed equations. □

Next we consider the case of $\alpha+\beta=0$. Denote $X_{\alpha,x,y}=\phi(L_\alpha x, L_{-\alpha}y)$.

Lemma 4.2. *For any $\alpha\in\Gamma$ and $x,y\in A$, $X_{\alpha,x,y}=X_{\alpha,1,xy}$.*

Proof. Using the Jacobi identity on the triple $(L_\alpha x, L_\beta y, L_{-\alpha-\beta}z)$, one has

$$(-2\beta-\alpha)X_{\alpha,x,yz}=(\beta-\alpha)X_{\alpha+\beta,xy,z}+(-2\alpha-\beta)X_{\beta,y,xz}. \tag{10}$$

Taking $\beta=\alpha$ in (10), we have $\alpha X_{\alpha,x,yz}=\alpha X_{\alpha,y,xz}$. Setting $y=1$ implies $\alpha X_{\alpha,x,z}=\alpha X_{\alpha,1,xz}$. Setting $\alpha=0$ and $y=1$ in (10), we get $X_{0,x,z}=0$. Hence $X_{\alpha,x,y}=X_{\alpha,1,xy}$ for any $\alpha\in\Gamma$. □

From Lemma 4.2, $X_{\alpha,x,y}$ only depends on α and xy. So we introduce a new notation from now on : $X_{\alpha,x}=\phi(L_\alpha, L_{-\alpha}x)$.

Theorem 4.1. *Suppose $\{x_i\,|\,i\in I\}$ is a basis of A. The second cohomology group $H^2(\mathscr{W}_A(\Gamma),\ \mathbb{F})=\prod_{k\in I}\mathbb{F}\bar{\phi_k}$ is the direct product of all $\mathbb{F}\bar{\phi_k}$, $k\in I$, where $\bar{\phi_k}$ is the cohomology class of ϕ_k defined by $\phi_k(L_\alpha x_i, L_\beta)=\delta_{\alpha+\beta,0}\delta_{i,k}\frac{\alpha^3-\alpha}{12}$ for $i,k\in I$, $\alpha,\beta\in\Gamma$ and $x,y\in A$.*

Proof. We obtain by a direct computation that $X_{\alpha,x}=\frac{1}{6}\alpha(\alpha-1)(\alpha+1)X_{2,x}$. Then $\phi=2\Sigma_{i\in I}X_{2,x_i}\phi_i$ (possibly infinite sum). Suppose $\Sigma_{i\in I}c_i\phi_i=\psi_g$ (possibly infinite sum in the left side) for some linear function on $\mathscr{W}_A(\Gamma)$. Applying this equation on the pair $(L_\alpha x_i, L_{-\alpha})$, we obtain $\frac{1}{12}c_i(\alpha^3-\alpha)=-2\alpha g(L_0x_i)$ for $\alpha\in\Gamma$. So $g=0$ and then $c_i=0$. □

Acknowledgement

This work is supported by NSF grant no. 10825101, 11001200 and 11101269 of China and the Fundamental Research Funds for the Central Universities.

References

1. D. Dokovic and K. Zhao, *Tran. Ams. Math.Soc.* **2350**, 643 (1998).
2. Y. Su, X. Xu and H. Zhang, *J. Algebra* **233**, 642 (2000).
3. R. Farnsteiner, *J. Algebra* **118**, 33 (1988).
4. Y. Su, *Comm. Algebra* **30**, 763 (2002).
5. Y. Su, *Science in China A* **46**, 346 (2003).
6. Y. Su and K. Zhao, *J. Algebra* **252**, 1 (2002).
7. H. Wu, S. Wang and X. Yue, *Structures of generalized loop Virasoro algebras, Comm. Alg,* **submitted**.
8. V. Kac, *Infinite-dimensional Lie algebras*, 3rd edn. (Cambridge University Press, Cambridg, 1990).

PARALLEL SESSION
Applications in Particle and Nuclear Physics

22 & 24 AUGUST 2012

NUCLEON-PAIR APPROXIMATION OF THE SHELL MODEL WITH ISOSPIN

G. J. FU and Y. M. ZHAO

INPAC, Department of Physics and Shanghai Key Lab for Particle Physics and Cosmology, Shanghai Jiao Tong University, Shanghai, 200240, China

In this paper we present a study of the nucleon-pair approximation with the isospin symmetry. In this model the building blocks are collective nucleon-pairs with given spin and isospin. We exemplify our work by using the ^{96}Cd nucleus.

Keywords: nucleon-pair approximation; isospin; spin-aligned proton-neutron pair.

1. Introduction

The nuclear shell model is a very good framework to study the low-lying excitations in atomic nuclei. As the number of single-particle orbits increases with the atomic mass, however, the configuration space of the shell model becomes too gigantic to handle, even for the best computers. Truncations of the shell model space are therefore necessary. Among various truncation schemes, the nucleon pair approximation is one convenient and efficient approach.

Many formulations of the nucleon-pair approximation have hitherto been developed. Important examples include the seniority scheme or quasispin scheme,[1] the Bardeen-Cooper-Schrieffer (BCS) theory,[2] the broken pair approximation,[3] the Ginocchio model (which was later further developed and called the fermion dynamical symmetry model).[4] There have been many studies of the microscopic foundation of the interacting boson models (IBM),[5] along which the validity of pair approximation has been examined in many cases.[6,7] These models have been widely used in studies of low-lying states in nuclei.

Most previous studies of nucleon pair approximation focused on nucleon pairs with isospin $T = 1$, namely, proton-proton or neutron-neutron pairs.

A number of phenomena, for instance, spin-zero ground states of even-even nuclei, large one-nucleon separation energy of even-mass nuclei, large energies of the first 2^+ states for semi-magic nuclei, are explained in terms of such pairing correlation. On the other hand, much attention has been also paid to the proton-neutron pairing correlation, e.g., the isospin generalized BCS theory and the isospin generalized HFB theory,[8] the generalized seniority scheme with isospin,[9] the cranked mean-field model,[10] the SO(5) and SO(8) models,[11–16] the BCS-like model,[17] the particle number projected BCS model[18,19] and the quartet condensation model.[20] These models have been applied to calculate $N \approx Z$ nuclei, for which contributions from proton-neutron pairs are important. However, various assumptions have been made in these studies. See Refs.[21,22] for comprehensive reviews.

In this talk, we introduce a formulation of the nucleon-pair approximation of the shell model (NPA) with the isospin symmetry. This formulation provides us with an appropriate tool in studies of proton-neutron pairing correlation. We present preliminary results for the ^{96}Cd nucleus, for which spin-aligned proton-neutron pairs were found to be dominant in the low-lying states.[23–25]

2. Theoretical framework

In 1997 Chen suggested a formulation of nucleon-pair approximation of the shell model.[6] The key technique of this formulation is to compute one- and two-body matrix elements via recursions, called "Wick theorem for the coupled clusters". Many examples of the NPA calculations in pair-truncated spaces can be found in Reference.[7]

In the NPA with the isospin symmetry, the building blocks of the configuration space are nucleon-pairs with given spin and isospin. A pair with spin J and isospin T is defined by

$$A^{(JT)^\dagger} = \sum_{ab} y(abJT) \left(C_a^\dagger \times C_b^\dagger \right)^{(JT)} ,$$

where $y(abJT)$ are called pair-structure coefficients of collective pair with spin J and isospin T. If there are $2N$ nucleons, our nucleon-pair model space with total spin L_N and isospin S_N is constructed by stepwise coupling of nucleon-pairs, i.e.,

$$A^{(L_N S_N)^\dagger}(J_1 T_1 \cdots J_N T_N, L_1 S_1 \cdots L_N S_N)|0\rangle \ .$$

Here

$$A^{(L_N S_N)\dagger}(J_1T_1\cdots J_NT_N, L_1S_1\cdots L_NS_N)$$
$$= \left(\cdots\left(\left(A^{(J_1T_1)\dagger}\times A^{(J_2T_2)\dagger}\right)^{(L_2S_2)}\times A^{(J_3T_3)\dagger}\right)^{(L_3S_3)}\cdots\times A^{(J_NT_N)\dagger}\right)^{(L_NS_N)}.$$

In Ref.[26] we generalized the technique of Refs.[6] with further consideration of the isospin symmetry. The overlaps between pair basis $A^{(L_NS_N)\dagger}(J_1T_1\cdots J_NT_N, L_1S_1\cdots L_NS_N)|0\rangle$, matrix elements of one-body and two-body operators in the pair basis, are derived.

3. Calculated results for ^{96}Cd

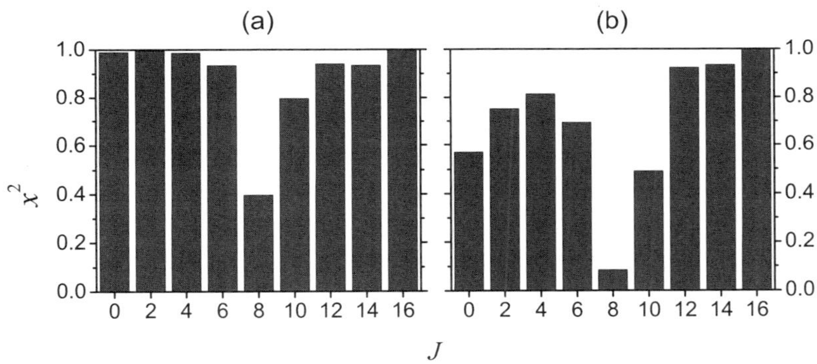

Fig. 1. Overlap squared (x^2) for a few yrast spin-J states, between the wave functions calculated in the nucleon-pair ($J = 9$ and $T = 0$) subspace and the two shell model spaces: (a) the single-j $1g_{9/2}$ shell, (b) the $p_{1/2}p_{3/2}f_{5/2}g_{9/2}$ shell. The same interaction parameters (JUN-45, taken from Ref.[27]), are assumed in these calculations.

Recent measurements of ^{92}Pd showed[23] that the low-lying states of this nucleus are dominated by spin-aligned proton-neutron pairs of the $1g_{9/2}$ shell. The importance of the spin-aligned proton-neutron pairs in ^{96}Cd, ^{94}Ag and ^{92}Pd was subsequently studied.[24,25,28–31]

Our NPA with the isospin symmetry is an ideal tool to study the relevance of the above picture. Below we present our preliminary results of our numerical calculations in the following model spaces:

(a) the configuration space constructed by using nucleon pairs with $J = 9$ and $T = 0$ in the $1g_{9/2}$ orbit,
(b) the shell model space constructed by the single-j $1g_{9/2}$ orbit,
(c) the shell model space constructed by the $p_{1/2}p_{3/2}f_{5/2}g_{9/2}$ orbits.

The one- and two-body interaction parameters that we used are the effective interactions for the $2p_{1/2}2p_{3/2}1f_{5/2}1g_{9/2}$ shell, given in Ref.[27] We investigate the wave functions obtained in the above configuration spaces.

In Fig. 1 we show the squares (denoted by x^2) of overlaps between the wave functions obtained in the nucleon-pair ($J = 9$ and $T = 0$) subspace and the shell model spaces (the single-j $g_{9/2}$ shell or the $p_{1/2}p_{3/2}f_{5/2}g_{9/2}$ shell), for a few yrast states with different spin. One sees that the spin-aligned proton-neutron pairs with $J = 9$ and $T = 0$ play a dominate role in these states, in accordance with previous results in Refs.[24,25,28,30,31]

4. Summary

In this paper we have discussed the framework of nucleon pair approximation of the shell model. We give an introduction to our new development along this line, namely, we consider the isospin symmetry in pair approximation. The building blocks of our new version are nucleon pairs with given spin and isospin. A computer code of the pair approximation with both the spin and isospin symmetry is now available.

We present preliminary results by applying our new version of pair approximation. We discuss the relevance of spin-aligned nucleon pairs with $J = 9$ and $T = 0$ (in the $1g_{9/2}$ orbit) for low-lying yrast states of the ^{96}Cd nucleus. Our results show that the spin-aligned pairs consisting of nucleons in the $1g_{9/2}$ orbit provide us with dominant configurations for most of these states, as suggested in many recent works.

Acknowledgments

We thank the National Natural Science Foundation of China (Grants No. 11225524, 10975096 and 11145005), the 973 Program of China (Grant No. 2013CB834401), and the Science & Technology Committee of Shanghai city (Grant No. 11DZ2260700) for financial support.

References

1. G. Racah, *Phys. Rev.* **62**, 438 (1942); I. Talmi, *Nucl. Phys.* **172**, 1 (1971).
2. J. Bardeen, L. N. Cooper, and J. R. Schrieffer, *Phys. Rev.* **106**, 162 (1957); *ibid.* **108**, 1175 (1957).

3. K. Allart, E. Boeker, G. Bonsignori, M. Saroia, and Y. K. Gambir, *Phys. Rep.* **169**, 209 (1988).
4. J. N. Ginocchio, *Ann. Phys.* **126**, 234 (1980); C. L. Wu, D. H. Feng, and M. Guidry, *Adv. Nucl. Phys.* **21**, 227 (1994).
5. A. Arima and F. Iachello, *Phys. Rev. Lett.* **35**, 1069 (1975); F. Iachello and A. Arima, *the Interacting Boson Model*, (Cambridge University Press, 1987).
6. Jin-Quan Chen, *Nucl. Phys. A* **626**, 686 (1997); Y. M. Zhao, N. Yoshinaga, S. Yamaji, J. Q. Chen, and A. Arima, *Phys. Rev. C* **62**, 014304 (2000).
7. Y. M. Zhao and A. Arima, preprint (submitted to *Physics Reports*).
8. A. L. Goodman, Adv. Nucl. Phys. **11**, 263 (1979); Phys. Rev. C **60**, 014311 (1999).
9. I. Talmi, *Simple Models of Complex Nuclei*, (Harwood Academic Publishers, 1993); I. Talmi, *Nucl. Phys. A* **686**, 217 (2001).
10. W. Satula and R. Wyss, Phys. Lett. B **393**, 1 (1997).
11. J. Engel, K. Langanke, and P. Vogel, Phys. Lett. B **389**, 211 (1996).
12. J. Engel, S. Pittel, M. Stoitsov, P. Vogel, and J. Dukelsky, Phys. Rev. C **55**, 1781 (1997).
13. J. Dobeš and S. Pittel, Phys. Rev. C **57**, 688 (1998).
14. O. Juillet and S. Josse, Eur. Phys. J. A **8**, 291 (2000).
15. S. Lerma H., B. Errea, J. Dukelsky, and W. Satula, Phys. Rev. Lett. **99**, 032501 (2007).
16. J. Dukelsky, B. Brrea, S. Lerma H., G. G. Dussel, C. Esebbag, and N. Sandulescu, Int. J. Mod. Phys. E **17**, 2155 (2008).
17. R. A. Senkov and V. G. Zelevinsky, Phys. Atom. Nucl. **74**, 1267 (2011).
18. N. Sandulescu and G. F. Bertsch, Phys. Rev. C **78**, 064318 (2008).
19. N. Sandulescu, B. Errea, and J. Dukelsky, Phys. Rev. C **80**, 044335 (2009).
20. N. Sandulescu, D. Negrea, J. Dukelsky, and C. W. Johnson, Phys. Rev. C **85**, 061303(R) (2012).
21. D. D. Warner, M. A. Bentley, and P. Van Isacker, Nature Physics 2, 311 (2006).
22. A. V. Afanasjev, arXiv:1205.2134 (2012).
23. B. Cederwall et al., *Nature* **469**, 68 (2011).
24. C. Qi, J. Blomqvist, T. Back, B. Cederwall, A. Johnson, R. J. Liotta, and R. Wyss, *Phys. Rev. C* **84**, 021301 (2011).
25. S. Zerguine and P. Van Isacker, *Phys. Rev. C* **83**, 064314 (2011).
26. G. J. Fu, Y. Lei, Y. M. Zhao, S. Pittel, and A. Arima, to be published.
27. M. Honma, T. Otsuka, T. Mizusaki, and M. Hjorth-Jensen, Phys. Rev. C **80**, 064323 (2009).
28. C. Qi, J. Blomqvist, T. Bäck, B. Cederwall, A. Johnson, R. J. Liotta, and R. Wyss, *Phys. Scr.* **T150**, 014031 (2012).
29. P. Van Isacker, *Phys. Scr.* **T150**, 014042 (2012).
30. Z. X. Xu, C. Qi, J. Blomqvist, R. J. Liotta, and R. Wyss, *Nucl. Phys. A* **877**, 51 (2012).
31. L. Coraggio, A. Covello, A. Gargano, and N. Itaco, *Phys. Rev. C* **85**, 034335 (2012).

RELATIVISTIC SYMMETRIES IN NUCLEI

JOSEPH N. GINOCCHIO

MS 283, Los Alamos National Laboratory
Los Alamos, NM, 87545, USA
E-mail: gino@lanl.gov

We show that pseudospin symmetry is a symmetry of the Dirac Hamiltonian for which the sum of the scalar and vector potentials are a constant. In this paper we discuss some of the implications of this relativistic symmetry and the experimental data that support these predictions. We show that pseudo-U(3) symmetry is a symmetry of the Dirac Hamiltonian for which the sum of harmonic oscillator vector and scalar potentials are equal to a constant, and we give the generators of pseudo-U(3) symmetry. We also show that pseudospin for nuclei implies spin symmetry for anti-nucleons moving in a nuclear environment.

Keywords: pseudospin symmetry, relativistic mean field, Dirac Hamiltonian, SU(3) symmetry, anti-nucleons.

1. Introduction

More than forty years a quasi-degeneracy in the states $(n\,\ell_j, n'\ell'_{j'})$ of spherical nuclei was discovered,[1,2] where $n' = n-1, \ell' = \ell+2, j' = j+\frac{1}{2}$ and n, ℓ, j are the radial, orbital angular momentum, and total angular momentum quantum numbers, respectively. These quasi-degeneracies persist in recent measurements in nuclei far from stability.[3] The authors realized that, if they define the average of the orbital angular momenta as a pseudo-orbital angular momentum $(\tilde{\ell})$ and then couple a pseudospin $(\tilde{s} = \frac{1}{2})$ to the pseudo-orbital angular momentum, they will get the total angular momenta $(j = \tilde{\ell} \pm \frac{1}{2})$. For example, for the $(1s_{\frac{1}{2}}, 0d_{\frac{3}{2}})$ orbits, $\tilde{\ell} = 1$, which gives the total angular momenta $j = \frac{1}{2}, \frac{3}{2}$. Subsequently pseudospin doublets in deformed nuclei were discovered.[4] Pseudospin symmetry was later revealed to be a symmetry of the Dirac Hamiltonian.[5,6]

2. Symmetries of the Dirac Hamiltonian

The Dirac Hamiltonian with a Lorentz scalar potential, $V_S(\vec{r})$, and a potential which is the fourth component of a Lorentz vector potential, $V_V(\vec{r})$, is

$$H = \vec{\alpha} \cdot \vec{p} + \beta(V_S(\vec{r}) + M) + V_V(\vec{r}), \tag{1}$$

where $\vec{\alpha}$, β are the Dirac matrices, $\vec{p}$ is the momentum, M is the mass, $\vec{r}$ is the radial coordinate, and the velocity of light is set equal to unity, c =1. The Dirac Hamiltonian has spin symmetry when the difference of the vector and scalar potentials in the Dirac Hamiltonian is a constant, $V_S(\vec{r}) - V_V(\vec{r}) = C_s$.[7] Hadrons[8] and anti-nucleons in a nuclear environment have spin symmetry.[9] These are relativistic systems and normally, in such systems, we would expect large spin-orbit splittings, but, in this limit, spin doublets are degenerate.

2.1. *Pseudospin Symmetry: A Symmetry of the Dirac Hamiltonian*

Another SU(2) symmetry of the Dirac Hamiltonian was discovered to occur when the sum of the vector and scalar potentials in the Dirac Hamiltonian is a constant, $V_S(\vec{r}) + V_V(\vec{r}) = C_{ps}$.[7] The generators for this SU(2) algebra $\vec{S}$, which commute with the Dirac Hamiltonian with any potential $V(\vec{r})$, spherical[1,2] or deformed,[4] $[\,H\,,\,\vec{S}\,] = 0$, are given by[10]

$$\vec{\tilde{S}} = \begin{pmatrix} U_p\;\vec{s}\;U_p & 0 \\ 0 & \vec{s} \end{pmatrix}, \tag{2}$$

where $\vec{s} = \vec{\sigma}/2$ are the usual spin generators, $\vec{\sigma}$ the Pauli matrices, and $U_p = \frac{\boldsymbol{\vec{\sigma}\cdot\vec{p}}}{p}$ is the helicity unitary operator introduced in.[11] The generators are four by four matrices as appropriate for the Dirac Hamiltonian.

This symmetry was shown to be pseudospin symmetry.[5] The eigenfunctions of the Dirac Hamiltonian in this limit will have degenerate doublets of states, one of which has pseudospin aligned and the other with pseudospin unaligned. The "upper" matrix of the pseudospin generators in Eq (2), $U_p\;\vec{s}\;U_p$, have the spin intertwined with the momentum which enables the generators to connect the states in the doublet, which differ by two units of angular momentum. The approximate equality in magnitude of the vector and scalar fields in nuclei and their opposite sign have been confirmed in relativisitc mean field theories[6] and in QCD sum rules.[6,12]

3. Consequences of Relativistic Pseudospin Symmetry

One immediate consequence of pseudospin symmetry as a relativistic symmetry is that the "lower" matrix of the pseudospin generators in Eq (2), $\vec{s}$, does not change the radial wavefunction of the "lower" component of the Dirac eigenfunctions. Hence this symmetry predicts that the radial wavefunctions of the "lower" component is the same for the two states in the doublet. Previous to this discovery many relativistic mean field calculations of nuclear properties had been made. Hence this prediction was tested with existing calculations and, indeed, these wavefunctions are very similar for both spherical[13,14] and deformed nuclei.[15,16] Because of the momentum dependence of the "upper" matrix of the generators the relationship between the "upper" components involves a differential equation and these have also been tested in spherical[17] and deformed nuclei[16] with success.

Magnetic dipole and Gamow-Teller transitions between the two states in pseuodspin doublets are forbidden non-relativistically because the states differ by two units of angular momentum. However, they are not forbidden relativistically which means that they are proportional to the lower component of the Dirac eigenfunction. This leads to a condition between the magnetic moments of the states and the magnetic dipole transtion between them because the radial amplitudes of the lower components of the two states in a pseudospin doublet are equal. Therefore the magnetic dipole transition between the two states in the doublet can be predicted if the magnetic moments of the states are known.[6,18] Likewise pseudospin symmetry also predicts Gamow-Teller transitions between a state in a parent nucleus to the partner state in the daughter nucleus if the Gamow-Teller transition to the same states in the parent and daughter nucleus is known. We do not have space to discuss these relationships in detail but one example occurs the mirror nuclei ${}^{39}_{19}\mathrm{K}_{20}$ and ${}^{39}_{20}\mathrm{Ca}_{19}$. The ground state and first excited state of ${}^{39}_{19}\mathrm{K}_{20}$ are interpreted as a $0d_{3/2}$ and $1s_{1/2}$ proton hole respectively, while the ground state and first excited state of ${}^{39}_{20}\mathrm{Ca}_{19}$ are interpreted as a $0d_{3/2}$ and $1s_{1/2}$ neutron hole respectively. These states are members of the $\tilde{n}_r = 1, \tilde{\ell} = 1$ pseudospin doublet. Using the magnetic moment of ${}^{39}\mathrm{Ca}$ a transition rate is calculated which is only about 37 % larger than the measured. However, the two states in the doublet are not pure single-particle states. A modification of these relations has been derived which take into account the fact that these states are not pure single particle states.[6,19] The modified relations give a transition rate that agrees with the measured value to within experimental error. Again using the mass 39 nuclei, the Gamow - Teller transitions from the ground state of ${}^{39}\mathrm{Ca}$ to the ground and first excited state of ^{39}K are known, which is enough information to predict the transition from the ground state to the excited state. In the non-relativistic shell model an effective tensor term g_{eff} $[Y_2\sigma]^{(1)}$ is added to the magnetic dipole operator and the Gamow-Teller operator to produce a transition, where g_{eff} is a calculated effective coupling constant. However, the magnetic dipole transition calculated between the same states is an order of magnitude lower than the experimental transition[20] although the calculated Gamow-Teller agrees with the experimental value within the limits of experimental and theoretical uncertainty. This inconsistency has been a puzzle for the non-relativistic shell model. On the other hand the relativistic single-nucleon model gives a consistent description of both of these transitions. A global prediction of magnetic dipole transitions throughout the periodic table has had reasonable success as well.[6,19] However, a global prediction of Gamow-Teller transitions have not been done yet. Pseuodspin symmetry can also be used to relate quadrupole transitions between multiplets.[6]

4. The Dirac Hamiltonian with Pseudo-U(3) Symmetry

It is well known that the non relativistic spherical harmonic oscillator potential has a U(3) symmetry.[21] The energies depend linearly on the total number of harmonic oscillator quanta, $N = 2n + \ell$, and the generators of the U(3) symmetry are the angular momentum operator, a quadrupole operator, and a monopole operator which commute with the non-relativistic Hamiltonian. We know that the Dirac Hamiltonian for which the difference of the scalar and vector potentials is a constant has spin symmetry. But if, in addition, the potentials are spherical harmonic oscillators, the Dirac Hamiltonian can be solved analytically[22] and has an U(3) symmetry.[23] The energies depend on the same harmonic quantum number N as the non-relativistic harmonic oscillator, and, therefore has the same degeneracies. However, the energy dependence on N depends on the ratio of the M to the strength of the potential. For M large, the dependence is linear just as in the non-relativistic oscillator. For M small, the energy goes like $N^{\frac{2}{3}}$, which means that, in the relativistic limit, the oscillator is not harmonic. Hence even a relativistic system can have U(3) symmetry.

If, on the other hand, the sum of the vector and scalar potentials is a constant and are spherical harmonic oscillators, the Dirac Hamiltonian, $\tilde{H}$, has a pseudospin symmetry and a pseudo-U(3) symmetry. The energy depends on the total number of pseudo-quanta, $\tilde{N} = 2\tilde{n} + \tilde{\ell}$. The quantum numbers, $\tilde{N}, \tilde{n}, \tilde{\ell}$, refer to the "lower" component of the Dirac eigenfunction.

The energy eigenvalues are[6]

$$E_{\tilde{N}} = \frac{M}{3}\left[3B(A_{\tilde{N}}) - 1 + \frac{4}{3\ B(A_{\tilde{N}})}\right], \tag{3}$$

where $B(A_{\tilde{N}}) = \left[\frac{A_{\tilde{N}}+\sqrt{A_{\tilde{N}}^2+\frac{32}{27}}}{2}\right]^{\frac{2}{3}}$, and $A_{\tilde{N}} = \frac{\sqrt{2}\hbar\omega}{M}(\tilde{N}+\frac{3}{2})$. The spectrum goes like $\tilde{N}$ for M small and $\tilde{N}^{\frac{2}{3}}$ for M large; i.e., the relativistic harmonic oscillator is not harmonic.

The pseudo-U(3) generators which commute with the Dirac Hamiltonian, are the pseudo-orbital angular momentum and the quadrupole operator which is given by

$$\tilde{Q}_m = \sqrt{\frac{3}{M\omega^2\hbar^2(\tilde{H}-M)}}\left(M\omega^2\ (\tilde{H}-M)\ [rr]^{(2)}_m + [pp]^{(2)}_m\right), \tag{4}$$

along with the pseudo-quantum number operator

$$\tilde{\mathcal{N}} = \frac{\sqrt{\tilde{H}-M}(\tilde{H}+M)}{\hbar\sqrt{2M\omega^2}} - \frac{3}{2}. \tag{5}$$

However, the eigenfunctions of $\tilde{H}$ are not valence Dirac eigenfunctions, but hole Dirac eigenfunctions or, in common parlance, "negative" energy states. The energies are not negative, though, since the harmonic oscillator is not finite. But the nodal structure of the eigenfunctions is that of hole states.[6] There are no bound valence states of $\tilde{H}$ and hence these solutions do not correspond to realistic nuclei. Of course, for real nuclei, the scalar and vector potentials are not exactly equal in magnitude, and so pseudospin is approximately conserved since the ratio of the sum of the potentials to their difference is approximately .03. For this reason we are attempting to solve analytically the Dirac Hamiltonian for general scalar and vector harmonic oscillator potentials.

5. Anit-nucleon in a Nuclear Environment

Charge conjugation changes a nucleon into an anti-nucleon. Under charge conjugation the scalar potential of a nucleon remains invariant while the its vector potential changes sign. Hence the pseudospin condition $V_S(\vec{r}) + V_V(\vec{r}) \approx C_{ps}$ becomes $\bar{V}_S(\vec{r}) - \bar{V}_V(\vec{r}) \approx C_{ps} = \bar{C}_s$. Hence approximate pseudospin for nucleons predicts approximate spin symmetry for anti-nucleons in a nuclear environment. Since the potentials are also very deep we would expect approximate $U(3)$ symmetry as well. The sparse data on the scattering of polarized anti-nucleons on nuclei supports this prediction.[24] Perhaps more data on the scattering of polarized anti-nucleons on nuclei will be forthcoming from GSI.

Acknowledgments

This research was supported by the US Department of Energy, Contract No. W-7405-ENG-36.

References

1. A. Arima, M. Harvey, and K. Shimizu, *Phys. Lett. B* **30**, 517 (1969).
2. K. T. Hecht, and A. Adler, *Nucl. Phys. A* **137**, 129 (1969).
3. O. Sorlin, and M. G. Porquet, *Prog. Part. Nucl. Phys.* **61**, 602 (2008).
4. A. Bohr, I. Hamamoto, and B. R. Mottelson, *Phys. Scr.* **26**, 267 (1982).
5. J. N. Ginocchio, *Phys.Rev. Lett.* **78**, 436 (1997).
6. J. N. Ginocchio, *Phys.Rep.* **414**, 165 (2005).
7. J. S. Bell, and H. Ruegg *Nucl.Phys. B* **98** 151 (1997).
8. P. R. Page, T. Goldman, and J. N. Ginocchio, *Phys.Rev. Lett.* **86**, 204 (2001).
9. J. N. Ginocchio *Phys.Rep.* **315**, 231 (1999).
10. J. N. Ginocchio, and A. Leviatan, *Phys. Lett. B* **425**, 1 (1998).
11. A. L. Blokhin, C. Bahri, and J. P. Draayer *Phys.Rev. Lett.* **74**, 4149 (1995).
12. T. D. Cohen, *et al*, *Prog. Part. Nucl. Phys.* **35**, 221 (1995).
13. J. N. Ginocchio, D. G. Madland, *Phys. Rev. C* **57**, 1167 (1998).
14. J. Meng, K. Sugawara-Tanabe, S. Yamaji, P. Ring, A. Arima, *Phys. Rev. C* **58**, R628 (1998).
15. K. Sugawara-Tanabe, S. Yamaji, and A. Arima, *Phys. Rev. C* **65**, 054313 (2002).
16. J. N. Ginocchio, A. Leviatan, J. Meng, and Shan-Gui Zhou, *Phys. Rev. C* **69** , 03430 (2004).
17. J. N. Ginocchio, *Phys. Rev. C* **66**, 064312 (2002).
18. J. N. Ginocchio, *Phys. Rev. C* **59**, 2487 (1999).
19. P. von Neumann-Cosel and J. N. Ginocchio, *Phys. Rev. C* **62** 014308 (2000).
20. E. Hagberg, et al, *Nucl. Phys. A* **571**, (1994) 555.
21. J. P. Elliott, *Proc. R. Soc. London, Ser. A* **245**, 128 (1958).
22. J. N. Ginocchio, *Phys. Rev. C* **69**, 034318 (2004).
23. J. N. Ginocchio, *Phys. Rev. Lett.* **95**, 252501 (2005).
24. D. Garetta, et al., *Phys. Lett. B* **151**, 473 (1985).

THE HEINE-STIELTJES CORRESPONDENCE AND THE POLYNOMIAL APPROACH TO THE SU(1,1) AND SU(2) GAUDIN MODELS

F. PAN*, X. GUAN, Y. YU

Department of Physics, Liaoning Normal University, Dalian 116029, China
** E-mail: daipan@dlut.edu.cn*
www.lnnu.edu.cn

K. D. LAUNEY, J. P. DRAAYER

Department of Physics and Astronomy, Louisiana State University, Baton Rouge, LA 70803-4001, USA
www.phys.lsu.edu

The Heine-Stieltjes correspondence is extended and applied to solve Bethe ansatz equations of the SU(1,1) and SU(2) Gaudin models, from which the extended Heine-Stieltjes polynomial approach to these models is proposed. As examples for the application, exact solutions of the standard two-site Bose-Hubbard model and the standard pairing model are formulated from the corresponding polynomials.

Keywords: Exactly solvable models; the Heine-Stieltjes correspondence; the polynomial approach.

1. Introduction

In the application of the Bethe ansatz method, Gaudin and Richardson independently found exact solutions for the spin[1] or nuclear pairing many-body problems,[2] which are referred to as the Gaudin(-Richardson) models. In this talk, we report our recent work on relation of solutions of the Bethe ansatz equations (BAE) for the SU(1,1) and SU(2) Gaudin models and zeros of the extended Heine-Stieltjes polynomials.

There is a large class of polynomials, called Heine-Stieltjes polynomials $y(x)$, which satisfy a second-order Fuchsian equation

$$A(x)\frac{d^2y(x)}{dx^2} + B(x)\frac{dy(x)}{dx} + V(x)y(x) = 0, \tag{1}$$

where $A(x)$ is a polynomial of degree m with $A(x) = \prod_{i=1}^{m}(x - a_i)$, $B(x)$ is a polynomial of degree $m - 1$ such that for a set of real positive parameters $\{\gamma_i\}$ and another set of real parameters $\{a_i\}$ is given as $B(x)/A(x) = \sum_{i=1}^{m} \gamma_i/(x - a_i)$, and $V(x)$ is an unknown, but to be determined polynomial of degree $m - 2$ that is allowed to depend on the solution $y(x)$. The latter are often called Van Vleck polynomials.[3] If $y(x)$ is a polynomial of degree k with simple zeros $\{x_1, x_2, \cdots, x_k\}$, one may write $y(x) = \prod_{j=1}^{k}(x - x_j)$. Hence, at any zero x_i, $y(x)$ satisfies

$$\frac{y''(x_i)}{y'(x_i)} = \sum_{j \neq i} \frac{2}{x_i - x_j}. \tag{2}$$

Combining (1) and (2), one obtains the following important relations among the zeros:

$$\sum_{j \neq i}^{k} \frac{2}{x_i - x_j} + \sum_{\mu=1}^{m} \frac{\gamma_\mu}{x_i - a_\mu} = 0 \tag{3}$$

for $i = 1, 2, \cdots, k$. It should be noted that the BAEs frequently appearing in search for exact solutions of quantum many-body problems, such as those associated with Gaudin type systems, are similar to the relations shown in (3). Roots of the BAEs (3) simultaneously determine the eigenenergies and eigenstates of the corresponding quantum many-body problem. As an alternative, roots of the BAEs may also be calculated as zeros of the corresponding polynomial $y(x)$. In this way, a link between solutions of the Gaudin type for quantum many-body problems and the corresponding polynomials is established.

2. The SU(1,1) Case

As a simple extension of the Stieltjes correspondence, we revisit the SU(1,1) Bethe ansatz solutions for the standard two-site Bose-Hubbard model,[4] for which the Hamiltonian is

$$\hat{H}_{\rm BH} = -t(c^\dagger d + d^\dagger c) + U(c^\dagger c c^\dagger c + d^\dagger d d^\dagger d), \tag{4}$$

where $c^\dagger$ (c) and $d^\dagger$ (d) are boson creation (annihilation) operators. The parameters t and U in the Bose-Hubbard model are related to the Josephson coupling and the charging energy, respectively. We use the unitary transformation for the boson operators with $c = (a - ib)/\sqrt{2}$ and $d = (a + ib)/\sqrt{2}$. Then, the Hamiltonian (4) can be rewritten in terms of a- and b-boson operators as

$$\hat{H}_{\rm BH} = t(b^\dagger b - a^\dagger a) - \frac{1}{2} U B^+(0) B^-(0) + U\hat{n}^2, \tag{5}$$

where $\hat{n} = a^\dagger a + b^\dagger b$ is the operator for the total number of bosons in the system, and $B^+(0) = b^{\dagger 2} + a^{\dagger 2}$, $B^-(0) = b^2 + a^2$ are boson pairing operators. The algebraic Bethe ansatz implies that eigenvectors of (5) may be expressed as

$$|n,\ \zeta, \nu_1, \nu_2\rangle = B^+(x_1^{(\zeta)}) B^+(x_2^{(\zeta)}) \cdots B^+(x_k^{(\zeta)}) |\nu_1, \nu_2\rangle \tag{6}$$

with $n = 2k + \nu_1 + \nu_2$, and $B^+(x_i^{(\zeta)}) = a^{\dagger 2}/(x_i^{(\zeta)} + 1) + b^{\dagger 2}/(x_i^{(\zeta)} - 1)$ and $|\nu_1, \nu_2\rangle = a^{\dagger \nu_1} b^{\dagger \nu_2} |0\rangle$ with $\nu_{1,2} = 0$ or 1, in which $x_i^{(\zeta)}$ $(i = 1, 2, \cdots, k)$ are spectral parameters to be determined, and ζ is an additional quantum number for distinguishing different eigenvectors with the same quantum number k. It can then be verified by using the corresponding eigenequation that (6) is a solution when the spectral parameters $x_i^{(\zeta)}$ $(i = 1, 2, \cdots, k)$ satisfy the following set of BAEs:

$$U\left(\frac{2\nu_1 + 1}{x_i^{(\zeta)} + 1} + \frac{2\nu_2 + 1}{x_i^{(\zeta)} - 1}\right) + 2t + 4U \sum_{j(\neq i)} \frac{1}{x_i^{(\zeta)} - x_j^{(\zeta)}} = 0 \tag{7}$$

for $i = 1, 2, \cdots, k$, with the corresponding eigenenergy given by $E_n^{(\zeta)} = 2t \sum_{i=1}^{k} x_i^{(\zeta)} + t(\nu_2 - \nu_1) + Un^2$.

In order to compare to the Heine-Stieltjes correspondence, we assume $t > 0$ and $U \neq 0$, and set $\alpha = \nu_2 + 1/2$, $\beta = \nu_1 + 1/2$, and $\gamma = t/U$ Then, the BAEs (7) become

$$\left(\frac{\alpha}{x_i^{(\zeta)} - 1} + \frac{\beta}{x_i^{(\zeta)} + 1}\right) + \gamma + \sum_{j(\neq i)} \frac{2}{x_i^{(\zeta)} - x_j^{(\zeta)}} = 0. \tag{8}$$

Although the $\gamma = 0$ result is trivial, according to the Stieltjes correspondence, the polynomial corresponding to (8) in this case is the Jacobi polynomial $P_k^{(\alpha-1,\beta-1)}(x)$. In this case, the Van Vleck polynomial $V(x)$ is trivially a k-dependent constant. Hence, there is only one set of zeros of $P_k^{(\alpha-1,\beta-1)}(x_i)$ with $i = 1, 2, \cdots, k$, satisfying the Bethe ansatz equation (8). The case with $\gamma \neq 0$ is non-trivial. According to (1)-(3), we write a differential equation corresponding to the Bethe ansatz equation (8) as

$$\frac{d^2y(x)}{dx^2} + \left(\frac{\alpha}{x-1} + \frac{\beta}{x+1} + \gamma\right)\frac{dy(x)}{dx} + \frac{V(x)}{x^2-1}y(x) = 0 \qquad (9)$$

with the corresponding polynomials $A(x) = x^2 - 1$ and $B(x) = \gamma x^2 + (\alpha + \beta)x + \alpha - \beta - \gamma$ shown in (1). In contrast to the Heine-Stieltjes equation (1), however, in this case the polynomial $B(x)$ is of the same degree as that of $A(x)$. Therefore, the polynomials $y(x)$ determined by (9) should be similar to but different from those of Heine-Stieltjes type.

In the search for polynomial solutions of (9), we write $y_k(x) = \sum_{j=0}^{k} b_j x^j$. Substitution of it into (9) determines the corresponding polynomial $V(x) = -\gamma kx + f$, where f is an undetermined constant depending on k, and together with the unknown expansion coefficients b_j satisfy the following four-term relations:

$$\begin{aligned}(j(\alpha + \beta + j - 1) + f)\, b_j &= (j+2)(j+1)b_{j+2} + \\ &(j+1)(\beta - \alpha + \gamma)b_{j+1} + \gamma(k - j + 1)b_{j-1}\end{aligned} \qquad (10)$$

with $b_j = 0$ for $j \leq -1$ or $j \geq k+1$. It is clear that (10) can be represented as a matrix with $k+1$ eigenvalues labeled as $f^{(\zeta)}$ $(\zeta = 1, 2, \cdots, k+1)$ and with the corresponding eigenvectors $\{b_1^{(\zeta)}, \cdots, b_k^{(\zeta)}\}$, which determine the polynomial $y_k^{(\zeta)}(x)$. The number of solutions for f is indeed the same as that of the eigenstates of the standard two-site Bose-Hubbard Hamiltonian (4) when $\gamma \neq 0$. The polynomials $y_k(x) = \sum_{j=0}^{k} b_j x^j$ determined by (10) with $\gamma \neq 0$ are called the extended Heine-Stieltjes polynomials.[5]

3. The SU(2) Case

The Hamiltonian of the standard pairing model is given by

$$\hat{H}_{\text{SP}} = \sum_{j=j_1}^{j_n} \epsilon_j \hat{n}_j - G \sum_{jj'} S_j^+ S_{j'}^-, \qquad (11)$$

where n is the total number of levels considered, $G > 0$ is the overall pairing strength, $\{\epsilon_j\}$ are unequal single-particle energies, $\hat{n}_j = \sum_m a_{jm}^\dagger a_{jm}$

is the number operator for the valence particles in the j-th level, and $S_j^+ = \sum_{m>0}(-)^{j-m}a_{jm}^\dagger a_{j\,-m}^\dagger$ ($S_j^- = (S_j^+)^\dagger$) are pair creation (annihilation) operators generating the SU(2) quasi-spin algebras. Since the formalism for even-odd systems is similar, in the following, we only focus on the even-even seniority zero case. According to the Gaudin-Richardson method, k-pair eigenstates of (11) can be written as

$$|k;x\rangle = S^+(x_1)S^+(x_2)\cdots S^+(x_k)|0\rangle, \tag{12}$$

where $S^+(x_i) = \sum_j S_j^+/(2\epsilon_j - x_i)$, in which x_i $(i = 1, 2, \cdots, k)$ are spectral parameters to be determined. It can then be verified by using the corresponding eigenequation that (12) is the eigenstates of (11) only when the spectral parameters x_i $(i = 1, 2, \cdots, k)$ satisfy the following set of BAEs:

$$1 - 2G\sum_j \frac{\rho_j}{x_i - 2\epsilon_j} - 2G\sum_{i'(\neq i)} \frac{1}{x_i - x_{i'}} = 0, \tag{13}$$

where $\rho_j = -(j + 1/2)/2$, with the corresponding eigenenergy given by $E_{n,\ k} = \sum_{i=1}^k x_i$.

According to Heine-Stieltjes correspondence, for nonzero pairing strength G, $A(x)$ and $B(x)$ of the Fuchsian equation (1) in this case are given by

$$A(x) = \prod_{j=j_1}^{j_n}(x - 2\epsilon_j), \quad B(x)/A(x) = \sum_{j=j_1}^{j_n}\frac{2\rho_j}{x - 2\epsilon_j} - \frac{1}{G}\,. \tag{14}$$

The corresponding Van Vleck polynomial $V(x)$ in this case is of degree $n-1$, which needs to be determined according to Eq. (1). In the original electrostatic analogue considered by Heine and Stieltjes, the parameters $\{\rho_j\}$ acting as fixed charges should all be positive with no external electrostatic field, $1/G \to 0$. Therefore, the polynomials $y(x)$ satisfying Eq. (1) with negative $\{\rho_j\}$ and $1/G \neq 0$ in this case are called the extended Heine-Stieltjes polynomials, which tend to be the original Heine-Stieltjes polynomials with negative $\{\rho_j\}$ in the $G \to \infty$ limit.

In the search for polynomial solutions of (1) for this case, we write[6]

$$y(x) = \sum_{j=0}^{k} a_j x^j, \quad V(x) = \sum_{j=0}^{n-1} b_j x^j, \tag{15}$$

where $\{a_j\}$ and $\{b_j\}$ are expansion coefficients to be determined. Substitution of (15) into (1) yields two matrix equations. Namely, the condition that the coefficients in front of x^i $(i = 0, \cdots, k)$ must be zero generates a $(k+1) \times (k+1)$ matrix $\mathbf{F}$ with $\mathbf{Fv} = b_0\mathbf{v}$, where the eigenvector $\mathbf{v}$ of $\mathbf{F}$ is given by the expansion coefficients $\mathbf{v} = \{a_0, \cdots, a_k\}$, while the condition that the coefficients in front of x^i $(i = k+1, \cdots, n+k-1)$ must be zero generates another $(n-1) \times (k+1)$ upper-triangular matrix $\mathbf{P}$ with $\mathbf{Pv} = 0$, which provides a unique solution for $\{b_1, b_2, \cdots, b_{n-1}\}$ in terms of $\{a_j\}$. Entries of the two matrices are all linear in the coefficients $\{b_1, b_2, \cdots, b_{n-1}\}$. Matrices $\mathbf{F}$ and $\mathbf{P}$ can be easily constructed, for which a simple MATHEMATICA code is now available.[7]

4. Summary

In this talk, the Heine-Stieltjes correspondence is extended and applied to solve Bethe ansatz equations of the SU(1,1) and SU(2) Gaudin models, from which the extended Heine-Stieltjes polynomial approach to the Gaudin-Richardson models is proposed. Solutions of the standard two-site Bose-Hubbard model and the standard pairing model for nuclei are formulated from the corresponding extended Heine-Stieltjes polynomials. The results by extension clearly show a new link between Bethe ansatz type solutions of a large class of Gaudin type quantum many-body problems and a class of extended Heine-Stieltjes polynomials satisfying second-order differential equations, which, in turn, opens a new way to find solutions of these BAEs from the zeros of the polynomials.

Support from the U.S. National Science Foundation (OCI-0904874), the U. S. Department of Energy (DE-SC0005248), the Southeastern Universities Research Association, the National Natural Science Foundation of China (11175078), the Doctoral Program Foundation of State Education Ministry of China (20102136110002), and the LSU–LNNU joint research program (9961) is acknowledged.

References

1. M. Gaudin, *J. Phys. France* **37**, 1087 (1976).
2. R. W. Richardson, *J. Math. Phys.* **6**, 1034 (1965).
3. E. B. Van Vleck, *Bull. Amer. Math. Soc.* **4**, 426 (1898).
4. F. Pan and J. P. Draayer, *Phys. Lett.* A **339**, 403 (2005).
5. F. Pan, L. Bao, L. Zhai, X. Cui, and J. P. Draayer, *J. Phys. A* **44**, 395305 (2011).
6. X. Guan, K. D. Launey, M. Xie, L. Bao, F. Pan, and J. P. Draayer, *Phys. Rev.* C **86**, 024313 (2012).
7. X. Guan and K. D. Launey, *MATHEMATICA code for polynomial solutions to the exact pairing problem*, 2012, unpublished.

THE GROUP THEORETIC METHOD IN THE MULTI-QUARK SYSTEM STUDY

JIALUN PING*

School of Physical Science and Technology, Nanjing Normal University, Nanjing, 210023, P.R. China
**E-mail: jlping@njnu.edu.cn*

FAN WANG

School of Physics, Nanjing University, Nanjing, 210097, P. R. China

A group theoretic method for the systematic study of multi-quark states is developed. The calculation of matrix elements of many-body Hamiltonians is simplified by transforming the physical bases (quark cluster bases) to symmetry bases (group chain classified bases), where the fractional parentage expansion method can be used. The technique is applied to 5- and 6-quark systems.

Keywords: multi-quark system; physical basis; symmetry basis; fractional parentage expansion technique.

1. Introduction

Symmetry plays a important role in exploring and understanding the properties of fundamental particles. A remarkable example is Gell-Mann's prediction of baryon Ω^- in 1962,[1] which based on flavor SU(3) symmetry. The particle was discovered in 1964.[2] Group theory is the mathematical theory of symmetry. The application of group theoretic method in physics can be categorized into three classes. (1) In exploring the new research field, the guidance of setting up theory can be provided by the symmetry of the system. (2) Various states of the system can be classified according to its symmetry. (3) The group theoretic method can be used to construct wavefunctions of many-body system and to simplify the calculation of matrix elements. The topic of the present discussion is to explore the application of group theoretic method in the multi-quark study.

The multi-quark system is an exotic system which beyond the naive picture of hadrons: baryon-qqq and meson-$q\bar{q}$. Although QCD-inspired models and their variants predicted the existence of the multi-quark states, even QCD does not deny their existence, no multi-quark state has been confirmed experimentally so far. However, recent experiments reported a lot of states: XYZ particles,[3] Θ^+,[4] etc. which are difficult to fit into the conventional hadron picture. Various explanations, molecule, multi-quark state or hybrid states and so on, are proposed.[3] The isoscalar ABC structure in the pd-scattering can be accounted for by introducing the 6-quark system: dibaryon.[5,6] The "proton spin crisis" and the strange magnetic moment of proton call for 5-quark components.[7,8] So the further study of multi-quark system is needed.

The calculation of multi-quark system is genuinely a many-body, many channel coupling one, which is tedious and difficult to perform. A powerful method is needed. In the present work, two bases, physical basis and symmetry basis, are used to construct the wavefunctions of 5- and 6-quark systems. The transformation between two bases are derived through the isoscalar factors of $SU(mn) \supset SU(m) \times SU(n)$. The matrix elements of operators are calculated by using fractional parentage (fp) expansion technique, which has been widely used in atomic and nuclear physics.

2. Physical basis, symmetry basis and fractional parentage expansion technique

There are two types of bases for the multi-quark system. One is called physical basis (quark cluster basis), where the p-particle system is separated to two sub-clusters, every sub-cluster has definite symmetry properties and relate to physical particles in most cases. The bases of the system are obtained by antisymmetrizing the coupled bases of two sub-clusters. The advantage of the physical basis is that it has a clear physical content. The other is called symmetry bases (group chain classification bases), or the irreducible bases of the symmetry group chain of the system. The symmetry group chain for the p-quark system within the u, d, s world is

$$\begin{aligned} &\overset{[1^p]}{\mathrm{SU}(36)} \supset \overset{[\nu]}{\mathrm{SU}^x(2)} \times \overset{[\tilde{\nu}]}{\mathrm{SU}(18)} \supset \overset{[\sigma]}{\mathrm{SU}^c(3)} \times \overset{[\mu]}{\mathrm{SU}(6)} \\ &\qquad \supset ((\overset{[f]}{\mathrm{SU}^f(3)} \supset \overset{I}{\mathrm{SU}^\tau(2)} \times \overset{Y}{\mathrm{U}^Y(1)}) \times \overset{J}{\mathrm{SU}^\sigma(2)}) \end{aligned} \tag{1}$$

where $[\nu]$ etc. are the Young diagrams describing the permutational and SU(n) symmetries. Although the symmetry basis do not shown us the clear physical contents, it has the advantage that the fp expansion technique can be used to simplify the calculation of the matrix elements. So to have a clear physical contents of the state and to make use of fp technique, the transformation between the physical basis and the symmetry basis is needed. In the following, we illustrate the procedure for 5- and 6-quark systems. The details can be found in Refs. 11–13.

2.1. *Six-quark system*

For 6-quark system, it is naturally to decompose the system to two 3-quark clusters, baryons if there are color singlet. Then the physical basis can be written as,

$$\begin{aligned}\Psi_{\alpha\kappa}(B_1B_2) &= \mathcal{A}\left[\psi_B(q_1q_2q_3)\psi_B(q_4q_5q_6)\right]^{[\sigma]IJ}_{WM_IM_J} \\ &= \mathcal{A}\left[\left|\begin{array}{c}[\nu_1]\\ [\sigma_1][\mu_1][f_1]Y_1I_1J_1\end{array}\right\rangle\left|\begin{array}{c}[\nu_2]\\ [\sigma_2][\mu_2][f_2]Y_2I_2J_2\end{array}\right\rangle\right]^{[\sigma]IJ}_{WM_IM_J},\end{aligned} \tag{2}$$

where $[\ ]^{[\sigma]IJ}_{WM_IM_J}$ means the couplings in terms of the $SU^c(3)$, $SU^\tau(2)$ and $SU^\sigma(2)$ Clebsch-Gordan coefficients (CGC) so that it has the total color symmetry $[\sigma]W$, isospin IM_I and spin JM_J. $\mathcal{A}$ is a normalized antisymmetric operator. $\alpha = (YIJ)$ with $Y = Y_1 + Y_2$, κ represents the quantum numbers $\nu_i, \sigma_i, \mu_i, f_i, Y_i, I_i, J_i$ (i=1,2).

The symmetry basis is the group chain (Eq.1) classification basis

$$\Phi_{\alpha K}(q^6) = \left|\begin{array}{c}[\nu]l^3r^3\\ [\sigma]W[\mu]\beta[f]YIJM_IM_J\end{array}\right\rangle \tag{3}$$

Physical and symmetry bases both form a complete set in a truncated Hilbert space, and are related by a unitary transformation,[9,11]

$$\begin{aligned}\Psi_{\alpha\kappa}(B_1B_2) = \sum_{\tilde{\nu}\mu\beta f\gamma} & C^{[\tilde{\nu}][\sigma][\mu]}_{[\tilde{\nu}_1][\sigma_1][\mu_1],[\tilde{\nu}_2][\sigma_2][\mu_2]} C^{[\mu]\beta[f]\gamma J}_{[\mu_1][f_1]J_1,[\mu_2][f_2]J_2} \\ & C^{[f]\gamma YI}_{[f_1]Y_1I_1,[f_2]Y_2I_2}\Phi_{\alpha K}(q^6)\end{aligned} \tag{4}$$

A physical six quark state with quantum number $\alpha = (YIJ)$ is the linear combination of the physical bases,

$$\Psi_\alpha = \sum_\kappa C_\kappa \Psi_{\alpha\kappa}(B_1B_2). \tag{5}$$

The coefficients C_κ is determined by diagonalizing the six quark Hamiltonian as usual. The calculation of six quark Hamiltonian matrix elements in the physical basis is difficult and tedious. To take the advantage of the fp expansion technique, the symmetry basis has to be used. In fp expansion method, The matrix elements $\langle \Phi_{\alpha K} |H| \Phi_{\alpha K'} \rangle$ can be written as,

$$\langle \Phi_{\alpha K} |H| \Phi_{\alpha K'} \rangle = \sum \binom{6}{2} \langle \Phi_{\alpha K} |\alpha_1 K_1, \alpha_2 K_2 \rangle \langle \alpha_1' K_1', \alpha_2' K_2' | \Phi_{\alpha K'} \rangle \langle \alpha_1 K_1 | \alpha_1' K_1' \rangle \langle \alpha_2 K_2 | H_{56} | \alpha_2' K_2' \rangle. \quad (6)$$

Here $\langle \alpha_1 K_1 | \alpha_1' K_1' \rangle$ is the four quark overlap, $\langle \alpha_2 K_2 | H_{56} | \alpha_2' K_2' \rangle$ is the two body matrix element and H_{56} represents two-body operator for the last pair. $\binom{6}{2} = 15$ is the interacting pair number. $\langle \Phi_{\alpha K} |\alpha_1 K_1, \alpha_2 K_2 \rangle$ and $\langle \alpha_1' K_1', \alpha_2' K_2' | \Phi_{\alpha K'} \rangle$ are the total CGC of the group chain Eq.(1). They are calculated as follows,

$$\langle \Phi_{\alpha K} |\alpha_1 K_1, \alpha_2 K_2 \rangle = \sum_{\gamma} C^{[\sigma]W}_{[\sigma_1]W_1^c,[\sigma_2]W_2^c} C^{IM_I}_{I_1 M_{I_1},I_2 M_{I_2}} C^{JM_J}_{J_1 M_{J_1},J_2 M_{J_2}} C^{[\nu]l^3 r^3}_{[\nu_1]W_1^x,[\nu_2]W_2^x} C^{[1^6][\nu][\tilde{\nu}]}_{[1^4][\nu_1][\tilde{\nu}_1],[1^2][\nu_2][\tilde{\nu}_2]} C^{[\tilde{\nu}][\sigma][\mu]}_{[\tilde{\nu}_1][\sigma_1][\mu_1],[\tilde{\nu}_2][\sigma_2][\mu_2]} C^{[\mu]\beta[f]\gamma J}_{[\mu_1][f_1]J_1,[\mu_2][f_2]J_2} C^{[f]\gamma Y I}_{[f_1]Y_1 I_1,[f_2]Y_2 I_2}. \quad (7)$$

The first four C's are the $SU^c(3)$, $SU^\tau(2)$, $SU^\sigma(2)$ and $SU^x(2)$ CGC, the next three C's are the $SU(36) \supset SU^x(2) \times SU(18)$, $SU(18) \supset SU^c(3) \times SU(6)$, $SU(6) \supset SU^f(3) \times SU^\sigma(2)$ isoscalar factors, the last one is the $SU^f(3) \supset SU^\tau(2) \times U^Y(1)$ isoscalar factor. All these isoscalar factors (for particle number ≤ 6) can be found in Ref.10.

The matrix elements in symmetry basis are related to the matrix elements in physical basis through the unitary transformation Eq.(4),

$$H_{\kappa\kappa'} = \sum_{K,K'} C_{\kappa K} C_{\kappa K'} \langle \Phi_{\alpha K} |H| \Phi_{\alpha K'} \rangle. \quad (8)$$

2.2. *Five-quark system*

The physical basis (meson-baryon basis) and the symmetry basis for 5-quark system are defined as

$$\Psi_{\alpha k}(q^4\bar{q}) = \mathcal{A}\left[\psi_B(q_1 q_2 q_3)\psi_M(q_4 \bar{q}_5)\right]^{[c]IJ}_{WM_I M_J}, \quad (9)$$

$$\Phi_{\alpha K}(q^4\bar{q}) = \left[\Phi_{\alpha_4 K_4}(q^4))\psi_{[\bar{c}]\bar{I}\bar{J}}(\bar{q}_5)\right]^{[c]IJ}_{WM_I M_J}. \quad (10)$$

The unitary transformation between two bases can be setup after a series of CG re-coupling,

$$\Psi_{\alpha k}(q^4\bar{q}) = \sum_K C_{kK} \Phi_{\alpha K}(q^4\bar{q}) = \sum_{\tilde{\nu}_4 \mu_4 f_4} \{UUU\}\{CCC\} \Phi_{\alpha K}(q^4\bar{q}). \quad (11)$$

where $\{UUU\}$ and $\{CCC\}$ are Racah coefficients and isoscalar factors,

$$\{UUU\}_{\rm Racah} = U(c_3c_1cc_{\bar{1}};c_4c_2)U(I_3I_1II_{\bar{1}};I_4I_2)U(J_3J_1JJ_{\bar{1}};J_4J_2) \quad (12)$$

$$\{CCC\}_{\rm ISF} = C^{[\tilde{\nu}_4][c_4][\mu_4]}_{[\tilde{\nu}_3][c_3][\mu_3],[\tilde{\nu}_1][c_1][\mu_1]} C^{[\mu_4][f_4][J_4]}_{[\mu_3][f_3][J_3],[\mu_1][f_1][J_1]} C^{[f_4]Y_4I_4}_{[f_3]Y_3I_3,[f_1]Y_1I_1} \quad (13)$$

As before, the matrix elements in physical basis are related to the matrix elements in symmetry basis through the above transformation. Because of the antiquark, different fp expansions for qq interaction and $q\bar{q}$ interaction have to be used in calculating the matrix elements in symmetry basis. For qq interaction, $4 \to 2+2$ is used,

$$\begin{aligned}\langle\Phi_{\alpha K}\,|H_{34}|\,\Phi_{\alpha K'}\rangle = &\sum_{1,2} C^{[1^4][\nu_4][\tilde{\nu}_4]}_{[1^2][\nu_1][\tilde{\nu}_1],[1^2][\nu_2][\tilde{\nu}_2]} C^{[\tilde{\nu}_4][c_4][\mu_4]}_{[\tilde{\nu}_1][c_1][\mu_1],[\tilde{\nu}_2][c_2][\mu_2]} \\ &C^{[\mu_4][f_4]J_4}_{[\mu_1][f_1]J_1,[\mu_2][f_2]J_2} C^{[f_4]Y_4I_4}_{[f_1]Y_1I_1,[f_2]Y_2I_2} C^{[1^4][\nu_4][\tilde{\nu}_4]}_{[1^2][\nu'_1][\tilde{\nu}'_1],[1^2][\nu'_2][\tilde{\nu}'_2]} C^{[\tilde{\nu}_4][c_4][\mu_4]}_{[\tilde{\nu}'_1][c'_1][\mu'_1],[\tilde{\nu}'_2][c'_2][\mu'_2]} \\ &C^{[\mu_4][f_4]J_4}_{[\mu'_1][f'_1]J'_1,[\mu'_2][f'_2]J'_2} C^{[f_4]Y_4I_4}_{[f'_1]Y'_1I'_1,[f'_2]Y'_2I'_2} C^{[\nu_4]W_{x_4}}_{[\nu_1]W_{x_1},[\nu_2]W_{x_2}} C^{[\nu_4]W_{x'_4}}_{[\nu'_1]W_{x'_1},[\nu'_2]W_{x'_2}} \\ &\langle\alpha_1K_1|\alpha'_1K'_1\rangle\langle\alpha_2K_2|H_{34}|\alpha'_2K'_2\rangle. \end{aligned} \quad (14)$$

For $q\bar{q}$ interaction, $4 \to 3+1$ is used,

$$\begin{aligned}\langle\Phi_{\alpha K}\,|H_{15}|\,\Phi_{\alpha K'}\rangle = &\sum_{3,1} C^{[1^4][\nu_4][\tilde{\nu}_4]}_{[1^3][\nu_3][\tilde{\nu}_3],[1][\nu_1][\tilde{\nu}_1]} C^{[\tilde{\nu}_4][c_4][\mu_4]}_{[\tilde{\nu}_3][c_3][\mu_3],[\tilde{\nu}_1][c_1][\mu_1]} \\ &C^{[\mu_4][f_4]J_4}_{[\mu_3][f_3]J_3,[\mu_1][f_1]J_1} C^{[f_4]Y_4I_4}_{[f_3]Y_3I_3,[f_1]Y_1I_1} C^{[1^4][\nu_4][\tilde{\nu}_4]}_{[1^3][\nu'_3][\tilde{\nu}'_3],[1][\nu'_1][\tilde{\nu}'_1]} C^{[\tilde{\nu}_4][c_4][\mu_4]}_{[\tilde{\nu}'_3][c'_3][\mu'_3],[\tilde{\nu}'_1][c'_1][\mu'_1]} \\ &C^{[\mu_4][f_4]J_4}_{[\mu'_3][f'_3]J'_3,[\mu'_1][f'_1]J'_1} C^{[f_4]Y_4I_4}_{[f'_3]Y'_3I'_3,[f'_1]Y'_1I'_1} U(c_3c_1cc_{\bar{1}};c_4c_2)U(I_3I_1II_{\bar{1}};I_4I_2) \\ &U(J_3J_1JJ_{\bar{1}};J_4J_2)U(c'_3c'_1cc'_{\bar{1}};c_4c'_2)U(I'_3I'_1II'_{\bar{1}};I_4I'_2)U(J'_3J'_1JJ'_{\bar{1}};J_4J'_2) \\ &C^{[\nu_4]W_{x_4}}_{[\nu_3]W_{x_3},[\nu_1]W_{x_1}} C^{[\nu_4]W_{x'_4}}_{[\nu'_3]W_{x'_3},[\nu'_1]W_{x'_1}} \langle\alpha_3K_3|\alpha'_3K'_3\rangle\langle\alpha_2K_2|H_{45}|\alpha'_2K'_2\rangle. \end{aligned} \quad (15)$$

At last the matrix elements of 5-body Hamiltonian can be obtained,

$$\langle\Phi_{\alpha K}|H_5|\Phi_{\alpha K'}\rangle = 6\langle\Phi_{\alpha K}|H_{12}|\Phi_{\alpha K'}\rangle + 4\langle\Phi_{\alpha K}|H_{15}|\Phi_{\alpha K'}\rangle \quad (16)$$

For 5-quark system, there is another configuration, $qq-qq-\bar{q}$. We define the physical basis (quark cluster basis) as

$$\Psi_{\alpha k}(q^4\bar{q}) = \left[\mathcal{A}\left[\psi_2(q_1q_2)\psi_2(q_3q_4)\right]^{[c_4]I_4J_4}_{W_{c_4}M_{I_4}M_{J_4}} \psi_{[\bar{c}]\bar{I}\bar{J}}(\bar{q}_5)\right]^{[c]IJ}_{WM_IM_J}. \quad (17)$$

where $\psi_2(q^2)$ is the group chain classification basis with $p = 2$.

The physical basis in this case is related to the symmetry basis through the following transformation,

$$\Psi_{\alpha k}(q^4\bar{q}) = \sum_K C_{kK}\Phi_{\alpha K}(q^4\bar{q}) \tag{18}$$

$$C_{kK} = \sum_{\tilde{\nu}_4\mu_4 f_4} C^{[\tilde{\nu}_4][c_4][\mu_4]}_{[\tilde{\nu}_2][c_2][\mu_2],[\tilde{\nu}'_2][c'_2][\mu'_2]} C^{[\mu_4][f_4][J_4]}_{[\mu_2][f_2][J_2],[\mu'_2][f'_2][J'_2]} C^{[f_4]Y_4 I_4}_{[f_2]Y_2 I_2,[f'_2]Y'_2 I'_2}$$

The above technique has been used in the 5- and 6-quark systems quark model calculation. The transformation coefficients between physical basis and symmetry basis for 5- and 6-quark systems have been tabulated in Ref.11–13. The obtained physics results are also given in Ref.12–14. Applying to systems with more quarks is in progress.

3. Summary

There are two ways to construct the wavefunctions for multiquark system: physical bases and symmetry bases. The calculation is generally performed in the physical basis because of its clear physical contents. However, the calculation of matrix elements in physical basis is difficult and tedious. So the symmetry basis is used. With the help of fractional parentage expansion technique, the matrix elements in the symmetry basis can be reduced to two-body matrix elements. A systematic study of multiquark system is possible with the help of the transformation between two bases and the fp expansion technique.

References

1. M. Gell-Mann, *Phy. Rev.* **125**, 1067 (1962).
2. V. E. Barnes, *et al.*, *Phy. Rev. Lett.* **12**, 204 (1964).
3. S. L. Olson, *Nucl. Phy. A* **827**, 53c (2009).
4. T. Nakano, *et al.*, *Phy. Rev. Lett.* **91**, 012002 (2003).
5. M. Bashkanov, *et al.* [CELSIUS/WASA Collaboration], *Phys. Rev. Lett.* **102** 052301 (2009).
6. J. L. Ping, H. X. Huang, H. R. Pang, *et al.*, *Phys. Rev. C* **79** 024001 (2009).
7. D. Qing, X. S. Chen and F. Wang, *Phys. Rev. D* **58** 114032 (1998).
8. B. S. Zou and D. O. Riska, *Phys. Rev. Lett.* **95** 072001 (2005).
9. M. Harvey, *Nucl. Phys. A* **A352** 301 (1981).
10. J. Q. Chen, *et al. Tables of the SU(mn)⊃ SU(m)×SU(n) Coefficients of Fractional Parentage*, (World Scientific, Singapore, 1991).
11. F. Wang, J. L. Ping and T. Goldman, *Phy. Rev. C* **51**, 1648 (1995).
12. H. X. Huang, J. L. Ping and F. Wang, *Phy. Rev. C* **77**, 025201 (2007).
13. J. L. Ping, *et al.*, *Phy. Rev. C* **79**, 065203 (2009).
14. F. Wang, J. L. Ping and T. Goldman, *Phy. Rev. C* **51**, 3411 (1995).

ANALYTIC STUDY OF THE PARTIAL SENIORITY CONSERVATION

CHONG QI

Royal Institute of Technology (KTH), Alba Nova University Center, SE-10691 Stockholm, Sweden
E-mail: chongq@kth.se

In this contribution I present an analytic proof for the partial conservation of the seniority symmetry in $j = 9/2$ shells. It is found that all non-diagonal (and the relevant diagonal) matrix elements can be re-expressed in closed forms and are proportional to certain one-particle cfp's. This remarkable occurrence of partial dynamic symmetry is the consequence of the peculiar property of the $j = 9/2$ shell, where all $v = 3$ and 5 states are uniquely defined.

Keywords: Seniority symmetry; Partial conservation; Coefficients of fractional parentage (cfp's).

The seniority quantum number refers to the number of unpaired particles in a single-j shell for a given configuration $|j^n; I\rangle$ where I is the total angular momentum. This concept is nowadays been applied in a variety of fields, including nuclear physics, atomic and molecular physics and quantum chemistry. Seniority is a good quantum number for systems with identical fermions in shells with $j \leq 7/2$ and breaks in shells with $j \geq 9/2$. Efforts have been made to find cases for which that symmetry is partially fulfilled. It was thus found that the rotationally-invariant interaction has to satisfy a number of constraints in order to conserve seniority.[1] However, it was noted that in $j = 9/2$ shell two special eigenstates with $I = 4$ and 6 have good seniority for an arbitrary interaction.[2,3] They exhibit partial dynamic symmetry and the solvability property.[2–7]

For a system with $n = 4$ identical fermions in a $j = 9/2$ shell, the $I = 4$ (and $I = 6$) states may be constructed so that one state has seniority $v = 2$ and the other two have seniority $v = 4$ (denoted as $|j^4, v = 4, \alpha_1 I\rangle$ and $|j^4, v = 4, \alpha_2 I\rangle$ where the index α symbolizes any additional quantum number needed). The Hamiltonian matrix elements can be written as linear combinations of the interaction terms $V_J = \langle j^2; J|\hat{V}|j^2; J\rangle$ as

$$H^I_{\gamma\gamma'} = n(n-1)/2 \sum_J M^I_{\gamma\gamma'}(J) V_J, \quad (1)$$

where γ denote the basis states. The elements of the symmetric matrices $M^I(J)$ are $M^I_{\gamma\gamma'}(J) = 1/2\langle\gamma|((a^\dagger a^\dagger)^J \times (aa)^J)_0|\gamma'\rangle$. The matrices H^I and $M^I(J)$ can be constructed with the help of coefficients of fractional parentage (cfp's). The consequences of the vanishing of non-diagonal matrix elements was examined in Ref.[5] for the cases of $I = J$. It was thus shown that this is due to a special relation of certain one-particle cfp's

$$\frac{[j^4(v{=}4,\alpha_1 I)jI_5|\}j^5, v{=}5, I_5{=}j]}{[j^4(v{=}4,\alpha_2 I)jI_5|\}j^5, v{=}5, I_5{=}j]} = \frac{[j^3(v_3 = 3, j)jI|\}j^4, v{=}4, \alpha_1 I]}{[j^3(v_3{=}3, j)jI|\}j^4, v{=}4, \alpha_2 I]}, \quad (2)$$

where the $v = 3$ and 5 states are uniquely specified. An analytic proof can be derived for Eq. (2) through the principal-parent procedure.[6] Closed-form expressions for the vanishing non-diagonal matrix elements of $M^I(J)$, which are valid for all J values, are given in Ref.[7] The basic idea behind Ref.[7] is briefly described below.

The $v \to v-1$ one-particle cfp's can be factorized as[6–8]

$$[j^{n-1}(v{-}1, \alpha_1, I_1)jI|\}j^n v, \alpha, I] = \sqrt{\frac{v(2j{+}3{-}n{-}v)}{n(2j{+}3{-}2v)}} R(j, v{-}1, \alpha_1 I_1; v\alpha I).$$

The factor R in this equation can be constructed using the principal-parent procedure as

$$R(j, v-1, \alpha_1 I_1; v\alpha I) = \sum_{\alpha'_1 I'_1} c_{\alpha'_1 I'_1} R'(j, v-1, \alpha_1 I_1; v(\alpha'_1 I'_1)I), \quad (3)$$

where $\alpha'_1 I'_1$ denote the principal parents. The isoscalar factor R' can be calculated using the recursion relation given by[8]

$$R'(j, v-1, \alpha_1 I_1; v(\alpha'_1 I'_1)I) = \frac{P(\alpha'_1 I'_1 \alpha_1 I_1 I)}{\sqrt{v P(\alpha'_1 I'_1 \alpha'_1 I'_1 I)}}. \quad (4)$$

from which one finds

$$R'(j, v-1, \alpha_1 I_1; v(\alpha'_1 I'_1)I) R'(j, v-1, \alpha'_1 I'_1; v(\alpha'_1 I'_1)I\alpha_2,) = \frac{P(\alpha'_1 I'_1 \alpha_1 I_1 I)}{v}.$$

For the special case of $n = v$ of concern in this work, we simply have

$$[j^{v-1}(v-1, \alpha_1, I_1)jI|\}j^v v, \alpha, I] = R(j, v-1, \alpha_1 I_1; v\alpha I). \quad (5)$$

Moreover, the state $|j^n v(\alpha'_1 I'_1)I\rangle$ can be written as,

$$[j^{n-1}(v', \alpha_1, I_1)jI|\}j^n v(\alpha'_1 I'_1)I] = \sum_\beta x_\beta [j^{n-1}(v', \alpha_1, I_1)jI|\}j^n v, \beta, I], \quad (6)$$

For the cfp of the special case $\alpha_1' = \alpha_1$ and $I_1' = I_1$, one gets

$$[j^{n-1}(v', \alpha_1', I_1')jI|\}j^n v, \beta, I] = x_\beta [j^{n-1}(v', \alpha_1', I_1')jI|\}j^n v(\alpha_1' I_1')I], \quad (7)$$

and

$$\begin{aligned} & v[j^{v-1}(v-1, \alpha_1, I_1)jI|\}j^v v(\alpha_1' I_1')I][j^{v-1}(v-1, \alpha_1', I_1')jI|\}j^v v(\alpha_1' I_1')I] \\ & = v\sum_\beta [j^{v-1}(v-1, \alpha_1, I_1)jI|\}j^v v, \beta, I][j^{v-1}(v-1, \alpha_1', I_1')jI|\}j^v v\beta I]. \quad (8) \end{aligned}$$

One can apply these relations to derive a simple proof of Eq. (2).[7] As in Ref.,[6] we start from the unique state $|j^5, v=5, I=j\rangle$. It can be easily constructed as the tensor product of any $n=4$ state times the single particle state j. For the principal parent one can take $|\alpha_1' I_1'\rangle = |j^4, v=4, I_1'=0\rangle$. We have

$$\begin{aligned} & 5[j^4(v=4, \alpha_1, I_1)jI|\}j^5, v=5, j][j^4(v_1'=4, I_1'=0)jI|\}j^5, v=5, j] \\ & = \left[\delta_{\alpha_1'\alpha_1}\delta_{I_1' I_1} - \frac{3\sqrt{2I_1+1}}{5}\right][j^3(v_3=3, I_3=j)jI|\}j^4, \alpha_1, v=4, I_1], \quad (9) \end{aligned}$$

which is valid for all the $v=4$ states $|\alpha_1 I_1\rangle$ in $j=9/2$ shells. For $I_1 = I$, it is equivalent to the special relation of Eq. (2).

A four-particle state $|j^4; I\rangle$ can be written in terms of three-particle states with the help of one-particle cfp's as

$$|j^4, v, \alpha, I\rangle = \sum_{v_3\alpha_3 I_3} [j^3(v_3\alpha_3 I_3)jI|\}j^4, v, \alpha, I]|j^3(v_3\alpha_3 I_3)j; I\rangle. \quad (10)$$

The Hamiltonian matrix element can be calculated as

$$M^I_{\gamma\gamma'}(J) = \sum [j^3(v_3\alpha_3 I_3)jI|\}j^4 v\alpha I][j^3(v_3'\alpha_3' I_3)jI|\}j^4, v'\alpha' I]M^{I_3}_{\gamma_3\gamma_3'}(J),$$

where γ_3 denote the intermediate three-particle states. For the $j=9/2$ shell of concern, the index α_3 can be omitted since all the three-particle states can be uniquely specified.

The non-diagonal matrix element between the $v=2$ and 4 states is

$$\begin{aligned} M^I_{2\alpha}(J) = & \sum_{I_3} [j^3(v_3=3, I_3)jI|\}j^4 v=2I][j^3(v_3'=3, I_3)jI|\}j^4, v=4, \alpha, I] \\ & \times [j^2(J)jI_3|\}j^3, v_3=3, I_3][j^2(J)jI_3|\}j^3, v_3'=3, I_3] \quad (11a) \\ & + [j^3(v_3=1, I_3=j)jI|\}j^4, v=2, I] \\ & \times [j^3(v_3'=3, I_3=j)jI|\}j^4, v=4, \alpha, I]M^{I_3=j}_{13}(J), \quad (11b) \end{aligned}$$

We can rewrite the first part of $M^I_{2\alpha}(J)$ as,

$$M^I_{2\alpha}(J;a) = \sqrt{\frac{2J+1}{2I+1}}\sum_{I_3}[j^2(I)jI_3|\}j^3, v_3=3, I_3][j^2(J)jI_3|\}j^3, v_3=3, I_3]$$
$$\times[j^3(v_3=3, I_3)jJ|\}j^4, v=2, J][j^3(v_3=3, I_3)jI|\}j^4, v=4, \alpha I].$$

The second part of $M^I_{2\alpha}(J)$ can be rewritten as

$$M^I_{2\alpha}(J;b) = [j^2(v=2, J)jI_3|\}j^3, v_3=3, j][j^3(v_3=3, j)jI|\}j^4 v=4, \alpha I]$$
$$\times\sqrt{\frac{4(2J+1)}{3(2j+1)(2j-1)}}[j^3(v_3=1, I_3=j)jI|\}j^4, v=2, I].$$

Finally we have

$$M^I_{2\alpha}(J) = \left[\frac{2}{3}\sqrt{\frac{2J+1}{10}}[j^2(J)jI_3|\}j^3, v_3=3, I_3=j]\right]$$
$$\times[j^3(v_3=3, I_3=j)jI|\}j^4, v=4, \alpha, I]. \quad (12)$$

The term in the bracket defines the seniority conservation condition.[1,4,9,10]

For the matrix elements involving the two $v=4$ states one has

$$M^I_{\alpha_1\alpha_2}(J) = \sum_{I_3}[j^3(v_3 I_3)jI|\}j^4, v=4, \alpha_1 I][j^3(v_3 I_3)jI|\}j^4, v=4, \alpha_2 I]$$
$$\times\left[\frac{1}{3}+\frac{2}{3}(2J+1)\begin{Bmatrix} I_3 & j & J \\ j & j & J \end{Bmatrix}\right] \quad (13a)$$
$$-[j^3(v_3, j)jI|\}j^4, v=4, \alpha_1 I][j^3(v_3, j)jI|\}j^4, v=4, \alpha_2 I]$$
$$\times\frac{4(2J+1)}{3(2j+1)(2j-1)}. \quad (13b)$$

For the special case $I=J$, the matrix elements become

$$M^I_{\alpha_1\alpha_2}(J=I) = \frac{(2I+1)}{6}[j^3(v_3=3, I_3=j)jI|\}j^4, v=4, \alpha_1, I]$$
$$\times[j^3(v_3=3, I_3=j)jI|\}j^4, v=4, \alpha_2, I] + \frac{1}{6}\delta_{\alpha_1\alpha_2}. \quad (14)$$

For cases with $I=J$ the matrix elements $M^I_{\alpha_1\alpha_2}(J)$ are proportional to $[j^3(v_3=3, I_3=j)jI|\}j^4, v=4, \alpha, I]$.

One may write the special $v=4$ state of concern (denoted as $|a\rangle$) as a combination of an arbitrary set of $v=4$ states as

$$|j^4, v=4, a, I\rangle = \alpha|j^4, v=4, \alpha_1, I\rangle + \beta|j^4, v=4, \alpha_2, I\rangle, \quad (15)$$

It is straightforward to construct a special $v = 4$ state that satisfies $H^I_{2a} = M^I_{2a}(J) = 0$. The diagonal matrix elements can be rewritten as

$$M^I_{bb}(J) = M^I_{aa}(J) + C_J[j^3(v_3 = 3, I_3 = j)jI|\}j^4, b, v = 4, I]^2, \quad (16)$$

where $C_J = -13/6$, $3/2$, $13/6$ and $-3/2$ for $J = 2$, 4, 6 and 8, respectively. We have

$$M^I_{\alpha_1\alpha_1}(J) = M^I_{aa}(J) + C_J[j^3(v_3 = 3, I_3 = j)jI|\}j^4, v = 4, \alpha_1, I]^2, \quad (17)$$

and

$$M^I_{\alpha_1\alpha_2}(J) = C_J[j^3(v_3 = 3, I_3 = j)jI|\}j^4, v = 4, \alpha_1, I] \\ [j^3(v_3 = 3, I_3 = j)jI|\}j^4, v = 4, \alpha_2, I]. \quad (18)$$

The Hamiltonian matrix elements may be calculated in a more straightforward way in terms of two-particle cfp's as[1,4,11,12]

$$M^I_{\alpha\beta}(J) = \sum [j^{n-2}(\alpha_{n-2}I_{n-2})j^2(J)I|\}j^n\alpha I][j^{n-2}(\alpha_{n-2}I_{n-2})j^2(J)I|\}j^n\beta I],$$

where the summation runs over all possible j^{n-2} states.[1,4,7] For four identical nucleons in a single-j shell, the state can be written as the tensor product of two-particle states as $|j^2(J_\alpha)j^2(J_\beta); I\rangle$. The overlap between such states is[12,13]

$$A^j_I(J_\alpha J_\beta; J'_\alpha J'_\beta) = \langle j^2(J_\alpha)j^2(J_\beta); I|j^2(J'_\alpha)j^2(J'_\beta); I\rangle$$

$$= \delta_{J_\alpha J'_\alpha}\delta_{J_\beta J'_\beta} + (-1)^I\delta_{J_\alpha J'_\beta}\delta_{J_\beta J'_\alpha} - 4\hat{J}_\alpha\hat{J}_\beta\hat{J}'_\alpha\hat{J}'_\beta \begin{Bmatrix} j & j & J_\alpha \\ j & j & J_\beta \\ J'_\alpha & J'_\beta & I \end{Bmatrix}, \quad (19)$$

where $\hat{J} = \sqrt{2J+1}$. One of the seniority $v = 4$ states can be written as,[7]

$$[j^2(K)j^2(K')I|\}j^4[J_\alpha J_\beta], v = 4, I]$$

$$= \sqrt{\frac{2}{n(n-1)}}\mathcal{N}_{J_\alpha J_\beta}\left[A^j_I(KK'; J_\alpha J_\beta) - \mathcal{N}^2_{0I}A^j_I(J_\alpha J_\beta; 0I)A^j_I(KK'; 0I)\right],$$

where J_α and J_β are the so-called principal parents. For $I = 4$, one can construct the $v = 2$ state as $|j^4[J_\alpha = 0 J_\beta = 4], v = 2, I\rangle$ and the first $v = 4$ state as $|j^4[J_\alpha = 2 J'_\beta = 2], v = 4, I\rangle$. One can get

$$\frac{M^I_{2\alpha_1}(J)}{M^I_{2\alpha_2}(J)} = -\frac{\beta}{\alpha} = \frac{\sqrt{182}}{10\sqrt{510}}, \quad (20)$$

and

$$\frac{M^I_{\alpha_1\alpha_1}(J) - M^I_{\alpha_2\alpha_2}(J)}{M^I_{\alpha_1\alpha_2}(J)} = \left[\frac{\alpha}{\beta} - \frac{\beta}{\alpha}\right] = -\frac{25409}{10\sqrt{23205}}. \quad (21)$$

These two relations are valid for all J values of concern.

For $I = 6$, one can construct the $v = 2$ state as $|j^4[J_\alpha = 0 J_\beta = 6], v = 2, I\rangle$ and the first $v = 4$ state as $|j^4[J_\alpha = 2 J'_\beta = 4], v = 4, I\rangle$. We have

$$\frac{M^I_{2\alpha_1}(J)}{M^I_{2\alpha_2}(J)} = -\frac{\beta}{\alpha} = -\frac{5\sqrt{15}}{6\sqrt{2261}}, \tag{22}$$

and

$$\frac{M^I_{\alpha_1\alpha_1}(J) - M^I_{\alpha_2\alpha_2}(J)}{M^I_{\alpha_1\alpha_2}(J)} = \left[\frac{\alpha}{\beta} - \frac{\beta}{\alpha}\right] = \frac{27007}{10\sqrt{33915}}. \tag{23}$$

The special $I = 4$ and 6 states of concern have very large overlap with the states $|j^4[J_\alpha = 2 J_\beta = 2], v = 4, I\rangle$ and $|j^4[J_\alpha = 2 J_\beta = 2], v = 4, I\rangle$. The overlaps are calculated to be $\alpha = 10\sqrt{255}/\sqrt{25591} = 0.998220$ and $2\sqrt{6783}/\sqrt{27257} = 0.997704$, respectively.

In summary, the partial conservation of seniority in $j = 9/2$ shells is studied within a fully analytic framework with the help of the one-particle and two-particle cfp's. By using the variety of recursion relations,[1,7] it is shown that all the relevant diagonal and non-diagonal interaction matrix elements can be expressed in very simple ways in terms of certain one-particle cfp's. This is related to the fact that all $v = 3$ and 5 states in $j = 9/2$ shells are uniquely defined.

The author acknowledges the supports from the Swedish Research Council (VR) under grant No. 621-2010-4723.

References

1. I. Talmi, *Simple Models of Complex Nuclei* (Harwood Academic Publishers, Chur, Switzerland, 1993).
2. A. Escuderos and L. Zamick, *Phys. Rev. C* **73** 044302 (2006).
3. L. Zamick, *Phys. Rev. C* **75**, 064305 (2007).
4. P. Van Isacker and S. Heinze, *Phys. Rev. Lett.* **100**, 052501 (2008).
5. L. Zamick and P. Van Isacker, *Phys. Rev. C* **78**, 044327 (2008).
6. C. Qi, *Phys. Rev. C* **83**, 014307 (2011).
7. C. Qi, Z.X. Xu, R.J. Liotta, *Nucl. Phys. A* **884-885**, 21 (2012).
8. H.Z. Sun, Q.Z. Han, M. Zhang, G.L. Long, *Commun. Theor. Phys.* **11**, 441 (1989).
9. C. Qi, X.B. Wang, Z.X. Xu, R.J. Liotta, R. Wyss, and F.R. Xu, *Phys. Rev. C* **82**, 014304 (2010) .
10. G. Rosensteel and D.J. Rowe, *Phys. Rev. C* **67**, 014303 (2003).
11. I. Talmi, *Nucl. Phys. A* **846**, 31 (2010).
12. C. Qi, *Phys. Rev. C* **81**, 034318 (2010).
13. Y. M. Zhao, A. Arima, J. N. Ginocchio, and N. Yoshinaga, *Phys. Rev. C* **68**, 044320 (2003).

GROUP REPRESENTATION VERSUS QUANTUM REPRESENTATION AND A NEW APPROACH TO GROUP REPRESENTATION CALCULATIONS

FAN WANG*

Department of physics, Nanjing University,
Nanjing, 210093, China
**E-mail: fgwang@chenwang.nju.edu.cn*

We developed a group representation theory and calculation method based on quantum representation theory. All the calculations, such as Character, Irreducible bases and Matrices, Projection operators, Clebsch-Gordan series and coefficients are reduced to solve a set of eigen-equations of a complete set of commuting operators consisting of the class operators of the group. This approach had been successfully applied to all of the finite groups and compact Lie groups and various useful coefficients for physical application of group representation theory had been calculated and tabulated.

Keywords: group representation; quantum representation; complete set of commuting operators; eigen-function method.

1. Introduction

Through many decades efforts of mathematicians and physicists the group theoretical methods are widely used in many fields of physics now. However the application of group theoretical methods in physics is still not as harmony as calculus and differential equations case. We suppose this is due to the fact that the latter was developed closely related to the development of physics and so the mathematical concepts and methods matched the physical application very well but the group theoretical methods were developed independently to the physical applications and the concepts and methods are strange for physicists. So even after the establishment of quantum mechanics, which is the main engine for the application of group theoretical methods in physics, many renowned physicists were still reluctant to apply group theoretical methods. Another reason might be due to that physicists prefer to have an universal group theoretical calculation method which can be used to deal with all groups used in physics but the traditional or standard group representation theory uses quite different methods to deal with different groups which is inconvenient for physical applications.

One can get all of the results of the rotation group through the solution of the eigen-equations of three commuting operators J^2, J_3,$\bar{J}_3$, which are the square, the third component and the third component in the intrinsic frame of the angular momentum operator respectively in quantum mechanics. In group representation theory they are Casimir operators, the infinitesimal operators of the first and second parameter group of SO(3). These operators are the complete set of commuting operators(CSCO) of SO(3) in the group parameter space, i.e., the regular representation space, since SO(3) is an order three and rank one compact Lie group. These concepts are parallel to the complete set of commuting operators introduced to the quantum representation theory by P.A.M. Dirac.

These motivated J.Q. Chen and me to initiate a new approach to the group representation theory completely based on the quantum representation theory of P.A.M. Dirac, i.e., to find the CSCO in the group representation space, to solve the eigen-equations in the group representation space to find the characters, the irreducible bases(IRB) and irreducible matrices(IRM), even the Clebsch-Gordan series(CGS) and coefficients(CGC) both for finite and compact Lie groups. These results were summarized in a book "Group Representation Theory for Physicists" co-authored by J.Q. Chen, J.L. Ping and Fan Wang, published by World Scientific, Singapore in 2002. Some useful results of the permutation group, SU(n) group, $SU(mn) \supset SU(m) \otimes SU(n)$ group are tabulated in books: J.Q. Chen and M.J. Gao, *Reduction Coefficients of the Permutation Group and their Applications* (Beijing Science Press, Beijing, 1981); J.Q. Chen, P.N. Wang, Z.M. Lü and X.B. Wu, *Tables of the Clebsch-Gordan, Racah, and Subduction Coefficients of SU(n) groups* (World Scientific, Singapore, 1987); J.Q. Chen, X.B Wu and M.J. Gao, *Tables of the* $SU(mn) \supset SU(m) \otimes SU(n)$ *Coefficients of Fractional Parentage* (World Scientific, Singapore, 1990).

In the following subsections I will give a schematic review of the main concepts and methods of this new group representation approach based on quantum representation theory..

2. Eigen function method of complete set of commuting operators for group representation calculations

P.A.M. Dirac introduced the concept of complete set of commuting operators of a quantum system. He called a set of observables (operators in quantum mechanics) which all commute each other and for which there is only one simultaneous eigen-state corresponding to any set of eigen-values (no degeneracy) as the complete set of commuting operators(CSCO). For a quantum system with n degree of freedom the complete set consists of n mutual commuting operators.

For an order n rank l compact Lie group the Cartan-Weyl theory already shows that one needs a set of l commuting operators with whose eigen-value set to specify the irreducible representation(IRR) of this Lie group. Furthermore one needs additional (n-l) commuting operators from its canonical subgroup chain with those eigen-value set to specify the IRB and IRM from the decomposition of the regular representation. A typical example is the SO(3), where the eigen-value of operator J^2 specifies the IRR of SO(3). The eigen-value set of operators J_3 and $\bar{J}_3$ provides the additional quantum numbers to specify the IRB and IRM from the decomposition of the regular representation.

We first extend this concepts and methods to finite group. The class operators of group algebra provide the natural candidates of the CSCO in the group representation space. It is well known that the class operators commute each other. We proved that the eigen-value $\lambda_i^{(\nu)}$ of class operator C_i in the IRR (ν) is simply related to the character $\chi_i^{(\nu)}$ as $\chi_i^{(\nu)} = (h_\nu/g_i)\lambda_i^{(\nu)}$. Here h_ν and g_i are the dimension of the IRR and the number of elements in the class (i). Since a set of characters specify the IRR of a group completely, the eigen-value set of class operators specify the IRR completely as well. It is well known that the class operators of a group algebra is linear independent. We found they are functionally correlated. For example, the 3-cycle class operator C_3 of permutation group S_3 is functionally correlated to the 2-cycle class operator C_2 as $C_3 = 1/3(C_2)^2 - e$. Here e is the identity element. Therefore we only need the eigen-value set of one or few class operators to specify an IRR of a group. For permutation group up to S_5, one 2-cycle class operator is already enough.

A set of commuting operators (class operators of a group) whose eigen-value set specifies the IRR completely is called the CSCO of this group or CSCO-I of this group and simply denoted by C.

To specify the IRB and IRM of a group needs additional mutual commuting operators. We proved the CSCO of the canonical subgroup chain $G \subset G(S_1) \subset G(s_2) \cdots$ will provide these additional operators which is called CSCO-II of a group and denoted simply as C,C(s). However they are not enough to specify the IRB obtained from the regular representation of a group because the well known fact that the h_ν dimensional IRR will appear h_ν times in the decomposition of the regular representation of this group. Here one needs more additional operators which provide the more additional eigen-value set to specify the h_ν set IRB appeared in the decomposition of the regular representation. In the SO(3) case we found in addition to J_3 one needs $\bar{J}_3$ which is the infinitesimal operator of the second parameter group. In essence it is a group induced by the right multiplication. We accordingly introduce a right multiplication operator $\bar{R}$, which is defined as

$$\bar{R}S = SR \quad for\ \ all\ \ S \in G. \tag{1}$$

These operators form a group $\bar{G}$ which is anti-isomophic and commute with the original group G. The CSCO of the subgroup chain $\bar{G} \subset G(\bar{s}_1) \subset G(\bar{S}_2) \cdots$ will provide the additional operators, which we simply denoted as $\bar{C}(s)$, whose eigen-values will provide the necessary additional quantum numbers to specify the IRB appeared in the decomposition of regular representation.

The operators C, C(s),$\bar{C}(s)$ form the CSCO for the regular representation of a group. We call them as CSCO-III of a group. Their simultaneous eigen equations

$$CP_m^{(\nu)k} = \nu P_m^{(\nu)k}, \tag{2}$$

$$C(s)P_m^{(\nu)k} = mP_m^{(\nu)k}, \tag{3}$$

$$\bar{C}(s)P_m^{(\nu)k} = kP_m^{(\nu)k}, \tag{4}$$

$$P_m^{(\nu)k} = \sum_{a=1}^{g} u_{\nu mk,a} R_a. \tag{5}$$

will fix the character ν, IRB and IRM without any letup. Here ν,m,k are eigen-value set of operator set C, C(s),$\bar{C}(s)$, $P_m^{(\nu)k}$ are eigenvectors in the regular representation, $u_{\nu mk,a}$ are coefficients to be determined by the eigen-equations above.

With suitable normalization, it has been proved that the eigenvectors are nothing else but the projection operators for the IRB $\psi_m^{(\nu)k}$,

$$R_a P_m^{(\nu)k} = \sum_j D_{jm}^{(\nu)k}(R_a) P_j^{(\nu)k}, \tag{6}$$

$$\bar{R}_a P_m^{(\nu)k} = \sum_j D_{jk}^{(\nu)}(\bar{R}_a) P_m^{(\nu)j} \tag{7}$$

$$P_j^{(\nu)i} P_m^{(\mu)k} = \sqrt{g/h_\nu}\,\delta_{\nu\mu}\delta_{im} P_j^{(\nu)k}. \tag{8}$$

The IRM $D_{jm}^{(\nu)}(R_a)$ and $D_{jk}^{(\nu)}(\bar{R}_a)$ are simply related to the coefficients $u_{\nu jm,a}$ of the eigen-vector as

$$D_{jm}^{(\nu)}(R_a) = D_{mj}^{(\nu)}(\bar{R}_a) = \sqrt{g/h_\nu}\, u^*_{\nu jm,a}. \tag{9}$$

This eigen function method can be directly extended to calculate the CGS and CGC in the product space of two irreducible representations of a group. The only difference is now one has to solve the eigen equations of the CSCO-III in the product space instead of the regular representation space. All of the CGS and CGC of the point group and permutation group had been calculated by this method and tabulated.

This eigen function method can also be extended to the compact Lie group. The SO(3) group representation had been recalculated by this method. SO(3) is an infinite group and it has infinite numbers of class operators. All rotations with a fixed rotation angle ϕ but with respect to all possible directions $(\theta'\phi')$ in the 3-dimensional Euclidean space form a class operator,

$$C(\phi) = \int_0^{2\pi} d\phi' \int_0^{\pi} \sin\theta' d\theta' \exp(-i(\vec{n}\cdot\vec{J})\phi). \tag{10}$$

This can be integrated with the coherent state technique and the result is,

$$C(\phi) = 4\pi \frac{\sin(\frac{1}{2}\phi\widehat{t})}{\widehat{t}\sin(\frac{1}{2}\phi)}. \tag{11}$$

Here $\widehat{t} = \sqrt{4J^2+1}$. For small ϕ its Taylor expansion is

$$C(\phi) = 4\pi[1 - \frac{1}{3!}J^2(\phi)^2 + \frac{1}{5!}J^2(J^2 - \frac{1}{3})(\phi)^4 + \cdots]. \tag{12}$$

One can see that different class operators are all determined by one operator J^2 which is just the square of the angular momentum operator in quantum mechanics and the Casimir operator in the Cardan-Weyl theory. J^2, J_3 and $\bar{J}_3$ form the CSCO-III of SO(3) group. Calculating these infinitesimal operators in the group parameter (usually Euler angles) space one get exactly the same partial differential equations and their solutions as Wigner obtained in 1930's.

However to directly apply this method to higher order Lie groups is not convenient either. For U(n) and SU(n) groups we found that the Gelfand invariants used to determine the Gelfand IRB and IRM are functions of the class operators *C(f)* of state permutation group S_f so one can get Gelfand IRB of U(n) and SU(n) groups through solving the eigen equations of the class operators *C(f)* of the state permutation group. Operators C, C(s) and *C(f)* form the CSCO-III in the n^f dimensional tensor representation space of U(n) and SU(n) group.

There is Littlewood rule for the outer-product of permutation group $S_{n_1} \otimes S_{n_2}$ and it is also well known that the CGS of the product representation of two IRR of U(n) and SU(n) can be obtained from the same

Littlewood rule. However there is never a discussion of the out-product reduction coefficients or the induction coefficients for these induction representations. We defined these outer-product reduction coefficients and calculated them with the eigen function method and tabulated. Moreover we proved that the CGC of U(n) and SU(n) groups for any n can be obtained from these outer-product reduction coefficients through assimilation process. We thereout proved that the CGC of U(n) and SU(n) are independent of the dimension n but only dependent on the IRR ν_1, ν_2 and ν of the inner product $\nu_1 \otimes \nu_2 = \sum_\nu (\nu_1\nu_2\nu)(\nu)$. Also this makes it possible to tabulate the CGC of U(n) and SU(n) of any n with one table.

We proved further that the IRB of the $SU(mn) \supset SU(m) \otimes SU(n)$ group chain can be obtained by using the CGC of permutation group to couple the SU(m) and SU(n) IRB together. The IRB of $SU(m+n) \supset SU(m) \otimes SU(n)$ can be obtained by using the out-product reduction coefficients to couple the SU(m) and SU(n) IRB together. The Racah coefficients and 9ν coefficients also can be obtained from the calculation of the non-standard IRB of permutation group.

This eigen function method had been extended to SO(n), SP(n), graded Lie group and quantum group by F. Pan and others. It had been widely used in large-molecular physics, nuclear physics and multi-quark hadron physics by J.L. Ping and others.

3. Few remarks

The new approach of group representation theory based on the quantum representation theory makes the group representation theory easier to be understood for physicists. If the group representation theory were developed specially for the application to quantum mechanics it might have taken this version. The new approach provides an universal calculation method to do group representation calculations, i.e., the eigen function method, which is powerful and easy to be used and so it will facilitate the application of group representation theory in physics. In the computer popular era it has the advantage to be easy to be computerized.

Usually physicists use mathematics to study physics. Here we use physical method to solve the mathematical problems. Such an approach might be useful for other mathematical problems.

J.Q. Chen made the dominant contribution to this new approach. Unfortunately he passed away untimely. This paper is dedicated to his lifelong efforts in this new approach.

E(2) DYNAMICAL SYMMETRY

Y. ZHANG*, Y. ZHAO and Y. WANG
Department of Physics, Liaoning Normal University,
Dalian, Liaoning, China
** E-mail: dlzhangyu@yahoo.com.cn*

The E(2) dynamical symmetry is realized with both the differential operators and boson operators. It is shown that the discrete spectrum of the E(2) dynamical symmetry just correspond to the solutions of the infinite square well Hamiltonian in two dimensions, of which the eigenfunction can be related with the Bessel equation of integral order.

Keywords: E(2) symmetry, Bessel function, infinite square well.

1. Introduction

E(2) group can be defined as follow.[1] Suppose V_2 is a real Euclidean space in two dimensions. One can find the transformation δ which keep the distance between any point in V_2 unchanged, i.e.

$$|\delta\vec{x} - \delta\vec{y}| = |\vec{x} - \vec{y}| \,. \tag{1}$$

δ is called a length preserving transformation. All these transformation form a group E(2) known the Euclidean group in two dimensions. The simplest elements in E(2) are translations T_a with

$$T_a\vec{x} = \vec{x} + \vec{a} \,. \tag{2}$$

All the translations T_a in two-dimensional Euclidean space form a Abelian subgroup T(2) of the E(2) group since

$$T_aT_b = T_bT_a = T_{a+b} \,. \tag{3}$$

Another length preserving transformation is just the rotation in two dimension with

$$|R_\theta\vec{x}| = |\vec{x}'| \,. \tag{4}$$

The rotation defines the subgroup SO(2) of the E(2) group. As a result, the E(2) group is the semiirect product of the translation group and the rotation group, namely $E(2) = T_2 \otimes_s SO(2)$. The E(2) dynamical symmetry is that the situation in which the Hamiltonian operator, H, can written in terms of Casimir operators C_i of the E(2) group and its subgroup.

2. The E(2) dynamical symmetry in the square well

The free Hamiltonian is, naturally, invariant under both translations and rotations in two-dimensional Euclidean space and is recognized as the Casimir operator of E(2) algebra[2]

$$C_2[E(2)] = \hat{p}^2. \tag{5}$$

The eigenvalues of the Casimir operators provide good quantum numbers for the eigenstates. The algebras just described form the subalgebra chain $E(2) \supset SO(2)$. The free Hamiltonian describes a dynamical symmetry situation for this algebra chain, as $H = C_2[E(2)]$. The square well Hamiltonian also describes the E(2) dynamical symmetry, provided that all operators are restricted to the bounded domain defined by $r \leq r_W$, which indicates

$$V(r) = \begin{cases} 0\,, & r \leq r_W \\ \infty\,, & r > r_W \end{cases} \tag{6}$$

with r_W being the width of the square well.

As well-known, the dynamics in a 2D infinite square well can be described by a Bessel equation with Bessel function solutions, that is, it is exactly solvable. Indeed, in this way one finds that an infinite square well in two dimensions has an E(2) dynamical symmetry, which differs from the E(5) dynamical symmetry[3] in the interacting boson model or the E(3) dynamical symmetry[4] in the U(4) vibron model only in dimension. And by exploiting this E(2) dynamical symmetry, physical quantities of the system can be calculated exactly as done in the E(5) and E(3) cases.[3,4] The 2D infinite square well has been also used to model the critical point in the transition from vibrational to rotational regimes in the pairing phase in Ref.[5]

The dynamical structure of the Hamiltonian with the E(2) dynamical symmetry can be characterized by group chain $E(2) \supset SO(2)$ with eigenstates

$$\left| \begin{matrix} E(2) \supset SO(2) \\ \xi \quad\quad L \end{matrix} \right\rangle, \tag{7}$$

where ξ and L are the quantum numbers of the E(2) and the SO(2), respectively. Accordingly, the eigenfunctions of the 2D infinite square well in the polar coordinates can be written as $\psi(r,\varphi) = R_L(r)e^{iL\varphi}$, where $R_L(r)$ is the solution of radial equation

$$\{-\frac{\hbar^2}{2M}[\frac{d^2}{dr^2} + \frac{1}{r}\frac{d}{dr} - \frac{L^2}{r^2}] + V(r)\}R = ER\,, \tag{8}$$

$V(r)$ is defined in Eq. (6), and M is a mass parameter. After setting $\rho = kr$ with $k = \sqrt{\frac{2ME}{\hbar^2}}$, Eq. (8) can be rewritten as

$$[\frac{d^2}{d\rho^2} + \frac{1}{\rho}\frac{d}{d\rho} + (1 - \frac{L^2}{\rho^2})]R = 0\,, \tag{9}$$

which is just the Bessel equation. Then, it is easy to find $R_L(r) = C_{k,L}J_L(kr)$, where J_L is a Bessel function of order L, and the normalization constants $C_{k,L}$ can be determined by imposing the condition

$$\int_0^{r_W} \mid C_{k,L}j_L(kr) \mid^2 rdr = 1\,. \tag{10}$$

The energy spectrum can be determined as $E_{L_\xi} = \frac{\hbar^2}{2M}(\frac{y_{\xi,L}}{r_W})^2$, where $y_{\xi,L}$ is the ξth zero of the Bessel function $J_L(y)$. It should be noted that the E(2) symmetry corresponds to an integer order Bessel equation, while the Bessel equations in the E(5) and E(3) symmetries are of half-integer orders.

Table 1. Excitation energies of the E(2) dynamical symmetry.

	$\xi = 1$	$\xi = 2$	$\xi = 3$
$L = 0$	0	2.77	7.77
$L = 1$	1	4.88	10.98
$L = 2$	2.31	7.31	14.52
$L = 3$	3.92	10.06	18.39

Table 1 shows excitation energies of some low-lying states in the E(2) symmetry, where the energy of ground state is set to zero and all energies are normalized to the energy of the first excited state. Some typical energy ratios E_{2_1}/E_{1_1}, E_{3_1}/E_{1_1} can be obtained by using results shown in Table 1, where the state is labeled by L_ξ. $E1$ transition rates in the E(2) model can be calculated by taking matrix elements of the dipole operator $\hat{T}(E1) = \alpha re^{\pm i\theta}$, where α is the effective charge. Since, up to an overall scale, $E1$ transition rates are determined uniquely by symmetry property of the corresponding states, we are more interested in $B(E1)$ ratios.

3. Algebraic realization of the E(2) dynamical symmetry

The above analysis shows that the 2D infinite square well problem can be described naturally within a frame of differential equations corresponding to the geometrical description of the E(2) dynamical symmetry. In the following, we will turn to constructing an algebraic model with the E(2) symmetry.

In the two dimensional Euclidean space, one can define creation and annihilation operators of a boson τ with coordinates x, y and the momentum operators $p_x = -i\partial/\partial x, p_y = -i\partial/\partial y$ as

$$\tau_x^\dagger = \frac{x - ip_x}{\sqrt{2}}, \ \tau_x = \frac{x + ip_x}{\sqrt{2}}, \ \tau_y^\dagger = \frac{y - ip_y}{\sqrt{2}}, \ \tau_y = \frac{y + ip_y}{\sqrt{2}}, \tag{11}$$

where $\hbar$ is set to 1. One can prove that $[\tau_i, \ \tau_j^\dagger] = \delta_{i,j}$ with $i, \ j = x, \ y$ according to the commutation relations between the coordinates and momentum operators. It is easy to find that

$$C_2[E(2)] = p^2 = p_x^2 + p_y^2 = \tau_x^\dagger\tau_x + \tau_y^\dagger\tau_y + 1 - \frac{1}{2}[\tau_x^2 + (\tau_x^\dagger)^2 + \tau_y^2 + (\tau_y^\dagger)^2]\,. \tag{12}$$

As mentioned before, the E(2) group is the semidirect product of the rotation group – a circle – and translations, then the E(2) algebra is the semidirect sum of the corresponding subalgebras. One can thus define the Lie algebra as

$$\xi_1 = ip_x = \frac{1}{\sqrt{2}}(\tau_x - \tau_x^\dagger) \tag{13}$$

$$\xi_2 = ip_y = \frac{1}{\sqrt{2}}(\tau_y - \tau_y^\dagger)\,, \tag{14}$$

$$\xi_3 = i(xp_y - yp_x) = \tau_y\tau_x^\dagger - \tau_y^\dagger\tau_x\,, \tag{15}$$

which satisfy the following commutation relations of the E(2) algebra:

$$[\xi_1, \xi_2] = 0, \quad [\xi_1, \xi_3] = \xi_2, \quad [\xi_3, \xi_2] = \xi_1, \quad [\xi_i, C_2[E(2)]] = 0 \tag{16}$$

for $i = 1, 2, 3$. Thus, the Casimir operator of the E(2) algebra can also be expressed as

$$C_2[E(2)] = -\xi_1^2 - \xi_2^2\,. \tag{17}$$

With the definition

$$\tau_\pm^\dagger = \mp\frac{\tau_x^\dagger \pm i\tau_y^\dagger}{\sqrt{2}}, \ \tau_\pm = \mp\frac{\tau_x \mp i\tau_y}{\sqrt{2}}, \tag{18}$$

Table 2. Typical quantities in the different models

	E(2)	F(2)	U(2)	SO(3)
E_{2_1}/E_{1_1}	2.31	2.31	2	4
E_{3_1}/E_{1_1}	3.92	3.92	3	9
E_{0_2}/E_{1_1}	2.77	2.77	2	$4N$-2
$\frac{B(E1;\ 2_1\to 1_1)}{B(E1;\ 1_1\to 0_1)}$	0.92	0.92	$\frac{6(N-1)}{5N}$	$\frac{3(N-1)(N+2)}{5N(N+1)}$
$\frac{B(E1;\ 0_2\to 1_1)}{B(E1;\ 1_1\to 0_1)}$	1.59	1.59	$\frac{3(N-1)}{N}$	0

the Casimir operator of $E(2)$ algebra can be rewritten as

$$C_2[E(2)] = \hat{n} + 1 - (\hat{P} + \hat{P}^\dagger)\,, \tag{19}$$

where $\hat{n} = \tau_+^\dagger \tau_+ + \tau_-^\dagger \tau_-$, $\hat{P} = -\tau_+\tau_-$, and $\hat{P}^\dagger = -\tau_+^\dagger \tau_-^\dagger$. Thus, we can get the algebraic E(2) model with $\hat{H}_{E(2)} = C_2[E(2)]$.

As we know,[6] the E(2) dynamical symmetry may be approximately to describe the dynamics at the critical point of the shape phase transition in the U(3) vibron model. The U(3) vibron model involves two dynamical limit, namely U(2) limit and SO(2) limit, and all the dynamical operators including the Hamiltonian and transition operators can be constructed by the two types of boson, namely the scalar s boson and the 2D vector τ boson. If we take the τ boson operators of the algebraic E(2) model to be the same operators as those in the U(3) viborn model, the Hamiltonian $\hat{H}_{E(2)}$ can be then diagonalized within the subspace spanned by all irreducible representations of the subgroup $U(2)$ of the $U(3)$ vibron model for fixed N, of which the corresponding model is referred to as the F(2) model. It follows that the F(2) model reduces to the E(2) model in the large N limit. It should be noted that both the Hamiltonian of the F(2) model and that of the E(2) model in the infinite well can be written as $\hat{H} = C_2[E(2)]$, but the former is realized by the boson operators and the latter is realized in differential form of the coordinate r. As is well known, the square well radius r_W only changes values of energy levels but does not affect structure of the eigenstates. Therefore, the boundary of the infinite square well in the E(2) model, which guarantees spectrum of the E(2) model is discrete, can be naturally realized in the F(2) model. More importantly, the results in Table 2 indicates that all the energy ratios and $B(E1)$ ratios obtained from the E(2) model, which lie between the values of the U(2) and SO(3) limits, can be exactly reproduced by the F(2) model in the large N limit, which further confirms that the $F(2)$ model is just the algebraic description of the E(2) dynamical symmetry.

4. Summary

In this work, we have analyzed the dynamics of the E(2) symmetry in both geometric and algebraic settings. The results indicate that the bound state condition in the geometric scheme is naturally involved in the algebraic realization of the E(2) dynamical symmetry. which is similar to the second quantization of the 2D harmonic oscillator problem. And the resulting spectrum in the algebraic scheme is also consistent with those solved from the Hamiltonian involving only the kinetic energy term under the bound state condition, which is actually the dynamics in a square well.

5. Acknowledgments and Appendices

Support from the Natural Science Foundation of China (11005056) is acknowledged.

References

1. J. Q. Chen, *Group representation theory for physicists* (World Scientific, Singapore, 1988).
2. M. A. Caprio and F. Iachello, *Nucl. Phys. A* **781**, 26 (2007).
3. F. Iachello, *Phys. Rev. Lett.* **85**, 3580 (2000).
4. Y. Zhang, Z. F. Hou, H. Chen, Y. X. Liu and H. Wei, *Phys. Rev. C* **78**, 024314 (2008).
5. R. M. Clark, A. O. Macchiavelli, L. Fortunato and R. Krucken, *Phys. Rev. Lett.* **96**, 032501 (2006).
6. Y. Zhang, F. Pan, Y. X. Liu and J. P. Draayer, *J. Phys. B* **43**, 225101 (2010).

THE SO(3) SYMMETRY IN THE NUCLEAR SHELL MODEL

Y. M. ZHAO*

INPAC, Department of Physics and Shanghai Key Lab for Particle Physics and Cosmology, Shanghai Jiao Tong University, Shanghai, 200240, China
** E-mail: ymzhao@sjtu.edu.cn*

In this contribution I would like to discuss the SO(3) symmetry in the nuclear shell model and its applications, based on our recent works. We shall focus on the enumeration of number of states with given spin, J-th pairing interaction, and sum rules of angular momentum coupling coefficients such as six-j and nine-j symbols.

Keywords: SO(3) symmetry; number of states; sum-rules of angular momentum coupling coefficients.

1. Introduction

It is a great pleasure for me to be here and to talk with many friends. I stayed here (Nankai University) in 1985-1989, for my undergraduate period. These days I saw great change in this beautiful campus.

This contribution starts from three interesting problems as below.

- It is well known that the enumeration of states of the nuclear shell model configurations with given spin (and isospin) is a fundamental practice in the shell model theory. Usually this is done via iterations by computer. Can we have the number of states with given spin by simple formulas [without computer enumerations]? My answer is yes, but sometimes.
- In nuclear structure theory, the seniority scheme simplifies greatly the problem for single-j systems with only monopole pairing interaction. Can we generalize this for other cases [non-monopole pairing interactions]? My answer is yes, but sometimes.
- Can one imagine, somehow, matrix elements of J-pairing interaction, sum rules of angular momentum coupling coefficients, and number of states in the shell model configurations, are related closely ? My answer is again yes, but sometimes!

In this talk I would like to discuss the above three questions, including what these "sometimes" mean.

2. Number of given-spin states for a single-j shell

We came to this problem accidentally. In summer of 2001, Prof. Arima gave a talk in American-Japanese joint nuclear physics symposium. His talk was based on collaborated work about spin-zero ground state dominance in the presence of random interactions.[1] In that conference Prof. Arima showed a figure of the spin-zero ground state probability $P(0)$ of four fermions in a single-j shell versus j. $P(0) \simeq 60\%$ when $j = 9/2, 15/2, 21/2, 27/2$, and 20-30% when $j = 13/2, 19/2, 25/2, 31/2$. In the autumn of 2001, Joe Ginocchio wrote to prof. Arima by fax, in which Joe pointed out an interesting coincidence : $P(0)$ comes to a peak if the number of states with spin zero ($I = 0$) increases by one. At that time we actually had known this fact (also Zelevinsky and collaborators knew this coincidence). However, I did not know the Ginocchio-Haxton formula:[2] The number of states with spin $I = 0$ for four fermions in a single-j shell is $[(2j+3)/6]$, where "[ ]" means to take the largest integer inside.

For $n = 2$, the number of states is 1 for even I and 0 for odd I. Thus one might be able to construct simple formulas for $n = 3, 4, 5$, for single-j shells. A hand-waving idea would be to get any hint simply by looking at the systematics of those numbers. I tried this practice carefully and found the empirical formulas.[3] I could not prove them at that time, however, I readily confirmed them by numerical experiments with computers, as much as possible. Therefore the formulas that we obtained "should" be correct: indeed they were later proved by Talmi,[4] Zhao and Arima.[5]

Let us denote the number of spin I states by D_I. For a single-j shell, one usually obtains D_I by subtracting the combinatorial number of angular momentum projection $M = I + 1$ from that with $M = I$. More specifically, D_I equals to the combinatorial number of $M = I$ subtracted by that of $M = I + 1$, where $M = m_1 + m_2 + \cdots + m_n$, with the requirement that $m_1 \geq m_2 \geq \cdots \geq m_n$ for bosons and $m_1 > m_2 > \cdots > m_n$ for fermions, where n is the number of particles. We call this enumeration method process A. There is an equivalent method, called process B, as follows. Let $\mathcal{P}(n, I_0)$ be the number of partitions of $I_0 = i_1 + i_2 + \cdots i_n$, with $0 \leq i_1 \leq i_2 \leq \cdots \leq i_n \leq 2j+1-n$ for fermions or $0 \leq i_1 \leq i_2 \leq \cdots \leq i_n \leq 2l$ for spin-l bosons. Here $I_{\max} = nj - \frac{n(n-1)}{2}$ for fermions in a single-j shell, and $I_{\max} = nl$ for bosons with spin l. One defines $\mathcal{P}(n, 0) = D_{I=I_{\max}} = 1$ for $I_0 = 0$. Then one has $D_{I=I_{\max}-I_0} = \mathcal{P}(n, I_0) - \mathcal{P}(n, I_0 - 1)$.

Let us consider D_I of $\bar{n} = 2l$ spin-$L = \frac{n}{2}$ "bosons" or $\bar{n} = 2j+1-n$ for fermions. $I_{\max}$ of these $\bar{n}$ "bosons" with spin L equals that of n bosons with spin l or that of n fermions in a single-j shell. Furthermore, $\mathcal{P}(n, I_0)$ of $I_0 = i_1 + i_2 + \cdots i_{\bar{n}}$ with the requirement $0 \le i_1 \le i_2 \le \cdots \le i_{\bar{n}} \le 2L = n$, always equals that of $I_0 = i_1 + i_2 + \cdots i_n$ with the requirement that $0 \le i_1 \le i_2 \le \cdots \le i_n \le 2j+1-n$ for n fermions or $0 \le i_1 \le i_2 \le \cdots \le i_n \le 2l$ for n bosons. This result can be explained from the fact as follows. The $\mathcal{P}(n, I_0)$ of $\bar{n}$ "bosons" with spin L corresponds to Young diagrams up to n rows, and $2l$ columns for bosons or $2j+1-n$ columns for fermions. The conjugates of these Young diagrams are those up to $2l$ rows for bosons or $2j+1-n$ rows for fermions, and up to n columns, which correspond to partitions in Process B for n fermions in a single-j shell or bosons with spin l. Therefore, Process B for $\bar{n}$ bosons with spin $L = n/2$ provides us with an alternative to construct D_I for n bosons with spin l or n fermions in a single-j shell. For $l = (2j+1-n)/2$ (n is even), i.e., $I_{\max}$ of bosons equals that of fermions, then D_I for bosons equals that of fermions. In Process A, j-shell fermions or spin-l bosons involve of the SU$(2j+1)$ or SU$(2l+1)$ symmetry. In Process B, one needs to study the SU$(n+1)$ symmetry, which is *independent* of j and l. For example, one can derive the number of spin-I states for $n = 4$ of a single-j shell by studying the number of states for d bosons, which follows the SU(5) symmetry. This was exemplified in Ref. 5.

We note that the the Ginocchio-Haxton formula, $D_{I=0} = [(2j+3)/6]$, for $n = 4$, was also studied by Zamick and Escuderos.[6] Talmi proposed recursive formulas of D_I for different particle number n.[4] Extension to boson systems was studied in Ref. [7].

3. J-pairing interaction

The J-pairing interaction which couples two fermions to an angular momentum J is

$$H_J = G_J \sum_{M=-J}^{J} A_M^{J\dagger}\, A_M^J, \quad A_M^{J\dagger} = \frac{1}{\sqrt{2}}\left[a_j^\dagger \times a_j^\dagger\right]^J,$$

$$A_M^J = -(-1)^M \frac{1}{\sqrt{2}}\left[\tilde{a}_j \times \tilde{a}_j\right]_{-M}^J, \quad \tilde{A}^J = -\frac{1}{\sqrt{2}}\left[\tilde{a}_j \times \tilde{a}_j\right]^J. \qquad (1)$$

where $[\;]_M^J$ means coupled to angular momentum J and projection M. We take $G_J = -1$.

For $n = 3$, there is at most one non-zero eigenvalue for each spin I with $H = H_J$. The non-zero eigenvalue equals $\langle j^3 K' : I | H_J | j^3 K : I\rangle$, are zero except

$$\langle |j^3[jJ]I|H_J|j^3[jJ]I\rangle = 1 + 2(2J+1)\begin{Bmatrix} J & j & I \\ J & j & j \end{Bmatrix}. \tag{2}$$

This fact gives new sum rules. The summation of all eigenvalues with given I is equal to $\frac{n(n-1)}{2}$ times $D_I(n=3)$. $D_I(n=3)$ was obtained in Ref. [3] and proved in Ref. [4]. For a half integer j, we have

$$\sum_{J=\text{even}} 2(2J+1)\begin{Bmatrix} j & I & J \\ j & j & J \end{Bmatrix} = \begin{cases} 3\left[\frac{2I+3}{6}\right] - I - \frac{1}{2} & \text{if } I \leq j; \\ 3\left[\frac{3j-3-I}{6}\right] + 3\delta_I^j - \left[\frac{3j+1-I}{2}\right] & \text{if } I \geq j\ . \end{cases} \tag{3}$$

where

$$\delta_I^j = \begin{cases} 0 \text{ if } (3j-3-I) \bmod 6 = 1 \\ 1 \text{ otherwise}\ . \end{cases}$$

Similar sum rules of six-j symbols can be obtained if one replace j by integer l, see appendix of Ref. [8]. One can similarly derive many sum rules of nine-j symbols, see Ref. [9].

For $J = J_{\max} = 2j-1$, we observe an interesting phenomenon. Let us compare the eigenvalues of two system, one with $n=3$ and the other with $n=4$. For $j=31/2$ and $n=4$ the total number of states is 790. The number of states with non-zero eigenvalues is 380. Within a precision 10^{-2}, the eigenvalues of these 308 states are located at the eigenvalues of $n=3$, and 21 states have eigenvalues closely at -2. We note that almost all the "non-integer" eigenvalues of $n=4$ can be very accurately given by one of three-particle clusters with $I \sim I_{\max}$ coupled to a single-j particle. This picture continues to work for $n=5$ and 6, see Ref. [8] for details.

As is well known, the existence of systematic degeneracy indicates that the Hamiltonian has certain symmetry. The degeneracy for the $J_{\max}$ pairing interaction, however, is not exact. It would be interesting to explore the broken symmetry hidden in the $J_{\max}$ pairing interaction discussed in this paper. It would be also interesting to discuss the modification of the $J_{\max}$ pairing interaction in order to recover the exact degeneracy.

4. An application

We generalize the enumeration of D_I for identical particles in a single-j shell to nucleons with isospin degree of freedom. In this way we obtain the formulas of number of states with given spin I and isospin T (denoted by

$D_{I,T}$). Let us first define our single-j shell model Hamiltonian,

$$H_{JT'} = \sum_{M=-J}^{J} A^{(JT')\dagger}_{MM_{T'}} A^{(JT')}_{MM_{T'}},$$

$$A^{(JT')\dagger}_{MM_{T'}} = \frac{1}{\sqrt{2}} \left[a^\dagger_{jt} a^\dagger_{jt}\right]^{(JT')}_{MM_{T'}}, A^{(JT')}_{M} = -(-1)^{M+M_{T'}} \frac{1}{\sqrt{2}} \left[\tilde{a}_{jt}\tilde{a}_{jt}\right]^{(JT')}_{-M-M_{T'}},$$

where $[\,]^{(JT')}_{MM_{T'}}$ means an operator in which two nucleons are coupled to spin J and isospin T with spin projection M and isospin projection M_T. Here we exemplify our results by using $n = 4$; in this case we have

$$\begin{aligned} 6D_{I,T} &= \sum_{JT'} \sum_{\alpha} \langle j^4 \alpha IT | H_{JT'} | j^4 \alpha IT \rangle \\ &= \sum_{JK\ T_2T_2'} \langle 0| \left[A^{(JT_2)} A^{(KT_2')}\right]^{(IT)}_{MM_T} \left[A^{(JT_2)\dagger} A^{(KT_2')\dagger}\right]^{(IT)}_{MM_T} |0\rangle . \end{aligned} \quad (4)$$

Now we exemplify this formula by using $T = 1$. We have

$$\begin{aligned} &6D_{I,T=1} \\ &= \sum_{\text{even } J \text{even K}} (1-(-)^I \delta_{JK} - 36(2J+1)(2K+1) \begin{Bmatrix} j & j & J \\ j & j & K \\ J & K & I \end{Bmatrix} \begin{Bmatrix} \frac{1}{2} & \frac{1}{2} & 1 \\ \frac{1}{2} & \frac{1}{2} & 1 \\ 1 & 1 & 1 \end{Bmatrix}) \\ &+ \sum_{\text{odd } J \text{even K}} (1-(-)^I \delta_{JK} - 12(2J+1)(2K+1) \begin{Bmatrix} j & j & J \\ j & j & K \\ J & K & I \end{Bmatrix} \begin{Bmatrix} \frac{1}{2} & \frac{1}{2} & 1 \\ \frac{1}{2} & \frac{1}{2} & 0 \\ 1 & 0 & 1 \end{Bmatrix}) \\ &+ \sum_{\text{even } J \text{odd K}} (1-(-)^I \delta_{JK} - 12(2J+1)(2K+1) \begin{Bmatrix} j & j & J \\ j & j & K \\ J & K & I \end{Bmatrix} \begin{Bmatrix} \frac{1}{2} & \frac{1}{2} & 0 \\ \frac{1}{2} & \frac{1}{2} & 1 \\ 0 & 1 & 1 \end{Bmatrix}) \\ &= \sum_{\text{even } J \text{ even K}} (1-(-)^I \delta_{JK}) + 2 \sum_{\text{even } J \text{ odd K}} (1-(-)^I \delta_{JK}) \\ &\quad -\frac{1}{2} S(j^4, \text{even } J \text{ even K}). \end{aligned} \quad (5)$$

For $n = 4$ and $T = 1$, $I_{\max}$ equals $4j - 3$ (odd value). When $I \geq 2j$, let us define $I = I_{\max} - 2I_0$ for odd I and $I_{\max} - 2I_0 - 1$ for even I. Using Eq. (5) and Eq. (22) of Ref. 8, we obtain

$$D_{I,T=1} = \left(\left[\frac{I_0}{2}\right] + 1\right) \left(\left[\frac{I_0}{2}\right] + 1 + (I_0 \bmod 2)\right) ;$$

When $I \leq 2j$, we use Eq. (5) and Eq. (21) of Ref. 8, and obtain

$$D_{I,T=1} = (I_0+1)j - \left(1+4\left[\frac{I_0}{2}\right]+6\left[\frac{I_0}{2}\right]^2+(I_0 \bmod 2)(6\left[\frac{I_0}{2}\right]+3)\right)/2\,,$$

where $I_0 = (I-1)/2$, and $I \leq 2j$.

$D_{I,T=0}$ for $n = 4$ and $D_{I,T=1/2}$ for $n = 3$ can be obtained in the same way.[10] Zamick and Escuderos studied certain relations[11] of $D_{I,T}$.

5. Summary

In this paper, I have given a brief introduction to our efforts on a few topics related to the SO(3) symmetry in the nuclear shell model theory. We discuss a few simple formulas of number of spin I states for a single-j shell, some features of J-pairing interaction, sum rules of angular momentum coupling coefficients. It is very interesting to note once more that these seemingly different issues can be related to each other. The $J_{\max}$-pairing interaction favors clustering picture, with many asymptotic degeneracy. We have found many new sum rules of angular momentum coupling coefficients for a single-j shell.

I thank the local organizers at Nankai University, for their hospitality and generosity of providing accommodation, food and transportation, and for their great patience and efforts in organizing this symposium. I thank Prof. Akito Arima for his collaboration and constant encouragements in these works. I would like to thank the National Natural Science Foundation of China for supporting this work under Grants No. 10975096 and 11145005.

References

1. Y. M. Zhao, A. Arima, and N. Yoshinaga, Phys. Rep. 400, 1 (2004).
2. J. N. Ginocchio and W. C. Haxton, in *Symmetries in Science VI*, edited by B. Gruber and M. Ramek (Plenum, New York, 1993), p. 398.
3. Y. M. Zhao and A. Arima, Phys. Rev. C68, 044310 (2003).
4. I. Talmi, Phys. Rev. C72, 037302 (2005).
5. Y. M. Zhao and A. Arima, Phys. Rev. C71, 047304 (2005).
6. L. Zamick and A. Escuderos, Phys. Rev. C71, 054308 (2005).
7. L. H. Zhang, Y. M. Zhao, L. Y. Jia, and A. Arima, Phys. Rev. C77, 014301 (2008).
8. Y. M. Zhao and A. Arima, Phys. Rev. C70, 034306 (2004).
9. Y. M. Zhao and A. Arima, Phys. Rev. C72, 054307 (2005).
10. Y. M. Zhao and A. Arima, Phys. Rev. C72, 064333 (2005).
11. L. Zamick and A. Escuderos, Phys. Rev. C72, 044317 (2005).

PARALLEL SESSION

General Quantum Mechanics and its Mathematical Methods Involving Group Theory

20 & 21 & 22 & 23 AUGUST 2012

PARTIAL INNER PRODUCT SPACES, A UNIFYING FRAMEWORK FOR QUANTUM MECHANICS

J-P. ANTOINE*

Institut de Recherche en Mathématique et Physique (IRMP)
Université catholique de Louvain
B-1348 Louvain-la-Neuve, Belgium
**E-mail: jean-pierre.antoine@uclouvain.be*

Keywords: Partial inner product spaces, operators, group representations.

1. Introduction

Many function spaces that play a central role in analysis come in the form of families, indexed by one or several parameters that characterize the behavior of functions (smoothness, behavior at infinity, ...). The typical structure is a *chain of Hilbert or (reflexive) Banach spaces*. Let us give two familiar examples.

(i) The Lebesgue L^p spaces on $[0,1]$, $\mathcal{I} = \{L^p([0,1], \mathrm{d}x), 1 \leqslant p \leqslant \infty\}$.
(ii) The scale of Hilbert spaces $\{\mathcal{H}_n\}_{n\in\mathbb{Z}}$ built on the powers of a positive self-adjoint operator $A \geq 1$ in a Hilbert space $\mathcal{H}_0$. Here $\mathcal{H}_n = D(A^n)$, with the graph norm $\|f\|_n = \|A^n f\|$, $f \in D(A^n)$, for $n \in \mathbb{N}$ or $n \in \mathbb{R}^+$, and $\mathcal{H}_{-n} = \mathcal{H}_n^\times$ (conjugate dual).

In both cases, the inner product of the central Hilbert space cannot be extended to the largest space of the chain, but only on *compatible* pairs. Thus one gets a *partial inner product space* (PIP-space). In this paper, we shall give a quick overview of PIP-spaces and operators on them. A complete information may be found in our monograph,[1] including references to the original work.

2. Partial inner product spaces

A formal way of introducing PIP-spaces is given by the idea of *linear compatibility*,[1] by which we mean a symmetric binary relation # on V which preserves linearity:

$$f\#g \iff g\#f,\ \forall\, f,g \in V,$$
$$f\#g, f\#h \implies f\#(\alpha g + \beta h),\ \forall\, f,g,h \in V,\ \forall\, \alpha,\beta \in \mathbb{C}.$$

Then, for any subset $S \subset V$, the set $S^{\#} = \{g \in V : g\#f, \forall\, f \in S\}$ is a vector subspace of V and one has $S^{\#\#} = (S^{\#})^{\#} \supseteq S$, $S^{\#\#\#} = S^{\#}$. Thus,

$$f\#g \iff f \in \{g\}^{\#} \iff \{f\}^{\#\#} \subseteq \{g\}^{\#}. \tag{1}$$

From now on, we will call *assaying subspace* of V a subspace S such that $S^{\#\#} = S$ and denote by $\mathcal{F}(V,\#)$ the family of all assaying subsets of V, ordered by inclusion. Assaying subsets will be denoted $V_r, V_q, \ldots$ and the index set by F. By definition, $q \leq r$ if and only if $V_q \subseteqq V_r$. We also write $V_{\overline{r}} = V_r^{\#}$, $r \in F$. Thus the relations (1) mean that $f\#g$ if and only if there is an index $r \in F$ such that $f \in V_r$, $g \in V_{\overline{r}}$. In other words, assaying subspaces are the building blocks of the whole structure.

Theorem 1. *The family $\mathcal{F}(V,\#) = \{V_r, r \in F\}$ of all assaying subspaces, ordered by inclusion, is a complete involutive lattice, under the following operations, arbitrarily iterated:*

. *involution:* $V_r \leftrightarrow V_{\overline{r}} := (V_r)^{\#}$,
. *infimum:* $V_{p\wedge q} := V_p \wedge V_q = V_p \cap V_q$, $\quad (p,q,r \in F)$
. *supremum:* $V_{p\vee q} := V_p \vee V_q = (V_p + V_q)^{\#\#}$.

The smallest element of $\mathcal{F}(V,\#)$ is $V^{\#} = \bigcap_r V_r$ and the greatest element is $V = \bigcup_r V_r$. We also note the following relations:

$$(V_{p\wedge q})^{\#} = V_{\overline{p\wedge q}} = V_{\overline{p}\vee\overline{q}} = V_{\overline{p}} \vee V_{\overline{q}}.$$

A *partial inner product* on $(V, \#)$ is a hermitian form $\langle\cdot|\cdot\rangle$, not necessarily positive definite, defined exactly on compatible pairs of vectors. A *partial inner product space* (PIP-space) is a vector space V equipped with a linear compatibility and a partial inner product.

We require the PIP-space $(V, \#)$ to be *nondegenerate*, i.e., $\langle f|g\rangle = 0$ for all $f \in V^{\#}$ implies $g = 0$. Then $(V^{\#}, V)$ is a dual pair in the sense of topological vector spaces,[3] and so is every couple $(V_r, V_{\overline{r}})$, $r \in F$. In the sequel, we also assume that the partial inner product is positive definite.

Next we assume that every V_r carries its *Mackey topology* $\tau(V_r, V_{\overline{r}})$. As a consequence, if V_r is a Hilbert space or a reflexive Banach space, then $\tau(V_r, V_{\overline{r}})$ coincides with the norm topology. Next, $r < s$ implies $V_r \subset V_s$, and the embedding operator $E_{sr} : V_r \to V_s$ is continuous and has dense range. In particular, $V^{\#}$ is dense in every V_r.

Let us give two simple examples.

(1) $V = \omega$, the space of *all* complex sequences $x = (x_n)$, with the compatibility relation $x\#y \Leftrightarrow \sum_{n=1}^{\infty} |x_n\, y_n| < \infty$, and the partial inner product $\langle x|y\rangle = \sum_{n=1}^{\infty} \overline{x_n}\, y_n$, for $x\#y$.

(2) $V = L^1_{\rm loc}(\mathbb{R}, \mathrm{d}x)$, the space of Lebesgue measurable, locally integrable functions, with the compatibility relation $f\#g \Leftrightarrow \int_{\mathbb{R}} |f(x)g(x)|\, \mathrm{d}x < \infty$ and the partial inner product $\langle f|g\rangle = \int_{\mathbb{R}} \overline{f(x)}g(x)\, \mathrm{d}x$, for $f\#g$.

The previous examples show that $\mathcal{F}(V,\#)$ is a huge lattice (it is complete!) and that assaying subspaces may be complicated, such as Fréchet spaces, non metrizable spaces, etc. This situation suggests to choose an involutive sublattice $\mathcal{I}$ of $\mathcal{F}$, indexed by I, such that

(i) $\mathcal{I}$ is *generating*, that is, $f\#g$ iff $\exists\, r \in I$ such that $f \in V_r, g \in V_{\overline{r}}$;

(ii) every $V_r, r \in I$, is a Hilbert space or a reflexive Banach space;

(iii) $\mathcal{I}$ contains a unique self-dual Hilbert space $V_o = V_{\overline{o}}$.

In that case, the structure $V_I := (V, \mathcal{I}, \langle\cdot|\cdot\rangle)$ is called, respectively, a *lattice of Hilbert spaces* (LHS) or a *lattice of Banach spaces* (LBS). Note that $V^{\#}, V$ themselves usually do *not* belong to the family $\{V_r,\, r \in I\}$, but they can be recovered as $V^{\#} = \bigcap_{r\in I} V_r$, $V = \sum_{r\in I} V_r$.

In the LBS case, the lattice structure takes the following form

- $V_{p\wedge q} = V_p \cap V_q$, with the *projective* norm $\|f\|_{p\wedge q} = \|f\|_p + \|f\|_q$;
- $V_{p\vee q} = V_p + V_q$, with the *inductive* norm

$$\|f\|_{p\vee q} = \inf_{f=g+h} \left(\|g\|_p + \|h\|_q\right), g \in V_p, f \in V_q.$$

These norms are usual in interpolation theory.[2] In the LHS case, one takes similar definitions with squared norms, in order to get Hilbert norms.

We list a series of concrete examples of LBS/LHS (in one dimension, for simplicity). The first two are those described in the Introduction.

(i) Scales of Hilbert or Banach spaces

(a) The Lebesgue L^p spaces on $[0,1]$, $\mathcal{I} = \{L^p([0,1], \mathrm{d}x),\ 1 \leqslant p \leqslant \infty\}$.

(b) The scale of Hilbert spaces built on the powers of a positive self-adjoint operator $A \geq 1$ in a Hilbert space $\mathcal{H}_0$. For instance, in $\mathcal{H}_0 = L^2(\mathbb{R}, \mathrm{d}x)$:

- $(A_{\rm m} f)(x) = (1 - \frac{d^2}{dx^2})^{1/2} f(x)$;
- $(A_{\rm osc} f)(x) = (1 + x^2 - \frac{d^2}{dx^2}) f(x)$.

In the case of $A_{\rm m}$, the intermediate spaces are the Sobolev spaces $H^s(\mathbb{R})$, $s \in \mathbb{Z}$ or $\mathbb{R}$. Note that $\mathcal{H}_\infty(A_{\rm osc})$ coincides with the Schwartz space $\mathcal{S}(\mathbb{R})$ of smooth functions of fast decay, and $\mathcal{H}_{-\infty}(A_{\rm osc})$ with the space $\mathcal{S}^\times(\mathbb{R})$ of tempered distributions.

(ii) Sequence spaces
In ω, we may take the lattice $\mathcal{I} = \{\ell^2(r)\}$ of the weighted Hilbert spaces $\ell^2(r) = \{(x_n) : \sum_{n=1}^{\infty} |x_n|^2 r_n^{-1} < \infty\}$, $r = (r_n)$, $r_n > 0$.

(iii) Spaces of locally integrable functions
In $L^1_{\rm loc}(\mathbb{R}, {\rm d}x)$, we may take the lattice $\mathcal{I} = \{L^2(r)\}$ of the weighted Hilbert spaces $L^2(r) = \{f \in L^1_{\rm loc}(\mathbb{R}, dx) : \int_{\mathbb{R}} |f(x)|^2 r(x)^{-1} \, {\rm d}x < \infty\}$, with $r, r^{-1} \in L^2_{\rm loc}(\mathbb{R}, dx), r(x) > 0$ a.e.

(iv) The spaces $L^p(\mathbb{R}, {\rm d}x)$, $1 < p < \infty$
These spaces do not constitute a scale, since one has only the inclusions $L^p \cap L^q \subset L^s$, $p < s < q$. Thus one has to consider the lattice they generate, which is a genuine lattice of Banach spaces, $L^p \cap L^q$ and $L^p + L^q$, reflexive for $1 < p, q < \infty$.

3. Operators on pip-spaces

According to the general philosophy, an operator on a PIP-space should be defined in terms of assaying subspaces V_r only. Thus we state:

Given a LHS or LBS $V_I = \{V_r, \, r \in I\}$, an *operator* on V_I is a map A from a subset $\mathcal{D} \subseteq V$ into V, where

(i) $\mathcal{D}$ is a nonempty union of assaying subsets of V_I;

(ii) for every assaying subset V_q contained in $\mathcal{D}$, there exists a $p \in I$ such that the restriction A_{pq} of A to V_q is linear and continuous into V_p;

(iii) A has no proper extension satisfying (i) and (ii), i.e., it is maximal.

The linear bounded operator $A_{pq} : V_q \to V_p$ is called a *representative* of A. In terms of the latter, the operator A may be characterized by the set $\mathsf{j}(A) := \{(q,p) \in I \times I : A_{pq}$ exists$\}$. Thus the operator A may be identified with the (coherent) collection of its representatives,

$$A \simeq \{A_{pq} : V_q \to V_p : (q,p) \in \mathsf{j}(A)\}.$$

We denote by $\mathrm{Op}(V_I)$ the set of all operators on V_I. A similar definition may be given for operators $A : V_I \to Y_K$ between two LHSs or LBSs.

Algebraic operations may be defined for operators on a PIP-space:

(i) *Adjoint $A^\times$:* given $A \in \mathrm{Op}(V_I)$ with $(r, s) \in \mathsf{j}(A)$, it has a unique adjoint $A^\times \in \mathrm{Op}(V_I)$ defined by

$$\langle A^\times x | y \rangle = \langle x | Ay \rangle, \text{ for } y \in V_r \text{ and } x \in V_{\overline{s}},$$

that is, $(A^\times)_{\overline{rs}} = (A_{sr})^*$ (usual Hilbert/Banach space adjoint).

It follows that $A^{\times\times} = A$, for every $A \in \mathrm{Op}(V_I)$: no extension is allowed, by the maximality condition (iii) of the definition.

(ii) *Partial multiplication:* AB is defined whenever there is a continuous factorization through some V_q :

$$V_r \xrightarrow{B} V_q \xrightarrow{A} V_s, \quad \text{i.e.,} \quad (AB)_{sr} = A_{sq}B_{qr} .$$

The concept of PIP-space operator allows to treat on the same footing all kinds of operators, from bounded ones to very singular ones. Take, for instance,

$$V_r \subset V_o \simeq V_{\overline{o}} \subset V_s \quad (V_o = \text{Hilbert space}).$$

Then three cases may arise:

- if A_{oo} exists, then A corresponds to a *bounded* operator $V_o \to V_o$;
- if A_{oo} does not exist, but only $A_{or} : V_r \to V_o$, $r < o$, then A corresponds to an *unbounded* operator, with Hilbert space domain containing V_r;
- if no A_{or} exists, but only $A_{sr} : V_r \to V_s$, $r < o < s$, then A corresponds to a *singular* operator, with domain in V_o possibly $\{0\}$.

A nice application is a rigorous analysis of singular quantum Hamiltonians, e.g., the Kronig–Penney crystal model or δ interactions (Ref. 1, Sec.7.1.3).

4. Special classes of operators on PIP-spaces

(1) Homomorphisms

An operator $A \in \text{Op}(V_I, Y_K)$ is called a *homomorphism* if

(i) $\forall\, r \in I$ there exists $u \in K$ such that both A_{ur} and $A_{\overline{u}\overline{r}}$ exist;

(ii) $\forall\, u \in K$, there exists $r \in I$ such that both A_{ur} and $A_{\overline{u}\overline{r}}$ exist.

We denote by $\text{Hom}(V_I, Y_K)$ the set of all homomorphisms between the two LHS V_I, Y_K. Clearly $A \in \text{Hom}(V_I, Y_K)$ if and only if $A^\times \in \text{Hom}(Y_K, V_I)$.

A homomorphism $A \in \text{Hom}(V_I, Y_K)$ is an *isomorphism* if there exists $B \in \text{Hom}(Y_K, V_I)$ such that $BA = 1_V, AB = 1_Y$ (identity operators).

An operator U is *unitary* if $U^\times U$ and $UU^\times$ are defined and $U^\times U = 1_V, UU^\times = 1_Y$. However, unitary operators need *not* be homomorphisms! This implies that the natural setting for group representations in a LHS is that of *unitary isomorphisms* (Ref. 1, Sec.3.3.4; see also (4) below) .

(2) Symmetric operators

An operator A is *symmetric* if $A^\times = A$. Symmetric operators satisfy a generalized KLMN theorem, stating when a symmetric operator has a self-adjoint restriction to the central Hilbert space V_o (Ref. 1, Sec.3.3.5).

(3) Orthogonal projections

An *orthogonal projection* on a nondegenerate PIP-space V is a homomorphism that satisfies the relations $P^2 = P = P^\times$. On the other hand, there is a natural notion of PIP-subspace, called *orthocomplemented*, which guarantees that such a subspace W of V is again a nondegenerate PIP-space with the induced compatibility relation and the restriction of the partial inner product. Then the basic theorem about projections states that a PIP-subspace W of V is orthocomplemented if and only if W is the range of an orthogonal projection $P : W = PV$. Then $V = W \oplus Z$, where $Z = W^{\perp}$ is another orthocomplemented PIP-subspace.

(4) Representations of Lie groups and Lie algebras

Given a strongly continuous unitary representation U_{00} of a Lie group G in a Hilbert space $\mathcal{H}_0$, define the Gårding domain $\mathcal{H}_0^G$, which consists of C^∞ vectors for U_{00}. The domain $\mathcal{H}_0^G$ is dense in $\mathcal{H}_0$ and is stable under $U_{00}(g)$, $\forall\, g \in G$, and also under the representatives of all elements of the Lie algebra $\mathfrak{g}$ of G. Define the Nelson operator $\Delta = \sum_{j=1}^n X_j^2$, where $\{X_j, j = 1, \ldots, n\}$ are the representatives under U_{00} of a basis of $\mathfrak{g}$. The operator Δ is essentially self-adjoint on $\mathcal{H}_0^G$ and $\overline{\Delta}$ is self-adjoint and positive. Define the canonical scale of Hilbert spaces $V_I := \{\mathcal{H}_n, n \in \mathbb{Z}\}$ generated by the powers of $(\overline{\Delta} + 1)$:

$$V^\# := \mathcal{D}^\infty(\overline{\Delta}) \subset \ldots \subset \mathcal{H}_n \subset \ldots \subset \mathcal{H}_0 \subset \ldots \subset \mathcal{H}_{-n} \subset \ldots \subset V := \mathcal{D}_{\overline{\infty}}(\overline{\Delta}).$$

Then $D^\infty(\overline{\Delta}) = \bigcap_{n=1}^\infty D(\overline{\Delta}^n) = \mathcal{H}_0^\infty = \{C^\infty\text{-vectors of } U_{00}\}$. For every $g \in G$ and every $n = 0, 1, 2, \ldots$, $U_{00}(g) : \mathcal{H}_n \to \mathcal{H}_n$ continuously. By transposition, the same holds for $U_{nn}(g^{-1}) : \mathcal{H}_{-n} \to \mathcal{H}_{-n}$, $n = 1, 2, \ldots$. Thus U_{00} extends to a unitary, totally regular representation U in the LHS V_I (Ref. 1, Sec.7.7.4). Therefore, the scale V_I is the natural tool for the representation of $\mathfrak{g}$ or $\mathfrak{U}(\mathfrak{g})$:

(i) Given $x \in \mathfrak{g}$ or $L \in \mathfrak{U}(\mathfrak{g})$, the representative $U(x)$, resp. $U(L)$, originally defined on $\mathcal{H}_0^G$, extends to a regular operator on V_I;

(ii) $\mathcal{H}_0^\infty = D^\infty(\overline{\Delta})$ is a domain of essential self-adjointness for many operators of the form $U(L), L \in \mathfrak{U}(\mathfrak{g})$, e.g. the generators $\{X_j, j = 1, \ldots, n\}$ or elliptic operators, such as the Nelson operator Δ of G or of $H \subset G$.

Thus, if G is the symmetry group of a quantum system, this analysis yields many candidates for physically meaningful observables.

References

1. J-P. Antoine and C. Trapani, *Partial Inner Product Spaces: Theory and Applications*, vol. 1986 of *Lecture Notes in Mathematics* (Springer, Berlin, 2009).
2. J. Bergh and J. Löfström, *Interpolation Spaces* (Springer, Berlin, 1976).
3. H. H. Schaefer, *Topological Vector Spaces* (Springer, Berlin, 1971).

ENHANCED QUANTIZATION: THE PARTICLE ON THE CIRCLE

J. BEN GELOUN
Perimeter Institute, 31 Caroline Str N, Waterloo, ON N2L 2Y5, Canada
E-mail: jbengeloun@perimeterinstitute.ca

Enhanced quantization is an improved program for overcoming difficulties which may arise during an ordinary canonical quantization procedure. We review here how this program applies for a particle on circle.

Keywords: Quantization, coherent states, periodic coordinates.
pi-mathphys-302

1. Introduction

Conventional canonical quantization works very well for many systems but it has also notorious drawbacks. Enhanced Quantization[1] is an improved quantization procedure which has matured through years.[2] It is only recently that the name of "Enhanced Quantization" (EQ) has been found. This procedure yields a new interpretation of the very process of quantization that encompasses the usual canonical formalism and offers additional features as well.

EQ yields a new sense to and improves major points which have been discussed for a long time in the ordinary quantization version. For instance,[3] a) the invariance of the theory under canonical transformations is ensured under EQ; b) EQ may remove singularities in classical solutions (Hydrogen atom and quantum toy models[3] and other simple cosmological models[4]); c) the triviality of certain quantum field models is now traded for a non trivial behavior (pseudo-free theories) and d) EQ preserves the metric positivity in quantum gravity kinematics.

Let us review now the main ingredients of the EQ program.[3]

(i) From canonically conjugate pairs P and Q of self-adjoint operators (a stronger condition than Hermiticity), one generates unitary operators acting on a fiducial state $|\eta\rangle$. This provides a set of coherent states

$$|p,q\rangle = e^{-\frac{i}{\hbar}qP}e^{\frac{i}{\hbar}pQ}|\eta\rangle \tag{1}$$

spanning the Hilbert space $\mathfrak{H}$.

(ii) Referring to the action principle, arbitrary variations of $|\psi(t)\rangle$ for a microscopic system are not accessible to a macroscopic observer who can only change the velocity or position of the microscopic system. Hence, the only state she/he could make are those represented by $|p(t),q(t)\rangle$. Thus, restricting the quantum action functional

$$A_Q = \int_0^T \langle\psi(t)|\left[i\hbar\partial_t - \mathcal{H}\right]|\psi(t)\rangle\, dt \tag{2}$$

to only coherent states yields

$$\begin{aligned} A_{Q(R)} &= \int_0^T \langle p(t),q(t)|\left[i\hbar\partial_t - \mathcal{H}\right]|p(t),q(t)\rangle\, dt \\ &= \int_0^T \Big[p(t)\dot{q}(t) - H(p(t),q(t))\Big]\, dt \end{aligned} \tag{3}$$

The restricted action (3) may be called enhanced classical action since is of a classical form but, because still $\hbar > 0$, it includes certain quantum modifications. The usual classical action is given by

$$A_C = \int_0^T \Big[p(t)\dot{q}(t) - H_c(p(t),q(t))\Big]\, dt\,, \tag{4}$$

where $H_c(p(t),q(t)) = \lim_{\hbar\to 0} H(p(t),q(t))$.

According to this point of view, enhanced classical theory forms a subset of quantum theory, and they both co-exist just like they do in the real world where $\hbar > 0$.[1]

In this paper, we review the EQ program for a classical particle moving on a circle of finite radius.[5] The particle motion on the circle has been studied in different ways[6–13] but none of previous contributions addresses the issue of the relationship between classical and quantum actions which is our main concern here. The present study leads to yet another set of coherent states which serves to unite the classical and quantum theories for such a system. At the quantum level, we find that the non trivial topology induces a possibly shifted momentum easily reabsorbed by a canonical transformation.

2. Enhanced quantization on the circle

Consider a particle on the circle S^1 parametrized by $\theta \in [-\pi, \pi)$. At the quantum level, position and momentum operators Q and P, respectively, obey the commutation relation:

$$[Q, P] = i\hbar \,. \tag{5}$$

The spectrum of the operator Q is bounded in $[-\pi, \pi)$ modulo 2π. The EQ program starts with two self-adjoint operators P and Q. Let us investigate the self-adjoint possibilities of the operator P.

Self-adjoint extension of P. Let us review,[14] the properties of $P = -i\hbar\partial_\theta$ acting on $L^2([-\pi, \pi), d\theta)$.

Call $D(P)$ the domain of P. Consider the inner product for any two functions $\psi, \varphi, \psi', \varphi' \in L^2([-\pi, \pi), d\theta)$. If we adopt the boundary conditions $\varphi(\pm\pi) = 0$ and make no restriction on ψ, then $D(P) = \big\{\varphi;\ \varphi, \varphi' \in L^2([-\pi, \pi), d\theta);\quad \varphi(\pi) = \varphi(-\pi) = 0\big\}$ and $(\psi, P\varphi) = (P\psi, \varphi)$ holds on $D(P)$. Hence P is symmetric, i.e. $P^\dagger = P$ on $D(P)$. However, the domain of $P^\dagger$ is larger, $D(P^\dagger) = \big\{\varphi;\ \varphi, \varphi' \in L^2([-\pi, \pi), d\theta)\big\} \supset D(P)$.

Imposing the boundary condition $\varphi(\pi) = e^{2\pi i\alpha}\varphi(-\pi)$, for a given $\alpha \in [0, 1)$, enlarges the domain of P and reduces the domain of $P^\dagger$ so that

$$\widetilde{D}(P_\alpha) = \Big\{\varphi;\ \varphi, \varphi' \in L^2([-\pi, \pi), d\theta);\ \varphi(\pi) = e^{2\pi i\alpha}\varphi(-\pi)\Big\} = \widetilde{D}(P_\alpha^\dagger)\,. \tag{6}$$

Coherent states. One denotes P by P_α, where $\alpha \in [0, 1)$ labels the different inequivalent representations of the momentum operator. We work in units such that Q is dimensionless and so the dimension of P is that of $\hbar$. Define eigenvectors $|\theta\rangle$ for the operator Q, obeying $\langle\theta|\theta'\rangle = \delta_{S^1}(\theta - \theta')$, where δ_{S^1} is periodic on S^1, as well as eigenvectors $|n, \alpha\rangle$ of P_α, satisfying $\langle n, \alpha|m, \alpha\rangle = \delta_{n,m}$, such that

$$Q|\theta\rangle = \theta|\theta\rangle\,, \quad P_\alpha|n, \alpha\rangle = p_{n,\alpha}|n, \alpha\rangle\,. \tag{7}$$

One has $\langle\theta|P_\alpha|n, \alpha\rangle = (-i\hbar)\partial_\theta\langle\theta|n, \alpha\rangle = p_{n,\alpha}\langle\theta|n, \alpha\rangle$. The spectrum of P_α on the circle is such that $p_{n,\alpha} = \hbar(n + \alpha)$, $(n, \alpha) \in \mathbb{Z} \times [0, 1)$ and corresponds to the normalized wave functions

$$\langle\theta|n, \alpha\rangle = \frac{1}{\sqrt{2\pi}} e^{i(n+\alpha)\theta}\,. \tag{8}$$

The unitary operators $e^{-\frac{i}{\hbar}qP_\alpha}$ and $e^{-\frac{i}{\hbar}pQ}$, where $(q, p) \in S^1 \times \mathbb{R}$ allow us to define a set of states

$$|p, q\rangle = e^{-\frac{i}{\hbar}qP_\alpha} e^{\frac{i}{\hbar}pQ}\,|\eta_\alpha\rangle\,, \tag{9}$$

where $|\eta_\alpha\rangle$ is called the fiducial state. One proves that the states (9) are normalized $\langle p,q|p,q\rangle = \langle\eta_\alpha|\eta_\alpha\rangle = 1$, and they satisfy a resolution of unity:

$$\int_{\mathbb{R}\times S^1} |p,q\rangle\langle p,q| \,\frac{dpdq}{2\pi\hbar} = I_{\mathfrak{H}} \,. \tag{10}$$

The set of states $\{|p,q\rangle\}$ forms an overcomplete family of normalized states which can be called coherent states.

We can now discuss the dynamics associated with such states by introducing a general quantum Hamiltonian of the form $\mathcal{H}(P, e^{iQ}, e^{-iQ})$. Consider the restricted quantum action associated with $|\psi(t)\rangle \to |p(t),q(t)\rangle$ which leads to

$$A_{Q(R)} = \int_0^T \langle p(t),q(t)|\Big[i\hbar\partial_t \; - \; \mathcal{H}\Big]|p(t),q(t)\rangle \; dt \,. \tag{11}$$

A class of fiducial vectors $|\eta_\alpha\rangle$ is choosen in the domain of both Q and P_α such that these states satisfy

$$\langle\eta_\alpha|Q|\eta_\alpha\rangle = 0 \qquad \text{and} \qquad \langle\eta_\alpha|P_\alpha|\eta_\alpha\rangle = \hbar\alpha \,. \tag{12}$$

Hence $|\eta_\alpha\rangle$ obeying the condition (12) can be considered analogs of the "physically centered" fiducial vectors $|\eta\rangle$ for the canonical case.

A direct evaluation using (12) yields

$$\langle p(t),q(t)|\Big[i\hbar\partial_t\Big]|p(t),q(t)\rangle = (\hbar\alpha + p)\dot{q} \,. \tag{13}$$

Moreover, one proves that

$$H_\alpha(p(t),q(t)) = \langle\eta_\alpha|\mathcal{H}(P_\alpha + p, e^{i(Q+q)}, e^{-i(Q+q)})|\eta_\alpha\rangle \,. \tag{14}$$

Equations (13) and (14) imply that the restricted quantum action is of the form

$$A_{Q(R)} = \int_0^T \Big\{\Big[p\dot{q} - H_\alpha(p(t),q(t))\Big] + \hbar\,\alpha\dot{q}\Big\}dt \,. \tag{15}$$

One notices that the part $\widetilde{A}_C = \int_0^T [p\dot{q} - H_\alpha(p(t),q(t))]dt$, as in the ordinary situation[1] can be related to a classical action A_C up to $\hbar$ corrections using

$$H_\alpha(p,q) = H_c(p,q) + O(\hbar; p, q) \,. \tag{16}$$

$H_c(p,q)$ is viewed as the usual classical Hamiltonian. Note that the quantum parameter α induces a surface term $\hbar\alpha\dot{q}$ in $A_{Q(R)}$ which makes no influence on the enhanced classical equations of motion whatsoever. Thus, one has

$$A_{Q(R)} = A_C + O(\hbar) \,. \tag{17}$$

Let us apply the above formalism to the particular instance of a particle governed by the following (periodic) dynamics

$$\mathcal{H}(P_\alpha, e^{iQ}, e^{-iQ}) = P_\alpha^2 + V(e^{iQ}, e^{-iQ}), \tag{18}$$
$$V(e^{iQ}, e^{-iQ}) = a_0 + \sum_{n=1}^{m} \left[a_n \cos nQ + b_n \sin nQ\right],$$

in mass units such that $1/2\mu = 1$ and with m a positive integer. As a fiducial vector, we consider

$$\eta_\alpha(\theta) := \langle\theta|\eta_\alpha\rangle = N e^{(r/\hbar)(\cos\theta - 1) + i\alpha\theta},$$
$$N = \left[2\pi e^{-2r/\hbar}\, I_0\!\left(\frac{2r}{\hbar}\right)\right]^{-1/2}, \tag{19}$$

where $r/\hbar > 0$, $I_0(z)$ stands for a modified Bessel funtion,[15] and N is a normalization fixed by $\langle\eta_\alpha|\eta_\alpha\rangle = 1$. Among its properties, $|\eta_\alpha(\theta)|$ is even, periodic and fulfills (12). For large $r/\hbar \gg 1$ and for $|\theta| \le \pi$, one makes the approximation

$$|\eta_\alpha(\theta)|^2 = N^2 e^{\frac{2r}{\hbar}[\cos\theta - 1]} \lesssim K N^2 e^{-\frac{r}{\hbar}\theta^2}, \tag{20}$$

for a large constant K impying that $|\theta| \lesssim \sqrt{\hbar/r}$ is small. Therefore, $\eta_\alpha(\theta)$ acts as a large θ-value cut-off. Calculating the diagonal coherent state matrix elements of $\mathcal{H}$, and denoting $\alpha' = \hbar\alpha$, we come to the restricted quantum action given by

$$A_{Q(R)} = \int_0^T \left\{(p + \alpha')\dot{q} - \left[(p + \alpha')^2 + V(e^{iq}, e^{-iq})\right] + O(\hbar)\right\} dt. \tag{21}$$

Therefore, up to constants and a canonical shift in momentum ($p \to p - \alpha'$),

$$A_{Q(R)} = \int_0^T \left\{p\dot{q} - \left[p^2 + V(e^{iq}, e^{-iq})\right] + O(\hbar)\right\} dt$$
$$= A_C + O(\hbar). \tag{22}$$

During the evaluation of $H_\alpha(p, q)$ in (11) yielding then (21), the term $\langle\eta_\alpha|(P_\alpha + p)^2|\eta_\alpha\rangle$ contains some terms proportional to α' of order $O(\hbar)$ plus other higher order $\hbar$-corrections. We have preferred to remove these terms in a unified way by making a canonical shift of momenta at the last stage (22). Other terms contributing to $O(\hbar)$ in (21) come from the potential expectation value. Note that, strictly speaking, the quantity $\langle\eta_\alpha|\, V(e^{i(Q+q)}, e^{-i(Q+q)})\, |\eta_\alpha\rangle$ contains at the first order of approximation, terms of order $O(\hbar/r)$. These are subtleties[5] without any consequence on the fact that (22) holds.

Acknowledgements

The organizers of the XXIXth International Colloquium on Group-Theoretical Methods in Physics, Nankai, China, are warmly thank for their welcome and hospitality. Discussions with John R. Klauder are gratefully aknowledged. Research at Perimeter Institute is supported by the Government of Canada through Industry Canada and by the Province of Ontario through the Ministry of Research and Innovation.

References

1. J. R. Klauder, *J. Phys. A* **45** 285304 (2012).
2. J. R. Klauder, *A Modern Approach to Functional Integration*, (Birkhäuser, Boston, MA, 2010).
3. J. R. Klauder, *Enhanced Quantum Procedures that Resolve Difficult Problems*, arXiv:1206.4017.
4. M. Fanuel and S. Zonetti, *Affine Quantization and the Initial Cosmological Singularity*, *Europhys. Lett.* **101**, 10001 (2013) .
5. J. Ben Geloun and J. R. Klauder, *Enhanced quantization on the circle*, *Phys. Scr.* **87**, 035006 (2013).
6. J. Govaerts and V. M. Villanueva, *Int. J. Mod. Phys. A* **15**, 4903 (2000).
7. J. Govaerts and F. Payen, *Mod. Phys. Lett. A* **22**, 119 (2007).
8. W. Chagas-Filho, *2T Physics and Quantum Mechanics*, arXiv:0802.2840 [hep-th].
9. S. T. Ali, J.-P. Antoine and J.-P. Gazeau, *Coherent States, Wavelets, and their Generalizations*, (Springer-Verlag, Berlin, 2000).
10. J.-P. Gazeau, *Coherent states in quantum physics*, (Wiley-VCH, Weinheim, 2009).
11. K. Kowalski, J. Rembielinski and L. C. Papaloucas, *J. Phys. A* **29**, 4149 (1996).
12. J. A. Gonzalez and M. A. del Olmo, *J. Phys. A* **31**, 8841 (1998).
13. G. Chadzitaskos, P. Luft and J. Tolar, *J. Phys.: Conf. Ser.* **284**, 012016 (2011).
14. M. Reed and B. Simon, *Methods of Modern Mathematical Physics. 2. Fourier Analysis, Selfadjointness*, (Accademic Press, NY, 1975).
15. *Handbook of Mathematical Functions*, 10th edition, Appl. Math. Ser. **55**, Section 9 p. 374, editors A. Abramowitz and I. A. Stegun (Dover, NY, 1972).

GROUPS OR SEMIGROUPS FOR THE DYNAMICAL EVOLUTION IN QUANTUM PHYSICS

A. BOHM* and H. V. BUI

Department of Physics, University of Texas at Austin,
Austin, Texas 78712, USA
** E-mail: bohm@physics.utexas.edu*

According to a story told to me by Larry Schulman Wigner was going around Princeton in the 1960s asking people: Tell me, what is the $2\,P^{1/2}$ (decaying) state of the H-atom? Wigner knew from his work on the Weisskopf-Wigner approximation (1930) and on the Breit-Wigner lineshape (1936) that there could not be an exponentially decaying state (Gamow 1926) vector or a Breit-Wigner (Lorentzian) resonance state within the mathematical framework of quantum mechanics at his time. We shall discuss how new mathematical tools, Hardy spaces predicting semigroup time evolution, can overcome this problem and lead to a unified description of resonance and decay phenomena.

1. Distinction between States and Observables

In quantum physics, one strictly distinguishes between states and observables. This is documented in experiments and it must be reflected in the theory which describes the experiment. An example is the scattering experiment depicted in Fig.1 with explanation in the caption.

The Born probabilities to measure the observable $A(t)$ in the state ρ or in the pure state ϕ^+ is calculated in the theory as:

$$\begin{aligned}\text{Born probability} = \mathcal{P}_\rho(A(t)) &= \mathrm{Tr}(A(t)\,\rho) = \mathrm{Tr}(A(t)\;|\phi^+\rangle\langle\phi^+\,|) \\ &= \langle\phi^+|A(t)\,|\phi^+\rangle = \langle\phi^+\,|\;A(t)\;|\phi^+\rangle\,.\end{aligned} \tag{1}$$

For the special case $A(t) = |\psi^-(t)\rangle\langle\psi^-(t)\,|$, this probability is given by:

$$\mathcal{P}_{\phi^+}(|\psi^-(t)\rangle\langle\psi^-(t)\,| = \underbrace{|\langle\psi^-(t)\,|\phi^+\rangle|^2}_{\text{in Heisenberg picture}} = \underbrace{|\langle\psi^-\,|\phi^+(t)\rangle|^2}_{\text{in Schrödinger picture}}\,. \tag{2}$$

To test the theory by experiments, one compares:

the theoretically calculated Born probabilities	with	measured ratio of large number of detector counts
$\mathcal{P}_\rho(A(t)) = \mathrm{Tr}(A(t)\rho) = \mathrm{Tr}(A\rho(t))$	$\approx$	$N(t)/N$

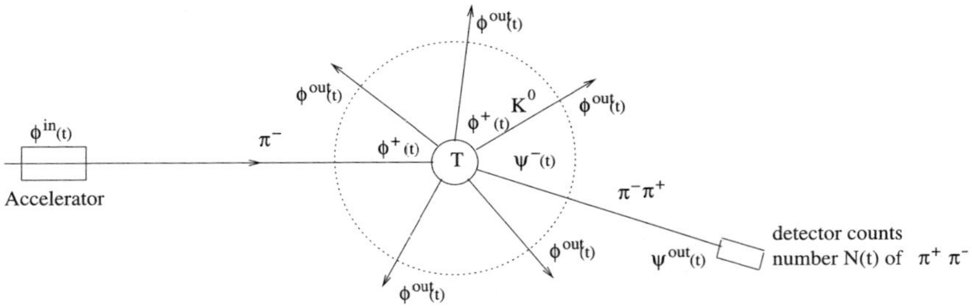

Fig. 1. The accelerator (preparation apparatus) prepares an in-state $\pi^- p$ which due to the interaction at the target T results into an uncontrolled mixture of out-states $\phi^+(t) \to \phi^{out}(t)$, of all kind of particles etc. One result of these reactions is $\pi^- p \to \Lambda K^0 \to \Lambda \pi^- \pi^+$. The detector is built to register a specific observable described by $A = A^\dagger$ or, in a special case (if $A = |\psi^-\rangle\langle\psi^-|$) by $\psi^-(t)$. The detector defines the observable, in this example it detect $|\pi^-\pi^+\rangle\langle\pi^-\pi^+| = |\psi^-\rangle\langle\psi^-|$ and is placed outside the interaction region of T: $\psi^-(t) \to \psi^{out}(t)$. It counts the number $N(t) = N(t; \Delta\Omega, E^{out})$ of π^+, π^- as function of the travel time from the target T to the decay vertex of $K^0 \to \pi^-\pi^+$ which provides a means to measure the lifetime of the K^0 as a function of the travel time of the K^0 from T to the $(\pi^+\pi^-)$-decay vertex.

It is clear that the probability $|\langle\psi^-(t)|\phi^+\rangle|^2$ to detect an observable $A(t) = |\psi^-(t)\rangle\langle\psi^-(t)|$ in the state ϕ^+ can not be measured for all t: $-\infty < t < +\infty$, because the clicks of the detector can not be registered at times $t < t_0$, where t_0 is the time at which the state ϕ^+ is prepared. A detector in the experiment of Fig.1 can not detect anything relevant to the interaction before the time t_0, at which the accelerator will be turned on, because this would contradict the causality principle:

> A state ϕ^+ needs to be prepared first, by a time t_0, *before* an observable $|\psi^-(t)\rangle\langle\psi^-(t)|$ can be measured in that state $\phi^+(t_0)$ by the detector clicks, $N(t)/N$ for times $t \geq t_0$. (3)

This causality requires that the dynamical equations have solutions which predict the Born probabilities:

$$\mathcal{P}_{\phi^+}(|\psi^-(t)\rangle\langle\psi^-(t)|) = |\langle\psi^-(t)|\phi^+\rangle|^2 = |\langle\psi^-|\phi^+(t)\rangle|^2 \qquad \text{for } t > t_0\,. \quad (4)$$

The time t_0 is the finite time at which the state is prepared and after which the observable can be registered in the state. A quantum mechanics ensemble consists of a large number N of micro-systems and t_0 is thus measured as an ensemble $t_0^{(i)}$ of the beginnings of time for the i-th member of this ensemble. Such "beginnings of time" $t_0^{(i)}$ have been observed as the on-set times of dark periods in the experiments with single ions in a Paul trap.[1] The i-th individual excited state lives for a duration $\Delta t^{(i)} = t_1^{(i)} - t_0^{(i)}$

decaying at $t_1^{(i)}$ and the reproducible quantity is the ensemble average of the $\Delta t^{(i)}$, which gives the lifetime τ of the excited state.[2]

2. Solving the Dynamical Equations under Different Boundary Conditions

In quantum theory, the Born probabilities that are measured by the detector counts, $N(t)/N$, are calculated by solving the dynamical equations of quantum mechanics with Hamiltonian H for the quantum system, either in the Schrödinger picture or in the Heisenberg picture.

Heisenberg equations for the observables $A(t)$ or for $\psi^-(t)$ are:

$$i\hbar\frac{\partial A(t)}{\partial t} = -[H, A(t)]\,, \qquad i\hbar\frac{\partial}{\partial t}\psi^-(t) = -H\psi^-(t)\,. \tag{5a}$$

Or, the von Neumann or Schrödinger equation for a state described by the state operator $\rho(t)$ ("density" operator) or for state vector $\phi^+(t)$ are:

$$i\hbar\frac{\partial\rho(t)}{\partial t} = [H, \rho(t)]\,, \qquad i\hbar\frac{\partial}{\partial t}\phi^+(t) = H\phi^+(t)\,. \tag{5b}$$

To solve differential equations (like Eqs. (5a) or (5b)) one needs to choose **Boundary Conditions**, that means one need to select the mathematical spaces which the solutions of the dynamical equations, (i.e ϕ^+ of the Schrödinger or ψ^- of the Heisenberg picture), must belong to. In the past, one had chosen for the boundary condition: the Hilbert space $\mathcal{H}$ (Lebesgue square integrable wave function) or the Schwartz space Φ.

$$\text{Set of state } \{\phi^+\} = \text{Set of observables } \{\psi^-\} = \Phi \text{ or } \mathcal{H}\,. \tag{6}$$

From either of the boundary conditions (6) follows, (by the Stone-von Neumann theorem[3] for Hilbert space $\mathcal{H}$ and by a similar theorem for Schwartz space Φ[4,5]), that the time evolution is given by the (*unitary*) group evolution. The solutions of the Schrödinger equation are given by:

$$\begin{array}{l}\phi^+(t) = U^\dagger(t)\,\phi^+ = e^{-iHt/\hbar}\,\phi^+\,, \quad \text{for } -\infty < t < +\infty \\ \text{in the Schrödinger picture for } \phi^+ \in \mathcal{H} \text{ (as well as for } \phi^+ \in \Phi)\,.\end{array} \tag{7a}$$

The solutions for the Heisenberg equation are given by:

$$\begin{array}{l}\psi^-(t) = U(t)\,\psi^- = e^{iHt/\hbar}\,\psi^-, \quad \text{for } -\infty < t < +\infty \\ \text{in the Heisenberg picture for } \psi^- \in \mathcal{H} \text{ (as well as for } \psi^- \in \Phi)\,.\end{array} \tag{7b}$$

This leads to predictions that the Born probabilities of an observable $A(t) = |\psi^-(t)\rangle\langle\psi^-(t)|$ in a state ρ given by

$$\mathcal{P}_\rho(A(t)) = Tr(\rho\,A(t)) = Tr(\rho(t)\,A) = |\langle\psi^-(t)\,|\phi^+\rangle|^2 = |\langle\psi^-\,|\phi^+(t)\rangle|^2\,, \tag{8}$$

are predicted for all $-\infty < t < +\infty$. But this prediction for Eq. (8) violates the causality condition $t > t_0$ in Eqs. (3) and (4).

The way out is the following: There is no compelling theoretical reason to solve the dynamical differential equations under Hilbert space or under the Schwartz space boundary conditions (6), (the Schwartz space provides the mathematical justification of the Dirac kets.)

In order to obtain a causal time evolution as required by (3) and (4), one needs to solve the dynamical Eqs. (5a) or (5b) under different boundary condition than (6). In the Lax-Phillip scattering theory, Hardy spaces were used in applications to the scattering of classical wave.[6] Hardy spaces have been used more recently in the quantum theory of scattering.[7] They are also needed to obtain a quantum theory that unifies resonance and decay phenomena.[8–10] They are the only mathematical tools that allow an unification of resonance scattering and of decay.[11]

3. The Hardy Space Axiom from *S*-Matrix Poles

What should one choose for the mathematical space of the in-states $\{\phi^+\}$ representing the preparation apparatus and for the mathematical space of the observables $\{\psi^-\}$ representing the detector in Fig. 1?

In order to arrive at the Hardy space axiom, we start with the S-matrix pole at $z_R = E_R - i\Gamma/2$ second sheet of the j-th partial S-matrix element $S_j(E)$ as the definition of a resonance.[7] From the pole term $S_j^{P.T}(E) = 2ia_j^{BW}(E) + 1$ of the S-matrix of angular momentum j at z_R, we obtain a Breit-Wigner resonance amplitude:

$$a_j^{BW}(E) = \frac{R_{-1}}{E - (E_R - i\Gamma/2)} \quad \Longleftrightarrow \quad |z_R, j, j_3, \eta^-\rangle\sqrt{2\pi\Gamma}\,, \tag{9a}$$

and associate to it a ket with Breit-Wigner energy distribution:

$$|z_R, j, j_3, \eta^-\rangle\sqrt{2\pi\Gamma} = \int_{-\infty_{II}}^{+\infty} dE\ |E, j, j_3, \eta^-\rangle \frac{i\sqrt{\frac{\Gamma}{2\pi}}}{E - (E_R - i\Gamma/2)}\,. \tag{9b}$$

The ket with Breit-Wigner energy wave function (9b) should be an eigenket of the Hamiltonian H with a discrete complex eigenvalue $z_R = E_R - i\Gamma/2$:

$$\langle\psi^- \mid H^\times\ |z_R, j, j_3, \eta^-\rangle = z_R\langle\psi^-\ |z_R, j, j_3, \eta^-\rangle \quad \text{for all } \psi^- \in \Phi_+\,. \tag{10a}$$

It should also have the property that it exponentially decays:

$$\begin{aligned}\langle e^{iHt/\hbar}\psi_\eta^-|z_R, j, j_3, \eta^-\rangle &= \langle\psi_\eta^-|e^{-iH^\times t/\hbar}|z_R, j, j_3, \eta^-\rangle \\ &= e^{-iE_R t/\hbar}e^{-(\Gamma/2)t/\hbar}\langle\psi_\eta^-|z_R, j, j_3, \eta^-\rangle \end{aligned}\tag{10b}$$

but, because of causality (4), *only* for $t \geq 0$.

Then a resonance state with Breit-Wigner energy distribution (9b) would be also an exponentially decaying Gamow ket (10b) with lifetime $\tau = \hbar/\Gamma$.

In order to obtain all these properties (9a), (9b), (10a) and (10b), one has to make assumptions about the energy wave functions of the states $\phi^+(E) = \langle^+E, j, j_3\,|\phi^+\rangle$ and of the observables $\psi^-(E) = \langle^-E, j, j_3\,|\psi^-\rangle$ in the Dirac basis vector expansion:

$$\phi^+ = \sum_{j,j_3,\eta} \int_0^\infty dE\,|E, j, j_3, \eta^+\rangle\langle^+E, j, j_3, \eta\,|\phi^+\rangle = \int_0^\infty dE\,|E^+\rangle\phi^+(E) \quad (11a)$$

$$\psi^- = \sum_{j,j_3,\eta} \int_0^\infty dE'|E', j, j_3, \eta^-\rangle\langle^-E', j, j_3, \eta|\psi^-\rangle = \int_0^\infty dE'|E'^-\rangle\psi^-(E') \quad (11b)$$

These assumptions turned out to be identical with the requirement:

$$\phi^+(E) = \langle^+E\,|\phi^+\rangle \in (\mathcal{H}^2_- \cap \mathcal{S})_{\mathbb{R}_+} = \text{smooth Hardy functions on } \mathbb{C}_-. \quad (12a)$$
$$\psi^-(E) = \langle^-E\,|\psi^-\rangle \in (\mathcal{H}^2_+ \cap \mathcal{S})_{\mathbb{R}_+} = \text{smooth Hardy functions on } \mathbb{C}_+. \quad (12b)$$

$\mathbb{C}_\pm$ are upper and lower complex energy plane second sheet of the S-matrix and the integrals extend over $\mathbb{R}_+$ along the cut of the S-matrix.
This is the new Hardy space axiom for the solutions of Eqs. (5a) and (5b):

$$\text{Set of prepared (in-) states } \{\phi^+\} = \Phi_- \subset \mathcal{H} \subset \Phi_-^\times. \quad (13a)$$
$$\text{Set of registered out- observables } \{\psi^-\} = \Phi_+ \subset \mathcal{H} \subset \Phi_+^\times. \quad (13b)$$

The Hardy spaces Φ_- and Φ_+ provide physicists with a new quantum theory that unifies resonance and decay phenomena:

a Breit-Wigner Resonance $a_j^{BW}(E)$ of width Γ in Eq. (9a) represents the same quantum object as a decaying state in Eqs. (10a), (10b) with lifetime τ; and these are the exact inverse of each other, $\tau = \frac{\hbar}{\Gamma}$.

4. Conclusion

In the same way as the mathematical theorem of Stone and von Neumann[3] for the Hilbert space $\mathcal{H}$ and a similar theorem[4] for the Schwartz space Φ leads to group evolution in $\mathcal{H}$ or Φ, the Paley-Wiener theorem[12] for Hardy spaces axiom (13a) (13b) provides a causal solution of the Heisenberg Eq. (5a) and the Schrödinger Eq. (5b): The time evolution of the observables ψ^- in the Heisenberg picture is given by the semigroup evolution:

$$\psi^-(t) = \mathrm{e}^{-iH(t-t_0)/\hbar}\,\psi^-(t_0)\,, \qquad t_0 < t < +\infty\,, \quad (14a)$$

or time evolution for the in-state ϕ^+ (Schrödinger picture) is given by:

$$\phi^+(t) = \mathrm{e}^{-iH(t-t_0)/\hbar}\,\phi^+(t_0)\,, \qquad t_0 < t < +\infty\,. \quad (14b)$$

The same Hardy space property of the energy wave functions (12a) (12b) that unifies resonance and decay phenomena also predicts -by the Paley-Wiener theorem[12]- the semigroup time evolution (14a) (14b) and therewith a finite beginning of time t_0 for quantum systems.

As a consequence, the Born probabilities to detect the observable ψ^- in the state ϕ^+ representing ensemble of microsystems fulfill:

$$\mathcal{P}_{\phi^+}(\psi^-(t)) = |\langle\psi^-(t)\,|\phi^+\rangle|^2\,, \qquad \text{only for} \quad t \geq t_0\,(=0)\,, \tag{15a}$$

or similar for

$$\mathcal{P}_\rho(A(t)) = \mathrm{Tr}(A(t)\,\rho)\,, \qquad \text{only for} \quad t \geq t_0\,(=0). \tag{15b}$$

Here t_0 represents an ensemble times $t_0^{(i)}$ for the ensemble of individual microsystems, observed in[1] and as discussed in[2] .

References

1. H. Dehmelt, W. Nagourney and J. Sandberg, *Phys. Rev. Lett.* **56**, 2797 (1986); X. Zhao, N. Yu, H. Dehmelt and W. Nagourney, *Phys. Rev.* A **51**, 4483 (1995); Th. Sauter, W. Neuhauser, R. Blatt and P. E. Toschek, *Phys. Rev. Lett.* **57**, 1696 (1986); J. C. Bergquist, Randall G. Hulet, Wayne M. Itano and D. J. Wineland, *Phys. Rev. Lett.* **57**, 1699 (1986); E. Peik, G. Hollemann and H. Walther, *Phys. Rev.* A **49**, 402 (1994).
2. A. Bohm, P. Kielanowski and S. Wickramasekara, *Ann. Phys.* **32**, 2299 (2006).
3. M. H. Stone, *Ann. Math.* **33**(3), 643 (1932); J. von Neumann, *Ann. Math.* **33**(3), 567 (1932).
4. A. Bohm and M. Gadella, *Dirac Kets, Gamow Vectors and Gelfand Triplet*, Lecture Note in Physics, vol. 348, p. 88 Proposition II (Springer, Berlin, 1989).
5. S. Wickramasekara and A. Bohm, *J. Phys. A* **35**, 807 (2002).
6. P. D. Lax and R. S. Phillips, *Scattering theory*, 1st edition (Academic Press, New York, 1967).
7. A. Bohm, H. Uncu And S. Komy, *Rep. Math. Phys.* **65**, 5 (2009).
8. A. Bohm, *Quantum Mechanics: Foundations and Applications*, 1st edition (1979), 2nd edition (1986), 3rd edition (1993) or Paperback edition (Springer, New York, 2001), upon a suggestion by H. Baumgartel.
9. Y. Strauss and L. P. Horwitz and E. Eisenberg, *J. Math. Phys.* **41**, No. 12, 8050 (2000); Y. Strauss, *Int. J. Theor. Phys.* **42**, 2285 (2003); Y. Strauss, L. P. Horwitz, and A. Volovick, *J. Math. Phys.* **47**, 123505 (2006).
10. H. Baumgartel, *Int. J. Theor. Phys.* **46**, 1959 (2007); H. Baumgartel, *J. Math. Phys.* **51**, 113508 (2010); H. Baumgartel, *Rev. Math. Phys.* **18**, 61 (2006).
11. O. Civitarese and M. Gadella, *Physics Report*, **396**, 41-43 (2004).
12. R. Paley and N. Wiener, *Fourier Transform in the Complex Domain* (American Mathematical Society, New York, 1934).

ON GENERALIZED OSCILLATOR ALGEBRAS AND THEIR ASSOCIATED COHERENT STATES

J. D. BUKWELI-KYEMBA† and M. N. HOUNKONNOU*

International Chair in Mathematical Physics and Applications
(ICMPA - UNESCO Chair), University of Abomey-Calavi
Cotonou, 072 B.P.: 50, Republic of Benin
† *E-mail:desbuk@gmail.com*
* *E-mail: norbert.hounkonnou@cipma.uac.bj; hounkonnou@yahoo.fr*

A unified method of calculating structure functions from commutation relations of deformed single-mode oscillator algebras is presented. A natural approach to building coherent states associated to deformed algebras is then deduced.

Keywords: coherent states; structure function; deformed algebra; Heisenberg algebra.

1. Introduction

Generalization of Heisenberg-Weyl algebra was suggested long before the discovery of quantum groups by Heisenberg to achieve regularization for (nonrenormalizable) nonlinear spinor field theory. The issue was considered as small additions to the canonical commutation relations.[1,2] Snyder,[3] investigating the infrared catastrophe of soft photons in the Compton scattering, raised this issue and built a noncommutative Lorentz invariant space-time where the noncommutativity of space operators is proportional to nonlinear combinations of phase space operators.

Further modifications of the oscillator algebra and their possible physical interpretations as spectrum generating algebras of non standard statistics were made since the earlier work of Snyder. As matter of citation, let us mention the q-oscillator algebras by Coon and coworkers,[4,5] Kuryshkin,[6] Jannussis and collaborators.[7]

With the development of quantum groups, new aspects of q-oscillators were identified[8–10] as a tool for providing a boson realization of quantum algebra $su_q(2)$ using a q-analogue of the harmonic oscillator and the Jordan-Schwinger mapping, and then generalizations in view of unifying or extending different existing q-deformed algebras were elaborated[11,12]

Despite all useful properties and applications motivated by various one-parameter deformed algebras, the multi-parameter deformations aroused much interest because of their flexibility when dealing with concrete physical models. See for example[13,14] and references therein.

Coherent states (CS) have practically followed the same trend as the quantum algebras. They were invented by Schrödinger in 1926 in the context of the quantum harmonic oscillator, and defined as minimum-uncertainty states that exhibit a classical behavior.[15]

In 1963, coherent states were simultaneous rediscovered by Glauber,[16,17] Klauder[18,19] and Sudarshan[20] in quantum optics of coherent light beams emitted by lasers.

The vast field covered by coherent states motivated their generalizations to other families of states deducible from noncanonical operators and satisfying not necessarily all properties of CS.[21,22]

The more general class is essentially based on the overcompleteness property of coherent states. This property was the *raison d'être* of the mathematically oriented construction of generalized coherent states by Ali *et al*[21]or of the ones with physical orientations.[23,24] Numerous publications continue to appear using this property. See for example[25,26]and references therein.

In this paper we give a method of computing the so-called structure function which is the basis of coherent state construction from a given algebra.

2. Unified deformed single-mode oscillator algebras and associated CS

The Heisenberg-Weyl algebra is a Lie algebra generated by the position Q, the momentum P and the unity $\mathbf{1}$ operators satisfying:

$$[Q,\ P] = i\hbar\mathbf{1}, \quad [Q,\ \mathbf{1}] = 0, \quad [P,\ \mathbf{1}] = 0. \tag{1}$$

Defining the annihilation and creation operators by $b := \frac{Q+iP}{\sqrt{2\hbar}}$ and $b^\dagger := \frac{Q-iP}{\sqrt{2\hbar}}$, the Weyl-Heisenberg algebra is generated by $\{b,\ b^\dagger,\ \mathbf{1}\}$ satisfying:

$$[b,\ b^\dagger] = \mathbf{1}, \quad [b,\ \mathbf{1}] = 0, \quad [b^\dagger,\ \mathbf{1}] = 0. \tag{2}$$

From (2), one defines the operator $N := b^\dagger b$, also called *number operator*, with the properties:

$$[N,\ b] = -b, \quad [N,\ b^\dagger] = b^\dagger, \quad [N,\ \mathbf{1}] = 0. \tag{3}$$

The canonical coherent states (CS) are normalized states $|z\rangle$ satisfying one of the following three equivalent conditions:[15,16,18,20]

$$(i)\quad (\Delta Q)(\Delta P) = \frac{\hbar}{2},$$
$$(ii)\quad b|z\rangle = z|z\rangle;$$
$$(iii)\quad |z\rangle = e^{zb^\dagger - \bar{z}b}|0\rangle = e^{-|z|^2/2}e^{zb^\dagger}|0\rangle = e^{-|z|^2/2}\sum_{n=0}^{\infty}\frac{z^n}{\sqrt{n!}}|n\rangle,$$

where $(\Delta A)^2 := \langle z|A^2 - \langle A\rangle^2|z\rangle$, $\langle A\rangle := \langle z|A|z\rangle$, $b|0\rangle = 0$ and $\langle 0|0\rangle = 1$.

They belong to the Fock space $\mathcal{F} = \text{span}\{|n\rangle \mid n \in \mathbb{N}\cup\{0\}\}$, where the states

$$|n\rangle = \frac{1}{\sqrt{n!}}(b^\dagger)^n|0\rangle, \quad n = 1,\ 2,\ \cdots \tag{4}$$

satisfy the orthogonality and completeness conditions:

$$\langle m|n\rangle = \delta_{m,n}, \qquad \sum_{n=0}^{\infty}|n\rangle\langle n| = \mathbf{1}. \tag{5}$$

The important feature of these coherent states resides in the partition (resolution) of identity:

$$\int_{\mathbb{C}}\frac{[d^2z]}{\pi}|z\rangle\langle z| = \mathbf{1}, \tag{6}$$

where we have put $[d^2z] = d(Rez)d(Imz)$ for simplicity.

Definition 2.1. We call deformed Heisenberg algebra, an associative algebra generated by the set of operators $\{\mathbf{1},\ a,\ a^\dagger,\ N\}$ satisfying the relations

$$[N,\ a^\dagger] = a^\dagger, \qquad [N,\ a] = -a, \tag{7}$$

such that there exists a non-negative analytic function f, called the structure function, defining the operator products $a^\dagger a$ and $aa^\dagger$ in the following way:

$$a^\dagger a := f(N), \qquad aa^\dagger := f(N+\mathbf{1}), \tag{8}$$

where N is a self-adjoint operator, a and its Hermitian conjugate $a^\dagger$ denote the deformed annihilation and creation operators, respectively.

The deformed Heisenberg algebras have a common property characterized by the existence of a self-adjoint number operator N, a lowering operator a and its formal adjoint, called raising operator, $a^\dagger$, and differ by the expression of the structure function f.

The associated Fock space $\mathcal{F}$ is now spanned by the orthonormalized eigenstates of the number operator N given by:

$$|n\rangle = \frac{1}{\sqrt{f(n)!}}(a^\dagger)^n|0\rangle, \quad n \in \mathbb{N} \cup \{0\}, \tag{9}$$

where $f(n)! = f(n)f(n-1)...f(1)$ with $f(0) = 0$. Moreover,

$$a|n\rangle = \sqrt{f(n)}|n-1\rangle, \quad a^\dagger|n\rangle = \sqrt{f(n+1)}|n+1\rangle. \tag{10}$$

In this context, the structure function f appears as a key unifying methods of coherent state construction corresponding to deformed algebras.

Denote $\mathbf{D}_f := \{z \in \mathbb{C} : |z|^2 < R_f\}$, where R_f is the radius of convergence of the series (the *deformed exponential function*):

$$\mathcal{N}_f(x) := \sum_{n=0}^{\infty} \frac{x^n}{[f(n)]!}. \tag{11}$$

Then, the states

$$|z, f\rangle := (\mathcal{N}_f(|z|^2))^{-1/2} \sum_{n=0}^{\infty} \frac{z^n}{\sqrt{[f(n)]!}}|n\rangle, \quad z \in \mathbf{D}_f, \tag{12}$$

are normalized eigenstates of the raising operator a with the eigenvalues z. They are not orthogonal to each other. Moreover, the map $z \mapsto |z, f\rangle$ from $\mathbf{D}_f \subset \mathbb{C}$ to the Fock space $\mathcal{F}$ is continuous.

The set $\{|z, f\rangle : z \in \mathbf{D}_f\}$ will be called *family of coherent states* whether there exists a positive measure μ_f such that[27]

$$\int_{\mathbf{D}_f} d\mu_f(\bar{z}, z)|z, f\rangle\langle z, f| = \sum_{n=0}^{\infty} |n\rangle\langle n| = \mathbf{1}. \tag{13}$$

Passing to polar coordinates, $z = \sqrt{x}e^{i\theta}$, where $0 \leq \theta \leq 2\pi$, $0 < x < R_f$, and $d\mu(\bar{z}, z) = d\omega_f(x)d\theta$, this corresponds to the following classical power-moment problem:[28,29]

$$\int_0^{R_f} x^n \frac{2\pi \, d\omega_f(x)}{\mathcal{N}_f(x)} = [f(n)]!, \quad n = 0,\ 1,\ 2,\ ... \tag{14}$$

Note immediately that not all deformed algebras lead to coherent states in this construction formalism because the moment problem (14) does not always have a solution.[28,29]

Many techniques have been proposed in the literature to determine the structure function f corresponding to a given algebra.[11–13]

Meljanac *et al*[12] defined the generalized q-deformed single-mode oscillator as an algebra generated by the identity operator $\mathbf{1}$, a self-adjoint number operator N, a lowering operator a and an operator $\bar{a}$ which is not necessarily conjugate to a satisfying

$$[N,\ a] = -a, \quad [N,\ \bar{a}] = \bar{a}, \tag{15}$$

$$a\bar{a} - F(N)\bar{a}a = G(N) \tag{16}$$

where F and G are arbitrary complex analytic functions.

Such an algebra furnishes an appropriate approach for the unification of classes of deformed algebras known in the literature.

For the purpose, let us start from the relations (15) to get $[N,\ a\bar{a}] = 0 = [N,\ \bar{a}a]$ implying the existence of a complex analytic function φ such that

$$\bar{a}a = \varphi(N) \quad \text{and} \quad a\bar{a} = \varphi(N+1). \tag{17}$$

Eqs (16) and (17) give

$$\varphi(N+1) - F(N)\varphi(N) = G(N). \tag{18}$$

Denote now $a^\dagger$ the Hermitian conjugate of the operator a. Then,

$$[N,\ a^\dagger] = a^\dagger, \quad \text{and} \quad \bar{a} = c(N)a^\dagger, \tag{19}$$

where $c(N)$ is a complex function. For convenience take $c(N) = e^{i \arg \varphi(N)}$. Therefore, from (17) and the fact that $a^\dagger a$ and $aa^\dagger$ are Hermitian operators we necessarily have

$$a^\dagger a = |\varphi(N)| \quad \text{and} \quad aa^\dagger = |\varphi(N+1)|. \tag{20}$$

We now assume the existence of a "vacuum state" $|0\rangle$ such that

$$N|0\rangle = 0,\ a|0\rangle = 0 \text{ and } \langle 0|0\rangle = 1, \tag{21}$$

and construct the non normalized eigenvectors, $(a^\dagger)^n|0\rangle$ of the operator N. Then, applying (18) to these vectors we obtain

$$\varphi(n) = [F(n-1)]! \sum_{k=0}^{n-1} \frac{G(k)}{[F(k)]!}, \quad n \geq 1, \tag{22}$$

unless the initial condition $\varphi(0) = 0$ is satisfied, and where

$$[F(k)]! = \begin{cases} F(k)F(k-1)\cdots F(1) & \text{if } k \geq 1 \\ 1 & \text{if } k = 0 \end{cases}. \tag{23}$$

Hence, the structure function characterizing a given deformation is defined as follows: $f(n) := \varphi(n)$ if $\varphi(n) \geq 0$, and $f(n) := |\varphi(n)|$, in general. Provided the function f, the above formalism can be displayed to construct generalized CS.

Acknowledgments

This work is partially supported by the Abdus Salam International Centre for Theoretical Physics (ICTP, Trieste, Italy) through the Office of External Activities (OEA) - Prj-15. The ICMPA is in partnership with the Daniel Iagolnitzer Foundation (DIF), France.

References

1. S. M. Dancoff, *Phys. Rev.* **78** (1950).
2. I. E. Tamm, *J. Phys. UdSSR* **9** (1945).
3. H. S. Snyder, *Phys. Rev.* **71** (1947).
4. M. Arik and D. D. Coon, *J. Math. Phys.* **17** (1976).
5. S. Yu, M. Baker and D. D. Coon, *Phys. Rev. D* **5** (1972).
6. V. V. Kuryshkin, *Annales de la Fondation Louis de Broglie* **5** (1980).
7. A. Jannussis, G. Bbodimas, D. Sourlas, L. Papaloucas, P. Siafaricas and K. Vlachos, *Hadronic J.* **6** (1983).
8. L. C. Biedenharn, *J. Phys. A* **22** (1989).
9. A. J. Macfarlane, *J. Phys. A* **22** (1989).
10. C.-P. Sun and H.-C. Fu, *J. Phys. A: Math. Gen.* **22** (1989).
11. I. M. Burban, *Phys. Let. A* **42** (2009).
12. S. Meljanac, M. Milekovi and S. Pallua, *Phys. Lett. B* **328** (1994).
13. E. Baloïtcha, M. N. Hounkonnou and E. B. Ngompe-Nkouankam, *J. Math. Phys.* **53** (2012).
14. M. N. Hounkonnou and J. D. Bukweli-Kyemba, *J. Math. Phys.* **51** (2010).
15. E. Schrödinger, *Naturwissensehaften* **14** (1926).
16. R. J. Glauber, *Phys. Rev. Letters* **10** (1963).
17. R. J. Glauber, *Phys. Rev.* **131** (1963).
18. R. J. Klauder, *J. Math. Phys.* **4** (1963).
19. R. J. Klauder, *J. Math. Phys.* **4** (1963).
20. E. C. G. Sudarshan, *Phys. Rev. Letters* **10** (1963).
21. S. T. Ali, J. P. Antoine and J. P. Gazeau, *Coherent States, Wavelets and their Generalizations* (Springer-Verlag, New-York, 1999).
22. A. Perelomov, *Generalized Coherent States and their Applications* (Springer-Verlag, Berlin Heidelberg, 1986).
23. J.-P. Gazeau, *Coherent States in Quantum Physics* (WILEY-VCH Verlag GmbH & Co. KGaA, Weinheim, 2009).
24. J. R. Klauder, K. A. Penson and J. M. Sixderniers, *Phys. Rev. A* **64** (2001).
25. M. Daoud, *Phys. Letters A* **305** (2002).
26. M. N. Hounkonnou and E. B. Ngompe-Nkouankam, *J. Phys. A: Math. Theor.* **40** (2007).
27. J. R. Klauder and B. S. Skagerstam, *Coherent states: Applications in Physics and Mathematical Physics* (World Scientific,Singapore, 1985).
28. N. I. Akhiezer, *The Classical Moment Problem and Some Related Questions in Analysis* (Olivier and Boyd, London,, 1965).
29. J. Shohat and J. Tamarkin, *The Problem of Moments* (APS, New York, 1943).

QUANTUM MEASUREMENTS CONSTRAINED BY SYMMETRIES

P. BUSCH

Department of Mathematics, University of York,
York YO10 5DD, UK
E-mail: paul.busch@york.ac.uk

L. D. LOVERIDGE

Department of Physics and Astronomy,
University of British Columbia,
Vancouver, BC V6T 1Z1 Canada
E-mail: Leon@phas.ubc.ca

We revisit the theorem of Wigner, Araki and Yanase (WAY) describing limitations to repeatable quantum measurements that arise from the presence of conservation laws. We will review a strengthening of this theorem by exhibiting and discussing a condition that has hitherto not been identified as a relevant factor. We will also show that an extension of the theorem to continuous variables such as position and momentum can be obtained if the degree of repeatability is suitably quantified.

Keywords: Quantum measurement; conservation laws; WAY theorem.

1. Introduction

Measurements are physical processes and as such are subject to the laws of physics. In particular, the interaction between a measurement device or probe and an object system needed to establish suitable correlations between the quantity being measured and a pointer observable are constrained by fundamental symmetries, or equivalently the associated conservation laws. In quantum mechanics this has been found to lead to restrictions of the measurability of certain physical quantities, as demonstrated first by Wigner[1] and then in more generality by Araki and Yanase.[2]

Here we will review a strengthening and an an extension of these results in two respects. First we show that the presence of an additive conservation law not only precludes *repeatable, sharp* measurements of a discrete quantity not commuting with the system part of the conserved quantity, but in addition a further property of a measurement must be violated: the so-called Yanase condition, which states that the pointer observable commutes with the conserved quantity (Sec. 3). Second, a WAY-type limitation is shown to hold for position measurements obeying momentum conservation, thus extending the WAY theorem to continuous quantities (Sec. 4). We begin with a brief account of the formal theory of quantum measurement as far as required for the formulation of these results.

2. Quantum Measurements

In the standard formulation of quantum mechanics, a physical system is associated with a Hilbert space $\mathcal{H}$ over $\mathbb{C}$; its states are represented as density operators ρ, with pure states as normalized vectors $\psi \in \mathcal{H}$. Observables are given as positive operator valued measures (POVMs) E from a σ-algebra Σ of subsets of the set of measurement outcomes to the set of effects (positive operators), $X \mapsto \mathsf{E}(X)$. The probability that a measurement of E in a state ρ gives an outcome in the set X then is $\mathsf{p}_\rho^{\mathsf{E}}(X) = \mathrm{tr}\big[\rho\, \mathsf{E}(X)\big]$.

As an example, we refer to an *unsharp*, or *smeared* position observable, defined as follows:

$$\mathsf{Q}^e(X) = \int_{\mathbb{R}} e(x)\mathsf{E}^Q(x+X)dx = \chi_X * e(Q), \tag{1}$$

$$\langle\varphi|\mathsf{Q}^e(X)\varphi\rangle = \mathsf{p}_\varphi^{\mathsf{Q}^e}(X) = \int dx\, |\varphi(x)|^2\, \chi_X * e(x), \tag{2}$$

where E^Q denotes the spectral measure of the position operator Q, χ_X a characteristic (set indicator) function, $*$ denotes convolution, and e is a distribution function representing the measurement inaccuracy. Such POVMs occur in a variety of models of approximate position measurements.

The time evolution of an isolated system is given by a unitary group U_t via $\rho_t = U_t\rho_0 U_t^*$, $U_t = \exp(-iHt/\hbar)$, or the associated Schrödinger equation, $i\hbar\, d\psi_t/dt = H\psi_t$.

A measurement is modelled as an interaction between the object system (represented by $\mathcal{H}$) and an apparatus or probe system (represented by $\mathcal{K}$), such that the time evolution U acts on the tensor product $\mathcal{H}\otimes\mathcal{K}$. Thus, a measurement $\mathcal{M}$ of an observable E is summarized as a quintuple $\langle\mathcal{K}, U, \phi, Z, h\rangle$, where ϕ is the initial probe state, Z the pointer observable (normally modelled as a selfadjoint operator with associated spectral measure E^Z), and h a pointer scaling function mapping the pointer values to the values of the measured observable. The latter is then determined by the *probability reproducibility condition*, according to which the coupling U sending the initially uncorrelated system-probe state to an entangled state,

$$\Psi_i \equiv \varphi \otimes \phi \to \Psi_f \equiv U(\varphi \otimes \phi) \in \mathcal{H} \otimes \mathcal{K},$$

serves as a measurement of the observable E if the pointer probabilities reproduce the probabilities of E:

$$\left\langle \Psi_f | \mathbb{1} \otimes \mathsf{E}^Z(h^{-1}(X))\Psi_f \right\rangle \equiv \langle \varphi | \mathsf{E}(X)\varphi \rangle . \tag{3}$$

A measurement is called *repeatable* if

$$\left\langle \Psi_f | \mathsf{E}(X) \otimes \mathsf{E}^Z(h^{-1}(X))\Psi_f \right\rangle = \langle \varphi | \mathsf{E}(X)\varphi \rangle . \tag{4}$$

The system state conditional on an outcome (in) X is defined via

$$\rho_X = \mathrm{tr}_{\mathcal{K}} \left[\mathbb{1} \otimes \mathsf{E}^Z \left(h^{-1}(X) \right) |\Psi_f\rangle\langle\Psi_f| \right] ; \quad \hat{\rho}_X = \frac{\rho_X}{\mathrm{tr}[\rho_X]}.$$

Then the probability reproducibility and repeatability conditions are compactly given as $\mathrm{tr}[\rho_X] = \langle \varphi | \mathsf{E}(X)\varphi \rangle$ and $\mathrm{tr}[\hat{\rho}_X \, \mathsf{E}(X)] = 1$, respectively.

It is a historical curiosity that until rather recently the term measurement was understood by many as including the property of repeatability. Only very few authors, such as Pauli, Landau and Peierls in the 1930s and Wigner, allowed for the possibility of measurements that are not necessarily repeatable.

3. The Theorem of Wigner, Araki and Yanase

In 1952, Wigner[1] considered a spin-$\frac{1}{2}$ system and showed that a repeatable measurement of the spin component s_x violates conservation of (say) the z-component of the total angular momentum shared between system and probe, $s_z + J_z$. He also showed that if the probe's angular momentum J_z is allowed to have a large spread in it's initial state ϕ, then an arbitrarily accurate and repeatable measurement of s_x becomes possible. Indeed, with a measurement coupling:

$$\varphi_+ \otimes \phi \longrightarrow \varphi_+ \otimes \phi_+ + \varphi_- \otimes \eta, \quad \varphi_- \otimes \phi \longrightarrow \varphi_- \otimes \phi_- - \varphi_+ \otimes \eta$$

(and $\langle \eta | \phi_\pm \rangle = \langle \phi_+ | \phi_- \rangle = 0$), the measured observable has *three* effects:

$$E_\pm = (1 - \|\eta\|^2)P[\varphi_\pm], \quad E_? = \|\eta\|^2 \, \mathbb{1}.$$

Here $\varphi_\pm$ are the eigenstates of the measured spin component s_x and $\phi_\pm, \eta$ are pointer eigenstates. The third effect is trivial: it provides no information as its expectation value does not depend on the object state. The associated outcome is 'uncertain'.

Wigner noted that the conservation law can be satisfied in the above dynamics while $\|\eta\|^2 = 1/(2n-1)$, or even $\|\eta\|^2 \sim \frac{1}{n^2}$, if $\phi = \sum_{\nu=1}^{n} \phi_\nu$ (where ϕ_ν are J_z-eigenvectors). For large n, one obtains $E_\pm \simeq P[\varphi_\pm]$, $E_? \simeq \mathbb{O}$.

Araki and Yanase[2] generalized Wigner's result to the effect that any observable that allows a repeatable measurement in the presence of an additive conserved quantity must commute with the object part of that conserved quantity. This statement is subject to the assumption that the measured observable is discrete (thus allowing repeatable measurements) and at least the object part of the conserved quantity is bounded. Yanase[3] constructed a model in which, in analogy to Wigner's observation, arbitrary accuracy and good repeatability are obtained if large variances in the probe part of the conserved quantity are allowed. This followed even under an additional assumption, now known as the *Yanase condition*: that the pointer observable commutes with the probe part of the conserved quantity. This can be seen as a pre-emptive assumption that ensures that the WAY measurement obstruction does not reappear in the measurement of the pointer.

It was shown recently that besides repeatability, the Yanase condition cannot be satisfied in a sharp measurement of an observable that does not commute with an additive conserved quantity.[4] This is the strengthened version of the WAY theorem.

4. Position measurement vs. momentum conservation

It has remained an open question for a long time whether the WAY theorem extends to object observables with continuous spectra. In that case repeatability cannot be obtained since according to a fundamental theorem due to Ozawa,[5] the existence of repeatable measurements requires the observable to be discrete. However, it is known that approximate repeatability is still available in the general case.

One can thus investigate whether momentum conservation entails an obstruction to approximately repeatable position measurements. This was denied by Ozawa[6] in 1991, who presented a model of a momentum-conserving position measurement. Closer scrutiny of this model (which satisfies the Yanase condition) has revealed that good accuracy and repeatability come at the expense of a large variance of the probe's momentum. Indeed, the measured observable is of the form (1), and it can be seen that the function e can be made narrow only by increasing the momentum spread in a state corresponding to the preparation of the apparatus. This can be seen as a quantitative version of the WAY theorem, and it has been shown by the present authors[7] that there is a general trade-off relation between the quality of repeatability and the variance of the probe momentum under the satisfaction of the Yanase condition. This will be briefly reviewed here.

A common way of quantifying the difference between the observable one aims to measure approximately, represented by selfadjoint operator M, and the actually measured observable is by means of the *noise operator*, $N = Z_f - M$, where $Z_f = U^* \mathbb{1} \otimes ZU$ and Z is the selfadjoint pointer observable. The *noise* $\epsilon(\varphi)$ is then given by $\epsilon(\varphi)^2 := \langle \varphi \otimes \phi | N^2 \varphi \otimes \phi \rangle \equiv \langle N^2 \rangle$. Noting that $\langle N^2 \rangle \geq (\Delta N)^2$ and using the Cauchy-Schwarz inequality, Ozawa[8] obtained the following inequality:

$$\epsilon(\varphi)^2 \geq \frac{1}{4} \frac{|\langle [Z_f - M, L_1 + L_2] \rangle|^2}{(\Delta_\varphi L_1)^2 + (\Delta_\phi L_2)^2}. \tag{5}$$

Here $L_1 + L_2$ is the additive conserved quantity. Stipulating the Yanase condition, the numerator becomes $|\langle [M, L_1] \rangle|^2$. We can then estimate the *global error*, $\epsilon := \sup_\varphi \epsilon(\varphi)$ and obtain:

$$\epsilon^2 \geq \sup_\varphi \frac{1}{4} \frac{|\langle [M, L_1] \rangle|^2}{(\Delta_\varphi L_1)^2 + (\Delta_\phi L_2)^2}. \tag{6}$$

This can be applied to position measurements obeying momentum conservation, where $M = Q$ and $L_1 = P$, $L_2 = P_{\mathcal{A}}$, giving

$$\epsilon^2 \geq \frac{\hbar^2}{4(\Delta_\phi P_{\mathcal{A}})^2}. \tag{7}$$

Similarly one can define a measure of the degree of repeatability by replacing M in the definition of the noise operator by the Heisenberg-evolved $M_f = U^* M \otimes \mathbb{1} U$: $\mu(\varphi)^2 := \langle \varphi \otimes \phi | (M_f - Z_f)^2 \varphi \otimes \phi \rangle$. For $\mu := \sup \mu(\varphi)$ one obtains:

$$\mu^2 \geq \sup_\varphi \frac{1}{4} \frac{|\langle [M(\tau) - Z(\tau), L_1 + L_2] \rangle|^2}{(\Delta_\varphi L_1)^2 + (\Delta_\phi L_2)^2}. \tag{8}$$

With the Yanase condition the numerator becomes $|\langle [M_f, L_1 + L_2] \rangle|^2$. In the case of position measurements the above inequality then gives

$$\mu^2 \geq \frac{\hbar^2}{4(\Delta_\phi P_{\mathcal{A}})^2}. \tag{9}$$

A first conclusion to be drawn is that under the Yanase condition, momentum conserving position measurements can only be performed with good accuracy and repeatability if the probe system is 'large', in the sense that its momentum variance must be large. This limitation is confirmed with Ozawa's position measurement model.[7]

The case of position and momentum allows for a particular way of violating the Yanase condition, so that the numerators of inequalities (5) and (8) become zero: this happens when the pointer is chosen to be the position $Q_{\mathcal{A}}$. In this case there is no size constraint on the probe that needs to be fulfilled in order that ϵ and μ can be made small, as can be confirmed by another position measurement model.[7] However, as noted above, the problem of measuring position in view of momentum conservation has now been shifted to the probe system. Thus a second conclusion is that information transfer about the position of an object to a probe system is perfectly possible if one relinquishes the Yanase condition, but whether this can be used for measurements depends on the measurability of the probe position under the constraint of momentum conservation.

Finally we note that if instead of position one seeks to measure the relative position between object and probe $M = Q - Q_{\mathcal{A}}$, then the numerators in (5) and (8) vanish if the Yanase condition holds since the relative position commutes with the total momentum. Closer inspection of measurement models shows that it is not primarily the large variance of the probe momentum that one needs to seek to achieve to obtain good measurement of the object position alone. It is rather the fact that the relative position effectively becomes indistinguishable from the object position if the probe position is prepared with low inaccuracy; but this entails large variance of the probe momentum by virtue of the uncertainty relation.[9]

Acknowledgments

This work was supported by a doctoral training grant from EPSRC (UK).

References

1. E. Wigner, *Z. Phys.* **133**, 101 (1952). English translation: arXiv:1012.4372.
2. H. Araki, M.M. Yanase, *Phys. Rev.* **120**, 622 (1960).
3. M.M. Yanase, *Phys. Rev.* **123**, 666 (1961).
4. L.D. Loveridge, P. Busch, *Eur. Phys. J. D* **62**, 297 (2011).
5. M. Ozawa, *J. Math. Phys.* **25**, 79 (1984).
6. M. Ozawa, *Phys. Rev. Lett.* **67**, 1956 (1991).
7. P. Busch, L.D. Loveridge, *Phys. Rev. Lett.* **106**, 110406 (2011).
8. M. Ozawa, *Phys. Rev. Lett.* **88**, 050402 (2002).
9. L.D. Loveridge, *Quantum Measurements in the Presence of Symmetry*, Doctoral Thesis, University of York, 2012 (http://etheses.whiterose.ac.uk/2670).

GEOMETRIC REALISATION OF A CLASS OF UNITARY RIEMANN SURFACE BRAID GROUP REPRESENTATIONS

M. SPERA*

Dipartimento di Matematica e Fisica "Niccolò Tartaglia"
Università Cattolica del Sacro Cuore
Via dei Musei 41, 25121 Brescia, Italia
**E-mail: mauro.spera@unicatt.it*

In this note, we report on ongoing research concerning geometric realisations of the simplest unitary Riemann surface braid group representations. We discuss in particular the genus one case by means of noncommutative geometric techniques.

Keywords: Riemann surface braid groups, stable bundles, Laughlin wave functions, noncommutative geometry.

1. Introduction

In this note, based on work in progress[1] motivated by topological quantum computing issues,[2] we show that searching for the simplest unitary Riemann surface braid group representations naturally leads to consideration of vector valued wave functions. These wave functions obey in general *fractional* statistics and can be represented by holomorphic sections of (semi)stable[3] vector bundles. The latter are holomorphic vector bundles admitting a Hermitian-Einstein structure – namely a (unique) hermitian connection with central constant curvature – in view of the Donaldson-Uhlenbeck-Yau theorem.[4–6] Stable vector bundles already emerge in the study of quantum Hall effects.[7–10] In our framework *they arise directly from a braid group theoretic perspective.* Our crucial remark is that *braiding is inextricably tied to curvature.* In the $g = 1$ case everything can be made quite explicit by resorting to A. Connes' noncommutative geometric setting[11,12] and to notion of noncommutative theta vectors introduced by A. Schwarz.[13] As a consequence of the general discussion, one may easily construct Laughlin-type wave functions.[7,14–16]

2. The Riemann surface braid group and its simplest unitary representations

The braid group on n strands $B(X, n)$ pertaining to a topological space X is the fundamental group of the associated configuration space $C_n(X)$ consisting of all n-point subsets of X. In our case $X = \Sigma_g$, a closed orientable surface of genus $g \geq 1$ (actually, a Riemann surface). Its associated braid group $B(\Sigma_g, n)$ admits, among others, the following presentation due to P. Bellingeri.[17] The generators are $\sigma_1, \ldots, \sigma_{n-1}; a_1, \ldots, a_g, b_1, \ldots, b_g$ (the latter have a simple geometric interpretation, in terms of the natural dissection of the surface by means of a $4g$-gon). In the presentation, in addition to the ordinary braid relations, one has "mixed" relations

$$
\begin{array}{lll}
(R1): & a_r\sigma_i = \sigma_i a_r, & 1 \leq r \leq g, \quad i \neq 1 \\
 & b_r\sigma_i = \sigma_i b_r, & 1 \leq r \leq g, \quad i \neq 1 \\
(R2): & \sigma_1^{-1}a_r\sigma_1^{-1}a_r = a_r\sigma_1^{-1}a_r\sigma_1^{-1}, & 1 \leq r \leq g, \\
 & \sigma_1^{-1}b_r\sigma_1^{-1}b_r = b_r\sigma_1^{-1}b_r\sigma_1^{-1}, & 1 \leq r \leq g, \\
(R3): & \sigma_1^{-1}a_s\sigma_1 a_r = a_r\sigma_1^{-1}a_s\sigma_1, & s < r \\
 & \sigma_1^{-1}b_s\sigma_1 b_r = b_r\sigma_1^{-1}b_s\sigma_1, & s < r \\
 & \sigma_1^{-1}a_s\sigma_1 b_r = b_r\sigma_1^{-1}a_s\sigma_1, & s < r \\
 & \sigma_1^{-1}b_s\sigma_1 a_r = a_r\sigma_1^{-1}b_s\sigma_1, & s < r \\
(R4): & \sigma_1^{-1}a_r\sigma_1^{-1}b_r = b_r\sigma_1^{-1}a_r\sigma_1, & 1 \leq r \leq g \\
(TR): & [a_1, b_1^{-1}]\cdots[a_g, b_g^{-1}] = \sigma_1\sigma_2\cdots\sigma_{n-1}^2\cdots\sigma_2\sigma_1 &
\end{array}
$$

with the usual convention $[a, b] = aba^{-1}b^{-1}$. We shall restrict ourselves to *unitary* representations, with $\sigma_j = \sigma\, I$, $j = 1, \ldots, n$ (I being the identity operator on the representation Hilbert space). One writes

$$\sigma = e^{2\pi i\theta}$$

and calls θ (a priori defined up to integers) the *statistics parameter.* The relations R1 and R2 are automatically fulfilled, R3 become:

$$[a_s, a_r] = [b_s, b_r] = I, \qquad r, s = 1, \ldots, g$$

whereas R4 yields:

$$[a_r, b_r] = \sigma^2 I, \qquad r = 1, \ldots, g$$

and TR gives, in turn, after checking that $[a_r, b_r^{-1}] = \sigma^{-2}I$,

$$\sigma^{2(n-1+g)} = 1$$

furnishing (for $n - 1 + g \neq 0$)

$$\theta = \frac{k}{2(n-1+g)}, \qquad k \in \mathbb{Z}$$

that is, *fractional statistics*, in general. Notice that if $\sigma^2 = 1$, that is $\theta \in 1/2 \cdot \mathbb{Z}$ (slight abuse of notation) we recover ordinary Fermi-Bose statistics.

We introduce the following tensor product Hilbert space:

$$\mathcal{H} := H_1 \otimes H_2$$

with H_1 carrying a representation ρ_1 of the *Weyl-Heisenberg Canonical Commutation Relations (CCR)*[18] in view of R3 and R4, whereas H_2 is one-dimensional. Upon setting (obvious notation)

$$\begin{aligned} &\rho(a_r) = \rho_1(a_r) \otimes I_{H_2}, \quad \rho(b_r) = \rho_1(b_r) \otimes I_{H_2}, \qquad r = 1, \ldots, g \\ &\rho(\sigma_j) = I_{H_1} \otimes \sigma\, I_{H_2}, \qquad j = 1, \ldots, n \end{aligned}$$

it is immediately checked that ρ *is indeed a unitary representation of the Riemann surface braid group* $B(\Sigma_g, n)$ *on* $\mathcal{H}$. The above setting is naturally tailored to the treatment of generalised Laughlin wave functions as we shall see below.

3. Geometric realisation of $\mathcal{H}$

We shall now outline a geometric construction of the Hilbert spaces H_j, $j = 1, 2$ and of the representation ρ.

3.1. *Stable bundles*

Concerning the space H_1, and strongly influenced by the author's geometric picture,[19] *we are naturally led to look for hermitian complex vector bundles over the Jacobian* $J(\Sigma_g)$ *of the Riemann surface in question, equipped with a hermitian connection of constant central curvature equal to* $2\pi i \cdot 2\theta$ *(Hermitian-Einstein connection). This is tantamount to considering holomorphic* stable *bundles over Jacobians (in physical applications, they represent the spectral (Brillouin) manifolds) with* slope (i.e. degree/rank)

$$\mu(\mathcal{E}) := \frac{\deg(\mathcal{E})}{\mathrm{rk}(\mathcal{E})} = 2\theta$$

in view of the Donaldson-Uhlenbeck-Yau theorem.[3–6] The space of L^2- sections of such a bundle yields the Hilbert space H_1. Moreover, *the generators* a_r *and* b_s *can be realised as suitable* parallel transport operators *pertaining to the constant central curvature hermitian connection, and actually will yield a representation of the Weyl-Heisenberg Commutation Relations.* We shall discuss the genus one case in some detail below. Here we just notice in passing that a stable bundle with statistical parameter $2\theta = -1/(n+g-1)$ can be easily manufactured via the Fourier-Mukai transform.[1,9]

3.2. *The prime form*

In this subsection we give an explicit construction of the one-dimensional space H_2 in terms of the so-called *prime form* on a Riemann surface.[20] The prime form $E(x,y) = -E(y,x)$ (for $x,\ y \in X = \Sigma_g$) on $X \times X$ is a holomorphic section of a suitable holomorphic line bundle or, alternatively, a differential form of weight $(-1/2, -1/2)$ on its universal cover $\widetilde{X} \times \widetilde{X}$. Its construction is meant to provide a generalization to its genus zero counterpart

$$E_0(x,y) = \frac{x-y}{\sqrt{dx}\,\sqrt{dy}}$$

to which it reduces, up to a third order term, upon working in local coordinates. We wish then to consider the following 1-forms

$$\omega_{ij} := \omega_{x_j - x_0}(x_i) + \omega_{x_i - x_0}(x_j)$$

(with $x_0 \in X$ fixed), where $\omega_{a-b}(x) := d_x \log(E(x,a)/E(x,b))$ is a *differential of the third kind* on X, namely the unique differential 1-form on X with zero A-periods and single poles at a and b with residues ± 1, respectively[20] . The quotient of prime forms above is well defined as a holomorphic function on $\widetilde{X}$. Notice that on $\mathbb{P}^1$, taking $x_0 = \infty$, one retrieves the standard (Arnol'd) logarithmic forms $\omega_{ij} = d\log(z_i - z_j)$. So we may introduce the analogue of the *Knizhnik-Zamolodchikov-Kohno (KZK)-connection*[21] on the configuration space $C_n(X)$

$$\omega := \theta \sum_{i<j} \omega_{ij}$$

whose *parallel transport* on a trivial line bundle yields a scalar representation of the subroup of the braid group generated by the σ_j's. We may thus take $H_2 = \langle \Psi \rangle$, with

$$\Psi(x_1, \ldots, x_n) = \prod_{i<j}^{\otimes} \frac{E(x_i, x_j)^{2\theta}}{E(x_i, x_0)^{\theta} E(x_j, x_0)^{\theta}}$$

which turns out to be well defined and enjoys the correct transformation law under the exchange of two points $x_i \leftrightarrow x_j$:

$$\Psi \mapsto (-1)^{2\theta}\, \Psi = \sigma\, \Psi$$

The upshot is that we may devise generalised "ground state" Laughlin wave functions[14–16] in $\mathcal{H}$ as follows:

$$\Phi = \xi \cdot \Psi$$

with ξ a holomorphic section of the stable bundle entering the construction (one works with a sort of "centre of mass" coordinate). Notice that they are not invariant under the action of the full braid group, since parallel transport does not preserve the holomorphic structure, in general. In this way we generalise the geometric treatment given for the standard (i.e. complex plane) braid group by A. Besana and the author.[22]

4. Stable bundles via noncommutative geometry

In this section we treat the torus case $g = 1$, and we use noncommutative geometry to sketch a construction of "classical" stable bundles thereon. Everything is ultimately based on A. Connes' seminal paper[11] (see also[12]), to which we refer for complete details.

If $\mathcal{A}_\vartheta$ denotes the smooth subalgebra pertaining to the noncommutative torus A_ϑ, $\vartheta \in \mathbb{R}/\mathbb{Z}$ (deformation of the standard commutative algebra of smooth functions on a torus - via Fourier theory) one has a natural (hermitian) $\mathcal{A}_\theta$-right module

$$\mathcal{E}_{p,q} = \mathcal{S}(\mathbb{R}, \mathbb{C}^q)$$

(vector valued Schwartz functions), p and q positive integers, g.c.d.$(p, q) = 1$, having "rank" (à la Murray-von Neumann)

$$\mathrm{rk}(\mathcal{E}_{p,q}) := \tau_{\mathrm{End}(\mathcal{E}_{p,q})}(I) = p - \vartheta q$$

(the trace on the endomorphism algebra being involved).

The *Connes connection*[11,12] (slight modifications) reads:

$$(\nabla_1 \xi)(s) = \xi'(s), \qquad (\nabla_2 \xi)(s) = 2\pi i \, \frac{s}{\epsilon} \, \xi(s), \qquad \epsilon := \frac{p}{q} - \vartheta.$$

It actually fulfils the appropriate version of the Leibniz rule and has constant curvature

$$\Omega := [\nabla_1, \nabla_2] \, e_1 \wedge e_2 = 2\pi i \, \frac{q}{p - \vartheta q} \, e_1 \wedge e_2$$

and *first Chern class* $c_1(\mathcal{E}_{p,q})$ given by the usual expression

$$c_1(\mathcal{E}_{p,q}) = \frac{1}{2\pi i} \, \tau_{\mathrm{End}(\mathcal{E}_{p,q})}(I)(2\pi i \, \frac{q}{p - \vartheta q}) = q.$$

We now introduce a *holomorphic structure*[23–25] on $\mathcal{E}_{p,q}$ via the $\overline{\nabla}$ operator as

$$\overline{\nabla} := \nabla_1 + i\nabla_2$$

The kernel of the $\overline{\nabla}$ operator is called the space of *noncommutative theta vectors*[13] (see[19] as well) and it easily seen to be q-dimensional. The modules $\mathcal{E}_{p,q}$ are actually independent of p. Upon taking $p = n$ and $\vartheta = 0$, we obtain a "classical" hermitian holomorphic vector bundle over a complex torus, which, in view of the above discussion, is indeed stable, with slope = statistical parameter = $1/(2\pi i)\cdot$ curvature = q/n.

Acknowledgements

The author wishes to express his deepest gratitude to all Organisers of the XXIX Colloquium on Group Theoretical Methods in Physics - and to their collaborators at all levels - for their kind invitation and financial support, superb organisation, warm hospitality and excellent scientific atmosphere. He is also indebted to MIUR for partial financial support (ex 60% funds).

References

1. M. Spera, in preparation.
2. Z. Wang, *Topological Quantum Computation* (American Mathematical Society, Providence, 2010).
3. D. Mumford, J. Fogarty and F. Kirwan, *Geometric Invariant Theory* (Springer, Berlin, 1994).
4. S. K. Donaldson, *J. Diff. Geom.* **18**, 269 (1983).
5. K. Uhlenbeck and S. T. Yau, *Commun. Pure Appl. Math.* Vol. XXXIX, Suppl., 257 (1986) and *Commun. Pure Appl. Math.* Vol. XLII, 703 (1989).
6. S. Kobayashi, *Differential geometry of complex vector bundles* (Iwanami Shoten Publishers, Tokyo – Princeton University Press, Princeton, 1987).
7. T. Chakraborty and P. Pietiläinen, *The Quantum Hall Effects - Fractional and Integral* (Springer, Berlin, 1995).
8. J. M. Guilarte, J. M. Mufioz Porras and M. de la Torre Mayado, *J. Geom. Phys.* **27**, 297 (1998).
9. C. Tejero Prieto, *Commun. Math. Phys.* **265**, 373 (2006).
10. R. Varnhagen, *Nucl. Phys. B* **443**, 501 (1995).
11. A. Connes, *C.R. Acad. Sc. Paris* **290**, 599 (1980).
12. A. Connes, *Noncommutative geometry* (Academic Press, London, 1994).
13. A. Schwarz, *Lett. Math. Phys.* **58**, 81 (2001).
14. R. B. Laughlin, *Phys. Rev. Lett.* **50**, 1395 (1983).
15. F.D.M. Haldane and E. H. Rezayi, *Phys. Rev. B* **31**, 2529 (1985).
16. R. Iengo and D. Li, *Nucl. Phys. B* **413**, 735 (1994).
17. P. Bellingeri, *J. Algebra* **274**, 543 (2004).
18. A. Perelomov, *Generalized Coherent States and Their Applications* (Springer, Berlin, 1986).
19. M. Spera, *Rend. Sem. Mat. Univers. Politecn. Torino* **44**, 383 (1986).
20. D. Mumford, *Tata Lectures on Theta II* (Birkhäuser, Basel, 1984).
21. T. Kohno, *Conformal Field Theory and Topology* (American Mathematical Society, Providence, 2002).
22. A. Besana and M. Spera, *J. Knot Theory Ram.* **15**, 883 (2006).
23. M. Spera, *Rend. Sem. Fac. Scienze Univ. Cagliari Suppl.* **58**, 409 (1988).
24. M. Spera, *Can. J. Math.* **44**, 368 (1992).
25. A. Polishchuck and A. Schwarz, *Commun. Math. Phys.* **236**, 135 (2003).

ON THE CONTRACTION OF $so(4)$ TO $iso(3)$

E. M. SUBAG[1] and E. M. BARUCH[2]
Department of Mathematics, Technion-Israel Institute of Technology, Haifa 32000, Israel
[1] *E-mail: esubag@tx.technion.ac.il*
[2] *E-mail: embaruch@math.technion.ac.il*

J. L. BIRMAN
Department of Physics, The City College of the City University of New York, New York N.Y. 10031, USA
E-mail: birman@sci.ccny.cuny.edu

A. MANN
Department of Physics, Technion-Israel Institute of Technology, Haifa 32000, Israel
E-mail: ady@physics.technion.ac.il

For any skew-Hermitian irreducible infinite dimensional representation η of $iso(3)$, we find a sequence of (finite dimensional) irreducible representations ρ_n of $so(4)$ which contract to η.

Keywords: Contractions, Lie algebras, Representation theory.

1. Introduction

One of the first known examples of contraction of Lie algebra representations, given in the early work of İnönü and Wigner[1], is the contraction of the representations of the Lie algebra $so(3)$ to those of $iso(2)$. Starting from a sequence $\{\rho_j\}_{j=1}^{\infty}$ of finite dimensional representations of $so(3)$ with increasing dimension, they obtained an infinite dimensional representation η_q of $iso(2)$. They proved the contraction of the representations by the following type of convergence of matrix elements:

$$\lim_{j \longrightarrow \infty} \left\langle {}^{j}_{m'} | \rho_j(X) |{}^{j}_{m} \right\rangle = \langle {}^{q}_{m'} | \eta_q(X) |{}^{q}_{m} \rangle \tag{1}$$

where $|{}^{j}_{m}\rangle$ (respectively $|{}^{q}_{m}\rangle$) is an element in an orthonormal basis of ρ_j (respectively η_q).

In this paper we show that the same type of convergence of matrix elements as in (1), holds for the contraction of the finite dimensional irreducible representations of $so(4)$ to infinite dimensional irreducible representations of $iso(3)$.

In section 2 we describe the representation theory of $so(4)$ and $iso(3)$. In section 3 we recall the contraction of the algebra $so(4)$ to $iso(3)$ and prove the convergence of the appropriate matrix elements.

2. Representation theory of $so(4)$ and $iso(3)$.

Recall that the Lie algebra $so(4)$ is the direct sum of two copies of the Lie algebra $so(3)$. Moreover every irreducible representation of $so(4)$ is a tensor product of two irreducible representations of $so(3)$. From the work of Weimar-Woods[2,3] we know all the contractions of representations of $so(3)$ and hence we also know all the contractions of representations of $so(4)$ that respect the decomposition $so(4) = so(3) \oplus so(3)$. As noted in Ref. 2, the contraction of $so(4)$ to $iso(3)$ does not respect this decomposition. Hence we use another description[4] of the representations of $so(4)$. The resemblance of the representations of $so(4)$ and $iso(3)$ in this description is more convenient for the contraction procedure.

The Lie algebras $so(4)$ and $iso(3)$ can be defined by the bases $\{M_1, M_2, M_3, N_1, N_2, N_3\}$ and $\{J_1, J_2, J_3, P_1, P_2, P_3\}$ respectively that satisfy the following commutation relations:

$$\begin{aligned} [M_i, M_j]_{so(4)} &= i\epsilon_{ijk} M_k \quad &, [J_i, J_j]_{iso(3)} &= i\epsilon_{ijk} J_k \\ [N_i, N_j]_{so(4)} &= i\epsilon_{ijk} M_k \quad &, [P_i, P_j]_{iso(3)} &= 0 \\ [M_i, N_j]_{so(4)} &= i\epsilon_{ijk} N_k \quad &, [J_i, P_j]_{iso(3)} &= i\epsilon_{ijk} P_k \end{aligned} \tag{2}$$

We will describe the irreducible representations of $so(4)$ and $iso(3)$ in terms of other bases which are $\{M_+, M_-, M_3, N_+, N_-, N_3\}$ and $\{J_+, J_-, J_3, P_+, P_-, P_3\}$, where $M_\pm = M_1 \pm iM_2, N_\pm = N_1 \pm iN_2$, $J_\pm = J_1 \pm iJ_2$, $P_\pm = P_1 \pm iP_2$.

$so(4)$ has two independent invariants (Casimir operators) $\vec{M} \cdot \vec{N}$ and $\frac{1}{2}(M^2 + N^2)$ that in each irreducible representation act as scalar operators with scalars that we denote by G and F respectively. These scalars determine uniquely (up to an isomorphism) the irreducible representation of $so(4)$. We denote the irreducible representation of $so(4)$ with F and G by $\rho_{F,G} : so(4) \longrightarrow gl(V_{F,G})$. The representation space $V_{F,G}$ has an orthonormal basis of the form $\left\{|^{F,G}_{j,m_j}\rangle : j \in \{j_0, j_0+1, ..., n\}, m_j \in \{-j, -j+1, ..., j\}\right\}$, where $j_0, n \in$

$\{0, \frac{1}{2}, 1, \frac{3}{2}, 2, ...\}$ and they satisfy $G^2 = j_0^2(n+1)^2$, $2F = j_0^2 + (n+1)^2 - 1$. The dimension of $V_{F,G}$ is given by $\sum_{i=j_0}^{n}(2i+1) = (n+1)^2 - j_0^2$. The representation $\rho_{F,G}$ is given by:

$$\rho_{F,G}(M_3)|_{j,m_j}^{F,G}\rangle = m_j|_{j,m_j}^{F,G}\rangle \tag{3}$$

$$\rho_{F,G}(M_\pm)|_{j,m_j}^{F,G}\rangle = \sqrt{(j \mp m_j)(j \pm m_j + 1)}|_{j,m_j\pm 1}^{F,G}\rangle \tag{4}$$

$$\begin{aligned}\rho_{F,G}(N_3)|_{j,m_j}^{F,G}\rangle = {} & \alpha_j^{F,G}\sqrt{(j+m_j)(j-m_j)}|_{j-1,m_j}^{F,G}\rangle \\ & +\beta_j^{F,G} m_j|_{j,m_j}^{F,G}\rangle + \\ & \alpha_{j+1}^{F,G}\sqrt{(j+m_j+1)(j-m_j+1)}|_{j+1,m_j}^{F,G}\rangle\end{aligned} \tag{5}$$

$$\begin{aligned}\rho_{F,G}(N_\pm)|_{j,m_j}^{F,G}\rangle = {} & \pm\alpha_j^{F,G}\sqrt{(j \mp m_j)(j \mp m_j - 1)}|_{j-1,m_j\pm 1}^{F,G}\rangle \\ & +\beta_j^{F,G}\sqrt{(j \mp m_j)(j \pm m_j + 1)}|_{j,m_j\pm 1}^{F,G}\rangle \\ & \mp\alpha_{j+1}^{F,G}\sqrt{(j \pm m_j + 1)(j \pm m_j + 2)}|_{j+1,m_j\pm 1}^{F,G}\rangle\end{aligned} \tag{6}$$

where $\alpha_j^{F,G} = \sqrt{\frac{2F+1-j^2-\frac{G^2}{j^2})}{(2j+1)(2j-1)}}$, $\beta_j^{F,G} = \frac{G}{j(j+1)}$. Similarly, $iso(3)$ has two independent invariants (Casimir operators) P^2 and $\vec{J} \cdot \vec{P}$ that in each irreducible representation act as scalar operators with scalars that we denote by p^2 and C respectively. These scalars determine uniquely (up to an isomorphism) the irreducible representation of $iso(3)$. We denote the irreducible representation of $iso(3)$ with given p^2 and C by $\eta_{p^2,C} : iso(3) \longrightarrow gl(W_{p^2,C})$. The representation space $W_{p^2,C}$ has an orthonormal basis of the form $\left\{|_{j,m_j}^{p^2,C}\rangle : j \in \{j_0, j_0+1, ...\}, m_j \in \{-j, -j+1, ..., j\}\right\}$, where $j_0 \in \{0, \frac{1}{2}, 1, \frac{3}{2}, 2, ...\}$ and they satisfy $C^2 = j_0^2 p^2$. All the $W_{p^2,C}$ are infinite dimensional. The representation $\eta_{p^2,C}$ is given by:

$$\eta_{p^2,C}(J_3)|_{j,m_j}^{p^2,C}\rangle = m_j|_{j,m_j}^{p^2,C}\rangle \tag{7}$$

$$\eta_{p^2,C}(J_\pm)|_{j,m_j}^{p^2,C}\rangle = \sqrt{(j \mp m_j)(j \pm m_j + 1)}|_{j,m_j\pm 1}^{p^2,C}\rangle \tag{8}$$

$$\begin{aligned}\eta_{p^2,C}(P_3)|_{j,m_j}^{p^2,C}\rangle = {} & \tilde{\alpha}_j^{p^2,C}\sqrt{(j+m_j)(j-m_j)}|_{j-1,m_j}^{p^2,C}\rangle \\ & +\tilde{\beta}_j^{p^2,C} m_j|_{j,m_j}^{p^2,C}\rangle + \\ & \tilde{\alpha}_{j+1}^{p^2,C}\sqrt{(j+m_j+1)(j-m_j+1)}|_{j+1,m_j}^{p^2,C}\rangle\end{aligned} \tag{9}$$

$$\eta_{p^2,C}(P_\pm)|_{j,m_j}^{p^2,C}\rangle = \pm\tilde{\alpha}_j^{p^2,C}\sqrt{(j\mp m_j)(j\mp m_j-1)}|_{j-1,m_j\pm1}^{p^2,C}\rangle + \tilde{\beta}_j^{p^2,C}\sqrt{(j\mp m_j)(j\pm m_j+1)}|_{j,m_j\pm1}^{p^2,C}\rangle \mp\tilde{\alpha}_{j+1}^{p^2,C}\sqrt{(j\pm m_j+1)(j\pm m_j+2)}|_{j+1,m_j\pm1}^{p^2,C}\rangle \quad (10)$$

where $\tilde{\alpha}_j^{p^2,C} = \sqrt{\frac{p^2-\frac{C^2}{j^2}}{(2j+1)(2j-1)}}$, $\tilde{\beta}_j^{p^2,C} = \frac{C}{j(j+1)}$.

3. Contraction of the representations

In this section, for each of the representations $\eta_{p^2,C}$ we specify a suitable sequence of the representations $\rho_{F(n),G(n)}$ such that we obtain the desired convergence of matrix elements. We will not address the question of contraction of the group representations which was solved in a different approach by Dooley and Rice[5] and was considered by others[6–8]. A contraction of the representations of $so(3,1)$ to those of $iso(3)$ was done by Weimar-Woods[9].

We realize the contraction[1,10,11] of $so(4)$ to $iso(3)$ by the family of maps $t_\epsilon : iso(3) \longrightarrow so(4)$ which for every $\epsilon \in (0,1]$ is given by $t_\epsilon(J_i) = M_i$, $t_\epsilon(P_i) = \epsilon N_i$ for every $i \in \{1,2,3\}$. Explicitly, one can see that for any $X, Y \in iso(3)$

$$[X,Y]_{iso(3)} = \lim_{\epsilon\longrightarrow 0^+} t_\epsilon^{-1}[t_\epsilon(X), t_\epsilon(Y)]_{so(4)} \quad (11)$$

We fix a representation $\eta_{p_1^2,C_1}$ of $iso(3)$ and define the sequence of $so(4)$ representations which consists of those $\rho_{(F,G)}$ that satisfy $j_0 = |C_1/p_1|$ and $sgn(G) = sgn(C_1)$. One can show that this sequence is given by $\{(\rho_{(F(n),G(n)}, V_{F(n),G(n)})\}_{n\in I}$ where

$$G(n) = \mathrm{sgn}\,(C_1)\left|\frac{C_1}{p_1}\right|(n+1), \qquad F(n) = \frac{(\frac{C_1}{p_1})^2 + (n+1)^2 - 1}{2} \quad (12)$$

and $I = \{|C_1/p_1|, |C_1/p_1|+1, ...\}$. Now we can state our main result (for a proof see arxiv.org/abs/1210.6023)

Theorem 3.1. *Let* $\epsilon_n = \frac{|p_1|}{\sqrt{2F(n)}}$ *and* $t_n = t(\epsilon_n)$. *Then for any* $|_{j,m_j}^{F(n),G(n)}\rangle$, $|_{\tilde{j},\tilde{m}_{\tilde{j}}}^{F(n),G(n)}\rangle \in V_{F(n),G(n)}$ *and any* $X \in iso(3)$

$$\lim_{n\to\infty}\left\langle {}_{\tilde{j},\tilde{m}_{\tilde{j}}}^{F(n),G(n)}|\rho_{F(n),G(n)}(t_n(X))|_{j,m_j}^{F(n),G(n)}\right\rangle = \left\langle {}_{\tilde{j},\tilde{m}_{\tilde{j}}}^{p_1^2,C_1}|\eta_{p_1^2,C_1}(X)|_{j,m_j}^{p_1^2,C_1}\right\rangle \quad (13)$$

Graphical representation of the contraction process
In figure 1 each point with coordinates (k, l) represents the irreducible

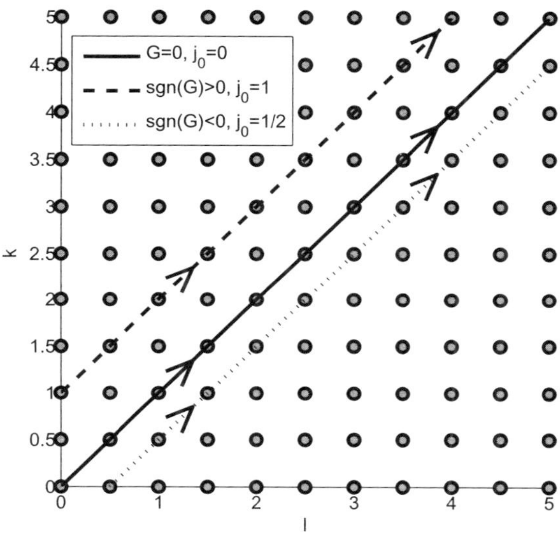

Fig. 1. Grid of $so(4)$ finite dimensional irreducible representations and some of their contractions.

representation of $so(4)$ which is a tensor product of two irreducible representations of $so(3)$ with Casimir values of k and l . In each "diagonal" line, j_0 is constant and equal to the value of $|k - l|$. Going along each diagonal line in the direction of the arrow (which is equivalent to taking ϵ_n to zero) we are increasing the value of n by one unit at each step, and this is the picture of the contraction. The solid, dashed and dotted diagonal lines correspond to contractions toward different representations $\eta_{p^2,C}$.

4. Discussion

The four-dimensional rotation group, $SO(4)$, occurs as a symmetry group of a physical system. The best known example is the Hydrogen atom. The group of isometries of the three-dimensional space, $\mathbb{R}^3$, i.e., the Euclidean group $ISO(3)$ is another group that is naturally related to many physical systems. Among others, $ISO(3)$ is a subgroup of both Poincaŕe group and Galilei group. The relation between $SO(4)$ and $ISO(3)$ was only partially studied, e.g., in Refs. 12–14.

In other work[15,16] we give a definition for contraction of Lie algebra representations using the notion of direct limit. We also show there that the convergence of matrix elements implies the convergence in norm of the sequence of operators. This shows that the contraction we obtained here is also a contraction according to the definition in Ref. 15.

Acknowledgments

AM is grateful to Prof. Weimar-Woods for a helpful discussion.
EMS would like to thank Mr. ShengQuan Zhou for sharing his notes on the contractions of $so(4)$.
The research of the 2nd author was supported by the center of excellence of the Israel Science Foundation grant no. 1438/06.
JLB thanks the Department of Physics, Technion, for its warm hospitality and support during visits while this work was being carried out, and the FRAP-PSC-CUNY for some support.

References

1. E. Inonu and E. P. Wigner, *Proceedings of the National Academy of Science* **39**, 510 (1953).
2. E. Weimar-Woods, *Journal of Mathematical Physics* **32**, 2028 (1991).
3. E. Weimar-Woods, *Journal of Mathematical Physics* **32**, 2660 (1991).
4. W. Pauli, Continuous groups in quantum mechanics, Springer Tracts in Modern Physics Vol. 37 (Springer Berlin / Heidelberg, 1965) pp. 85–104.
5. A. H. Dooley and J. W. Rice, *Mathematical Proceedings of the Cambridge Philosophical Society* **94**, p. 509 (1983).
6. K. B. Wolf, *Journal of Mathematical Physics* **12**, 197 (1971).
7. M. K. F. Wong and H.-Y. Yeh, *Journal of Mathematical Physics* **21**, 1 (1980).
8. M. K. F. Wong and H.-Y. Yeh, *Journal of Mathematical Physics* **22**, 1559 (1981).
9. E. Weimar-Woods, *Reviews in Mathematical Physics* **18**, 655 (2006).
10. E. J. Saletan, *Journal of Mathematical Physics* **2**, 1 (1961).
11. R. Gilmore, *Lie groups, Lie algebras, and some of their applications* (Wiley New York, 1974).
12. M. Levy-Nahas and R. Seneor, *Communications in Mathematical Physics* **9**, 242 (1968).
13. W. J. Holman, III, *Annals of Physics* **52**, 176 (1969).
14. A. I. Bobenko, *Functional Analysis and Its Applications* **20**, 53 (1986).
15. E. M. Subag, E. M. Baruch, J. L. Birman and A. Mann, *Journal of Physics: Conference Series* **343**, p. 012116 (2012).
16. E. M. Subag, E. M. Baruch, J. L. Birman and A. Mann, *Journal of Physics A: Mathematical and Theoretical* **45**, p. 265206 (2012).

POSTER SESSION

20 & 22 AUGUST 2012

VERTEX OPERATOR REALIZATION OF JACK SYMMETRIC FUNCTIONS

TOMMY WUXING CAI

School of Sciences
South China University of Technology
Guangzhou, Guangdong 510640, CHINA
E-mail: caiwx@scut.edu.cn

NAIHUAN JING

Department of Mathematics
North Carolina State University
Raleigh, NC 27695-8205, USA
E-mail: jing@math.ncsu.edu

We summarize the recent development of realizing arbitrary Jack symmetric functions by a sequence of vertex operators associated with rectangular Young tableaux.

Keywords: Jack polynomials; vertex operators; vertex operator realization.

1. Introduction

Jack polynomials[5] are a special family of orthogonal symmetric functions generalizing Schur functions and zonal spherical polynomials. They figure very importantly in the Calegero-Sutherland model in statistical mechanics and also appear as extremal vectors in the Fock representation of the Virasoro algebra[1,9] . Macdonald and Stanley first proved the existence by using a generalized Laplace-Beltrami operator[8,10] (see also[2]).

Vertex operators have long been used to construct symmetric functions. Schur functions were constructed as Bernstein operators[12] at the very beginning of vertex representations of affine Lie algebras. Later Schur Q-functions[6] , Hall-Littlewood functions[7] are realized by vertex operators. However one only knows a few examples for Macdonald functions[11] . In our recent work[3] we find a vertex operator realization of Jack symmetric functions based on the vertex operator realization of the rectangular Jack functions[4,9] (cf.[1]) .

In this paper we describe this new approach to Jack symmetric functions using a sequence of rectangular Jack vertex operators.

2. Jack symmetric functions

We first recall some basic notions and terminologies. A partition λ of n is a sequence of weakly decreasing positive integers $\lambda = (\lambda_1, \cdots, \lambda_l)$, with weight $|\lambda| := \lambda_1 + \cdots + \lambda_l$ equal to n. Here $l(\lambda) := l$ is called the length of λ. For convenience, one sets $\lambda_i = 0$ if $i > l(\lambda)$. The multiplicity of i in λ is $m_i(\lambda) := \sharp\{j|\lambda_j = i\}$. Partition λ of n *dominates* another partition μ of n, denoted $\lambda \geq \mu$, if the corresponding partial sums satisfy $\sum_{i=1}^k \lambda_i \geq \sum_{i=1}^k \mu_i$. If the multiplicities $m_i(\mu) \leq m_i(\lambda)$ for all i, we say that μ is multiplicity-contained in λ and denoted as $\mu \subset' \lambda$. In this case we define $\lambda\backslash\mu$ by $m_i(\lambda\backslash\mu) = m_i(\lambda) - m_i(\mu)$. One usually visualizes λ by its Young diagram consisting of $l(\lambda)$ rows of left justified boxes, with λ_i boxes on the ith row. Thus a partition of the form $(k, k, \cdots, k) = (k^s)$ is said to be of rectangular shape. The dual partition λ' of λ corresponds to the transpose of the Young diagram along the diagonal. A box at the ith row and jth column in the Young diagram is denoted as $s = (i, j) \in \lambda$. For an indeterminate α, the (Jack) lower/upper hook-lengths of s are defined as $h_*^\lambda(s) = \alpha(\lambda_i - j) + (\lambda'_j - i + 1)$ and $h^*_\lambda(s) = \alpha(\lambda_i - j + 1) + (\lambda'_j - i)$ respectively.

Let Λ be the ring of symmetric functions in infinitely many variables x_n, $n \in \mathbb{N}$ over the field $F = \mathbb{Q}(\alpha)$. As a vector space Λ has well-known linear bases such as monomial symmetric functions m_λ and power sum symmetric functions p_λ indexed by partitions. Here $m_\lambda = m_\lambda(x_1, x_2, \cdots)$ is the formal sum of distinct monomials $x_{i_1}^{\lambda_1} x_{i_2}^{\lambda_2} \cdots$, and $p_\lambda = p_{\lambda_1} p_{\lambda_2} \cdots$ with $p_n = m_{(n)}$ ($p_0 = 1$).

The inner product of Λ is defined by requiring that p_λ's are orthogonal[8,10] and

$$\langle p_\lambda, p_\lambda \rangle = \alpha^{l(\lambda)} \prod_{i=1}^{l(\lambda)} i^{m_i(\lambda)} m_i(\lambda)!. \tag{1}$$

Jack symmetric functions[5] $P_\lambda(\alpha)$ are one-parameter deformation of Schur functions. They form another orthogonal basis of Λ and satisfy

$$P_\lambda(\alpha) = \sum_{\lambda \geq \mu} c_{\lambda\mu}(\alpha) m_\mu,$$

where $c_{\lambda\mu}(\alpha) \in F$ ($\lambda, \mu \in \mathcal{P}$) and $c_{\lambda\lambda}(\alpha) = 1$. The dual Jack functions with respect to the inner product are denoted as $Q_\lambda(\alpha)$.

3. Vertex operator realization

Any symmetric function f acts on Λ as a multiplication operator. The dual operator f^* is defined by $\langle fu, v\rangle = \langle u, f^*\cdot v\rangle$ for any $u, v \in \Lambda$. Let $\mathbb{Q}[\frac{1}{2}\mathbb{Z}h]$ be the group algebra of the one-dimensional lattice spanned by e^{mh}, $m \in \frac{1}{2}\mathbb{Z}$. We introduce the vertex operator X_n acting on $\mathbb{Q}[\frac{1}{2}\mathbb{Z}h] \otimes \Lambda$ by

$$\begin{aligned} X(z) &= e^h z^{2\alpha^{-1}\partial_h} \otimes \exp\Big(\sum_{n=1}^{\infty} \frac{z^n}{n\alpha} p_n\Big) exp\Big(-\sum_{n=1}^{\infty} 2z^{-n}\frac{\partial}{\partial p_n}\Big) \\ &= \sum_{n\in\mathbb{Z}} X_n z^{-n-\alpha^{-1}}. \end{aligned}$$

The following result provides a realization of Jack functions of rectangular shapes. The first part was proved by Mimachi-Yamada[9] for rectangular Jack polynomials, on which our general result is based.

Theorem 3.1.[1] *Assume that $k, s, \alpha^{-1} \in \mathbb{Z}_{>0}$, for partition $\lambda = (k^s)$ one has that*

$$\begin{aligned} &(X_{-k})^s.(e^{-sh/2} \otimes 1) = c_s(\alpha) Q_\lambda(\alpha) \\ &= c_s(\alpha) \otimes \sum_{\mu} \frac{(-1)^{\alpha^{-1}s(s-1)/2}(\alpha^{-1}!)^s}{(s\alpha^{-1})!(-2)^{l(\mu)}} \sum_{\underline{\mu},\underline{\nu}} \prod_{i=1}^{s} \frac{(-2\alpha^{-1})^{l(\nu^i)}}{z_{\nu^i}} \binom{m(\mu^{i-1})}{m(\mu^i\backslash\nu^i)} p_\mu, \end{aligned}$$

where $c_s(1) = (-1)^{s(s-1)/2}s!e^{sh/2}$. The second sum is over sequences of partitions $\underline{\mu} = (\mu^1, \mu^2, \cdots, \mu^s)$ and $\underline{\nu} = (\nu^1, \nu^2, \cdots, \nu^s)$ such that $\mu^0 = (0)$, $\mu^s = \mu$, $\nu^i \subset' \mu^i$, $|\mu^i| = i(k + (s-i)\alpha^{-1})$, and $\mu^i\backslash\nu^i \subset' \mu^{i-1}$.

The following result proves Stanley's positivity conjecture on the Littlewood-Richardson coefficients of Jack functions in the case of rectangular λ. If a partition ν satisfies $\nu_1 \le r$, $l(\nu) \le s$, we define partition $\mu = (r^s) -' \nu$ by $\mu_i = r - \nu_{s-i+1}$. Geometrically, it is the *upside down* complement of ν in λ.

Theorem 3.2.[3] *For a rectangular partition $\lambda = (r^s)$, $\langle J_\mu J_\nu, J_\lambda\rangle \neq 0$ if and only if $\mu = \lambda -' \nu$. And in this case we have*

$$\begin{aligned} \langle J_\mu J_\nu, J_\lambda\rangle &= \prod_{s\in\mu} h_*^\mu(s) \prod_{s\in\nu} h_\nu^*(s) \qquad (2) \\ &\times \prod_{(i,j)\in\mu} [(\mu_j' - i) + \alpha(r-j+1)] \cdot \prod_{(i,j)\in\nu} [(s-i+1) + \alpha(j-1)]. \end{aligned}$$

Let λ be a partition of multiplicity type $(m_1, \cdots, m_s)$, i.e. , where the non-vanishing multiplicities are $m_i = m_{a_i}(\lambda)$, where $a_1 > a_2 > \cdots > a_s > 0$

are distinguished parts of λ. Then the conjugate partition λ' has multiplicity type $(p_1, p_2, \cdots, p_s)$, where $p_i = a_{s+1-i} - a_{s+2-i}$ with $a_{s+1} = 0$. We associate λ with a sequence of rectangular partitions $R_\lambda = (R_1, R_2, \cdots, R_s)$ such that λ will be the alternating set union of $R_i's$. The associated filtration of partitions is given explicitly by: $R_1 = (p_1 + p_2 + \cdots + p_s)^{m_1+m_2+\cdots+m_s}$, $R_2 = (p_2 + \cdots + p_s)^{m_2+\cdots+m_s}$, $R_3 = (p_2 + \cdots + p_{s-1})^{m_2+\cdots+m_{s-1}}$, $R_4 = (p_3 + \cdots + p_{s-1})^{m_3+\cdots+m_{s-1}}$, $R_5 = (p_3 + \cdots + p_{s-2})^{m_3+\cdots+m_{s-2}}$, $\cdots$, $R_s = (p_{s'}^{n_{s'}})$, where $s' = [s/2]+1$ and $[s/2]$ is the integer part of $s/2$.

Theorem 3.3.[3] *Let a partition λ and its filtration $R_\lambda = (R_1, R_2, \cdots, R_s)$ be defined as above. Assume that $f_i = Q_{R_i}(\alpha)$ is the vertex realization of the rectangular R_i, then we have*

$$Q_\lambda(\alpha) = c'_\lambda(\alpha)(\cdots(((f_s^* \cdot f_{s-1})^* \cdot f_{s-2})^* \cdot f_{s-3})^* \cdot \cdots \cdot f_2)^* \cdot f_1,$$
$$= c'_\lambda(\alpha) \sum_{\underline{\xi}} \prod_{i=1}^{s} g_{R_i,\xi^i}(\alpha) \prod_{i=1}^{s-1} z_{\dot{\xi}^i} \alpha^{l(\dot{\xi}^i)} \binom{m(\xi^i)}{m(\dot{\xi}^i)} p_{\dot{\xi}^0}$$

where $c'_\lambda(\alpha) \neq 0$, $g_{R_i,\xi^i}(\alpha)$ is the coefficient of p_μ in Q_{R_i} given by (3.1), and the sum is over sequences of partitions $\underline{\xi} = (\xi^0, \xi^1, \cdots, \xi^s)$ which is associated with $\underline{\dot{\xi}} = (\dot{\xi}^0, \dot{\xi}^1, \cdots, \dot{\xi}^s)$ such that $\dot{\xi}^i = \xi^{i+1} \backslash \dot{\xi}^{i+1} \subseteq' \xi^i$ for $i = s-1, \cdots, 2, 1, 0$ with $\dot{\xi}^s = 0$ and $\xi^0 = \xi^1$.

Acknowledgments

We thank the support of NSFC during this work.

References

1. W. Cai, N. Jing, *J. Alg. Combin.* **32**, 579 (2010).
2. W. Cai, N. Jing, *European J. Combin.* **33**, 556 (2012).
3. W. Cai, N. Jing, *Jack vertex operators and realization of Jack functions, J. Alg. Combin.*, to appear. (arXiv:1108.2007)
4. B. Feigin, E. Feigin, *Adv. Math.* **206**, 307 (2006).
5. H. Jack, *Proc. Roy. Soc. Edinburgh Sect. A* **69**, 1 (1970/1971).
6. N. Jing, *J. Algebra* **138**, 340 (1991).
7. N. Jing, *Adv. Math.* **87**, 226 (1991).
8. I. G. Macdonald, *Symmetric functions and Hall polynomials*, 2nd edition (Oxford Univ. Press, New York, 1995).
9. K. Mimachi, Y. Yamada, *Comm. Math. Phy.*, **174** (1995).
10. R. Stanley, *Adv. Math.* **77**, 76 (1989).
11. M. Zabrocki, *Electron. J. Combin.* **5**, 45 (1998).
12. A. Zelevinsky, *Representations of finite classical groups*, (Springer-Verlag, Berlin-New York, 1981).

ON THE RACAH COEFFICIENTS OF $sl_{-1}(2)$ AND BANNAI–ITO POLYNOMIALS

VINCENT X. GENEST* and LUC VINET
Centre de Recherches Mathématiques, Université de Montréal,
Montréal, Québec, Canada, H3C 3J7
**E-mail: genestvi@crm.umontreal.ca*
www.crm.umontreal.ca

ALEXEI ZHEDANOV
Donetsk Institute for Physics and Technology
Donetsk 83114, Ukraine
E-mail: zhedanov@fti.dn.ua

The Bannai–Ito (BI) polynomials are shown to arise as Racah coefficients of the Hopf algebra $sl_{-1}(2)$.

Keywords: Bannai–Ito polynomials, Racah coefficients, sl_{-1} algebra.

1. Introduction

The $sl_{-1}(2)$ algebra was introduced recently[1] as a deformation of the classical Lie algebra $sl(2)$; it has four generators, one of which is an involution, that obey both commutation and anticommutation relations. This algebra can also be obtained from the quantum algebra $sl_q(2)$ by taking the limit $q \to -1$. We here consider the Racah problem for this algebra.

Recently,[2,4] a series of orthogonal polynomials (OPs) corresponding to limits $q \to -1$ of q-polynomials of the Askey scheme have been discovered. These OPs are eigenfunctions of operators of Dunkl type, which involve the reflection[5] operator. Interestingly, they have also been related to Jordan anticommutator[6] algebras. At the top of the discrete branch of this $q = -1$ class of polynomials lie the Bannai–Ito (BI) polynomials, which depend on four parameters and are expressible in terms of Wilson polynomials.

It is known[1] that the Clebsch-Gordan coefficients of $sl_{-1}(2)$ are given in terms of the dual -1 Hahn polynomials. Here it is shown that the Racah coefficients of $sl_{-1}(2)$ are expressed in terms of the Bannai–Ito polynomials.

Our approach consist in constructing the Jordan algebra of the intermediary Casimir operators that appear in the coproduct of three $sl_{-1}(2)$ algebras. This anticommutator algebra coincides with the Bannai–Ito algebra,[2] an observation which suffices to obtain the Racah coefficients in terms of the BI polynomials.[7]

2. The $sl_{-1}(2)$ algebra

The Hopf algebra $sl_{-1}(2)$[1] is generated by four operators J_0, J_+, J_- and R satisfying the relations

$$[J_0, J_\pm] = \pm J_\pm, \quad [J_0, R] = 0, \quad \{J_+, J_-\} = 2J_0, \quad \{J_\pm, R\} = 0, \tag{1}$$

where $[x, y] = xy - yx$ and $\{x, y\} = xy + yx$. The operator R is an involution, which means that it satisfies the property $R^2 = \mathfrak{id}$, where $\mathfrak{id}$ is the identity. The Casimir operator is given by

$$\mathcal{Q} = J_+J_-R - J_0R + R/2. \tag{2}$$

Let $\epsilon = \pm 1$ and $\mu \geqslant 0$ be two parameters; we denote by (ϵ, μ) the infinite-dimensional vector space spanned by the basis $e_n^{(\epsilon,\mu)}$, $n \in \mathbb{N}$, endowed with the actions

$$\begin{aligned} J_0 e_n^{(\epsilon\mu)} &= (n + \mu + 1/2) e_n^{(\epsilon\mu)}, & R e_n^{(\epsilon,\mu)} &= \epsilon(-1)^n e_n^{(\epsilon,\mu)}, \\ J_+ e_n^{(\epsilon,\mu)} &= \sqrt{[n+1]_\mu}\, e_{n+1}^{(\epsilon,\mu)}, & J_- e_n^{(\epsilon,\mu)} &= \sqrt{[n]_\mu}\, e_{n-1}^{(\epsilon,\mu)}, \end{aligned} \tag{3}$$

where $[n]_\mu$ denotes the μ-number $[n]_\mu = n + \mu(1 - (-1)^n)$. With the actions (3), the vector space (ϵ, μ) forms an irreducible $sl_{-1}(2)$-module. On this module, the Casimir operator acts as a multiple of the identity

$$\mathcal{Q} e_n^{(\epsilon,\mu)} = -\epsilon\, \mu e_n^{(\epsilon,\mu)}. \tag{4}$$

The algebra $sl_{-1}(2)$ admits a non-trivial addition rule, or coproduct. Let $J_0^{(i)}$, $J_\pm^{(i)}$ and $R^{(i)}$ be two mutually commuting sets of $sl_{-1}(2)$ generators. A third algebra, denoted symbolically $3 = 1 \oplus 2$, is obtained by defining

$$J_0^{(3)} = J_0^{(1)} + J_0^{(2)}, \quad J_\pm^{(3)} = J_\pm^{(1)} R^{(2)} + J_\pm^{(2)}, \quad R^{(3)} = R^{(1)} R^{(2)}. \tag{5}$$

It is easily verified that the generators $J_0^{(3)}$, $J_\pm^{(3)}$ and $R^{(3)}$ satisfy the defining relations of $sl_{-1}(2)$ given in (1). The Casimir operator for the third algebra, denoted by $\mathcal{Q}_{12}$, takes the form

$$\mathcal{Q}_{12} = \left(J_-^{(1)} J_+^{(2)} - J_+^{(1)} J_-^{(2)}\right) R^{(1)} - R^{(1)} R^{(2)}/2 + \mathcal{Q}_1 R^{(2)} + \mathcal{Q}_2 R^{(1)}. \tag{6}$$

3. Bannai–Ito polynomials

The Bannai–Ito polynomials were introduced in 1984 in the complete classification of orthogonal polynomials satisfying the Leonard duality property.[8] The monic BI polynomials satisfy the recurrence relation

$$P_{n+1}(x) + (\rho_1 - A_n - C_n)P_n(x) + A_{n-1}C_nP_{n-1}(x) = xP_n(x), \tag{7}$$

where

$$A_n = \begin{cases} \frac{(n+1+2\rho_1-2r_1)(n+1+2\rho_1-2r_2)}{4(n+1-r_1-r_2+\rho_1+\rho_2)}, & n \text{ even}, \\ \frac{(n+1-2r_1-2r_2+2\rho_1+2\rho_2)(n+1+2\rho_1+2\rho_2)}{4(n+1-r_1-r_2+\rho_1+\rho_2)}, & n \text{ odd}, \end{cases} \tag{8}$$

$$C_n = \begin{cases} -\frac{n(n-2r_1-2r_2)}{4(n-r_1-r_2+\rho_1+\rho_2)}, & n \text{ even}, \\ -\frac{(n-2r_2+2\rho_2)(n-2r_1+2\rho_2)}{4(n-r_1-r_2+\rho_1+\rho_2)}, & n \text{ odd}. \end{cases} \tag{9}$$

For the BI polynomials, the positive-definiteness condition $U_n = A_{n-1}C_n > 0$ cannot be satisfied for all n. However, if $U_{N+1} = 0$ and $U_n > 0$ for $n \in \{0, \ldots, N\}$, the BI polynomials form a finite system of orthogonal polynomials $P_0(x), \ldots, P_N(x)$. For example, if we consider the following parametrization:

$$\begin{aligned} 2\rho_1 &= (\beta + \gamma), \qquad 2\rho_2 = -(\beta + \gamma + N + 1), \\ 2r_1 &= (\gamma - \beta), \qquad 2r_2 = -(2\alpha + \beta + \gamma + N + 1), \end{aligned} \tag{10}$$

where N is an odd integer and where α, β and γ are positive parameters, the truncation and positivity conditions are satisfied and the BI polynomials obey the orthogonality relation

$$\sum_{\ell=0}^{N} \Omega_\ell P_n(x_\ell)P_m(x_\ell) = \Phi_n\delta_{nm}. \tag{11}$$

The grid is given by $x_\ell = (1/2)[(-1)^\ell(\ell + \gamma + \beta + 1/2) - 1/2]$; the BI polynomials also have a hypergeometric representation in terms of ${}_4F_3$'s; more details can be found in ref. [4].

4. Racah problem

The addition rule (5) possesses an associativity property when three $sl_{-1}(2)$ algebras are added. We consider three mutually commuting sets of $sl_{-1}(2)$ generators $J_0^{(j)}$, $J_\pm^{(j)}$ and $R^{(j)}$ with $j \in \{1, 2, 3\}$. The Racah problem consists in finding the overlap between the respective eigenstates of the intermediary Casimir operators $\mathcal{Q}_{12}$ and $\mathcal{Q}_{23}$. The Racah coefficients are defined by

$$e_m^{(q_{12},\vec{\lambda})} = \sum_{q_{23}} R_{q_{12},q_{23}}^{\vec{\lambda}} e_m^{(q_{23},\vec{\lambda})}, \tag{12}$$

where $\vec{\lambda} = (\epsilon_1\mu_1, \ldots, \epsilon_4\mu_4)$ and where we have

$$\mathcal{Q}_{12}e_m^{(q_{12},\vec{\lambda})} = q_{12}e_m^{(q_{12},\vec{\lambda})} \qquad \mathcal{Q}_{23}e_m^{(q_{23},\vec{\lambda})} = q_{23}e_m^{(q_{23},\vec{\lambda})} \tag{13}$$

The vectors $e_m^{(q_{ij},\vec{\lambda})}$ are eigenvectors of the local Casimir operators $\mathcal{Q}_i$, $i = 1, \ldots, 4$, where $\mathcal{Q}_4$ is the Casimir operator of the three added algebras $4 = 1 \oplus 2 \oplus 3$. We define $K_1 = \mathcal{Q}_{12}$ and $K_2 = \mathcal{Q}_{23}$, where the $\mathcal{Q}_{ij}$ are given by the formula (6). The Racah problem is non-trivial since the operators K_1 and K_2 do not commute. A direct computation shows that

$$\{K_1, K_2\} = K_3 + \alpha_3, \quad \{K_2, K_3\} = K_1 + \alpha_1, \quad \{K_1, K_3\} = K_2 + \alpha_2, \tag{14}$$

with

$$\alpha_1 = -2(\lambda_1\lambda_2 + \lambda_3\lambda_4),\ \alpha_2 = -2(\lambda_1\lambda_4 + \lambda_2\lambda_3),\ \alpha_3 = 2(\lambda_1\lambda_3 + \lambda_2\lambda_4), \tag{15}$$

and where K_3 is *defined* by the first relation of (14). The algebra (14) is known as the Bannai–Ito algebra. This algebra is related to the Bannai–Ito polynomials through its irreducible representations. Considering an $N+1$-dimensional irreducible representation of (14), it is possible to obtain

$$R_{q_{12}q_{23}}^{\vec{\lambda}} = \sqrt{\frac{\Omega_\ell(x_\ell)}{\Phi_{N,n}}}P_n(\rho_1, \rho_2, r_1, r_2; x_\ell), \tag{16}$$

with $x_\ell = -\frac{1}{2}\left((-1)^\ell(\mu_2 + \mu_3 + 1/2 + \ell) - 1/2\right)$,

$$\ell = |q_{23}| - \mu_2 - \mu_3 - 1/2 \qquad n = |q_{12}| - \mu_1 - \mu_2 - 1/2, \tag{17}$$

$\epsilon_i = 1$, N even and where we have the following identification:

$$\rho_1 = \tfrac{1}{2}(\mu_2 + \mu_3),\ \rho_2 = \tfrac{1}{2}(\mu_1 + \mu_4),\ r_1 = \tfrac{1}{2}(\mu_3 - \mu_2),\ r_2 = \tfrac{1}{2}(\mu_4 - \mu_1). \tag{18}$$

Acknowledgments V.X.G. holds a scholarship from FRQ-NT (Québec). The research of L.V. is supported by NSERC (Canada).

References

1. S. Tsujimoto, L. Vinet and A. Zhedanov, *S.I.G.M.A.* **7** (2011).
2. S. Tsujimoto, L. Vinet and A. Zhedanov, *Adv. Math.* **229**, 2123 (2012).
3. L. Vinet and A. Zhedanov, *S.I.G.M.A.* **7** (2011).
4. S. Tsujimoto, L. Vinet and A. Zhedanov, *ArXiV:1108.0132* (2011).
5. C. Dunkl, *Comm. Math. Phys* **197**, 451 (1998).
6. S. Tsujimoto, L. Vinet and A. Zhedanov, *J. Math. Phs.* **52**, p. 103512 (2011).
7. V.X. Genest, L. Vinet and A. Zhedanov, *ArXiv:1205.4215* (2012).
8. E. Bannai and T. Ito, *Algebraic Combinatorics I: Associaion Schemes* (1984).

DISCORD DERIVED FROM TSALLIS ENTROPY FOR WERNER AND ISOTROPIC STATES

JACEK JURKOWSKI

Institute of Physics, Nicolaus Copernicus University
Grudziadzka 5, 87–100 Toruń, Poland

Due to some ambiguity in the definition of mutual Tsallis entropy in classical probability theory, its generalization to quantum theory is discussed. We define a q-discord derived from the q-expectation value of mutual Tsallis entropy. We provide analytical expressions for q-discord for two-qubit Werner states as well as for isotropic states and we show that both are positive, at least for states under investigation, for all $q > 0$.

1. Introduction

There is a growing interest in quantifying and measuring correlations in composite quantum systems (for a recent review see[1]). Such notions as entanglement, quantum discord and other entropic measures of quantumness proved to be useful in these investigations.[2–5] While entanglement from the very definition reflects by itself some specific structure of the density matrix representing the state, discord measures the amount of quantum correlations between subsystems in terms of mutual von Neumann entropy. It is the strong subadditivity property of von Neumann entropy that underlies the nonnegativity of quantum discord and, hence, provides the most fundamental interpretation of quantum discord as a measure of correlations.

On the other hand, for many years the nonadditive generalizations of Shannon and von Neumann entropy functions, such as Renyi entropy,[6] Tsallis entropy[7] and other ones,[5,8] have been used to describe some aspects of statistical and thermodynamic behaviour of systems far from equilibrium. These families of entropy functions parametrized by $q > 0$ seem to incorporate in a natural manner some correlations present in composite systems. In this paper we focus on Tsallis entropy which, except additivity, displays similar properties (nonnegativity, concavity) as von Neumann entropy.[12] Therefore, it can be used in most applications in the field of quantum information as well.[9–11,13] In particular, an observation due to Abe[14] that using Tsallis entropy one can detect more separable two-qubit Werner states than using Shannon entropy seems very promising.

Let (Ω, P) be a discrete probability space. For any given random variable $f = (f_1, \ldots, f_n)$ and probability $P = (p_1, \ldots, p_n)$, instead of ordinary expectation value $\langle f \rangle = \sum_j p_j f_j$, we define q-expectation value with respect to P as

$$\langle f \rangle_q = \frac{\sum_j p_j^q f_j}{\sum_j p_j^q} \tag{1}$$

which satisfies $\langle 1 \rangle_q = 1$. Let us denote by $p_{ij} = P(A_i, B_j)$ joint probabilities with marginals p_i^A and p_j^B. Now, let $p_{i|j} := P(A_i, B_j)/P(B_j) = p_{ij}/p_j^B$ be conditional probabilities. Recall that Tsallis entropy is defined as[7]

$$T_q(P) = \frac{1}{q-1}\Big(1 - \sum_j p_j^q\Big).$$

Hence, Tsallis entropy of the conditional probability distribution is $T_q(A|B_j) := \big(1 - \sum_i p_{i|j}^q\big)/(q-1)$. As a consequence conditional Tsallis entropy $T_q(A|B)$ can be defined as q-expectation value[15] (1) with respect to marginal probablity p_j^B

$$T_q(A|B) = \langle T_q(A|B_j)\rangle_q^{(B)} = \sum_j \Big(\frac{p_j^q(A)}{\sum_i p_i^q(A)}\Big) T_q(A|B_j), \tag{2}$$

in analogy to the case of Shannon entropy S, where $S(A|B) = \langle S(A|B_j)\rangle^{(B)}$. Here $\langle \cdot \rangle^{(B)}$ means the usual expectation value with respect to marginal probability $P(B)$. Using Tsallis entropy of conditional probability and the relation $p_{ij} = p_j^B p_{i|j}$ we can rewrite (2) as

$$T_q(A|B) = \frac{T_q(A,B) - T_q(B)}{1 + (1-q)T_q(B)}. \tag{3}$$

Note, that another possibility is to define $T_q(A|B) = \sum_j (p_j^B)^q T_q(A|B_j)$, i.e., without normalization of $(p_j^B)^q$. This leads to the traditional form of mutual entropy

$$\widetilde{\mathcal{I}}_q(A:B) = T_q(A) + T_q(B) - T_q(AB) \tag{4}$$

well known from the case of Shannon entropy. But, in what follows, we consequently define mutual information as $\mathcal{I}_q(A:B) = T_q(A) - T_q(A|B)$. Hence, from (3) we obtain

$$\mathcal{I}_q(A:B) = \frac{T_q(A) + T_q(B) - T_q(A,B) + (1-q)T_q(A)T_q(B)}{1 + (1-q)T_q(B)}. \tag{5}$$

2. Quantum Tsallis entropy

In the framework of composite quantum systems quantum Tsallis entropy can be defined in perfect analogy with the classical case as

$$T_q(A,B) \;:=\; T_q(\rho_{AB}) \;=\; \frac{1}{q-1}\Big(1-\mathrm{Tr}\,(\rho_{AB}^q)\Big)\,,$$

where ρ_{AB} is a density matrix of the composite system.

In quantum theory all the objects occurring in (5) are already well-defined. Hence, we define the *quantum mutual Tsallis entropy*, which measures total correlations between subsystems A and B by (5), where $T_q(A) = T_q(\rho_A)$, $T_q(B) = T_q(\rho_B)$ and ρ_A, ρ_B are reduced density matrices. On the other hand, due to[3] we can describe classical correlations present in the state ρ_{AB} by considering a complete, projective measurement $\{\Pi_k\}$ on subsystem B and states $\rho_k = (I\otimes\Pi_k)\rho_{AB}(I\otimes\Pi_k)^\dagger/p_k$, after measurements, where $p_k = \mathrm{Tr}\,(I\otimes\Pi_k)\rho_{AB}$. Next, define quantum conditional Tsallis entropy with respect to this measurement by

$$T_q(\rho_{AB}|\{\Pi_k\}) \;:=\; \langle T_q(\rho_k)\rangle_q^{(\Pi)} \;=\; \frac{\sum_k p_k^q T_q(\rho_k)}{\sum_k p_k^q}\,. \tag{6}$$

In fact, (6) is fully characterized by an ensemble $\{\rho_k, p_k\}$. Now, the following quantity encodes classical correlations: $\mathcal{C}_q(A:B) = \sup_{\{\Pi_k\}}\Big(T_q(A) - T_q(\rho_{AB}|\{\Pi_k\})\Big)$. Finally, we define the q-discord by $D_q(A:B) = \mathcal{I}_q(A:B) - \mathcal{C}_q(A:B)$. In general, q-discord is not positive (see the discussion in[16] for the case (4)).

3. q-discord for Werner and isotropic states

Let us consider two-qubit Werner states $\rho_W(\lambda) = \frac{1}{6}[(2-\lambda)I_4 + (2\lambda-1)\mathbb{F}]$, parametrized by $0\le\lambda\le 1$, where $\mathbb{F}$ is the flip operator and I_4 is 4×4 identity matrix. Straightforward algebra yields q-discord for Werner states (see Fig. 1),

$$D_q(A:B) = \frac{1}{q-1}\Big[\frac{1}{2}(1-\lambda)^q + \frac{1}{2}\Big(\frac{1+\lambda}{3}\Big)^q - \Big(\frac{2-\lambda}{3}\Big)^q\Big]\,.$$

Similar calculations can be performed also for $2\otimes 2$ isotropic states $\rho_{\mathrm{iso}}(\lambda) = \lambda\Pi_+ + \frac{1-\lambda}{3}(I_4 - \Pi_+)$, $0\le\lambda\le 1$, where $|\psi_+\rangle = 1/\sqrt{2}(|00\rangle + |11\rangle)$ and the projector $\Pi_+ = |\psi_+\rangle\langle\psi_+|$. After some algebra q-discord for isotropic states reads

$$D_q(A:B) = \frac{1}{q-1}\Big[\frac{1}{2}\Big(\frac{2-2\lambda}{3}\Big)^q + \frac{1}{2}(2\lambda)^q - \Big(\frac{1+2\lambda}{3}\Big)^q\Big]\,.$$

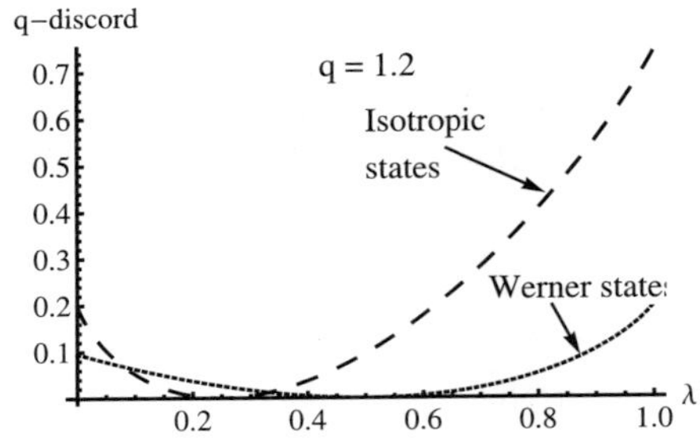

Figure 1. q-Discord for Werner and isotropic states

We can prove analytically that the q-discords $D_q(A:B)$ for $2 \otimes 2$ isotropic and Werner states are nonnegative for every $q > 0$ (see Fig. 1).

Bibliography

1. L. C. Céleri, J. Maziero, and R. M. Serra, Int. J. Quant. Inf. **9**, 1837 (2011) and other articles in a special issue of Int. J. Quant. Inf. 9 Nos. 7 & 8 (2011).
2. M. A. Nielsen and I. L. Chuang, *Quantum computation and quantum information*, Cambridge University Press, Cambridge, 2000; R. Horodecki, P. Horodecki, M. Horodecki, K. Horodecki, Rev. Mod. Phys. **81**, 865–942 (2009).
3. H. Ollivier and W.H. Żurek, Phys. Rev. Lett. 88, 017901 (2002); L. Henderson and V. Vedral, J. Phys. A 34, 6899 (2001).
4. M. Horodecki, P. Horodecki, R. Horodecki, J. Oppenheim, A. Sen, U. Sen, and B. Synak-Radtke, Phys. Rev. A 71, 062307 (2005).
5. R. Rossignoli, N. Canosa, L. Ciliberti, Phys. Rev. A **82**, 052342 (2010).
6. A. Rényi, *Probability Theory*, North-Holland, Amsterdam, 1970.
7. C. Tsallis, J. Stat. Phys. **52**, 479 (1988).
8. M. Masi, Phys. Lett. A **338**, 217 (2005).
9. T. Yamano, Phys. Rev. E **63**, 046105 (2001).
10. O. Gühne and M. Lewenstein, Phys. Rev. A **70**, 022316 (2004).
11. J. S. Kim, Phys. Rev. A 81, 062328 (2010).
12. S. Furuichi, K. Yanagi, K. Kuriyama, J. Math. Phys. 45 4868–4877 (2004); S. Furuichi, J. Math. Phys. **47**, 023302 (2006); X. Hu, Z. Ye, J. Math. Phys. **47**, 023502 (2006).
13. S. Abe and A. K. Rajagopal, Phys. Rev. E 60, 3461 (1999).
14. S. Abe and A.K. Rajagopal, Physica A **289**, 157–164 (2001).
15. C. Tsallis, R. S. Mendes, A. R. Plastino, Physica A **261**, 534 (1998).
16. P. J. Coles, *Non-negative discord strengthens the subadditivity of quantum entropy functions*, arXiv: 1101.1717 [quant-ph].

GENERALIZED ASSOCIATIVITY IN THE CAYLEY NUMBERS

S. KUWATA

Department of Frontier Sciences, Hiroshima City University,
Asaminami-ku, Hiroshima 731-3194, Japan
E-mail: kuwata@hiroshima-cu.ac.jp

The Cayley number, which is closely related to the exceptional Lie groups, is somewhat difficult to deal with, due to the lack of its associativity, where the associativity can be represented by the commutativity between the left and right multiplications L_x and R_y, respectively. We obtain a relation between L_x and R_y by generalizing the commutator $[L_x, R_y]$. However, the resultant relation depends upon the geometrical configuration between x and y.

Keywords: Cayley number; associativity; alternativity; Moufang identity.

1. Introduction

The Cayley number $\mathbb{O}$, which is an eight-dimensional alternative division algebra over the real numbers $\mathbb{R}$, is closely related to the exceptional Lie groups,[1] and has been applied to Lorentz geometry and supersymmetry,[2] non-bijective quadratic transformations,[3] and some other physical models.[4] Due to the lack of the associativity of the $\mathbb{O}$, the commutativity between the left multiplication L_x and the right multiplication R_y (for $x, y \in \mathbb{O}$) does not hold in general. However, owing to the alternativity of the $\mathbb{O}$, we can obtain a relation between L_x and R_y by generalizing the commutator $[L_x, R_y]$. Finally, we give a simple application of the resultant relation between L_x and R_y.

2. Cayley numbers

For the Cayley number $\mathbb{O} \ni x = \sum_{i=0}^{7} x_i e_i$ $(x_i \in \mathbb{R})$, the e_i's satisfy $e_0 = 1$, $e_i^2 = -1$ for $i = 1, 2, \ldots, 7$, and $e_i e_j e_k = -1$ for $(i, j, k) \in W$, where $W = \{(1,2,3), (1,4,5), (2,4,6), (3,4,7), (2,5,7), (6,1,7), (5,3,6)\}$. For the conjugation $\bar{x} = 2x_0 - x$, it follows that $\overline{xy} = \bar{y}\bar{x}$ for all $x, y \in \mathbb{O}$. The

Euclidean norm and inner product in $\mathbb{R}^8$ are given by

$$\|x\|^2 = x\bar{x} = \bar{x}x, \quad \langle x, y\rangle = \frac{1}{2}(x\bar{y} + y\bar{x}),$$

respectively. The Cayley number is a composition algebra: $\|xy\| = \|x\|\|y\|$ for all $x, y \in \mathbb{O}$. It it known that the alternative laws $x^2y = x(xy)$ and $yx^2 = (yx)x$ lead to the flexible law $(xy)x = x(yx)$ and the Moufang identities[5]

$$(xax)y = x[a(xy)], \quad y(xax) = [(yx)a]x, \quad (xy)(ax) = x(ya)x. \tag{1}$$

For later analysis, it is convenient to introduce the left and right multiplications L_a, R_a of an algebra $\mathfrak{A}$ (for $a \in \mathfrak{A}$) by $L_a : x \mapsto ax, R_a : x \mapsto xa$ for all $x \in \mathfrak{A}$. If $\mathfrak{A}$ is associative, then the associativity can be written as a commutativity of the left and right multiplications:

$$\forall x, y, z \in \mathfrak{A},\ (xz)y - x(zy) = 0 \Longleftrightarrow \forall x, y \in \mathfrak{A},\ [L_x, R_y] = 0,$$

where $[A, B] = AB - BA$. It should be remarked here that the linear operators R_x and L_y act on an element $z \in \mathfrak{A}$ from the right, so that, for example, $L_xR_y : z \mapsto (xz)y$. This can be easily understood by regarding $z \in \mathfrak{A}$ as a row vector $\langle z|$, so that $\langle z|L_xR_y = \langle xz|R_y = \langle (xz)y|$.

For the Cayley numbers, the alternative laws $x^2y = x(xy)$ and $zx^2 = (zx)x$ can be rewritten after the linearization $x \to x + z$ and $x \to x + y$ as

$$[L_x, R_y] = R_{xy} - R_xR_y, \tag{2}$$

$$R_{xy} + R_{yx} = R_xR_y + R_yR_x, \tag{3}$$

respectively. Due to the lack of the associativity of the $\mathbb{O}$, the commutativity $[L_x, R_y] = 0$ does not hold in general. However, owing to the alternativity of the $\mathbb{O}$, we can obtain a relation between L_x and R_y by generalizing the commutator $[L_x, R_y]$, as will be shown as follows.

Before going further, it should be noticed that $[L_x, R_y]$ is irrelevant to the real part of x and y, and that $L_{\lambda x} = \lambda L_x, R_{\mu y} = \mu R_y$ for all $\lambda, \mu \in \mathbb{R}$. Bearing these in mind, one may find it useful to introduce a subset of $\mathbb{O}$ as

$$\hat{\mathbb{O}} := \{x \in \mathbb{O} | x + \bar{x} = 0, \|x\| = 1\}.$$

For $\alpha, \beta \in \hat{\mathbb{O}}$, let r, s, and t denote

$$r := L_\alpha R_\beta, \quad s := R_{\alpha\beta}, \quad t := R_\alpha R_\beta.$$

Then the relations of (2) and (3) can be rephrased as

$$r - r^{-1} = s - t = t^{-1} - s^{-1}, \tag{4}$$

where r^{-1} is defined as such that $r^{-1}r = rr^{-1} = 1$; and s^{-1} and t^{-1} are defined in an analogous way. Explicitly, we have $r^{-1} = R_\beta L_\alpha$, $s^{-1} = R_{\beta\alpha}$, and $t^{-1} = R_\beta R_\alpha$.

In what follows, we will rewrite $r^n - r^{-n}$ (for $n \in \mathbb{N}$) using s and t. For this purpose, we first prepare the following relations:

$$[s, t] = 0, \quad r^2 = s^{-1}t. \tag{5}$$

To show $[s, t] = 0$, it may be useful to employ the identity $[A, BC] = [A, B]_+C - B[A, C]_+$ (where $[A, B]_+ = AB + BA$), together with (3) and the Moufang identity $R_{\alpha\beta\alpha} = R_\alpha R_\beta R_\alpha$. To show $r^2 = s^{-1}t$, it may be useful to decompose $r^2 = L_\alpha r^{-1} R_\beta = -L_\alpha(r^{-1}R_\alpha)t$, and to employ another Moufang identity $r^{-1}R_\alpha = L_\alpha s^{-1}$. Once we have prepared (5), we can rewrite $r^n - r^{-n}$ in terms of s and t. For $n = 1$, we have already obtained the first relation of (4). For $n = 2$, from (5) and the second relation of (4), we obtain $r^2 - r^{-2} = s^{-1}t - t^{-1}s = t^2 - s^2 - [s, t] = t^2 - s^2$. For $n \geq 3$, we can calculate $r^n - r^{-n}$ inductively through the recursion identity $r^n - r^{-n} = (r^{n-2} - r^{-(n-2)})((r - r^{-1})^2 + 2) - (r^{n-4} - r^{-(n-4)})$. To summarize,

$$r^n - r^{-n} = (-1)^n(t^n - s^n) \quad (\text{for } n \in \mathbb{N}). \tag{6}$$

In a general setting, the angle between α and β (for $\alpha, \beta \in \hat{\mathbb{O}}$) can be chosen as arbitrary. Denote the angle by $\theta_{\underset{\smile}{\alpha\beta}}$. Then β can be written using α and $\theta_{\underset{\smile}{\alpha\beta}}$ as $\beta = \cos\theta_{\underset{\smile}{\alpha\beta}} \cdot \alpha + \sin\theta_{\underset{\smile}{\alpha\beta}} \cdot \alpha_\perp$, where $\alpha_\perp \in \hat{\mathbb{O}}$ is so chosen that $\langle \alpha, \alpha_\perp \rangle = 0$. In this case, $t^n - s^n$ can be rewritten as $t^n - s^n = (-1)^n \sin n\theta_{\underset{\smile}{\alpha\beta}} \cdot (R_{\alpha\alpha_\perp} - R_\alpha R_{\alpha_\perp})$ (for all $n \in \mathbb{N}$), which, together with (6), leads to $r^n - r^{-n} = \sin n\theta_{\underset{\smile}{\alpha\beta}} \cdot (R_{\alpha\alpha_\perp} - R_\alpha R_{\alpha_\perp})$. Notice that n in this relation can be chosen as a real number, if r^ν and $r^{-\nu}$ (for $\nu \in \mathbb{R}$) are defined as a power series of r and r^{-1}, respectively. As a consequence, it is found that for $\alpha, \beta \in \hat{\mathbb{O}}$

$$(L_\alpha R_\beta)^\nu - (R_\beta L_\alpha)^\nu = 0 \quad (\text{for } \nu \text{ such that } \sin\nu\theta_{\underset{\smile}{\alpha\beta}} = 0). \tag{7}$$

For $\alpha \parallel \beta$, where $\theta_{\underset{\smile}{\alpha\beta}} = 0, \pi$, we can choose $\nu = 1$. In this case, the relation of (7) simply indicates the flexible law $[L_\alpha, R_\alpha] = 0$.

3. Conclusion

We have obtained the relation of (7) by using the alternativity of the Cayley numbers. Some may point out that other alternative algebras might satisfy the same relation as (7). Consider, for example, the split-octonion $\mathbb{O}'$, which is also an alternative (and composition) algebra. However, the norm $\|x\|$ for

$x \in \mathbb{O}'$ is not positive definite, so that the angle between $\alpha, \beta \in \mathbb{O}'$ cannot be well defined for $\|\alpha\| = 0$ or $\|\beta\| = 0$ even if $\alpha, \beta \neq 0$. Thus in this case, the value of ν in (7) is not well defined.

As a simple application of (7), consider a double-sided Fourier transform of an octonion-valued function. For $f : \mathbb{R}^2 \to \mathbb{O}$ with $f \in L^2(\mathbb{R}^2)$, let $\mathcal{F}_{LR}f$ and $\mathcal{F}_{RL}f$ be defined by (double-sided Fourier transform was first proposed in [6] for a quaternion-valued function)

$$(\mathcal{F}_{LR}f)(x,y) := \iint_{\mathbb{R}^2} \left(e^{2\pi ikx}\, f(k,\ell)\right) e^{2\pi j\ell y}\, dk\, d\ell,$$

$$(\mathcal{F}_{RL}f)(x,y) := \iint_{\mathbb{R}^2} e^{2\pi ikx} \left(f(k,\ell) e^{2\pi j\ell y}\right) dk\, d\ell,$$

where $i = e_1 \in \mathbb{O}$ and $j = e_2 \in \mathbb{O}$, with satisfying $\langle i, j\rangle = 0$. Due to the lack of the associativity of $\mathbb{O}$, it follows that $\mathcal{F}_{LR}f \neq \mathcal{F}_{RL}f$ in general. However, owing to $(L_iR_j)^2 = (R_jL_i)^2$ by (7), it is found that

$$(\mathcal{F}_{LR})^2 f = (\mathcal{F}_{RL})^2 f \tag{8}$$

as follows. To show (8), it is convenient to decompose f as $f = f_{++} + f_{+-} + f_{-+} + f_{--}$, where $f_{\sigma\tau} \in \Phi_{\sigma\tau} := \{\phi(x,y) \in L^2(\mathbb{R}^2) | \sigma\phi(-x,y) = \phi(x,y) = \tau\phi(x,-y)\}$ for $\sigma, \tau \in \{+,-\}$. Noticing that under the Fourier transform of ϕ, the even-odd property of ϕ is preserved, that is, $\mathcal{F}_{LR}\phi, \mathcal{F}_{RL}\phi \in \Phi_{\sigma\tau}$ for $\phi \in \Phi_{\sigma\tau}$, and using $\mathcal{F}_{LR}\phi = \mathcal{F}_{RL}\phi$ for $\phi \in \{\Phi_{++}, \Phi_{+-}, \Phi_{-+}\}$, and $(\mathcal{F}_{LR})^2\phi = (\mathcal{F}_{RL})^2\phi$ for $\phi \in \Phi_{--}$ by $(L_iR_j)^2 = (R_jL_i)^2$, we eventually obtain the relation of (8).

Acknowledgments

The author is indebted to K. Terada for invaluable discussion and suggestion.

References

1. N. Jacobson, *Lie Algebras* (Wiley, New York, 1962).
2. J. C. Baez, *Bull. Amer. Math. Soc.* **42**, 229 (2005).
3. D. Lambert and M. Kibler, *J. Phys. A: Math. Gen.* **21**, 307 (1988).
4. S. Okubo, *Introduction to Octonion and Other Non-Associative Algebras in Physics* (Cambrridge, Cambridge, 1995).
5. R. D. Schafer, *An Introduction to NonAssociative Algebras* (Dover, New York, 1995).
6. T. A. Ell, Quaternion-Fourier Transforms for Analysis of Two-Dimensional Linear Time-Invariant Partial Differential Systems, in *Proceedings of the 32nd Conference on Decision and Control*, (San Antonio, Texas, 1993).

ON A ROTATIONAL INVARIANT IN QUANTUM FOUR-BODY SYSTEMS

ZHONG-QI MA

Institute of High Energy Physics, Chinese Academy of Sciences, Beijing 100049, China and Institute for Advanced Study, Tsinghua University, Beijing, 100084, China
** E-mail: mazq@ihep.ac.cn*

ZONG-CHAO YAN

Department of Physics, University of New Brunswick, Fredericton, New Brunswick, Canada E3B 5A3 and Center for Cold Atom Physics, Chinese Academy of Sciences, Wuhan 430071, China
E-mail: zyan@unb.ca

In this paper the rotational invariants constructed by the products of three spherical harmonic polynomials are expressed generally as homogeneous polynomials with respect to three coordinate vectors, where the coefficients are calculated explicitly in compact forms. The details can be found in arXiv 1203.6702.

Keywords: four-body system; rotational invariants; spherical harmonic polynomial.

After separating the center-of-mass motion there are three Jacobi coordinate vectors $\mathbf{r}_j$ in a four-body system. A measurable physical quantity in the four-body system is generally proportional to

$$T^{jk\ell}_{\mu\nu\rho} = \int (d\xi)(dR)G(\xi)\mathcal{Y}^j_\mu(\mathbf{r}_1)\mathcal{Y}^k_\nu(\mathbf{r}_2)\mathcal{Y}^\ell_\rho(\mathbf{r}_3). \tag{1}$$

where $(d\xi)$ is the integral element for the six internal variables, $G(\xi)$ is a rotational invariant depending on the internal variables, and $(dR) = \sin\beta d\alpha d\beta d\gamma/(8\pi^2)$ is the integral element for the Euler angles. From group theory,[1] the Euler angles can be separated as follows

$$T^{jk\ell}_{\mu\nu\rho} = \begin{pmatrix} j & k & \ell \\ \mu & \nu & \rho \end{pmatrix} \int (d\xi)G(\xi)\left\{\sum_{\mu,\nu,\rho}\begin{pmatrix} j & k & \ell \\ \mu & \nu & \rho \end{pmatrix}\mathcal{Y}^j_\mu(\mathbf{r}_1)\mathcal{Y}^k_\nu(\mathbf{r}_2)\mathcal{Y}^\ell_\rho(\mathbf{r}_3)\right\}. \tag{2}$$

It is therefore desirable to reexpress the sum in the curly brackets of (2) in terms of a polynomial of the internal variables, which have widespread applications in many branches of physics.

Weyl (see p.53 in [2]) first established a theorem on the important structure for rotational invariants. In their famous Encyclopedia of Mathematics Vol. 9 on the Racah-Wigner algebra in quantum theory,[3] Biedenharn and Louck studied the important case $n = 3$ of the general theorem in some detail (§6.17 in [3]), and defined the even invariant (see (6.153) in [3]) as

$$\begin{aligned} I_{j,k,\ell}(\mathbf{r}_1,\mathbf{r}_2,\mathbf{r}_3) &= \sqrt{\frac{(4\pi)^3}{(2j+1)(2k+1)(2\ell+1)}} \\ &\times \sum_{\mu,\nu,\rho} \begin{pmatrix} j & k & \ell \\ \mu & \nu & \rho \end{pmatrix} \mathcal{Y}^j_\mu(\mathbf{r}_1)\mathcal{Y}^k_\nu(\mathbf{r}_2)\mathcal{Y}^\ell_\rho(\mathbf{r}_3) = \sum_{(\alpha)} A_{(\alpha)} \prod_{a\le b} (\mathbf{r}_a\cdot\mathbf{r}_b)^{\alpha_{ab}}, \end{aligned} \tag{3}$$

where $\begin{pmatrix} j & k & \ell \\ \mu & \nu & \rho \end{pmatrix}$ is the Wigner 3-j symbol, $\mathcal{Y}^j_\mu(\mathbf{r})$ denotes the spherical harmonic polynomial, and $\sum \alpha_{ab} = (j+k+\ell)/2$. The odd invariant $I_{j,k,\ell}$ is proportional to an even invariant multiplied by a factor $(\mathbf{r}_1 \times \mathbf{r}_2)\cdot\mathbf{r}_3$. However, Biedenharn and Louck pointed out in their book (p.308 in [3]) that: "Unfortunately, the expression for the general coefficients (6.157) has not been given in the literature, and one has had to work out these invariant polynomials from the definition, Eq. (6.153)", where the coefficients (6.157) mean $A_{(\alpha)}$. The purpose of this paper is to present an independent calculation of the coefficients $A_{(\alpha)}$ in (3) for both even and odd rotational invariants $I_{j,k,\ell}$ generally in terms of the internal variables in compact forms using group theoretical method and the mathematical inductions. The results are as follows.

$$\begin{aligned} I_{j,k,j+k-2n}(\mathbf{r}_1,\mathbf{r}_2,\mathbf{r}_3) &= \begin{pmatrix} j & k & j+k-2n \\ 0 & 0 & 0 \end{pmatrix} P^{-1}_{j,k,j+k-2n} \sum_{a=0}^{[j/2]} \sum_{b=0}^{j-2a} \\ &\times \sum_{c=\max\{0,n-a-b\}}^{[(k-b)/2]} A_{abc}\xi_1^a\xi_2^c\xi_3^{a+b+c-n}\eta_1^{k-2c-b}\eta_2^{j-2a-b}\eta_3^b, \end{aligned}$$

where $j \le k$, $P_{j,k,j+k-2n} = \sum_a\sum_b\sum_c A_{abc}$, $[m]$ denotes the largest integer equal to or less than the non-negative real number m, $\xi_a = \mathbf{r}_a\cdot\mathbf{r}_a$, $\eta_1 = \mathbf{r}_2\cdot\mathbf{r}_3$ and so on.

$$\begin{aligned} A_{ab(n-a-b)} &= G_{a,b}\frac{(j-n)!(k-n)!}{(j-2a-b)!(k-2n+2a+b)!} \\ &\times \prod_{m=1}^{\lambda} 2m(2j+2k-4n-2m+1), \\ 0 \le a \le n, &\quad 0 \le b \le n-a, \end{aligned}$$

$$G_{a,b} = \sum_{r=\max\{0,2a+b-n\}}^{a} \frac{(-1)^{a+b+r} n!(j-2a-b+r)!}{2^{a-r} r!(a-r)!b!(n-2a-b+r)!(j-n)!}$$
$$\times \frac{(k-2n+2a+b)!(2j-2a-1)!!(2k-2n+4a+2b-2r-1)!!}{(k-2n+2a+b-r)!(2j-2n-1)!!(2k-2n-1)!!},$$
$$G_{a,b} = 0 \quad \text{if } a<0, \text{ or } b<0, \text{ or } a+b>n.$$

$$A_{ab(n-a-b+c)} = \sum_{s=0}^{\min\{c,b\}} \sum_{r=\max\{s,c-a\}}^{c} G_{a-c+r,b-s} \frac{(-1)^c 2^s (c!)}{(c-r)!(r-s)!s!(j-2a-b)!}$$
$$\times \frac{(j-n)!(k-n)!}{(k-2n+2a+b-2c)!} \prod_{m=c+1}^{\lambda} 2m(2j+2k-4n-2m+1),$$
$$0 \le a \le n, \quad 0 \le b \le n-a, \text{ and } 0 \le c \le [(k+b)/2]+a-n.$$

$$A_{a(n-a+b)c} = \sum_{s=\max\{0,b-a\}}^{\min\{b+c,n-a+b\}} \sum_{r=\max\{s,b+c-a\}}^{\min\{s+c,b+c\}} G_{a-b-c+r,n-a+b-s} \frac{(-1)^{b+c} 2^s (b+c)!}{(r-s)!s!(b+c-r)!}$$
$$\times \frac{(j-n)!(k-n)!}{(j-n-a-b)!(k-n+a-b-2c)!} \prod_{m=b+c+1}^{\lambda} 2m(2j+2k-4n-2m+1),$$
$$0 \le a \le n, \quad 0 \le b \le j-n-a, \text{ and } 0 \le c \le [(k-n+a-b)/2].$$

$$A_{(n+a)bc} = \sum_{s=\max\{0,b-n\}}^{b} \sum_{r=\max\{s,b+c-n\}}^{s+c} G_{n-b-c+r,b-s} \frac{(-1)^{a+b+c} 2^s (a+b+c)!}{(r-s)!s!(a+b+c-r)!}$$
$$\times \frac{(j-n)!(k-n)!}{(j-2n-2a-b)!(k-2c-b)!} \prod_{m=a+b+c+1}^{\lambda} 2m(2j+2k-4n-2m+1),$$
$$0 \le a \le [j/2]-n, \quad 0 \le b \le j-2n-2a, \text{ and } 0 \le c \le [(k-b)/2].$$

Similarly, the odd invariant can be written more explicitly in the form

$$I_{j+1,k+1,j+k-2n+1}(\mathbf{r}_1,\mathbf{r}_2,\mathbf{r}_3) = \frac{\mathrm{i}\zeta}{2}\sqrt{\frac{(j+2)!(k+2)!}{j!k!}}$$
$$\times \begin{pmatrix} j+1 & k+1 & j+k-2n+1 \\ 1 & -1 & 0 \end{pmatrix} Q^{-1}_{j,k,j+k-2n}$$
$$\times \sum_{a=0}^{[j/2]} \sum_{b=0}^{j-2a} \sum_{c=\max\{0,n-a-b\}}^{[(k-b)/2]} B_{abc} \xi_1^a \xi_2^c \xi_3^{a+b+c-n} \eta_1^{k-2c-b} \eta_2^{j-2a-b} \eta_3^b,$$
$$Q_{j,k,j+k-2n} = \sum_a \sum_b \sum_c B_{abc}, \qquad j \le k. \tag{4}$$

The expression for B_{abc} is the same as that for A_{abc} except for the products replaced by $\prod_m 2m(2j+2k-4n-2m+3)$, and $G_{a,b}$ replaced by $F_{a,b}$,

$$F_{a,b} = \sum_{r=\max\{0,2a+b-n\}}^{a} \frac{(-1)^{a+b+r} n!}{2^{a-r} r!(a-r)!b!}$$
$$\times \frac{(j-2a-b+r)!(k-2n+2a+b)!}{(n-2a-b+r)!(j-n)!(k-2n+2a+b-r)!}$$
$$\times \frac{(2j-2a+1)!!(2k-2n+4a+2b-2r+1)!!}{(2j-2n+1)!!(2k-2n+1)!!},$$
$$F_{a,b} = 0 \text{ if } a<0, \text{ or } b<0, \text{ or } a+b>n.$$

The rotational invariants $I_{j,k,\ell}(\mathbf{r}_1, \mathbf{r}_2, \mathbf{r}_3)$ constructed by three spherical harmonic polynomials are the homogeneous polynomials of orders j, k, and ℓ with respect to the three coordinate vectors $\mathbf{r}_1$, $\mathbf{r}_2$, and $\mathbf{r}_3$, respectively. We have rewritten the definitions for the invariants given by Biedenharn and Louck more explicitly, derived the recursive relations for the coefficients A_{abc} and B_{abc} using the Laplace's equations, defined two key functions $G_{a,b}$ and $F_{a,b}$, and calculated analytically the expressions for A_{abc} and B_{abc} in compact forms by mathematical induction. Therefore, we have completely solved the problem raised by Biedenharn and Louck (p.308 in [3]). The present method can in principle be generalized to the rotational invariants constructed by four or more spherical harmonic polynomials although the definition for the invariants depends on the order of coupling.

Acknowledgments

ZQM would like to thank Professor Fu-Chun Zhang for the warm hospitality during his visit at the University of Hong Kong where part of this work was completed. This work was partly supported by the Natural Science Foundation of China under grant No. 11174099 and No.11075014. ZCY was supported by NSERC of Canada.

References

1. Zhong-Qi Ma, *Group Theory for Physicists*, (World Scientific, Singapore, 2007).
2. H. Weyl, in *The Classical Groups: Their Invariants and Representations*, (Princeton Univ. Press, Princeton, 1946).
3. L. C. Biedenharn and J. D. Louck, in *The Racah-Wigner algebra in Quantum Theory*, Encyclopedia of Mathematics and its Application, Vol. 9, Ed. G.-C. Rota, (Addison-Wesley, Massachusetts, 1981).

EXACTLY SOLVABLE, NON-HERMITIAN BCS HAMILTONIAN

AMIR MOGHADDAM*, JON LINKS, and YAO-ZHONG ZHANG

Centre for Mathematical Physics, School of Mathematics and Physics, The University of Queensland, Brisbane, QLD 4072, Australia
**E-mail: a.moghaddam1@uq.edu.au*
www.maths.uq.edu.au/cmp

Understanding the nature of the roots of the Bethe ansatz equations is central to understanding the mathematical physics underpinning quantum integrable and exactly solvable models. Here we analyse an exactly solvable, non-hermitian BCS pairing Hamiltonian dependent on two real-valued coupling parameters. For particular constraints on these prameters, we show that a class of solutions for the Bethe ansatz equations can be obtained analytically. A discussion is given on the physical implications of these solutions.

Keywords: Non-hermitian Hamiltonian; BCS pairing Hamiltonian; Bethe ansatz.

1. Introduction

Quantum phase transition and quantum critical points are crucial foundations of modern statistical mechanics and condensed-matter theory. The predominant formulation is through the Landau–Ginzburg–Wilson theory for which the "order parameter" is a central concept. More recently, Senthil *et al.* identified specific examples of quantum phase transitions that are beyond the Landau–Ginzburg–Wilson paradigm.[1] The key concept in these examples is not the order parameter, but degrees of freedom specific to the critical point.

Here we analyse a recently introduced non-hermitian variant of the reduced Bardeen–Cooper–Schrieffer (BCS)[2] pairing model, which admits an exact solution via Bethe ansatz methods.[3] It will be shown that for particular constraints on the coupling parameters, a class of solutions for the Bethe ansatz equations can be obtained analytically. We will briefly discuss some properties of this class of solutions, and how they relate to the identification of quantum critical points in the spirit of Senthil *et al.*[1] More detail can be found in Ref. 3.

2. A non-hermitian BCS Hamiltonian

Originally proposed in 1957,[2] the reduced BCS Hamiltonian has the general form

$$H_{BCS} = \sum_{j=1}^{L} \epsilon_j n_j - \sum_{j,k=1}^{L} G_{jk} c_{k+}^{\dagger} c_{k-}^{\dagger} c_{j-} c_{j+}. \tag{1}$$

where $j = 1, \cdots, L$, labels a shell of doubly degenerate energy levels with energy ϵ_j, $c_{j\pm}$,$c_{j\pm}^{\dagger}$ are the annihilation and creation operators for fermions at level j, $\pm$ refer to time–reversed pairs, $n_j = c_{j+}^{\dagger} c_{j+} + c_{j-}^{\dagger} c_{j-}$ is the fermion number operator for level j, and G_{jk} are the coupling variables. The first exact solution of the BCS Hamiltonian was presented by Richardson in 1963.[4] In the Richardson model the coupling variables G_{jk} are real and equal. We will set G_{jk} as follows:

$$G_{jk} = \begin{cases} G_+ & j < k \\ \dfrac{G_+ + G_-}{2} & j = k \\ G_- & j > k \end{cases} \tag{2}$$

In 2004 LeClair *et al.*[5] put forward the Russian Doll model, for which $G_+ = G_-^*$ above, as an example of a quantum mechanical system which admits a cyclic renormalisation group map. This choice gives a hermitian Hamiltonian. It was shown by Dunning and Links[6] that this model is exactly solvable by the algebraic Bethe ansatz. Here we study a version of the BCS Hamiltonian for which G_+, G_- are real coupling variables. In this case the Hamiltonian is not hermitian, and consequently not guaranteed to have a real eigenvalue spectrum.

Below, we will choose the ϵ_j to be uniformly and symmetrically distributed around zero

$$\epsilon_j = \left(j - \frac{L+1}{2} \right) \delta \tag{3}$$

where $\delta > 0$ provides the level spacing.

3. Bethe ansatz solution of the Hamiltonian

It is convenient to work with the parameterisation obtained by introducing new variables α and η such that

$$\alpha = \frac{1}{2} \ln \left(\frac{G_+}{G_-} \right), \qquad \eta = \frac{1}{2} \left(G_+ - G_- \right).$$

These relations can be inverted to give

$$G_+ = \frac{2\eta e^{\alpha}}{e^{\alpha} - e^{-\alpha}}, \qquad G_- = \frac{2\eta e^{-\alpha}}{e^{\alpha} - e^{-\alpha}}.$$

In terms of these variables the Bethe ansatz solution for the energy eigenvalues, in the sector with no unpaired fermions and M Cooper pairs, is[3]

$$E = 2\sum_{j=1}^{M} v_j.$$

such that $\{v_j\}$ is a solution of the Bethe ansatz equations

$$e^{2\alpha}P(v_k+\eta)\prod_{j\neq k}^{M}(v_k-v_j-\eta) = P(v_k)\prod_{j\neq k}^{M}(v_k-v_j+\eta), \qquad k=1,\ldots,M \quad (4)$$

where $P(u) = \prod_{j=1}^{L}(u - \epsilon_j - \eta/2)$.

Typically, solutions of the Eqs. 4 cannot be found analytically. However there is a special case in which a class of solutions can be found in a simple, closed form. This occurs when we set $\eta = \pm\delta$, which corresponds to

$$G_+ - G_- = \pm 2\delta. \quad (5)$$

With this restriction the Bethe ansatz Eqs. 4 can be expressed in the factorised form

$$Q(v_k)\left(e^{2\alpha}\left(v_k + \frac{\eta L}{2}\right)\prod_{j\neq k}^{M}(v_k - v_j - \eta) - \left(v_k - \frac{\eta L}{2}\right)\prod_{j\neq k}^{M}(v_k - v_j + \eta)\right) = 0 \qquad k=1,\ldots,M$$

where

$$Q(u) = \prod_{l=1}^{L-1}\left(u - \eta\left(l - \frac{L}{2}\right)\right).$$

In this form it is clear that we obtain a solution set by choosing $v_k \in \mathcal{S}$, $k = 1, \ldots, M$, for

$$\mathcal{S} = \{\eta(k - L/2) : k = 1, \ldots, L-1\}.$$

For these analytic solutions a free quasiparticle interpretation is appropriate, since a solution set for M Cooper pairs is simply a union of one-pair solutions chosen from $\mathcal{S}$.

4. Concluding comments

To conclude, we mention the following distinguishing properties[3] of the analytic solution sets which are found on the lines given by Eqs. 5.

1. Numerical diagonalisation of the Hamiltonian Eq. 1, subject to Eqs. 2,3, shows that between the lines given by Eq. 5 the spectrum is real. Outside of these boundary lines complex eigenvalues can arise.
2. The analytic solutions taken from $\mathcal{S}$ are independent of the variable α, in stark contrast to generic solutions of Eqs. 4.
3. The analytic solutions taken from $\mathcal{S}$ are found to obey generalised exclusion statistics in the sense proposed by Haldane.[7] In particular we can only choose $v_j, v_k \in \mathcal{S}$ whenever $|v_j - v_k| > \delta$. For given L and $M \leq L/2$ the total number of analytic solutions of this type is
$$n(L, M) = \frac{(L-M)!}{(L-2M)!M!}.$$

In this manner we identify the lines given by Eqs. 5 as lines of critical points, in a similar vein proposed by Senthil *et al.*[1] In the above BCS model the transition is of the type described as broken PT-symmetry found in other non-hermitian Hamiltonians.[8–10]

Acknowledgments

Jon Links and Yao-Zhong Zhang are supported by the Australian Research Council through Discovery Projects DP110101414 and DP110103434 respectively. Amir Moghaddam is supported by IPR and UQR Scholarships.

References

1. T. Senthil, A. Vishwanath, L. Balents, S. Sachdev, and M.P.A. Fisher, *Science* **303**, 1490 (2004); T. Senthil, L. Balents, S. Sachdev, A. Vishwanath, and M.P.A. Fisher, *Phys. Rev. B* **70**, 144407 (2004).
2. J. Bardeen, L.N. Cooper, and J.R. Schrieffer, *Phys. Rev.* **108**, 1175 (1957).
3. J. Links, A. Moghaddam and Y.-Z. Zhang, *J. Phys. A: Math. Theor.* in press, arXiv:1206.2687v2.
4. R.W. Richardson, *Phys. Lett.* **3**, 277 (1963).
5. A. LeClair, J.M. Roman, and G. Sierra, *hys. Rev. B* **69**, 020505(R) (2004).
6. C. Dunning and J. Links, *Nucl. Phys. B* **702**, 481 (2004).
7. F.D.M. Haldane, *Phys. Rev. Lett.* **67**, 937 (1991).
8. C.M. Bender, *Rep. Prog. Phys.* **70**, 947 (2007).
9. K. Jones-Smith and H. Mathur, *Phys. Rev. A* **82**, 042101 (2010).
10. Y.C. Hu and T.L. Hughes, *Phys. Rev. B* **84**, 153101 (2011).

QUANTUM INTERFEROMETERS, EULER ANGLES, UNITARY REPRESENTATIONS OF SU(2)

CLAUDIO PARMEGGIANI

Università di Milano (Italy)
E-mail: clparm@infinito.it

The "one particle" and the "two particles" interferometers can be (quantum mechanically) described in terms of Hilbert spaces of states and scattering operators. In this case the scattering operators realize a unitary representation of SU(2) and the Euler angles of the SU(2) group are related to the interferometers parameters (transmission coefficients, phase shifts).

1. Introduction. Quantum Interferometers

A quantum interferometer can be described, as any physical system, by a Hilbert space of "states" and, if we are interested to *transition probabilities*, by a *scattering operator* (see,[1] for example).

For a (quantum) *one particle* interferometer, the describing Hilbert space, $\mathcal{H}$, is the direct sum of two isometric closed subspaces, $\mathcal{H}_1$ and $\mathcal{H}_2$; the interferometer scattering operator is realized by mean of a (fixed) self-adjoint isometry from $\mathcal{H}_1$ to $\mathcal{H}_2$, J; then, if I is the identity operator of $\mathcal{H}$, $J.J = I$.

In the next sections we shall examine the so called Mach-Zehnder interferometer and we shall see as the scattering operator realizes a unitary representation of the SU(2) group, on the Hilbert space $\mathcal{H}$.

In a similar way it is possible to describe a (quantum) *two particles* interferometer, for example the EPR (Einstein, Podolski and Rosen) one. In this case the Hilbert space of the physical system is the tensor product of two copies of the same Hilbert space; that is $\mathcal{H} = \mathcal{H} \otimes \mathcal{H}$ where $\mathcal{H}$ is the one particle space. Now, for a suitable choice of the "initial conditions", we shall obtain the well know interference and correlation (entanglement) effects.

2. Beam splitters, phase shifter

2.1. *The mathematical beam splitter*

The *unitary* operator ($E_\pm = \frac{I \pm J}{2}$):

$$\begin{aligned} S(\alpha) &= \cos(\tfrac{\alpha}{2}).I + i.\sin(\tfrac{\alpha}{2}).J \\ &= \exp(+i\tfrac{\alpha}{2}).E_+ + \exp(-i\tfrac{\alpha}{2}).E_- \end{aligned} \tag{1}$$

is the *scattering operator* of a *beam splitter* (a semitransparent mirror, a perfect crystal[1]); $\cos(\frac{\alpha}{2})$ is the transmission coefficient of the splitter; then if $\cos(\frac{\alpha}{2}) = 0$ (and $\sin(\frac{\alpha}{2}) = \pm 1$) we have a true, reflecting mirror.

A splitter because, if the ongoing unit vector ϕ_1 belong to $\mathcal{H}_1$, the outgoing vector:

$$S(\alpha).\phi_1 = \cos(\tfrac{\alpha}{2}).\phi_1 + i.\sin(\tfrac{\alpha}{2}).\phi_2 \tag{2}$$

(where $\phi_2 = J.\phi_1$) is *split* between the closed orthogonal subspaces $\mathcal{H}_1$ and $\mathcal{H}_2$.

For a physical splitter $\cos(\frac{\alpha}{2})$ is always > 0, but we can obtain a mathematical splitter combining a physical one and a mirror (as $S(\alpha).S(\beta) = S(\alpha + \beta)$).

2.2. *The phase shifter*

If E_1 and E_2 are, respectively, the orthogonal projectors on the closed subspaces $\mathcal{H}_1$ and $\mathcal{H}_2$, the unitary operator ($K = E_1 - E_2$):

$$\begin{aligned} T(\vartheta) &= \exp(+i\tfrac{\vartheta}{2}).E_1 + \exp(-i\tfrac{\vartheta}{2}).E_2 \\ &= \cos(\tfrac{\vartheta}{2}).I + i.\sin(\tfrac{\vartheta}{2}).K \end{aligned} \tag{3}$$

describes, quantum mechanically, a *phase shifter* (the shift is generally a consequence of strong, electromagnetic or gravitational interactions[1]); ϑ is the phase difference generated by the shifter.

Observe that $J.K = -K.J$ and $J.T(\vartheta).J = T(-\vartheta)$ so:

$$E_\pm.K.E_\pm = 0 \tag{4a}$$

$$E_\pm.K.E_\mp = E_\pm.K. \tag{4b}$$

3. The Mach-Zehnder interferometer

3.1. *The scattering operator. The Euler angles*

An one particle, *Mach-Zehnder* interferometer is described by the *unitary scattering operator*:

$$U(\alpha, \beta, \vartheta) = S(\alpha).T(\vartheta).S(\beta), \tag{5}$$

that is splitter, shifter, splitter (plus mirror).

Now we can easily verify that U is a unitary representation of $SU(2)$ (on the Hilbert space $\mathcal{H}$); α, β, ϑ are the corresponding *Euler angles.* In fact, let consider the $SU(2)$ matrix:

$$\begin{pmatrix} a & b \\ -b^* & a^* \end{pmatrix} \tag{6}$$

where, as usual:[2]

$$a = \cos(\tfrac{\vartheta}{2}).\exp(+i\tfrac{\alpha+\beta}{2}) \tag{7a}$$

$$b = i.\sin(\tfrac{\vartheta}{2}).\exp(+i\tfrac{\alpha-\beta}{2}); \tag{7b}$$

hence we shall obtain, remembering the definitions of U, S and T (Section 2) and the relations (4a, 4b):

$$U(\alpha,\beta,\vartheta) = V(a,b) =$$

$$\left(e^{i\frac{\alpha}{2}}E_+ + e^{-i\frac{\alpha}{2}}E_-\right)\left(\cos(\tfrac{\vartheta}{2})I + i.\sin(\tfrac{\vartheta}{2})K\right)\left(e^{i\frac{\beta}{2}}E_+ + e^{-i\frac{\beta}{2}}E_-\right) = \tag{8}$$

$$a.E_+ + a^*.E_- + b.E_+.K - b^*.E_-.K$$

Then we can directly verify the multiplication formula of the $SU(2)$ group:

$$V(a,b).V(c,d) = (ac - bd^*).E_+ + (a^*c^* - b^*d).E_- + \tag{9}$$
$$+(ad + bc^*).E_+.K + (-a^*d^* - cb^*).E_-.K$$

3.2. *Transition Probabilities. Interferences*

If the unit vectors ϕ_1 and ϕ_2 ($= J.\phi_1$) belong, respectively, to $\mathcal{H}_1$ and $\mathcal{H}_2$, the *Mach-Zehnder* interferometer transition probabilities are (remembering the definition of the transition probability, in a quantum contest):

$$\text{Prob}(\phi_1,\phi_1) = |(\phi_1|U(\alpha,\beta,\vartheta).\phi_1)|^2 = A(\alpha,\beta) - 2.C(\alpha,\beta).\cos(\vartheta) \tag{10a}$$

$$\text{Prob}(\phi_1,\phi_2) = |(\phi_1|U(\alpha,\beta,\vartheta).\phi_2)|^2 = B(\alpha,\beta) + 2.C(\alpha,\beta).\cos(\vartheta) \tag{10b}$$

Here A, B, C are polynomials of the sin and cos of α and β, $A + B = 1$. The $\cos(\vartheta)$ term, coming from the phase shifter, is clearly responsible of the interference effects.

4. The EPR Interferometer

4.1. *The scattering operator*

Now the Hilbert space of the physical system (the *EPR, Einstein, Podolski and Rosen* interferometer) is the tensor product (eventually symmetric or antisymmetric) $\mathcal{H} \otimes \mathcal{H}$ of two copies of the same Hilbert space $\mathcal{H}$ (the one particle space); for every particle there is a phase shifter and a beam splitter. The splitters have two different transmission coefficients, $\cos(\frac{\alpha}{2})$ and $\cos(\frac{\beta}{2})$; the shifter have two different phase differences, ϑ_A and ϑ_B.

Then, if the S and T operators are defined as in the previous sections, the interferometer scattering operator is:

$$(S(\alpha) \otimes S(\beta)).(T(\vartheta_A) \otimes T(\vartheta_B)) = (S(\alpha).T(\vartheta_A)) \otimes (S(\beta).T(\vartheta_B)) \quad (11)$$

4.2. *Transition Probabilities. Entanglement*

Now, for a suitable choice of the ongoing unit vector, Φ, we shall obtain the well know interference and correlation (entanglement) effects. In fact, let be:

$$\Phi = c_1.(\phi_1 \otimes \phi_1) + c_2.(\phi_2 \otimes \phi_2) \quad (12)$$

where ($\phi_2 = J.\phi_1$): $\phi_1 \in \mathcal{H}_1$, $\phi_2 \in \mathcal{H}_2$ and $(c_1)^2 + (c_2)^2 = 1$, c_1 and c_2 are real numbers.

Then we have, repeating the calculations of the previous section:

$$\begin{aligned} \text{Prob}(\phi_1 \otimes \phi_1, \Phi) &= (c_1)^2 A_1 + (c_2)^2 B_1 - 2c_1c_2.C.\cos(\vartheta) \\ \text{Prob}(\phi_1 \otimes \phi_2, \Phi) &= (c_1)^2 A_2 + (c_2)^2 B_2 + 2c_1c_2.C.\cos(\vartheta) \end{aligned} \quad (13)$$

Here A_1, A_2, B_1, B_2 are, again, polynomials of the sin and cos of α and β, $C(\alpha, \beta) = \frac{1}{4}\sin(\alpha).\sin(\beta)$, $D(\alpha, \beta) = \frac{1}{4}\cos(\alpha).\cos(\beta)$, $\vartheta = \vartheta_A + \vartheta_B$.

If $c_1 = 0$ or $c_2 = 0$ (*disentanglement*) we lose the $\cos(\vartheta)$ interference term. But if $c_1 = \pm c_2$ the *entanglement* is maximal and we have the "correlation" formulae:

$$\text{Prob}(\phi_1 \otimes \phi_1, \Phi) = \text{Prob}(\phi_2 \otimes \phi_2, \Phi) = \tfrac{1}{4} + D \mp C.\cos(\vartheta) \quad (14)$$

$$\text{Prob}(\phi_1 \otimes \phi_2, \Phi) = \text{Prob}(\phi_2 \otimes \phi_1, \Phi) = \tfrac{1}{4} - D \pm C.\cos(\vartheta) \quad (15)$$

and, eventually, $\text{Prob}(\phi_1 \otimes \phi_1, \Phi) = 0$ or $\text{Prob}(\phi_1 \otimes \phi_2, \Phi) = 0$.

References

1. H. Rauch, *Reality in Neutron Interference Experiments*, in J. Ellis, D. Amati (Ed.), *Quantum Reflections*, Cambridge UP, 2000.
2. A.R. Edmonds, *Angular momentum in Quantum Mechanics*, Princeton UP, 1996.

ON t-DESIGNS AND GENERALISED COHERENT STATES

P. S. TURNER

Department of Physics, University of Tokyo
7-3-1 Hongo, Bukyo-ku, Tokyo, 113-0033 Japan
E-mail: turner@phys.s.u-tokyo.ac.jp

In this brief report we connect the notion of quantum (state) t-designs, motivated by the theory of measurement in quantum information theory, to generalised coherent states, specifically involving finite group actions on finite dimensional Hilbert space. We use this to show that the Clifford group, which plays an important role in many areas, can give only 3-designs in the case of one qubit.

Keywords: t-designs; Coherent states; Group theory; Quantum measurement.

1. Introduction

The rapid growth of quantum information science has, among many other things, increasingly driven research into the theory of quantum measurement. In order to realise the potential of quantum information protocols, the need for finer coherent control of quantum systems has placed a high demand on techniques that exhaustively measure such systems. One of the most elegant and useful theoretical tools for considering such measurements are quantum t-designs.[1] As conceptual objects they are found in various branches of mathematics,[2] and they happen to subsume two of the most famous mathematical problems in quantum theory — both symmetric informationally complete positive operator valued measures (SICPOVMs),[3] as well as mutually unbiased bases (MUBs),[4] are examples of 2-designs. It has also been shown that for finite dimensions, measurements corresponding to 2-designs are optimal for quantum state and process tomography,[5] the most exhaustive characterisation of a quantum system possible. Applications with $t > 2$ are less abundant but also known.[6] Designs are therefore of both significant theoretical as well as practical interest.

We will briefly review t-designs in finite dimensions and show how coherent states of finite groups give a natural construction. We will use this to show how the Clifford group of one qubit can give 3-designs but must fail to give 4-designs, a fact of interest in measurement based computation.[7] We will be interested here only in state (vector) designs, as opposed to so-called unitary (matrix) designs.[8]

2. Background

Loosely speaking, a t-design is a subensemble of pure states that 'mimics' the uniformly distributed ensemble of all pure states, in that averages of any degree t polynomial over the t-design equal those over the entire uniform ensemble. The latter is defined as the set of all pure states distributed according to the natural unitary Haar measure. One also says that a t-design has the same t^{th} moments as the uniform ensemble.

More precisely, let $\mathcal{H}$ be a finite d dimensional Hilbert space upon which $\mathrm{U}(d)$ acts via the defining irreducible representation U with Young diagram (1), let Ψ denote the set of all pure states, and let $\mathrm{d}\psi$ be the Haar measure such that $\mathrm{d}(U\psi) = \mathrm{d}\psi$ for all $U \in \mathrm{U}(d)$. A finite set $\Phi := \{|\phi\rangle\}_\phi$ distributed uniformly with probabilities $\frac{1}{|\Phi|}$ is a t-design iff

$$\int_\Psi \mathrm{d}\psi \langle\psi|^{\otimes t}\hat{X}|\psi\rangle^{\otimes t} = \sum_\Phi \frac{1}{|\Phi|}\langle\phi|^{\otimes t}\hat{X}|\phi\rangle^{\otimes t} \tag{1}$$

$$\Leftrightarrow \mathrm{tr}\left[\hat{X}\int_\Psi \mathrm{d}\psi|\psi\rangle\langle\psi|^{\otimes t}\right] = \mathrm{tr}\left[\hat{X}\sum_\Phi \frac{1}{|\Phi|}|\phi\rangle\langle\phi|^{\otimes t}\right], \tag{2}$$

for all operators $\hat{X} \in \mathcal{B}(\mathcal{H}^{\otimes t})$, since they map onto the set of degree t polynomials in the components of the state vectors. Here we've assumed the t-design is finite, which gives an indication of a design's utility: for finite d, it's a finite set that mimics an infinite set as far as t^{th} moments are concerned.

The representation $(1)^{\otimes t} = (t) \oplus \cdots$, and by Schur-Weyl duality (t) is multiplicity free and irreducible on $\mathcal{H}_{(t)} \subseteq \mathcal{H}^{\otimes t}$, the totally symmetric subspace under tensor factor permutations in S_t, which contains $|\psi\rangle^{\otimes t}$ for any $|\psi\rangle$. By Schur's lemma we have that

$$\int_\Psi \mathrm{d}\psi|\psi\rangle\langle\psi|^{\otimes t} \propto \mathbb{1}_{(t)}, \tag{3}$$

the projector onto the symmetric subspace. Since Eq. (2) is true for all $\hat{X}$, an equivalent definition of a t-design is

$$\sum_\Phi \frac{1}{|\Phi|}|\phi\rangle\langle\phi|^{\otimes t} \propto \mathbb{1}_{(t)}. \tag{4}$$

Equation (4) hints at a natural construction for designs, as it was arrived at using the known irreducibility of the unitary group on $\mathcal{H}_{(t)}$. Schur's lemma ensures that *any* group with a Haar measure that is so irreducible will also satisfy Eq. (4) if the left hand side is well defined, and we can go about mining character tables for candidates. Following this line of reasoning, we write the set of all pure states as $\Psi = \{U|\psi_0\rangle | U \in \mathrm{U}(d)\}$, where ψ_0 denotes any fiducial state. We are thus interested in designs of the form

$$\Phi = \{U_G(g)|\phi_0\rangle | g \in G\}, \tag{5}$$

where, in order to guarantee Haar measureability and convergence, we restrict our attention to finite groups G with a unitary irrep U_G on $\mathcal{H}$[a]. The set Φ can therefore be considered as a set of generalised coherent states.[10,11] The usefulness of this viewpoint depends on the nature of the players involved, namely G, U_G and $|\phi_0\rangle$. The following is an example.

3. The Clifford group on one qubit

Consider the case $\mathcal{H} = \mathbb{C}^2 = \mathrm{span}\{|0\rangle, |1\rangle\}$, a *qubit*, and the group generated by the operators

$$H = \frac{1}{\sqrt{2}}\begin{bmatrix}1 & 1\\ 1 & -1\end{bmatrix}, \quad P = \begin{bmatrix}1 & 0\\ 0 & i\end{bmatrix}. \tag{6}$$

This is known as the Clifford group in quantum information, $C := \langle H, P\rangle$. GAP[12] calls it (SL(2,3):C4):C2, referring to the special linear group of 2×2 matrices over $\mathbb{Z}_3$ and the cyclic groups C4, C2. It has 192 elements, and it can be shown without too much difficulty (*e.g.* using a computer algebra package) that for any fiducial state $|\phi_0\rangle = \cos\theta|0\rangle + e^{i\varphi}\sin\theta|1\rangle$ it generates a 3-design in the sense of Eq. (5). As is usual with coherent states, it is a coset of the group and not the entire group that is necessary for the design to hold, where the coset depends upon the isotropy subgroup of the fiducial state. For example, in the basis of Eq. (6), choosing $|0\rangle \leftrightarrow \begin{bmatrix}1 & 0\end{bmatrix}^{\mathrm{T}}$ to be the fiducial state gives six coherent states, which coincide with the set of mutually unbiased bases for qubits.

GAP tells us that C has 8, 12, 8 and 4 inequivalent irreps of dimension 1, 2, 3, and 4, respectively. We immediately see that C cannot generate a 4-design, because $\dim\mathbb{C}^2_{(4)} = 5$ which exceeds the maximum 4, and so C cannot be irreducible on the symmetric subspace of four qubits, violating

[a]There is a clear analogy for continuous groups, and the topic presented at this conference was the case of Gaussian 2-designs.[9]

Eq. (4). We conclude that the Clifford group for qubits can generate at most 3-designs, and that indeed it does so for any fiducial state.

4. Conclusion

We have given a very brief outline of (state) t-designs in quantum theory, and have connected a very natural group theoretical construction to the theory of generalised coherent states. We used this to show quite straightforwardly that no Clifford group construction can give a 4-design for qubits.

The example worked out here was quite simple, but obviously this line of reasoning can be used to paint a much broader picture. A similar approach has been taken in the context of matrix designs,[13] but to our knowledge the connection to generalised coherent states was not made.

The ubiquitous nature of the Clifford group and it's generalisations in the theory of SICPOVMs and MUBs motivated the specific Clifford group example here, which shows that an appropriate group structure is sufficient for generating designs. Knowing the extent to which such a group structure is also *necessary* for a design structure would be an important step forward in solving these notorious problems.

References

1. J. Renes, R. Blume-Kohout, A. Scott and C. Caves, *Journal of Mathematical Physics* **45**, p. 2171 (2004).
2. C. Colbourn and J. Dinitz (eds.), *Handbook of Combinatorial Designs, Second Edition* (Chapman & Hall/CRC, 2006).
3. D. Appleby, *Journal of Mathematical Physics* **46**, p. 052107 (2005).
4. W. Wootters and B. Fields, *Annals of Physics* **191**, 363 (1989).
5. A. Scott, *Journal of Physics A: Mathematical and Theoretical* **41**, p. 055308 (2008).
6. W. Matthews, S. Wehner and A. Winter, *Communications in Mathematical Physics* **291**, 813 (2009).
7. T.-C. Wei, I. Affleck and R. Raussendorf, *Physical Review Letters* **106**, p. 070501 (2011).
8. C. Dankert, R. Cleve, J. Emerson and E. Livine, *Physical Review A* **80**, p. 012304 (2009).
9. R. Blume-Kohout and P. Turner, `arXiv:1110.1042` (2011).
10. A. Perelomov, *Generalized coherent states and their applications* (Springer-Verlag New York Inc., New York, NY, 1986).
11. S. Ali, J.-P. Antoine and J.-P. Gazeau, *Coherent States, Wavelets and their Generalizations* (Springer, New York, 2000).
12. `http://www.gap-system.org/`.
13. D. Gross, K. Audenaert and J. Eisert, *Journal of Mathematical Physics* **48**, p. 052104 (2007).

REDUCTIONS OF CONTROLLED HAMILTONIAN SYSTEMS WITH SYMMETRY

HONG WANG

School of Mathematical Sciences and LPMC,
Nankai University, Tianjin 300071, P.R.China
E-mail: hongwang@nankai.edu.cn

This article introduces briefly some recent developments of reduction theory of controlled Hamiltonian systems with symmetry.

Keywords: Controlled Hamiltonian system; Regular and singular reduction; Optimal reduction; Poisson reduction.

1. Introduction

The theory of mechanical control systems is a very important subject, following the theoretical development of geometric mechanics, a lot of important problems about this subject are being explored and studied. In this article, we introduce briefly some recent developments of reduction theory of controlled Hamiltonian (CH) systems with symmetry. Our main idea is that we regard the set of Hamiltonian systems on T^*Q as a subset of the set of CH systems on T^*Q, and study regular, singular, and optimal reduction theory of CH systems with symplectic (resp. Poisson) structure and symmetry, as an extension of the corresponding regular, singular, and optimal reduction theory of Hamiltonian systems under controlled Hamiltonian equivalence conditions. One can show from Marsden et al,[1] Wang and Zhang,[2] Wang,[3] Ratiu and Wang[4] for more details.

2. Regular Reductions of RCH Systems with Symmetry

At first, we define a regular point reducible RCH (i.e. regular controlled Hamiltonian) system with symplectic structure and symmetry, by using moment map and the associated reduced symplectic forms. Let Q be a smooth manifold and T^*Q its cotangent bundle with symplectic form ω. Let $\Phi : G\times Q \to Q$ be a smooth left action of the Lie group G on Q, which is free and proper. Then the cotangent lifted left action $\Phi^{T^*}: G\times T^*Q\to T^*Q$ is symplectic, free and proper, and admits an Ad^*-equivariant moment map $\mathbf{J}: T^*Q\to\mathfrak{g}^*$, where $\mathfrak{g}$ is the Lie algebra of G and $\mathfrak{g}^*$ is the dual of $\mathfrak{g}$. Let $\mu\in\mathfrak{g}^*$ be a regular value of $\mathbf{J}$ and denote by G_μ the isotropy subgroup of the coadjoint G-action at $\mu\in\mathfrak{g}^*$. It follows that G_μ acts also freely and properly on $\mathbf{J}^{-1}(\mu)$, so that $(T^*Q)_\mu=\mathbf{J}^{-1}(\mu)/G_\mu$ is a symplectic manifold with symplectic form ω_μ uniquely characterized by the relation $\pi_\mu^*\omega_\mu=i_\mu^*\omega$. The pair $((T^*Q)_\mu,\omega_\mu)$ is called the symplectic point reduced space of (T^*Q,ω) at μ. If $H: T^*Q\to\mathbb{R}$ is a G-invariant Hamiltonian, the associated regular point reduced Hamiltonian function $h_\mu:(T^*Q)_\mu\to\mathbb{R}$ is defined by $h_\mu\cdot\pi_\mu=H\cdot i_\mu$, and the Hamiltonian vector fields X_H and X_{h_μ} are π_μ-related. Moreover, if considering the external force and control, we can introduce a regular point reducible RCH systems as follows.

Definition 2.1. (Regular Point Reducible RCH System) A 6-tuple (T^*Q,G,ω,H,F,W), where the Hamiltonian $H: T^*Q\to\mathbb{R}$, the fiber-preserving map $F: T^*Q\to T^*Q$ and the fiber submanifold W of T^*Q are all G-invariant, is called a regular point reducible RCH system if there exists a point $\mu\in\mathfrak{g}^*$ which is a regular value of the moment map $\mathbf{J}$, such that the regular point reduced system, that is, the 5-tuple $((T^*Q)_\mu,\omega_\mu,h_\mu,f_\mu,W_\mu)$, where $(T^*Q)_\mu=\mathbf{J}^{-1}(\mu)/G_\mu$, $\pi_\mu^*\omega_\mu=i_\mu^*\omega$, $h_\mu\cdot\pi_\mu=H\cdot i_\mu$, $f_\mu\cdot\pi_\mu=\pi_\mu\cdot F\cdot i_\mu$, $W\subset\mathbf{J}^{-1}(\mu)$, $W_\mu=\pi_\mu(W)$, is a RCH system, which is simply written as R_P-reduced RCH system.

Next, we note that when a RCH system is given, the force map F is determined, but the feedback control law $u: T^*Q\to W$ could be chosen. In order to describe the feedback control law to modify the structure of regular point reducible RCH system, we need to give the controlled Hamiltonian matching conditions and RpCH-equivalence. By using the notations for vertical lifts along fiber, we can give an expression of the dynamical vector field $X_{(T^*Q,\omega,H,F,u)}$ of the RCH system (T^*Q,ω,H,F,W) with a control law u as follows $X_{(T^*Q,G,\omega,H,F,u)}=(\mathbf{d}H)^\sharp+\mathrm{vlift}(F)+\mathrm{vlift}(u)$. Thus, we can define the following RpCH-equivalence.

Definition 2.2. (RpCH-equivalence) Suppose that we have two regular point reducible RCH systems $(T^*Q_i,G_i,\omega_i,H_i,F_i,W_i)$, $i=1,2$, we say them to be RpCH-equivalent, or simply, $(T^*Q_1,G_1,\omega_1,H_1,F_1,W_1)\overset{RpCH}{\sim}(T^*Q_2,G_2,\omega_2,H_2,F_2,W_2)$, if there exists a diffeomorphism $\varphi: Q_1\to Q_2$ such that the following controlled Hamiltonian matching conditions hold:
RpHM-1: The cotangent lift map $\varphi^*: T^*Q_2\to T^*Q_1$ is symplectic.

RpHM-2: For $\mu_i \in \mathfrak{g}_i^*$, the regular reducible points of RCH systems $(T^*Q_i, G_i, \omega_i, H_i, F_i, W_i)$, $i = 1, 2$, the map $\varphi_\mu^* = i_{\mu_1}^{-1} \cdot \varphi^* \cdot i_{\mu_2} : \mathbf{J}_2^{-1}(\mu_2) \to \mathbf{J}_1^{-1}(\mu_1)$ is $(G_{2\mu_2}, G_{1\mu_1})$-equivariant and $W_1 = \varphi_\mu^*(W_2)$, where $\mu = (\mu_1, \mu_2)$, and denote by $i_{\mu_1}^{-1}(S)$ the preimage of a subset $S \subset T^*Q_1$ for the map $i_{\mu_1} : \mathbf{J}_1^{-1}(\mu_1) \to T^*Q_1$.
RpHM-3: $Im[(\mathbf{d}H_1)^\sharp + \text{vlift}(F_1) - ((\varphi_*)^*\mathbf{d}H_2)^\sharp - \text{vlift}(\varphi^* F_2 \varphi_*)] \subset \text{vlift}(W_1)$, where Im means the pointwise image of the map in brackets.

Then, we can prove the regular point reduction theorem for RCH system, which can be regarded as an extension of regular point reduction theorem of Hamiltonian systems under controlled Hamiltonian equivalence conditions.

Theorem 2.1. *Two regular point reducible RCH systems* $(T^*Q_i, G_i, \omega_i, H_i, F_i, W_i)$, $i = 1, 2$, *are RpCH-equivalent if and only if the associated* R_P*-reduced RCH systems* $((T^*Q_i)_{\mu_i}, \omega_{i\mu_i}, h_{i\mu_i}, f_{i\mu_i}, W_{i\mu_i}), i = 1, 2$, *are RCH-equivalent.*

For $\mu \in \mathfrak{g}^*$, a regular value of the moment map $\mathbf{J}$, let $\mathcal{O}_\mu = G \cdot \mu \subset \mathfrak{g}^*$ be a orbit of the coadjoint G-action through the point μ. Then the quotient space $(T^*Q)_{\mathcal{O}_\mu} = \mathbf{J}^{-1}(\mathcal{O}_\mu)/G$ is a symplectic manifold with symplectic form $\omega_{\mathcal{O}_\mu}$ given by $i_{\mathcal{O}_\mu}^* \omega = \pi_{\mathcal{O}_\mu}^* \omega_{\mathcal{O}_\mu} + \mathbf{J}_{\mathcal{O}_\mu}^* \omega_{\mathcal{O}_\mu}^+$. The pair $((T^*Q)_{\mathcal{O}_\mu}, \omega_{\mathcal{O}_\mu})$ is the symplectic orbit reduced space of (T^*Q, ω). Thus, in the same way we can also introduce a kind of regular orbit reducible RCH systems and its RoCH-equivalence, and prove the regular orbit reduction theorem for RCH system. See Marsden et al[1] for more details.

3. Singular Reductions of RCH Systems with Symmetry

If Lie group G acts only properly on Q, does not act freely, then the reduced space $(T^*Q)_\mu = \mathbf{J}^{-1}(\mu)/G_\mu$ (resp. $(T^*Q)_{\mathcal{O}_\mu} = \mathbf{J}^{-1}(\mathcal{O}_\mu)/G$) is not necessarily smooth manifold, but just quotient topological space, and it is a symplectic Whitney stratified space in the sense that the strata is symplectic manifold in a very natural way. From Sjamaar's principle, we know that the singular symplectic strata are symplectic diffeomorphic to regular reduced spaces. In consequence, we can also define a singular point (resp. orbit) reducible RCH systems by using moment map and the associated singular reduced symplectic forms, as well as its SpCH-equivalence (resp. SoCH-equivalence), and prove the singular point (resp. orbit) reduction theorem for RCH system. See Wang[3] for more details.

4. Optimal Reductions of CH Systems with Symmetry

It is a natural problem what and how we could do, if we define a controlled Hamiltonian system on the cotangent bundle T^*Q by using a Poisson structure, and if symplectic reduction procedure given by Marsden et al[1] does not work or is not efficient enough. In Wang and Zhang,[2] we study the optimal reduction theory of CH systems with Poisson structure and symmetry by using the optimal moment map and reduced Poisson tensor (resp. reduced symplectic form), and prove the optimal point, optimal orbit, and regular Poisson reduction theorems for CH systems. Moreover, in Ratiu and Wang[4] we consider the singular Poisson reduction and SPR-CH-equivalence for CH system, and prove the singular Poisson reduction theorem.

5. Poisson Reductions by Controllability Distribution

We note that when the phase space of CH system is a Poisson manifold, and there is no moment map of Lie group action, in this case we cannot use the above reduction procedure, but we can consider the Poisson reduction by distribution. For a symmetric CH system, its control subset $W \subset T^*Q$ is a G-invariant fiber submanifold. If we assume that $D \subset TT^*Q|_W$ is a G-invariant controllability distribution of symmetric CH system, then we give Poisson reducible conditions of CH system by controllability distribution, and prove the Poisson reducible property for CH systems to keep invariant under the CH-equivalence. Moreover, we study the relationship between Poisson reduction for regular (resp. singular) Poisson reducible CH systems by G-invariant controllability distributions and that for associated reduced CH systems by reduced controllability distributions. See Ratiu and Wang[4] for more details.

Acknowledgments: The author would like to thank Prof. M.L. Ge and Prof. C.M. Bai for their understanding, support and help. The research was partially supported by LPMC, Ministry of Education, China.

References

1. J.E.Marsden, H.Wang and Z.X.Zhang, *Regular reduction of controlled Hamiltonian system with symplectic structure and symmetry* (arXiv: 1202.3564).
2. H. Wang and Z.X. Zhang, *Optimal reduction of controlled Hamiltonian system with Poisson structure and symmetry*, *J. Geom. Phys.* 62(5) (2012) 953-975.
3. H. Wang, *Singular reduction of controlled Hamiltonian system with symplectic structure and symmetry* (2012).
4. T.S. Ratiu and H. Wang, *Poisson reduction of controlled Hamiltonian system by controllability distribution* (2012) submitted.

AUTHOR INDEX